The IMA Volumes
in Mathematics
and its Applications

Volume 65

Series Editors
Avner Friedman   Willard Miller, Jr.

Springer

New York
Berlin
Heidelberg
Barcelona
Budapest
Hong Kong
London
Milan
Paris
Santa Clara
Singapore
Tokyo

Institute for Mathematics and
its Applications
IMA

The **Institute for Mathematics and its Applications** was established by a grant from the National Science Foundation to the University of Minnesota in 1982. The IMA seeks to encourage the development and study of fresh mathematical concepts and questions of concern to the other sciences by bringing together mathematicians and scientists from diverse fields in an atmosphere that will stimulate discussion and collaboration.

The IMA Volumes are intended to involve the broader scientific community in this process.

Avner Friedman, Director
Willard Miller, Jr., Associate Director

* * * * * * * * * *

IMA ANNUAL PROGRAMS

1982–1983	**Statistical and Continuum Approaches to Phase Transition**
1983–1984	**Mathematical Models for the Economics of Decentralized Resource Allocation**
1984–1985	**Continuum Physics and Partial Differential Equations**
1985–1986	**Stochastic Differential Equations and Their Applications**
1986–1987	**Scientific Computation**
1987–1988	**Applied Combinatorics**
1988–1989	**Nonlinear Waves**
1989–1990	**Dynamical Systems and Their Applications**
1990–1991	**Phase Transitions and Free Boundaries**
1991–1992	**Applied Linear Algebra**
1992–1993	**Control Theory and its Applications**
1993–1994	**Emerging Applications of Probability**
1994–1995	**Waves and Scattering**
1995–1996	**Mathematical Methods in Material Science**

IMA SUMMER PROGRAMS

1987	**Robotics**
1988	**Signal Processing**
1989	**Robustness, Diagnostics, Computing and Graphics in Statistics**
1990	**Radar and Sonar (June 18 - June 29)**
	New Directions in Time Series Analysis (July 2 - July 27)
1991	**Semiconductors**
1992	**Environmental Studies: Mathematical, Computational, and Statistical Analysis**
1993	**Modeling, Mesh Generation, and Adaptive Numerical Methods for Partial Differential Equations**
1994	**Molecular Biology**

* * * * * * * * * *

SPRINGER LECTURE NOTES FROM THE IMA:

The Mathematics and Physics of Disordered Media

Editors: Barry Hughes and Barry Ninham
(Lecture Notes in Math., Volume 1035, 1983)

Orienting Polymers

Editor: J.L. Ericksen
(Lecture Notes in Math., Volume 1063, 1984)

New Perspectives in Thermodynamics

Editor: James Serrin
(Springer-Verlag, 1986)

Models of Economic Dynamics

Editor: Hugo Sonnenschein
(Lecture Notes in Econ., Volume 264, 1986)

Mark H.A. Davis   Darrell Duffie
Wendell H. Fleming   Steven E. Shreve
Editors

Mathematical Finance

Springer

Mark H.A. Davis
Department of Electrical and
  Electronic Engineering
Imperial College
London, SW7 2BT, England

Darrell Duffie
Graduate School of Business
Stanford University
Stanford, CA 94305 USA

Wendell H. Fleming
Division of Applied Mathematics
Brown University
Providence, RI 02912 USA

Steven E. Shreve
Department of Mathematics
Carnegie Mellon University
Pittsburgh, PA 15213 USA

Mathematics Subject Classifications (1991): 93E20, 90A09, 60G44, 60H10

Library of Congress Cataloging-in-Publication Data
Mathematical finance / Mark H.A. Davis . . . [et al.], editors.
        p.   cm. — (The IMA volumes in mathematics and its
    applications ; v. 65)
      Includes bibliographical references.
      ISBN 0-387-94439-7
      1. Investments—Mathematical models.   2. Securities—Mathematical
    models.   3. Investments—Mathematics.   I. Davis, M.H.A.
    II. Series.
    HG4515.3.M38   1995
    332.6'01'5118—dc20   94-44431

Printed on acid-free paper.

Production managed by Hal Henglein; manufacturing supervised by Genieve Shaw.
Camera-ready copy prepared by the IMA.
Printed and bound by Braun-Brumfield, Ann Arbor, MI.
Printed in the United States of America.

9 8 7 6 5 4 3 2

ISBN 0-387-94439-7 Springer-Verlag New York Berlin Heidelberg   SPIN 10536702

The IMA Volumes
in Mathematics and its Applications

Current Volumes:

Volume 1: Homogenization and Effective Moduli of Materials and Media
Editors: Jerry Ericksen, David Kinderlehrer, Robert Kohn, and
J.-L. Lions

Volume 2: Oscillation Theory, Computation, and Methods of
Compensated Compactness
Editors: Constantine Dafermos, Jerry Ericksen,
David Kinderlehrer, and Marshall Slemrod

Volume 3: Metastability and Incompletely Posed Problems
Editors: Stuart Antman, Jerry Ericksen, David Kinderlehrer, and
Ingo Muller

Volume 4: Dynamical Problems in Continuum Physics
Editors: Jerry Bona, Constantine Dafermos, Jerry Ericksen, and
David Kinderlehrer

Volume 5: Theory and Applications of Liquid Crystals
Editors: Jerry Ericksen and David Kinderlehrer

Volume 6: Amorphous Polymers and Non-Newtonian Fluids
Editors: Constantine Dafermos, Jerry Ericksen, and
David Kinderlehrer

Volume 7: Random Media
Editor: George Papanicolaou

Volume 8: Percolation Theory and Ergodic Theory of Infinite Particle
Systems
Editor: Harry Kesten

Volume 9: Hydrodynamic Behavior and Interacting Particle Systems
Editor: George Papanicolaou

Volume 10: Stochastic Differential Systems, Stochastic Control Theory,
and Applications
Editors: Wendell Fleming and Pierre-Louis Lions

Volume 11: Numerical Simulation in Oil Recovery
Editor: Mary Fanett Wheeler

Volume 12: Computational Fluid Dynamics and Reacting Gas Flows
Editors: Bjorn Engquist, M. Luskin, and Andrew Majda

Volume 13: Numerical Algorithms for Parallel Computer Architectures
Editor: Martin H. Schultz

Volume 14: Mathematical Aspects of Scientific Software
Editor: J.R. Rice

Volume 15: Mathematical Frontiers in Computational Chemical Physics
Editor: D. Truhlar

Volume 16: Mathematics in Industrial Problems
by Avner Friedman

Volume 17: Applications of Combinatorics and Graph Theory to the
Biological and Social Sciences
Editor: Fred Roberts

Volume 18: q-Series and Partitions
Editor: Dennis Stanton

Volume 19: Invariant Theory and Tableaux
Editor: Dennis Stanton

Volume 20: Coding Theory and Design Theory Part I: Coding Theory
Editor: Dijen Ray-Chaudhuri

Volume 21: Coding Theory and Design Theory Part II: Design Theory
Editor: Dijen Ray-Chaudhuri

Volume 22: Signal Processing: Part I - Signal Processing Theory
Editors: L. Auslander, F.A. Grünbaum, J.W. Helton, T. Kailath,
P. Khargonekar, and S. Mitter

Volume 23: Signal Processing: Part II - Control Theory and Applications
of Signal Processing
Editors: L. Auslander, F.A. Grünbaum, J.W. Helton, T. Kailath,
P. Khargonekar, and S. Mitter

Volume 24: Mathematics in Industrial Problems, Part 2
by Avner Friedman

Volume 25: Solitons in Physics, Mathematics, and Nonlinear Optics
Editors: Peter J. Olver and David H. Sattinger

Volume 26: Two Phase Flows and Waves
    Editors: Daniel D. Joseph and David G. Schaeffer

Volume 27: Nonlinear Evolution Equations that Change Type
    Editors: Barbara Lee Keyfitz and Michael Shearer

Volume 28: Computer Aided Proofs in Analysis
    Editors: Kenneth Meyer and Dieter Schmidt

Volume 29: Multidimensional Hyperbolic Problems and Computations
    Editors: Andrew Majda and Jim Glimm

Volume 30: Microlocal Analysis and Nonlinear Waves
    Editors: Michael Beals, R. Melrose, and J. Rauch

Volume 31: Mathematics in Industrial Problems, Part 3
    by Avner Friedman

Volume 32: Radar and Sonar, Part I
    by Richard Blahut, Willard Miller, Jr., and Calvin Wilcox

Volume 33: Directions in Robust Statistics and Diagnostics: Part I
    Editors: Werner A. Stahel and Sanford Weisberg

Volume 34: Directions in Robust Statistics and Diagnostics: Part II
    Editors: Werner A. Stahel and Sanford Weisberg

Volume 35: Dynamical Issues in Combustion Theory
    Editors: P. Fife, A. Liñán, and F.A. Williams

Volume 36: Computing and Graphics in Statistics
    Editors: Andreas Buja and Paul Tukey

Volume 37: Patterns and Dynamics in Reactive Media
    Editors: Harry Swinney, Gus Aris, and Don Aronson

Volume 38: Mathematics in Industrial Problems, Part 4
    by Avner Friedman

Volume 39: Radar and Sonar, Part II
    Editors: F. Alberto Grünbaum, Marvin Bernfeld, and
    Richard E. Blahut

Volume 40: Nonlinear Phenomena in Atmospheric and Oceanic Sciences
    Editors: George F. Carnevale and Raymond T. Pierrehumbert

Volume 41: Chaotic Processes in the Geological Sciences
        Editor: David A. Yuen

Volume 42: Partial Differential Equations with Minimal Smoothness and
        Applications
        Editors: B. Dahlberg, E. Fabes, R. Fefferman, D. Jerison, C. Kenig,
        and J. Pipher

Volume 43: On the Evolution of Phase Boundaries
        Editors: Morton E. Gurtin and Geoffrey B. McFadden

Volume 44: Twist Mappings and Their Applications
        Editors: Richard McGehee and Kenneth R. Meyer

Volume 45: New Directions in Time Series Analysis, Part I
        Editors: David Brillinger, Peter Caines, John Geweke,
        Emanuel Parzen, Murray Rosenblatt, and Murad S. Taqqu

Volume 46: New Directions in Time Series Analysis, Part II
        Editors: David Brillinger, Peter Caines, John Geweke,
        Emanuel Parzen, Murray Rosenblatt, and Murad S. Taqqu

Volume 47: Degenerate Diffusions
        Editors: Wei-Ming Ni, L.A. Peletier, and J.-L. Vazquez

Volume 48: Linear Algebra, Markov Chains, and Queueing Models
        Editors: Carl D. Meyer and Robert J. Plemmons

Volume 49: Mathematics in Industrial Problems, Part 5
        by Avner Friedman

Volume 50: Combinatorial and Graph-Theoretic Problems in Linear
        Algebra
        Editors: Richard A. Brualdi, Shmuel Friedland, and Victor Klee

Volume 51: Statistical Thermodynamics and Differential Geometry of
        Microstructured Materials
        Editors: H. Ted Davis and Johannes C.C. Nitsche

Volume 52: Shock Induced Transitions and Phase Structures in General
        Media
        Editors: J.E. Dunn, Roger Fosdick, and Marshall Slemrod

Volume 53: Variational and Free Boundary Problems
        Editors: Avner Friedman and Joel Spruck

Volume 54: Microstructure and Phase Transitions
Editors: David Kinderlehrer, Richard James, Mitchell Luskin, and Jerry L. Ericksen

Volume 55: Turbulence in Fluid Flows: A Dynamical Systems Approach
Editors: George R. Sell, Ciprian Foias, and Roger Temam

Volume 56: Graph Theory and Sparse Matrix Computation
Editors: Alan George, John R. Gilbert, and Joseph W.H. Liu

Volume 57: Mathematics in Industrial Problems, Part 6
by Avner Friedman

Volume 58: Semiconductors, Part I
Editors: W.M. Coughran, Jr., Julian Cole, Peter Lloyd, and Jacob White

Volume 59: Semiconductors, Part II
Editors: W.M. Coughran, Jr., Julian Cole, Peter Lloyd, and Jacob White

Volume 60: Recent Advances in Iterative Methods
Editors: Gene Golub, Anne Greenbaum, and Mitchell Luskin

Volume 61: Free Boundaries in Viscous Flows
Editors: Robert A. Brown and Stephen H. Davis

Volume 62: Linear Algebra for Control Theory
Editors: Paul Van Dooren and Bostwick Wyman

Volume 63: Hamiltonian Dynamical Systems: History, Theory, and Applications
Editors: H.S. Dumas, K.R. Meyer, and D.S. Schmidt

Volume 64: Systems and Control Theory for Power Systems
Editors: Joe H. Chow, Petar V. Kokotovic, and Robert J. Thomas

Volume 65: Mathematical Finance
Editors: Mark H.A. Davis, Darrell Duffie, Wendell H. Fleming, and Steven E. Shreve

Forthcoming Volumes:

Mathematics in Industrial Problems, Part 7

1991-1992: *Applied Linear Algebra*

Linear Algebra for Signal Processing

1992 Summer Program: *Environmental Studies*

1992–1993: *Control Theory*

> Control and Optimal Design of Distributed Parameter Systems
>
> Flow Control
>
> Robotics
>
> Nonsmooth Analysis & Geometric Methods in Deterministic Optimal Control
>
> Adaptive Control, Filtering and Signal Processing
>
> Discrete Event Systems, Manufacturing, Systems, and Communication Networks

1993 Summer Program: *Modeling, Mesh Generation, and Adaptive Numerical Methods for Partial Differential Equations*

1993-1994: *Emerging Applications of Probability*

> Probability and Algorithms
>
> Finite Markov Chain Renaissance
>
> Random Discrete Structures
>
> Mathematical Population Genetics
>
> Stochastic Networks
>
> Stochastic Problems for Nonlinear Partial Differential Equations
>
> Image Models (and their Speech Model Cousins)
>
> Stochastic Models in Geosystems
>
> Classical and Modern Branching Processes

FOREWORD

This IMA Volume in Mathematics and its Applications

MATHEMATICAL FINANCE

is based on the proceedings of a workshop that was an integral part of the 1992–93 IMA program on "Control Theory." We thank Mark H.A. Davis, Darrell Duffie, Wendell H. Fleming, and Steven E. Shreve for editing the proceedings. We also take this opportunity to thank the National Science Foundation and the Army Research Office, whose financial support made the workshop possible.

Avner Friedman

Willard Miller, Jr.

PREFACE

This volume is the Proceedings of the Workshop on Mathematical Finance held at the Institute for Mathematics and its Applications, June 14–18, 1993. The organizers for this workshop were Mark Davis, Darrell Duffie, Wendell Fleming, and Steven Shreve. This workshop was part of the 1992–1993 Year on Control at IMA, for which the organizing committee was Wendell Fleming, Pramod Khargonekar, P.R. Kumar, David Russell, Steven Shreve, and Héctor Sussmann (chairman). The Workshop on Mathematical Finance was preceded by a one-week tutorial on the same subject, given by Darrell Duffie and Ioannis Karatzas.

A workshop on mathematical finance can be held only because of two revolutions that have taken place on Wall Street in the latter half of the twentieth century. Both these revolutions began at universities and have led inexorably to an escalation in the level of mathematics, probability, and statistics used in financial practice.

The first revolution, which was the introduction of quantitative methods to the black art of equity fund management, began with the 1952 publication by Harry Markowitz of his Ph.D. dissertation "Portfolio Selection." This work showed how to understand and quantify the trade-offs between risk and return inherent in a portfolio. The vehicle for this so-called "mean-variance" analysis of portfolios is linear regression. Once this analysis is complete, one can pose and solve the optimization problem of choosing the portfolio with the largest mean return subject to keeping the risk below an acceptable level.

The second revolution in finance began with the 1973 publication by Fischer Black and Myron Scholes (in consultation with Robert Merton) of the solution to the option pricing problem. The Black-Scholes formula brought to the finance industry the modern methodology of martingales and stochastic calculus, methodology that enables investment banks to produce, price, and hedge an endless variety of "derivative securities."

These two revolutions in finance have created a stream of practical problems whose solutions require the expertise of research mathematicians. This workshop and the preceding tutorial addressed a number of these problems. These events brought together finance practitioners, finance faculty, and mathematics faculty. The proceedings are necessarily an incomplete record of all that transpired during these two weeks, and indeed, the proceedings are not even a complete record of all the talks presented. Nonetheless, even an imperfect record can communicate the spirit and many of the particulars of a meeting.

We would like to thank Professor Avner Friedman for promoting the idea of these meetings and Professor Willard Miller, Jr. for his assistance in bringing them to fruition. We are grateful to Kathy Boyer, Paul Ewing, Joan Felton, Ceil Mcaree, John Pliam, Kathi Polley, Pam Rech, and Mary Saunders of the IMA staff for their flawless logistical support. To Patricia V. Brick, Ruth Capp, Stephan J. Skogerboe, and Kaye Smith we express our gratitude for collecting and editing the contents of these proceedings.

Mark H.A. Davis
Darrell Duffie
Wendell H. Fleming
Steven E. Shreve

CONTENTS

Foreword xi

Preface xiii

Continuous trading with asymmetric information and
imperfect competition... 1
  Kerry Back

Contingent claim valuation and hedging with constrained portfolios... 13
  Jakša Cvitanić and Ioannis Karatzas

On portfolio optimization under "drawdown" constraints............. 35
  Jakša Cvitanić and Ioannis Karatzas

American options and transaction fees 47
  Mark H.A. Davis and Thaleia Zariphopoulou

The optimal stopping problem for a general American put-option..... 63
  Nicole El Karoui and Ioannis Karatzas

Optimal investment models and risk sensitive stochastic control 75
  Wendell H. Fleming

Arbitrage and free lunch in a general financial market model; the
fundamental theorem of asset pricing................................ 89
  Marco Frittelli and Peter Lakner

Which model for term-structure of interest rates should one use? 93
  L.C.G. Rogers

Liquidity premium for capital asset pricing with transaction costs ... 117
  S.E. Shreve

CONTINUOUS TRADING WITH ASYMMETRIC INFORMATION AND IMPERFECT COMPETITION

KERRY BACK*

Abstract. A review is given of the continuous-time equilibrium model of Kyle [11] and Back [1], in which there is a single risk-neutral informed trader, liquidity traders, and risk-neutral competitive market makers. Pricing is done by the market makers, who make inferences from the order flow about the private information. The informed trader is concerned with the effects of his trades on prices, which is a form of transaction cost. Some approaches to analyzing a risk averse informed trader are explored.

Kyle [11] introduced a model of asset pricing with asymmetric information and imperfect competition that has become the standard model for the study of market microstructure. In addition to single-period and discrete-time multiperiod models, Kyle analyzed a continuous-time version. He did so by a direct, but heuristic, analysis and also by taking the limit of discrete-time models. The continuous-time model was formalized and extended by Back [1,2,3].

Briefly, the model is as follows. Trading of a single risky asset and a risk-free asset occurs continuously on [0, 1]. Some information is to be released to the market after the close of trading at time 1. One trader has better knowledge than the rest of the market regarding the content of this information release. Other traders try to infer this information from orders. Their inference problem is complicated by the existence of non-information-based trades, which cannot be distinguished from informed orders. The motive for the non-information-based trades, called "liquidity trades," is assumed to be portfolio rebalancing, but this is not explicitly modeled. It is assumed that orders are handled by risk-neutral market makers, who compete among themselves and thereby drive each transaction price to equal the discounted conditional expectation of the post-information-release price.

Unlike prior models of asset pricing with asymmetric information (see Grossman [6]), the informed trader in Kyle's model is assumed to be aware that his trades move prices. This avoids what Hellwig [8] calls the "schizophrenia" problem of competitive rational-expectations models. This element of imperfect competition introduces complications that do not arise when analyzing the portfolio choice problem of a "small" trader, as in Merton [12]. The effect of trades on transaction prices is a form of transaction cost, but it is a different form than studied by Davis and Norman [4] and others.

In the following section, I describe a general version of the model. In §2, I summarize and expand upon the results in Back [1]. In § 3, I present

---

* Olin School of Business, Washington University in St. Louis, St. Louis, MO 63130. phone: 314-935-7321. email: back@backfin.wustl.edu. I thank Saikat Nandi for helpful comments.

some thoughts on a model with risk aversion, which is currently unsolved.

1. The model. Let $(\Omega, \mathcal{F}, P)$ denote a probability space with filtration $\mathbf{F} = (\mathcal{F}_t)$ satisfying "the usual conditions," where $\mathcal{F}_1 \subset \mathcal{F}$ with the inclusion being possibly strict. Let Z be a semimartingale on this space, and let v be a random variable that is independent of Z (and that may or may not be measurable with respect to any particular $\mathcal{F}_t$). The random variable v denotes the price at which the risky asset will trade after the information release. The process Z denotes the cumulative liquidity trades. In other words, Z_t is the total number of shares owned by liquidity traders at time t (assuming, for convenience, that they own no shares at time 0). The filtration $\mathbf{F}$ represents the information of the informed trader. One could question whether it should be assumed that the informed trader sees the liquidity trades rather than just the series of transaction prices, but this is without loss of generality if the asset price is a strictly monotone function of the orders, as will be true in equilibrium. Denote the number of shares initially owned by the informed trader by x_0. A trading strategy of the informed agent is taken to be a semimartingale X on $(\Omega, \mathbf{F}, P)$ satisfying $X_{0-} = x_0$. The interpretation of X_t is that it is the number of shares owned by the informed trader at time t, so $X_t - x_0$ denotes net purchases. A trading strategy X defines a total order process Y^X according to $Y_t^X(\omega) = X_t(\omega) - x_0 + Z_t(\omega)$. Let $\mathbf{F}(Y^X) = (\mathcal{F}_t(Y^X))$ denote the filtration generated by Y^X.

Let $\mathbf{G} = (\mathcal{G}_t)$ denote a filtration with $(\forall t)\ \mathcal{G}_t \subset \mathcal{F}_t$. This represents exogenous public information. In addition to this information, market makers observe the order flow, which is an element of $D[0, 1]$, the space of cadlag functions. A pricing rule is a function $S : [0, 1] \times \Omega \times D[0, 1] \to \Re$ such that, for each t, $S_t(\cdot)$ is measurable with respect to the product of $\mathcal{G}_t$ with the σ-field on $D[0, 1]$ generated by the projection maps $(y \mapsto y_s, 0 \leq s \leq t)$. We will need to assume enough regularity of $S(\cdot)$ and X to ensure the process $S(Y^X)$ is a semimartingale. This issue will be addressed in the context of the example in the next section.

A trading strategy X^* and pricing rule S^* constitute an equilibrium if two conditions are satisfied. The first is that transaction prices equal conditional expectations of v, given the market maker's information. I will call this the "market efficiency condition." It can be expressed as:

(1.1) $(\forall t)\ S_t^*(\omega, Y^{X^*}(\omega)) = E[v | \mathcal{G}_t \vee \mathcal{F}_t(Y^{X^*})](\omega)$ a.s.

The second equilibrium condition is that the informed trader optimizes, given his information and the pricing rule S^*. Given a semimartingale (S_t) on $(\Omega, \mathbf{F}, P)$ and a trading strategy X, the trader's final wealth is the random variable

(1.2) $$W(X, S) = (v - S_1)X_1 + \int_{[0,1]} X_{t-}\, dS_t.$$

Here, we are normalizing the initial wealth to zero. The first term in the above represents the capital gain following the information release. The second term is the usual sum over time of the instantaneous capital gains, as in Harrison and Pliska [7]. The stochastic integral exists because the process (X_{t-}), being left-continuous, is locally bounded. The process (X_{t-}) is a trading strategy in the terminology of Harrison and Pliska, who allow more general strategies than this.

One reason for assuming Z and X are semimartingales is that this model is a continuous-time version of a "market order" model, in which traders submit orders to the market and then market makers set the price at which they are cleared. The orders are the changes in the positions of the liquidity and informed traders, which here would be represented by "dZ" and "dX." In order to make any sense at all of this, we need Z and X to be semimartingales. The semimartingale assumption allows us to integrate by parts in (1.2) to obtain

$$(1.3) \qquad W(X, S) = \int_{[0,1]} (v - S_{t-}) \, dX_t - [S, X]_1,$$

which is useful in the analysis of the informed trader's optimization problem.

Assume the informed trader has a von Neumann-Morgenstern utility function U. Then the optimality condition, which is the second part of the equilibrium condition, is

$$(1.4) \qquad X^* \in \operatorname{argmax}_X E[U(W(X, S^*(Y^X)))].$$

This condition reflects the assumption that the informed trader is not a price-taking agent. There are two ways in which this could conceivably be understood, which need to be carefully distinguished. One way is that the informed trader could expect the pricing rule to depend on his trading strategy. It is true that the market efficiency condition (1.1) establishes a relationship between the pricing rule and X^*. However, this does not mean that the informed trader is able to directly affect the pricing rule. If he tried to do so by changing his trading strategy, he would have to make the revised strategy known to market makers, and there is no credible way of communicating such a revision. What is meant, instead, is that the informed trader takes the *pricing rule* S^* as given, understanding that this implies *prices* are not given, but rather depend on orders Y^X.

An additional point to note is that, given an equilibrium (S^*, X^*), if the informed trader deviates from his equilibrium strategy and uses some other strategy X, then prices will be given by

$$S_t^*(Y^X) \neq E[v|\mathcal{F}_t(Y^X)].$$

This means that market makers, being unaware of the deviation, will incorrectly compute the conditional expectation of v. The equilibrium concept requires that it not be profitable for the informed trader to create such pricing errors—the deviation X cannot be preferred to X^*.

2. An example. This model has been solved in some special cases. The example considered by Back [1], which is a generalization of that studied by Kyle [11], is as follows. The informed trader is risk neutral; i.e., $U(w) = w$. The risk-free rate is normalized to zero. The liquidity trade process Z is a Brownian motion with zero drift, variance σ_z^2, and $Z_0 = 0$. It is straightforward to generalize to the case of deterministically varying parameters ($dZ_t = \mu(t)\,dt + \sigma_z(t)\,dB_t$ for a Wiener process B). The σ-field $\mathcal{F}_0$ is independent of Z. The filtration $\mathbf{F}$ is the minimal P-complete, right-continuous filtration containing $\mathcal{F}_0$ with respect to which Z is adapted. There is no exogenous public information: $(\forall t)\ \mathcal{G}_t = \{\emptyset, \Omega\}$. Let $\hat{v} = E[v|\mathcal{F}_0]$. Assume the random variable $\hat{v}$ has a continuous distribution function F that is strictly increasing on $\{a|0 < F(a) < 1\}$. The expectation of (1.3) will be unchanged if v is replaced by $\hat{v}$, and the expectation in (1.4) also, so there is no reason to distinguish beween v and $\hat{v}$. Hence, the "hat" on v will be dropped in the remainder of this section.

Because there is no exogenous public information, S must be a function of (t, Y) alone. In fact, in this setting, there is an equilibrium in which S_t^* depends only on Y_t, so we can take S^* to be a function on $[0, 1] \times \Re$. Let $\pi(y, \cdot, \tau)$ denote the normal density function with mean y and variance $\sigma_z^2 \tau$. The actual equilibrium pricing rule is

$$(2.1) \qquad S^*(t, y) \stackrel{\text{def}}{=} \int s(u)\pi(y, u, 1-t)\,du,$$

where $s = F^{-1} \circ N$, and N denotes the normal $(0, \sigma_z^2)$ distribution function. Because S^* is smooth, $S^*(Y^X)$ is a semimartingale for each semimartingale X. We will give two examples and then address the informed trader's optimization problem, taking (2.1) as given. It will then be shown that (2.1) is an equilibrium pricing rule.

Example 2.1. If v is normally distributed with mean μ and variance σ_v^2, then $s(y) = \mu + \lambda y$, where $\lambda = \sigma_v/\sigma_z$. This implies

$$(2.2) \qquad S^*(t, y) = \mu + \lambda y.$$

Example 2.2. If $\log v$ is normally distributed with mean μ and variance σ_v^2, then $s(y) = \exp(\mu + \lambda y)$, where as before $\lambda = \sigma_v/\sigma_z$. This implies

$$(2.3) \qquad S^*(t, y) = \exp(\mu + \lambda y + \lambda^2 \sigma_z^2 (1-t)/2).$$

The informed trader's optimization problem can be analyzed by dynamic programming. Because the price at each time t is an invertible function of cumulative orders, we could use either the price or cumulative orders as the state variable. We will use cumulative orders in most of the following. We need a constraint on X analogous to the "no doubling strategies" condition needed in competitive models; otherwise, infinite wealth can be obtained a.s. This condition is that

$$E\left[\int_0^1 S^*(t, Y_t^X)^2\, dt\right] < \infty.$$

It has a straightforward economic interpretation, which is discussed in Back [1].

Back [1] shows that jumps in X are strictly suboptimal, so to simplify the notation, we will assume here that the informed trader is constrained to use a continuous strategy. Given this simplification, Itô's formula applied to $S(t, Y_t)$ and (1.3) yield

$$
\begin{aligned}
W(X, S^*(Y^X)) &= \int_0^1 \Big\{ (v - S^*(t, Y_t^X))\, dX_t - \tfrac{\partial}{\partial y} S^*(t, Y_t^X)\, d[X, X]_t \\
&\quad - \tfrac{\partial}{\partial y} S^*(t, Y_t^X)\, d[X, Z]_t \Big\}.
\end{aligned}
$$

The term involving $d[X, X]$ would not appear in a competitive model. This term "punishes" infinite-variation strategies, and in fact, all optimal strategies have continuous finite-variation paths.

If we assume that $dX_t = \alpha_t\, dt$ for some adapted process α, then (1.3) simplifies further to

$$W(X, S^*(Y^X)) = \int_0^1 (v - S^*(t, Y_t^X))\alpha_t\, dt.$$

We can view $(v - S^*(t, y))\alpha_t$ as the instantaneous profit at time t and define the remaining value as

$$J(v, t, y) = \sup_\alpha E\left[\int_t^1 (v - S^*(u, Y_u^X))\alpha_u\, du \,|\, v, Y_t = y\right].$$

It is shown in Back [1] that

$$(2.4) \qquad J(v, y, t) = \int j(v, u)\pi(y, u, 1 - t)\, du,$$

where

$$(2.5) \qquad j(v, y) = \int_y^{s^{-1}(v)} (v - s(x))\, dx.$$

Moreover, this is the value when general semimartingale strategies are allowed.

Example 2.1. If v is normal (μ, σ_v^2), then

$$J(v, y, t) = \frac{(\mu + \lambda y - v)^2}{2\lambda} + \frac{\lambda \sigma_z^2 (1 - t)}{2}.$$

Example 2.2. In the lognormal case, the value function is simplified somewhat by making the change of variables suggested by (2.3): $p = \exp(\mu + \lambda y + \sigma_z^2(1 - t)/2)$. The value at time t is

$$\frac{v \log v - v \log p + p - v}{\lambda} + \frac{\lambda^2 \sigma_z^2 v(1 - t)}{2}.$$

The function $j(v, y)$ gives the area in price-quantity space circumscribed by the "supply curve" $p = S^*(1, q)$, the line $p = v$, and the lines $q = y$ and $q = s^{-1}(v)$. This is the profit that a perfect monopsonist would obtain if faced by this supply curve. The value function J is the expected value of this profit, assuming the only trades that arrive between times t and 1 are liquidity trades. Trading against the time-1 supply curve in this way is not actually possible; nevertheless, this construction gives the actual value function.

In each of the two examples, and in general, the value is the sum of two terms. The first term is the profit a perfect monopsonist would obtain by trading against the time-t supply curve, pushing the price to v, and the second is the expected profit attributable to liquidity-trade-induced mispricing between times t and 1.

The verification theorem is proven in Back [1]. This shows that any strategy with continuous finite-variation paths that implies $S^*(1, Y_1^X) = v$ a.s. is optimal. The multiplicity of optima can be seen by examining the Bellman equation. Both the state dynamics $dY_t = \alpha \, dt + dZ_t$ and the instantaneous profit $(v - S(t, y))\alpha$ are linear in the control α. Given the pricing rule (2.1), the coefficient on the control in the Bellman equation is zero, so any control is locally optimal. Global optimality requires the boundary condition $S^*(1, Y_1^X) = v$ to be satisfied. Jumps and local martingale components are punished by the quadratic term $d[X, S^*(Y^X)]$ and hence are strictly suboptimal. More details regarding the Bellman equation are given in the next section.

A particular optimal strategy is

$$(2.6) \qquad X_t^* \overset{\text{def}}{=} ts^{-1}(v) + (1 - t) \int_0^t \frac{-Z_s}{(1 - s)^2} \, ds.$$

The second term in (2.6) when added to Z_t gives a Brownian bridge beginning at zero and ending at zero. Thus, the informed trades offset the liquidity trades. The first term represents trading on information.

The pricing rule S^* and trading strategy X^* constitute an equilibrium. What has not yet been discussed is the market efficiency condition (1.1). The price (2.1) can be understood as the conditional expectation of $s(Y_1^{X^*})$ at time t, assuming $(Y_u^{X^*})_{t \le u \le 1}$ is a Brownian motion with the same drift and variance as $(Z_u)_{t \le u \le 1}$. Taking the conditional expectation of $s(Y_1^{X^*})$ is the correct thing to do, because $s(Y_1^{X^*}) = v$ a.s. Treating $(Y_u^{X^*})_{t \le u \le 1}$ as being distributed like $(Z_u)_{t \le u \le 1}$ is also the correct thing to do, because Y^{X^*} is a Brownian bridge on $\mathbf{F}$ with instantaneous variance σ_z^2, and the $\mathcal{F}_0(Y^{X^*})$-conditional distribution of the terminal value $s^{-1}(v)$ is normal with mean zero and variance σ_z^2; hence $(Y^{X^*}, \mathbf{F}(Y^{X^*}))$ is a Brownian motion.

Another way to establish that $(Y^{X^*}, \mathbf{F}(Y^{X^*}))$ is a Brownian motion, which may have more general applicability than the above argument, is to use the Kalman-Bucy filter. If X has continuous finite-variation paths, and $(Y^X, \mathbf{F}(Y^X))$ is a martingale, then by a well-known theorem of P. Levy (Dellacherie and Meyer [5, VIII.59]), $(Y^X, \mathbf{F}(Y^X))$ is a Brownian motion with variance σ_z^2. This reasoning can be used to compute X^*. For an arbitrary X, take the observation process to be (Y_t^X) and the signal (state) process to be (ξ_t), where $\xi_t \overset{\text{def}}{=} Y_t^X - Y_1^X$. Suppose now that $dX_t = \theta(t)\xi_t$ for some deterministic function θ. Set

$$\hat{\xi}_t = E[\xi_t | \mathcal{F}_t(Y^X)].$$

The process $(Y^X, \mathbf{F}(Y^X))$ is a martingale iff $\hat{\xi} \equiv 0$. It is immediate from the definition of $s(\cdot)$ that $\hat{\xi}_0 = 0$, so we need $d\hat{\xi}_t \equiv 0$. The Kalman-Bucy filtering equations (Kallianpur, [10, 10.3.11a and 10.3.15]) show that this will be the case iff $\theta(t) = -1/(1 - t)$, which is the same as (2.6).

An interesting aspect of this model is the "constancy" of price pressure over time. In the normal-distribution example, the price dynamics are $dS^*(t, Y_t^{X^*}) = \lambda \, dY_t^{X^*}$, so an order of one unit always moves the price by λ, regardless of when it is submitted. In the lognormal example, $dS^*(t, Y_t^{X^*})/S^*(t, Y_t^{X^*}) = \lambda \, dY_t^{X^*}$, so the percent change is in constant proportion to orders. More generally, $dS^*(t, Y_t^{X^*}) = S_y^*(t, Y_t^{X^*}) \, dY_t^{X^*}$ and the price-pressure process $S_y^*(Y^{X^*})$ is a martingale relative to the filtration $\mathbf{F}(Y^{X^*})$. Notice that in the lognormal example, the equilibrium price process is a geometric Brownian motion relative to the filtration it generates, as is commonly assumed in financial analysis.

3. Risk aversion. An unrealistic aspect of the example in the preceding section is that the informed trader is assumed to be risk neutral. It is reasonable to assume market makers are risk neutral because the risk about which there is private information should be idiosyncratic and market making firms hold diversified portfolios. However, the informed trader will have an undiversified position. A related assumption, which is also unrealistic, is that the wealth of the informed trader is allowed to go negative.

In this section, I will retain the assumption that wealth is uncon-strained but add risk aversion. Specifically, assume the utility function is of the negative exponential variety: $U(w) = -e^{-w}$. Keep the informa-tional assumptions of the preceding section and suppose in addition that v is $\mathcal{F}_0$-measurable. In the following subsections, I address the possibility of an equilibrium with, respectively, Y_t as the state variable and price as the state variable. Neither exploration is successful.

3.1. Cumulative orders as the state variable. Fix a pricing rule of the form $S(t, Y_t)$. I will consider only continuous finite-variation trading strategies. For any such strategy X, define the wealth process

$$W_t^X = \int_0^t (v - S(u, Y_u^X))\, dX_u.$$

The value function

$$J(v, t, w, y) \overset{\text{def}}{=} \sup_X E[-\exp(-W_1^X)|v, Y_t^X = y, W_t^X = w]$$

is easily seen to satisfy

$$J(v, t, w, y) = e^{-w} J(v, t, 0, y).$$

Set $\hat{J}(v, t, y) = J(v, t, 0, y)$. Then

$$dJ(v, t, W_t^X, Y_t^X) = \exp(-W_t^X)(\hat{J}_t\, dt + \hat{J}_y\, dY_t^X - \hat{J}dW_t^X + (1/2)\sigma_z^2\hat{J}_{yy}\, dt),$$

where the subscripts on $\hat{J}$ denote partial derivatives. Considering strategies of the form $dX_t = \alpha_t\, dt$, the Bellman equation is

$$\sup_\alpha\ (\hat{J}_t + \alpha\hat{J}_y - (v - S)\alpha\hat{J} + (1/2)\sigma_z^2\hat{J}_{yy}) = 0.$$

Because α is unconstrained, the above supremum can be finite only if the coefficient on α is zero. Thus, we must have

(3.1) $\hat{J}_y = (v - S)\hat{J},$

and

(3.2) $\hat{J}_t + (1/2)\sigma_z^2\hat{J}_{yy} = 0.$

What seems to be the natural boundary condition is

(3.3) $J(v, 1, w, y) \geq -e^{-w}$, with equality iff $S(1, y) = v.$

If these could be solved, then the verification theorem could be proven as in Back [1], yielding the result that any continuous finite-variation strategy satisfying $S(1, Y_1^X) = v$ is optimal.

Equation (3.2) suggests that $\hat{J}$ may be of the form

$$\hat{J}(v,t,y) = \int j(v,u)\pi(y,u,1-t)\,du,$$

for some function j, where π is the transition density of Z, as in the preceding section. A natural candidate to try for j is

$$j(v,y) = -\exp\left(-\int_y^{s^{-1}(v)} (v-s(x))\,dx\right),$$

where $s(y) = S(1,y)$. Making this choice, and assuming we can interchange expectation and differentiation, we have

$$\begin{aligned}
\hat{J}_y(v,t,y) &= \int j_u(v,u)\pi(y,u,1-t)\,du \\
&= \int j(v,u)(v-s(u))\pi(y,u,1-t)\,du \\
&= \hat{J}(v,t,y)v - \int j(v,u)s(u)\pi(y,u,1-t)\,du.
\end{aligned}$$

Hence, (3.1) will be satisfied if

$$S(t,y) = \frac{\int j(v,u)s(u)\pi(y,u,1-t)\,du}{\int j(v,u)\pi(y,u,1-t)\,du}.$$

We can write this in a form similar to (2.1), namely

(3.4) $$S(t,y) = \int s(u)\rho(v,y,u,1-t)\,du$$

where

$$\rho(v,y,u,\tau) = \frac{j(v,u)\pi(y,u,\tau)}{\int j(v,u)\pi(y,u,\tau)\,du}.$$

This will satisfy the market efficiency condition if $\rho(v,\cdot)$ is the transition density of $(Y^X, \mathbf{F}(Y^X))$.

The problem is the apparent dependence of ρ on v, whereas the left-hand side of (3.4) does not depend on v. The function ρ will not depend on v if $j(v,y)$ is multiplicatively separable in v and y, equivalently, if

$$h(v,y) \stackrel{\text{def}}{=} \int_y^{s^{-1}(v)} (v-s(x))\,dx$$

is additively separable in v and y. However, for any function $s(\cdot)$, the cross partial of h is -1, so it is not additively separable.

3.2. Price as the state variable. Constancy of price pressure is not compatible with an informed trader who knows the post-information-release price and is risk averse. The only risk such a trader faces is that liquidity trades will move the asset price closer to its true value, thereby reducing his profit opportunities. If price pressure were constant over time, such a trader would move the price to the true value in an arbitrarily small amount of time (actually, there would be no optimal strategy). In conjunction with the market efficiency condition, this suggests that there should be no price pressure after the first instant of trading, which contradicts the hypothesis of constant price pressure. This is similar to what happens when two identically informed risk-neutral traders compete (Holden and Subrahmanyam [9]).

A reasonable alternative to explore is the possibility of deterministically varying (increasing) price pressure (though it should be noted that price pressure is deterministic in the risk-neutral case only when the asset value is normally distributed). Fix a pricing rule of the form

$$S(t, Y^X) = S(0, Y^X) + \int_0^t \lambda(u)\, dY_u^X$$

for a deterministic function λ. The informed trader's optimization problem is a Markovian control problem with state variables $S(t, Y^X)$ and

$$W_t^X = \int_0^t (v - S(u, Y^X))\, dX_u.$$

Define the value function

$$J(v, t, w, p) = \sup_X E[-\exp(-W_1^X)|v, S(t, Y^X) = p, W_t^X = w].$$

As before, this satisfies

$$J(v, t, w, p) = e^{-w} J(v, t, 0, p).$$

Set $\hat{J}(v, t, p) = J(v, t, 0, p)$. Considering strategies of the form $dX_t = \alpha_t\, dt$, the Bellman equation is

$$\sup_\alpha (\hat{J}_t + \lambda \alpha \hat{J}_p - (v - p)\alpha \hat{J} + (1/2)\lambda^2 \sigma_z^2 \hat{J}_{pp}) = 0.$$

Reasoning as before, we must have

(3.5) $$\lambda \hat{J}_p = (v - p)\hat{J},$$

and

(3.6) $$\hat{J}_t + (1/2)\lambda^2 \sigma_z^2 \hat{J}_{pp} = 0.$$

The natural boundary condition is

(3.7) $$J(v, 1, w, p) \geq -e^{-w}, \text{ with equality iff } p = v.$$

Equation (3.6) suggests that $\hat{J}$ may be of the form

$$(3.8) \qquad \hat{J}(v, t, p) = \int j(v, q)\phi(p, q, 1 - t)\, dq,$$

for some function j, where $\phi(y, \cdot, 1-t)$ is the normal $(0, \int_t^1 \lambda^2(u)\, du)$ density function. As before, we can compute the wealth attainable by trading against the time-1 pricing rule and try

$$j(v, y) = -\exp\left(-\frac{(v - p)^2}{2\lambda(1)}\right).$$

Making this substitution, (3.8) can be computed by evaluating the moment generating function for a chi-square random variable. After some manipulation, (3.5) reduces to

$$\lambda(t) = (1 + \lambda(1)) \int_t^1 \lambda^2(u)\, du.$$

However, the only solution of this is $\lambda \equiv 0$, which is clearly inconsistent with equilibrium.

This suggests, but obviously does not prove, that there is no equilibrium with deterministically varying price pressure. It is possible that the price pressure depends on the contemporaneous price (as in the risk-neutral case when the asset value is not normally distributed), or, what seems more likely to me, it may depend on the entire history of the process.

4. Conclusion. The analysis in the preceding section was obviously preliminary, but it may provide some insight into the problem. To solve a model of this sort requires judicious guessing about the form of the equilibrium price function. The guesses in the preceding section were apparently insufficiently inspired.

Apart from the equilibrium problem, it seems interesting to study the optimization problem of a possibly uninformed trader who faces price pressure. This type of transaction cost is different from the proportional transaction cost that has been studied before. Proportional transaction costs penalize reversals (buying and then selling an asset, or the opposite), so the optimum involves a no-transaction region (Davis and Norman [4]). On the other hand, the type of price pressure studied in this paper penalizes large trades. This reflects the real phenomenon that better execution is obtained if an order is split into small pieces before being sent to the floor of the exchange. Reversals actually are costless: the quadratic variation term is zero for any continuous finite-variation strategy. Both types of transaction costs render local martingale components suboptimal, but they do so for different reasons. In particular, there is no reason to have a no-transaction region in the price-pressure problem.

One issue that would need to be addressed when modeling price pressure is whether the price pressure is temporary or permanent. Efficient markets theory suggests it should be modeled as temporary. This is indeed the nature of the price pressure imposed by the liquidity traders in the model discussed here—the Brownian bridge component of the informed trader's strategy eventually undoes the liquidity trades. This device could be included in a study of the optimal trading of an uninformed agent facing price pressure.

REFERENCES

[1] K.BACK, *Insider trading in continuous time*, Review of Financial Studies, **5** (1992), 387–409.

[2] K.BACK, *Asymmetric information and options*, Review of Financial Studies, **6** (1993), 435–472.

[3] K.BACK, *Time-varying liquidity trading, price pressure, and volatility*, Washington University in St. Louis (*working paper*) 1993.

[4] M.H.A.DAVIS, A.R.NORMAN, *Portfolio selection with transaction costs*, Mathematics of Operations Research **15** (1990), 676–713.

[5] C.DELLACHERIE, P.-A.MEYER, *Probabilities and potential B*, North-Holland, Amsterdam 1982.

[6] S.J.GROSSMAN, *An introduction to the theory of rational expectations under asymmetric information*, Review of Economic Studies **48** (1981), 541–559.

[7] J.M.HARRISON, S.R.PLISKA, *Martingales and stochastic integrals in the theory of continuous trading*, Stochastic Processes and Their Applications **11** (1981), 215–260.

[8] M.F.HELLWIG, *On the aggregation of information in competitive markets*, Journal of Economic Theory **22** (1980), 477–498.

[9] C.W.HOLDEN, A.SUBRAHMANYAM, *Long-lived private information and imperfect competition*, Journal of Finance **47** (1992), 247–270.

[10] G.KALLIANPUR, Stochastic Filtering Theory, Springer-Verlag, New York 1980.

[11] A.S.KYLE, *Continuous auctions and insider trading*, Econometrica **53** (1985), 1315–1335.

[12] R.C.MERTON, *Lifetime portfolio selection under uncertainty: the continuous-time case*, Review of Economics and Statistics **51** (1969), 247–57.

CONTINGENT CLAIM VALUATION AND HEDGING WITH CONSTRAINED PORTFOLIOS*

JAKŠA CVITANIĆ[†] AND IOANNIS KARATZAS[‡]

Abstract. We use a stochastic control approach to study the question of hedging contingent claims by portfolios constrained to take values in a given closed, convex subset of $\mathcal{R}^d$. In the framework of our work on utility maximization with constrained portfolios, we extend results of El Karoui & Quenez [12] on incomplete markets and treat the case of different interest rates for borrowing and lending. The latter turns out to be an example of an "explicitly" solvable Backwards SDE in the sense of Pardoux and Peng [26]. This paper is an abbreviated and slightly modified version of Cvitanić & Karatzas [8], which should be consulted for detailed proofs.

1. Introduction and summary. The celebrated papers of Black & Scholes [5] and Merton [23] paved the way for pricing options on stocks, based on the following principle: In a *complete market* (such as the one in section 2 of this article) every contingent claim can be attained exactly by investing in the market and starting with a large enough initial capital; thus, the "fair price" of the claim is taken to be the minimal such capital. This turns out to be equal to the expectation of the discounted value of the claim, under an equivalent, so called "risk-neutral", probability measure ([16], [15], [6]). The argument that leads to this result, and to the associated "valuation formulae", is by now standard (e.g. [20], [19]); it is based on the martingale representation and Girsanov theorems from stochastic analysis.

The above argument fails, however, in an *incomplete market*, a prototypical example of which is a market in which claims can depend on stocks that are not available for investment. The option pricing problem under incompleteness of this type has been studied, among others, by Föllmer & Schweizer [14] who adopt a risk-minimization approach, and by Ansel & Stricker [1], Jacka [17] and, most notably, El Karoui & Quenez [12]. Using a stochastic control approach, similar to that of the latter paper, we attack here a more general problem: *the hedging of contingent claims with portfolios constrained to take values in a given closed, convex set K.* The model employed is a by now standard generalization of that in Merton [23], [24]. The methodologies of Cvitanić & Karatzas [7], in which we studied the constrained portfolio optimization problem, are essential in obtaining the main results. These can be summarized as follows: under appropriate conditions, it is possible to replicate contingent claims even

* Research supported in part by the National Science Foundation under Grant NSF-DMS-90-22188. We are indebted to Prof. Mark Brown for bringing to our attention the well-known fact 7.4 for the European call option. We are also grateful to M.C. Quenez for sending us the preprint [11], which inspired Proposition 6.10 and elicited Remark 9.
† IMA, University of Minnesota, Minneapolis, MN 55455.
email: cvitanic@ima.umn.edu. On leave from Columbia University.
‡ Department of Statistics, Columbia University, New York, NY 10027.
email: ik@stat.columbia.edu

with constrained portfolios, albeit some additional consumption may be necessary; the minimal initial capital which makes this replication possible is equal to the supremum of the expected discounted values of the claim under new probability measures in a suitably large family (Theorem 6.4); replication without extra consumption is possible only if the Black-Scholes replicating portfolio happens to take values in K (Theorems 6.6, 6.7); and the associated wealth process is the minimal adapted solution of a backward stochastic differential equation with convex constraints. The main mathematical tool—namely, the martingale approach to stochastic control—is adapted from Davis & Varaiya [9] and Elliott [10].

The paper is organized as follows: the ingredients of the model are laid out in sections 2–5. In section 6 we state and prove the main results. Section 7 deals with some special cases, in which more explicit results can be obtained. We discuss some possible extensions and ramifications in section 8. Finally, we show in section 9 how to apply this same approach in the (unconstrained) case of a market with different interest rates for borrowing and lending. It turns out that in such a market a large class of contingent claims, including European call options, are attainable. The results of this section extend those in the recent preprint by Korn [21]; compare also with Bergman [4] and Jouini & Kallal [18]. These results also provide a concrete example, with "explicit" solution, of an *adapted* solution to a *nonlinear, backwards stochastic differential equation* in the spirit of Pardoux & Peng [26]; cf. Remark 9 and Example 9.

The detailed proofs of the results outlined here can be found in Cvitanić & Karatzas [8], of which the present paper is an abbreviated and slightly modified version.

2. The model. We consider a financial market $\mathcal{M}$ which consists of one *bond* and several (d) *stocks*. The prices $P_0(t), \{P_i(t)\}_{1 \le i \le d}$ of these financial instruments evolve according to the equations

$$(2.1) \qquad dP_0(t) = P_0(t)r(t)dt , \qquad P_0(0) = 1$$

$$(2.2) \qquad dP_i(t) = P_i(t)[b_i(t)dt + \textstyle\sum_{j=1}^{d} \sigma_{ij}(t)dW^{(j)}(t)] ,$$
$$P_i(0) = p_i \in (0, \infty) ; \quad i = 1, \ldots, d .$$

Here $W = (W^{(1)}, \ldots, W^{(d)})^*$ is a standard Brownian motion in $\mathcal{R}^d$, defined on a complete probability space $(\Omega, \mathcal{F}, \mathbf{P})$, and we shall denote by $\{\mathcal{F}_t\}$ the $\mathbf{P}$-augmentation of the filtration $\mathcal{F}_t^W = \sigma(W(s); \ 0 \le s \le t)$ generated by W. The *coëfficients* of $\mathcal{M}$ - i.e., the processes $r(t)$ (scalar interest rate), $b(t) = (b_1(t), \ldots, b_d(t))^*$ (vector of appreciation rates) and $\sigma(t) = \{\sigma_{ij}(t)\}_{1 \le i,j \le d}$ (volatility matrix)—are assumed to be progressively measurable with respect to $\{\mathcal{F}_t\}$ and *bounded* uniformly in $(t, \omega) \in [0, T] \times \Omega$. We shall also impose the following strong non-degeneracy condition on the matrix $a(t) \overset{\triangle}{=} \sigma(t)\sigma^*(t)$:

$$(2.3) \qquad \xi^* a(t)\xi \geq \varepsilon||\xi||^2, \quad \forall \ (t,\xi) \ \in \ [0,T] \times \mathcal{R}^d$$

almost surely, for a given real constant $\varepsilon > 0$. All processes encountered throughout the paper will be defined on the fixed, finite horizon $[0, T]$.

We introduce also the bounded "relative risk" process

$$(2.4) \qquad \theta(t) \overset{\triangle}{=} \sigma^{-1}(t)[b(t) - r(t)\mathbf{1}]$$

where $\mathbf{1} = (1, \ldots, 1)^*$, the exponential martingale

$$(2.5) \qquad Z_0(t) \overset{\triangle}{=} \exp[-\int_0^t \theta^*(s)dW(s) - \frac{1}{2}\int_0^t ||\theta(s)||^2 ds] \ ,$$

and the discount process

$$(2.6) \qquad \gamma_0(t) \overset{\triangle}{=} \exp\{-\int_0^t r(s)ds\} \ .$$

3. Portfolio, consumption and wealth processes. Consider now an economic agent whose actions cannot affect market prices, and who can decide, at any time $t \in [0, T]$, what proportion $\pi_i(t)$ of his wealth $X(t)$ to invest in the ith stock $(1 \leq i \leq d)$, and what amount of money $c(t + h) - c(t) \geq 0$ to withdraw for consumption during the interval $(t, t + h]$, $h > 0$. These decisions can only be based on the current information $\mathcal{F}_t$, without anticipation of the future. With $\pi(t) = (\pi_1(t), \ldots, \pi_d(t))^*$ chosen, the amount $X(t)[1 - \sum_{i=1}^d \pi_i(t)]$ is invested in the bond. Thus, in accordance with the model set forth in 2.1, 2.2, the wealth process $X(t)$ satisfies the linear stochastic equation

$$(3.1) \qquad \begin{aligned} dX(t) &= \sum_{i=1}^d \pi_i(t)X(t)\{b_i(t)dt + \sum_{j=1}^d \sigma_{ij}(t)dW^{(j)}(t)\} \\ &\quad + \{1 - \sum_{i=1}^d \pi_i(t)\}X(t)r(t)dt - dc(t) \\ &= r(t)X(t)dt - dc(t) + X(t)\pi^*(t)\sigma(t)dW_0(t) \ ; \quad X(0) = x > 0 \ , \end{aligned}$$

where $c(t)$ is the cumulative consumption up to time t, and

$$(3.2) \qquad W_0(t) \overset{\triangle}{=} W(t) + \int_0^t \theta(s)ds \ , \quad 0 \leq t \leq T \ .$$

We formalize the above discussion as follows.

DEFINITION 3.1.

(i) An $\mathcal{R}^d$-valued, $\{\mathcal{F}_t\}$-progressively measurable process $\pi = \{\pi(t), 0 \leq t \leq T\}$ with $\int_0^T ||\pi(t)||^2 dt < \infty$, a.s., will be called a portfolio process.

(ii) A nonnegative, nondecreasing, $\{\mathcal{F}_t\}$-progressively measurable process $c = \{c(t), 0 \leq t \leq T\}$ with RCLL paths, $c(0) = 0$ and $c(T) < \infty$, a.s., will be called a cumulative consumption process.

(iii) Given a pair (π, c) as above, the solution $X \equiv X^{x,\pi,c}$ of the equation 3.1 will be called the wealth process corresponding to the portfolio/consumption pair (π, c) and initial capital $x \in (0, \infty)$.

DEFINITION 3.2. A portfolio/consumption process pair (π, c) is called admissible for the initial capital $x \in (0, \infty)$, if

$$(3.3) \qquad X^{x,\pi,c}(t) \geq 0, \qquad \forall\, 0 \leq t \leq T$$

holds almost surely. The set of admissible pairs (π, c) will be denoted by $\mathcal{A}_0(x)$.   ◇

In the notation of 2.5, 2.6, the equation 3.1 leads to

$$(3.4) \qquad \begin{aligned} M_0(t) &\overset{\Delta}{=} \gamma_0(t)X(t) + \int_0^t \gamma_0(s)dc(s) \\ &= x + \int_0^t \gamma_0(s)X(s)\pi^*(s)\sigma(s)dW_0(s). \end{aligned}$$

In particular, the process $M_0(\cdot)$ of 3.4 is seen to be a continuous local martingale under the so-called "risk-neutral equivalent martingale measure"

$$(3.5) \qquad \mathbf{P}^0(A) \overset{\Delta}{=} E[Z_0(T)1_A], \quad A \in \mathcal{F}_T\ .$$

If $(\pi, c) \in \mathcal{A}_0(x)$, the $\mathbf{P}^0$-local martingale $M_0(\cdot)$ of 3.4 is also nonnegative, thus a supermartingale, and this gives

$$(3.6) \quad E^0[\gamma_0(T)X^{x,\pi,c}(T) + \int_0^T \gamma_0(t)dc(t)] \leq x, \quad \forall\, (\pi, c) \in \mathcal{A}_0(x)\ .$$

Here, E^0 denotes the expectation operator with respect to $\mathbf{P}^0$; under this measure, the process W_0 of 3.2 is standard Brownian motion by the Girsanov theorem (e.g. [20], §3.5) and the discounted stock prices $\gamma_0(\cdot)P_i(\cdot)$ are martingales, since

$$(3.7) \qquad \begin{aligned} dP_i(t) &= P_i(t)[r(t)dt + \textstyle\sum_{j=1}^d \sigma_{ij}(t)dW_0^{(j)}(t)]\ , \\ P_i(0) &= p_i;\ i = 1, \ldots, d\ . \end{aligned}$$

4. Hedging without constraints. Let us suppose now that an agent promises to pay a random amount $B(\omega) \geq 0$ at time $t = T$. *What is the value of this promise at time $t = 0$?* In other words, how much should the agent charge for selling a contractual obligation that entitles its holder to a payment of size $B(\omega)$ at $t = T$? For instance, suppose that this obligation stipulates selling one share of the first stock at a contractually specified price q. If at time $t = T$ the price $P_1(T, \omega)$ of the stock is below q, the contract is worthless to its holder; if not, the holder can purchase the stock at the price q per share and then sell it at price $P_1(T, \omega)$, thus making a profit of $P_1(T, \omega) - q$. In other words, this contract entitles its holder to a payment of $B(\omega) = (P_1(T, \omega) - q)^+$ at time $t = T$; it is called a (European) *call option* with "exercise price" q and "maturity date" T.

To answer the question of the first paragraph, one argues as follows. Suppose the agent sets aside an amount $x > 0$ at time $t = 0$; he invests in the market $\mathcal{M}$ according to some portfolio $\pi(\cdot)$ and withdraws (possibly) funds according to a cumulative consumption process $c(\cdot)$, but wants to be certain that at time $t = T$ he will be able to *cover his obligation*, i.e., that $X^{x,\pi,c}(T) \geq B$ will hold almost surely. What is the smallest value of $x > 0$ for which such "hedging" is possible? This smallest value will then be the "price" of the contract at time $t = 0$.

DEFINITION 4.1. A Contingent Claim is a nonnegative, $\mathcal{F}_T$-measurable random variable B that satisfies

$$(4.1) \qquad 0 < E^0[\gamma_0(T)B] < \infty \ .$$

The hedging price of this contingent claim is defined by

$$(4.2) \quad u^{(0)} \triangleq \inf\{x > 0; \ \exists(\pi, c) \in \mathcal{A}_0(x) \ \text{s.t.} \ X^{x,\pi,c}(T) \geq B \ \text{a.s.}\} \ .$$

The following "classical" result identifies $u^{(0)}$ as the expectation, under the risk-neutral probability measure in 3.5, of the claim's discounted value.

PROPOSITION 4.2. *The infimum in 4.2 is attained, and we have*

$$(4.3) \qquad u^{(0)} = E^0[\gamma_0(T)B] \ .$$

Furthermore, there exists a portfolio $\pi_0(\cdot)$ such that $X_0(\cdot) \equiv X^{u^{(0)},\pi_0,0}(\cdot)$ is given by

$$(4.4) \qquad X_0(t) = \frac{1}{\gamma_0(t)} E^0[\gamma_0(T)B|\mathcal{F}_t] \ , \quad 0 \leq t \leq T \ .$$

In the sequel, we shall refer to $u^{(0)}, X_0(\cdot), \pi_0(\cdot)$ as the unconstrained hedging price, price process, and portfolio, respectively. It should be noticed that

$$(4.5) \qquad X_0(T) = X_0^{u^{(0)},\pi_0,0}(T) = B \ , \quad a.s.$$

in Theorem 4.2; we express this by saying that the contingent claim is *attainable* (with initial capital $u^{(0)}$, portfolio π_0, and zero consumption).

Example 4.3. Constant $r(\cdot) \equiv r > 0, \sigma(\cdot) \equiv \sigma$ nonsingular. In this case, the solution $P(t) = (P_1(t), \ldots, P_d(t))^*$ is given by $P_i(t) = h_i(t -$

$s, P(s), \sigma(W_0(t) - W_0(s)))$, $0 \le s \le t$ where $h : [0, \infty) \times \mathcal{R}_+^d \times \mathcal{R}^d \to \mathcal{R}_+^d$ is the function defined by

$$(4.6) \qquad h_i(t, p, y; r) \stackrel{\triangle}{=} p_i \exp[(r - \frac{1}{2}a_{ii})t + y_i] , \quad i = 1, \ldots, d.$$

Consider now a contingent claim of the type $B = \varphi(P(T))$, where $\varphi : \mathcal{R}_+^d \to [0, \infty)$ is a given continuous function, that satisfies polynomial growth conditions in both $\|p\|$ and $1/\|p\|$. Then it is rather straightforward, using the Feynman-Kac theorem (e.g. [20], p.366) and Itô's rule, to see that the processes $X_0(\cdot), \pi_0(\cdot)$ of Proposition 4.2 are given as

$$(4.7) \qquad X_0(t) = e^{-r(T-t)} E^0[\varphi(P(T)) | \mathcal{F}_t] = U(T - t, P(t))$$

$$(4.8) \qquad \pi_{0i}(t) = \frac{P_i(t) \cdot \frac{\partial}{\partial p_i} U(T - t, P(t))}{U(T - t, P(t))} , \quad i = 1, \ldots, d$$

respectively, where

$$U(t, p)$$
$$(4.9) \qquad \stackrel{\triangle}{=} \left\{ \begin{array}{ll} e^{-rt} \int_{\mathcal{R}^d} \varphi(h(t, p, \sigma z; r)) \frac{e^{-\|z\|^2/2t}}{(2\pi t)^{d/2}} dz & ; \; t > 0, \; p \in \mathcal{R}_+^d \\ \varphi(p) & ; \; t = 0, \; p \in \mathcal{R}_+^d \end{array} \right\}.$$

In particular, the unconstrained hedging price $u^{(0)}$ of 4.3 is given, in terms of the function U of 4.9, by

$$(4.10) \qquad u^{(0)} = X_0(0) = U(T, P(0)) .$$

A very explicit computation for the function U is possible for $d = 1$ in the case $\varphi(p) = (p - q)^+$ of a call option: with $\sigma = \sigma_{11} > 0$, exercise price $q > 0, \Phi(z) = \frac{1}{\sqrt{2\pi}} \int_{-\infty}^z e^{-u^2/2} du$ and $\nu_\pm(t, p) \stackrel{\triangle}{=} \frac{1}{\sigma\sqrt{t}} \left[\log(\frac{p}{q}) + (r \pm \frac{\sigma^2}{2})t\right]$, we have the famous Black & Scholes [5] formula

$$U(t, p)$$
$$(4.11) \qquad = \left\{ \begin{array}{ll} p\Phi(\nu_+(t, p)) - qe^{-rt}\Phi(\nu_-(t, p)) & ; \; t > 0, \; p \in (0, \infty) \\ (p - q)^+ & ; \; t = 0, \; p \in (0, \infty) \end{array} \right\}.$$

5. Convex sets and constrained portfolios. We shall fix throughout a nonempty, closed, convex set K in $\mathcal{R}^d$, and denote by

$$(5.1) \qquad \delta(x) \equiv \delta(x|K) \stackrel{\triangle}{=} \sup_{\pi \in K} (-\pi^* x) \; : \; \mathcal{R}^d \to \mathcal{R} \cup \{+\infty\}$$

the support function of the convex set $-K$. This is a closed, positively homogeneous, proper convex function on $\mathcal{R}^d$ ([27], p.114). It is finite on its *effective domain*

$$
(5.2) \quad \begin{aligned}
\tilde{K} &\triangleq \{x \in \mathcal{R}^d; \; \delta(x|K) < \infty\} \\
&= \{x \in \mathcal{R}^d \; ; \;\; \exists \; \beta \in \mathcal{R} \; s.t. \; -\pi^* x \le \beta, \; \forall \, \pi \in K\} \, ,
\end{aligned}
$$

which is a convex cone (called the "barrier cone" of $-K$). It will be assumed throughout that the function

$$
(5.3) \qquad\qquad \delta(\cdot|K) \quad \text{is continuous on} \quad \tilde{K}
$$

and bounded from below on $\mathcal{R}^d$:

$$
(5.4) \qquad \delta(x|K) \ge \delta_0, \quad \forall \, x \in \mathcal{R}^d \quad \text{for some} \quad \delta_0 \in \mathcal{R} \, .
$$

Condition 5.4 is obviously satisfied (with $\delta_0 = 0$) if K contains the origin. On the other hand, Theorem 10.2, p.84 in Rockafellar [27] guarantees that 5.3 is satisfied, in particular, if $\tilde{K}$ is locally simplicial.

Example 5.1. The rôle of the closed, convex set K that we just introduced, is to model reasonable constraints on portfolio choice. One may, for instance, consider the following examples, all of which satisfy the conditions 5.3 and 5.4.

(i) *Unconstrained case:* $K = \mathcal{R}^d$. Then $\tilde{K} = \{0\}$, and $\delta \equiv 0$ on $\tilde{K}$.

(ii) *Prohibition of short-selling:* $K = [0,\infty)^d$. Then $\tilde{K} = K$, and $\delta \equiv 0$ on $\tilde{K}$.

(iii) *Incomplete Market:* $K = \{\pi \in \mathcal{R}^d; \pi_i = 0, \, \forall \, i = m+1,\ldots,d\}$ for some fixed $m \in \{1,\ldots,d-1\}$. Then $\tilde{K} = \{x \in \mathcal{R}^d; \; x_i = 0, \, \forall \, i = 1,\ldots,m\}$ and $\delta \equiv 0$ on $\tilde{K}$.

(iv) *Incomplete Market with prohibition of short-selling:* $K = \{\pi \in \mathcal{R}^d; \; \pi_i \ge 0, \forall \, i = 1,\ldots,m$ and $\pi_i = 0, \, \forall \, i = m+1,\ldots,d\}$ with m as in (iii). Then $\tilde{K} = \{x \in \mathcal{R}^d; \; x_i \ge 0, \, \forall \, i = 1,\ldots,m\}$ and $\delta \equiv 0$ on $\tilde{K}$.

(v) K *is a closed, convex cone in* $\mathcal{R}^d$. Then $\tilde{K} = \{x \in \mathcal{R}^d; \; \pi^* x \ge 0, \, \forall \, \pi \in K\}$ is the polar cone of $-K$, and $\delta \equiv 0$ on $\tilde{K}$. This case obviously generalizes (i)–(iv).

(vi) *Prohibition of borrowing:* $K = \{\pi \in \mathcal{R}^d; \sum_{i=1}^d \pi_i \le 1\}$. Then $\tilde{K} = \{x \in \mathcal{R}^d; \; x_1 = \ldots = x_d \le 0\}$, and $\delta(x) = -x_1$ on $\tilde{K}$.

(vii) *Rectangular constraints:* $K = \times_{i=1}^d I_i$, $I_i = [\alpha_i, \beta_i]$ for some fixed numbers $-\infty \le \alpha_i \le 0 \le \beta_i \le \infty$, with the understanding that the interval I_i is open to the right (left) if $b_i = \infty$ (respectively, if $\alpha_i = -\infty$). Then $\delta(x) = \sum_{i=1}^d (\beta_i x_i^- - \alpha_i x_i^+)$ and $\tilde{K} = \mathcal{R}^d$ if all the $\alpha_i's$, $\beta_i s$ are real. In general, $\tilde{K} = \{x \in \mathcal{R}^d; x_i \ge 0, \, \forall \, i \in \mathcal{S}_+$ and $x_j \le 0, \, \forall \, j \in \mathcal{S}_-\}$ where $\mathcal{S}_+ \triangleq \{i = 1,\ldots,d \; / \; \beta_i = \infty\}$, $\mathcal{S}_- \triangleq \{i = 1,\ldots,d \; / \; \alpha_i = -\infty\}$.   ◇

From now on, we consider only portfolios that take values in the given, convex, closed set $K \subset \mathcal{R}^d$, i.e., we replace the set of admissible policies $\mathcal{A}_0(x)$ with

$$(5.5)\ \mathcal{A}'(x) \overset{\Delta}{=} \{(\pi, c) \in \mathcal{A}_0(x); \ \pi(t, \omega) \in K, \quad \text{for} \quad \ell \otimes \mathbf{P} - a.e. \ (t, \omega)\} \ .$$

As in Cvitanić & Karatzas [7], hereafter abbreviated as [CK], consider the class $\mathcal{H}$ of $\tilde{K}$-valued, $\{\mathcal{F}_t\}$-progressively measurable processes $\nu = \{\nu(t), 0 \le t \le T\}$ which satisfy $E \int_0^T \|\nu(t)\|^2 dt + E \int_0^T \delta(\nu(t)) dt < \infty$, and introduce for every $\nu \in \mathcal{H}$ the analogues

$$(5.6) \qquad \begin{aligned} \theta_\nu(t) &\overset{\Delta}{=} \theta(t) + \sigma^{-1}(t)\nu(t) \ , \\ \gamma_\nu(t) &\overset{\Delta}{=} \exp[-\int_0^t \{r(s) + \delta(\nu(s))\} ds] \ , \end{aligned}$$

$$(5.7) \qquad Z_\nu(t) \overset{\Delta}{=} \exp[-\int_0^t \theta_\nu^*(s) dW(s) - \frac{1}{2}\int_0^t \|\theta_\nu(s)\|^2 ds] \ ,$$

$$(5.8) \qquad W_\nu(t) \overset{\Delta}{=} W(t) + \int_0^t \theta_\nu(s) ds \ ,$$

of the processes in 2.4–2.6, 3.2, as well as the measure

$$(5.9) \qquad \mathbf{P}^\nu(A) \overset{\Delta}{=} E[Z_\nu(T) 1_A] = E^\nu[1_A], \quad A \in \mathcal{F}_T$$

by analogy with 3.5. Finally, denote by $\mathcal{D}$ the subset consisting of the processes $\nu \in \mathcal{H}$ for which the exponential local martingale $Z_\nu(\cdot)$ of 5.7 is actually a martingale. Thus, for every $\nu \in \mathcal{D}$, the measure $\mathbf{P}^\nu$ of 5.9 is a probability measure and the process $W_\nu(\cdot)$ of 5.8 is a $\mathbf{P}^\nu$−Brownian motion.

Remark 5.2. As we saw in Proposition 4.2, in the case $K = \mathcal{R}^d$ the number $u^{(0)} = E^0[\gamma_0(T)B]$ is the "unconstrained" hedging price for the contingent claim B. In the framework of [CK] [7], the number $u_\nu \overset{\Delta}{=} E^\nu[\gamma_\nu(T)B] = E[\gamma_\nu(T)Z_\nu(T)B]$ is the unconstrained hedging price for B in an auxiliary market $\mathcal{M}_\nu$; this consists of a bond with interest rate $r^{(\nu)}(t) \overset{\Delta}{=} r(t) + \delta(\nu(t))$ and d stocks, with the same volatility matrix $\{\sigma_{ij}(t)\}_{1 \le i,j \le d}$ as before and appreciation rates $b_i^{(\nu)}(t) \overset{\Delta}{=} b_i(t) + \nu_i(t) + \delta(\nu(t))$, $1 \le i \le d$, for any given $\nu \in \mathcal{D}$. Thus, the prices of these instruments in $\mathcal{M}_\nu$ are given, by analogy with 2.1, 2.2, as

$$(5.10) \qquad dP_0^{(\nu)}(t) = P_0^{(\nu)}(t)[r(t) + \delta(\nu(t))]dt \ , \ P_0^{(\nu)}(0) = 1$$

$$(5.11) \qquad \begin{aligned} dP_i^{(\nu)}(t) = P_i^{(\nu)}(t) \\ \left[\{b_i(t) + \nu_i(t) + \delta(\nu(t))\}dt + \sum_{j=1}^d \sigma_{ij}(t) dW^{(j)}(t)\right] \ , \end{aligned}$$

$$P_i^{(\nu)}(0) = p_i \in (0, \infty); \; i = 1, \ldots, d.$$

We shall see in the next section that the price for hedging B with a constrained portfolio in the market $\mathcal{M}$, is given by the supremum of the unconstrained hedging prices $u_\nu = E^\nu[\gamma_\nu(T)B]$ in these auxiliary markets $\mathcal{M}_\nu$, $\nu \in \mathcal{D}$, namely by

$$(5.12) \qquad\qquad V(0) \stackrel{\triangle}{=} \sup_{\nu \in \mathcal{D}} E^\nu[\gamma_\nu(T)B].$$

Remark 5.3. In terms of the $\mathbf{P}^\nu$-Brownian motion $W_\nu(\cdot)$ of 5.8, the stock price equations 2.2 can be re-written for any given $\nu \in \mathcal{D}$ as

$$(5.13) \qquad dP_i(t) = P_i(t)\left[(r(t) - \nu_i(t))dt + \sum_{j=1}^d \sigma_{ij}(t)dW_\nu^{(j)}(t)\right],$$
$$i = 1, \ldots, d \;.$$

6. Hedging under constraints. We introduce in this section the "hedging price" of a contingent claim B, with portfolios constrained to take values in the set K of sections 4 and 5, and show that this price coincides with the number $V(0)$ of 5.12.

DEFINITION 6.1. The hedging price with K-constrained portfolios of a contingent claim B is defined by

$$h(0)$$
$$(6.1) \stackrel{\triangle}{=} \left\{ \begin{array}{c} \inf\{x \in (0,\infty); \exists(\pi,c) \in A'(x), \; s.t. \; X^{x,\pi,c}(T) \geq B \; a.s.\} \\ \infty \qquad , \text{ if the above set is empty} \end{array} \right\}.$$

If $h(0) < \infty$, we say that B is K-hedgeable. $\qquad\qquad\qquad\qquad \diamond$

Let us denote by $\mathcal{S}$ the set of all $\{\mathcal{F}_t\}$-stopping times τ with values in $[0,T]$, and by $\mathcal{S}_{\rho,\sigma}$ the subset of $\mathcal{S}$ consisting of stopping times τ s.t. $\rho(\omega) \leq \tau(\omega) \leq \sigma(\omega)$, $\forall \, \omega \in \Omega$, for any two $\rho \in \mathcal{S}, \sigma \in \mathcal{S}$ such that $\rho \leq \sigma$, a.s. Denote also by $\mathcal{D}_{\rho,\sigma}$ the restriction of $\mathcal{D}$ to the stochastic interval $[\![\rho,\sigma]\!]$. For every $\tau \in \mathcal{S}$ consider the $\mathcal{F}_\tau$-measurable random variable

$$(6.2) \qquad V(\tau) \stackrel{\triangle}{=} ess\sup_{\nu \in \mathcal{D}} E^\nu[B\gamma_0(T)\exp\{-\int_\tau^T \delta(\nu(s))ds\}|\mathcal{F}_\tau] \,,$$

and notice the notational agreement with the definition 5.12.

PROPOSITION 6.2. For any contingent claim B for which $V(0) < \infty$, the family 6.2 of random variables $\{V(\tau)\}_{\tau \in \mathcal{S}}$ satisfies the equation of Dynamic Programming

$$(6.3) \qquad V(\tau) = ess\sup_{\nu \in \mathcal{D}_{\tau,\theta}} E^\nu[V(\theta)\exp\{-\int_\tau^\theta \delta(\nu(u))du\}|\mathcal{F}_\tau] \,;$$
$$\forall \; \theta \in \mathcal{S}_{\tau,T} \;.$$

PROPOSITION 6.3. *The process* $V = \{V(t), \mathcal{F}_t; 0 \leq t \leq T\}$ *of Proposition 6.2 can be considered in its RCLL modification and, for every* $\nu \in \mathcal{D}$,

$$(6.4) \qquad \left\{ \begin{array}{l} Q_\nu(t) \stackrel{\triangle}{=} V(t)e^{-\int_0^t \delta(\nu(u))du}, \mathcal{F}_t; \ 0 \leq t \leq T \\ \\ \text{is a } \mathbf{P}^\nu\text{-supermartingale with RCLL paths} \end{array} \right\}.$$

Furthermore, V *is the smallest adapted, RCLL process that satisfies 6.4 as well as*

$$(6.5) \qquad V(T) = B\gamma_0(T), \quad a.s. \qquad\qquad \diamond$$

The following can be regarded as the main result of the paper.

THEOREM 6.4. *For an arbitrary contingent claim* B, *we have* $h(0) = V(0)$. *Furthermore, if* $V(0) < \infty$, *then* B *is* K−*hedgeable and there exists a pair* $(\hat{\pi}, \hat{c}) \in \mathcal{A}'(V(0))$ *such that* $X^{V(o), \hat{\pi}, \hat{c}}(T) = B$, *a.s.*

Proof. (sketch). We first argue that $h(0) \leq V(0)$. Clearly, we may assume $V(0) < \infty$. From 6.4, the martingale representation theorem and the Doob-Meyer decomposition (e.g. [20], §1.4), we have for every $\nu \in \mathcal{D}$:

$$(6.6) \qquad Q_\nu(t) = V(0) + \int_0^t \psi_\nu^*(s)dW_\nu(s) - A_\nu(t), \quad 0 \leq t \leq T,$$

where $\psi_\nu(\cdot)$ is an $\mathcal{R}^d$-valued, $\{\mathcal{F}_t\}$-progressively measurable and a.s. square-integrable process and $A_\nu(\cdot)$ is adapted with increasing, RCLL paths and $A_\nu(0) = 0, A_\nu(T) < \infty$ a.s. The idea then is to consider the positive, adapted, RCLL process

$$(6.7) \qquad \hat{X}(t) \stackrel{\triangle}{=} \frac{V(t)}{\gamma_0(t)} = \frac{Q_\nu(t)}{\gamma_\nu(t)}, \quad 0 \leq t \leq T \quad (\forall \ \nu \in \mathcal{D})$$

with $\hat{X}(0) = V(0), \hat{X}(T) = B$ a.s., and to find a pair $(\hat{\pi}, \hat{c}) \in \mathcal{A}'(V(0))$ such that $\hat{X}(\cdot) = X^{V(0), \hat{\pi}, \hat{c}}(\cdot)$; this will prove that $h(0) \leq V(0)$. Comparing $Q_\mu(\cdot)$ with $Q_\nu(\cdot)$, $\mu, \nu \in \mathcal{D}$, we get $\psi_\nu^*(t) \ e^{\int_0^t \delta(\nu(s))ds} = \psi_\mu^*(t) \ e^{\int_0^t \delta(\mu(s))ds}$ and hence that this expression is independent of $\nu \in \mathcal{D}$:

$$(6.8) \quad \psi_\nu^*(t) \ e^{\int_0^t \delta(\nu(s))ds} = \hat{X}(t)\gamma_0(t)\hat{\pi}^*(t)\sigma(t); \quad \forall \ 0 \leq t \leq T, \ \nu \in \mathcal{D},$$

for some adapted, $\mathcal{R}^d$-valued, a.s. square integrable process $\hat{\pi}$ (we do not know yet that $\hat{\pi}$ takes values in K). Similarly, we have

$$e^{\int_0^t \delta(\nu_s)ds} dA_\nu(t) - \gamma_0(t)\hat{X}(t)[\delta(\nu_t) + \hat{\pi}_t^*\nu_t]dt$$

$$= e^{\int_0^t \delta(\mu_s)ds} dA_\mu(t) - \gamma_0(t)\hat{X}(t)[\delta(\mu_t) + \hat{\pi}_t^*\mu_t]dt$$

and hence this expression is also independent of $\nu \in \mathcal{D}$:

$$(6.9) \quad \hat{c}(t) \triangleq \int_0^t \gamma_\nu^{-1}(s)dA_\nu(s) - \int_0^t \hat{X}(s)[\delta(\nu(s)) + \nu^*(s)\hat{\pi}(s)]ds ,$$

for every $0 \le t \le T, \nu \in \mathcal{D}$. From 6.9 with $\nu \equiv 0$, we obtain $\hat{c}(t) = \int_0^t \gamma_0^{-1}(s)dA_0(s), \quad 0 \le t \le T$ and hence

$$(6.10) \quad \left\{ \begin{array}{c} \hat{c}(\cdot) \text{ is an increasing, adapted, RCLL process} \\ \text{with } \hat{c}(0) = 0 \quad \text{and} \quad \hat{c}(T) < \infty, a.s. \end{array} \right\}.$$

Next, 6.9 and the nonnegativity of A_μ, for all $\mu \in \mathcal{D}$, imply

$$(6.11) \qquad \delta(\nu(t,\omega)) + \nu^*(t,\omega)\hat{\pi}(t,\omega) \ge 0, \quad \ell \otimes \mathbf{P} - a.e.$$

holds for every $\nu \in \mathcal{D}$. Then the arguments of [CK] ([7], Theorem 9.1) lead to the fact that

$$(6.12) \qquad \hat{\pi}(t,\omega) \in K \quad \text{holds} \quad \ell \otimes \mathbf{P} - a.e. \text{ on } [0,T] \times \Omega$$

(these arguments need the continuity condition 5.3, and the assumption that the set K is closed). It is now straightforward to show that $\hat{X}(\cdot) \equiv X^{V(0),\hat{\pi},\hat{c}}(\cdot)$, and hence $h(0) \le V(0) < \infty$.

The reverse inequality, $h(0) \ge V(0)$, follows from the supermartingale property of the processes

$$M_\nu(u) \triangleq \gamma_\nu(u)X^{x,\pi,c}(u) + \int_0^u \gamma_\nu(t)dc(t) + \int_0^u \gamma_\nu(t)X^{x,\pi,c}(t)\{\delta(\nu(t)) + \nu^*(t)\pi(t)\}dt], \ \nu \in \mathcal{D}.$$

$\square$

DEFINITION 6.5. We say that a K-hedgeable contingent claim B is $K-$attainable, if there exists a portfolio process π with values in K such that $(\pi, 0) \in \mathcal{A}'(V(0))$ and $X^{V(0),\pi,0}(T) = B$, a.s.

THEOREM 6.6. *For a given contingent claim B with $V(0) < \infty$, and any given $\lambda \in \mathcal{D}$, the conditions*

$$(6.13) \ \{Q_\lambda(t) = V(t)e^{-\int_0^t \delta(\lambda(u))du}, \mathcal{F}_t; \ 0 \le t \le T\} \quad \text{is a } \mathbf{P}^\lambda\text{-martingale}$$

$$(6.14) \qquad \lambda \ \text{achieves the supremum in} \ V(0) = \sup_{\nu \in \mathcal{D}} E^\nu[B\gamma_\nu(T)]$$

$$(6.15) \quad \left\{ \begin{array}{c} B \text{ is } K\text{-attainable (by a portfolio } \pi\text{), and the} \\ \text{corresponding } \gamma_\lambda(\cdot)X^{V(0),\pi,0}(\cdot) \text{ is a } \mathbf{P}^\lambda\text{-martingale} \end{array} \right\}$$

are equivalent, and imply

(6.16) $\hat{c}(t,\omega) = 0,\ \delta(\lambda(t,\omega)) + \lambda^*(t,\omega)\hat{\pi}(t,\omega) = 0;\ \ \ell \otimes P - a.e.$

for the pair $(\hat{\pi}, \hat{c}) \in \mathcal{A}'(V(0))$ of Theorem 6.4.

Proof. The $\mathbf{P}^\lambda$-supermartingale $Q_\lambda(\cdot)$ is a $\mathbf{P}^\lambda$-martingale, if and only if $Q_\lambda(0) = E^\lambda Q_\lambda(T) \Leftrightarrow V(0) = E^\lambda[B\gamma_\lambda(T)] \Leftrightarrow 6.14$. On the other hand, 6.13 implies $A_\lambda(\cdot) \equiv 0$, and so from 6.9: $\hat{c}(t) = -\int_0^t \hat{X}(s)[\delta(\lambda(s))+\lambda^*(s)\hat{\pi}(s)]ds$. Now 6.16 follows from the increase of $\hat{c}(\cdot)$ and the nonnegativity of $\delta(\lambda) + \lambda^*\hat{\pi}$, since $\hat{\pi}$ takes values in K. From 6.14 (and its consequences 6.13, 6.16), the process $\hat{X}(\cdot)$ of 6.7 coincides with $X^{V(0),\hat{\pi},0}(\cdot)$, and we have: $\hat{X}(T) = B$ almost surely, $\gamma_\lambda(\cdot)\hat{X}(\cdot)$ is a $\mathbf{P}^\lambda$-martingale; thus 6.15 is satisfied with $\pi \equiv \hat{\pi}$. On the other hand, suppose that 6.15 holds; then $V(0) = E^\lambda[B\gamma_\lambda(T)]$, so 6.14 holds.   □

The following result is also quite straightforward.

THEOREM 6.7. Let B be a contingent claim with $V(0) < \infty$. Suppose that, for any $\nu \in \mathcal{D}$ with $\delta(\nu) + \nu^*\hat{\pi} \equiv 0$,

(6.17) $Q_\nu(\cdot)$ in 6.4 is of class $D[0,T]$, under $\mathbf{P}^\nu$.

Then, for any given $\lambda \in \mathcal{D}$, the conditions 6.13, 6.14, 6.16 are equivalent, and imply

(6.15)° $\left\{ \begin{array}{l} B\ \text{is}\ K\text{-attainable (by a portfolio}\ \pi),\ \text{and the} \\ \text{corresponding}\ \gamma_0(\cdot)X^{V(0),\pi,0}(\cdot)\ \text{is a}\ \mathbf{P}^0\text{-martingale} \end{array} \right\}.$

Conversely, if 6.15° holds, then the conditions 6.13, 6.14, 6.16 are satisfied by any $\lambda \in \mathcal{D}$ with $\delta(\lambda) + \lambda^*\hat{\pi} \equiv 0$. Furthermore, if the portfolio $\pi_o(\cdot)$ of Proposition 4.2 satisfies $\pi_o \in K, \ell \otimes \mathbf{P}-$ a.e., these conditions are then satisfied, with $(\hat{\pi}, \hat{c}) \equiv (\pi_o, 0)$, by $\lambda \equiv 0$ (and thus also by any $\lambda \in \mathcal{D}$ with $\delta(\lambda) + \lambda^*\hat{\pi} \equiv 0$), and the corresponding wealth process X_o of 4.4 is equal to $\hat{X}$ of 6.7.

Remark 6.8. The conditions $V(0) < \infty$ and 6.17 are satisfied (the latter, in fact, for every $\nu \in \mathcal{D}$) in the case of the simple European call option $B = (P_1(T) - q)^+$, provided

(6.18) the function $x \mapsto \delta(x) + x_1$ is bounded from below on $\tilde{K}$.

The same is true for any contingent claim B that satisfies $B \leq \alpha P_1(T)$ a.s., for some $\alpha \in (0, \infty)$. It should be noted that the condition 6.18 is indeed satisfied, if

(6.19)   the convex set K contains both the origin and the point

$(1, 0, \ldots, 0)$

(and thus also the line-segment adjoining these points); for then $x_1 + \delta(x) \geq x_1 + \sup_{0 \leq \alpha \leq 1}(-\alpha x_1) = x_1^+ \geq 0$, $\forall x \in \tilde{K}$. This is the case in the Examples 5 (i)–(iv), (vi), and (vii) with $1 \leq \beta_1 \leq \infty$.

Remark 6.9. If the condition 6.18 is not satisfied, we have $V(0) = \infty$ for the European call option $B = (P_1(T) - q)^+$ with $\delta(\cdot) \geq 0$, $r(\cdot) \geq 0$. In other words, such constraints make impossible the hedging of this contingent claim, starting with a finite initial capital.

The conditions 6.18, 6.19 fail, for instance, in the case of *rectangular constraints* $K = \times_{i=1}^{d}[\alpha_i, \beta_i]$ of Example 5(vii) with $\beta_1 < 1$. They also fail in the case of an incomplete market in which investment in the first stock is prohibited, say with $K = \{\pi \in \mathcal{R}^d; \pi_1 = \ldots = \pi_m = 0\}$ for some $1 \leq m \leq d$; then $\tilde{K} = \{x \in \mathcal{R}^d; x_{m+1} = \ldots = x_d = 0\}$, and $\delta \equiv 0$ on $\tilde{K}$. (However, the "option with a ceiling" $B = \min\{(P_1(T) - q)^+, L\}$ for some real $L > 0$, is bounded, thus hedgeable, for any constraint set K).

The next result characterizes the process $\hat{X}(\cdot)$ of 6.7 as the *minimal* solution of a certain *Backwards Stochastic Differential Equation* (BSDE) *with Convex Constraints*.

PROPOSITION 6.10. *Suppose we have $V(0) < \infty$, and let (X, π, c) be any triple of $\mathcal{R} \times \mathcal{R}^d \times [0, \infty)$–valued, adapted processes, such that $c(\cdot)$ has increasing, RCLL paths, $c(T) + \int_0^T \|\pi(s)\|^2 ds < \infty$ a.s., and such that the BSDE*

$$(6.20) \quad X(t) = B + (c(T) - c(t)) - \int_t^T X(s)[r(s)ds + \pi^*(s)\sigma(s)dW_0(s)] ,$$
$$0 \leq t \leq T$$

and the Convex Constraint

$$(6.21) \qquad (X(t), \pi(t)) \in [0, \infty) \times K , \quad 0 \leq t \leq T$$

are satisfied almost surely. Then the triple $(\hat{X}, \hat{\pi}, \hat{c})$ of the Theorem 6.4 solves the problem 6.20, 6.21, and we have $\hat{X}(\cdot) \leq X(\cdot)$, a.s.

7. Examples. Let us now illustrate the results of section 6 by means of some simple examples.

Example 7.1. (no short-selling): In the case $d = 1$, $K = [0, \infty)$ of Example 5(ii) and with $r, \sigma \equiv \sigma_{11}$ positive constants, we have $\tilde{K} = K$, $\delta(x) = \left\{ \begin{matrix} 0; & x \geq 0 \\ \infty; & x < 0 \end{matrix} \right\}$ and so $x + \delta(x) = x \geq 0$ on $\tilde{K}$. Now take $B = \varphi(P_1(T))$, where $\varphi : \mathcal{R}^+ \to [0, \infty)$ is continuous, increasing, piecewise continuously differentiable, and satisfies $\varphi(p) \leq \alpha p$, for some real $\alpha > 0$. Then we have from Remark 6 that $V(0) < \infty$, and that condition 6.17 is satisfied. In fact, from the "classical" theory of section 4 (cf. Example 4), we know that

$$(7.1) \qquad X(t) = e^{rt}V(t) = e^{-r(T-t)}U(T - t, P_1(t)),$$

$$(7.2) \qquad \hat{\pi}(t) = \frac{Q(T-t, P_1(t))}{U(T-t, P_1(t))} = P_1(t) \frac{\frac{\partial}{\partial p} U(T-t, P_1(t))}{U(T-t, P_1(t))} \geq 0 \ ,$$

where

$$(7.3) \qquad \begin{aligned} U(t, p) &\triangleq \int\limits_{-\infty}^{\infty} \varphi(pe^{\sigma(\xi + \delta t)}) \frac{e^{-\xi^2/2t}}{\sqrt{2\pi t}} d\xi, \\ Q(t, p) &\triangleq \int\limits_{-\infty}^{\infty} \psi(pe^{\sigma(\xi + \delta t)}) \frac{e^{-\xi^2/2t}}{\sqrt{2\pi t}} d\xi = p \cdot \frac{\partial}{\partial p} U(t, p) \end{aligned}$$

with $\delta = \frac{r}{\sigma} - \frac{\sigma}{2}$ and $\psi(p) \triangleq p\varphi'(p) \geq 0$ (hence $Q(t, p) \geq 0$). If $p\varphi'(p) \geq \varphi(p)$ holds everywhere on $\mathcal{R}_+$, and $p\varphi'(p) > \varphi(p)$ holds on a set of positive Lebesgue measure, then $p \cdot \frac{\partial}{\partial p} U(t, p) > U(t, p)$, whence

$$(7.4) \qquad \hat{\pi}(t) > 1, \quad 0 \leq t \leq T.$$

For instance, this is the case of the European call option $\varphi(p) = (p - q)^+$ with exercise price $q > 0$. Thus, the unconstrained hedging portfolio does not require short-selling, and the constraint $K = [0, \infty)$ makes no difference. In particular, the supremum $V(0)$ of 5.12 is achieved by $\nu \equiv 0$, and is equal to the unconstrained hedging price $u^{(0)}$ of the option.

Example 7.2. (no borrowing): Let $d = 1, K = (-\infty, 1]$ as in Example 5(vi); then $\hat{K} = (-\infty, 0]$, $\delta(\nu) = -\nu$, and consider the contingent claim $B = (P_1(T) - q)^+$. From 5.13 the process $e^{\int_0^t \nu(s)ds} \gamma_0(t)P_1(t)$ is a $\mathbf{P}^\nu$-martingale, for every $\nu \in \mathcal{D}$. Consequently,

$$(7.5) \qquad \begin{aligned} V(t) &\leq \operatorname{ess\,sup}_{\nu \in \mathcal{D}} e^{-\int_0^t \nu(s)ds} E^\nu[e^{\int_0^T \nu(s)ds} \gamma_0(T)P_1(T)|\mathcal{F}_t] \\ &= \gamma_0(t)P_1(t), \ 0 \leq t \leq T. \end{aligned}$$

On the other hand, with $\mathcal{D}_d$ denoting the class of non-random functions in $\mathcal{D}$, we have by Jensen's inequality

$$V(t) \geq \operatorname{ess\,sup}_{\nu \in \mathcal{D}} \left(e^{-\int_0^t \nu(s)ds} E^\nu[e^{\int_0^T \nu(s)ds} \gamma_0(T)P_1(T)|\mathcal{F}_t] \right.$$

$$\left. - E^\nu[e^{\int_t^T \nu(s)ds} \gamma_0(T)q|\mathcal{F}_t] \right)^+$$

$$\geq \operatorname{ess\,sup}_{\nu \in \mathcal{D}_d} \left(\gamma_0(t)P_1(t) - e^{\int_t^T \nu(s)ds} q E^\nu[\gamma_0(T)|\mathcal{F}_t] \right)^+ = \gamma_0(t)P_1(t),$$

(7.6)
for $0 \leq t < T$. The inequalities 7.5, 7.6 imply

$$(7.7) \qquad V(t) = \left\{ \begin{array}{ll} \gamma_0(t)P_1(t); & 0 \leq t < T \\ \gamma_0(T)(P_1(T) - q)^+; & t = T \end{array} \right\},$$

or equivalently

(7.8) $$dV(t) = \gamma_0(t)P_1(t)\sigma(t)dW_0(t) - dA_0(t),$$

where

(7.9) $$A_0(t) = \left\{ \begin{array}{ccc} 0 & ; & 0 \le t < T \\ \gamma_0(T)[P_1(T) - (P_1(T) - q)^+] & ; & t = T \end{array} \right\}.$$

In particular, 7.8 implies $X \triangleq V/\gamma_0 \equiv X^{V(0),\hat{\pi},\hat{c}}$ with $\hat{\pi}(t) \equiv 1$, $\hat{c}(t) = \int_0^t \gamma_0^{-1}(s)dA_0(s)$. In other words, in order to replicate $B = (P_1(T) - q)^+$ without borrowing, one has to invest all the wealth in the stock, not consume before the expiration date T, and consume at time $t = T$ the amount $\hat{c}(T) = P_1(T) - (P_1(T) - q)^+ = \min(P_1(T), q)$.

This example shows that, *in general, the supremum of 5.12 is not attained*; indeed, one has to let $\nu \equiv -\infty$ in order to achieve equality in 7.6.

Example 7.3. (Option with a ceiling on a stock that cannot be traded): Let $K = \{x \in \mathcal{R}^d; x_1 = 0\}$, $B = (P_1(T) - q)^+ \wedge L$, for some real $q > 0, L > 0$. Then $\tilde{K} = \{x \in \mathcal{R}^d; x_2 = x_3 = \ldots = x_d = 0\}$, and $\delta \equiv 0$ on $\tilde{K}$. Assume deterministic market coefficients. We want to verify

(7.10) $$V(0) = \gamma_0(T)L,$$

by first showing $V(0) \ge \gamma_0(T)L$, and then proving the reverse inequality by constructing a consumption process c such that the wealth process corresponding to the triple $(\gamma_0(T)L, 0, c)$ satisfies $X(T) = B$, a.s. In the notation of 7.6, we have

(7.11) $$V(0) \ge \gamma_0(T)L \; ess \sup_{\nu \in \mathcal{D}_d} E^\nu 1_{\{P_1(T) - q > L\}}.$$

Define an $\mathcal{R}_+^d$-valued process $\tilde{P}^{(\nu)}(\cdot) = \{\tilde{P}_i^\nu(\cdot)\}_{i=1}^d$ by

(7.12) $$d\tilde{P}_i^{(\nu)}(t) = \tilde{P}_i^{(\nu)}(t)[r(t) - \nu_i(t)]dt + \tilde{P}_i^{(\nu)}(t)\sum_{j=1}^d \sigma_{ij}(t)dW_0^{(j)}(t),$$
$$\tilde{P}_i^{(\nu)}(0) = P_i(0)$$

for $i = 1, \ldots, d$ and $\nu \in \mathcal{D}_d$. Then a comparison of 7.12 with 5.13 shows that $\tilde{P}^{(\nu)}(\cdot)$ has the same distribution under $\mathbf{P}^0$, as $P(\cdot) = \{P_i(\cdot)\}_{i=1}^d$ has under $\mathbf{P}^\nu$, and thus

(7.13) $$E^\nu 1_{\{P_1(T) - q > L\}} = E^0 1_{\{\tilde{P}_1^{(\nu)}(T) - q > L\}}.$$

Letting $\nu \to -\infty$, 7.11–7.13 imply

(7.14) $$V(0) \ge \gamma_0(T)L.$$

Next, define a consumption process c by

$$c(t) = \left\{ \begin{array}{ccc} 0 & ; & t < T \ or \ t = T, P_1(T) - q > L \\ L - (P_1(T) - q)^+ & ; & t = T, P_1(T) - q \leq L \end{array} \right\}.$$

Then the wealth process $X(\cdot)$ associated with the policy $(\gamma_0(T)L, 0, c)$ is given by $X(t) = \frac{\gamma_0(T)}{\gamma_0(t)}L$ for $t < T$, and by $X(T) = L - c(T) = B$ for $t = T$. This implies $V(0) \leq \gamma_0(T)L$ by Theorem 6.4 and, in conjunction with 7.14, leads to 7.10.

Consequently, the way to hedge a bounded option on a stock that is not available for investment, is to replicate the upper bound of the option by investing in the bond only, and then to consume the difference at the expiration date.

8. Extensions and ramifications

1. As in section 16 of [CK] [7], we can let the constraint set K depend on $(t, \omega) \in [0, T] \times \Omega$ in a non-anticipative way.
2. The hedging price $V(0)$ can be regarded as an upper bound for the "fair price" of the contingent claim; or, in the terminology of El Karoui & Quenez [11], it can be called the *selling price*. As in that paper, we could also consider a lower bound, or the *purchase price*, by replacing *sup* operator by *inf* operator.
3. As for numerical calculations, we refer again to El Karoui & Quenez [11]. It is shown in that paper, in the special case of "incompleteness" constraints and constant r, σ that $V(0) = Q(0, p)$, where $Q(t, p)$ is the pointwise limit $Q(t, p) = \lim_{n \to \infty} Q^n(t, p)$. Here we are considering a contingent claim of the form $B = \varphi(P(T))$, for an appropriate continuous function $\varphi : \mathcal{R}_+^d \longrightarrow \mathcal{R}_+$ of the vector $P(t) \triangleq (P_1(t), \ldots, P_d(t))^*$ of stock prices at the terminal time $t = T$, and define

$$(8.1) \quad Q^n(t, p) \triangleq \sup_{\nu \in \mathcal{D}_n} E^\nu[\varphi(P(T)) \frac{\gamma_\nu(T)}{\gamma_\nu(t)} \mid P(t) = p];$$
$$0 \leq t \leq T, \ p \in \mathcal{R}_+^d,$$

with $\mathcal{D}_n \triangleq \{\nu \in \mathcal{D}; \|\nu(t, \omega)\| \leq n, \text{ for } \ell \otimes \mathbf{P} - a.e. (t, \omega)\}$, $n \in \mathbf{N}$. Moreover, from 8.1 and the dynamics 5.13 of the process $P(\cdot)$ under $\mathbf{P}^\nu$, the value function Q^n of 8.1 can be characterized in terms of the following Cauchy problem for the associated HJB equation (cf.

[13]):

$$\frac{\partial Q^n}{\partial t} + \frac{1}{2} \sum_{i=1}^{d} \sum_{j=1}^{d} a_{ij} p_i p_j \frac{\partial^2 Q^n}{\partial p_i \partial p_j} + r \left(\sum_{i=1}^{d} p_i \frac{\partial Q^n}{\partial p_i} - Q^n \right)$$

(8.2) $$\quad + \max_{\nu \in \tilde{K}; \|\nu\| \leq n} \left(-\sum_{i=1}^{d} \nu_i p_i \frac{\partial Q^n}{\partial p_i} - \delta(\nu) Q^n \right) = 0;$$

$$0 \leq t < T, \ p \in \mathcal{R}_+^d$$

$$Q^n(T, p) = \varphi(p); \quad t = T, \ p \in \mathcal{R}_+^d \ .$$

9. Hedging claims with higher interest rate for borrowing. We have studied so far a model in which one is allowed to borrow money, at an interest rate $R(\cdot)$ equal to the bond rate $r(\cdot)$. In this section we consider the more general case of a financial market $\mathcal{M}^*$ in which $R(\cdot) \geq r(\cdot)$, without constraints on portfolio choice. We assume that the progressively measurable process $R(\cdot)$ is also bounded. As it is not reasonable in this model to borrow money and to invest money in the bond at the same time, we restrict ourselves to policies for which the relative amount borrowed at time t is equal to $\left(1 - \sum_{i=1}^{d} \pi_i(t)\right)^-$. Then, as in section 18 of [CK] [7], the wealth process $X = X^{x,\pi,c}$ corresponding to initial capital $x > 0$ and portfolio/cumulative consumption pair (π, c) as in Definition 3, satisfies

$$dX(t) = r(t)X(t)dt - dc(t)$$

$$+X(t) \left[\pi^*(t)\sigma(t)dW_0(t) - (R(t) - r(t))\left(1 - \sum_{i=1}^{d} \pi_i(t)\right)^- dt \right] \ .$$

(9.1)
We set $\delta(\nu(t)) = -\nu_1(t)$ for $\nu \in \mathcal{D}$, where

(9.2) $$\mathcal{D} \stackrel{\triangle}{=} \{\nu; \ \nu \text{ progressively measurable, } \mathcal{R}^d - \text{valued process with}$$

$$r - R \leq \nu_1 = \ldots = \nu_d \leq 0, \ \ell \otimes \mathbf{P} - a.e.\} \ .$$

With this notation, *the theory of the previous sections goes then through with only minor changes.* For instance, in Theorem 6.6 we have to replace 6.16 by

(9.3) $$\hat{c}(t, \omega) \equiv 0, \ \Psi^{\lambda, \hat{\pi}}(t, \omega) = 0; \quad \ell \otimes \mathbf{P} - a.e.,$$

where $(\hat{\pi}, \hat{c})$ is the portfolio/consumption process pair of Theorem 6.4, and

(9.4) $$\Psi^{\nu, \pi}(t) \stackrel{\triangle}{=} [R(t) - r(t) + \nu_1(t)]\left(1 - \sum_{i=1}^{d} \pi_i(t)\right)^-$$

$$-\nu_1(t)\left(1 - \sum_{i=1}^{d} \pi_i(t)\right)^+, \quad 0 \leq t \leq T$$

is a nonnegative process. Similarly, Theorem 6.7 now takes the following form:

THEOREM 9.1. *Let $\hat{\pi}$ be the portfolio process of Theorem 6.4, and suppose that 6.17 holds for every $\nu \in \mathcal{D}$ that satisfies $\Psi^{\nu,\hat{\pi}} \equiv 0$. Then, for any given $\lambda \in \mathcal{D}$, the conditions 6.13, 6.14, 9.3 are equivalent, and imply*

$$(9.5) \quad \left\{ \begin{array}{c} B \text{ is attainable (by a portfolio } \pi \text{) and the} \\ \text{corresponding process } \gamma_{\hat{\lambda}} X^{V(0),\pi,0}(\cdot) \text{ is a} \\ \mathbf{P}^{\hat{\lambda}} - \text{martingale} \end{array} \right\}$$

for the process $\hat{\lambda} \in \mathcal{D}$ given by 9.6 below. Conversely, if 9.5 holds, then the conditions 6.13, 6.14, 9.3 are satisfied for some $\lambda \in \mathcal{D}$, and in particular for

$$(9.6) \quad \hat{\lambda}(t) = \hat{\lambda}_1(t)\mathbf{1}, \quad \hat{\lambda}_1(t) \triangleq [r(t) - R(t)] \, 1_{\{\sum_{i=1}^d \hat{\pi}_i(t) > 1\}}. \qquad \diamond$$

Actually, in this case, due to the boundedness of $\mathcal{D}$ we have the following *existence result* under a condition analogous to 6.17.

THEOREM 9.2. *If the process $Q_{\hat{\lambda}}(\cdot)$ of 6.13 is of class $D[0, T]$ under $\mathbf{P}^{\hat{\lambda}}$, with $\hat{\lambda}$ as in 9.6, then $\hat{\lambda}$ is optimal; namely, 6.14 holds for $\lambda = \hat{\lambda}$.*

Theorem 9.2 implies that, under its conditions, the contingent claim B is attainable in the market $\mathcal{M}^*$ with different interest rates for borrowing and lending. In the case $d = 1$, $B = \varphi(P_1(T))$ with $\varphi : \mathcal{R}_+ \to [0, \infty)$ as in Example 7, and with constant $R > r$, the condition of Theorem 9.2 is actually satisfied (cf. Remark 6). If $p\varphi'(p) \geq \varphi(p)$ holds everywhere on $\mathcal{R}_+$ and strictly on a set of positive measure, then we may take $\hat{\lambda} \equiv r - R$, and the Black-Scholes formulae 7.1, 7.3 remain valid if we replace in them r by R. This follows from 7.4, which can be shown, in the present context, either directly, or as in the following example.

Example 9.3. Let us consider the case of constant coefficients $r, R, \{\sigma_{ij}\} = \sigma$. Then the vector $P(t) = (P_1(t), \ldots, P_d(t))$ of stock price processes satisfies the equations

$$(9.7) \quad \begin{aligned} dP_i(t) &= P_i(t)[b_i(t)dt + \sum_{i=1}^d \sigma_{ij}dW^{(j)}(t)] \\ &= P_i(t)[(r - \nu_1(t))dt + \sum_{i=1}^d \sigma_{ij}dW_\nu^{(j)}(t)], \quad 1 \leq i \leq d, \end{aligned}$$

for every $\nu \in \mathcal{D}$ (recall 2.2 and 5.13). Consider now a contingent claim of the form $B = \varphi(P(T))$, for a given continuous function $\varphi : \mathcal{R}_+^d \to [0, \infty)$ that satisfies a polynomial growth condition, as well as the value function

$$(9.8) \quad Q(t, p) \triangleq \sup_{\nu \in \mathcal{D}} E^\nu [\varphi(P(T))e^{-\int_t^T (r - \nu_1(s))ds} | P(t) = p]$$

on $[0, T] \times \mathcal{R}_+^d$. Clearly, the processes $\hat{X}, V$ of 6.7, 6.2 are given as

$$\hat{X}(t) = Q(t, P(t)), \quad V(t) = e^{-rt}\hat{X}(t); \quad 0 \leq t \leq T,$$

where Q solves the semilinear parabolic partial differential equation of the Hamilton-Jacobi-Bellman (HJB) type

$$\frac{\partial Q}{\partial t} + \frac{1}{2}\sum_i\sum_j a_{ij}p_ip_j\frac{\partial^2 Q}{\partial p_i\,\partial p_j} + \max_{r-R\leq\nu_1\leq 0}\left[(r-\nu_1)\{\sum_i p_i\frac{\partial Q}{\partial p_i} - Q\}\right]$$

$$= 0 \; ; \quad 0 \leq t < T, \; p \in \mathcal{R}_+^d,$$

(9.9)

$$Q(T, p) = \varphi(p); \; p \in \mathcal{R}_+^d$$

associated with the control problem of 9.8 and the dynamics 9.7 (cf. [22] for the basic theory of such equations, and [13] for the connections with stochastic control). Clearly, the maximization in 9.9 is achieved by $\nu_1^* = -(R-r).1_{\{\sum_i p_i\frac{\partial Q}{\partial p_i}\geq Q\}}$; the portfolio $\hat{\pi}(\cdot)$ of Theorem 9.1, and the process $\hat{\lambda}_1(\cdot)$ of 9.6, are then given, respectively, by

(9.10) $$\hat{\pi}_i(t) = \frac{P_i(t)\cdot\frac{\partial}{\partial p_i}Q(t, P(t))}{Q(t, P(t))}, \qquad i = 1,\ldots,d$$

and

(9.11) $$\hat{\lambda}_1(t) = (r-R)1_{\{\sum_i\hat{\pi}_i(t)\geq 1\}}.$$

Suppose now that the function φ satisfies $\sum_i p_i\frac{\partial\varphi(p)}{\partial p_i} \geq \varphi(p)$, $\forall\, p \in \mathcal{R}_+^d$. Then the solution Q of the equation 9.9 also satisfies this inequality, namely

(9.12) $$\sum_i p_i\frac{\partial Q(t, p)}{\partial p_i} \geq Q(t, p), \qquad 0 \leq t \leq T$$

for all $p \in \mathcal{R}_+^d$, and is actually given explicitly as

$$Q(t, p) = E^{(r-R)\mathbf{1}}[e^{-R(T-t)}\varphi(P(T))|P(t) = p]$$

$$= \left\{\begin{array}{l} e^{-R(T-t)}\int_{R^d}\varphi(h(T-t,p,\sigma z; R))(2\pi t)^{-d/2}\exp(-\frac{\|z\|^2}{2t})dz \; ; \\ \qquad 0 \leq t < T, \; p \in \mathcal{R}_+^d \\ \varphi(p) \; ; \quad t = T, \; p \in R_+^d \end{array}\right\}$$

(9.13)

in the notation of 4.7; recall Example 4. Indeed, it is straightforward to check that, in this case, the function of 9.13 satisfies the inequality 9.12, as well as the linear PDE

$$\frac{\partial Q}{\partial t} + \frac{1}{2} \sum_i \sum_j p_i p_j a_{ij} \frac{\partial^2 Q}{\partial p_i \partial p_j}$$

$$+ R\left(\sum_i p_i \frac{\partial Q}{\partial p_i} - Q\right) = 0; \quad 0 \le t < T, \quad p \in \mathcal{R}_+^d,$$

$$Q(T, p) = \varphi(p); \quad p \in \mathcal{R}_+^d,$$

and thus also the nonlinear equation 9.9. In this case the portfolio $\hat{\pi}(\cdot)$ always borrows: $\sum_{i=1}^d \hat{\pi}_i(t) \ge 1$, $0 \le t \le T$ (a.s.), and thus $\hat{\lambda}_1(t) = r - R$, $0 \le t \le T$.

Remark 9.4. The pair of adapted processes $(\hat{X}, \hat{\pi})$, with $\hat{X}(\cdot) \equiv X^{x,\hat{\pi},0}(\cdot)$ as in Theorem 9.1, satisfies $\hat{X}(T) = B$ and

$$d\hat{X}(t) = \hat{X}(t)[r(t) - (R(t) - r(t))(1 - \sum_{i=1}^d \hat{\pi}_i(t))^-]dt$$

$$+ \hat{X}(t)\hat{\pi}^*(t)\sigma(t)dW_0(t), \quad 0 \le t \le T$$

almost surely. This is a Nonlinear Backwards Stochastic Differential Equation in the spirit of Pardoux & Peng [26]; it admits an "explicit" solution in the context of Example 9. In fact, El Karoui, Peng & Quenez [11] observe that the more general equation of the form

$$dX(t) = f(t, X_t, \pi_t)dt + \pi_t dW(t), \quad X(T) = B$$

with $f(t, \cdot, \cdot)$ convex admits, under certain conditions, as solution the process

$$X(t) = esssup_{(\mu,\nu) \in D} E^\nu \left[\gamma_\mu(t, T)B + \int_t^T \gamma_\mu(t, s)F(s, \mu_s, \nu_s)ds \ \bigg| \ \mathcal{F}_t \right],$$

$$0 \le t \le T$$

where $F(\cdot, \mu, \nu)$ is the dual function of $f(\cdot, x, \pi)$, D is the effective domain of F, $\gamma_\mu(t, s) = \exp[\int_t^s \mu(u)du]$, E^ν is the expectation under the measure defined by Z_ν as in 5.7, with θ_ν replaced by ν.

Remark 9.5. We can also study the combined problem of hedging under constraints and with higher interest rate for borrowing then for lending. In that case, the hedging price can be shown to be equal to $\sup_{(\nu,\mu) \in \mathcal{D}_1 \times \mathcal{D}_2} E[H_{\nu,\mu}(T)B]$, where $\mathcal{D}_1$ is the set $\mathcal{D}$ of section 5, $\mathcal{D}_2$ is the set of 9.2 in this section, $\theta_{\nu,\mu}(t) \triangleq \theta(t) + \sigma^{-1}(t)[\nu(t) + \mu(t)]$, and

$$H_{\nu,\mu}(t) \triangleq \exp\left[-\int_0^t \{r(s) + \delta(\nu(s)) - \mu_1(s)\}ds \right.$$

$$\left. -\int_0^t \theta_{\nu,\mu}^*(s)dW(s) - \frac{1}{2}\int_0^t \|\theta_{\nu,\mu}(s)\|^2 ds \right].$$

REFERENCES

[1] J.P.ANSEL, C.STRICKER, Couverture des actifs contingents (preprint) 1992.

[2] E.N.BARRON, R.JENSEN, A stochastic control approach to the pricing of options, Mathem. Operations Research 15 (1990), 49–79.

[3] V.E.BENEŠ, Existence of optimal stochastic control laws, SIAM J. Control 8 (1971), 446–475.

[4] Y.A.BERGMAN, Option pricing with different borrowing and lending rates (preprint) 1991.

[5] F.BLACK, M.SCHOLES, The pricing of options and corporate liabilities, J. Polit. Economy 81 (1973), 637–659.

[6] J.COX, S.ROSS, The valuation of options for alternative stochastic processes, J. Financial Econ. 3 (1976), 145–166.

[7] J.CVITANIĆ, I.KARATZAS, Convex duality in constrained portfolio optimization, Annals of Applied Probability 2 (1992), 767–818.

[8] J.CVITANIĆ, I.KARATZAS, Hedging contingent claims with constrained portfolios, Annals of Applied Probability 3 (1993), 652–681.

[9] M.H.A.DAVIS, P.P.VARAIYA, Dynamic programming conditions for partially-observable stochastic systems, SIAM J. Control 11 (1973), 226–261.

[10] R.J.ELLIOTT Stochastic Calculus and Applications, Springer-Verlag, New York 1982.

[11] N.EL KAROUI, S.PENG, M.C.QUENEZ, Backwards stochastic differential equations in finance and optimization (preprint).

[12] N.EL KAROUI, M.C.QUENEZ, Dynamic programming and pricing of contingent claims in an incomplete market, SIAM J. Control & Optimization, (to appear).

[13] W.H.FLEMING, R.W.RISHEL, Deterministic and Stochastic Optimal Control, Springer-Verlag, New York 1975.

[14] H.FÖLLMER, M.SCHWEIZER, Hedging of contingent claims under incomplete information, Applied Stochastic Analysis 5 (1991), 389–414.

[15] J.M.HARRISON, D.M.KREPS, Martingales and arbitrage in multi-period security markets, J. Econ. Theory 20 (1979), 381–408.

[16] J.M.HARRISON, S.R.PLISKA, Martingales and stochastic integrals in the theory of continuous trading, Stoch. Processes & Appl. 11 (1981), 215–260.

[17] S.D.JACKA, A martingale representation result and an application to incomplete financial markets, Mathematical Finance 2 (1992), 239–250.

[18] E.JOUINI, H.KALLAL, Arbitrage in securities markets with short sales constraints, (preprint) 1993.

[19] I.KARATZAS, Optimization problems in the theory of continuous trading, SIAM J. Control & Optimization 27 (1989), 1221–1259.

[20] I.KARATZAS, S.E.SHREVE, Brownian Motion and Stochastic Calculus, Springer-Verlag, New York 1988.

[21] R.KORN, Option pricing in a model with a higher interest rate for borrowing than for lending (preprint) 1992.

[22] O.A.LADYŽENSKAJA, V.A.SOLONNIKOV, N.N.URAL'TSEVA, Linear and quasilinear equations of parabolic type, Translations of Mathematical Monographs 23, American Math. Society, Providence, R.I. 1968.

[23] R.C.MERTON, Lifetime portfolio selection under uncertainty: the continuous-time model, Rev. Econom. Stat. 51 (1969), 247–257.

[24] R.C.MERTON, Theory of rational option pricing, Bell J. Econ. Manag. Sci. 4 (1973), 141–183.

[25] J.NEVEU, Discrete-Parameter Martingales. (english translation) North-Holland & Elsevier, Amsterdam & New York 1975.

[26] E.PARDOUX, S.G.PENG, Adapted solution of a backward stochastic differential equation, Systems & Control Letters 14 (1990), 55–61.

[27] R.T.ROCKAFELLAR, Convex Analysis, Princeton University Press, Princeton 1970.

[28] G.L.XU, A duality method for optimal consumption and investment under short-selling prohibition, (doctoral dissertation) Carnegie-Mellon University 1990.

ON PORTFOLIO OPTIMIZATION UNDER "DRAWDOWN" CONSTRAINTS*

JAKŠA CVITANIĆ[†] AND IOANNIS KARATZAS[‡]

Abstract. We study the problem of portfolio optimization under the "drawdown constraint" that the wealth process never falls below a fixed fraction of its maximum-to-date, and one strives to maximize the long-term growth rate of its expected utility. This problem was introduced and solved explicitly by Grossman and Zhou; we present an approach which simplifies and extends their results.

1. Introduction and summary. In a very interesting recent article, Grossman & Zhou [1] consider the classical portfolio optimization problem of Merton [5] under the "drawdown constraint" that the wealth process $X^\pi(\cdot)$ satisfy:

$$(1.1) \qquad X^\pi(t) > \alpha \max_{0 \le s \le t} X^\pi(s), \quad \forall 0 \le t < \infty$$

almost surely. In other words, one admits only those portfolios $\pi(\cdot)$ for which the corresponding wealth process $X^\pi(\cdot)$ never falls below $100\alpha\%$ of its maximum-to-date, for some given constant $\alpha \in (0,1)$. The objective is then to maximize the long-term growth rate

$$(1.2) \qquad \mathcal{R}(\pi) := \varlimsup_{T \to \infty} \frac{1}{T} \log E(X^\pi(T))^\delta$$

of expected utility, for some power $\delta \in (0,1)$, over portfolio rules $\pi(\cdot)$ that satisfy (1.1).

Using a mixture of analytical and probabilistic arguments, Grossman and Zhou provide an explicit solution to this problem, when investment is between a bond and one stock (modeled by geometric Brownian motion with constant coefficients). They show that the optimal portfolio $\hat{\pi}(\cdot)$ always invests a *constant* proportion of the difference $X^{\hat{\pi}}(t) - \alpha \cdot \max_{0 \le s \le t} X^{\hat{\pi}}(t)$, $0 \le t < \infty$, in the risky asset. Their insights are impressive, but the arguments are rather lengthy and detailed.

We present in this paper an approach to the above problem, which simplifies the results of Grossman & Zhou [1]—and extends them to the case of several stocks with general deterministic coefficients. The model and the problem are introduced in sections 2 and 3, respectively. The approach is based on an auxiliary finite-horizon stochastic control problem,

---

* Research supported in part by the National Science Foundation under Grant NSF-DMS-93-19816.

† On leave from Columbia University. IMA, University of Minnesota, Minneapolis, MN 55455. email: cvitanic@ima.umn.edu.

‡ Department of Statistics, Columbia University, New York, NY 10027. email: ik@stat.columbia.edu.

formulated in section 4 (Problem 4.1, Remark 4.1), in terms of the positive process $N_\alpha^\pi(\cdot)$ in (4.1), which is "close" to the wealth process $X^\pi(\cdot)$ (in the sense of equality (5.6)), and which has the local martingale property under appropriate discounting (see (4.4)). Precisely because of this property, the auxiliary problem admits an explicit optimal portfolio $\hat{\pi}(\cdot)$, which, in fact, turns out to be independent of the time-horizon $T \in (0, \infty)$ and can be found using "classical" martingale and duality arguments. It is then a relatively straightforward matter to show that this portfolio $\hat{\pi}(\cdot)$ is also optimal for the problem of maximizing (1.2); this is carried out in section 5. In section 6 we find the optimal portfolio for the case of logarithmic utility function, and in section 7 we show that the same portfolio maximizes the long-term growth rate almost surely, not just in expectation. Moreover, this portfolio is optimal even if we allow random (adapted) market coefficients.

2. The model. Let us consider the following standard model of a financial market $\mathcal{M}$ with one riskless asset ("bond", price $P_0(t)$ at time t) and a risky assets ("stocks"; prices $P_i(t)$ at time t, $1 \le i \le d$), modeled by the stochastic equations

$$(2.1) \quad dP_0(t) \; = \; P_0(t)r(t)dt, \; P_0(0) = 1$$

$$(2.2) \quad dP_i(t) \; = \; P_i(t)\left[b_i(t)dt + \sum_{j=1}^{d} \sigma_{ij}(t)dW_j(t)\right], \; P_i(0) = p_i > 0,$$

for $i = 1, \ldots, d$. These equations are driven by the d-dimensional Brownian motion $W = (W_i, \ldots, W_d)'$, which generates all the randomness in the model – in the sense that the interest rate $r(\cdot)$, the vector of appreciation rates $b(\cdot) = (b_1(\cdot), \ldots, b_d(\cdot))'$ and the volatility matrix $\sigma(\cdot) = \{\sigma_{ij}(\cdot)\}_{1 \le i,j \le d}$ are bounded measurable processes, adapted to the augmentation $\mathbb{F} = \{\mathcal{F}(t)\}_{0 \le t < \infty}$ of the filtration $\mathcal{F}^W(t) = \sigma(W(s), 0 \le s \le t)$, $0 \le t < \infty$ generated by W. We shall assume throughout that the matrix $\sigma(\cdot)$ is invertible, and the "relative risk" process $\theta(t) := \sigma^{-1}(t)[b(t) - r(t)\mathbb{I}]$, $0 \le t < \infty$ is bounded as well, where $\mathbb{I}$ is a d-dimensional vector with all entries equal to one. We shall also denote by

$$(2.3) \qquad \beta(t) := \frac{1}{P_0(t)} = \exp\left(-\int_0^t r(s)ds\right), \quad 0 \le t < \infty$$

the "discount process" of this model.

Consider now an economic agent who invests in this market according to a portfolio rule $\pi(\cdot) = (\pi_1(\cdot), \ldots, \pi_d(\cdot))'$ in such a way that his corresponding wealth process $X^\pi(\cdot)$ is governed by the equation

$$dX^\pi(t) = \sum_{i=1}^{d} \pi_i(t) \left(X^\pi(t) - \frac{\alpha M^\pi(t)}{\beta(t)} \right) \left[b_i(t)dt + \sum_{j=1}^{d} \sigma_{ij}(t)dW_j(t) \right]$$

$$+ \left[\left(1 - \sum_{i=1}^{d} \pi_i(t) \right) \left(X^\pi(t) - \frac{\alpha M^\pi(t)}{\beta(t)} \right) + \alpha \frac{M^\pi(t)}{\beta(t)} \right] r(t)dt$$

(2.4)

$$= r(t)X^\pi(t)dt + \left(X^\pi(t) - \frac{\alpha M^\pi(t)}{\beta(t)} \right)$$

$$\pi'(t) \left[(b(t) - r(t)\mathbb{1})dt + \sigma(t)dW(t) \right],$$

$$X^\pi(0) = x,$$

and satisfies the "drawdown constraint"

(2.5) $$P\left[\beta(t)X^\pi(t) > \alpha M^\pi(t), \quad \forall 0 \le t < \infty \right] = 1.$$

Here $\alpha \in (0,1)$ is a given constant, and

(2.6) $$M^\pi(t) := \max_{0 \le s \le t} (\beta(s)X^\pi(s)).$$

The interpretation is this: the agent does not tolerate the "drawdown $1 - \frac{\beta(t)X^\pi(t)}{M^\pi(t)}$ of his discounted wealth, from its maximum-to-date", to be greater than or equal to the constant $1 - \alpha$, *at any time* $t \ge 0$; thus, he imposes the (almost sure) constraint (2.5). He invests a proportion $\pi_i(t)$ of the difference $X^\pi(t) - \alpha \frac{M^\pi(t)}{\beta(t)} > 0$ in the ith stock, $i = 1, \ldots, d$, and invests the remainder $\left(1 - \sum_{i=1}^{d} \pi_i(t) \right) \left(X^\pi(t) - \alpha \frac{M^\pi(t)}{\beta(t)} \right) + \alpha \frac{M^\pi(t)}{\beta(t)}$ of his wealth in the bond.

With this interpretation in mind, we set up the formal model as follows.

DEFINITION 2.1. For a given initial capital $x > 0$, let $\mathcal{A}_\alpha(x)$ denote the class of measurable, $\mathbb{F}$-adapted processes $\pi : [0, \infty) \times \Omega \to \mathbb{R}^d$ which satisfy

(2.7) $$\int_0^T \|\pi'(t)\sigma(t)\|^2 dt < \infty, \quad \text{a.s.}$$

for any given $T \in (0, \infty)$, and for which the stochastic functional/differential equation (2.4), (2.6) has a unique $\mathbb{F}$-adapted solution $X^\pi(\cdot)$ that obeys the constraint (2.5).

The elements of $\mathcal{A}_\alpha(x)$ will be called "admissible portfolio processes".

The class of Definition 2.1 is non-empty. In fact, it is shown in the Appendix that

(2.8) $$\left\{ \begin{array}{c} \text{for any measurable, locally square-integrable (a.s.),} \\ \text{and } \mathbb{F}\text{-adapted process} \\ \rho : [0, \infty) \times \Omega \to \mathbb{R}^d, \ \hat{\pi} = (\rho'\sigma^{-1})' \text{ is an admissible portfolio} \\ \text{in } \mathcal{A}_\alpha(x), \quad \text{for any } x > 0 \end{array} \right\}.$$

3. The Grossman-Zhou problem. Let $U : (0, \infty) \to \mathbb{R}$ be a *utility function*: a strictly increasing, strictly concave function of class C^1 with $U'(0+) = \infty, U'(\infty) = 0$ and $U(0+) \geq -\infty$. The convex dual of this function is given by

$$(3.1) \qquad \tilde{U}(y) := \max_{x>0}[U(x) - xy] = U(I(y)) - yI(y), y > 0$$

where $I(\cdot)$ is the inverse of $U'(\cdot)$.

3.1. Problem. *(Grossman & Zhou [1])*: For some given $0 < \delta < 1$, maximize the long-term rate of growth

$$(3.2) \qquad \mathcal{R}(\pi) := \varlimsup_{T \to \infty} \frac{1}{T} \log E(X^\pi(T))^\delta$$

of expected power-utility, over $\pi \in \mathcal{A}_\alpha(x)$. In particular, compute

$$(3.3) \qquad v(\alpha) := \sup_{\pi \in \mathcal{A}_\alpha(x)} \mathcal{R}(\pi)$$

and find $\hat{\pi} \in \mathcal{A}_\alpha(x)$, for which the limit $\lim_{T \to \infty} \frac{1}{T} \log E(X^{\hat{\pi}}(T))^\delta = \mathcal{R}(\hat{\pi})$ exists and achieves the supremum in (3.3).

Grossman and Zhou solved Problem 3.1 for $d = 1$ and constant r, b_1, σ_{11}, using rather lengthy analytical and probabilistic techniques. We present in section 5 a simple solution to this problem that allows general $d \geq 1$ and deterministic coefficients $r(\cdot), b(\cdot), \sigma(\cdot)$ in (2.1), (2.2).

4. An auxiliary process and problem. For any portfolio process $\pi \in \mathcal{A}_\alpha(x)$ as in Definition (2.1), consider the auxiliary process

$$(4.1) \qquad N_\alpha^\pi(t) := \left(X^\pi(t) - \alpha \frac{M^\pi(t)}{\beta(t)} \right) (M^\pi(t))^{\frac{\alpha}{1-\alpha}}, \quad 0 \leq t < \infty.$$

Because the increasing process $M^\pi(\cdot)$ of (2.6) is flat off the set $\{t \geq 0; \beta(t)X^\pi(t) = M^\pi(t)\}$, we have from (2.4), (2.3), (4.1):

$$(4.2) \qquad \begin{aligned} d(\beta(t)N_\alpha^\pi(t)) &= (\beta(t)N_\alpha^\pi(t))\pi'(t)\sigma(t)dW_0(t), \\ W_0(t) &:= W(t) + \int_0^t \theta(s)ds. \end{aligned}$$

Consider also the processes

$$(4.3) \qquad \begin{aligned} Z(t) &:= \exp\left\{ -\int_0^t \theta'(s)dW(s) - \tfrac{1}{2}\int_0^t \|\theta(s)\|^2 ds \right\}, \\ H(t) &:= \beta(t)Z(t). \end{aligned}$$

From the product rule $d(H(t)N_\alpha^\pi(t)) = \beta(t)N_\alpha^\pi(t)dZ(t) + Z(t)d(\beta(t)N_\alpha^\pi(t)) + d\langle Z, \beta N_\alpha^\pi\rangle(t)$ and (4.2), (4.3) we obtain: $d(H(t)N_\alpha^\pi(t))$

$= H(t)N_\alpha^\pi(t)(\pi'(t)\sigma(t) - \theta'(t))dW(t)$. In other words, for any $\pi \in \mathcal{A}_\alpha(x)$ the process

$$H(t)N_\alpha^\pi(t)$$
$$= (1 - \alpha)x^{\frac{1}{1-\alpha}} \exp\left\{\int_0^t (\pi'\sigma - \theta')(s)dW(s) - \tfrac{1}{2}\int_0^t \|\pi'\sigma - \theta'\|^2(s)ds\right\}$$

(4.4)

is a positive local martingale, hence supermartingale, which thus satisfies

$$(4.5) \qquad E\left[H(T)N_\alpha^\pi(T)\right] \leq (1 - \alpha)x^{\frac{1}{1-\alpha}}, \quad \forall T \in (0, \infty).$$

We now pose an auxiliary stochastic control problem, involving the process $N_\alpha^\pi(\cdot)$ of (4.1).

4.1. An auxiliary, finite-horizon, control problem. *For a given* $T \in (0, \infty)$ *and utility function* $U : (0, \infty) \to \mathbb{R}$, *denote by* $\mathcal{A}_\alpha(x, T)$ *the class of portfolios* $\pi(\cdot)$ *that satisfy the requirements of Definition 2.1 on the finite horizon* $[0, T]$, *and find* $\hat\pi(\cdot) \in \mathcal{A}_\alpha(x, T)$ *that achieves*

$$(4.6) \qquad V(\alpha; T, x) := \sup_{\pi \in \mathcal{A}_\alpha(x, T)} EU(N_\alpha^\pi(T)).$$

There is a fairly straightforward solution to this problem along the lines of Karatzas, Lehoczky & Shreve [3], as follows: For any $y > 0$, $\pi \in \mathcal{A}_\alpha(x, T)$ we have from (3.1), (4.5):

$$(4.7) \qquad EU(N_\alpha^\pi(T)) \leq E\tilde{U}(yH(T)) + yE(H(T)N_\alpha^\pi(T)) \leq E\tilde{U}(yH(T))$$
$$+y(1 - \alpha)x^{\frac{1}{1-\alpha}}$$

The inequalities in (4.7) are equalities, if and only if $y = \hat{y}$ and $\pi(\cdot) = \hat\pi(\cdot)$ are such that both

$$(4.8) \qquad\qquad N_\alpha^{\hat\pi}(T) \quad = \quad I(\hat{y}H(T)),$$
$$(4.9) \qquad E[H(T)I(\hat{y}H(T))] \quad = \quad (1 - \alpha)x^{1/1-\alpha}$$

hold. Now $\hat{y}$ is uniquely determined from (4.9), and a portfolio $\hat\pi \in \mathcal{A}_\alpha(x)$ satisfying (4.8) can be found by introducing the positive martingale

$$\mathcal{Q}(t) := E[H(T)I(\hat{y}H(T))|\mathcal{F}(t)] = (1 - \alpha)x^{\frac{1}{1-\alpha}} + \int_0^t \mathcal{Q}(s)\varphi'(s)dW(s),$$
$$0 \leq t \leq T.$$

(4.10)

The second equality follows from the representation theorem for Brownian martingales as stochastic integrals with respect to the Brownian motion W (e.g. Karatzas & Shreve [4], §3.4), and $\varphi : [0, T] \times \Omega \to \mathbb{R}^d$ is a measurable, $\mathbb{F}$-adapted process with $\int_0^T \|\varphi(s)\|^2 ds < \infty$, a.s. Comparing (4.10) with (4.4) and recalling (2.8), we see that

$$(4.11) \quad \hat\pi(\cdot) = \left((\theta' + \varphi')\sigma^{-1}\right)'(\cdot) \in \mathcal{A}_\alpha(x, T), \quad H(\cdot)N^{\hat\pi}(\cdot) = \mathcal{Q}(\cdot), a.s.$$

In particular, (4.8) follows, and

$$(4.12) \qquad V(\alpha;T,x) = E[(U \circ I)(\hat{y}H(T))].$$

Remark 4.1. Let us consider now Problem 4.1 with utility function

$$(4.13) \qquad U(x) = \frac{1}{\gamma}x^{\gamma} \quad \text{for } \gamma := \delta(1-\alpha), \quad 0 < \delta < 1.$$

Then, with $\mu := \frac{\gamma}{1-\gamma}$, the formulae (4.9), (4.12) become

$$(4.14) \qquad \begin{aligned} \hat{y}^{-\frac{1}{1-\gamma}} &= \frac{(1-\alpha)x^{\frac{1}{1-\alpha}}}{E[(H(T))^{-\mu}]}, \\ V(\alpha;T,x) &= \frac{1}{\gamma}\left((1-\alpha)x^{\frac{1}{1-\alpha}}(E(H(T))^{-\mu})^{1/\mu}\right)^{\gamma}. \end{aligned}$$

If, in addition, the coefficients $r(\cdot)$, $b(\cdot)$, $\sigma(\cdot)$ are deterministic, then

$$\begin{aligned} (H(t))^{-\mu} &= \exp\left[\mu \int_0^t \theta'(s)dW(s) - \frac{\mu^2}{2}\int_0^t \|\theta(s)\|^2 ds\right] \\ &\quad \exp\left\{\mu \int_0^t \left(r(s) + \frac{1+\mu}{2}\|\theta(s)\|^2\right)ds\right\} \end{aligned}$$

and (4.10), (4.11), (4.14) give

$$(4.15) \qquad \begin{aligned} Q(t) &= (1-\alpha)x^{\frac{1}{1-\alpha}}\exp\left\{\mu \int_0^t \theta'(s)dW(s) - \frac{\mu^2}{2}\int_0^t \|\theta(s)\|^2 ds\right\}, \\ \varphi(t) &= \mu\theta(t) \end{aligned}$$

$$(4.16) \qquad \begin{aligned} \hat{\pi}'(t)\sigma(t) &= (1+\mu)\theta'(t) = \frac{1}{1-\delta(1-\alpha)}\theta'(t), \\ &\quad \text{independent of } T, \end{aligned}$$

$$(4.17) \qquad \begin{aligned} &V(\alpha;T,x) \\ &= \frac{1}{\gamma}\left((1-\alpha)x^{\frac{1}{1-\alpha}}\exp\left\{\int_0^T \left(r(t) + \frac{1+\mu}{2}\|\theta(t)\|^2\right)dt\right\}\right)^{\gamma}. \end{aligned}$$

Clearly, the portfolio $\hat{\pi}(\cdot)$ of (4.16) is well-defined for all $0 \le t < \infty$; it belongs to $\mathcal{A}_\alpha(x)$ of Definition 2.1 for any $x \in (0,\infty)$, by (2.8).

5. Solution of the Grossman-Zhou problem. We shall assume in this section that

$$(5.1) \qquad \left\{ \begin{array}{c} \text{the coefficients } r(\cdot), b(\cdot), \sigma(\cdot) \text{ in the model of (2.1), (2.2)} \\ \text{are deterministic,} \\ \text{and that } r_* := \lim_{T\to\infty}\frac{1}{T}\int_0^T r(s)ds, \\ \|\theta_*\|^2 := \lim_{T\to\infty}\frac{1}{T}\int_0^T \|\theta(s)\|^2 ds \\ \text{exist and are finite.} \end{array} \right\}$$

THEOREM 5.1.   *Under the assumption (5.1), the portfolio $\hat{\pi}(\cdot)$ of (4.16) is optimal for the Problem 3.1. In fact, in the notation of (3.2), (3.3), (4.17) and (5.1) we have*

$$(5.2) \qquad \lim_{T \to \infty} \frac{1}{T} \log E(X^{\hat{\pi}}(T))^{\delta} = \mathcal{R}(\hat{\pi}) = v(\alpha) = V(\alpha) + \alpha \delta r_*,$$

where

$$(5.3) \qquad \begin{aligned} V(\alpha) \quad &:= \quad \lim_{T \to \infty} \tfrac{1}{T} \log V(\alpha; T, x) = \gamma r_* + \tfrac{\gamma}{2}(1+\mu)\|\theta_*\|^2 \\ &= \quad \delta(1-\alpha)\left[r_* + \tfrac{\|\theta_*\|^2}{2}\tfrac{1}{1-\delta(1-\alpha)}\right]. \end{aligned}$$

In order to establish this result, it will be helpful to consider the auxiliary problem

$$(5.4) \quad \bar{v}(\alpha) := \sup_{\pi \in \mathcal{A}_\alpha(x)} \bar{\mathcal{R}}_\alpha(\pi), \quad \bar{\mathcal{R}}_\alpha(\pi) := \overline{\lim_{T \to \infty}} \frac{1}{T} \log E(N^\pi(T))^{\delta(1-\alpha)}.$$

From the fact that the portfolio $\hat{\pi}(\cdot)$ of (4.16) does not depend on the horizon $T \in (0, \infty)$, it is clear that

$$(5.5) \qquad \lim_{T \to \infty} \frac{1}{T} \log E(N_\alpha^{\hat{\pi}}(T))^{\delta(1-\alpha)} = \bar{\mathcal{R}}_\alpha(\hat{\pi}) = \bar{v}(\alpha) = V(\alpha).$$

It will also be helpful to note from (4.1) that

$$(5.6) \qquad (N_\alpha^\pi(t))^{\delta(1-\alpha)} = (\beta(t))^{\alpha\delta}(X^\pi(t))^\delta \left(f_\alpha\left(\frac{\alpha M^\pi(t)}{\beta(t)X^\pi(t)}\right)\right)^\delta,$$

where the function $f_\alpha(x) := \left(\frac{x}{\alpha}\right)^\alpha (1-x)^{1-\alpha}$, $0 \le x \le 1$ is strictly increasing on $(0, \alpha)$ and strictly decreasing on $(\alpha, 1)$.

5.1. Proof of Theorem 5.1. From (5.6) we obtain

$$(5.7) \qquad E(N_\alpha^\pi(T))^{\delta(1-\alpha)} \le (\beta(T))^{\alpha\delta}(1-\alpha)^{\delta(1-\alpha)}E(X^\pi(T))^\delta,$$

whence

$$(5.8) \qquad \bar{\mathcal{R}}_\alpha(\pi) \le \mathcal{R}(\pi) - \alpha \delta r_* \le v(\alpha) - \alpha \delta r_*, \quad \forall \pi \in \mathcal{A}_\alpha(x)$$

and therefore $V(\alpha) \le v(\alpha) - \alpha \delta r_*$. In order to establish the reverse inequality, take $\eta \in (0, \alpha)$ close enough to α so that $f_\eta(\eta) \ge f_\eta(\eta/\alpha)$, and observe from (5.6) that for an arbitrary $\pi \in \mathcal{A}_\alpha(x)$ ($\subseteq \mathcal{A}_\eta(x)$) we have

$$(5.9) \qquad \begin{aligned} E(N_\eta^\pi(T))^{\delta(1-\eta)} &\ge (\beta(T))^{\eta\delta}(f_\eta(\eta/\alpha))^\delta E(X^\pi(T))^\delta \\ &= (\beta(T))^{\eta\delta}\left(\alpha^{-\eta}(1-\eta/\alpha)^{1-\eta}\right)^\delta E(X^\pi(T))^\delta. \end{aligned}$$

Consequently

$$V(\eta) \geq \bar{\mathcal{R}}_\eta(\pi) \geq \mathcal{R}(\pi) - \eta \delta r_*, \quad \forall \pi \in \mathcal{A}_\alpha(x),$$

whence $V(\eta) \geq v(\alpha) - \eta \delta r_*$; letting $\eta \uparrow \alpha$ and invoking *the continuity of the function* $V(\cdot)$ *in (5.3)*, we obtain $V(\alpha) \geq v(\alpha) - \alpha \delta r_*$ and thus the third equality of (5.2):

$$(5.10) \qquad v(\alpha) = V(\alpha) + \alpha \delta r_* = \delta r_* + \frac{\|\theta_*\|^2}{2} \frac{\delta(1-\alpha)}{1 - \delta(1-\alpha)}.$$

To obtain the second equality in (5.2), it suffices to observe that (5.8), (5.5), (5.10) imply

$$v(\alpha) \geq \mathcal{R}(\hat{\pi}) \geq \bar{\mathcal{R}}_\alpha(\hat{\pi}) + \alpha \delta r_* = V(\alpha) + \alpha \delta r_* = v(\alpha).$$

Finally, the first equality in (5.2), i.e., the existence of the indicated limit, follows from the double inequality

$$-\frac{\delta(1-\alpha)}{T} \log(1-\alpha) + \frac{\alpha \delta}{T} \int_0^T r(s)ds + \frac{1}{T} \log E(N^{\hat{\pi}}(T))^{\delta(1-\alpha)}$$

$$(5.11) \quad \leq \tfrac{1}{T} \log E(X^{\hat{\pi}}(T))^\delta \leq -\tfrac{\delta}{T} \log(\alpha^{-\eta}(1 - \tfrac{\eta}{\alpha})^{1-\eta}) + \tfrac{\eta \delta}{T} \int_0^T r(s)ds$$

$$+\tfrac{1}{T} \log E(N^{\hat{\pi}}(T))^{\delta(1-\eta)}$$

(a consequence of (5.7), (5.9)) in conjunction with (5.5), by passing to the limit as $T \to \infty$ and then letting $\eta \uparrow \alpha$.

Remark 5.2. Formally setting $\alpha = 0$ in (4.16), we recover the well-known optimal portfolio $\hat{\pi}'(t)\sigma(t) = \frac{\theta'(t)}{1-\delta}$ for the investment problem without the constraint (2.5), with utility function $U(x) = \frac{1}{\delta}x^\delta$ from wealth and deterministic coefficients.

6. Maximizing long-term rate of expected logarithmic utility.
The methods of sections 4 and 5 can also be used to show that the portfolio

$$(6.1) \qquad\qquad \pi^*(t) = (\theta'(t)\sigma^{-1}(t))', \quad 0 \leq t < \infty$$

is optimal for the problem of *maximizing the long-term rate of expected logarithmic utility* under the constraint (2.5):

$$(6.2) \quad \begin{aligned} &\varlimsup_{T\to\infty} \tfrac{1}{T} E(\log X^\pi(T)) \leq \lim_{T\to\infty} \tfrac{1}{T} E(\log X^{\pi_*}(T)) \\ &= (1-\alpha)\left(\bar{r} + \tfrac{\|\bar{\theta}\|^2}{2}\right) + \alpha \bar{r}, \quad \forall \pi \in \mathcal{A}_\alpha(x). \end{aligned}$$

This problem was also considered by Grossman & Zhou [1] in their setting. It turns out that (6.2) holds for *general random*, $\mathbb{F}$-*adapted coefficients* $r(\cdot)$, $b(\cdot)$, $\sigma(\cdot)$, for which the conditions of section 2 are satisfied and the limits

$$(6.3) \quad \bar{r} := \lim_{T\to\infty} \frac{1}{T} \int_0^T Er(t)dt, \quad \|\bar{\theta}\|^2 := \lim_{T\to\infty} \frac{1}{T} \int_0^T E\|\theta(t)\|^2 dt$$

exist and are finite.

Indeed, consider $u(\alpha) = \sup_{\pi \in \mathcal{A}_\alpha(x)} \mathcal{P}(\pi)$, $\mathcal{P}(\pi) := \varlimsup_{T \to \infty} \frac{1}{T} E(\log X^\pi$
$(T))$ and $\bar{u}(\alpha) = \sup_{\pi \in \mathcal{A}_\alpha(x)} \bar{\mathcal{P}}_\alpha(\pi)$, $\bar{\mathcal{P}}_\alpha(\pi) := \varlimsup_{T \to \infty} \frac{1-\alpha}{T} E(\log N_\alpha^\pi(T))$ instead of the quantities in (3.3), (3.2) and (5.4), respectively. Solving the
Problem 4.1 with $U(x) = (1-\alpha)\log x$ leads to $\mathcal{Q}(\cdot) \equiv (1-\alpha)x^{1/1-\alpha}$, $\varphi(\cdot) \equiv$
0 in (4.10), and thus the optimal portfolio of (4.11) takes the form $\pi^*(\cdot)$
in (6.1), *independent of the finite horizon* $T > 0$; furthermore, (4.14) be-
comes $V(\alpha; T, x) = \log x + \log(1-\alpha)^{1-\alpha} + (1-\alpha)E \int_0^T \left(r(s) + \frac{1}{2}\|\theta(s)\|^2\right) ds$,
whence

$$\bar{u}(\alpha) = \lim_{T \to \infty} \frac{V(\alpha; T, x)}{T} = (1-\alpha)\left(\bar{r} + \frac{\|\bar{\theta}\|^2}{2}\right).$$

Now one writes (5.6) with $\delta = 1$, and uses exactly the same methodology
as in section 5 to prove that $\lim_{T \to \infty} \frac{1}{T} E(\log X^{\pi^*}(T)) = \mathcal{P}(\pi^*) = u(\alpha) =$
$\bar{u}(\alpha) + \alpha\bar{r}$, thus establishing (6.2).

7. Maximization of long-term growth rate from investment.
More important than the optimality property (6.2), however, is the fact
that the portfolio $\pi^*(\cdot)$ of (6.1) *maximizes the long-term growth rate from
investment*

$$(7.1) \quad \begin{aligned} \mathcal{S}(\pi) &:= \varlimsup_{T \to \infty} \frac{1}{T} \log X^\pi(T) \le \lim_{T \to \infty} \frac{1}{T} \log X^{\pi^*}(T) \\ &= (1-\alpha)\left(r^* + \frac{\|\theta^*\|^2}{2}\right) + \alpha r^*, \quad a.s. \end{aligned}$$

over all $\pi \in \mathcal{A}_\alpha(x)$. Again, this comparison is valid for *general random,*
$\mathbb{F}$-*adapted coefficients* as in section 2, under the proviso that the limits

$$(7.2) \quad r^* := \lim_{T \to \infty} \frac{1}{T} \int_0^T r(t)dt, \quad \|\theta^*\|^2 := \lim_{T \to \infty} \frac{1}{T} \int_0^T \|\theta(t)\|^2 dt$$

exist and are finite, almost surely.

In order to prove (7.1), let us start by noticing that $\Lambda(t) := N_\alpha^\pi(t)/$
$N_\alpha^{\pi^*}(t)$, $0 \le t < \infty$ satisfies the stochastic equation

$$d\Lambda(t) = \Lambda(t)(\pi'(t)\sigma(t) - \theta'(t))dW(t), \quad \Lambda(0) = 1$$

from (4.2) and Itô's rule, and is thus a positive supermartingale, for any
$\pi \in \mathcal{A}_\alpha(x)$. It follows readily from this (e.g. Karatzas [2], p. 1243) that
$\varlimsup_{t \to \infty} \frac{1}{t} \log \Lambda(t) \le 0$, or equivalently

$$\begin{aligned} \bar{\mathcal{S}}_\alpha(\pi) &:= \varlimsup_{T \to \infty} \frac{1}{T} \log(N_\alpha^\pi(T))^{1-\alpha} \le \lim_{T \to \infty} \frac{1}{T} \log(N^{\pi^*}(T))^{1-\alpha} = \\ &= (1-\alpha)\left(r^* + \frac{\|\theta^*\|^2}{2}\right) =: \bar{s}(\alpha), \quad a.s. \end{aligned}$$

(7.3)

The existence of this last limit, and its value, follow from (4.4) (which gives $N_\alpha^{\pi^*}(t) = (1 - \alpha)x^{\frac{1}{1-\alpha}}H^{-1}(t)$) and (7.2). On the other hand, the inequality (5.6) with $\delta = 1$ leads (as in (5.11)) to

$$(7.4) \quad \frac{1}{T}\log(N_\alpha^\pi(T))^{1-\alpha} + \frac{\alpha}{T}\int_0^T r(s)ds - \frac{1-\alpha}{T}\log(1-\alpha) \leq \frac{1}{T}\log X^\pi(T)$$

$$\leq \frac{1}{T}\log(N_\eta^\pi(T))^{1-\eta} + \frac{\eta}{T}\int_0^T r(s)ds - \frac{1}{T}\log(\alpha^{-\eta}(1-\eta/\alpha)^{1-\eta})$$

almost surely, for any $\pi \in \mathcal{A}_\alpha(x) \subseteq \mathcal{A}_\eta(x)$ and any $\eta \in (0, \alpha)$ sufficiently close to α. In particular, (7.4) gives

$$(7.5) \qquad \bar{S}_\alpha(\pi) + \alpha r^* \leq s(\alpha) := \operatorname*{esssup}_{\pi \in \mathcal{A}_\alpha(x)} \mathcal{S}(\pi), \text{ a.s.}$$

in the notation of (7.1)-(7.3), whence $\bar{s}(\alpha) + \alpha r^* \leq s(\alpha)$, a.s.; similarly,

$$\mathcal{S}(\pi) - \eta r^* \leq \bar{S}_\eta(\pi) \leq \bar{s}(\eta), \text{ whence } s(\alpha) - \eta r^* \leq \bar{s}(\eta)$$

and in the limit as $\eta \uparrow \alpha$: $s(\alpha) - \alpha r^* \leq \bar{s}(\alpha)$, a.s. It develops that $s(\alpha) = \bar{s}(\alpha) + \alpha r^* = (1 - \alpha)\left(r^* + \frac{1}{2}\|\theta^*\|^2\right) + \alpha r^*$, and it remains to show the existence of the limit and the equality in (7.1). But both of these follow by writing the double inequality (7.4) with $\pi \equiv \pi^*$, letting $T \to \infty$ to obtain in conjunction with (7.3)

$$s(\alpha) = \bar{s}(\alpha) + \alpha r^* \leq \varliminf_{T\to\infty} \frac{1}{T}\log X^{\pi^*}(T) \leq \varlimsup_{T\to\infty} \frac{1}{T}\log X^{\pi^*}(T) \leq s(\eta),$$

and then letting $\eta \uparrow \alpha$ to conclude $\lim_{T\to\infty} \frac{1}{T}\log X^{\pi^*}(T) = s(\alpha)$, almost surely.

A. Appendix. We devote this section to the proof of the claim (2.8). Clearly, it suffices to show that for any $x \in (0, \infty)$ and $\rho(\cdot)$ as in (2.8), the stochastic equation

$$(A.1) \qquad \begin{aligned} d\hat{X}(t) &= (\hat{X}(t) - \alpha\hat{M}(t))\rho'(t)dW_0(t), \\ \hat{M}(t) &= \max_{0\leq s\leq t}\hat{X}(s); \quad \hat{X}(0) = x \end{aligned}$$

admits a unique $\mathbb{F}$-adapted solution that satisfies a.s.

$$(A.2) \qquad \hat{X}(t) > \alpha\hat{M}(t), \quad \forall 0 \leq t < \infty.$$

For then $\hat{X}(\cdot)/\beta(\cdot)$ coincides with $X^{\hat{\pi}}(\cdot)$, the wealth process corresponding to the portfolio $\hat{\pi} = (\rho'\sigma^{-1})'$ according to (2.4), and (2.5) is satisfied; and vice-versa.

Suppose that $\hat{X}(\cdot)$ is an $\mathbb{F}$-adapted process that satisfies (A.1), (A.2). Following Grossman & Zhou (1993), p. 269, observe that

$$d\left(\frac{\hat{X}(t)}{\hat{M}(t)} - \alpha\right) = \left(\frac{\hat{X}(t)}{\hat{M}(t)} - \alpha\right)\rho'(t)dW_0(t) - \frac{d\hat{M}(t)}{\hat{M}(t)}, \text{ whence}$$

(A.3) $d\left(\log\left(\frac{\hat{X}(t)}{\hat{M}(t)} - \alpha\right)\right) = d\xi(t) - \frac{1}{1-\alpha}\frac{d\hat{M}(t)}{\hat{M}(t)},$

$$\xi(t) := \int_0^t \rho'(s)dW_0(s) - \frac{1}{2}\int_0^t \|\rho(s)\|^2 ds.$$

Therefore,

(A.4) $0 \leq R(t) := \log(1-\alpha) - \log\left(\frac{\hat{X}(t)}{\hat{M}(t)} - \alpha\right) = -\xi(t) + \log\left(\frac{\hat{M}(t)}{x}\right)^{\frac{1}{1-\alpha}}.$

Clearly, the continuous increasing process $K(t) := \log\left(\frac{\hat{M}(t)}{x}\right)^{\frac{1}{1-\alpha}}$ is flat away from the set $\left\{t \geq 0/\hat{X}(t) = \hat{M}(t)\right\}$, i.e., away from the zero-set of the continuous nonnegative process $R(\cdot)$ of (A.3). From the theory of the Skorohod equation (e.g. Karatzas & Shreve [4], §3.6) we have then $K(t) = \max_{0 \leq s \leq t} \xi(s)$, and from this and (A.3):

(A.5) $\qquad \hat{M}(t) \equiv \tilde{M}(t) := x\exp\left\{(1-\alpha)\max_{0\leq s\leq t}\xi(s)\right\},$

(A.6)
$$\hat{X}(t) \equiv \tilde{X}(t) := x\exp\left\{(1-\alpha)\max_{0\leq s\leq t}\xi(s)\right\}$$
$$\left[\alpha + (1-\alpha)\exp\left\{\xi(t) - \max_{0\leq s\leq t}\xi(s)\right\}\right].$$

It is straightforward to check that $\tilde{X}(\cdot)$ satisfies (A.1), (A.2). Notice also (from (A.4), (A.5) and (4.1)) that

(A.7) $\beta(t)N_\alpha^{\hat{\pi}}(t) = (\hat{X}(t) - \alpha\hat{M}(t))(\hat{M}(t))^{\frac{\alpha}{1-\alpha}} = (1-\alpha)x^{\frac{1}{1-\alpha}}e^{\xi(t)}.$

REFERENCES

[1] S.J. GROSSMAN AND Z. ZHOU, *Optimal investment strategies for controlling drawdowns*, Math. Finance **3** (3), (1993), pp. 241–276.

[2] I. KARATZAS, *Optimization problems in the theory of continuous trading*, SIAM J. Control & Optimization **27**, (1989) pp. 1221-1259.

[3] I. KARATZAS, J.P. LEHOCZKY AND S.E. SHREVE, *Optimal portfolio and consumption decisions for a "small investor" on a finite horizon*, SIAM J. Control & Optimization **25**, (1987), pp. 1157–1586.

[4] I. KARATZAS AND S.E. SHREVE, *Brownian Motion and Stochastic Calculus* (2nd edition), Springer-Verlag, New York (1991).

[5] R.C. MERTON, *Optimum consumption and portfolio rules in a continuous time model*, J. Econom. Theory **3**, (1971), pp. 373–413, Erratum: Ibid **6** (1973), pp. 213–214.

AMERICAN OPTIONS AND TRANSACTION FEES

MARK H.A. DAVIS* AND THALEIA ZARIPHOPOULOU[†]

Abstract. We consider the problem of pricing American options in a market model similar to the Black-Scholes one except that proportional transaction charges are levied on all sales and purchases of stock. "Perfect replication" is no longer possible and holding an option involves an essential element of risk. We obtain a definition of the option price by computing the maximum price at which a utility-maximizing investor would include the option in his portfolio. This definition reduces to the Black-Scholes value for American options when the transaction costs are removed. Computing the price involves solving a combination of stochastic control problems with singular policies and optimal stopping.

Key words. Option pricing, Black-Scholes formula, transaction costs, utility maximization, stochastic control, free boundary problem, variational inequalities, constrained viscosity solutions

AMS(MOS) subject classifications. Primary 35R35, 90A16, 93E20; Secondary 35R45, 49B60, 90A09

1. Introduction. The classic Black-Scholes option pricing theory relies on ideas of arbitrage and "perfect replication": the price of an option is the endowment required to form a portfolio whose value at any time exactly coincides with that of the option. The advantage of this approach is that option valuation is *preference independent*: the writer's or buyer's attitude to risk is irrelevant since all risk is "hedged away". However, the theory is unsatisfying on at least two counts:

(a) perfect hedging is an idealization that can never be realized in practice; and

(b) the theory does not explain why options exist. (Why not just form the hedging portfolio directly?)

If we abandon perfect hedging then preference-independent valuation formulas are no longer possible, but in return we might hope to provide some answer to (b). This is a primary objective of the present paper, which continues the investigations begun by Hodges and Neuberger into option pricing in a continuous-time market model with proportional transaction costs using an approach based on utility maximization.

In our earlier paper [6] with V. G. Panas, we considered pricing of European call options from the option writer's point of view, comparing the two situations where (i) the writer forms a portfolio with the objective of maximizing utility of wealth at a given time T, and (ii) the same, except that the writer accepts a payment at time 0 and the obligation to fulfill

---

* Department of Electrical and Electronic Engineering, Imperial College, London SW7 2BT England.

† Partially supported by NSF Grant DMS 9204866. Department of Mathematics and Finance, University of Wisconsin, Madison, WI 53706 USA.

the terms of an option contract with exercise time T. The fair price was then defined as the payment required so that the maximum utility in the two cases is the same. This is the minimum price at which it is worth the writer's while to write the option.

This approach fails in the case of American options since it is the buyer and not the writer who determines the exercise strategy, and without knowing this the writer cannot make the above computation. In this paper we take a complementary approach to [6] and ask what is the maximum price a buyer would be prepared to pay for an option (which we *can* now take as American). Of course, the answer depends on what the buyer is trying to achieve, and we assume the objective is maximization of utility at time T. The option is one instrument available for investment, and will be purchased if doing so augments utility; this addresses point (b) above.

The problem is stated formally in section 2 below. We suppose that options on some number of shares are purchased at time zero and either held to maturity or exercised at some intermediate time τ. (It would be better also to allow trading of the options, but this is beyond our scope at present.) In section 3 we show that calculating the price involves solving two singular stochastic control problems in which the solution of the first acts as boundary data for the second. The main technical results of the paper are in section 3, showing that the value functions of these control problems are the unique viscosity solutions of the corresponding Bellman equations. This means (see [1], [7] or [12, Chapter 9]) that these solutions can be computed by standard discretization methods (at the time of writing we have not yet done this.) The second problem is interesting from a stochastic control point of view in that it combines *singular control* and *optimal stopping*. Such problems have, to our knowledge, never been considered in the literature except for the preliminary exploration of Davis and Zervos [7] which uncovered the fact that optimal strategies take some interesting and perhaps unexpected forms. This whole area undoubtedly merits further investigation.

2. American option pricing via utility maximization. In this section, we give a general definition of the fair price for American options based on utility maximization. This is a generalization of the results on pricing European options by the authors and V. Panas in [6]. Many of the key ideas of the utility maximization approach were introduced by Hodges and Neuberger [9]. The main result of this section, which demonstrates that our approach is well founded, is Theorem 2.1, which shows that if the transaction costs vanish, then our price reduces to the Black and Scholes option price in the framework of markets with no friction.

We consider a time interval $[0, T]$ and a market, which consists of a stock whose price $S(t)$ is assumed to be a stochastic process on a given probability space $(\Omega, \mathcal{F}, \mathcal{F}_t, P)$. Investors can also keep their funds in cash, i.e. a risk-free asset, denoted by B. We wish to give a price applicable at time 0 for an American option with exercise time τ in $[0, T]$ on the stock $S(t)$.

Let $\mathcal{T}(B)$ denote the set of *admissible trading strategies* available to an investor who starts at time 0 with an amount x in cash and no holding in stock. We identify an element $\pi \in \mathcal{T}(B)$ with a vector stochastic process $(z^\pi(t), y^\pi(t))$, $t \in [0, T]$, where $z^\pi(t)$ denotes the *amount held in a bond account with interest rate* $r > 0$ and $y^\pi(t)$ the *number of shares* of stock held, over $[0, T]$ (this may or may not be constrained to an integral number). There may be costs associated with transactions. In particular, $c(y, S)$ is the liquidated cash value of the portfolio component y, i.e. the residual cash value when long positions $(y > 0)$ are sold and short positions $(y < 0)$ are closed. We assume that $c(0, S) = 0$.

An American option on stock $S(t)$ is defined by an adapted process X_t and it is the claim to the payoff X_τ at a stopping (exercise) time τ chosen by the holder of the security. The classic example is the case of a put option, when $X_t = (K - S(t))^+$, where K is the prespecified exercise price. For simplicity of exposition, we discuss only the American option in this paper, although the approach is equally applicable to other types of options.

We suppose that the buyer of the option starts with wealth x in cash. At time $t = 0$, he splits his wealth into amounts x_1 and $x_2 = x - x_1$ and starts the following funds: he uses the amount x_1 to buy, at $t = 0$, $\frac{x_1}{p}$ shares of an American option written on $S(t)$, with p being the price of the option, to be determined. He also uses the rest of the cash x_2 to form a portfolio $\pi \in \mathcal{T}(x_2)$ using the bond and the stock.

Let $\mathcal{U} : \mathbb{R} \to \mathbb{R}$, the buyer's utility function, be a concave increasing function such that $\mathcal{U}(0) = 0$. Note that $\mathcal{U}(x)$ is defined for both positive and negative wealth. The use of the second fund is to maximize the investor's utility of terminal wealth given by
(2.1)
$$V(t, S, y, z) = \sup_{\mathcal{A}} E[\mathcal{U}(z(T) + c(y(T), S(T)) \mid z(t) = z, y(t) = y, S(t) = S]$$

where $0 \leq t \leq T$ and $\mathcal{A}$ is the set of admissible policies to be defined in detail later on.

At time τ, the buyer decides to exercise the option and transfers the proceeds to the portfolio fund. In other words, he receives the amount of $\frac{Kx_1}{p}$ dollars and surrenders to the writer $\frac{x_1}{p}$ shares of the stock whose value is, of course, $\frac{x_1 S(\tau)}{p}$. If $(y(\tau), z(\tau))$ is the composition of his portfolio at time τ, after the transfers are made the new portfolio components at time τ^+ are

(2.2)
$$\left(z(\tau) + \frac{Kx_1}{p}, y(\tau) - \frac{x_1}{p} \right)$$

Thus, let

(2.3) $V_1(t, S, y, z; x_1) = V\left(t, S, y - \frac{x_1}{p}, z + \frac{Kx_1}{p}\right)$

and define the value function
(2.4)
$U(t, S, y, z; x_1) = \sup_{\mathcal{A}, \tau} E[V_1(\tau, S(\tau), y(\tau), z(\tau); x_1) \mid S(t) = S, y(t) = y, z(t) = z]$

where τ is a stopping time in $[t, T]$.

Next, we look at the original amounts x_1 and x_2 and we define the functions $\alpha : \mathbb{R}_+^4 \to \mathbb{R}$, $\beta : \mathbb{R}_+^3 \to \mathbb{R}$ given by

(2.5) $\alpha(x_1, x_2, S, p) = U(0, S, 0, x_2; x_1)$

and

(2.6) $\beta(x, S, p) = \sup_{x_1 + x_2 = x} \alpha(x_1, x_2, S, p).$

Finally, assuming that β is well defined, we let

(2.7) $X^*(p, S, x) = \arg\max_{x_1} \alpha(x_1, x - x_1, S, p)$

and we define the writing price of the American option to be

(2.8) $p^*(S) = \sup_x \{p : X^*(p, S, x) > 0\}.$

Thus, the fair price is defined as *the maximum price at which a positive investment is made in the option at time $t = 0$.*

We conclude this section by giving a primary justification for this definition. We will show that it reduces to the Black-Scholes valuation price in the case where it is applicable.

THEOREM 2.1. *Suppose that a replicating portfolio exists and that there are no position limits. Then $p^*(S)$ is equal to the initial endowment of the replicating portfolio.*
Proof. Suppose that the option is offered at a price $p < p_0$, where p_0 is the initial endowment of the replicating portfolio; then the buyer has an arbitrage opportunity. Given initial wealth x, he can take x_1 and purchase x_1/p options, at the same time taking a short position in the hedging portfolio, whose value is $x_1 p_0/p$. He now has an excess fund of $x_1(p_0 - p)/p$ which can be invested to yield a riskless profit, and the value of the option given optimal exercise is always enough to close the short position. Thus the optimal x_1 in this case is $+\infty$. It is $-\infty$ if $p > p_0$. Thus $p^*(S) = p_0$.   □

3. Transaction costs: The Bellman equation for the value functions. It is now assumed that investors must pay transaction costs, which are proportional to the amount transferred from the stock to the bank. A market model, similar to that of Davis and Norman [5], is described. The main purpose of this section is to give a heuristic derivation of the fully nonlinear partial differential equation, actually a combination of variational inequalities with gradient and obstacle constraints, satisfied by all the value functions of the utility maximization problems stated in Section 2.

The cash value of a number of shares, $y(t)$, of the stock is

$$(3.1) \qquad c(y(t), S(t)) = \begin{cases} (1 + \lambda)y(t)S(t); & \text{if } y(t) < 0 \\ \\ (1 - \mu)y(t)S(t); & \text{if } y(t) \geq 0 \end{cases}$$

where λ and μ are the fraction of the traded amount in stock, which the investor pays in transaction costs when buying or selling stock respectively. The time horizon considered is $[0, T]$ and the market model equations are

$$(3.2) \qquad dz(t) = rz(t)dt - (1 + \lambda)S(t)dL(t) + (1 - \mu)S(t)dM(t)$$

$$(3.3) \qquad dy(t) = dL(t) - dM(t)$$

$$(3.4) \qquad dS(t) = S(t)(\alpha dt + \sigma dW(t))$$

where $L(t)$ and $M(t)$ are the cumulative number of shares bought or sold, respectively, over $[0, T]$, $W(t)$ is a P-Brownian motion which represents the single source of uncertainty, α and σ are non-random constants, and $\mathcal{F}_t$ denotes the natural filtration of $W(t)$. This system of equations describes a degenerate diffusion in $\mathbb{R}^3$.

DEFINITION 3.1. The set of trading strategies, $\mathcal{T}(B)$, consists of all the 2-dimensional, right-continuous, measurable processes, $(z^\pi(t), y^\pi(t))$, which are the solution of equations (3.2) to (3.4), corresponding to some pair of right-continuous, measurable, $\mathcal{F}_t$-adapted, increasing control processes, $(L(t), M(t))$, such that:

$$(3.5) \qquad (z^\pi(t), y^\pi(t), S(t)) \in \mathcal{S}_{\overline{K}}, \quad \forall t \in [0, T]$$

where $\overline{K}$ is a large constant and

$$(3.6) \qquad \mathcal{S}_{\overline{K}} = \{(z, y, S) \in \mathbb{R} \times \mathbb{R} \times \mathbb{R}^+ : z + c(y, S) > -\overline{K}\}.$$

By convention, $L(0-) = M(0-) = 0$.

REMARK: Investors may start with any combination of the two assets, at time $s \in [0, T]$, in the general utility maximization problem and in that case, the class of admissible trading strategies depends on the time s and the initial portfolio, which is characterized by the initial amount in the bank, z, an initial number of shares, y, and the initial value of the stock, S. The constraint (3.5) is required for technical reasons, in Section 4, and it only rules out strategies which are clearly non-optimal, as the objective is the maximization of the utility of final wealth. Also, either $L(0)$ or $M(0)$ may be positive, i.e. a jump at the initial time is possible. Finally, (3.2) and (3.3) imply that the trading strategies are self financing.

Next, we derive the Hamilton-Jacobi-Bellman equations of the two singular stochastic control problems, defined, respectively, by (2.1) which corresponds to the utility maximization by forming a portfolio and by (2.4) which correspond to employing the option. We start with the optimization problem given by (2.1) and we derive, at least in a formal way, the associated HJB equation. Consider, temporarily, a smaller class of trading strategies $\mathcal{T}'$, such that $L(t)$ and $M(t)$ are absolutely continuous processes, given by:

$$(3.7) \qquad L(t) = \int_s^t l(\xi)d\xi \ \text{ and } \ M(t) = \int_s^t m(\xi)d\xi$$

where $l(\xi)$ and $m(\xi)$ are non-negative and uniformly bounded by $k < \infty$. Then, (3.2)–(3.4) is a vector stochastic differential equation with controlled drift and the value function of this approximate problem denoted by V^k satisfies the HJB equation

$$\max_{0 \leq l, m \leq k} \left\{ \left(\frac{\partial V^k}{\partial y} - (1 + \lambda)S \frac{\partial V^k}{\partial z} \right) l - \left(\frac{\partial V^k}{\partial y} - (1 - \mu)S \frac{\partial V^k}{\partial z} \right) m \right\} +$$

$$(3.8) \qquad\qquad + \frac{\partial V^k}{\partial s} + rz \frac{\partial V^k}{\partial z} + \alpha S \frac{\partial V^k}{\partial S} + \frac{1}{2} \sigma^2 S^2 \frac{\partial^2 V^k}{\partial S^2} = 0,$$

$$V^k(T, z, y, S) = U(z + c(y, S)),$$

for $(s, z, y, S) \in [0, T] \times \mathbb{R} \times \mathbb{R} \times \mathbb{R}^+$. The optimal trading strategy is determined by considering the following three possible cases:

$$(3.9) \qquad\qquad (\text{i}) \quad \frac{\partial V^k}{\partial y} - (1 + \lambda)S \frac{\partial V^k}{\partial z} \geq 0,$$

where the maximum is achieved by $m = 0$ and buying at the maximum possible rate, $l = k$,

$$(3.10) \qquad\qquad (\text{ii}) \quad \frac{\partial V^k}{\partial y} - (1 - \mu)S \frac{\partial V^k}{\partial z} \leq 0,$$

where the maximum is achieved by $l = 0$ and selling at the maximum possible rate, $m = k$,

$$(3.11) \qquad \text{(iii)} \quad (1 - \mu)S\frac{\partial V^k}{\partial z} \leq \frac{\partial V^k}{\partial y} \leq (1 + \lambda)S\frac{\partial V^k}{\partial x},$$

where the maximum is achieved by doing nothing, that is $m = 0$ and $l = 0$. (Note that in this case the process $(z^\pi(t), y^\pi(t), S(t))$ becomes an uncontrolled diffusion, which drifts under the influence of the stock process only.) All the other permutations of inequalities are impossible, as all the value functions are increasing functions of z and y.

The above results suggest that the optimization problem is a free boundary problem, where if the value function is known in the 4-dimensional space, defined by the state of the investor (s, z, y, S), the optimal trading strategy is determined by the above inequalities. Also, the state space is divided into 3 regions called the Buy, Sell and No-Transaction regions, which are characterized by (3.9), (3.10) and (3.11), respectively. Clearly, the Buy and Sell regions do not intersect, as it is not optimal to buy and sell at the same time. The boundaries between the No-Transaction region and the Buy and Sell regions are denoted by ∂B and ∂S.

As $k \to \infty$, the class of admissible trading strategies becomes the class defined at the beginning of this section. It is conjectured that the state space remains divided into a Buy region, a Sell region and a No-Transaction region, and the optimal trading strategy mandates an immediate transaction to ∂B or ∂S if the State is in the Buy region or Sell region, followed by transactions of "local time" type at ∂B and ∂S. Therefore, each of the value functions satisfies the following set of equations:

(i)   In the Buy region, the value function remains constant along the path of the state, dictated by the optimal trading strategy and therefore

$$(3.12) \qquad V(s, z, y, S) = V(s, z - (1 + \lambda)S\delta y_b, y + \delta y_b, S)$$

where δy_b, the number of shares bought by the investor, can take any positive value up to the number required to take the state to ∂B. Allowing $\delta y_b \to 0$, equation (3.12) becomes

$$(3.13) \qquad \frac{\partial V}{\partial y} - (1 + \lambda)S\frac{\partial V}{\partial z} = 0.$$

(ii) Similarly, in the Sell region, the value function obeys the equation:

$$(3.14) \qquad V(s, z, y, S) = V(s, z + (1 - \mu)S\delta y_s, y - \delta y_s, S)$$

where δy_s, the number of shares sold by the investor, can take any positive value up to the number required to take the state to ∂S. Allowing $\delta y_s \to 0$, equation (3.14) becomes

$$(3.15) \qquad \frac{\partial V}{\partial y} - (1 - \mu)S\frac{\partial V}{\partial z} = 0.$$

(iii) In the No-Transaction region, the value function obeys the same set of equations obtained for the class of absolutely continuous trading strategies, and therefore the value function is given by

$$(3.16) \qquad \frac{\partial V}{\partial s} + rz\frac{\partial V}{\partial z} + \alpha S\frac{\partial V}{\partial S} + \frac{1}{2}\sigma^2 S^2 \frac{\partial^2 V}{\partial S^2} = 0$$

and the pair of inequalities, shown above in (3.11) also hold. Note that, due to the continuity of the value function, if the value function is known in the No-Transaction region, it can be determined in the Buy region and Sell region by equations (3.12) and (3.14), respectively.

In the Buy region, the left hand side of equation (3.15) is negative and in the Sell region, the left hand side of equation (3.13) is positive. Also, from the two pairs of inequalities (3.9) and (3.10), it is conjectured that the left hand side of equation (3.16) is negative in both the Buy and Sell regions stopping. Therefore, the above set of equations is condensed into the fully non-linear partial differential equation

$$\min\left\{ -\frac{\partial V}{\partial y} + (1+\lambda)S\frac{\partial V}{\partial z}, \quad \left(\frac{\partial V}{\partial y} - (1-\mu)S\frac{\partial V}{\partial z}\right), \right.$$

$$(3.17) \qquad \left. -\left(\frac{\partial V}{\partial s} + rz\frac{\partial V}{\partial z} + \alpha S\frac{\partial V}{\partial S} + \frac{1}{2}\sigma^2\frac{\partial^2 V}{\partial S^2}\right)\right\} = 0$$

for $(s, z, y, S) \in [0, T] \times \mathbb{R} \times \mathbb{R} \times \mathbb{R}^+$.

For the optimization problem given in (2.4) we observe that if we let the stopping time $\tau = t$, then this policy is, in general, suboptimal. Therefore

$$(3.18) \qquad U(t, S, y, z) \geq V_1(t, S, y, z) \quad \text{in } [0, T] \times \mathcal{S}_{\overline{K}}.$$

Using for U arguments similar to the ones we used above for V and taking into account (3.18), we get that, at least formally, U solves

$$\min\left\{ U - V_1, -\left(\frac{\partial U}{\partial s} + rz\frac{\partial U}{\partial z} + \alpha S\frac{\partial U}{\partial S} + \frac{1}{2}\sigma^2\frac{\partial^2 U}{\partial S^2}\right), \right.$$

$$(3.19) \qquad \left. (1+\lambda)S\frac{\partial U}{\partial z} - \frac{\partial U}{\partial y}, -(1-\mu)S\frac{\partial U}{\partial z} + \frac{\partial U}{\partial y}\right\} = 0.$$

Note that (3.19) is a degenerate Variational Inequality with gradient constraints, due to the presence of transaction costs as well as an obstacle constraint which is inherent from the possibility of early stopping.

4. Existence and uniqueness of the solutions of the (HJB) equations. In this section we characterize the value functions V and U given by (2.1) and (2.4) as constrained weak (viscosity) solutions of the

Variational Inequalities (3.17) and (3.19). Since the stochastic control problem whose value function is given by (2.1) is a special case, with different terminal conditions of problem (2.4), we only examine the problem with value function U. We next show that this value function is a constrained viscosity solution of (3.19) on $[0, T] \times \overline{\mathcal{S}_{\overline{K}}}$, where $\mathcal{S}_{\overline{K}}$ is defined by (3.6); the characterization of U as a constrained viscosity solution of (3.19) is natural due to the presence of the state constraints (3.5).

The notion of *viscosity solutions* was introduced by Crandall and Lions [3] for first order and by Lions [12] for second order equations. For a general overview of the theory we refer to the *User's Guide* by Crandall, Ishii and Lions [4] and the book of Fleming and Soner [8]. Next we recall the notion of constrained viscosity solutions, which was introduced by Soner [13] and Capuzzo-Dolcetta and Lions [2] for first order equations (see also Ishii and Lions [10] and Katsoulakis [11] for second-order equations). To this end, we consider a non-linear second order partial differential equation of the form

$$(4.1) \qquad F(X, W, DW, D^2W) = 0, \quad \text{in } [0, T] \times \mathcal{S}$$

where $\mathcal{S} \subseteq \mathbb{R}^n$, DW and D^2W stand for the gradient vector and the second derivative matrix of W and the function F is continuous in all its arguments and degenerate elliptic, meaning that

$$(4.2) \qquad F(X, p, q, A + N) \leq F(X, p, q, A), \quad \text{if } N \geq 0,$$

for A a nonnegative definite symmetric matrix.

DEFINITION 4.1. A continuous function $W : [0, T] \times \overline{\mathcal{S}} \to \mathbb{R}$ is a *constrained viscosity solution* of (4.1) if

(i) W is a *viscosity subsolution* of (4.1) on $[0, T] \times \overline{\mathcal{S}}$, that is if for any $\phi \in C^{1,2}([0, T] \times \overline{\mathcal{S}})$ and any local maximum point $X_0 \in [0, T] \times \overline{\mathcal{S}}$ of $W - \phi$

$$(4.3) \qquad V(X_0, W(X_0), D\phi(X_0), D^2\phi(X_0))) \leq 0$$

and

(ii) W is a *viscosity supersolution* of (4.1) on $[0, T] \times \mathcal{S}$, that is if for any $\phi \in C^{1,2}([0, T] \times \overline{\mathcal{S}})$ and any local minimum point $X_0 \in [0, T] \times \mathcal{S}$ of $W - \phi$

$$(4.4) \qquad F(X_0, W(X_0), D\phi(X_0), D^2\phi(X_0)) \geq 0.$$

THEOREM 4.1. *The value function $U(s, S, y, z)$ defined by (2.4), is a constrained viscosity solution of*

$$\min \left\{ W - V_1, -\frac{\partial W}{\partial y} + (1 + \lambda)S\frac{\partial W}{\partial z}, \frac{\partial W}{\partial y} - (1 - \mu)S\frac{\partial W}{\partial z}, \right.$$

$$(4.5) \qquad -\left(\frac{\partial W}{\partial s} + rz\frac{\partial W}{\partial z} + \alpha S\frac{\partial W}{\partial S} + \frac{1}{2}\sigma^2 S^2 \frac{\partial^2 W}{\partial S^2}\right)\Bigg\} = 0$$

on $[0,T] \times \overline{\mathcal{S}_{\overline{K}}}$.

Proof. In our case, the state X is (s,x), where $x = (z,y,S) \in \overline{\mathcal{S}_{\overline{K}}}$. Let $X_0 = (s_0, z_0, y_0, S_0) \in [0,T] \times \overline{\mathcal{S}_K}$: it follows, from a straightforward generalization of the results of Zhu [Th. iv.2.2, 14], that there exists an optimal trading strategy, dictated by the triple $(L^*(t), M^*(t), \tau^*(t))$, where $X_0^*(t) = (t, B_0^*(t), y_0^*(t), S_0^*(t))$ is the optimal trajectory, with $X_0^*(s_0) = X_0$.

(i) First we prove that U is a viscosity subsolution of (4.5) on $[0,T] \times \overline{\mathcal{S}_{\overline{K}}}$; for this we need to show that for all smooth functions, $\phi(X)$, such that $U_1(X) - \phi(X)$ has a local maximum at $X_0 \in [0,T] \times \overline{\mathcal{S}_{\overline{K}}}$, the following inequality holds

$$\min\left\{ U(X_0) - V_1(X_0), -\frac{\partial\phi(X_0)}{\partial y} + (1+\lambda)S_0 \frac{\partial\phi(X_0)}{\partial z}, \frac{\partial\phi(X_0)}{\partial y} \right.$$
$$-(1-\mu)S_0\frac{\partial\phi(X_0)}{\partial z},$$

$$(4.6)$$
$$\left. -\left(\frac{\partial\phi(X_0)}{\partial s} + rz_0\frac{\partial\phi(X_0)}{\partial z} + \alpha S_0\frac{\partial\phi(X_0)}{\partial S} + \frac{1}{2}\sigma^2 S_0^2 \frac{\partial^2\phi(X_0)}{\partial S^2}\right)\right\} \leq 0.$$

Without loss of generality we assume that $U(X_0) = \phi(X_0)$ and $U \leq \phi$ on $[0,T] \times \overline{\mathcal{S}_{\overline{K}}}$. We first observe that (4.6) follows if $\varphi(X_0) \leq V_1(X_0)$. Therefore, it remains to prove (4.6) when

$$(4.7) \qquad \varphi(X_0) - V_1(X_0) > 0.$$

We are going to argue by contradiction: assume that the rest of the arguments inside the minimum operator of equation (4.6) satisfy

$$(4.8) \qquad \frac{\partial\phi(X_0)}{\partial y} - (1+\lambda)S_0\frac{\partial\phi(X_0)}{\partial z} < 0,$$

$$(4.9) \qquad \frac{\partial\phi(X_0)}{\partial y} - (1-\mu)S_0\frac{\partial\phi(X_0)}{\partial z} > 0,$$

and that there exists $\theta > 0$, such that:

$$(4.10) \qquad \frac{\partial\phi(X_0)}{\partial s} + rz_0\frac{\partial\phi(X_0)}{\partial z} + \alpha S_0\frac{\partial\phi(X_0)}{\partial S} + \frac{1}{2}\sigma^2 S_0^2\frac{\partial^2\phi(X_0)}{\partial S^2} < -\theta.$$

From the fact that ϕ is smooth the above inequalities become

$$(4.11) \qquad \phi(X) - V_1(X) > 0$$

(4.12)
$$\frac{\partial \phi(X)}{\partial y} - (1 + \lambda) S \frac{\partial \phi(X)}{\partial z} < 0,$$

(4.13)
$$\frac{\partial \phi(X)}{\partial y} - (1 - \mu) S \frac{\partial \phi(X)}{\partial z} > 0,$$

and

(4.14)
$$\frac{\partial \phi(X)}{\partial s} + rz \frac{\partial \phi(X)}{\partial z} + \alpha S \frac{\partial \phi(X)}{\partial S} + \frac{1}{2} \sigma^2 S^2 \frac{\partial^2 \phi(X)}{\partial S^2} < -\theta,$$

where $X = (s, B, y, S) \in \mathcal{B}(X_0)$, a neighborhood of X_0.

We observe that (3.18) together with (4.7) and $U(X_0) = \varphi(X_0)$ imply that in a sufficiently small neighborhood, say $B(X_0)$, of X_0 it is not optimal to exercise a "stopping" policy, i.e. to have $X(t) = X(s)$ for $s < t \leq T$ where $X(s) = (s, x)$ with $(s, x) \in B(X_0)$. Next, we claim that $\tau(\omega)$ defined by

$$\tau(\omega) = \inf\{t \in [s_0, T] : X_0^*(t) \notin B(X_0)\}$$

is positive P – a.s. or, equivalently, that the optimal trajectory has no jumps along the direction $(0, -(1+\lambda)S_0, 1, 0)$ (or $(0, (1-\mu)S_0, -1, 0)$) at $t = s_0^+$.

The proof of this argument for the optimization problem (2.3) can be found in Lemma 1 of [6]). We next assume that the optimal trajectory $X_0^*(t)$ has a jump of size ϵ, along the direction $(0, -(1+\lambda)S_0, 1, 0)$ and we denote this event by $A(\omega)$. By the principle of dynamic programming

$U(s_0, z_0, y_0, S_0)$
$$= E\{\max(U(s_0, z_0 - (1+\lambda)S_0\epsilon, y_0 + \epsilon, S_0), V_1(s_0, z_0 -$$
$(1+\lambda)S_0\epsilon, y_0 + \epsilon, S_0)\} =$
$$= \int_{A(\omega)} \max(U(s_0, z_0 - (1+\lambda)S_0\epsilon, y_0 + \epsilon, S_0), V_1(s_0, z_0 -$$
$(1+\lambda)S_0\epsilon, y_0 + \epsilon, S_0))dP$
$$+ \int_{\Omega - A(\omega)} \max(U(s_0, z_0 - (1+\lambda)S_0\epsilon, y_0 + \epsilon, S_0), V_1(s_0, z_0 -$$
$(1+\lambda)S_0\epsilon, y_0 + \epsilon, S_0))dP.$

Therefore,

$$\int_{A(\omega)} (\varphi(s_0, z_0 - (1+\lambda)S_0\epsilon, y_0 + \epsilon, S_0) - \varphi(s_0, z_0, y_0, S_0))dP \geq 0$$

since $U(X_0) = \varphi(X_0)$, $U(X) \leq \varphi(X)$ for $X \neq X_0$ and $V_1 \leq U$ $\forall X$. Dividing by ϵ the above inequality, sending $\epsilon \to 0$ and using Fatou's Lemma we contradict (4.12). Working similarly we can also contradict (4.13). It remains to contradict (4.14). Applying Itô's formula to $\varphi(X^*(t))$ and taking into account (4.14) we get

$$E\varphi(X_0^*(\tau)) < \varphi(X_0) + E \int_{s_0}^{\tau(\omega)} \left(-(1+\lambda)S_0^*(t)\frac{\partial\varphi(X_0^*(t))}{\partial z} \right) dL^*(t)$$

$$+E \int_{s_0}^{\tau(\omega)} \left((1-\mu)S_0^*(t)\frac{\partial\varphi(X_0^*(t))}{\partial z} - \frac{\partial\varphi(X_0^*(t))}{\partial y} \right) dM^*(t) - \theta E[\tau(\omega)]$$

which, together with the optimality of $L^*(t)$ and $M^*(t)$, violates the dynamic programming principle. Therefore, at least one of the inequalities (4.12), (4.13) and (4.14) must not hold and therefore U is a viscosity subsolution of (4.5).

ii) In the second part of the proof, we show that U is a viscosity supersolution of (4.5) in $[0,T] \times S_{\overline{K}}$; for this we must show that, for all smooth functions $\varphi(X)$, such that $U(X) - \varphi(X)$ has a local minimum at $X_0 \in [0,T] \times S_{\overline{K}}$, the following inequality holds

$$\min\left\{ \varphi(X_0) - V_1(X_0), -\frac{\partial\varphi(X_0)}{\partial y} + (1+\lambda)S_0 \frac{\partial\varphi(X_0)}{\partial z}, \frac{\partial\varphi(X_0)}{\partial y} - (1-\mu)S_0 \frac{\partial\varphi(X_0)}{\partial z}, \right.$$

$$\left. -\frac{\partial\varphi(X_0)}{\partial s} + rz_0 \frac{\partial\varphi(X_0)}{\partial z} + \alpha S_0 \frac{\partial\varphi(X_0)}{\partial S} + \frac{1}{2}\sigma^2 S_0^2 \frac{\partial^2\varphi(X_0)}{\partial S^2} \right\} \geq 0$$

where, without loss of generality we have assumed that $U(X_0) = \varphi(X_0)$ and $U \geq \varphi$ on $[0,T] \times S_{\overline{K}}$. In this case we need to show that each argument of the above minimum operator is nonnegative. The nonnegativity of the first argument follows from (3.18). The proof for the rest of the arguments is similar to the one of Theorem 2 in [6] and therefore it is not presented here.   □

We conclude this section by showing that the value function V and U are the unique constrained viscosity solutions of (3.17) and (3.19). For simplicity of exposition, we assume that the interest rate $r = 0$; all the arguments are the same but notationally more cumbersome, when $r > 0$. Finally, since this uniqueness result will be mainly used for the convergence of numerical approximations which usually involve utility functions bounded from above, we will establish the uniqueness property in the class of bounded functions. To this end, we assume that the utility function $\mathcal{U}$ in (2.1) satisfies for $(z, y, S) \in S_{\overline{K}}$

(4.15) $$\mathcal{U}(z + c(y, s)) \leq M - e^{-\gamma(z+c(y,s))}$$

where M and γ are positive constants.

We first present a comparison result for constrained viscosity solutions of (4.5). Although the proof of the next Theorem is a slight modification of the proof of Theorem 3 of Davis, Panas and Zariphopoulou [6] we present the main steps for completeness of the presentation.

THEOREM 4.2. *Let u be a bounded upper semi-continuous viscosity subsolution of (4.5) on $[0,T] \times \overline{S_{\overline{K}}}$ and v be a lower semi-continuous function which is bounded from below, exhibits sublinear growth and is a viscosity supersolution of (4.5) in $[0,T] \times S_{\overline{K}}$ such that $u(T,x) \leq v(T,x)$, $\forall x \in \overline{S_{\overline{K}}}$. Then $u \leq v$ on $[0,T] \times \overline{S_{\overline{K}}}$.*

Proof. (Sketch) We first construct a positive strict supersolution of (4.5) in $[0, T] \times \mathcal{S}_{\overline{K}}$ when $\mathcal{U}$ satisfies (4.15). To this end, let $h : [0, T] \times \overline{\mathcal{S}_{\overline{K}}} \to \mathbb{R}^+$ be given by

$$h(t, z, y, S) = M - \exp(-\gamma(z + kyS)) + C_1(T - t) + C_2$$

where the constants k, C_1 and C_2 satisfy:

(4.16) $\quad 1 + \lambda > k > 1 - \mu, \quad C_1 > \dfrac{\alpha^2}{2\sigma^2} \exp(\gamma \overline{K})$ and $C_2 > \exp \overline{K} - 1$.

Then:

(4.17) $\quad H(X, h_t, Dh, D^2 h) =$

$$= \min\left\{ -\frac{\partial h}{\partial y} + (1+\lambda)S\frac{\partial h}{\partial z}, \frac{\partial h}{\partial y} - (1-\mu)S\frac{\partial h}{\partial z}, -\frac{\partial h}{\partial s} - \alpha S\frac{\partial h}{\partial S} - \frac{1}{2}\sigma^2 S^2 \frac{\partial^2 h}{\partial S^2} \right\} =$$

$= \exp(-\gamma(z+kyS)) \times \min\{\gamma S(1+\lambda-k), \gamma S(k-(1-\mu)), C_1 \exp(\gamma(z+ kyS)) + \frac{1}{2}\gamma^2 k^2 \sigma^2 (yS)^2 - \alpha k \gamma(yS)\}$.

Using (4.16) and the fact that the minimum value of the quadratic $D(\zeta) = \frac{1}{2}\gamma^2 k^2 \sigma^2 \zeta^2 - \alpha \gamma k \zeta$ is $-\frac{\alpha^2}{2\sigma^2}$, the above inequality yields:
(4.18)
$H(X, h_t, Dh, D^2 h) > \exp(-\gamma(z+kyS)) \min\{\gamma S(1+\lambda-k), \gamma S(k-(1-\mu)), K'\}$

in $[0, T] \times \overline{\mathcal{S}_{\overline{K}}}$, where $0 < K' < C_1 - \frac{\alpha^2}{2\sigma^2} \exp(\gamma k)$.

Therefore, h is a strict supersolution (4.5). The fact that $h > 0$ follows from the choice of the constant C_2.

To conclude the proof of the theorem we will need the following key Lemma. Its proof follows along the lines of Theorem vi.5 in Ishii and Lions [10] and therefore it is not presented here.

LEMMA 4.1. *Let u be a bounded upper semicontinuous viscosity subsolution of (4.5) on $[0, T) \times \overline{\mathcal{S}_{\overline{K}}}$ and v be a bounded from below uniformly continuous viscosity supersolution of (4.5) in $[0, T) \times \mathcal{S}_{\overline{K}}$ of the equation $H(X, v_t, Dv, D^2 v) - f(x) = 0$, where $f > 0$ in $\mathcal{S}_{\overline{K}}$, $u(T, x) \le v(T, x)$, $\forall x \in \overline{\mathcal{S}_{\overline{K}}}$. Then $u \le v$ on $[0, T] \times \overline{\mathcal{S}_{\overline{K}}}$.*

We now conclude the proof of the Theorem. To this end we first observe that because of the choice of k, C_1 and C_2

(4.19) $\qquad\qquad h(T, z, y, S) > \mathcal{U}(z + c(y, S)).$

Next, we define the function $w^\theta = \theta v + (1 - \theta)h$, where $0 < \theta < 1$ and using (4.19) we get

(4.20) $\qquad\qquad w^\theta(T, z, y, S) \ge u(T, z, y, S).$

We also observe that w^θ is a viscosity supersolution of $H - g = 0$, in $[0, T] \times \overline{\mathcal{S}_{\overline{K}}}$, where $g = (1 - \theta)f$.

Applying Lemma 5.2 to $u \leq w^\theta$, on $[0,T] \times \overline{\mathcal{S}_K}$ and sending $\theta \uparrow 1$ concludes the proof of the theorem. □

We conclude this section by showing that the obstacle problem given by (4.5) has a unique bounded constrained viscosity solution.

THEOREM 4.3. *Let u be a bounded upper semi-continuous viscosity sub-solution of (4.5) on $[0,T] \times \overline{\mathcal{S}_K}$ and v be a lower semi-continuous function which is bounded from below, exhibits sublinear growth and is a viscosity supersolution of (4.5) in $[0,T] \times \mathcal{S}_K$ such that $u(T,x) \leq v(T,x)$, $\forall x \in \overline{\mathcal{S}_K}$. Then $u \leq v$ on $[0,T] \times \overline{\mathcal{S}_K}$.*

Proof. (Sketch) We first obtain an upper bound for the value function U. For the simplicity of calculations we assume that the constant k in (4.16) satisfies $1 - \mu < k < 1$. From Theorem 4.2 we have

$$V(t,S,y,z) \leq h(t,S,y,z) \quad \text{on } [0,T] \times \overline{\mathcal{S}_K}$$

which together with the choice of k and the nonnegativity of X_1 implies

$$V_1(t,S,y,z) \leq h(t,S,y,z) \quad \text{on } [0,T] \times \overline{\mathcal{S}_K}.$$

□

We next define the function

$$h'(t,z,y,S) = h(t,z,y,S) + C_3$$

with $C_3 > 0$. It is straightforward to show that h' is a strictly positive supersolution of the obstacle problem (4.5). The rest of the proof follows along the arguments of Theorem 4.2.

5. Conclusions. As pointed out in the introduction, the significance of Theorems 4.2 and 4.3 is that the option price p^* defined by (2.8) can in principle be computed by standard discretization methods. Clearly, however, a substantial amount of computation is involved. We are now investigating "differential" variants on (2.8) which we hope will give a more readily evaluated price.

Acknowledgements. Work on this paper was initiated while the first author was at the IMA, University of Minnesota, supported by funds derived from an NSF grant. He is grateful to the IMA for this opportunity and to the NSF for financial support.

The second author is thankful to the IMA for the hospitality during the Workshop on Mathematical Finance.

REFERENCES

[1] Barles, G. and P.E. Souganidis (1991), *Convergence of Approximation Schemes for Fully Nonlinear Second Order Equations*, Asymptotic Analysis, **4**, 271–283.

[2] Capuzzo-Dolcetta, I. and P.-L. Lions (1987), *Hamilton-Jacobi Equations and State Constraints Problems*, IMA preprint series #342, University of Minnesota, USA.

[3] Crandall, M.G. and P.-L. Lions (1983), *Viscosity Solutions of Hamilton-Jacobi Equations*, Trans. Amer. Math. Soc., **277**, 1–42.

[4] Crandall, M.G., H. Ishii and P.-L. Lions (1992), *User's Guide to Viscosity Solutions of Second Order Partial Differential equations*, Bull. AMS, **27**, 1–67.

[5] Davis, M.H.A. and A.R. Norman (1990), *Portfolio Selection with Transaction Costs*, Mathematics of Operations Research, **15**, 676–713.

[6] Davis, M.H.A., V.G. Panas and T. Zariphopoulou (1993), *European option pricing with transaction costs*, SIAM J. Control and Optimizations, **31**, 470–493.

[7] Davis, M.H.A. and M. Zervos (1994), *A problem of singular stochastic control with discretionary stopping*, Ann. Appl. Prob. **4**, 226–240.

[8] Fleming, W.H. and H. M. Soner (1993), *Controlled Markov Processes and Viscosity Solutions*, Springer-Verlag, New York.

[9] Hodges, S. D. and A. Neuberger (1989), *Optimal Replication of Contingent Claims under Transaction Costs*, Review of Futures Markets, **8**, 222–239.

[10] Ishii, H. and P.-L. Lions (1990), *Viscosity Solutions of Fully Non-Linear Second-Order Elliptic Partial Differential Equations*, Journal of Differential Equations, **83**, 26–78.

[11] Katsoulakis, M. (1991), *State-Constraints Problems for Second Order Fully Nonlinear Degenerate Partial Differential Equations*, Ph.D. Thesis, Brown University, USA.

[12] Lions, P.-L. (1983), *Optimal Control of Diffusion Processes and Hamilton-Jacobi-Bellman Equations, Part 1 & 2*, Comm. Part. Diff. Eq., **8**, 1101–1174 & 1229–1276.

[13] Soner, H. M. (1986), *Optimal Control with State-Space Constraints*, SIAM J. Control & Optimization, **24**, 552–561.

[14] Zhu, H. (1991), *Characterization of Variational Inequalities in Singular Control*, Ph.D. Thesis, Brown University, USA.

THE OPTIMAL STOPPING PROBLEM FOR A GENERAL AMERICAN PUT-OPTION

NICOLE EL KAROUI* AND IOANNIS KARATZAS†

Abstract. We derive the representation $E\left[\int_t^T e^{-\int_t^u r(s)ds}\left(K - \underline{M}(t,u)\right)^+ r(u)du + e^{-\int_t^T r(s)ds}\left(K \wedge S(T) - \underline{M}(t,T-)\right)^+ \Big| \mathcal{F}(t)\right]$ for the "early exercise premium" $V(t;K) - P_e(t;K)$ of the American put-option $V(t;K) = \text{esssup}_{t \le \tau \le T} E\left[e^{-\int_t^\tau r(u)du}\left(K - S(\tau)\right)^+ \Big| \mathcal{F}(t)\right]$ on an asset with positive, continuous price process $S(\cdot)$). Here the supremum is ever all stopping times τ with values in $[t,T]$, $r(\cdot) \ge 0$ is the interest rate of the numéraire, $K > 0$ is the "strike price" of the option, $P_e(t;K) = E\left[e^{-\int_t^T r(u)du}\left(K - S(T)\right)^+ \Big| \mathcal{F}(t)\right]$ is the value of the corresponding European put-option, E denotes expectation under the so-called "risk-neutral equivalent martingale measure", and $\underline{M}(t,\theta) = \inf_{t \le u \le \theta} M(u)$, $t \le \theta < T$ is the lower envelope of the "index process" $M(t) = \inf\left\{K > 0 / V(t;K) = K - S(t)\right\}$, $0 \le t \le T$.

1. Introduction and summary. We offer in this paper a representation for the early exercise premium $V(t;K) - P_e(t;K)$ of an American put-option with given strike-price $K > 0$, on a finite time-horizon $[0,T]$ and on an asset with arbitrary continuous, strictly positive process $S(\cdot)$. Here

$$(1.1) \quad V(t;K) = \underset{\substack{t \le \tau \le T \\ \tau stop.time}}{\text{esssup}}\ E\left[e^{-\int_t^\tau r(u)du}\left(K - S(\tau)\right)^+ \Big| \mathcal{F}(t)\right]$$

is the value of the American put-option at time $t \epsilon [0,T]$,

$$(1.2) \quad P_e(t;K) = E\left[e^{-\int_t^T r(u)du}\left(K - S(T)\right)^+ \Big| \mathcal{F}(t)\right]$$

the value of the corresponding European put-option, $r(\cdot) \ge 0$ is the interest rate process for the prevailing pure discount bound (numéraire) in the economy, and E denotes expectation with respect to the so-called "risk-neutral" equivalent martingale measure. In terms of the Gittins-index-like process

$$(1.3) \quad M(t) = \inf\left\{K > 0 / V(t;K) = K - S(t)\right\}, \quad 0 \le t \le T$$

---

* Laboratoire de Probabilités, Université Pierre et Marie Curie, 4, place Jussieu-Tour 56, 75252 Paris Cedex 05, France.

† Department of Statistics, 619 Mathematics Building, Columbia University, New York, N Y 10027. Research supported in part by the National Science Foundation under Grant DMS-90-22188.

which gives at any time t the smallest value of the strike-price that makes immediate exercise of the option profitable, and its lower envelope $\underline{M}(t, \theta) = \inf_{t \leq u \leq \theta} M(u)$, $t \leq \theta < T$, our representation of the early exercise premium is

$$
(1.4) \quad V(t; K) - P_e(t; K) = E\left[\int_t^T e^{-\int_t^u r(s)ds} \left(K - \underline{M}(t, u)\right)^+ r(u)du \right.
$$
$$
\left. + e^{-\int_t^u r(s)ds} \left(K \wedge S(T) - \underline{M}(t, T-)\right)^+ \middle| \mathcal{F}(t) \right].
$$

It takes on the simpler, and more familiar, form

$$
(1.5) \quad V(t; K) - P_e(t; K) = E\left[\int_t^T e^{-\int_t^u r(s)ds} K 1_{\{M(u) \leq K\}} r(u)du \middle| \mathcal{F}(t) \right]
$$

in a special case where $e^{-\int_0^t r(s)ds} S(t)$ is a P-martingale (cf. Remarks 5.4); and leads to the representation

$$
(1.6) \quad V(t; K) = K - E\left[\int_t^T e^{-\int_t^u r(s)ds} \left(K \wedge \underline{M}(t, u)\right) r(u)du \right.
$$
$$
\left. + e^{-\int_t^T r(s)ds} \left(K \wedge S(T) \wedge \underline{M}(t, T-)\right) \middle| \mathcal{F}(t) \right]
$$

for the value of the American put-option as in (1.1).

The paper is organized as follows. Section 2 studies the optimal stopping problem of (1.1) in some detail, including an explicit representation for the right-hand derivative of the convex mapping $K \mapsto V(t; K)$, whereas section 4 introduces the index process $M(\cdot)$ of (1.3) via its lower envelope, as in El Karoui & Karatzas (1994). The connection of this optimal stopping problem with the pricing of the American put-option is made in section 3, using the by now standard framework of Bensoussan (1984). The representation (1.4) is then derived in section 5, using the formula for the derivative of $K \mapsto V(t; K)$ and properties of the lower envelope for the index process. Several consequences of the representation (1.4) are also discussed in section 5.

The paper was presaged by Jacka (1991), who obtained a special case of the representation (1.5), using very different methods. Our approach is fully probabilistic, and reminiscent of our recent work El Karoui & Karatzas (1994) on the continuous-time dynamic allocation or "multi-armed bandit" problem.

2. The optimal stopping problem. Consider a complete probability space $(\Omega, \mathcal{F}, P)$, and a filtration $\mathbf{F} = \{\mathcal{F}(t)\}_{0 \leq t \leq T}$ of sub-σ-fields of $\mathcal{F}$ which satisfies the "usual conditions" of right-continuity and augmentation by P-negligible sets, and is quasi-left-continuous. With $T > 0$ a fixed real constant, and any $0 \leq v \leq u \leq T$, we denote by $\mathcal{S}_{v,u}$ the class of all stopping times of $\mathbf{F}$ with values in $[v, u]$. Let $r(\cdot)$, $S(\cdot)$ be two $\mathbf{F}$-progressively measurable processes with values in $[0, \infty)$ and $(0, \infty)$ respectively, and assume that $S(\cdot)$ has continuous paths with $P(S(t) = x) = 0$, $\forall t \epsilon(0, T]$, $x \epsilon \mathbb{R}$.

Our object of interest in this paper is the *family of optimal stopping problems*

$$(2.1) \quad V(t; K) := \operatorname{esssup}_{\tau \epsilon \mathcal{S}_{t,T}} E\left[\exp\left\{ \int_t^\tau r(u)du \right\} (K - S(\tau))^+ \Big| \mathcal{F}(t) \right],$$
$$0 \leq t \leq T$$

parametrized by $K \epsilon [0, \infty)$. As we shall discuss in more detail in the next section, the interpretation for $S(\cdot)$ is that of the price-per-share of a certain asset, and for $V(t; K)$ that of the value of an *American put-option* on the asset (i.e., of a contract which confers to its holder to the right to sell one share of the asset at the specified "strike-price" K, and at any time during the interval $[t, T]$). In this context, the process

$$(2.2) \quad P_e(t; K) := E\left[\exp\left\{ - \int_t^T r(u)du \right\} (K - S(T))^+ \Big| \mathcal{F}(t) \right],$$
$$0 \leq t \leq T$$

has the interpretation of the value of the corresponding *European put-option* (i.e., of a similar contract as before, but in which the holder can exercise his right only at the terminal time T).

We shall assume throughout that the process $P_e(\cdot; K)$ is strictly positive on $[0, T)$:

$$(2.3) \qquad\qquad P_e(t; K) > 0; \quad \forall 0 \leq t < T, \ 0 < K < \infty.$$

It is also obvious that

$$(2.4) \quad V(t; K) \geq P_e(t; K) \vee (K - S(t))^+; \quad \forall 0 \leq t \leq T, \ 0 < K < \infty.$$

From standard theory on optimal stopping (e.g. Fakeev (1970), Bismut & Skalli (1977), El Karoui (1981), Karatzas (1993)) we know that, with K fixed and

$$(2.5) \qquad Y(t; K) := e^{- \int_0^t r(u)du} (K - S(t))^+; \quad 0 \leq t \leq T,$$

the process

$$(2.6) \quad \begin{aligned} Z(t; K) &:= e^{- \int_0^t r(u)du} V(t; K) \\ &= \operatorname{esssup}_{\tau \epsilon \mathcal{S}_{t,T}} E\left[Y(\tau; K) | \mathcal{F}(t) \right], \quad 0 \leq t \leq T \end{aligned}$$

is the *Snell envelope* of (i.e., the smallest supermartingale that dominates) $Y(\cdot\,; K)$. Clearly, $Z(T; K) = Y(T; K)$ a.s. The stopping time

$$(2.7) \qquad \begin{aligned} \sigma_t(K) &:= \inf\left\{\theta\epsilon[t,T]/Z(\theta; K) = Y(\theta; K)\right\} \\ &= \inf\left\{\theta\epsilon[t,T]/V(\theta; K) = K - S(\theta)\right\} \wedge T \end{aligned}$$

is optimal, i.e., achieves the supremum in (2.1), (2.6), and

(2.8) the process $\left\{Z(\theta \wedge \sigma_t(K); K), \mathcal{F}(\theta)\right\}_{t \le \theta \le T}$ is a martingale.

Furthermore, since $Y(\cdot\,; K)$ has continuous paths with values in $[0, K]$, the supermartingale $Z(\cdot\,; K)$ also takes values in $[0, K]$ and is *regular,* thus *quasi-left-continuous* thanks to the quasi-left-continuity of the filtration **F**.

Here are some basic properties of the random fields $(t, K, \omega) \mapsto \sigma_t(K, \omega)$, $(t, K, \omega) \mapsto V(t; K, \omega)$, considered in their measurable versions.

LEMMA 2.1. *For every $t\epsilon[0, T)$, the mapping*
(i) $K \mapsto V(t; K)$ *is convex, increasing, null at $K = 0$, and strictly positive,*
(ii) $K \mapsto K - V(t; K)$ *is concave, increasing, null at $K = 0$, and dominated by $K \wedge S(t)$,*
(iii) $K \mapsto \sigma_t(K)$ *is decreasing, right-continuous, with $\sigma_t(0+) = T$, almost surely.*

Proof.
(i) The convexity and increase follow from the facts that the mapping $K \mapsto (K - x)^+$ has these properties, and that we are then taking supremum over the class $\mathcal{S}_{t,T}$ of stopping times.
(ii) We have from (2.1)

$$K - V(t; K) = \operatorname{essinf}_{\tau\epsilon\mathcal{S}_{t,T}}$$

$$(2.9) \quad E\left[K\left(1 - e^{-\int_t^\tau r(u)du}\right) + e^{-\int_t^\tau r(u)du}\left(K \wedge S(\tau)\right)\middle|\mathcal{F}(t)\right]$$
$$\le K \wedge S(t).$$

The two functions of K inside the expectation are linear and concave, respectively, and both are increasing; since we are taking an infimum (over the class $\mathcal{S}_{t,T}$), these properties persist.
(iii) Introduce the nonnegative random field

$$(2.10) \qquad \begin{aligned} \varphi(t; K) &:= V(t; K) - K + S(t), \\ &0 \le t \le T, \quad K\epsilon(0, \infty) \end{aligned}$$

in terms of which we can re-write (2.7) as $\sigma_t(K) = \inf\left\{\theta\epsilon[t,T]/\varphi(\theta; K) = 0\right\} \wedge T$. For any fixed $t\epsilon[0, T)$, the mapping $K \mapsto \varphi(t; K)$ is continuous and decreasing (from (ii)); thus if $\{K_n\}_{n\epsilon\mathbf{N}} \subset (K, \infty)$ is a strictly decreasing sequence with $K = \lim_{n\to\infty} K_n$, we have $0 \le \varphi(\sigma_t(K_2); K_1) \le \varphi(\sigma_t(K_2); K_2) = 0$ so that $\sigma_t(K_1) \le \sigma_t(K_2)$, a.s. Therefore, $\sigma_* := \lim_{n\to\infty} \uparrow \sigma_t(K_n)$ exists and $\sigma_* \le \sigma_t(K)$.

More generally, $\varphi(\sigma_t(K_\ell); K_m) = 0$ for $\ell > m$; now let $\ell \uparrow \infty$ to obtain, from the quasi-left-continuity of $Z(\cdot; K)$ (and thus of $V(\cdot, K), \varphi(\cdot; K)$ as well): $\varphi(\sigma_*; K_m) = 0$, $\forall m \epsilon \mathbf{N}$. Finally, let $m \to \infty$ and exploit the continuity of $\varphi(t; \cdot)$, to obtain $\varphi(\sigma_*; K) = 0$, a.s. It follows that $\sigma_t(K) \leq \sigma_*$, and thus $K \mapsto \sigma_t(K)$ is right-continuous. $\square$

THEOREM 2.2. *For every* $t\epsilon[0, T)$, *the convex mapping* $K \mapsto V(t, K)$ *has right-hand derivative given by*

(2.11)
$$\frac{\partial^+}{\partial K} V(t; K)$$
$$= E\big[R^t_{\sigma_t(K)} 1_{\{\sigma_t(K)<T\}} + R^t_T 1_{\{S(T)\leq K, \ \sigma_t(K)=T\}}\big|\mathcal{F}(t)\big]$$
$$= E\big[R^t_{\sigma_t(K)} - R^t_T 1_{\{\sigma_t(K)=T, S(T)>K\}}\big|\mathcal{F}(t)\big],$$

with the notation

(2.12)
$$R^t_\theta := \exp\left\{ - \int_t^\theta r(s)ds \right\}, \quad 0 \leq t \leq \theta \leq T.$$

Proof. Fix $(t, K)\epsilon[0, T) \times (0, \infty)$, and for any given $\varepsilon > 0$ denote $\sigma^\varepsilon \equiv \sigma_t(K + \varepsilon)$, $\sigma^0 \equiv \sigma_t(K)$. Of course $\sigma^\varepsilon \leq \sigma^0$ a.s., and thus $Z(\cdot \wedge \sigma^\varepsilon; K)$ is a martingale (from (2.8)); therefore, $V(t; K) = E\big[R^t_{\sigma^\varepsilon} \cdot V(\sigma^\varepsilon; K)\big|\mathcal{F}(t)\big]$. On the other hand, $V(t; K + \varepsilon) = E\big[R^t_{\sigma^\varepsilon} \cdot (K + \varepsilon - S(\sigma^\varepsilon))^+ \big|\mathcal{F}(t)\big]$ from the optimality of σ^ε at $(t, K + \varepsilon)$, whence

(2.13)
$$V(t; K + \varepsilon) - V(t; K) =$$
$$E\left[R^t_{\sigma^\varepsilon}\left\{(K + \varepsilon - S(\sigma^\varepsilon))^+ - V(\sigma^\varepsilon; K)\right\}\Big|\mathcal{F}(t)\right].$$

But on $\{\sigma^\varepsilon < T\}$, we have: $K+\varepsilon-S(\sigma^\varepsilon) = V(\sigma^\varepsilon; K+\varepsilon) > 0$, $V(\sigma^\varepsilon; K) > K - S(\sigma^\varepsilon)$, thus $(K + \varepsilon - S(\sigma^\varepsilon))^+ - V(\sigma^\varepsilon; K) < \varepsilon$. Furthermore, on the event $\{\sigma^\varepsilon = T\}$ we have $V(\sigma^\varepsilon; K) = (K - S(T))^+$, and so

$$(K + \varepsilon - S(\sigma^\varepsilon))^+ - V(\sigma^\varepsilon; K) =$$
$$(K + \varepsilon - S(T))^+ - (K - S(T))^+ \leq \varepsilon 1_{\{K+\varepsilon\geq S(T)\}}.$$

Back into (2.13), these observations lead to the upper bound

$$\frac{V(t; K + \varepsilon) - V(T; K)}{\varepsilon}$$
$$\leq E\big[R^t_{\sigma^\varepsilon} 1_{\{\sigma^\varepsilon<T\}} + R^t_T 1_{\{\sigma^0=T\}\cap\{S(T)\leq K+\varepsilon\}}\big|\mathcal{F}(t)\big]$$
$$\leq E\big[R^t_{\sigma^0} 1_{\{\sigma^0<T\}} + R^t_T 1_{\{\sigma^0=T\}\cap\{S(T)\leq K\}}\big|\mathcal{F}(t)\big]+$$
$$+E\big[(R^t_{\sigma^\varepsilon} - R^t_{\sigma^0})1_{\{\sigma^0<T\}} + 1_{\{\sigma^\varepsilon<\sigma^0=T\}}$$
$$+1_{\{K<S(T)\leq K+\varepsilon\}}\big|\mathcal{F}(t)\big].$$

But now $\lim_{\varepsilon \downarrow 0} \uparrow \sigma^\varepsilon = \sigma^0$ a.s., and from the (conditional) monotone and bounded convergence theorems, we obtain that the last conditional expectation goes to zero as $\varepsilon \downarrow 0$, whence

$$
(2.14) \quad \overline{\lim}_{\varepsilon \downarrow 0} \frac{V(t; K + \varepsilon) - V(t; K)}{\varepsilon} \leq
$$
$$
E\left[R^t_{\sigma^0} 1_{\{\sigma^0 < T\}} + R^t_T 1_{\{\sigma^0 = T, \, S(T) \leq K\}} \big| \mathcal{F}(t)\right].
$$

To obtain a lower bound, recall the supermartingale property of $Z(\cdot; K + \varepsilon)$, which gives $V(t; K + \varepsilon) \geq E\left[R^t_{\sigma^0} V(\sigma^0; K + \varepsilon) \big| \mathcal{F}(t)\right]$, and in conjunction with $V(t; K) = E\left[R^t_{\sigma^0}(K - S(\sigma^0))^+ \big| \mathcal{F}(t)\right]$ get

$$
\begin{aligned}
(2.15) \quad & V(t; K + \varepsilon) - V(t; K) \\
& \geq E\left[R^t_{\sigma^0}\{V(\sigma^0; K + \varepsilon) - (K - S(\sigma^0))^+\} \big| \mathcal{F}(t)\right] \\
& \geq E\left[R^t_{\sigma^0}\{(K + \varepsilon - S(\sigma^0))^+ - (K - S(\sigma^0))^+\} \big| \mathcal{F}(t)\right].
\end{aligned}
$$

Now on $\{\sigma^0 < T\}$, we have $K - S(\sigma^0) = V(\sigma^0; K) > 0$, whence

$$
(K + \varepsilon - S(\sigma^0))^+ - (K - S(\sigma^0))^+ = \varepsilon.
$$

On the other hand, on $\{\sigma^0 = T\}$ the last expression in braces in (2.15) becomes

$$
\begin{aligned}
(K + \varepsilon - S(T))^+ & - (K - S(T))^+ \\
& = \{(K + \varepsilon - S(T)) - (K - S(T))\}1_{\{S(T) \leq K\}} \\
& \quad + (K + \varepsilon - S(T))^+ 1_{\{S(T) > K\}} \\
& \geq \varepsilon 1_{\{S(T) \leq K\}}.
\end{aligned}
$$

Back into (2.15), these considerations give

$$
\underline{\lim}_{\varepsilon \downarrow 0} \frac{V(t; K + \varepsilon) - V(t; K)}{\varepsilon}
$$
$$
\geq E\left[R^t_{\sigma^0} 1_{\{\sigma^0 < T\}} + R^t_T 1_{\{\sigma^0 = T, \, S(T) \leq K\}} \big| \mathcal{F}(t)\right],
$$

and therefore also (2.11) in conjunction with (2.14). $\square$

3. The American put-option. Suppose now that the filtration **F** is the augmentation of the natural filtration generated by a d-dimensional standard Brownian motion $W = (W_1, \ldots, W_d)'$ on some complete probability space $(\Omega, \mathcal{F}, P_0)$. Consider a financial market $\mathcal{M}$ with $d+1$ instruments (assets), one *bond* with price $S_0(\cdot)$ governed by

$$
(3.1) \qquad dS_0(t) = S_0(t)r(t)dt, \quad S_0(0) = 1,
$$

and d *stocks*, with prices-per-share $S_i(\cdot)$ which satisfy the stochastic equations

$$
(3.2) \qquad dS_i(t) = S_i(t)\left[b_i(t)dt + \sum_{j=1}^d \sigma_{ij}(t)dW_j(t)\right];
$$
$$
i = 1, \ldots, d, \quad 0 \leq t \leq T
$$

driven by the Brownian motion W. Here $r(\cdot)$, $b(\cdot) = \{b_i(\cdot)\}_{i=1}^{d}$, $\sigma(\cdot) = \{\sigma_{ij}(\cdot)\}_{i,j=1}^{d}$ are all bounded, $\mathbf{F}$-progressively measurable processes, and we suppose that the same properties hold for $\theta(t) = \sigma^{-1}(t)[b(t) - r(t)\mathbf{1}]$. We are also assuming here that $\sigma(t)$ is invertible for all $0 \leq t \leq T$, almost surely. Then

$$(3.3) \quad Z(t) = \exp\left[-\int_0^t \theta'(s)dW(s) - \frac{1}{2}\int_0^t \|\theta(s)\|^2 ds\right], \quad 0 \leq t \leq T$$

is an $(\mathbf{F}, P_0)$-martingale, and thus $P(A) := E_0[Z(T)1_A]$ defines a new probability measure on $(\Omega, \mathcal{F})$ which is *equivalent* to P_0. Under this new measure P, the process $\tilde{W}(t) = W(t) - \int_0^t \theta(s)ds$, $0 \leq t \leq T$ is standard Brownian motion and the discounted stock price processes $\frac{S_i(\cdot)}{S_0(\cdot)}$, $i = 1, \ldots, d$ are *martingales* —whence the terminology "risk-neutral equivalent martingale measure" for P.

Consider now an arbitrary "asset" in this market $\mathcal{M}$, with price-per-share process $S(\cdot)$ satisfying the conditions of section 2; in particular, we can take $S(\cdot) \equiv S_i(\cdot)$ for some $i = 1, \ldots, d$ but this is not necessary. Suppose that, at time $t = 0$, you sign a contract with another "agent", which gives you the right (but not the obligation) to sell to the agent one share of the asset, at the contractually specified price $K\epsilon(0, \infty)$ and at *any* time ρ in $[0, T]$. Such a contract is called an *American put-option* with horizon T and "strike price" K. (The corresponding contract with only one possible exercise time, namely $\rho = T$, is called a *European* put-option). The signing of such a contract effectively commits the agent to make to you a payment of $(K - S(\rho))^+$ at the exercise time ρ. *What is the "fair price", or "value" of the contract, that you should be charged at $t = 0$?*

The agent can of course invest in the instruments of the market $\mathcal{M}$, by committing an initial capital $x > 0$ and then selecting a *portfolio* process $\pi = (\pi_1, \ldots, \pi_d)'$ and an increasing *cumulative consumption* process C with $C(0) = 0$ (both $\mathbf{F}$-progressively measurable with $C(T) + \int_0^T \|\pi(t)\|^2 dt < \infty$, a.s.). His wealth-process $X(\cdot) = X^{x,\pi,C}(\cdot)$ is then determined by

$$(3.4) \quad \begin{aligned} dX(t) &= \sum_{i=1}^{d} \pi_i(t)[b_i(t)dt + \sum_{j=1}^{d} \sigma_{ij}(t)dW_j(t)] \\ &\quad + \left(X(t) - \sum_{i=1}^{d} \pi_i(t)\right)r(t)dt - dC(t), \quad X(0) = x. \end{aligned}$$

The agent should strive to cover his obligation by selecting x and (π, C) in such a way that

$$(3.5) \quad X^{x,\pi,c}(t) \geq (K - S(t))^+, \quad \forall 0 \leq t \leq T$$

holds almost surely. Now the fair price F_A for the American put-option, should be the smallest initial capital that allows the agent to achieve this, i.e.,

(3.6) $F_A := \inf \left\{ x > 0 / \exists (\pi, C) \text{ s.t. (3.5) holds a.s.} \right\}.$

It was shown by Bensoussan (1984), Karatzas (1988, 1989) (see also Karatzas & Shreve (1994)) that

(3.7) $F_A = V(0; K) \text{ as in (2.1)};$

in other words, the valuation of this contract is given in terms of the optimal stopping problem of (2.1). It is a far more straightforward matter to see that the value

(3.8) $F_E := \inf \left\{ x > 0 / \exists (\pi, C) \text{ s.t. } X^{x, \pi, C}(T) \geq (K - S(T))^+, \text{ a.s.} \right\}$

of the corresponding European put-option is given by the famous Black & Scholes formula

(3.9) $F_E = P_e(0; K) \text{ as in (2.2)}.$

The nonnegative number $F_A - F_E = V(0; K) - P_e(0; K)$ has then an obvious interpretation as *early exercise premium* for the American put-option. Furthermore, $\sigma_0(K) = \inf \left\{ \theta \epsilon [0, T) / V(\theta; K) = K - S(\theta) \right\} \wedge T$ as in (2.7) has the interpretation of *optimal exercise time* ρ for this problem.

Similarly, $V(t; K)$ (resp. $P_e(t; K)$) can be interpreted as the *value* for the American (respectively, European) put-option, and $V(t; K) - P_e(t; K)$ as an "early exercise premium", *at any time* $t \epsilon [0, T]$; see Karatzas & Shreve (1994) for the particulars of this interpretation.

Remarks 3.1: Marc Romano (Université de Paris-Dauphine) observes that the third expression in (2.11) leads to the bounds

(3.10) $0 \leq E\left[R^0_{\sigma_0(K)} \right] - \frac{\partial^+}{\partial K} V(0; K) \leq E(R^0_T) \underset{T \to \infty}{\longrightarrow} 0,$

where the last property holds, for example, if $\int_0^\infty r(u) du = \infty$, a.s.

4. Index processes. For every $t \epsilon [0, T)$, the set $\left\{ K > 0 / V(t; K) = K - S(t) \right\}$ is an interval of the form $\left[M(t), \infty \right)$, where $M(t)$ is an $\mathcal{F}(t)$-measurable random variable that satisfies

(4.1) $M(t) = \inf \left\{ K > 0 / V(t; K) = K - S(t) \right\} \geq S(t),$ a.s.

We also define

(4.2) $M(T) := S(T).$

Now let

$$\underline{M}(t,\theta) :=$$

$$\begin{cases} \sup\{K > 0 / \sigma_t(K) > \theta\} = \inf\{K > 0 / \sigma_t(K) \le \theta\}, & t \le \theta < T \\ \underline{M}(t, T-) \wedge S(T), & \theta = T \end{cases}$$

(4.3)

be the right-continuous inverse of the decreasing mapping $K \mapsto \sigma_t(K)$. We have the properties

$$(4.4) \qquad \sigma_t(K) > \theta \iff \underline{M}(t,\theta) > K \iff M(u) > K, \quad \forall u \epsilon[t,\theta]$$

for $0 \le t \le \theta < T$, so that $\underline{M}(t,\theta) = \inf_{t \le u \le \theta} M(u)$ is the "lower envelope" of $M(\cdot)$ on $[t,\theta]$. It is also clear that $\sigma_t(K) = \inf\{\theta \epsilon[t,T)/M(\theta) \le K\} \wedge T$.

Remarks 4.1: For every $t\epsilon[0,T)$, we have from (4.4):

$$\bigcap_{\varepsilon > 0}\{\sigma_t(K + \varepsilon) < T\}$$

$$= \bigcap_{\varepsilon > 0}\bigcup_{T-t \le \alpha \le T}\{\sigma_t(K + \varepsilon) \le T - \alpha\}$$

$$= \bigcap_{\varepsilon}\bigcup_{\alpha}\{\underline{M}(t, T - \alpha) \le K + \varepsilon\}$$

$$= \bigcap_{\varepsilon > 0}\{\underline{M}(t, T-) \le K + \varepsilon\} = \{\underline{M}(t, T-) \le K\}.$$

Therefore,

$$\underline{M}(t, T-) = \inf\{K > 0 / \sigma_t(K) < T\}$$

as well as

$$\{\underline{M}(t, T-) \le K\}$$
$$= \{\sigma_t(K) < T\} \cup \{\sigma_t(K) = T; \sigma_t(K + \varepsilon) < T, \quad \forall \varepsilon > 0\}.$$

We deduce

$$(4.5) \qquad \{\underline{M}(t, T-) \le K < S(T)\} = \{\sigma_t(K) < T, \; S(T) > K\},$$

because on $\{\sigma_t(K + \varepsilon) < T\}$ we have $S(\sigma_t(K + \varepsilon)) = K + \varepsilon - V(\sigma_t(K + \varepsilon)); K+\varepsilon) \le K+\varepsilon$ and thus on $\cap_{\varepsilon > 0}\{\sigma_t(K) = T, \sigma_t(K+\varepsilon) < T\} : S(T) = S(\sigma_t(K)) = S(\lim_{\varepsilon \downarrow 0} \sigma_t(K + \varepsilon)) = \lim_{\varepsilon \downarrow 0} S(\sigma_t(K + \varepsilon)) \le K.$ ⋄

The processes $M(\cdot)$ and $\underline{M}(t, \cdot)$ have obvious similarities with the *Gittins index process* and its lower envelope, which have proved very useful in the study of dynamic allocation (or "multi-armed bandit") problems; see, for example, El Karoui & Karatzas (1994). Clearly, in the context of section 3, $M(t)$ can be interpreted as the smallest value of the strike-price $K > 0$ that makes immediate exercise of the American put-option profitable at time t.

5. Early exercise premium in terms of indices. We can present now our main result, the expression (5.1) for the *early exercise premium* $V(t; K) - P_e(t; K)$ for the American option of section 3 in terms of the lower envelope $\underline{M}(t, \cdot)$ of the index process as in (4.3). This leads to a similar representation (5.3) for the value $V(t; K)$ itself.

THEOREM 5.1. *In terms of the lower envelope $\underline{M}(t, \cdot)$ of the index process $M(\cdot)$ in (4.2), (4.3), we have for every $t \epsilon [0, T)$ the following representation of the "early exercise premium":*

$$
(5.1) \quad
\begin{aligned}
V(t; K) - P_e(t; K) = E\Big[& \int_t^T R_u^t (K - \underline{M}(t, u))^+ r(u) du \\
& + R_T^t ((K \wedge S(T)) - \underline{M}(t, T-))^+ \Big| \mathcal{F}(t) \Big].
\end{aligned}
$$

Proof. It is quite easy to compute $\frac{\partial^+}{\partial K} P_e(t; K) = E[R_T^t 1_{\{S(T) \leq K\}} | \mathcal{F}(t)]$ from (2.2). Therefore, we obtain in conjunction with (2.11):

$$
(5.2) \quad
\begin{aligned}
\frac{\partial^+}{\partial K} & \Big[V(t; K) - P_e(t; K) \Big] \\
&= E\Big[1_{\{\sigma_t(K) < T\}} \big(R_{\sigma_t(K)}^t - R_T^t 1_{\{S(T) \leq K\}} \big) \Big| \mathcal{F}(t) \Big] \\
&= E\Big[\big(R_{\sigma_t(K)}^t - R_T^t \big) + R_T^t 1_{\{S(T) > K\} \cap \{\sigma_t(K) < T\}} \Big| \mathcal{F}(t) \Big] \\
&= E\Big[\int_t^T R_u^t 1_{\{\underline{M}(t, u) \leq K\}} r(u) du + R_T^t 1_{\{\underline{M}(t, T-) \leq K < S(T)\}} \Big| \mathcal{F}(t) \Big],
\end{aligned}
$$

using (4.5), (4.4) and its corollary $R_{\sigma_t(K)}^t - R_T^t = \int_t^T R_u^t 1_{\{\sigma_t(K) \leq u\}} r(u) du = \int_t^T R_u^t 1_{\{\underline{M}(t, u) \leq K\}} r(u) du$.

Integrating with respect to K in (5.2) over $[0, K]$, and using $V(t; 0) = P_e(t; 0) = 0$, we obtain (5.1) from the conditional Fubini theorem. $\square$

COROLLARY 5.2. *For every $t \epsilon [0, T)$, $K \epsilon (0, \infty)$ we have the a.s. representation*

$$
(5.3) \quad
\begin{aligned}
V(t; K) = K - \\
E\Big[\int_t^T R_u^t (K \wedge \underline{M}(t, u)) r(u) du + R_T^t (K \wedge \underline{M}(t, T)) \Big| \mathcal{F}(t) \Big];
\end{aligned}
$$

in particular,

$$
(5.4) \quad \lim_{K \to \infty} \big[K - V(t; K) \big] = E\Big[\int_t^T R_u^t \underline{M}(t, u) r(u) du + R_T^t \underline{M}(t, T) \Big| \mathcal{F}(t) \Big].
$$

Proof. We can write (2.2) equivalently as

$$K - P_e(t; K) = E\left[\int_t^T R_u^t Kr(u)du + R_T^t(K \wedge S(T))\Big|\mathcal{F}(t)\right].$$

Subtracting (5.1) memberwise from this equality, and recalling the definition of (4.3) for $t = T$, we obtain (5.3); (5.4) follows then by Monotone Convergence. □

Remarks 5.3: The process

$$(5.5) \quad R_t^0\left[\underline{M}(0,t) - V(t, \underline{M}(0,t))\right] + \int_0^t R_u^0 \underline{M}(0,u)r(u)du, \quad 0 \leq t < T$$

is an **F***-martingale.* Indeed, (5.3) gives

$$\underline{M}(0,t) - V\left(t, \underline{M}(0,t)\right) = E\left[\int_t^T R_u^t \underline{M}(0,u)r(u)du + R_T^t \underline{M}(0,T)\Big|\mathcal{F}(t)\right],$$

and the martingale property follows directly from this.

Remarks 5.4: In the *special setting of section 3 with* $d = 1$, $r > 0$, $\sigma = \sigma_{11} > 0$ *real constants and* $S(\cdot) \equiv S_1(\cdot)$, (5.1) takes the simpler form

$$(5.6) \quad \begin{aligned} V(t; K) - P_e(t; K) &= E\left[\int_t^T R_u^t K 1_{\{M(u) \leq K\}} r(u)du \Big|\mathcal{F}(t)\right] \\ &= E\left[\int_t^T R_u^t K 1_{\{V(u;K) = K - S(u)\}} r(u)du \Big|\mathcal{F}(t)\right]. \end{aligned}$$

This last expression was obtained by S. Jacka (1991). In fact, for this special case it can be easily verified that

$$(5.7) \quad V(t; K) = u(T - t, S(t); K), \quad M(t) = \frac{S(t)}{b_1(T-t)}; \quad 0 \leq t < T$$

where $(\theta, x, K) \mapsto u(\theta, x; K) : (0, \infty)^3 \to (0, \infty)$ and $(\theta, K) \mapsto b_K(\theta) : (0, \infty)^2 \to (0, \infty)$ are suitable functions with the scaling properties $u(\theta, x; K) = Ku(\theta, \frac{x}{K}; 1)$, $b_k(\theta) = Kb_1(\theta)$. Furthermore, $b_1(\cdot)$ is then continuous and decreasing with $b_1(0+) = 1$; it is the optimal exercise boundary for the American put-option corresponding to $K = 1$, in the sense that the optimal stopping time of (2.7) with $t = 0$, $K = 1$ takes the form $\sigma_0(1) = \inf\{t\epsilon[0,T)/S(t) \leq b_1(T-t)\}$; see Jacka (1991), Myneni (1992) or Karatzas

& Shreve (1994) for details. In this special case, the formulae (5.2), (5.1) become, respectively,

$$\frac{\partial^+}{\partial K}\left[V(t;K) - P_e(t;K)\right] = E\left[\int_t^T 1_{\{\inf_{t \le \theta \le u} \frac{S(\theta)}{b_t(T-\theta)} \le K\}} re^{-r(u-t)}du\right.$$
$$\left. + e^{-r(T-t)} 1_{\{\inf_{t \le \theta < T} \frac{S(\theta)}{b_1(T-\theta)} \le K < S(T)\}}\bigg| \mathcal{F}(t)\right]$$

(5.8)

$$V(t;K) - P_e(t;K) = E\left[\int_t^T re^{-r(u-t)}\left(K - \inf_{t \le \theta \le u} \frac{S(\theta)}{b_1(T-\theta)}\right)^+ du\right.$$
$$\left. + e^{-r(T-t)}\left(K \wedge S(T) - \inf_{t \le \theta < T} \frac{S(\theta)}{b_1(T-\theta)}\right)^+ \bigg| \mathcal{F}(t)\right]$$

(5.9)

This last representation is of some interest, as it involves compound European options of the path-dependent (or "look-back") type. It follows readily from (5.4) that

$$\lim_{K \to \infty}\left[K - V(t;K)\right] = E\left[\int_t^T re^{-r(u-t)} \inf_{t \le \theta \le u}\left(\frac{S(\theta)}{b_1(T-\theta)}\right)du\right.$$
(5.10)
$$\left. + e^{-r(T-t)} \inf_{t \le \theta \le T}\left(\frac{S(\theta)}{b_1(T-\theta)}\right)\bigg| \mathcal{F}(t)\right].$$

REFERENCES

[1] A.BENSOUSSAN, *On the theory of option pricing.*, Acta Appl. Math. 2 (1984), 139–158.
[2] J.M.BISMUT, B. SKALLI, *Temps d' arrêt optimal, théorie générale de processus et processus de Markov*, Z. Wahrsch. verw. Gebiete 39 (1977), 301–313.
[3] N.EL KAROUI, I.KARATZAS, *Dynamic allocation problems in continuous time*, Annals of Applied Probability, to appear (1994).
[4] N.EL KAROUI, *Les aspects probabilistes du contrôle stochastique*, Lecture Notes in Mathematics 876 (1981), 73–238 Springer-Verlag, Berlin.
[5] A.G.FAKEEV, *Optimal stopping rules for processes with continuous parameter*, Theory Probab. Appal. 15 (1970), 324–331.
[6] S.D.JACKA, *Optimal stopping and the American put*, Math. Finance 1 (2) (1991), 1–14.
[7] I.KARATZAS, *On the pricing of American options*, Appl. Math. Optimiz. 17 (1988), 37–60.
[8] I.KARATZAS, *Optimization problems in the theory of continuous trading*, SIAM J. Control & Optimization 27 (1989), 1221–1259.
[9] I.KARATZAS, *Lectures on the probabilistic aspects of optimal stopping*, Class notes, Columbia University, Spring (1993).
[10] I.KARATZAS, S.E.SHREVE, *Monograph on mathematical finance*, In preparation (1994).
[11] R.MYNENI, *The pricing of the American option*, Annals of Applied Probability 2 (1992), 1–23.

OPTIMAL INVESTMENT MODELS
AND RISK SENSITIVE
STOCHASTIC CONTROL

WENDELL H. FLEMING*

Abstract. Risk sensitive stochastic control theory is concerned with asymptotic results which hold for problems with exponential cost criteria. Two kinds of results are considered in the context of optimal investment models. The first kind concerns totally risk averse limits for problems on a fixed finite time horizon. The second kind is related to optimal long term growth of the expected utility of wealth.

1. Introduction. In stochastic control theory, disturbances in a system being controlled are modelled as stochastic processes (random noise.) However, there are interesting alternative approaches in which disturbances are modelled deterministically. One such approach is the robust control formulation of the so-called disturbance attenuation problem. With this approach an associated "soft-constrained" two-player, zero-sum differential game arises naturally. One player in this game is the controller in the original problem. The opposing player (corresponding to "unfriendly nature") chooses a disturbance. See Basar - Bernhard [3]. For linear system models and quadratic cost criteria, the robust control disturbance attenuation problem becomes one of H - infinity control theory. Rather explicit solutions are found in terms of matrix Riccati equations.

The theory of risk sensitive control provides a link between stochastic and deterministic approaches to disturbances in control systems. The stochastic counterpart to the linear-quadratic robust control disturbance attenuation model is a linear exponential quadratic regulator (LEQR) problem, introduced by Jacobson [10]. His analysis of the LEQR problem led to the same differential games obtained later via the deterministic robust control approach. Whittle [16] provides a comprehensive introduction to risk sensitive control theory.

For nonlinear systems, or nonquadratic cost criteria, large deviations ideas can be used to establish similar connections between stochastic and robust control approaches. For control problems on a finite time interval $0 \leq t \leq T$, Whittle [17] used Freidlin - Wentzell type "small noise" asymptotics to make such a connection in a formal way. In Fleming - McEneaney [6] McEneaney [13] and James [11], Whittle's formula for the optimal large deviations rate was obtained using logarithmic transformations of optimal cost functions and viscosity solution methods for nonlinear partial differential equations. Those papers treat the "complete state information case" in which the current state of the process being controlled is known. For problems with partial state information, see Whittle [16] [18], Baras - Elliott -

* Division of Applied Mathematics, Brown University, Providence, Rhode Island 02912, supported by NSF grant DMS-9000038.

James [1] and references cited there.

The purpose of the present paper is to explore (without detailed proofs) risk sensitive control ideas in the context of some optimal investment models. In Section 2, we formulate a class of finite-time horizon stochastic control problems, with state-linear dynamics. This class includes, in particular, Merton-type models without consumption in which expected utility of wealth at the terminal time is to be maximized. In Section 3, we apply (following Barron - Jensen [2]) a nonlinear transformation to the value function (or optimal cost function). This transformation is just the inverse of the utility function. Then in Section 4 we consider totally risk averse limits. Some technical difficulties are encountered in making rigorous the limit, and in providing a differential game interpretation of it. To avoid these difficulties we proceed in a somewhat different way in Section 5, using modified logarithmic transformations of value functions.

In taking totally risk averse limits, the noise intensity (thus, the stock price volatility) must tend to zero. In Section 6 we consider another class of risk sensitive stochastic control problems, which are perhaps of more interest for finance applications. The volatility coefficients and utility functions are now fixed. For each "reasonable" stationary investment control policy, the expected utility of wealth at terminal time T should grow exponentially as $T \to \infty$. The problem is to find a policy which maximizes the growth rate. Problems of kind have been considered by Taksar - Klass - Assaf [15] and Morton - Pliska [14] in case of logarithmic utility functions. HARA utility functions are used in Fleming - Grossman - Vila - Zariphopoulou [5]. In each of these papers, transactions costs are included in the model and the stochastic control problems are of singular type. In Section 6 we connect the problem of maximum rate of growth of expected utility with another stochastic control problem, with an average cost per unit time payoff criterion. The discussion in Section 6 is to considerable extent formal, and is a preliminary report of work in progress by S. - U. Park and the author. In particular, artificial upper bounds are imposed to avoid issues related to singular stochastic control.

2. State-linear stochastic control model. Let us consider stochastic control models of the following form. The state at time t is a vector $\vec{x}_t$ in n- dimensional space $\mathbb{R}^n$, and the control at time t is a vector $\vec{u}_t \in U$, where the "control space" U is a closed subset of $\mathbb{R}^m$. The state dynamics are governed by a state-linear stochastic differential equation of the form

$$(2.1) \qquad d\vec{x}_t = A(\vec{u}_t)\, \vec{x}_t\, dt + B(\vec{u}_t)\, \vec{x}_t\, dw_t, \quad 0 \leq t \leq T,$$

where w_t is a brownian motion which we take (for simplicity) to be of dimension 1. The $n \times m$ - matrix valued functions $A(\vec{u}), B(\vec{u})$ are continuous on the control space U. The controller is allowed complete state observations. A convenient way to express this idea technically is to allow the control process $\vec{u}$ to be any U - valued stochastic process which is progressively measurable with respect to some reference probability system. See Fleming - Soner [8, Chap. 4]. We consider (2.1) with the initial data

$$(2.2) \qquad \vec{x}_0 = \vec{x},$$

where the initial state $\vec{x}$ is a given vector. The objective is to maximize an expectation of the form $EF[h(\vec{x}_T)]$. We will consider this problem by methods of dynamic programming and partial differential equations (PDEs). In dynamic programming the value function (or optimal cost function) has a central role. To obtain various asymptotic results (e.g. for totally risk averse limits) it is convenient to consider the PDEs satisfied by certain nonlinear transforms of value functions rather than the value functions themselves.

In a finance context, $\vec{x}_t$ is a vector related to amounts invested in various assets, and $h(\vec{x}_t)$ represents the total wealth. In this paper, we will take $h(x^1, \cdots, x^n) = \Sigma x^i$. The vector $\vec{u}_t$ represents investment controls. Consumption is not considered in this model. The function $F(y)$ represents the utility of wealth y, and the goal is to maximize the terminal expected utility of wealth.

Example. Merton type models (1 dimensional). Let x_t be the wealth at time t, divided between a riskless and a risky asset; and let u_t be the *fraction* of wealth in the risky asset ($u_t \in U$). The stochastic differential equation for x_t is now

$$(2.3) \qquad dx_t = (A_0 + A_1 u_t) x_t dt + \sigma u_t x_t dw_t,$$

and $h(x) = x$. Moreover, A_0, A_1, σ are positive constants. In the original Merton model, there were no control constraints ($U = \mathbb{R}^1$). If no borrowing or short selling are allowed then $U = [0, 1]$. See Zariphopoulou [19]. If the interest rate for borrowing exceeds the rate of return on the riskless asset, then the term $A_1 u_t$ in (2.3) should be replaced by $A_1 u_{1t} - A_2 u_{2t}$, where u_{1t}, u_{2t} represent respectively the fraction of wealth in the risky asset and borrowed. In this case

$$U = \{(u_1, u_2) : u_1 \geq 0, u_2 \geq 0, u_2 \geq u_1 + 1\}.$$

See Fleming - Zariphopoulou [9].

If the utility function F is HARA, then dynamic programming and a simple homogeneity argument give an explicit solution for Merton - type models. We consider HARA functions of the form

$$(2.4) \qquad F(x) = \frac{1}{\gamma} x^\gamma, \gamma < 1 (\gamma \neq 0).$$

For non-HARA utility functions F, a change of variables linearizes the dynamic programming equation in case of the original Merton model (with

$U = I\!\!R^1$.) See Karatzas - Lehoczhy - Sethi - Shreve [12]. However, this device is no longer available otherwise; and one must resort to numerical solutions of the dynamic programming equation to obtain optimal investment policies.

3. Nonlinear transformations of value functions. For the state - linear stochastic control model formulated in Section 2, let $\psi(\vec{x}, T)$ denote the value function. Thus

$$(3.1) \qquad \psi(\vec{x}, T) = \sup_{\vec{u}} E_{\vec{x}} F[h(\vec{x}_T)],$$

where the subscript $\vec{x}$ indicates the initial data (2.2) and $\vec{u}$ is any admissible control process. We are going to apply certain nonlinear transformations to ψ. One such class are logarithmic transformations, which turn out to be of interest in connection with large deviations questions. See Fleming - Soner [8, Chap. 6]. In Section 5, we will use a modified version of that idea. In the present section, we use a different transformation, which is just the inverse F^{-1} of the utility function F. This transformation was previously used by Barron - Jensen [2].

For notational simplicity let us describe the PDE satisfied by the transformed function $V = F^{-1}(\psi)$ only for the 1 - dimensional version of (2.1), with x_t, u_t scalar - valued processes. We assume that $F \in C^2(0, \infty)$; i. e. the second derivative $F''(y)$ is continuous for $y > 0$. Moreover,

$$F'(y) > 0, \ F''(y) < 0.$$

The dynamic programming PDE for $\psi(x, T)$ is

$$(3.2) \qquad \psi_T = \max_{u \in U} \left[\frac{B^2(u)x^2}{2} \psi_{xx} + A(u)x\psi_x \right],$$

where the subscripts denote partial derivatives. The initial data for the PDE (3.2) are

$$(3.3) \qquad \psi(x, 0) = F[h(x)].$$

Since $\psi = F(V)$ the transformed function $V(x, T)$ satisfies (by an elementary calculation) the PDE

$$(3.4) \qquad V_T = \max_{u \in U} \left[\frac{B^2(u)x^2}{2} V_{xx} + A(u)x V_x \right.$$

$$\left. - \frac{B^2(u)x^2}{2} r(V)V_x^2 \right],$$

where

(3.5)
$$r(V) = \frac{|F''(V)|}{F'(V)}$$

is a coefficient of risk aversion. The initial data for (3.4) are

(3.6)
$$V(x,0) = h(x).$$

Let us consider two examples, both of Merton type. Thus, we now take

$$A(u) = A_0 + A_1 u, B(u) = \sigma u, h(x) = x.$$

Example 1. For a HARA utility function, $\psi = \gamma^{-1} V^\gamma$. In this case, $V(x,T)$ is linear in x. in fact,

(3.7)
$$V(x,T) = e^{\lambda T} x.$$

The risk aversion coefficient is

(3.8)
$$r(V) = \frac{1-\gamma}{V}.$$

The constant λ is found by substituting (3.7) into the PDE (3.4):

(3.9)
$$\lambda = \max_{u \in U} \left[A(u) - \frac{B^2(u)(1-\gamma)}{2} \right].$$

Example 2. Negative exponential utility function, namely,

(3.10)
$$F(x) = C - \exp\left(-\frac{x}{\epsilon}\right)$$

where $\epsilon > 0$ is a parameter. The additive constant C plays no role: and we may take $C = 0$. In this case, the risk aversion coefficient is constant:

(3.11)
$$r(V) = \epsilon^{-1}.$$

For this example, we will obtain explicitly $V(x,T)$ in the totally risk averse limit (Section 4) for the original Merton model.

4. Totally risk averse limits. We now consider limits as $r(V) \to \infty$. For the last term on the right side of (3.4) to make sense in the limit, we require that $B^2(u)$ tend to 0 in such a way that

(4.1)
$$B^2(u)r(V) \to \beta^2(u,V).$$

This procedure is similar to one followed by Barron-Jensen [2], and in case F is negative exponential by Whittle [17]. Proceeding in a purely formal way, we should get in the limit a function $V^0(x, T)$ satisfying the first order partial differential equation

$$(4.2) \qquad V_T^0 = \max_{u \in U} \left[-\frac{\beta^2(u, V^0)}{2} x^2 (V_x^0)^2 + A(u) x V_x^0 \right]$$

with the initial data (3.6). The solution should be considered in the viscosity sense, since one cannot expect V^0 to be a classical (smooth) solution. Moreover, one may expect $V^0(x, T)$ to have an interpretation as the value function for a suitably defined differential game. We do not have a proof of such results at this level of generality. However, in Section 5 similar results will be obtained using a somewhat different transformation of value functions.

For negative exponential utility functions there is a natural differential game interpretation for $V^0(x, T)$. Let us take

$$(4.3) \qquad F = F^\epsilon(y) = -\exp(-\epsilon^{-1} y), \, B^\epsilon(u) = \sqrt{\epsilon} \beta(u).$$

By (3.11) the left side of (4.1) is exactly $\beta^2(u)$, which does not depend on V. Since

$$(4.4) \qquad -\frac{\beta^2(u) x^2 p^2}{2} = \min_{v \in \mathbb{R}^1} \left[\beta(u) v x p + \frac{1}{2} v^2 \right],$$

we recognize upon taking $p = V_x^0$ that (4.2) is the Isaacs PDE for the lower value of the following differential game. Let ξ_t denote the state of the game at time t. The maximizing player chooses a control $u_t \in U$ and the minimizing player chooses a control $v_t \in \mathbb{R}^1$. The game dynamics are

$$(4.5) \qquad \frac{d\xi_t}{dt} = A(u_t) \xi_t + \beta(u_t) \xi_t v_t.$$

The game payoff is

$$(4.6) \qquad P = \frac{1}{2} \int_0^T v_t^2 \, dt + h(\xi_T).$$

Note that (4.5) is the deterministic analogue of (2.1) in which the "noise term" $\sqrt{\epsilon} dw_t/dt$ is replaced by a "disturbance" v_t. This is just what is expected in passing from the stochastic to deterministic (robust control) model via the totally risk averse limit.

For the original Merton model with negative exponential utility function, $A(u) = A_0 + A_1 u, \beta^2(u) = \sigma^2 u^2, U = \mathbb{R}^1$ and $h(x) = x$. In this case the PDE (4.2) becomes

$$(4.7) \qquad V_T^0 = \frac{A_1^2}{2\sigma^2} + A_0 x V_x^0.$$

The desired solution to (4.7) is

$$(4.8) \qquad V^0(x, T) = \frac{A_1^2 T}{2\sigma^2} + x e^{A_0 T}.$$

5. Modified logarithmic transformations. We now consider a different nonlinear transformation of value functions, which is a modification of a logarithmic transformation. The passage to a totally risk averse limit can then be treated rigorously by the methods of [6] [13]. Let $\epsilon > 0$ be a small parameter, such that $\epsilon \to 0$ corresponds to total risk aversion. We take $B^\epsilon(u) = \sqrt{\epsilon}\beta(u)$ as in (4.3), and

$$(5.1) \qquad F^\epsilon(y) = -\exp[\epsilon^{-1}g^\epsilon(y)]$$

where g^ϵ satisfies assumptions (5.11) below. Moreover, we take $h(x) = x$. The Ito rule gives

$$(5.2) \qquad g^\epsilon(x_T) = g^\epsilon(x_0) + \int_0^T (L^{u_t}g^\epsilon(x_t)dt$$

$$+\sqrt{\epsilon}\beta(u_t)x_tg_x^\epsilon(x_t)dw_t),$$

where

$$(5.3) \qquad L^u g = \frac{\epsilon\beta^2(u)}{2}x^2g_{xx} + A(u)xg_x.$$

A Girsanov transformation then gives

$$(5.4) \qquad E_{x_0}F^\epsilon(x_T) = F^\epsilon(x_0)\,\widetilde{E}_{x_0}\exp\frac{1}{\epsilon}\int_0^T \ell^\epsilon(x_t, u_t)dt,$$

where the expectation $\widetilde{E}$ is with respect to a probability measure $\widetilde{P}$ such that

$$(5.5) \qquad dx_t = \tilde{A}^\epsilon(x_t, u_t)x_tdt + \sqrt{\epsilon}\beta(u_t)x_td\,\widetilde{w}_t,$$

with $\widetilde{w}_t$ a $\widetilde{P}$ - brownian motion and

$$(5.6) \qquad \tilde{A}^\epsilon(x, u) = A(u) + \beta^2(u)xg_x^\epsilon(x)$$

$$(5.7) \qquad \ell^\epsilon(x, u) = L^u g^\epsilon(x) + \frac{\beta^2(u)}{2}x^2(g_x^\epsilon(x))^2.$$

Since we wish to maximize the left side of (5.4) and $F(x_0)$ does not depend on the control, it is equivalent to maximize the expectation on the right side of (5.4), for a controlled process governed by the stochastic differential equation (5.5). We write the value function ψ^ϵ as

$$(5.8) \qquad \psi^\epsilon(x, T) = F^\epsilon(x)\exp(\epsilon^{-1}W^\epsilon(x, T)).$$

By a routine calculation we obtain from (3.4) the following PDE for W^ϵ:

(5.9) $$W_T^\epsilon = \min_{u \in U}\left[\frac{\epsilon\beta^2(u)}{2}x^2 W_{xx}^\epsilon + \tilde{A}^\epsilon(x,u)x W_x^\epsilon\right.$$

$$\left. +\frac{\beta^2(u)}{2}x^2(W_x^\epsilon)^2 + \ell^\epsilon(x,u)\right],$$

with the initial data

(5.10) $$W^\epsilon(x,0) = 0.$$

We make the following assumptions:

        (a) U is compact :

(5.11) (b) xg_x^ϵ and $x^2 g_{xx}^\epsilon$ are uniformly bounded;

        (c) As $\epsilon \to 0$, g^ϵ, g_x^ϵ tend to g^0, g_x^0, uniformly
            on compact subsets of $(0,\infty)$.

THEOREM 5.1. *As $\epsilon \to 0$, $W^\epsilon \to W^0$ where $W^0(x,T)$ is the lower value of the differential game described by (5.12) - (5.14).*

The game dynamics are

(5.12) $$\frac{d\xi_t}{dt} = \tilde{A}^0(\xi_t, u_t)\xi_t + \beta(u_t)\xi_t v_t$$

with $\xi_0 = x$. The controls for the minimizing and maximizing players are chosen from the control sets U and $\mathbb{R}^1$:

(5.13) $$u_t \in U, \; v_t \in \mathbb{R}^1.$$

The game payoff is

(5.14) $$P = \int_0^T [\ell^0(\xi_t, u_t) - \frac{1}{2}v_t^2]dt.$$

The reader may note similarities and dissimilarities between this game and the differential game described by (4.5) - (4.6).

    Let us sketch a proof of the Theorem. It is convenient to make the transformation $z = \log x$,

$$\bar{W}^\epsilon(z, T) = W^\epsilon(x, T).$$

Then

$$xW_x^\epsilon = \bar{W}_z^\epsilon, \quad x^2 W_{xx}^\epsilon = \bar{W}_{zz}^\epsilon - \bar{W}_z^\epsilon.$$

If we let $\bar{g}^\epsilon(z) = g^\epsilon(x)$, then similar formulas hold relating partial derivatives of g^ϵ and $\bar{g}^\epsilon$. By (5.5) and the Ito rule, $z_t = \log x_t$ satisfies

$$dz_t = [\bar{A}^\epsilon(z_t, u_t) - \frac{\epsilon}{2}\beta^2(u_t)]dt + \sqrt{\epsilon}\beta(u_t)d\tilde{w}_t,$$

$$\bar{A}^\epsilon(z, u) = A(u) + \beta^2(u)\bar{g}_z^\epsilon(z).$$

Since $\ell^\epsilon \to \ell^0$ uniformly as $\epsilon \to 0$, it suffices in (5.9) to replace ℓ^ϵ by ℓ^0. Moreover, $\ell^0(x, u) = \bar{\ell}^0(z, u)$ where

$$\bar{\ell}^0(z, u) = A(u)\bar{g}_z(z) + \frac{\beta^2(u)}{2}(\bar{g}_z(z))^2.$$

Let $\hat{W}^\epsilon(z, T)$ denote the corresponding solution to the PDE

(5.15) $\qquad \hat{W}_T^\epsilon = \min_{u \in U}[\frac{\epsilon\beta^2(u)}{2}\hat{W}_{zz}^\epsilon + (\bar{A}^\epsilon(z, u) - \frac{\epsilon}{2}\beta^2(u))\hat{W}_z^\epsilon$

$$+ \frac{\beta^2(u)}{2}(\hat{W}_z^\epsilon)^2 + \bar{\ell}^0(z, u)]$$

(5.16) $\qquad\qquad\qquad\qquad \hat{W}^\epsilon(z, 0) = 0.$

Since $\pm|\bar{\ell}^0\|T$ are super and subsolutions of (5.15) - (5.16), the estimate $|\hat{W}^\epsilon(z, T)| \le \|\bar{\ell}^0\|T$ holds. Since $\bar{A}^\epsilon$ tends to $\bar{A}^0$ uniformly on compact subsets of $\mathbb{R}^1 \times U$, the Barles - Perthame method shows that $\hat{W}^\epsilon \to \bar{W}^0$ uniformly on compact sets, where $\bar{W}^0(z, T)$ is the unique viscosity solution to (5.15) - (5.16) when $\epsilon = 0$. See [8, Sec 7.8]. Then $\bar{W}^0(z, T) = W^0(x, T)$, where $W^0(x, T)$ is the lower value of the differential game.

Example. Let F^ϵ be a perturbation of a HARA utility function, as follows. Let $\epsilon = -\gamma^{-1}$; thus $\epsilon \to 0$ corresponds to $\gamma \to -\infty$. Then

$$\gamma^{-1}x^\gamma = -\epsilon\exp(-\epsilon^{-1}\log x)$$

Thus, we take $g^0(x) = -\log x$. Since $xg_x^0 = -1$, $x^2g_{xx}^0 = 1$, we get from (5.3), (5.6), (5.7) with $\epsilon = 0$:

$$\tilde{A}^0(u) = A(u) - \beta^2(u)$$

$$\ell^0(u) = -A(u) + \frac{1}{2}\beta^2(u).$$

Neither of these depend on the state x. In this case, the desired solution is

$$W^0(T) = T\min_{u \in U}\ell^0(u).$$

The Theorem gives the asymptotic formula

$$\max_u E_x F^\epsilon(x_T) \sim F^\epsilon(x)\exp(\epsilon^{-1}W^0(T)).$$

6. Maximum long-term growth rate. In Sections 4 and 5 we considered some asymptotic problems associated with taking totally risk averse limits. In the setting of large deviations theory, such asymptotic problems are associated with Freidlin - Wentzell large deviations for small random perturbations of dynamical systems. In the present section we consider a different kind of asymptotic result, in which the state - linear control system described by (2.1) and the utility function F are fixed (not depending on a parameter ϵ). We are concerned with the rate of growth of $E_{\overrightarrow{x}}F[h(\overrightarrow{x}_T)]$ as $T \to \infty$. We assume that F is HARA and that $h(\overrightarrow{x}) = \Sigma x^i, \overrightarrow{x} = (x^1, \cdots, x^n)$. Then the growth should be exponential. The goal is to find a stationary Markov control policy which maximizes the exponential rate of growth. The maximum growth rate Λ should correspond to the optimal cost in a stochastic control problem with average cost per unit time criterion, obtained by making a logarithmic transformation. The corresponding large deviations principles are of Donsker - Varadhan type [4], concerned with large deviations from ergodicity. This section represents a preliminary report of work in progress with S.- U. Park. We again assume that the control space U is compact. In case the problem arises from a finance model with transactions costs, this amounts to imposing artificial upper bounds on transactions rates. We do this to avoid the additional complication of considering "singular" stochastic controls. Let

$$(6.1) \qquad y_t = h(\overrightarrow{x}_t) = \sum_{i=1}^{n} x_t^i,$$

$$(6.2) \qquad \overrightarrow{z}_t = y_t^{-1} \overrightarrow{x}_t .$$

By summing over the components of (2.1) we obtain since h is a linear function

$$(6.3) \qquad dy_t = y_t\{h[A(\overrightarrow{u}_t)z_t]dt + h[B(\overrightarrow{u}_t)\,\overrightarrow{z}_t]dw_t\}.$$

An application of the Ito differential rule to $\log y_t$ shows that $y_t > 0$ for all $t > 0$ if $y_0 > 0$. Moreover, $\overrightarrow{z}_t$ satisfies a stochastic differential equation, with coefficients depending on $\overrightarrow{z}_t$ and $\overrightarrow{u}_t$ but *not* on y_t. Since $h(\overrightarrow{z}_t) = 1$, the process $\overrightarrow{z}_t$ lies on the hyperplane $H = \{\overrightarrow{z}: \Sigma z^i = 1\}$. If, in addition, $x_t^i \geq 0$ for $i = 1, \cdots, n$, then $\overrightarrow{z}_t$ lies in the standard unit simplex on H.

If $F(y) = \gamma^{-1}y^\gamma$, $0 < \gamma < 1$, then the value function ψ in (3.1) has the form

$$(6.4) \qquad \psi(\overrightarrow{x}, T) = y^\gamma \Psi(\overrightarrow{z}, T), y = \sum_{i=1}^{n} x^i.$$

By writing the dynamic programming PDE for ψ and using (6.4), a formal calculation leads us to expect that Ψ is the value function for another stochastic control problem, in which $\vec{z}_t$ is the state. The state dynamics have the form

$$(6.5) \qquad d\vec{z}_t = f(\vec{z}_t, \vec{u}_t)dt + g(\vec{z}_t, \vec{u}_t)d\tilde{w}_t$$

with $\tilde{w}_t$ a 1 - dimensional brownian motion. The criterion to be maximized takes the form

$$(6.6) \qquad \tilde{E}_{\vec{z}} \exp \int_0^T \ell(\vec{z}_t, \vec{u}_t)dt.$$

We will write out the explicit form of f, g and ℓ only for the transactions costs model below. We conjecture that

$$(6.7) \qquad \Psi(\vec{z}, T) \sim e^{\Lambda T} \phi(\vec{z}) \text{ as } T \to \infty.$$

The interpretation of Λ is as the optimal long term growth rate. By substituting the right side of (6.7) in the dynamic programming equation for Ψ, one can interpret ϕ as an eigenvalue and ϕ as a positive eigenfunction in a kind of nonlinear eigenvalue problem. The function $W = \log \phi$ satisfies the dynamic programming PDE for a stochastic control problem with average cost per unit time criterion. The control problem (6.5) - (6.6) fits formally into the framework of Fleming - McEneaney [7]. However, the assumptions made in [7] on f and g are different, and the results there are not directly applicable.

Transactions costs models. Let us illustrate the above ideas with the following 2-dimensional model with transactions costs, considered in [5]. Let S_t, B_t denote the amounts invested in a stock and a bond, respectively. Let

$$x_1 = (1 - s)S, \quad x_2 = B,$$

where s is a unit transaction cost for selling the stock, The wealth is $y_t = x_{1t} + x_{2t}$. The state equations have the form

$$(6.8) \qquad \begin{cases} dx_{1t} = [(A_1 - u_{1t})x_{1t} + u_{2t}(1 - s)x_{2t}]dt \\ \qquad\qquad +\sigma x_{1t}dw_t \\ dx_{2t} = [u_{1t}x_{1t} + (A_0 - (1 + b)u_{2t})x_{2t}]dt, \end{cases}$$

where b is a unit transaction cost for selling the bond. The respective transactions rates are $u_{1t}S_t$ and $u_{2t}B_t$. To avoid considering singular controls, we impose artificial upper bounds on transactions rates:

$$(6.9) \qquad 0 \leq u_{1t} \leq \bar{u}_1 < \infty, \ 0 \leq u_{2t} \leq \bar{u}_2 < \infty.$$

Thus we take $U = [0, \bar{u}_1] \times [0, \bar{u}_2]$. Without these artificial upper bounds, the differential equation satisfied by Λ and ϕ would be replaced by a set of linear inequalities, which were solved rather explicitly in [5].

In the notation above, we have $\vec{x} = (x^1, x^2), y = x^1 + x^2, z^i = x^i/y$. Since $z^1 + z^2 = 1$, we may consider the scalar valued process $z_t = z_t^1$, with $z_t^2 = 1 - z_t^1$. In place of (6.5) we have

$$(6.10) \qquad dz_t = f(z_t, u_{1t}, u_{2t})dt + \sigma z_t(1 - z_t)d\tilde{w}_t,$$

$$(6.11) \qquad f(z, u_1, u_2) = \sigma^2 z^2(\gamma - 1)(1 - z) + (A_1 - A_0)z(1 - z)$$

$$-zu_1 + [1 - s + (s + b)z](1 - z)u_2.$$

In (6.6) we have

$$(6.12) \qquad \ell(z, u_2) = \gamma[\frac{\sigma^2 z^2}{2}(\gamma - 1) + A_1 z + A_0(1 - z)]$$

$$-\gamma(s + b)(1 - z)u_2.$$

The differential equation satisfied by Λ and $\phi(z)$ now becomes

$$(6.13) \qquad \Lambda\phi = \frac{\sigma^2 z^2(1 - z)^2}{2}\phi_{zz}+$$

$$\max_{u_1, u_2}[f(z, u_1, u_2)\phi_z + \ell(z, u_2)\phi].$$

Optimal stationary control policies $u_1^*(z), u_2^*(z)$ are found by taking arg max in (6.13). Since f and ℓ are linear in the controls u_1, u_2, these should be bang-bang. Thus (as one anticipates) transactions occur either at the maximum rates $\bar{u}_1, \bar{u}_2$ or not at all. As mentioned earlier, the discussion above is formal since we have not proved the existence of Λ and $\phi(x)$ satisfying(6.13). If initially $S_0 > 0$ and $B_0 > 0$, then for "reasonable" investment policies (in particular, for the optimal policies) we anticipate that $x_{1t} > 0$ and $x_{2t} > 0$. Then $0 < z_t < 1$. The endpoints of the interval $0 < z < 1$ are inaccessible, which accounts for the lack of boundary conditions for ϕ. We anticipate that the optimal policies have the form

$$u_2^*(z) = \begin{cases} \bar{u}_2, 0 < z < z_b, \\ \\ 0, z_b < z < 1 \end{cases}$$

$$u_1^*(z) = \begin{cases} 0, 0 < z < z_s, \\ \\ \bar{u}_1, z_s < z < 1 \end{cases}$$

where $z_b < z_s$. The second derivative of $\phi(z)$ must remain continuous at z_b and z_s. Moreover, $\phi(z)$ is bounded on the interval $(0,1)$.

By making the logarithmic transformation $W = \log \phi$, another interpretation of Λ is obtained as an optimal average cost per unit time. By using (6.13) and a formula like (4.4), we get

$$(6.14) \qquad \Lambda = \frac{\sigma^2 z^2 (1-z)^2}{2} W_{zz} + \max_v [\sigma z (1-z) v W_z - \frac{1}{2} v^2]$$

$$+ \max_{u_1, u_2} [f(z, u_1, u_2) W_z + \ell(z, u)].$$

This is the average cost per unit time dynamic programming equation of a stochastic control problem with state $\zeta_t \in (0,1)$ and controls $(u_{1t}, u_{2t}) \in U$, $v_t \in \mathbb{R}^1$. The state dynamics are

$$(6.15) \qquad d\zeta_t = [f(\zeta_t, u_{1t}, u_{2t}) + \sigma \zeta_t (1 - \zeta_t) v_t] dt$$

$$+ \sigma \zeta_t (1 - \zeta_t) dw_t,$$

with w_t some brownian motion. The criterion to be maximized is

$$(6.16) \qquad J = \limsup_{T \to \infty} \frac{1}{T} E \int_0^T [\ell(\zeta_t, u_{2t}) - \frac{1}{2} v_t^2] dt.$$

The function $W(z)$ has the role of a cost potential. Unlike the situation in Sections 4 and 5, the controllers choosing (u_{1t}, u_{2t}) and v_t both wish to maximize J. In this case, the logarithmic transformation leads to another stochastic control problem rather than a stochastic differential game. [If F is HARA with $\gamma < 0$, then $\phi(z) < 0$. In that case, one should take $W = \log(-\phi)$ and a stochastic differential game is obtained.]

REFERENCES

[1] J. S. Baras, R. J. Elliott and M. R. James, "Risk-sensitive control and dynamic games for partially observed discrete-time nonlinear systems", submitted to IEEE Trans. Auto. Control.

[2] E. N. Barron and R. Jensen "Total risk aversion, stochastic optimal control and differential games", Appl. Math. and Optimiz 19 (1989) 313–327.

[3] T. Başar and P. Bernhard, "H^∞ - Optimal Control and Related Minimax Design Problems", Birkhauser, Boston 1991.

[4] M. D. Donsker and S. R. S. Varadhan, "Asymptotic evaluation of certain Markov process expectations for large time", I, II, III, Comm. Pure Appl. Math. 28 (1975) 1-45, 279- 301; 29 (1976) 389–461.

[5] W. H. Fleming, S. G. Grossman, J-L Vila and T. Zariphopoulou, "Optimal portfolio rebalancing with transaction costs", submitted to Econometrica.

[6] W. H. Fleming and W. M. McEneaney, "Risk sensitive control and differential games", Springer Lecture Notes in Control and Info. Sci. No. 184, 1992, 185–197.

[7] W. H. Fleming and W. M. McEneaney, "Risk sensitive control on an infinite time horizon", Submitted to SIAM J. on Control and Optimization.

[8] W. H. Fleming and H. M. Soner, "Controlled Markov Processes and Viscosity Solutions", Springer–Verlag, 1992.

[9] W. H. Fleming and T. Zariphopoulou, "An optimal investment/consumption model with borrowing", Math of Operations Res. $\underline{16}$ (1991) 802–822.

[10] D. H. Jacobson "Optimal stochastic linear systems with exponential criteria and their relation to deterministic differential games", IEEE Trans. Automat. Control AC - $\underline{18}$ (1973) 124–131.

[11] M. R. James "Asymptotic analysis of nonlinear stochastic risk-sensitive control and differential games", Math of Control, Signals and Systems (to appear).

[12] I. Karatzas, J. Lehoczhy, S. Sethi and S. Shreve, "Explicit solution of a general consumption - investment problem", Math. of Operations Res. $\underline{11}$ (1986) 261–294.

[13] W. M. McEneaney, "Connections between risk-sensitive stochastic control, differential games and H - infinity control: The nonlinear case", Brown University PhD Thesis, 1993.

[14] A. J. Morton and S. R. Pliska, " Optimal portfolio management with fixed transaction costs", preprint.

[15] M. Taksar, M. J. Klass and D. Assaf, "A diffusion model for optimal portfolio selection in the presence of brokerage fees", Math. of Operations Res. $\underline{13}$ (1988) 277–294.

[16] P. Whittle, "Risk Sensitive Optimal Control", Wiley, 1990.

[17] P. Whittle, "A risk sensitive maximum principle", Syst. Control Lett. $\underline{15}$ (1990) 183–192.

[18] P. Whittle, "A risk sensitive maximum principle: The case of imperfect state information", IEEE Trans. Auto. Control $\underline{36}$ (1991) 793–801.

[19] T. Zariphopoulou, "Optimal investment -consumption models with borrowing", Brown Univ. PhD Thesis, 1988.

ARBITRAGE AND FREE LUNCH IN A GENERAL FINANCIAL MARKET MODEL; THE FUNDAMENTAL THEOREM OF ASSET PRICING

MARCO FRITTELLI* AND PETER LAKNER†

Abstract. We shall examine a financial market with a finite or countably infinite class of securities available for investment. A portfolio which generates almost surely positive gains without risks is called an arbitrage opportunity. Clearly, if such opportunity exists then the market cannot be in equilibrium. The "Fundamental Theorem of Asset Pricing" asserts that no arbitrage opportunity exists if and only if there exists a probability measure Q, equivalent to our "reference measure", such that the discounted asset prices are martingales under Q. This loosely phrased statement will receive a rigorous mathematical treatment here under the most general conditions. The concept of arbitrage is replaced by a generalization called "free lunch". No artificial mathematical conditions, such as integrability, continuity or boundedness will be imposed on the discounted asset price processes. The main result is that an equivalent martingale measure exists if and only if there is no free lunch opportunity in the financial market.

1. Introduction. An initial result on our subject, later called the Fundamental Theorem of Asset Pricing, was achieved by Harrison & Pliska [7]. The authors assumed that Ω, representing the state of the world, is finite. Various generalizations are now available in the literature. There is a very satisfactory treatment of the case of finite index set in Dalang, Morton & Willinger [2] in which the authors show that arbitrage is absent in the market if and only if an equivalent martingale measure exists.

Unfortunately for discrete infinite or continuous index set the absence of arbitrage is not a sufficient condition for the existence of a martingale measure. A generalization of arbitrage, called free lunch, was introduced by Kreps [8]. There are differences in the definition of free lunch in the various papers but the essence of the matter is the following. The absence of arbitrage postulates that the set of achievable gains by certain admissible, simple trading strategies does not include a positive random variable. The absence of free lunch, roughly speaking, postulates the same of the topological closure of the same set.

Schachermayer [10] and Delbaen & Schachermayer [4] mainly deal with the case of discrete, infinite index set. The beauty of this treatment is that the topological closure is replaced by taking limits of convergent sequences. Continuous time models were studied by Ansel & Stricker [1], Stricker [11], Duffie & Huang [5], Delbaen [3], Lakner [9], and Frittelli & Lakner [6]. In the first three of the above mentioned papers it is assumed that the discounted stock prices are in $L^p(P)$ for some $1 \leq p < \infty$. This condition is

* Universita degli Studi di Urbino, Facolta di Economia e Commercio, Piazza della Repubblica 5, Urbino, Italy. e-mail: frittema@imiucca.csi.unimi.it phone: (0722) 4475.

† New York University, Statistics and Oper. Res. Dept., 44 W. 4th St., Rm. 861, New York, N Y 10012. e-mail: plakner@rnd.stern.nyu.edu phone: (212) 998-0476.

not invariant under the substitution of P by an equivalent probability measure. Consequently, the definitions of free lunch and equivalent martingale measure depend on the selection of the reference measure P. This is not satisfying if we view P as an arbitrary subjective probability measure, not having any particular significance versus any other subjective probability measure.

In Delbaen [3] the discounted asset prices are assumed to be bounded and continuous. In Lakner [9] the only assumption is that every discounted asset price at every stopping time is bounded below by a constant. This is natural because the prices are non–negative. Eventually, in Frittelli & Lakner [6] none of these assumptions are imposed. In the present paper we give only results and send the interested reader to Frittelli & Lakner [6] for proofs and further details.

2. Martingale measures and free lunch. Let our index set I be an arbitrary subset of $[0, \infty)$ such that $0 \in I$. With this generality we cover simultaneously the usual choices of a finite interval $[0, T]$, N, or the entire of $[0, \infty)$. We have a probability space $(\Omega, \mathcal{F}, P)$ and a filtration $\{\mathcal{F}_t; \ t \in I\}$. $\mathcal{F}_0$ is assumed to be trivial and complete.

Let $\mathcal{X}$ be a finite or countably infinite set of adapted processes. These represent the discounted prices of available assets. Trading is permitted at any point of I.

We introduce the following two linear spaces:

$$K \quad = \quad \text{Lin}\{h(X_t - X_s); \ t, s \in I, s \leq t, h \in L^\infty(\Omega, \mathcal{F}_s, P)\}$$

$$K_U \quad = \quad \text{Lin}\{g(X_T - X_S); \ T, S, \text{ are bounded stopping times,}$$
$$S \leq T, g \in L^\infty(\Omega, \mathcal{F}_S, P)\}$$

We shall call K (resp. K_U) the set of achievable gains by simple trading strategies with deterministic times (resp. the set of achievable gains by simple trading strategies with stopping times).

The usual "no arbitrage" condition postulates that K or K_U contains no positive random variables. However, in a continuous time setting this condition is not sufficiently restrictive. In order to formulate the "no free lunch" condition we include K and K_U in the following two linear spaces:

$$S = \text{Lin}(K \cup L^\infty)$$

$$S_U = \text{Lin}(K_U \cup L^\infty) \ .$$

To talk about closures of the sets K and K_U we need topologies on S and on S_U. These will be topologies compatible with certain dualities. Specifically, let us consider the following subsets of the algebraic dual of S and S_U:

$$S' = \{x \in L^1(P) : E[|xs|] < \infty, \text{ for all } s \in S\}$$

$$S'_U = \{x \in L^1(P) : E[|xs|] < \infty, \text{ for all } s \in S_U\}.$$

Elements of S' and S'_U represent linear functionals on S and S_U given by $s \mapsto E[xs]$. The sets S' and S'_U, regarded as sets of linear functionals, are invariant under the substitution of P by an equivalent probability measure. For example S', as a set of linear functionals, can be described as the set of all bounded signed measures $\mu \ll P$ on $\mathcal{F}$, such that $\int |s| d|\mu| < \infty$ for all $s \in S$. The linear functionals are now given by $s \mapsto \int s\, d\mu$.

Let τ be any locally convex topology on S compatible with the duality (S, S') and τ_U be any locally convex topology compatible with the duality (S_U, S'_U). We shall formulate two conditions, one using deterministic times and the other using stopping times:

(A) $\quad \overline{K - L^\infty_+} \cap L^\infty_+ = \{0\}$

(B) $\quad \overline{K_U - L^\infty_+} \cap L^\infty_+ = \{0\}$.

Naturally, the closure in (A) is the τ closure, whereas in (B) it is the τ_U closure. The notation L^∞_+ represents the set of almost surely non-negative and bounded random variables, and $K - L^\infty_+$ is the set $\{k - l;\ k \in K,\ l \in L^\infty_+\}$. Condition (A) will be called "no free lunch with deterministic times" and condition (B) will be called "no free lunch with stopping times". Condition (A) can be explained in the following way. Suppose that it fails. Then there exists a gain achieved by a simple trading strategy, which dominates a random variable in S arbitrarily close to a positive bounded random variable. The explanation of (B) is quite similar.

DEFINITION 2.1. We call a probability measure Q an absolutely continuous (resp. equivalent) martingale measure if $Q \ll P$ (resp. $Q \sim P$), and each process $X \in \mathcal{X}$ is a martingale under Q with respect to the filtration $\{\mathcal{F}_t;\ t \in I\}$.

To formulate our results we need the following condition:

(C) $\quad$ *there exists an equivalent martingale measure.*

The main result is the following

THEOREM 2.2. *There is no free lunch with deterministic times if and only if there exists an equivalent martingale measure, i.e., (A) and (C) are equivalent. Furthermore, if every $X \in \mathcal{X}$ is right–continuous, then (A), (B), and (C) are equivalent.*

For proof and further information on the subject we refer to Frittelli & Lakner [6]. However, there is an intermediate result worth mentioning here:

PROPOSITION 2.3. *For any sequence $(Q_n)_{n \in N}$ of absolutely continuous martingale measures there exists an absolutely continuous martingale measure Q such that $Q(A) = 0$ if and only if $Q_n(A) = 0$ for every $n \in N$.*

3. Conclusion. The present paper achieves a high level of generality since no mathematical conditions are imposed on the asset prices which would not be naturally required in the finance model. Indeed, the only requirement is that the processes are adapted which is natural because no asset price can depend on information available only in the future. To achieve this generality we had to use topologies defined by certain dualities. For an elaboration of the meaning of these topologies in the finance model we send the interested reader to Lakner [9].

REFERENCES

[1] J.P.ANSEL, C.STRICKER, *Quelques remarques sur un théorème de yan*, Sém. de Probabilités, Lecture Notes in Mathematics **XXIV** Springer-Verlag, New York, 1990, 226–274.

[2] R.C.DALANG, A.MORTON, W.WILLINGER, *Equivalent martingale measures and no-arbitrage in stochastic securities market models*, Stochastics and Stochastic Reports, **29** (1990), 185–201.

[3] F.DELBAEN, *Representing martingale measures when asset prices are continuous and bounded*, Mathematical Finance **2** (1992), 107–130.

[4] F.DELBAEN, W.SCHACHERMAYER, *A general version of the fundamental theorem of asset prices*, (*preprint*) 1993.

[5] D.DUFFIE, C-F.HUANG, *Multiperiod security markets with differential information*, Journal of Mathematical Economics **15** (1986), 283–303.

[6] M.FRITTELLI, P.LAKNER, *Almost sure characterization of martingales*, Stochastics and Stochastic Reports (*forthcoming*) 1993.

[7] J.M.HARRISON, S.R.PLISKA, *Martingales and stochastic integrals in the theory of stochastic trading*, Stochastic Processes and their Applications **11** (1981), 215–260.

[8] D.M.KREPS, *Arbitrage and equilibrium in economies with infinitely many commodities*, Journal of Mathematical Economics **8** (1981), 15–35.

[9] P.LAKNER, *Martingale measures for a class of right-continuous processes*, Mathematical Finance **3**(1) January (1993), 43–53.

[10] W.SCHACHERMAYER, *Martingale measures for discrete time processes with infinite horizon*, (*mimeo*) 1992.

[11] C.STRICKER, *Arbitrage et lois de martingale*, Ann. Inst. Henri Poincare **26** (1990), 451–460.

WHICH MODEL FOR TERM-STRUCTURE OF INTEREST RATES SHOULD ONE USE?[*]

L.C.G. ROGERS[†]

1. Introduction. There are presently many different models of the term structure of interest rates, but little agreement on any one natural one. This is perhaps not surprising in view of the nature and quality of data on term structure; prices of coupon-bearing and zero-coupon bonds, LIBOR rates, index-linked bonds, together with options and futures on such things, all provide information about interest rates, and have to be compatible with any successful model, either by being used to estimate parameters of the model, or by being consistent with the predictions of the (fitted) model. In addition, interest rate swaps provide information, but this is a different market, and should not be expected necessarily to fit the same model.

Much of the work done in this area describes models which one *could* use, rather than claiming strongly that one *should* use them; I have very little to say about what model(s) one *should* use, but will say a few things about what one should *not* use! To decide this, it helps to be clear about what the goal is; we want to build a model that practitioners may rely on. Now a practitioner wants a model which is:

  (**a**) *flexible* enough to cover most situations arising in practice;
  (**b**) *simple* enough that one can compute answers in reasonable time;
  (**c**) *well-specified*, in that required inputs can be observed or estimated;
  (**d**) *realistic*, in that the model will not do silly things.

Additionally, the practitioner shares the view of an econometrician who wants

  (**e**) *a good fit* of the model to data;

and a theoretical economist would also require

  (**f**) *an equilibrium derivation* of the model.

For the practitioner, (a)-(e) already constitute Nirvana!

The fundamental object of study for term-structure is the spot-rate process $(r_t)_{t \geq 0}$. This is a continuous-time process, often assumed to have continuous paths, though sometimes also modelled (more realistically) with jumps. The spot rate represents the instanteneous rate of riskless return

---

[*] supported by SERC grant GR/H52023.

[†] School of Mathematical Sciences, University of Bath, Claverton Down, Bath BA2 7AY, Great Britain.

at any time, so \$1 invested at time t will have grown by later time T to

$$\$\exp\left(\int_t^T r_u du\right).$$

Now if one assumes that r is a stochastic process, the problem of pricing a bond is non-trivial: how much should one pay at time t to receive \$1 at later time T? Arbitrage pricing theory says the price must be

$$P(t,T) \equiv E\left[\exp\left(-\int_t^T r_u du\right)|\mathcal{F}_t\right]$$

where the expectation is with respect to the "risk-neutral" measure. So the central question is how to model the law of r under the risk-neutral measure so as to achieve (a)-(e).

The plan of the rest of the paper is as follows. Section 2 discusses some very simple Gaussian models for r, and provides a bound for the error arising from the possibility that r can be negative. Section 3 presents some simple one-factor squared-Gaussian models, with an appendix summarizing the analysis of this class of models. Section 4 discusses some two-factor models which have been proposed. In particular, we take a very general class of multi-factor squared Gaussian models and analyse the term-structures which can arise, and in Section 5 we discuss some "whole-yield" models currently in vogue. Section 6 is a discussion of the relationship between equilibrium and no-arbitrage pricing; essentially, the two are the same. Finally, in Section 7, we briefly survey the questions of estimating and fitting with observed data. The relevant literature is surveyed in the corresponding Sections.

As a piece of notation, we define the *yield curve* at time t to be the function

$$T \mapsto -\frac{1}{T-t}\log P(t,T) \quad (T > t)$$

and the *forward rate* f_{tu} (for $0 \le t \le u$) to be given by

$$P(t,T) = \exp\left[-\int_t^T f_{tu} du\right],$$

provided the yield curve at time t is differentiable.

2. Simple Gaussian models. If $\alpha, \beta, \sigma : \mathbb{R}^+ \to \mathbb{R}^+$ are any locally bounded functions, then the stochastic differential equation

$$(2.1) \qquad dr_t = \sigma_t dW_t + (\alpha_t - \beta_t r_t)dt$$

for r has a unique solution, which is a Gaussian process. This is the model considered by Hull & White [27], generalising the models of Vasicek [42], in which the functions α, β, σ are constant, and of Merton [34] where also $\beta = 0$. The analysis of (2.1) is very simple. Taking

$$K_t \equiv \int_0^t \beta_u \, du,$$

and multiplying (2.1) by $\exp(K_t)$ we obtain

$$d(e^{K_t} r_t) = e^{K_t}(\sigma_t dW_t + \alpha_t dt)$$

from which

$$(2.2) \qquad r_t = e^{-K_t} \left\{ r_0 + \int_0^t e^{K_u}(\sigma_u dW_u + \alpha_u du) \right\}.$$

This is a Gaussian process, for which

$$(2.3) \qquad \mu_t \equiv E r_t = e^{-K_t} \left\{ r_0 + \int_0^t e^{-K_u} \alpha_u du \right\}$$

$$(2.4) \qquad \rho(s,t) \equiv cov(r_s, r_t) = e^{-K_s - K_t} \int_0^{s \wedge t} e^{2K_u} \sigma_u^2 du.$$

Thus

$$Z_t \equiv \int_0^t r_u \, du \sim N(m_t, v_t)$$

where

$$(2.5) \qquad m_t \equiv \int_0^t e^{-K_u} \left(r_0 + \int_0^u e^{K_s} \alpha_s ds \right) du,$$

$$(2.6) \qquad v_t \equiv 2 \int_0^t du \int_0^u ds \int_0^s dy \sigma_y^2 e^{-K_s - K_u + 2K_y},$$

and hence the bond price is

$$(2.7) \qquad E \exp \left(-\int_0^t r_u \, du \right) = \exp \left(-m_t + \frac{1}{2} v_t \right)$$

$$(2.8) \qquad\qquad = \exp\{-r_0 B(0,t) - A(0,t)\},$$

where for $0 \le t \le T$

$$(2.9) \quad B(t,T) \quad \equiv \quad \int_t^T \exp(-K_u + K_t)du,$$

$$(2.10) \; A(t,T) \quad \equiv \quad \int_t^T du \int_t^u ds \left\{ \alpha_s e^{K_s - K_u} - \int_t^s dy \sigma_y^2 e^{-K_s - K_u + 2K_y} \right\}.$$

Thus this model is extremely *simple*, and the log-Gaussian distribution of bond prices makes it very easy to price derivative securities. For example, the price at time t of a European call option to be exercised at time T with strike X on a zero-coupon bond of maturity $T' > T$ is simply

$$E\left[\exp\left(-\int_t^T r_u du \right) \{\exp(-r_T B(T,T') - A(T,T')) - X\}^+ \, |\mathcal{F}_t \right]$$

which can be evaluated in closed form. Similarly the value at time t of a futures contract, delivery date T, on a zero-coupon bond with maturity $T' > T$ is

$$E(P(T,T')|F_t) = E\left[\exp(-r_T B(T,T') - A(T,T'))|F_t \right]$$

which can again be evaluated in closed form.

Is the model *well-specified*? Hull & White argue that if one knows at some time $(0$, say$)$ the volatility of r, and the volatility of bonds of all maturities, then since

$$dP(t,T) = -P(t,T)d(r_t B(t,T) + A(t,T)) + \frac{1}{2}P(t,T)B(t,T)^2 d\langle r \rangle_t,$$

the volatility of the maturity-T bond is $\sigma_0 B(0,T) P(0,T)$; and, since this is known, one can deduce $B(0,T)$ for all T. Then, since $B(0,\cdot)$ is known, we can recover $A(0,\cdot)$ from (2.8). But now knowing $B(0,\cdot)$ we can find K and therefore β from (2.9); and then differentiating $A(0,\cdot)$ gives

$$e^{K_T} A'(0,T) = \int_0^T \left(\alpha_s e^{K_s} - \int_0^s \sigma_y^2 e^{2K_y - K_s} dy \right) ds,$$

differentiating once more gives

$$e^{K_T} \left(\beta_T A'(0,T) + A''(0,T) \right) = \alpha_T e^{K_T} - \int_0^T \sigma_y^2 e^{2K_y - K_T} dy,$$

and another derivative gives

$$e^{2K_T}\left(2\beta_T^2 A'(0,T) + 3\beta_T A''(0,T) + \beta_T' A'(0,T) + A'''(0,T)\right)$$
$$= \left(\alpha_T' + 2\beta_T\alpha_T - \sigma_T^2\right)e^{2K_T}.$$

This cannot be uniquely solved for α, σ, but it gives us an equation

$$\alpha_T' + 2\beta_T\alpha_T - \sigma_T^2 = \varphi(T)$$

for some known function φ, and this could be satisfied by taking, for example,

$$\sigma_T^2 = \varepsilon + \varphi(T)^-,$$
$$\alpha_T' + 2\beta_T\alpha_T = \varepsilon + \varphi(T)^+$$

for some $\varepsilon > 0$ fixed. This model is also clearly extremely *flexible*, in that any initial yield curve, and any initial term-structure of volatility can be fitted. However, the estimation of the model from data is not practical. Firstly, the yield curve is not some nice smooth curve known at all positive real points; in practice, it is only known at a limited set of maturities (typically 10-20), with dubious accuracy of measurement. Any procedure which requires repeated differentiation of this "curve" cannot be expected to work. Secondly, even if one could obtain estimates of the functions α, β, σ from the data, there is no reason why we should get consistent estimates if we performed the same analysis of the term-structure as it appears one week later!

The best we might hope for is to restrict (α, β, σ) to lie in some small parametric family, then estimate the parameters. The smallest interesting family we could consider is the Vasicek model, where the functions are constant, and the bond prices given by (2.7) simplify to

$$(2.11) \qquad m_t = \frac{\alpha}{\beta}t + \frac{1 - e^{-\beta t}}{\beta}\left(r_0 - \frac{\alpha}{\beta}\right)$$

$$(2.12) \qquad v_t = \frac{\sigma^2}{2\beta^3}\left[2\beta t - 3 + 4e^{-\beta t} - e^{-2\beta t}\right].$$

This special case is concrete enough for us to investigate the only undesirable feature of these Gaussian models, namely that the interest rates may be *negative*. Now in the Vasicek model, the limiting distribution of r is $N\left(\alpha/\beta, \sigma^2/2\beta\right)$, so if we choose (α/β) to be reasonably large compared with the standard deviation $\sigma/\sqrt{2\beta}$, we might imagine that negative interest rates will not be a problem. However, taking

$$\alpha = \frac{1}{3} \times 10^{-3}, \qquad \beta = \frac{1}{3} \times 10^{-2}, \qquad \sigma^2 = \frac{8}{3} \times 10^{-6}$$

we have

$$(i)\ \frac{\alpha}{\beta} = 5\frac{\sigma}{\sqrt{2\beta}} \qquad (ii)\ \frac{\alpha}{\beta} = 0.1 \qquad (iii)\ \frac{1}{t}\log P(0,t) \to 0.02.$$

The first tells us that in equilibrium the probability of a negative interest rate is very small (about 3×10^{-7} in fact), the second tells us that the mean value of r is 0.1, not an unreasonable annual rate – but the third tells us that the bond prices *grow* exponentially, which is absurd! Admittedly $P(0, t)$ will not climb back to 1 for quite a long time, but a model which can do this is a model which must either be rejected, or handled with caution. It is not enough simply to hope that the problem can be neglected.

We should consider the spot-rate process to be r_t^+ rather than r_t, but then the tractable Gaussian behaviour is gone. We can say the following, however.

PROPOSITION 2.1. *Suppose that r is a Gaussian process, $E r_t = \mu_t$, $cov(r_s, r_t) = \rho_{st}$. Then*

$$
\begin{aligned}
0 \;\le\; & E \exp\left(-\int_0^T r_u \, du\right) - E \exp\left(-\int_0^T r_u^+ \, du\right) \\
\le\; & E e^{-R_T}\left\{1 - \exp\left(-E\left(e^{-R_T}\int_0^T r_u^- \, du\right)\Big/ E e^{-R_T}\right)\right\}
\end{aligned}
$$

where $R_T \equiv \displaystyle\int_0^T r_u \, du$;

$$
(2.13) \qquad = E e^{-R_T}\left\{1 - \exp\left(-\int_0^T \sqrt{\rho_{ss}}\, G\left(\frac{\mu_s - \int_0^T \rho_{st} \, dt}{\sqrt{\rho_{ss}}}\right) ds\right)\right\},
$$

where

$$
(2.14) \qquad G(a) \equiv E(W_1 - a)^+ = \frac{1}{\sqrt{2\pi}} e^{-a^2/2} - a\overline{\Phi}(a)
$$

with $\overline{\Phi}$ the tail of the standard normal distribution. Thus

$$
\begin{aligned}
0 \;\le\; & 1 - E\left(\exp -\int_0^T r_u^+ \, du\right)\Big/ E\left(\exp -\int_0^T r_u \, du\right). \\
(2.15) \qquad \le\; & 1 - \exp\left[-\int_0^T \sqrt{\rho_{ss}}\, G\left(\frac{\mu_s - \int_0^T \rho_{st} \, dt}{\sqrt{\rho_{ss}}}\right) ds\right]
\end{aligned}
$$

REMARK 2.1.   Using the inequality

$$
(2.16) \qquad G(a) \le \frac{e^{-a^2/2}}{\sqrt{2\pi}(1 + a^2)} + a^-
$$

gives a good idea of the sizes of the quantities involved. For the example of the Vasicek model,

$$\mu_s = r_0 e^{-\beta s} + \frac{\alpha}{\beta}(1 - e^{-\beta s}),$$

$$\rho_{st} = \frac{\sigma^2}{2\beta}\left(e^{-\beta|s-t|} - e^{-\beta s - \beta t}\right),$$

so

$$\int_0^T \rho_{st}\,dt = \frac{\sigma^2}{\beta^2}\left\{1 - e^{-\beta s} - e^{-\beta T}\sinh\beta s\right\}.$$

Taking the earlier example with $\alpha = \frac{1}{3} \times 10^{-3}$, $\beta = 10\alpha$, $\sigma^2 = \frac{8}{3} \times 10^{-6}$, evaluating the bound for $T = 10$ gives 0.014, for $T = 20$ gives 0.041, for $T = 30$ gives 0.078 and for $T = 50$ gives 0.182.

Proof. The first statement is immediate, the second uses Jensen's inequality and for the third, we use the result that if X and Z are zero-mean Gaussians, $EX^2 = 1$, then for any θ

$$Ee^{-Z}(X + \theta)^{-} = \exp\left(\frac{1}{2}EZ^2\right)G(\theta - cov(X, Z)).$$

□

REMARK 2.2.   The estimate (2.13) is likely to be a good approximation to the difference between the bond prices using the Gaussian process r, and the positive part of r. This is because the only approximation used to reach (2.13) is $1 - e^{-x} \leq x$. For small $x > 0$, the difference $x - 1 + e^{-x}$ is $O(x^2)$, and for larger values of x it is $O(x)$, but in the expectation, the distribution of r_s^- will have a rapidly decreasing tail if we have chosen parameters which make the probability of negative spot rates small. Thus the approximation should be accurate.

3. Squared Gaussian models. We can escape negative interest rates if we modify the variance structure in (2.1) to give

(3.1) $dr_t = \sigma_t\sqrt{r_t}\,dW_t + (\alpha_t - \beta_r r_t)dt,$

where once again $\alpha, \beta, \sigma : \mathbb{R}^+ \to \mathbb{R}^+$ are any locally bounded functions[1]. The process r will remain non-negative if it starts non-negative, a great advantage over the Gaussian models of the last section. Cox, Ingersoll & Ross [13] introduced such processes as models for the spot rate (taking α, β, σ to be constant); the time-dependent version which we take here is a generalisation due to Hull & White [27].

---

[1] It can be shown, using the Yamada-Watanabe theorem and time-change, that (3.1) has a pathwise unique strong solution if $\beta\sigma^{-2}$ is locally bounded; see, for example, V.26, V.40 of Rogers & Williams [38].

These models are particularly tractable because, like the Vasicek model, the yield curve is *affine in the spot rate*. More precisely, we have the following formula for the bond price:

$$(3.2) \qquad P(t,T) = \exp\{-r_t B(t,T) - A(t,T)\} \quad (0 \le t \le T),$$

where B solves the Riccati equation

$$(3.3) \quad \dot{B}(t,T) - \frac{1}{2}\sigma_t^2 B(t,T)^2 - \beta_t B(t,T) + 1 = 0, \quad B(T,T) = 0,$$

and A solves the simple first-order equation

$$(3.4) \qquad \dot{A}(t,T) = -\alpha_t B(t,T), \quad A(T,T) = 0.$$

It should be emphasized that the functions A, B used above are *not* the same as the functions A, B in the previous section ((2.9), (2.10)). Indeed, no simple closed form is available in general for A, B, though in the case of constant α, β, σ we have the (Cox-Ingersoll-Ross) formulae

$$(3.5) \qquad A(t,T) \;=\; -\frac{2\alpha}{\sigma^2} \log\left[\frac{\gamma e^{\beta\tau/2}}{\gamma\cosh\gamma\tau + \frac{1}{2}\beta\sinh\gamma\tau}\right],$$

$$(3.6) \qquad B(t,T) \;=\; \frac{\sinh\gamma\tau}{\gamma\cosh\gamma\tau + \frac{1}{2}\beta\sinh\gamma\tau},$$

where $\tau = T - t$, $2\gamma \equiv (\beta^2 + 2\sigma^2)^{1/2}$.

Compared to the simple Gaussian models of the last section, these models are *more realistic*, in that the spot rate stays non-negative, but *not so simple*, because one has to solve the Riccati equation (3.3). It is well known that (3.3) can be reduced to a second-order linear equation, and a brisk treatment of this is given in Appendix A.

All of the objections raised to the estimation of the Gaussian models of the last section apply equally well here, in particular, the functions α, β, σ cannot be uniquely determined from the term-structure and term-structure of volatility, and any attempt to recover them from the data involves differentiating up to three times. Jamshidian [28] has studied this class of models, and finds that if one restricts α, β, σ by insisting that α/σ^2 is constant, then certain simplifications result, and for given $A(0, \cdot), B(0, \cdot)$ there are unique α, β, σ satisfying $\alpha\sigma^{-2} =$ constant. To see why this restriction is natural, let us approach the problem from the other end, starting with the constant-coefficient Cox-Ingersoll-Ross model, and see what perturbations we could make to it. This work is done with Wolfgang Stummer.

One thing we could do is to make a *deterministic C^2 time-change*, and thus represent the bond prices as

$$(3.7) \qquad P(t,T) = E\left[\exp -\int_{\tau(t)}^{\tau(T)} r_u\, du \,\Big|\, \mathcal{F}_{\tau(t)}\right],$$

where r is a standard Cox-Ingersoll-Ross process, solving (3.1) with constant α, β, σ. If $P_0(x, \tau)$ denotes the price of a bond of maturity τ when the initial spot rate is x and the spot-rate process is r, then P_0 is available in closed form (from (3.2), (3.6)) and from (3.7) we have immediately that

$$(3.8) \qquad P(t, T) \quad = \quad P_0(r(\tau_t), \tau_T - \tau_t)$$

$$= \quad E\left[\exp - \int_t^T \tau_u' r(\tau_u)du | \mathcal{F}_{\tau(t)}\right]$$

$$(3.9) \qquad\qquad\qquad = \quad P_0(\rho_t/\tau_t', \tau_T - \tau_t),$$

where $\rho_t \equiv \tau_t' r(\tau_t)$ is the spot rate process, relative to the given time scale. (To see this, observe that $-\frac{\partial}{\partial T}\log P(t, T)|_{T=t} \equiv f_{tt}$ is the spot rate, and now differentiate (3.8) with respect to T.)

Routine methods of stochastic calculus show that ρ solves a stochastic differential equation

$$(3.10) \qquad d\rho_t = \sigma\tau_t'\sqrt{\rho_t}dW_t + \left[\alpha(\tau_t')^2 - (\beta\tau_t' - \tau_t''/\tau_t')\rho_t\right]dt.$$

More important, however, is the fact that with this model, one may fit *any initial term structure exactly*. This is trivial; we know that for fixed t the function $T \mapsto P(t, T)$ decreases continuously to zero, and we use (3.8) and (3.9) to tell us what should be the function τ. (The initial slope τ_0' is indeterminate, and could be chosen to match up the volatility of the spot rate, for example.)

A second simple transformation of the basic CIR model that we could perform is to *multiply by some positive C^1 function g*. It is easy to check that if we define $\tilde{r}_t \equiv g_t\rho_t$, then $\tilde{r}$ solves the SDE (3.1) with

$$(3.11) \qquad\qquad \sigma_t \quad = \quad \sigma\tau_t'\sqrt{g_t},$$

$$(3.12) \qquad\qquad \alpha_t \quad = \quad \alpha\sigma_t^2/\sigma^2,$$

$$(3.13) \qquad\qquad \beta_t \quad = \quad \beta\tau_t' - \frac{\tau_t''}{\tau_t'} - \frac{g_t'}{g_t}.$$

Given any functions $\sigma_\bullet > 0$ and $\beta_\bullet$, we can always choose τ increasing and $g > 0$ to make (3.11) hold.[2] To summarize then, *the processes satisfying (3.1) with $\sigma_t^{-2}\alpha_t = $ constant are exactly those obtained from a standard Cox-Ingersoll-Ross process by deterministic C^2 time-change, and multiplication by a deterministic C^1 function.*

REMARK 3.1.   Suppose that X is a Gauss-Markov process in $\mathbb{R}^d$ solving

$$dX_t = \frac{1}{2}\sigma_t dW_t - \frac{1}{2}\beta_t X_t dt.$$

---

[2] Write $v = \tau' > 0$, so that $v^2 g = \sigma_t^2/\sigma^2$ is known, and, multiplying (3.13) by vg we find $\beta_t v_t g_t = \beta(v_t^2 g_t) - (v_t g_t)' = \beta\sigma_t^2/\sigma^2 - (v_t g_t)'$; now this is easily solved for $v_t g_t$, and hence we deduce v_t, g_t.

Then if $r_t \equiv |X_t|^2$, and $\sigma, \beta : \mathbb{R}^+ \to \mathbb{R}$, simple Itô calculus reveals that

$$(3.14) \qquad dr_t = \sigma_t \sqrt{r_t} dZ_t + \left(\frac{d}{4} \sigma_t^2 - \beta_t r_t \right) dt$$

where W is Brownian motion in $\mathbb{R}^d$, Z is Brownian motion in $\mathbb{R}$. Comparing (3.14) with (3.1) shows that $4\alpha_t/\sigma_t^2$ is the "dimension" of the Gaussian process and so it becomes less surprising that the condition $\alpha\sigma^2 = $ constant should make an appearance. It also explains the title of this section! For those who are familiar with such things, we are in the land of Bessel processes (for a recent survey, see Revuz & Yor [36]), which also appear to be connected with the Ray-Knight theorems on Brownian local time (see, for example, Rogers & Williams [38]) and diffusion limits of branching processes; these things are all very closely related.

The squared-Gaussian and Gaussian models have an affine yield, but what other processes are there with this desirable property? This question has been answered in various forms (see Cox, Ingersoll & Ross [14], Brown & Schaefer [6] for example). Following [14], we shall consider the spot-rate process to solve the SDE

$$(3.15) \qquad dr_t = \sigma(t, r, \xi) dW_t + \mu(t, r, \xi) dt$$

where $\xi_t = (\xi_i(t))_{i=1}^n$ is a vector of exponentially-weighted past values of r,

$$(3.16) \qquad \xi_i(t) = e^{-\lambda_i t} \int_{-\infty}^{t} \lambda_i e^{\lambda_i u} r_u du$$

so that

$$(3.17) \qquad d\xi_i = \lambda_i (r - \xi_i) dt.$$

If now we require that

$$y(t, T) \equiv -\log P(t, T) = A(t, T) + B(t, T) r_t + \sum_{i=1}^n C_i(t, T) \xi_i(t)$$

then applying Itô's formula to the martingale $\exp \left(- \int_0^t r_u du \right) P(t, T)$ gives

$$(3.18) \quad \frac{1}{2} B(t, T)^2 \sigma(t, r, \xi)^2 \quad - \quad \dot{A}(t, T) - r - \dot{B}(t, T) r - \mu(t, r, \xi) B(t, T)$$

$$- \sum_1^n \{ \dot{C}_i(t, T) \xi_i + \lambda_i C_i(t, T)(r - \xi_i) \} = 0$$

where, as before, a dot denotes differentiation with respect to t. Hence

$$\mu(t, r, \xi) = \frac{1}{2}B(t, T)\sigma(t, r, \xi)^2 \quad - \quad \frac{\dot{A}}{B}(t, T) - \frac{r}{B(t, T)} - r\frac{\dot{B}}{B}(t, T)$$

$$- \sum_1^n \left\{ \frac{\dot{C_i}}{B}(t, T)\xi_i + \lambda_i \frac{C_i}{B}(t, T)(r - \xi_i) \right\}.$$

Now the right-hand side of this is the same for all $T > t$, so taking the difference for two values of $T > t$ we deduce that $\sigma(t, r, \xi)^2$ must be of the form

(3.19) $$\sigma(t, r, \xi) = \sigma_0(t)^2 + \sigma(t)^2 r + \sum_1^n \sigma_i(t)^2 \xi_i,$$

and from this

(3.20) $$\mu(t, r, \xi) = \mu_0(t) + \mu(t)r + \sum_1^n \mu_i(t)\xi_i.$$

Returning these to (3.18) yields differential equations for A,B,C, which can always be solved numerically.

REMARK 3.2.   (i) The Gaussian models appear if we take $\sigma_i = 0$, $\sigma = 0$ in (3.19), and $\mu_i = 0$ in (3.20). The squared-Gaussian models arise when we take $\sigma_0 = 0 = \sigma_i$ in (3.19).

(ii) If we were to insist that $\sigma_0(\cdot)$, $\sigma(\cdot)$, $\sigma_i(\cdot)$, $\mu_0(\cdot)$, $\mu(\cdot)$, and $\mu_i(\cdot)$ were all constants, we have a wide range of possible models available to fit any given yield curve, though the analysis will not be particularly simple. Nonetheless, it may be preferable to use one of this class of models rather than one of the time-dependent versions.

(iii) All of these models are *single-factor* models, and can be criticized on these grounds. Other single-factor models have been proposed by Dothan [16], Brennan & Schwartz [4], Courtadon [11], and Cox [12] (see also Beckers [3]); none of these appears to be conclusively superior to the models discussed already, and all have lost the analytical tractability of the models discussed above.

4. Multi-factor models. The single-factor Markovian models discussed so far are often criticized on the grounds that the long rate is a deterministic function of the spot rate, and that the prices of bonds of different maturities are perfectly correlated. These defects might perhaps be forgiven if the models appeared to match observed prices, but, as we shall review in Section 7 the weight of empirical evidence suggests that multi-factor models do significantly better than single-factor models.

In the last few years, there has been a rash of papers all dealing with the same broad class of models, the higher-dimensional squared-Gauss-Markov processes. This title is longer (and less informative) than the SDE defining them:

$$(4.1) \qquad dX_t \; = \; \sigma_t dW_t + (a_t + C_t X_t)dt$$

$$(4.2) \qquad r_t \; = \; \frac{1}{2}|X_t|^2,$$

where W is a Brownian motion in $\mathbb{R}^n$, and $\sigma, C : \mathbb{R}^+ \to \mathbb{R}^n \otimes \mathbb{R}^n$, $a : \mathbb{R}^1 \to \mathbb{R}^n$ are deterministic locally bounded functions. As was remarked above, the Cox-Ingersoll-Ross single factor models are obtained in this way if their dimension is integer. The papers known to me which deal with this class of models are Beaglehole & Tenney [2], El Karoui, Myneni & Viswanathan [21], Duffie & Kan [17], Constantinides [10], Jamshidian [28], although the starting points do differ in these various papers. They all end up with essentially the same class of models, examples of which were already studied by Cox, Ingersoll & Ross [13], Richard [37], and Longstaff & Schwartz [33].

The attraction of this class is that there is a (semi-)explicit formula for the bond price: it is easy to prove that

$$(4.3) \qquad P(t,T) = \exp\left\{ -\frac{1}{2}X_t^T Q_t X_t + b_t^T X_t - \gamma_t \right\},$$

where Q solves the matrix Riccati equation

$$(4.4) \qquad I + QC + (QC)^T + \dot{Q} - Q\sigma\sigma^T Q^T = 0, \quad Q(T,T) = 0,$$

and b, γ solve

$$(4.5) \qquad \dot{b} \; - \; Qa - (Q\sigma\sigma^T - C^T)b = 0, \quad b(T,T) = 0,$$

$$(4.6) \qquad \dot{\gamma} \; = \; b^T a - \frac{1}{2}tr(\sigma^T Q\sigma) + \frac{1}{2}b^T \sigma\sigma^T b, \quad \gamma(T,T) = 0.$$

This class of processes, and the differential equations (4.4)-(4.6), have been around for a long time in the world of stochastic processes, going back at least to Cameron & Martin [7]; see also Feller [22], Liptser & Shiryaev [31], and, for recent work, Donati-Martin & Yor [15], Rogers & Shi [39], Chan, Dean, Jansons & Rogers [9].

Though the matrix Riccati equation (4.4) can easily be solved numerically, there are only closed-form analytic solutions when the problem is one-dimensional, or can be reduced to independent one-dimensional processes; in general, the analysis of this class of models is sticky. Interest in these models will undoubtedly continue for some time to come; Duffie & Ken [17] have formulated the most general affine yield multifactor model, which appears to have closed-form solutions only in the special case of independent CIR processes (or some equivalent situation). Nevertheless, numerical solution is always a possibility.

By passing to multi-factor models, one should get an improved fit, but there is a heavy price to pay; if one wants to calculate prices of, say, options on bonds, the PDE to be solved is higher-dimensional, and will thus be much slower. Perhaps even more importantly, the factors used have to correspond to some observable variables if the formulae are ever to be used. Cox, Ingersoll & Ross [13] and Richard [37] both take the spot rate together with the rate of inflation, Longstaff & Schwartz [33] use the spot rate together with the volatility of the spot rate, and Duffie & Kan [17] use the yields on a fixed set of bonds, for example.

Outside of this class of squared Gaussian models, there are many which have been proposed, but I highlight just two. The first of these is the model of Brennan & Schwartz [4], which takes as the variables of the (two-factor) model the spot rate r and the long rate. As a proxy for the long rate, Brennan & Schwartz use the reciprocal of the price of a consol, which is a traded asset. This is a nice idea, and clarifies the analysis, though their pricing equation still has to be solved numerically. They use the model to analyze Canadian Government bonds, and obtain impressive results. A variant of the basic model is subsequently studied by Schaefer & Schwartz [40], who take the spot rate and the spread as the variables. In this context, it is worth recalling a result of Dybvig, Ingersoll & Ross [19], who prove that the long rate is non-decreasing. This makes one a little wary about a model which supposes that the long rate moves as a diffusion, even if only in the form of its proxy, the reciprocal of the consol.

The final model is the model of Fong & Vasicek [23], who take the spot rate r to follow the Ornstein-Uhlenbeck SDE (2.1) with constant α, β, but with σ itself following an independent squared-Gaussian diffusion. The model has a simple affine yield structure, which is attractive, but also the risk of negative spot rates. Whether this model is any better than a two-factor squared Gaussian model cannot be decided at a theoretical level, and must be resolved by comparing results on real data.

5. Whole-yield models. The approach adopted in whole-yield models is to model directly the forward-rate processes $(f_{tT})_{0 \leq t \leq T}$ for each T. The earliest (discrete-time) appearance of this approach was due to Ho & Lee [26]. In view of the fact (proved by Dybvig [18] and Jamshidian [28]) that the continuous-time limit of the Ho & Lee model is

$$dr_t = \theta_t dt + \sigma dW_t$$

for some deterministic function θ, it has not been adopted unreservedly. This possibility of spectacularly negative spot rates, together with the single-factor nature of the model, has led to more refined models, the main one being the model of Heath, Jarrow & Morton [25]. The same model was developed independently and contemporaneously by Babbs [1]. The idea here is to model the forward-rate curve by

$$(5.1) \qquad f_{tT} = f_{0T} + \int_0^t \sigma(s,T)dW_s + \int_0^t \alpha(s,T)ds \quad (0 \le t \le T),$$

where W is an n-dimensional Brownian motion, and $(f_{0T})_{T \ge 0}$ is the initial forward-rate curve. The functions σ and α cannot be chosen unrestrictedly; indeed, Heath, Jarrow & Morton prove that in fact

$$(5.2) \qquad \alpha(t,T) = \sigma(t,T)\left\{ \varphi_t + \int_t^T \sigma(t,s)ds \right\},$$

where φ is some n-vector previsible process which is zero when working in the risk-neutral measure. In fact, this structure becomes almost obvious when looked at in the right way. Indeed, if we consider the martingale (with respect to the risk-neutral measure)

$$(5.3) \; M_t \equiv E\left[\exp\left(-\int_0^T r_u du \right) | \mathcal{F}_t \right] = \exp\left\{ -\int_0^t r_u du - \int_t^T f_{tu} du \right\}$$

we know that it can be represented in the form

$$(5.4) \qquad M_t = M_0 \exp\left\{ -\int_0^t \Sigma(s,T)dW_s - \frac{1}{2}\int_0^t |\Sigma(s,T)|^2 ds \right\}$$

for some previsible n-vector process $\Sigma(\cdot, T)$ (any martingale on the Brownian filtration can be represented as a stochastic integral, and hence easily any non-negative martingale on the Brownian filtration is representable as an exponential of a stochastic integral.) Taking the two expressions of $\log M_t$ from (5.3) and (5.4) and comparing the martingale parts shows that

$$\Sigma(t,T) = \int_t^T \sigma(t,u)du,$$

and now differentiating with respect to T gives (5.2), and also (5.1) when we remember that $\log M_0 = \int_0^t f_{0u} du$.

Of course, various regularity conditions are needed to justify this, and indeed further regularity conditions (which are not made explicit in [25]) are need to ensure that

$$\lim_{T \downarrow t} f_{tT} = r_t$$

exists, is continuous and defines a semimartingale (none of which need hold in general). However, such points are trivial in comparison to real objections to the use of the model in practice. Heath, Jarrow & Morton give two examples where there are simple formulae for bond prices, but where r is allowed to go negative; and they give an example where $r \geq 0$, but there are no simple formulae. It appears very difficult to obtain both of the desirable properties together, not least because this approach begins by trying model *derived* quantities (the forward rates) instead of the *fundamental* quantity (the spot rate) and thus loses control.

If one ever *could* find a specification of the model where $r \geq 0$ and where bond prices were given by a simple formula, then one could just as well obtain this by starting directly with the spot rate process r! So these whole-yield models appear to offer no advantage over the approach of modelling r, although the ability to input the initial yield curve directly and to vary the volatilities of different forward rates are attractions (partly shared by some of the models of the preceding sections.)

6. Equilibrium or arbitrage pricing? Cox, Ingersoll & Ross [13] in their paper on term structure of interest rates discussed (in Section 5) the comparison of their own equilibrium approach and the arbitrage approach. What they wrote there appears to have caused considerable confusion about the relation between the two; the aim of this section is to prove that the two are essentially equivalent.

To begin with, we suppose that $(S_t)_{0 \leq t \leq T} = ((S_t^0, \ldots, S_t^n))_{0 \leq t \leq T}$ is an $(n+1)$-vector continuous semimartingale of *financial asset* prices. The first component S^0 is non-decreasing, and represents the price of a riskless bond. There will be a single *productive asset* whose price at time t is ξ_t, where ξ is also a continuous semimartingale. Prices are in terms of the unique commodity of this economy. An investor starts with wealth x, and holds a self-financing portfolio (θ_t, H_t) in the financial assets and the productive asset respectively. He consumes at rate $C_t \geq 0$ at time t, so his wealth at time t, X_t, obeys the wealth equation

$$(6.1) \qquad dX_t \; = \; \theta_t dS_t + H_t d\xi_t - C_t dt,$$
$$(6.2) \qquad X_t \; = \; \theta_t S_t + H_t \xi_t, \quad X_0 = x.$$

The investor aims to choose (θ, H) so as to

$$(6.3) \qquad \max E \int_0^T u(C_s) ds \text{ subject to } X_t \geq 0 \text{ for all } t \in [0, T].$$

We shall say that the price processes (S, ξ) constitute an *equilibrium* for this model if under optimal play the investor invests nothing in the financial assets. This terminology bears the usual interpretation; we imagine a number of identical investors, all investing in the market. The financial assets only exist by virtue of some agents going short, some going long; the net supply of financial assets is zero. The productive asset has a physical existence, however.

We assume the utility function in (6.3) to be strictly increasing, C^2, strictly concave and unbounded above. Thus if (S, ξ) is an equilibrium, there must be no arbitrage (otherwise the investor would be able to make unlimited gain, and his utility-maximization problem would be ill-posed). It is a folk theorem of the subject that no arbitrage implies the existence of an equivalent martingale measure. Without pausing to examine the exact result (which, in any case, has not yet been formulated correctly), we concentrate on the converse.

THEOREM 6.1. *Write $\beta_t \equiv 1/S_t^0$, and suppose that there exists a measure $\tilde{P}$ equivalent to P such that under $\tilde{P}$ the process $\tilde{S} \equiv \beta S$ is a local martingale. Define the P-martingale Z by*

$$Z_t = \left. \frac{d\tilde{P}}{dP} \right|_{\mathcal{F}_t}.$$

If we now take $U(x) \equiv \log x$, $\xi_t \equiv 1/\beta_t Z_t$, then the optimal policy for the investor is

(6.4) $C_t^* = \frac{x}{T}\xi_t, \quad \theta_t^* = 0, \quad H_t^* = x\left(1 - \frac{t}{T}\right).$

REMARK 6.1. What this says is that if there exists an equivalent measure under which all the financial asset prices, when discounted, are local martingales, *then there is an economy which supports these as equilibrium prices*: The reason that the example of [13] does not violate this is as follows. Cox, Ingersoll & Ross are working in a Brownian framework, where a change of measure corresponds to introducing a drift into the Brownian motion. However, not *every* possible drift corresponds to a change of measure, and what they have done is to consider a drift which does not correspond to such a change of measure! To quote Cox, Ingersoll & Ross, "The difficulty, of course, is that there is no underlying equilibrium which would support the assumed premiums" – but, on the other hand, there is no risk-neutral measure either!

Proof. With the portfolio and consumption plan (6.4), the wealth process is

$$X_t = x(1 - t/T)\xi_t$$

and it is two lines of calculus to confirm that this solves the wealth equation (6.1). Reworking the wealth equation (6.1) for general portfolio/consumption,

(6.5) $d(\beta_t X_t) = \theta_t d\tilde{S}_t + H_t d\tilde{\xi}_t - \beta_t C_t dt$

(where $\tilde{\xi}_t \equiv \beta_t \xi_t \equiv 1/Z_t$) so that

$$0 \leq \beta_t X_t = x + \int_0^t (\theta_u d\tilde{S}_u + H_u d\tilde{\xi}_u) - \int_0^t \beta_u C_u du.$$

Now under $\tilde{P}$, both $\tilde{S}$ and $\tilde{\xi}$ are local martingales, so by Fatou's lemma

$$(6.6) \qquad x \geq \tilde{E}\left(\int_0^T \beta_u C_u du\right) = E\left(\int_0^T \beta_u Z_u C_u du\right)$$

for any feasible consumption plan. The proof is completed by a simple "Lagrangian sufficiency" argument. As a piece of notation, we set for $x > 0$

$$(6.7) \qquad U_*(x) = \sup\{U(c) - xc\}$$
$$(6.8) \qquad\qquad\quad = U(I(x)) - xI(x),$$

where I is the inverse to U' (in fact, the second line only holds if U' decreases from infinity at $0+$ to zero at infinity, which is certainly true when, as we assume here, $U = \log)^3$. Now for any $\lambda > 0$, we have for any feasible consumption process

$$(6.9)\, E\int_0^T U(C_s)ds \;\leq\; E\left[\int_0^T U(C_s)ds + \lambda\left(x - \int_0^T \beta_u Z_u C_u du\right)\right]$$

$$= \;\lambda x + E\int_0^T \{U(C_s) - \lambda\beta_s Z_s C_s\}ds$$

$$\leq \;\lambda x + E\int_0^T U_*(\lambda\beta_s Z_s)ds$$

$$= \;\lambda x + E\int_0^T \{U(I(\lambda\beta_s Z_s)) - \lambda\beta_s Z_s I(\lambda\beta_s Z_s)\}ds.$$

Now if we write $C_s^\lambda \equiv I(\lambda\beta_s Z_s)$, then provided λ were chosen so that the "budget constraint" (6.6) is satisfied with equality, that is

$$(6.10) \qquad x = E\int_0^T \beta_s Z_s C_s^\lambda ds \equiv E\int_0^T \beta_s Z_s I(\lambda\beta_s Z_s)ds,$$

---

[3] U_* is the convex dual of U, and $U(x) = \inf_y\{U_*(y) - xy\}$.

then (6.9) becomes simply

$$E \int_0^T U(C_s^\lambda)ds.$$

However, we *can* achieve (6.10) in general, and trivially in this case where $I(x) = 1/x$ simply by choosing $\lambda = T/x$! Assembling the inequality, we have to conclude that for any feasible consumption plan C

$$E \int_0^T U(C_s)ds \le E \int_0^T U(C_s^{T/x})ds \equiv E \int_0^T U(C_s^*)ds,$$

and the proof is complete. □

Dybvig & Ross [20] have already remarked on the essential equivalence of 'equilibrium' and 'arbitrage' pricing. I understand that Heston has a similar argument, reported in Exercise 9.3 of Duffie's book *Dynamic Pricing Theory*.

7. Empirical aspects. A recent search of a computer data base turned up 135 articles which referenced the fundamental paper [13] of Cox, Ingersoll & Ross; of these, about 75% were principally theoretical. Of the papers which are principally empirical, I discuss here only a few, but I think the conclusions reported give a good feel for what has been deduced from data so far; no strongly-preferred model emerges, but some models appear to be inadequate.

The first problem to be faced is that there are very few zero-coupon bonds traded; US Treasury bills seem to be the only ones for which much data is available, and the maturities only go out to about a year. One is then faced with the problem of estimating the yield curve from other (coupon-bearing) bonds, and possibly other information. Anyone who works on interest rates, be they practitioner or academic, has a way of making a yield curve from such data, or has a source of yield-curve data where this information has already been stripped out or "estimated" and we shall not discuss further how these are obtained.

Most of the papers have some model, or class of models, in mind, and proceed to test the model, or some feature of it. The most common class of models is the class of squared-Gaussian models with one or more factors, though Chan, Karolyi, Longstaff & Sanders [8] cast their net wider and take their class to be (single-factor) models where r solves an SDE of the form

$$dr_t = \sigma r_t^\gamma dW_t + (\alpha + \beta r_t)dt$$

where the parameters $\alpha, \beta, \gamma > \frac{1}{2}$, σ are to be estimated. This includes the CEV model of Cox [12] (see Beckers [3] for a description of the model), for example. The one-factor squared-Gaussian (Cox-Ingersoll-Ross) model is tested by Brown & Dybvig [5] using a least-squares fit, and by Pearson & Sun [35] using exact maximum likelihood. Both conclude that the one-factor model does not satisfactorily fit the data. On the other hand, Gibbons & Ramaswamy [24] find that it does perform quite well on the short-term Treasury bill data. Multifactor squared Gaussian models are tested by Stambaugh [41], Longstaff & Schwartz [33], and Litterman, Scheinkman & Weiss [32], who all conclude that introducing additional factors significantly improves the fit; in fact, using two factors appears to be satisfactory from the work of these quthors, although Pearson & Sun [35] do not find that two factors are sufficient.

The generalized method of moments is a popular approach to the estimation. Despite the arbitrariness of the procedure, it has some attractive features; the large-sample behaviour does not depend on specific distributional assumptions, needing only that the spot rate process be stationary and ergodic. A little care is needed here; Chan et at. [14] conclude that Dothan's [16] model, the Cox CEV [13] and the Cox-Ingersoll-Ross variable-rate model all do better than Vasicek or the standard Cox-Ingersoll-Ross squared-Gaussian, but these three diffusions are not ergodic, so the results must be interpreted with caution.

The outcome of the empirical studies seems to be that a two- or three-factor squared-Gaussian model is reasonably satisfactory, but there is one conclusion that most agree on, namely, that more work is needed here!

8. Conclusions: where now. Long before stochastic calculus hit the industry, bonds were being traded; practitioners had a good idea what prices to charge, and were exploiting their knowledge of the market, and hunches about the future, to guide them. It is futile to imagine that increasingly sophisticated mathematical models will replace or displace such skill, and what is needed now is not *more* (and more complicated) mathematical models, but rather a serious attempt to combine practitioner input with (probably extremely simple) mathematical models; no mathematical model based on assumptions of time-homogeneity can ever represent a world of elections, trade figures, summits and treaties.

One thing which does appear to be well worth doing (of a more academic nature) is to try to model index-linked bonds. Asking practitioners what they consider the main influence on term-structure, the response is that anticipated inflation is the most important effect. By removing that, through studying index-linked bonds, we may be able to see a more orderly pattern emerging, just as one does with share prices when the "operational time" effect is removed (see, for example, the working paper "Some statistics for testing the influence of the number of transactions on the distributions of returns" by S.E. Satchell & Y. Yoon.).

A. Appendix. (i) We take r to solve the SDE

(A.1) $$dr_t = \sigma_t \sqrt{r_t} dW_t + (\alpha_t - \beta_t r_t) dt$$

so changing variables to $z_t \equiv 2r_t/\sigma_t^2$, we find that

(A.2) $$dz_t = \sqrt{2z_t} dW_t + \left[\frac{2\alpha_t}{\sigma_t^2} - \beta_t z_t - 2\dot\sigma_t \sigma_t^{-1} z_t \right] dt.$$

For notational simplicity, set $a_t \equiv 2\alpha_t \sigma_t^{-2}$, $b_t \equiv \beta_t + 2\dot\sigma_t \sigma_t^{-1}$, so that the bond price $P(z,t,T)$ $(0 \le t \le T)$ satisfies for fixed T

(A.3) $$\dot P + zP'' + (a - bz)P' - \frac{1}{2}\sigma^2 zP = 0,$$

where a dot denotes differentiation with respect to t, and a dash denotes differentiation with respect to z. This PDE has a solution of the form

$$P(z,t,T) = \exp\left[-z_t B(t,T) - A(t,T) \right]$$

provided

(A.4) $$\dot B - B^2 - bB + \frac{1}{2}\sigma^2 = 0, \qquad B(T,T) = 0$$

(A.5) $$\dot A = -aB, \qquad A(T,T) = 0.$$

(Note that, because of the change of variables to z, what was denoted by $B(t,T)$ in Section 3 is here denoted $2B(t,T)\sigma_t^{-2}$; the change of notation should cause no confusion, and is particularly convenient for the purposes of this section.)

Writing

(A.6) $$B(t,T) = -\dot\psi(t,T)/\psi(t,T),$$

we can recast (A.4) as

(A.7) $$\ddot\psi - b\dot\psi - \frac{1}{2}\sigma^2\psi = 0.$$

Let us write $\psi_\pm$ for two linearly independent solutions of this second-order linear differential equations; we shall choose ψ_+ to be increasing non-negative, ψ_- to be decreasing non-negative, $\psi_+(0) = \psi_-(0) = 1$, $\psi_+'(0) = 0$.[4] In terms of these, if we set

(A.8) $$\psi(t,T) = \psi_-(t)\dot\psi_+(T) - \psi_+(t)\dot\psi_-(T)$$

---

[4] To see that such a choice is possible, if we let x be the solution to the SDE

$$dx_t = dW_t - b(x_t, T)\,sgn(x_t)dt$$

and if $H_a \equiv \inf\{t : x_t = a\}$, $\varphi_t = \int\limits_0^t \frac{1}{2}\sigma(x_u,T)^2 du$, then we can take $\psi_-(x) = E_x \exp\{-\varphi(H_0)\}$, $\psi_+(x) = 1/E^0 \exp\{-\varphi(H_x \wedge H_{-x})\}$.

then

(A.9) $$B(t,T) = -\frac{\dot{\psi}(t,T)}{\psi(t,T)} = -\frac{\partial}{\partial t}\log\psi(t,T)$$

is the solution to the Riccati equation, and the form of A is

(A.10) $$A(t,T) = \int_t^T a_u B(u,T)du.$$

(ii) It is of interest to consider the inverse problem, that is, if we are told the initial functions $B(0,\cdot)$, $A(0,\cdot)$, can we find some process of the form (A.1) which would give these A, B? This needs an understanding of how $B(t,T)$ varies with T. Now

$$
\begin{aligned}
\frac{\partial}{\partial T}B(t,T) &\equiv B_T(t,T) \\
&= -\frac{\partial^2}{\partial t\partial T}\log\psi(t,T) \\
&= -\frac{\partial}{\partial t}\frac{\psi_-(t)\dot{\psi}_+(T) - \psi_+(t)\dot{\psi}_-(T)}{\psi_-(t)\psi_+(T) - \psi_+(t)\psi_-(T)} \\
&= -\frac{\partial}{\partial t}\left[b_T + \frac{1}{2}\sigma_T^2\frac{\psi_-(t)\psi_+(T) - \psi_+(t)\psi_-(T)}{\psi_-(t)\dot{\psi}_+(T) - \psi_+(t)\dot{\psi}_-(T)}\right]
\end{aligned}
$$

(A.11) $$= \frac{1}{2}\sigma_T^2\frac{\xi_t\xi_T}{\psi(t,T)^2},$$

where

(A.12) $$\xi_t = \psi_-(t)\dot{\psi}_+(t) - \psi_+(t)\dot{\psi}_-(t) = \psi(t,t).$$

If we define $\gamma(t,T) \equiv \psi_-(t)\psi_+(T) - \psi_+(t)\psi_-(T)$, then we have after some calculations

(A.13) $$
\begin{aligned}
\frac{\partial}{\partial T}\log B_T(t,T) &= \frac{2\dot{\sigma}_T}{\sigma_T} - b_T - \sigma_T^2\frac{\gamma(t,T)}{\psi(t,T)} \\
&= -\beta_T - \sigma_T^2\frac{\gamma(t,T)}{\psi(t,T)}
\end{aligned}
$$

and differentiating once more, and rearranging, one gets after some calculations

(A.14) $$\frac{\partial^2}{\partial T^2}\log B_T(t,T) = -\beta_T' - \sigma_T^2 - \frac{1}{2}\beta_T^2 + \frac{1}{2}\left(\frac{\partial}{\partial T}\log B_T(t,T)\right)^2.$$

If we were in the special case where a was constant, then knowing $A(0,\cdot)$ would tell us, from (A.9), (A.10), the function

$$\log\psi(T,T) - \log\psi(0,T)$$

and if we differentiate this with respect to T, we get

$$-\frac{1}{2}\sigma_T^2\,\frac{\gamma(0,T)}{\psi(0,T)}.$$

Combining with (A.13) would tell us what the function β should be, and now returning this to (A.14), we deduce the function σ^2. But notice that to recover the coefficient functions σ^2, β, we have had to differentiate the bond prices three times.

REFERENCES

[1] S. BABBS, *A family of Itô process models for the term structure of interest rates*, University of Warwick, preprint 90/24, 1990.

[2] D.R. BEAGLEHOLE & M.S. TENNEY, *General solution of som e interest rate-contingent claim pricing equations*, J. Fixed Income 1, 1991, pp. 69- -83.

[3] S. BECKERS, *The constant elasticity of variance model an d its implications for option pricing*, J. Finance 35, 1980, pp. 661–673.

[4] M.J. BRENNAN & E.S. SCHWARTZ, *A continuous time approa ch to the pricing of bonds*, J. Banking Fin. 3, 1979, pp. 133–155.

[5] S.J. BROWN & P.H. DYBVIG, *The empirical implications o f the Cox-Ingersoll-Ross theory of the term structure of interest rates*, J. Finance, 41, 1986, pp. 617–630.

[6] R.H. BROWN & S.M. SCHAEFER, *Interest rate volatility and the term structure*, London Business School preprint, 1991.

[7] R.H. CAMERON & W.T. MARTIN, *Evaluation of various Wien er integrals by use of certain Sturm-Liouville differential equations*, Bull. Amer. Math. Soc. 51, 1945, pp. 73–90.

[8] K.C. CHAN, G.A. KAROLYI, F.A. LONGSTAFF, & A.B. SANDERS, *An empirical comparison of alternative models of the short-term interest rate*, J. Finance 47, 1992, pp. 1209–1227.

[9] T. CHAN, D. DEAN, K.M. JANSONS, & L.C.G. ROGERS, *On polymer conformations in elongational flows*, Comm. Math. Phys. 160, 1994, pp. 239–257.

[10] G.M. CONSTANTINIDES, *A theory of the nominal term stru cture of interest rates*, Rev. Fin. Studies 5, 1992, pp. 531–552.

[11] G. COURTADON, *The pricing of options on default-free bonds*, J. Fin. Quantit. Anal. 17, 1982, pp. 75–100.

[12] J.C. COX, *Notes on option pricing I: constant elasticity of variance diffusions*, Stanford University working paper, 1975.

[13] J.C. COX, J.E. INGERSOLL, & S.A. ROSS, *A theory of th e term structure of interest rates*, Econometrica 53, 1985, pp. 385–408.

[14] J.C. COX, J.E. INGERSOLL, & S.A. ROSS, *A re-examinat ion of traditional hypotheses about the term structure of interest rates*, J. Finance 36, 1981, pp. 769–798.

[15] C. DONATI-MARTIN & M. YOR, *Fubini's theorem for doubl e Wiener integrals and the variance of the Brownian path*, Ann. Inst. Henri Poincaré bf 27, 1991, pp. 181–200.

[16] L.U. DOTHAN, *On the term structure of interest rates*, J . Fin. Econ. 6, 1978, pp. 59–69.

[17] D. DUFFIE & R. KAN, *A yield factor model of interest rates*, GSB Stanford University preprint, 1993.

[18] P.H. DYBVIG, *Bond and bond option pricing based on the current term structure*, Washington University working paper, 1988.

[19] P.H. DYBVIG, J.E. INGERSOLL, & S.A. ROSS, *Long forwar d rates can never fall*, Washington University working paper, 1992.

[20] P.H. DYBVIG & S.A. ROSS, *Arbitrage: The new Palgrave: a Dictionary of Economics*, Stockton Press, New York, 1987, pp. 100–106.
[21] N. EL KAROUI, R. MYNENI, & R. VISWANATHAN, *Arbitrage pricing and hedging of interest claims with state variables I: Theory*, Preprint, 1992.
[22] W. FELLER, *Two singular diffusion problems*, Ann. Math. **54**, 1951, pp. 173–182.
[23] H.G. FONG & O.A. VASICEK, *Interest-rate volatility as a stochastic factor*, Gifford Fong Associates working paper, 1991.
[24] M.R. GIBBONS & K. RAMASWAMY, *The term structure of int erest rates: empirical evidence*, Stanford University working paper, 1986.
[25] D. HEATH, R. JARROW, & A. MORTON, *Bond pricing and th e term structure of interest rates: a new methodology for contingent claims valuation*, Econometrica **60**, 1992, pp. 77–105.
[26] T.S.Y. HO & S.-B. LEE, *Term structure movements and pr icing interest rate contingent claims*, J. Finance **41**, 1986, pp. 1011–1029.
[27] J. HULL & A. WHITE, *Pricing interest-rate derivative securities*, Rev. Fin. Studies, **3** 1990, pp. 573–592.
[28] F. JAMSHIDIAN, *The one-factor Gaussian interest rate m odel: Theory and implementation*, Working paper, Merrill Lynch Capital Markets, 1988.
[29] F. JAMSHIDIAN, *A simple class of square-root interest rate models*, Working paper, Fuji International Finance, 1993.
[30] F. JAMSHIDIAN, *Bond, futures and option evaluation in the quadratic interest rate model*, Working paper, Fuji International Finance, 1993.
[31] R.S. LIPTSER & A.N. SHIRYAEV, *Statistics of Random P rocesses I*, Springer, Berlin, 1977.
[32] R. LITTERMAN, J. SCHEINKMAN, & L. WEISS, *Volatility a nd the yield curve*, Working paper, Goldman Sachs & Co., New York, 1988.
[33] F.A. LONGSTAFF & E.S. SCHWARTZ, *Interest-rate volatili ty and the term structure: a two factor general equilibrium model*, J. Finance **47**, 19 91, pp. 1259–1282.
[34] R.C. MERTON, *Theory of rational option pricing*, Bell J. Econ. Man. Sci. **4**, 1973, pp. 141–183.
[35] N.D. PEARSON & T.-S. SUN, *A test of the Cox-Ingersoll- Ross model of the term structure of interest-rates using the method of maximum likelihood*, Working paper, MIT Sloan School, 1989.
[36] D. REVUZ & M. YOR, *Continuous Martingales and Brownian Motion*, Springer, 1991.
[37] S.F. RICHARD, *An arbitrage model of the term structur e of interest rates*, J. Fin. Econ. **6**, 1978, pp. 33–57.
[38] L.C.G. ROGERS & D. WILLIAMS, *Diffusion Markov Processe s and Martingales*, Vol. **2**, Wiley, Chichester, 1987.
[39] L.C.G. ROGERS & Z. SHI, *Quadratic functionals of Brown ian motion, optimal control, and the "Colditz" example*, Stochastics and Stochasti cs Reports **41**, 1992, pp. 201–218.
[40] S.M. SCHAEFER & E.S. SCHWARTZ, *Time-dependent variance and the pricing of bond options*, J. Finance **42**, 1987, pp. 1113–1128.
[41] R.F. STAMBAUGH, *The information in forward rates: implic ations for models of the term structure*, J. Fin. Econ. **21**, 1988, pp. 41–70.
[42] O.A. VASICEK, *An equilibrium characterization of the ter m structure*, J. Fin. Econ. **5**, 1977, pp. 177–188.

LIQUIDITY PREMIUM FOR CAPITAL ASSET PRICING WITH TRANSACTION COSTS*

S.E. SHREVE[†]

Abstract. An agent solves an infinite-horizon consumption-investment problem when the investment possibilities are a constant-interest-rate, risk-free asset and a stock, modelled as a geometric Brownian motion. There are proportional transaction costs associated with moving wealth between these two assets. The direct utility for consumption is of the form $\frac{1}{p}c^p$ for some $p \in (0,1)$. The sensitivity of the indirect utility function (or value function) to small transaction costs is found to be of the order of the transaction cost to the 2/3 power.

1. Introduction. In a series of landmark papers, Merton (1969, 1971, 1973a) initiated the study of optimal consumption and investment in continuous-time financial models. This use of continuous-time models subsequently led to the solution of the option pricing problem (Black & Scholes (1973), Merton (1973b)). The option pricing methodology was generalized by Kreps (1981), Harrison & Kreps (1979) and Harrison & Pliska (1981, 1983), so that we now have a quite satisfactory arbitrage pricing theory for frictionless, complete market models. For such markets, we also have a good understanding of optimal consumption and investment with additive utility (see Pliska (1986), Cox & Huang (1989), Karatzas, Lehoczky & Shreve (1987)). In frictionless, complete markets, both the arbitrage pricing theory and the theory for optimal consumption and investment depends on the construction of portfolio processes which represent certain martingales as stochastic integrals.

When transaction costs are present, neither the arbitrage pricing theory nor the theory of optimal consumption and investment has yet reached a satisfactory stage of development. In the presence of transaction costs, the portfolio policies mandated by the martingale representation theorem become prohibitively expensive. One is forced to use other, less easily constructed and analyzed portfolios, and is faced with the question of how much is given up in this process. In arbitrage pricing theory, one wants to construct the "hedging" portfolio which duplicates a contingent claim, and this may still be possible, albeit starting from a higher initial wealth (e.g., Gilster & Lee (1984), Leland (1985), Boyle & Vorst (1992), Henrotte (1993)). Alternatively, one could seek a portfolio which merely dominates the contingent claim (e.g., Bensaid, Lesne, Pagés & Scheinkman (1992)). Finally, one could introduce the contingent claim as another asset in an optimal consumption and investment model, and study prices which would

* Presented at the Workshop on Mathematical Finance, Institute for Mathematics and its Applications, Minneapolis, June 14-18, 1993.

† Department of Mathematics, Carnegie Mellon University, Pittsburgh, PA 15213. Work supported by the National Science Foundation under grant DMS–92–03360 and by the Army Research Office through the Center for Nonlinear Analysis.

make it attractive to investors (e.g., Hodges & Neuberger (1989), Edirisinghe, Naik & Uppal (1993), Davis, Panas & Zariphopoulou (1993), Constantinides (1993), Davis & Zariphopoulou (1994)). For this last approach, one needs to understand the kind of model considered in this paper.

A second reason to study the problem of optimal consumption and investment in the presence of transaction costs is to seek a model which has solid empirical support. The relation between consumption and risk premia implied by the frictionless, complete-market, capital asset pricing model has been rejected by some empirical studies (e.g., Hansen & Singleton (1982, 1983)). This motivated Grossman & Laroque (1990) to construct a model in which a transaction cost must be paid each time the level of consumption is altered. The empirically rejected relation between consumption and risk premia is not a conclusion of such a model.

The ultimate goal of this paper is to understand the size of the effect the introduction of transaction costs has in consumption/investment models. In the ensuing discussion, the transaction cost is assessed only on changes in portfolio and is always a percentage of the size of the transaction. One way of measuring its effect is to examine the marginal indirect utility as a function of transaction cost. Another is to estimate the "liquidity premium," i.e., the amount by which the mean rate of return of the stock would have to be increased to compensate the investor for the introduction of transaction costs. Constantinides (1986) computed an upper bound on the liquidity premium in a model allowing intermediate consumption, and found that for transaction costs in the range of 0.5% to 20%, this upper bound was roughly .14 times the transaction cost. Fleming, Grossman, Vila & Zariphopoulou (1990), considering the asymptotic behavior of a model with consumption only at the terminal time, obtained the indirect utility in closed form and found that the reduction in indirect utility was of the order of the transaction cost to the 2/3 power; in particular, the marginal indirect utility is infinite. (See also Taksar, Klass & Assaf (1988) and Dumas & Luciano (1991) for the solution of closely related models.) In contrast to the apparent implication of the numerical results of Constantinides (1986), the Fleming, et. al. (1990) result means that the ratio of the liquidity premium to transaction cost would increase to infinity as the transaction cost falls to zero. In this paper we return to the model of Constantinides (1986), trying to determine if the presence of intermediate consumption explains the apparent discrepancy. Although the model with intermediate consumption does not admit a closed form solution, we analytically obtain sharp upper and lower bounds on the indirect utility for small transaction cost values. These bounds confirm that for this model also, the reduction in indirect utility is of the order of the transaction cost to the 2/3 power and the ratio of the liquidity premium to transaction cost increases to infinity as the transaction cost falls to zero. However, the behavior we observe occurs only for small transaction costs, where "small" means less than the 0.5% level considered by Constantinides (1986). In particular, we obtain no contradiction to Constantinides' results; we show only that one cannot extrapolate from them.

This paper is organized as follows. Section 2 formulates the transaction cost model. This model was introduced in discrete time by Constantinides (1979) and in continuous time by Magill & Constantinides (1976). The continuous-time model was placed on a firm mathematical footing by Davis & Norman (1990). Section 3 reports on the solution of the model, including the work of Davis & Norman (1990). Sections 4 – 6 sketch a derivation of the results of Section 3, using the theory of a *viscosity solution* to the associated *Hamilton-Jacobi-Bellman (HJB)* equation. Section 7 reports on the sensitivity of the value function to small transaction costs.

For this model, the viscosity solution analysis was first provided by Fleming & Soner (1993) and elaborated by Shreve & Soner (1994). The details of the present work can be found in these references and in the forthcoming paper Shreve (1994). A noteworthy new result of the viscosity solution analysis is the observation that when the model without transaction costs leads an optimally behaving agent to borrow in order to buy stock, then in a model with the same parameters except that there are transaction costs, the agent will prefer strictly less leverage and will trade to realize this preference, even if her initial distribution of wealth between the assets is optimal for the model without transaction costs. This statement is shown by Shreve & Soner (1992) to hold for some parameter values, and there is strong evidence to support the conjecture that it holds for all parameter values in which leverage is optimal for the model without transaction costs.

The foundational work on viscosity solutions of HJB equations is due to Crandall & Lions (1983), Crandall, Evans & Lions (1984), and Lions (1982). The extension to second-order equations was obtained by Lions (1983a,b,c), Jensen (1988), and Ishii (1989). The reader may consult Crandall, Ishii & Lions (1992) for a full account of viscosity solution theory. The application of the theory to stochastic control is developed in Fleming & Soner (1993). The first application of the theory to finance is due to Zariphopoulou (1989), who also used viscosity solution theory to study a transaction cost model (Zariphopoulou (1992)).

2. The transaction cost model. We imagine an agent who can apportion her wealth between two assets. One of these is a risk-free *money market* paying constant *interest rate r*. The other is a *stock*, whose price is modelled as a geometric Brownian motion with constant *mean rate of return* α and constant *volatility* σ. We assume $\alpha > r > 0$ and $\sigma > 0$.

The agent can transfer money between these assets, but pays a *transaction cost* λ times the size of the transfer for doing so. We assume $0 \leq \lambda < 1$. The agent can also consume, with the consumption being deducted from the money market account. The agent receives *utility* $\frac{1}{p}c^p$ for consumption level c, where $0 < p < 1$.

The agent's objective is to manage her consumption and investment so as to maximize the expected, infinite-horizon utility of consumption, subject to the *solvency condition* that even when the transaction cost is taken into account, the agent can always cover any short position in one asset by liquidating a long position in the other.

Mathematically, we begin with a standard, one-dimension Brownian motion $\{W(t), \mathcal{F}(t); 0 \leq t < \infty\}$ to drive the stock process. The agent begins with initial position $X(0^-) = x_0$ in the money market and $Y(0^-) = y_0$ in the stock. These positions evolve according to the equations

$$(2.1) \qquad dX(t) = (rX(t) - C(t))dt - dL(t) + (1 - \lambda)dM(t),$$

$$(2.2) \qquad dY(t) = \alpha Y(t)dt + \sigma Y(t)dW(t) + (1 - \lambda)dL(t) - dM(t),$$

where the $\{\mathcal{F}(t)\}$-adapted, non-negative process $C(\cdot)$ records the rate of consumption and the $\{\mathcal{F}(t)\}$-adapted, nondecreasing, right-continuous processes $L(\cdot)$ and $M(\cdot)$ record the respective cumulative transfers of wealth from money market to stock and vice versa. We set $L(0^-) = M(0^-) = 0$.

The agent must choose the processes $C(\cdot)$, $L(\cdot)$, $M(\cdot)$ so that for all $t \geq 0$, $(X(t), Y(t))$ is in the closure $\overline{\mathcal{S}}$ of the *solvency region*

$$(2.3) \qquad \mathcal{S} \doteq \{(x, y); x + \frac{y}{1 - \lambda} > 0, x + (1 - \lambda)y > 0\}.$$

We assume that the initial position is in $\overline{\mathcal{S}}$. Subject to this solvency condition, the agent wishes to maximize $E \int_0^\infty e^{-\beta t} \frac{1}{p} C^p(t)dt$, where β is a positive *discount factor*. We denote by $v(x_0, y_0)$ the maximal attainable expected utility of consumption when the initial position is $X(0^-) = x_0$, $Y(0^-) = y_0$, i.e.,

$$(2.4) \qquad v(x_0, y_0) \doteq \max_{C,L,U} E \int_0^\infty e^{-\beta t} \frac{1}{p} C^p(t)dt.$$

3. Properties of the value function. The value function v for the stochastic control problem of the previous section is defined on the closed, convex set $\overline{\mathcal{S}}$. It can easily be seen that v is continuous and concave there, and takes the value 0 on the boundary of $\mathcal{S}$. Furthermore, it inherits from the utility function the *homotheticity property*

$$(3.1) \qquad v(\gamma x, \gamma y) = \gamma^p v(x, y), \forall (x, y) \in \overline{\mathcal{S}}, \gamma \geq 0.$$

When $\lambda = 0$, we have (Merton (1971))

$$(3.2) \qquad v(x, y) = \frac{A}{p}(x + y)^p,$$

where

$$(3.3) \qquad A \doteq \frac{\beta - rp}{1 - p} - \frac{p(\alpha - r)^2}{2\sigma^2(1 - p)^2}.$$

In this case, the positivity of A is a necessary and sufficient condition for the finiteness of v. Merton (1971) further showed that the optimal ratio of wealth invested in the stock to the total wealth is

$$(3.4) \qquad \theta_* \doteq \frac{\alpha - r}{\sigma^2(1-p)}.$$

When $\lambda > 0$, the positivity of A is sufficient but not necessary for v to be finite. If v is finite, then there are numbers $0 < \theta_1 < \theta_2 < 1/\lambda$ which determine three subregions of $\mathcal{S}$:

$$(3.5) \qquad SMM \doteq \{(x,y) \in \mathcal{S}; \frac{y}{x+y} < \theta_1\},$$

$$(3.6) \qquad NT \doteq \{(x,y) \in \mathcal{S}; \theta_1 < \frac{y}{x+y} < \theta_2\}$$

$$(3.7) \qquad SS \doteq \{(x,y) \in \mathcal{S}; \theta_2 < \frac{y}{x+y}\}.$$

When the investor's position is in SMM, she should sell money market and buy stock in order to immediately move to the boundary of NT where $\frac{y}{x+y} = \theta_1$. When the investor's position is in SS, she should sell stock and buy money market in order to immediately move to the boundary of NT where $\frac{y}{x+y} = \theta_2$. When the investor's position is in NT, she should not transact but she should consume. Indeed, the level of optimal consumption at $(x,y) \in NT$ is $[v_x(x,y)]^{\frac{1}{p-1}}$. When the investor's position reaches one of the boundaries of NT, she should transact just enough and in the right direction to prevent her position from exiting $\overline{NT}$. Except for a possible initial discontinuity to bring the position from outside $\overline{NT}$ to the boundary of NT, the optimal cumulative transaction processes L and M are continuous, but singularly so, i.e., they have zero derivative Lebesgue-almost everywhere, but are not constant. This singularly continuous behavior is required to overcome the stock price fluctuations, which would otherwise cause the position to exit $\overline{NT}$ once the boundary is reached.

The regions SMM, NT and SS, and hence the numbers θ_1 and θ_2, are characterized in terms of the second-order differential operator

$$(3.8) \qquad \mathcal{L}v \doteq \beta v - \frac{1}{2}\sigma^2 y^2 v_{yy} - \alpha y v_y - rx v_x$$

and the inequalities

$$(3.9) \qquad -(1-\lambda)v_x + v_y \geq 0,$$

$$(3.10) \qquad \mathcal{L}v - \frac{1-p}{p} v_x^{\frac{p}{p-1}} \geq 0,$$

(3.11) $v_x - (1-\lambda)v_y \geq 0.$

In all of SMM, (3.11) is an equality and (3.9), (3.10) are strict inequalities. In particular, in SMM the function v is constant in the direction the position moves when the agent sells money market to buy stock. In NT, (3.10) is an equality and (3.9), (3.11) are strict inequalities. In particular, any transaction will move the position to a place where v is strictly lower, and thus no transactions should take place. Finally, in SS, (3.9) is an equality and (3.10), (3.11) are strict inequalities.

Because in every region, one of the inequalities in (3.9) – (3.11) is an equality, we can combine the three inequalities into the single Hamilton-Jacobi-Bellman equation

(3.12) $\min\{-(1-\lambda)v_x + v_y, \mathcal{L}v - \dfrac{1-p}{p}v_x^{\frac{p}{p-1}}, v_x - (1-\lambda)v_y\} = 0$ in 4pt $\mathcal{S}$.

The function v is twice continuously differentiable in all of $\mathcal{S}$, except possibly on the positive y-axis, but even there v is once continuously differentiable and v_{yy} is defined and continuous. Thus, all the derivatives appearing in (3.8) – (3.11) are defined in the classical sense.

As mentioned earlier, Magill & Constantinides (1976) introduced this model. Moreover, they guessed the form of the optimal solution and obtained some numerical results on the values of θ_1 and θ_2. Davis & Norman (1990) formulated the model in terms of singular stochastic control and proved all the results reported above under the assumptions

(3.13) $A > 0, \theta_* < \dfrac{1}{2},$

where A and θ_* are defined by (3.3), (3.4). They showed that under these conditions,

(3.14) $\theta_1 < \theta_* < \theta_2,$

i.e., the optimal position for the problem with $\lambda = 0$ lies inside the no-transaction region NT for the problem with $\lambda > 0$.

Shreve & Soner (1992) removed both the conditions in (3.13), replacing them by the sole assumption that

(3.15) $v(x,y) < \infty \qquad \forall (x,y) \in \mathcal{S}.$

Sufficient conditions for (3.15) other than $A > 0$ were provided. One such condition is

(3.16) $(1-\lambda)^2 < \dfrac{\beta - \alpha p}{\beta - rp},$

a condition which can hold for σ near zero, when both conditions in (3.13) are violated. Shreve & Soner (1992) provided analytical bounds on θ_1 and

θ_2, showed that $\theta_2 = 1$ when $\theta_* = 1$, and showed that (3.14) can (and probably always does) fail when $\theta_* > 1$. This is the result, alluded to in the introduction, that when leverage is optimal in the model without transaction cost, then the introduction of a nonzero transaction cost will cause the agent to reduce her leverage, even if trading from the formerly optimal position is required to accomplish this.

In the remainder of this paper, we sketch the approach taken in Shreve & Soner (1992) and Shreve (1993) to the derivation of the results reported here. More precisely, we define the notion of viscosity solution to (3.12), sketch the proof that v is such a solution, and then outline how this fact can be exploited to derive substantive conclusions about the value function and the optimal policy.

4. Viscosity solutions. In this section, we sketch the proof that the value function v is a solution to the HJB equation (3.12). The complete proof can be found in Fleming & Soner (1993) and Shreve & Soner (1992).

DEFINITION 4.1. (Crandall, Evans, Lions) *We say that v is a* viscosity supersolution *(respectively,* viscosity subsolution*) of (3.12) if, for every* $(x_0, y_0) \in \mathcal{S}$ *and for every twice continuously differentiable function* $\phi :$ $\mathcal{S} \to R$ *satisfying* $\phi(x_0, y_0) = v(x_0, y_0)$ *and* $\phi \leq v$ *(respectively,* $\phi \geq v$*) on* $\mathcal{S}$*, we have*

$$\min\{-(1-\lambda)\phi_x(x_0,y_0) + \phi_y(x_0,y_0), \mathcal{L}\phi(x_0,y_0) - \frac{1-p}{p}(\phi_x(x_0,y_0))^{\frac{p}{1-p}},$$

$$\phi_x(x_0,y_0) - (1-\lambda)\phi_y(x_0,y_0)\} \geq 0.$$

(respectively, $\min\{\cdots\} \leq 0$*.). We say that v is a* viscosity solution *of (3.12) if it is both a viscosity supersolution and a viscosity subsolution.*

THEOREM 4.2. *The value function v of (2.4) is a viscosity solution of (3.12).*

SKETCH OF PROOF: We simplify the sketch but retain its essence by replacing the HJB equation (3.12) by the simpler equation

(4.1) $$\min_{c \geq 0}\{\mathcal{L}v - \frac{1}{p}c^p + cv_x\} = \mathcal{L}v - \frac{1-p}{p}v_x^{\frac{p}{p-1}} = 0.$$

The first-order expressions omitted by the passage from (3.12) to (4.1) are more easily handled than the second order term retained in $\mathcal{L}v$.

Let us first consider the supersolution property. Let $(x_0, y_0) \in \mathcal{S}$ be given. Let $\phi : \mathcal{S} \to R$ by a C^2 function satisfying $\phi \leq v$ on $\mathcal{S}$ and $\phi(x_0, y_0) = v(x_0, y_0)$. Let $C(\cdot)$ be a continuous consumption process. Let $X(0) = x_0$, $Y(0) = y_0$, and let $X(\cdot)$, $Y(\cdot)$ be given by (2.1), (2.2) with $L \equiv M \equiv 0$. Define

$$\tau_\epsilon \doteq \inf\{t \geq 0; (X(t) - x_0)^2 + (Y(t) - y_0)^2 \geq \epsilon^2\}.$$

According to the *Principle of Dynamic Programming*,

$$(4.2) \quad v(x_0, y_0) \geq E\left[\int_0^{\tau_\epsilon} e^{-\beta s}\frac{1}{p}C^p(s)ds + e^{-\beta \tau_\epsilon}v(X(\tau_\epsilon), Y(\tau_\epsilon))\right]$$

$$\geq E\left[\int_0^{\tau_\epsilon} e^{-\beta s}\frac{1}{p}C^p(s)ds + e^{-\beta \tau_\epsilon}\phi(X(\tau_\epsilon), Y(\tau_\epsilon))\right].$$

On the other hand, Itô's rule implies

$$Ee^{-\beta \tau_\epsilon}\phi(X(\tau_\epsilon), Y(\tau_\epsilon)) = \phi(x_0, y_0) - E\int_0^{\tau_\epsilon} e^{-\beta s}\left[\mathcal{L}\phi(X(s), Y(s))\right.$$

$$(4.3) \hspace{5cm} \left. + C(s)\phi_x(X(s), Y(s))\right]ds.$$

Adding these relations, we obtain

$$0 \geq -E\int_0^{\tau_\epsilon} e^{-\beta s}\left[\mathcal{L}\phi(X(s), Y(s)) - \frac{1}{p}C^p(s) + C(s)\phi_x(X(s), Y(s))\right]ds.$$

Letting $\epsilon \downarrow 0$, we see that

$$\mathcal{L}\phi(x_0, y_0) - \frac{1}{p}C^p(0) + C(0)\phi_x(x_0, y_0) \geq 0.$$

Minimizing over $C(0) \geq 0$, we obtain the desired inequality

$$\min_{c \geq 0}\{\mathcal{L}\phi(x_0, y_0) - \frac{1}{p}c^p + c\phi_x(x_0, y_0)\} \geq 0.$$

For the subsolution inequality, we begin with $(x_0, y_0) \in \mathcal{S}$ and ϕ : $\mathcal{S} \to R$ satisfying $\phi \geq v$ on $\mathcal{S}$ and $\phi(x_0, y_0) = v(x_0, y_0)$. Let $C(\cdot)$ be an optimal consumption process for the initial condition (x_0, y_0). (Of course, the existence of such a $C(\cdot)$ is not guaranteed. A rigorous proof would have to develop an argument based on the use of a nearly optimal $C(\cdot)$, and would also have to admit the possibility of transactions. Because we ignore the possibility of transactions, we are able to restrict our attention to the simplified HJB equation (4.1)). For an optimal $C(\cdot)$, the first inequality in (4.2) becomes an equality. The second inequality is reversed because now $\phi \geq v$. Proceeding as before, we conclude that

$$\min_{c \geq 0}\{\mathcal{L}\phi(x_0, y_0) - \frac{1}{p}c^p - c\phi_x(x_0, y_0)\} \leq 0.$$

5. Reduction to one variable.. According to the homotheticity property (3.1), $v(x, y) = (x+y)^p v(\frac{x}{x+y}, \frac{y}{x+y})$ for all $(x, y) \in \mathcal{S}$, so v is completely determined by its values on the line segment $\{(1 - z, z); -\frac{1-\lambda}{\lambda} \leq z \leq \frac{1}{\lambda}\}$. Therefore, we introduce the function

$$(5.1) \hspace{2cm} u(z) \doteq v(1 - z, z), \hspace{1cm} -\frac{1-\lambda}{\lambda} \leq z \leq \frac{1}{\lambda}.$$

Because v is concave and continuous on $\mathcal{S}$, u is concave and continuous on $[-\frac{1-\lambda}{\lambda}, \frac{1}{\lambda}]$. Furthermore, $u(-\frac{1-\lambda}{\lambda}) = u(\frac{1}{\lambda}) = 0$. The HJB equation (3.12) transforms into the HJB equation for u:

$$\min\{\lambda p u(z) + (1 - \lambda z)u'(z), (\beta - pd_1(z))u(z) - d_2(z)u'(z) - d_3(z)u''(z)$$

$$- \frac{1-p}{p}\left(pu(z) - zu'(z)\right)^{\frac{p}{p-1}}, \lambda p u(z) - (1 - \lambda + \lambda z)u'(z)\} = 0,$$

$$\text{(5.2)} \qquad\qquad\qquad\qquad\qquad\qquad -\frac{1-\lambda}{\lambda} < z < \frac{1}{\lambda},$$

where

$$\text{(5.3)} \qquad \beta - pd_1(z) \;=\; (1-p)A + \frac{1}{2}p(1-p)\sigma^2(z - \theta_*)^2,$$

$$\text{(5.4)} \qquad\qquad d_2(z) \;=\; -(1-p)\sigma^2 z(1-z)(z - \theta_*),$$

$$\text{(5.5)} \qquad\qquad d_3(z) \;=\; \frac{1}{2}\sigma^2 z^2(1-z)^2.$$

Because v is a viscosity solution of (3.12), u is a viscosity solution of (5.2).

The viscosity solution property permits an important first step in the study of the regularity of u.

THEOREM 5.1. *The function u is continuously differentiable on* $(-\frac{1-\lambda}{\lambda}, \frac{1}{\lambda})$.

SKETCH OF PROOF: Because u is concave, it is continuously differentiable if and only if its left-hand derivative agrees with its right-hand derivative everywhere. If $z_0 \in (-\frac{1-\lambda}{\lambda}, \frac{1}{\lambda})$ is a point where these derivatives disagree, the concavity of u implies $u'(z_0^-) > u'(z_0^+)$. Therefore, for each $K > 0$, we can construct a C^2 function $\phi : (-\frac{1-\lambda}{\lambda}, \frac{1}{\lambda}) \to R$ such that $\phi \geq u$, $\phi(z_0) = u(z_0)$, $u'(z_0^+) < \phi'(z_0) < u'(z_0^-)$, and $\phi''(z_0) \leq -K$. Because u is a viscosity subsolution of (5.2), we must have

$$\begin{aligned}
0 \;\geq\; & (\beta - pd_1(z_0))\phi(z_0) - d_2(z_0)\phi'(z_0) - d_3(z_0)\phi''(z_0) \\
& - \frac{1-p}{p}(p\phi(z_0) - z_0\phi'(z_0))^{\frac{p}{p-1}} \\
\geq\; & (\beta - pd_1(z_0))\phi(z_0) - d_2(z_0)\phi'(z_0) + d_3(z_0)K. \\
& - \frac{1-p}{p}(p\phi(z_0) - z_0\phi'(z_0))^{\frac{p}{p-1}} .
\end{aligned}$$

If $d_3(z_0) > 0$, this cannot hold for arbitrarily large K, and we conclude that u is of class C^1 except possibly at $z_0 = 0$ and $z_0 = 1$.

The cases $z_0 = 0$ and $z_0 = 1$ must be handled separately. Fleming & Soner (1993) and Shreve & Soner (1992) show how to treat the case $z_0 = 1$ by approaching this point from the left and right. Shreve & Soner (1992) show that $(1,0) \in SMM$, and therefore v is of class C^∞ in a neighborhood of $(1,0)$. It follows that u is of class C^∞ in a neighborhood of $z_0 = 0$. This concludes the sketch of the proof.

Because u' exists, we can define three regions:

$$smm \doteq \{z \in (-\frac{1-\lambda}{\lambda}, \frac{1}{\lambda}); \lambda pu(z) + (1 - \lambda z)u'(z) > 0,$$

(5.6)
$$\lambda pu(z) - (1 - \lambda + \lambda z)u'(z) = 0\},$$

$$nt \doteq \{z \in (-\frac{1-\lambda}{\lambda}, \frac{1}{\lambda}); \lambda pu(z) + (1 - \lambda z)u'(z) > 0,$$

(5.7)
$$\lambda pu(z) - (1 - \lambda + \lambda z)u'(z) > 0\},$$

$$ss \doteq \{z \in (-\frac{1-\lambda}{\lambda}, \frac{1}{\lambda}); \lambda pu(z) + (1 - \lambda z)u'(z) = 0,$$

(5.8)
$$\lambda pu(z) - (1 - \lambda + \lambda z)u'(z) > 0\}.$$

The concavity of v can be invoked to show that these regions take the form

(5.9)
$$smm = (-\frac{1-\lambda}{\lambda}, \theta_1), nt = (\theta_1, \theta_2), ss = (\theta_2, \frac{1}{\lambda})$$

for some constants $0 \le \theta_1 < \theta_2 < \frac{1}{\lambda}$, and a further argument (see Shreve & Soner (1992)) implies $\theta_1 > 0$. The first-order equation satisfied by u in *smm* implies that u takes the form, for some constant $B_1 > 0$,

(5.10)
$$u(z) = B_1 \left(\frac{1-\lambda+\lambda z}{1-\lambda}\right)^p, \quad -\frac{1-\lambda}{\lambda} < z < \theta_1.$$

Similarly,

(5.11)
$$u(z) = B_2(1 - \lambda z)^p, \theta_2 < z < \frac{1}{\lambda}.$$

Finally, the first-order inequalities defining nt, coupled with the HJB equation (5.2), imply

(5.12)
$$(\beta - pd_1(z))u(z) - d_2(z)u'(z) - d_3(z)u''(z) - \frac{1-p}{p}(pu(z) - zu'(z))^{\frac{p}{p-1}} = 0$$

in (θ_1, θ_2). This equation is initially known to hold only in the viscosity sense, but since u is C^1, the equation can be used to define $u''(z)$ as a continuous function, except possibly at $z = 1$ (0 is not in nt because $\theta_1 > 0$), and then it can be verified that u is C^2 on $(\theta_1, \theta_2)\backslash\{1\}$ and satisfies (5.12) in the classical sense on this set.

We have outlined the proof of the following theorem.

THEOREM 5.2. *There are numbers $0 < \theta_1 < \theta_2 < 1/\lambda$ such that u satisfies (5.10) and (5.11) for some positive constants B_1 and B_2. Furthermore, u is of class C^2 on $(\theta_1, \theta_2)\backslash\{1\}$ and satisfies (5.12) there.*

6. Return to two variables. By the definition (5.1) of u and the homotheticity (3.1) of v, we have

(6.1)
$$v(x, y) = (x + y)^p u\left(\frac{y}{x+y}\right) \quad \forall (x, y) \in \overline{\mathcal{S}}.$$

Theorems 5.1 and 5.2 imply that v is C^1 on $\mathcal{S}$, C^∞ on SMM given by (3.5) and on SS given by (3.7), and C^2 on NT given by (3.6), except possibly on the positive y-axis. But on the positive y-axis, we have $v(0, y) = y^p u(1)$, so $v_{yy}(0, y)$ is defined for all $y > 0$. It can be shown that v_{yy} is continuous on the positive y-axis, as well.

To show that v is C^2 across the boundaries $\frac{y}{x+y} = \theta_1$ and $\frac{y}{x+y} = \theta_2$, one assumes the existence of a discontinuity in the second derivative at some point (x_0, y_0) on one of these boundaries and argues to a contradiction. Given such a discontinuity, it is possible to construct a perturbed region NT' of NT so that, with NT replaced by NT', the policy described in Section 3 starting at initial condition (x_0, y_0) gives an expected utility of consumption strictly larger than $v(x_0, y_0)$. The details are in Shreve & Soner (1992). This contradiction completes the proof of the following theorem.

THEOREM 6.1. *The value function v of (2.4) is C^1 on $\mathcal{S}$ and C^2 on $\mathcal{S} \backslash \{(0, y); y > 0\}$. Moreover, v_{yy} exists and is continous on all of $\mathcal{S}$.*

The assertions made in Section 3 about the form of the optimal policy now follow from a standard verification argument.

7. Sensitivity to the transaction cost. Throughout this section, we assume

$$(7.1) \qquad\qquad\qquad 0 < \theta_* < 1.$$

For the sensitivity analysis, we need sharp lower and upper bounds on the value function v. To obtain these, we will use the following comparison theorem.

THEOREM 7.1. *If $w : [-\frac{1-\lambda}{\lambda}, \frac{1}{\lambda}] \to R$ is C^1, piecewise C^2, satisfies $w(-\frac{1-\lambda}{\lambda}) = w(\frac{1}{\lambda}) = 0$, and also satisfies*

$$
\begin{aligned}
(7.2) \qquad min\{ & \lambda p w(z) + (1 - \lambda z) w'(z), (\beta - p d_1(z)) w(z) - d_2(z) w'(z) \\
& - d_3(z) w''(z) - \frac{1-p}{p} (p w(z) - z w'(z))^{\frac{p}{p-1}}, \\
& \lambda p w(z) - (1 - \lambda + \lambda z) w'(z) \} \quad \geq \quad 0, \\
& \qquad\qquad\qquad -\frac{1-\lambda}{\lambda} < z < \frac{1}{\lambda},
\end{aligned}
$$

(respectively, $min\{\cdots\} \leq 0$), then $w \geq u$ (respectively, $w \leq u$) on $[-\frac{1-\lambda}{\lambda}, \frac{1}{\lambda}]$, where u is given by (5.1).

SKETCH OF PROOF: From Theorem 6.1 we see that u is C^2 except possibly at $z = 1$, and even at $z = 1$, u is C^1. To simplify the presentation, we assume u is C^2 at $z = 1$; if it is not, one can use the fact that $d_3(1) = 0$ to modify the following argument (see Shreve (1993)).

We consider a function w satisfying (7.2). (The argument for the reverse inequality is similar.) Let us suppose $w \geq u$ does not hold. Then $w - u$ assumes a negative minimum value at some point $z_0 \in (-\frac{1-\lambda}{\lambda}, \frac{1}{\lambda})$, i.e., $w(z_0) < u(z_0)$. Moreover, $w'(z_0) = u'(z_0)$ and $w''(z_0) \geq u''(z_0)$. (Here we replace $w''(z_0)$ by the one-sided derivatives $w''(z_0^{\pm})$ if $w''(z_0)$ is not defined.) From these inequalities and (7.2), we see that the left side of (5.2), evaluated at $z = z_0$, is strictly positive. This contradicts Theorem 5.2 and concludes the sketch of the proof.

It remains to construct piecewise C^2 functions satisfying (7.2) and its reverse inequality. Toward that end, we define

$$(7.3)\; w(z) = \begin{cases} \frac{1}{p}(A^{p-1} - \epsilon - \gamma_1\epsilon\delta_1^2)\left(\frac{1-\lambda+\lambda z}{1-\lambda+\lambda z_1}\right)^p, & -\frac{1-\lambda}{\lambda} \leq z \leq z_1, \\ \frac{1}{p}(A^{p-1} - \epsilon - \gamma_1\epsilon(z - \theta_*)^2), & z_1 \leq z \leq z_2, \\ \frac{1}{p}(A^{p-1} - \epsilon - \gamma_1\epsilon\delta_2^2)\left(\frac{1-\lambda z}{1-\lambda z_2}\right)^p, & z_2 \leq z \leq \frac{1}{\lambda}, \end{cases}$$

where γ_1 and γ_2 are positive constants, $\epsilon = \frac{\gamma_2}{\gamma_1}\lambda^{2/3}$, $z_1 = \theta_* - \delta_1$, $z_2 = \theta_* + \delta_2$, and δ_1 and δ_2 are positive numbers, depending on λ, chosen to make w be C^1. It is not difficult to verify that for sufficiently small $\lambda > 0$,

$$(7.4) \qquad \frac{m}{\gamma_2}\lambda^{1/3} \leq \delta_i \leq \frac{M}{\gamma_2}\lambda^{1/3}, \qquad i = 1, 2,$$

where $m \doteq pA^{p-1}/2^p$, $M \doteq pA^{p-1}/(\theta_* \wedge (1 - \theta_*))$. It can also be verified (Shreve (1993)) that if

$$(7.5)\; \gamma_1 > 2^{\frac{2-p}{1-p}}\frac{A}{\sigma^2\theta_*^2(1 - \theta_*^2)}, \qquad \gamma_2 < \left[2^{\frac{3-2p}{p-1}}p(1 - p)\sigma^2 m^2 A^{p-2}\gamma_1\right]^{1/3},$$

then for sufficiently small $\lambda > 0$, w is a supersolution of (5.1), i.e., w satisfies (7.2), and hence $w \geq u$. On the other hand, if

$$(7.6) \qquad 0 < \gamma_1 < 16A/\sigma^2, \qquad \gamma_2 > \left[\frac{8p(1 - p)A^{p-1}M^2}{(16A/\sigma^2) - \gamma_1}\right]^{1/3},$$

then for sufficiently small $\lambda > 0$, w is a subsolution of (5.1), i.e., w satisfies the reverse of inequality (7.2), and hence $w \leq u$. We have then the following theorem.

THEOREM 7.2. *There exist constants $k_2 \geq k_1 > 0$, depending on p, A and θ_*, but not depending on λ, such that for all $\lambda > 0$ sufficiently small,*

$$(7.7) \qquad \frac{1}{p}A^{p-1} - k_2\lambda^{2/3} \leq u(\theta_*) \leq \frac{1}{p}A^{p-1} - k_1\lambda^{2/3}.$$

For $q > 0$, we denote by $O(\lambda^q)$ any function satisfying $\limsup_{\lambda\downarrow 0} \lambda^{-q}|O(\lambda^q)| < \infty$. The conclusion of Theorem 7.2 may be restated as

$$(7.8) \qquad v(1 - \theta_*, \theta_*; \lambda) = u(\theta_*; \lambda) = u(\theta_*; 0) - O(\lambda^{2/3}),$$

where now we have explicitly indicated the dependence of v and u on the transaction cost parameter λ.

REMARK 7.3. We have a rigorous result that the value function decreases like $\lambda^{2/3}$ near $\lambda = 0$. The definition of z_1 and z_2 used in the construction of w in (7.3) suggests that $(\theta_2(\lambda) - \theta_1(\lambda)) = O(\lambda^{1/3})$ (see (7.4)), where we have explicitly indicated the dependence of θ_1 and θ_2 on λ. This is the order of the width of the no-transaction interval obtained by Fleming, et. al (1990) for their transaction cost problem without intermediate consumption. It is also consistent with the numerical results of Constantinides (1986) and Davis & Norman (1990), both of whom found a rapid opening of the no-transaction interval as λ increases from zero.

We turn now to the notion of *liquidity premium*. Let us indicate the dependence of v on both α, the mean rate of return of the stock, and on λ, by writing $v(x, y; \alpha, \lambda)$. With r, σ, p and β fixed, let $A(\alpha)$ and $\theta_*(\alpha)$ be defined by (3.3), (3.4), where we have indicated their dependence on α. For $\lambda > 0$, the *liquidity premium* is defined to be that positive number $\rho(\lambda)$ for which

$$v(1 - \theta_*(\alpha + \rho(\lambda)), \theta_*(\alpha + \rho(\lambda)); \alpha + \rho(\lambda), \lambda) = v(1 - \theta_*(\alpha), \theta_*(\alpha); \alpha, 0).$$

Using this definition, (7.8) and (3.2), we have

$$
\begin{aligned}
0 &= \Big[v\big(1 - \theta_*(\alpha + \rho(\lambda)), \theta_*(\alpha + \rho(\lambda)); \alpha + \rho(\lambda), \lambda\big) \\
&\qquad - v\big(1 - \theta_*(\alpha + \rho(\lambda)), \theta_*(\alpha + \rho(\lambda)); \alpha + \rho(\lambda), 0\big) \Big] \\
&\quad + \Big[v\big(1 - \theta_*(\alpha + \rho(\lambda)), \theta_*(\alpha + \rho(\lambda)); \alpha + \rho(\lambda), 0\big) \\
&\qquad - v\big(1 - \theta_*(\alpha), \theta_*(\alpha); \alpha, 0\big) \Big] \\
&= O(\lambda^{2/3}) + \frac{p}{2\sigma^2(1-p)^2} [2(\alpha - r)\rho(\lambda) + \rho^2(\lambda)],
\end{aligned}
$$

from which we conclude that $\rho(\lambda) = O(\lambda^{2/3})$. In particular, $\lim_{\lambda \downarrow 0} \frac{\rho(\lambda)}{\lambda} = \infty$.

REFERENCES

BENSAID, B., LESNE, J.-P., PAGÉS, H., and SCHEINKMAN, J. (1992), Derivative asset pricing with transaction costs, *Math. Finance* **2**, 63–86.

BLACK, F. and SCHOLES, M. (1973), The pricing of options and corporate liabilities, *J. Political Economy* **81**, 637–659.

BOYLE, P.P. and VORST, T. (1992), Option replication in discrete time with transaction costs, *J. Finance* **47**, 272–293.

CONSTANTINIDES, G.M. (1979), Multiperiod consumption and investment behavior with convex transaction costs, *Management Sci.* **25**, 1127–1137.

CONSTANTINIDES, G.M. (1986), Capital market equilibrium with transaction costs, *J. Political Economy* **94**, 842–862.

CONSTANTINIDES, G.M. (1993), Option pricing bounds with transaction costs, Working paper, Graduate School of Business, University of Chicago.

COX, J. and HUANG, C. (1989), Optimal consumption and portfolio policies when asset prices follow a diffusion process, *J. Economic Theory* **49**, 33–83.

CRANDALL, M.G. and LIONS, P.-L. (1983), Viscosity solutions of Hamilton-Jacobi equations, *Trans. Amer. Math. Soc.* **277**, 1–42.

CRANDALL, M.G. EVANS, L.C., and LIONS, P.-L. (1984), Some properties of viscosity solutions of Hamilton-Jacobi equations, *Trans. Amer. Math. Soc.* **282**, 487–502.

CRANDALL, M.G., ISHII, H. and LIONS, P.-L. (1992), User's guide to viscosity solutions of second-order partial differential equations, *Bull. Amer. Math. Soc.* **27**, 1–67.

DAVIS, M.H.A. and NORMAN, A. (1990), Portfolio selection with transaction costs, *Math. Operations Research* **15**, 676–713.

DAVIS, M.H.A., PANAS, V.G., and ZARIPHOPOULOU, T. (1993), European option pricing with transaction costs, *SIAM J. Control Optimization* **31**, 470–493.

DAVIS, M.H.A. and ZARIPHOPOULOU, T. (1994), American options and transaction fees, *this volume*, 47–62.

DUMAS, B. and LUCIANO, E. (1991), An exact solution to a dynamic portfolio choice problem under transaction costs, *J. Finance* **46**, 577–595.

EDIRISINGHE, C., NAIK, V. and UPPAL, R. (1993), Optimal replication of options with transactions costs and trading restrictions, *J. Financial and Quantitative Analysis* **28**, 117-138.

FLEMING, W.H., GROSSMAN, S. G., VILA, J.-L., and ZARIPHOPOULOU, T. (1990), Optimal portfolio rebalancing with transaction costs, Working paper, Division of Applied Mathematics, Brown University.

FLEMING, W.H. and SONER, H.M. (1993), *Controlled Markov Processes and Viscosity Solutions*, Springer-Verlag.

GILSTER, J. and LEE, W. (1984), The effects of transaction costs and different borrowing and lending rates on the option pricing model: a note, *J. Finance* **39**, 1215–1222.

GROSSMAN, S.J. and LAROQUE, G. (1990), Asset pricing and optimal portfolio choice in the presence of illiquid durable consumption goods, *Econometrica* **58**, 25–51.

HANSEN, L.P. and SINGLETON, K.J. (1982), Generalized instrumental variables estimation of nonlinear rational expectations models, *Econometrica* **50**, 1269–1286.

HANSEN, L.P. and SINGLETON, K.J. (1983), Stochastic consumption, risk aversion, and the temporal behavior of asset returns, *J. Political Economy* **91**, 249–265.

HARRISON, J.M. and KREPS, D.M. (1979), Martingales and arbitrage in multiperiod security markets, *J. Econom. Theory* **20**, 381–408.

HARRISON, J.M. and PLISKA, S.R. (1981), Martingales and stochastic integrals in the theory of continuous trading, *Stochastic Processes and Appl.* **11**, 215–260.

HARRISON, J.M. and PLISKA, S.R. (1983), A stochastic calculus model of continuous time trading: complete markets, *Stochastic Processes and Appl.* **15**, 313–316.

HENROTTE, P. (1993), Transaction costs and duplication strategies, Working paper, Graduate School of Business, Stanford University.

HODGES, S.D. and NEUBERGER, A. (1989), Optimal replication of contingent claims under transaction costs, *Rev. Futures Markets* **8**, 222–239.

ISHII, H. (1989), On uniqueness and existence of viscosity solutions of fully nonlinear second order elliptic partial differential equations, *Comm. Pure Appl. Math.* **42**, 15–45.

JENSEN, R. (1988), The maximum principle for viscosity solutions of second order fully nonlinear partial differential equations, *Arch. Rat. Mech. Anal.* **101**, 1–27.

KREPS, D.M. (1981), Arbitrage and equilibrium in economies with infinitely-many commodities, *J. Math. Econom.* **8**, 15–35.

KARATZAS, I., LEHOCZKY, J.P., and SHREVE, S.E. (1987), Optimal portfolio and consumption decisions for a "small investor" on a finite horizon, *SIAM J. Control Optim.* **25**, 1557–1586.

LELAND, H.E. (1985), Option pricing and replication with transaction costs *J. Finance* **40**, 1283–1301.

LIONS, P.-L. (1982), *Generalized solutions of Hamilton-Jacobi equations*, Pitman, Boston.

LIONS, P.-L. (1983a), Optimal control of diffusion processes and Hamilton-Jacobi-Bellman equations I, *Comm. Partial Diff. Equations* **8**, 596–608.

LIONS, P.-L. (1983b), Optimal control of diffusion processes and Hamilton-Jacobi-Bellman equations, Part II: viscosity solutions and uniqueness, *Comm. Partial Diff. Equations* **8**, 1229–1276.

LIONS, P.-L. (1983c), Optimal control of diffusion processes and Hamilton-Jacobi-Bellman equations, Part III: regularity of the optimal cost function, in *Nonlinear Partial Differential Equations and Applications*, College de France Seminar Vol. V, Pitman, Boston.

MAGILL, M.J.P. and CONSTANTINIDES, G.M. (1976), Portfolio selection with transaction costs, *J. Economic Theory* **13**, 245–263.

MERTON, R.C. (1969), Lifetime portfolio selection under uncertainty: the continuous-time case, *Rev. Econ. Statist.* **51**, 247–257.

MERTON, R.C. (1971), Optimum consumption and portfolio rules in a continuous-time case, *J. Economic Theory* **3**, 373–413. Erratum: ibid. **6** (1973), 213-214.

MERTON, R.C. (1973a), An intertemporal capital asset pricing model, *Econometrica* **41**, 867–887.

MERTON, R.C. (1973b), Theory of rational option pricing, *Bell J. Econom. Manag. Sci.* **4**, 141–183.

PLISKA, S.R. (1986), A stochastic calculus model of continuous trading: optimal portfolio, *Math. Operations Research* **11**, 371–382.

SHREVE, S.E. and SONER, H.M. (1994), Optimal investment and consumption with transaction costs, *Ann. Appl. Probab.*, to appear.

SHREVE, S.E. (1994), Sensitivity of the indirect utility to transaction costs in a consumption-based model, (appendix to Shreve & Soner (1994)), *Ann. Appl. Probab.*, to appear.

TAKSAR, M., KLASS, M.J. and ASSAF, D. (1988), A diffusion model for optimal portfolio selection in the presence of brokerage fees, *Math. Operations Research* **13**, 277–294.

ZARIPHOPOULOU, T. (1989), Optimal investment-consumption models with constraints, Ph.D. Thesis, Brown University, Providence, RI.

ZARIPHOPOULOU, T. (1992), Investment-consumption models with transaction fees and Markov-chain parameters, *SIAM J. Control Optimization* **30**, 613-636.

Springer Series in
MATERIALS SCIENCE

96

Springer Series in
MATERIALS SCIENCE

Editors: R. Hull   R. M. Osgood, Jr.   J. Parisi   H. Warlimont

The Springer Series in Materials Science covers the complete spectrum of materials physics, including fundamental principles, physical properties, materials theory and design. Recognizing the increasing importance of materials science in future device technologies, the book titles in this series reflect the state-of-the-art in understanding and controlling the structure and properties of all important classes of materials.

99 **Self-Organized Morphology in Nanostructured Materials**
Editors: K. Al-Shamery and J. Parisi

100 **Self Healing Materials**
An Alternative Approach
to 20 Centuries of Materials Science
Editor: S. van der Zwaag

101 **New Organic Nanostructures for Next Generation Devices**
Editors: K. Al-Shamery, H.-G. Rubahn, and H. Sitter

102 **Photonic Crystal Fibers**
Properties and Applications
By F. Poli, A. Cucinotta,
and S. Selleri

103 **Polarons in Advanced Materials**
Editor: A.S. Alexandrov

104 **Transparent Conductive Zinc Oxide**
Basics and Applications
in Thin Film Solar Cells
Editors: K. Ellmer, A. Klein, and B. Rech

105 **Dilute III-V Nitride Semiconductors and Material Systems**
Physics and Technology
Editor: A. Erol

106 **Into The Nano Era**
Moore's Law Beyond Planar Silicon CMOS
Editor: H.R. Huff

107 **Organic Semiconductors in Sensor Applications**
Editors: D.A. Bernards, R.M. Ownes, and G.G. Malliaras

108 **Evolution of Thin-Film Morphology**
Modeling and Simulations
By M. Pelliccione and T.-M. Lu

109 **Reactive Sputter Deposition**
Editors: D. Depla amd S. Mahieu

110 **The Physics of Organic Superconductors and Conductors**
Editor: A. Lebed

Volumes 50–98 are listed at the end of the book.

Rüdiger Quay

Gallium Nitride Electronics

Springer

Dr. Rüdiger Quay
Fraunhofer Institut für Angewandte Festkörperphysik (IAF)
Tullastr. 72, 79108, Freiburg, Germany

Series Editors:

Professor Robert Hull
University of Virginia
Dept. of Materials Science and Engineering
Thornton Hall
Charlottesville, VA 22903-2442, USA

Professor R. M. Osgood, Jr.
Microelectronics Science Laboratory
Department of Electrical Engineering
Columbia University
Seeley W. Mudd Building
New York, NY 10027, USA

Professor Jürgen Parisi
Universität Oldenburg, Fachbereich Physik
Abt. Energie- und Halbleiterforschung
Carl-von-Ossietzky-Strasse 9–11
26129 Oldenburg, Germany

Professor Hans Warlimont
Institut für Festkörper-
und Werkstofforschung,
Helmholtzstrasse 20
01069 Dresden, Germany

ISBN 978-3-540-71890-1   e-ISBN 978-3-540-71892-5

DOI 10.1007/978-3-540-71892-5

Springer Series in Materials Sciences ISSN 0933-033X

Library of Congress Control Number: 2008924620

Typesetting and production: le-tex publishing services oHG, Leipzig, Germany
Cover design: WMXDesign, Heidelberg, Germany

Printed on acid-free paper

9 8 7 6 5 4 3 2 1

springer.com

To our son, Jonathan Benedikt

In memoriam, Oliver Winterer (1970–2006)

Preface

Electronic RF-communication and sensing systems have dramatically changed our daily lives since the invention of the first electronic transistor in 1947. Advanced semiconductor devices are key components within electronic systems and ultimately determine their performance. In this never ending challenge, wide-bandgap nitride semiconductors and heterostructure devices are unique contenders for future leading-edge electronic systems due to their outstanding material properties with respect to speed, power, efficiency, linearity, and robustness. At the same time, their material properties are challenging compared with any other material system due to high growth temperatures and many other intrinsic properties.

Wide bandgap semiconductors have attracted a lot of attention in the last ten years due to their use in optoelectronic and electronic applications. The field is developing rapidly due to the high investments in US, Japanese, and increasing European research and development activities. Some of the knowledge acquired may not be available to the general public because of military or civil restrictions. However, this work compiles and systemizes the available knowledge and evaluates remaining issues. This book is of interest to graduate students of electrical engineering, communication engineering, and physics; to material, device, and circuit engineers in research and industry; and to scientists with general interest in advanced electronics.

The author specially thanks those people, without whom and without whose individual contributions such a challenging work would have been impossible. He owes special thanks to:

Prof. Dr. Günter Weimann, director of the Fraunhofer Institute of Applied Solid-State Physics (IAF), for his encouragement, his advice, and continuous support.

Prof. Dr. Joachim Wagner for the encouragement to start this project.

Prof. Dr. Siegfried Selberherr, Insitut für Mikroelektronik, TU Wien, for continuous encouragement and support.

Dr. Michael Schlechtweg, head of the RF-device and circuits department at Fraunhofer IAF, and *Dr. Michael Mikulla,* head of the technology department at Fraunhofer IAF, for their generous support.

Dr. Rudolf Kiefer for his outstanding contributions and careful advise on the technology chapter, for his kind understanding, and for valuable discussions.

Dipl.Phys. Stefan Müller and *Dr. Klaus Köhler* for their valuable contributions of the epitaxy chapter and for proof reading.

Dr. Friedbert van Raay for his proof reading, contributions to large-signal modeling and circuit design and for countless discussions on modeling, large-signal measurements, layout, and circuit design.

Dr. Michael Dammann, Dipl.Ing. Helmer Konstanzer, and *Andreas Michalov* for their contributions and their work on device reliability.

Dr. Wolfgang Bronner for his contributions to the development of technology and the SiC back-end process.

Dr. Wilfried Pletschen for his thorough proof reading and his kind advise on etching.

Dr. Matthias Seelmann-Eggebert for his inspired work on thermal simulations and large-signal modeling.

Dr. Patrick Waltereit for numerous fruitful discussions on epitaxial and process development.

Dipl.Ing. Daniel Krausse for his work on low-noise amplifiers.

Dr. Vassil Palankovski and *Dipl.Ing. Stanislav Vitanov,* TU Vienna, for proof reading and their valuable support on physical device simulation.

Dr. Axel Tessmann for his continuous good mood and valuable motivation for the development of mm-wave technology.

Markus Riesle and *Dr. Herbert Walcher* for their contributions to the MMIC module and device packaging.

Martin Zink and *Ronny Kolbe* for the patient dicing and picking of a numerous MMICs.

Dipl.Ing. Christoph Schwörer for valuable discussions on circuit design and for his contribution on the broadband amplifiers.

Dr. Lutz Kirste for his help on crystal structures.

Dipl.Ing. Michael Kuri , Dipl.Ing. Hermann Massler and the members of the RF-devices and circuits characterization group at Fraunhofer IAF for their support.

Dr. Arnulf Leuther for wise hints and good cooperation on process development.

Fouad Benkhelifa for his creative and careful development of processing technology and for active discussions.

Further, I would like to thank the technical staff in the Fraunhofer RF-devices-and-circuit and technology-departments, especially *Dr. Gundrun Kauffel* and *W. Fehrenbach.*

Dr. Hardy Sledzik, Dr. Patrick Schuh, Dr. Ralf Leberer, and *Dr. Martin Oppermann* at EADS DE in Ulm for good cooperation and for valuable discussions.

Dipl.Ing. Dirk Wiegner, Dr. Wolfgang Templ, and *Ulrich Seyfried* at the Research Center at Alcatel-Lucent/Stuttgart for outstanding cooperation on highly-linear power amplifiers.

Dr. Thomas Rödle and his team at NXP research center, Nimwegen, for good cooperation on power amplifier technology.

The Team at United Monolithic Semiconductors (UMS), Ulm for good cooperation on device technology.

Mark van Heijningen and *Dr. Frank van Vliet* at TNO Safety and Security, the Hague, Netherlands, for the good cooperation on high-power amplifiers at Ka-band.

Joyce Visne, Wien, for her patient and thorough language corrections and her kind support.

The Springer team, especially, *Adelheid Duhm* and *Dr. Claus Ascheron* for their great support and kind understanding.

The le-tex team, especially *Steffi Hohensee,* for their kind support.

I thank my wife Christine and my son Jonathan, to whom this book is dedicated, for their endless patience.

Freiburg i.Br.   *Rüdiger Quay*
Januar 2008

Contents

List of Symbols XVII

List of Acronyms XXV

1   Introduction 1

2   III-N Materials, and the State-of-the-Art of Devices
    and Circuits 3
    2.1   State-of-the-Art of Materials Research 3
          2.1.1   Binary Materials 4
          2.1.2   Material Limitations 20
          2.1.3   Thermal Properties and Limitations 21
          2.1.4   Ternary and Quaternary III-N Materials 23
    2.2   Polar Semiconductors for Electronics 28
          2.2.1   Spontaneous Polarization 28
          2.2.2   Piezoelectric Polarization 30
          2.2.3   Device Design Using Polarization-Induced Charges .. 32
          2.2.4   Analytical Calculation
                  of Channel Charge Concentrations 38
          2.2.5   Doping Issues 38
          2.2.6   Surfaces and Interfaces 40
          2.2.7   Transport Properties in Polarized Semiconductors ... 46
          2.2.8   Polarization-Based Devices
                  and Their Specific Properties 48
    2.3   Electrical and Thermal Limitations of Materials and Devices   48
          2.3.1   Physical Modeling of Devices 49
          2.3.2   Devices: Figures-of-Merit 51
          2.3.3   III-N Devices: Frequency Dispersion 52
    2.4   Substrates for Electronic Devices 55
          2.4.1   Criteria for Substrate Choice 55
          2.4.2   Silicon Carbide Substrates 56

        2.4.3   Sapphire Substrates 61
        2.4.4   Silicon Substrates 62
        2.4.5   GaN and AlN Substrates 62
    2.5 State-of-the-Art of Devices and Circuits 65
        2.5.1   Nitride-Based Diodes 65
        2.5.2   Power Electronics 66
        2.5.3   RF-Metal Semiconductor Field-Effect Transistors
                (MESFETs) 68
        2.5.4   Metal Insulator Semiconductor Field-Effect
                Transistors (MISFETs) 70
        2.5.5   High-Electron Mobility Transistors (HEMTs) 71
        2.5.6   Heterojunction Bipolar Transistors (HBTs) 84
        2.5.7   MMIC HEMT Technology 86
    2.6 Applications Issues 87
        2.6.1   Broadband Communication 88
        2.6.2   Radar Components 88
        2.6.3   Electronics in Harsh Environments 89
    2.7 Problems 90

3   Epitaxy for III-N-Based Electronic Devices 91
    3.1 The AlGaN/GaN Material System 92
        3.1.1   Metal Organic Chemical Vapor
                Deposition (MOCVD) 92
        3.1.2   Molecular Beam Epitaxy (MBE) 112
        3.1.3   MOCVD and MBE Growth
                on Alternative Substrates 121
        3.1.4   Epitaxial Lateral Overgrowth (ELO) 122
        3.1.5   Hydride Vapor Phase Epitaxy (HVPE) 122
    3.2 Indium-Based Compounds and Heterostructures 123
        3.2.1   MOCVD Growth of Indium-Based Layers 124
        3.2.2   MBE Growth of Indium-Based Layers 125
        3.2.3   Indium-Based Heterostructure Growth 126
    3.3 Doping and Defects 127
        3.3.1   MOCVD Growth 128
        3.3.2   MBE Growth 130
    3.4 Epitaxial Device Design 131
        3.4.1   Geometrical Considerations 131
        3.4.2   Growth of Cap Layers 133
        3.4.3   Doping....................................... 134
        3.4.4   AlN Interlayer 135
        3.4.5   Channel Concepts 137
        3.4.6   Epitaxial In-Situ Device Passivation 137
    3.5 Problems 138

4   Device Processing Technology 139
  4.1   Processing Issues 139
  4.2   Device Isolation 142
     4.2.1   Mesa Structures 143
     4.2.2   Ion Implantation for Isolation 143
  4.3   Contact Formation 144
     4.3.1   Ohmic Contacts 144
     4.3.2   Schottky Contacts 151
  4.4   Lithography 157
     4.4.1   Optical Lithography 157
     4.4.2   Electron Beam Lithography....................... 158
     4.4.3   Field Plates and Gate Extensions.................. 160
  4.5   Etching and Recess Processes 165
     4.5.1   Dry Etching 166
     4.5.2   Wet Etching 169
     4.5.3   Recess Processes 170
  4.6   Surface Engineering and Device Passivation 174
     4.6.1   Passivation of the Ungated Device Region 174
     4.6.2   Physical Trapping Mechanisms 177
     4.6.3   Trap Characterization 178
     4.6.4   Technological Measures:
           Surface Preparation and Dielectrics 182
     4.6.5   Epitaxial Measures:
           Surface Preparation and Dielectrics 188
  4.7   Gate Dielectrics 189
  4.8   Processing for High-Temperature Operation 191
  4.9   Backside Processing 192
     4.9.1   Thinning Technologies 192
     4.9.2   Viahole Etching and Drilling Technologies 193
     4.9.3   Viahole Metallization 195
  4.10 Problems 196

5   Device Characterization and Modeling 197
  5.1   Device Characteristics 197
     5.1.1   Compact FET Analysis 197
     5.1.2   Compact Bipolar Analysis 209
  5.2   Frequency Dispersion 211
     5.2.1   Dispersion Effects and Characterization 211
     5.2.2   Dispersion Characterization and Analysis........... 214
     5.2.3   Models for Frequency Dispersion in Devices........ 217
     5.2.4   Suppression of Frequency Dispersion 220
  5.3   Small-Signal Characterization, Analysis, and Modeling 220
     5.3.1   RF-Characterization and Invariants................ 220
     5.3.2   Common-Source HEMTs 222
     5.3.3   Dual-Gate HEMTs 227

	5.3.4	Pulsed-DC- and RF-Characteristics	227
	5.3.5	Small-Signal Modeling	230
5.4	Large-Signal Analysis and Modeling		235
	5.4.1	Large-Signal Characterization and Loadpull Results .	235
	5.4.2	Large-Signal Modeling	241
5.5	Linearity Analysis and Modeling		255
	5.5.1	Basic Understanding	256
	5.5.2	Nitride-Specific Linearity Analysis	259
5.6	Noise Analysis		262
	5.6.1	Low-Frequency Noise	262
	5.6.2	RF-Noise Analysis and Characterization	265
5.7	Problems		270

6 Circuit Considerations and III-N Examples 271

6.1	Passive Circuit Modeling		271
	6.1.1	Coplanar-Waveguide Transmission-Line Elements	271
	6.1.2	Microstrip-Transmission-Line Elements	273
6.2	High-Voltage High-Power Amplifiers		274
	6.2.1	Basic Principles of High-Voltage High-Power Operation	274
	6.2.2	General Design Considerations of III-N Amplifiers	278
	6.2.3	Mobile Communication Amplifiers Between 500 MHz and 6 GHz	279
	6.2.4	C-Frequency Band High-Power Amplifiers	285
	6.2.5	X-Band High-Power Amplifiers	287
	6.2.6	Design, Impedance Levels, and Matching Networks	294
	6.2.7	Broadband GaN Highly Linear Amplifiers	297
	6.2.8	GaN Mm-wave Power Amplifiers	298
6.3	Robust GaN Low-Noise Amplifiers		300
	6.3.1	State-of-the-Art of GaN Low-Noise Amplifiers	300
	6.3.2	Examples of GaN MMIC LNAs	301
6.4	Oscillators, Mixers, and Attenuators		304
	6.4.1	Oscillators	307
	6.4.2	GaN HEMT Mixer Circuits	307
	6.4.3	Attenuators and Switches	308
6.5	Problems		309

7 Reliability Aspects and High-Temperature Operation 311

7.1	An Overview of Device Testing and of Failure Mechanisms		311
	7.1.1	Description of Device Degradation	311
	7.1.2	Degradation Mechanisms in III-N FETs	314
	7.1.3	III-V HBT Device Degradation	316
7.2	Analysis of Nitride-Specific Degradation Mechanisms		317
	7.2.1	DC-Degradation	318
	7.2.2	RF-Degradation	322

7.3   Failure Analysis 324
      7.3.1   Failure Mechanisms 325
      7.3.2   Reliability Case Studies 327
7.4   Radiation Effects 331
7.5   High-Temperature Operation 332
7.6   Problems 336

8   Integration, Thermal Management, and Packaging 337
    8.1   Passive MMIC Technologies 337
          8.1.1   Passive Element Technologies 337
          8.1.2   Microstrip Backend Technology 340
    8.2   Integration Issues 341
    8.3   Thermal Management 343
          8.3.1   Thermal Analysis 343
          8.3.2   Thermal Material Selection and Modeling 345
          8.3.3   Basic Thermal Findings, Heat Sources,
                  and Thermal Resistances 349
          8.3.4   Backside Cooling 352
          8.3.5   Flip-Chip Integration 355
          8.3.6   Dynamic Thermal Effects 357
    8.4   Device and MMIC Packaging 358
          8.4.1   Dicing 358
          8.4.2   Die-Attach 359
          8.4.3   Package Technology Selection 361
          8.4.4   Thermal Management for Linear Applications 363
          8.4.5   Active Cooling 365
    8.5   Problems 366

9   Outlook 367

Appendix 369

References of Chapter 2 371

References of Chapter 3 395

References of Chapter 4 405

References of Chapter 5 419

References of Chapter 6 429

References of Chapter 7 439

References of Chapter 8 447

Index 455

List of Symbols

Δ	Step, difference, change
ΔE_V, ΔE_C	Discontinuity of the valence/conduction band at a heterointerface
ΔE_g	Total difference of the bandgaps at a heterointerface
Δf	Frequency interval
ΔV	Voltage drop over the depletion zone
$\Theta(T_0)$	Temperature-dependent thermal resistance
α, β, γ	General exponent
α	General temperature coefficient
α	Common base current gain
α_a, α_c	Coefficients of thermal expansion along a- and c-axis
α_AB	Temperature exponent of the ternary $\mathrm{A}_x\mathrm{B}_{1-x}$
α_H	Hooge parameter
α_n, α_p	Impact ionization parameter for electrons and holes
α_n	Fitting parameter in the LS-model
α_T	Base transmission factor
α_S	Fitting parameter in the LS-model
β	Common emitter current gain in bipolar devices
γ_i	Emitter injection efficiency
δ	The base recombination factor
ε	Permittivity
ϵ_{ij}	Dielectric tensor components
ϵ_0	Dielectric constant
ε_r	Relative permittivity

$\varepsilon_{\mathrm{r}}^{\mathrm{eff}}$	Effective permittivity
$\varepsilon_{\mathrm{r}}^{\mathrm{inf}}$	Relative permittivity for $\omega \to \infty$
η_{d}	Drain efficiency
κ_{L}	Lattice thermal conductivity
$\kappa_{\mathrm{L}}(T)$	Thermal conductivity as a function of temperature
$\kappa_{300}^{\mathrm{A}}$	Thermal conductivity at $300\,\mathrm{K}$ of material A
λ_{n}	Fitting parameter in the LS-model
μ_{ν}	Mobility of carrier type ν
μ_{AB}	Mobility of the ternary semiconductor $A_x B_{1-x} N$
τ	Phase term of the transconductance g_{m}
τ_{thermal}	Thermal time constant
ϕ_{B}	Schottky barrier potential
ψ	Electrostatic potential
ψ_i	Polynomials of the potential
ρ_{s}	Semiconductor resistivity
ρ_{c}	Metal resistivity
σ_{B}	Total interface charge at the boundary
$\sigma_{\mathrm{B,SP}}$	Interface charge due to spontaneous polarization
$\sigma_{\mathrm{B,PZ}}$	Interface charge due to piezoelectric polarization
σ_A	Electric drift-region-conductance per unit area in the semiconductor
τ_{D}	The delay due to the extension of the depletion zone to the drain
τ	Small-signal phase constant
τ_{RC}	Channel-charge RC-delay
τ_{T}	Total delay
$\tau_{\mathrm{e}}, \tau_{\mathrm{b}}, \tau_{\mathrm{c}}$	Emitter, base, collector delay
τ_{TR}	Transistor delay
τ_{thermal}	Thermal time constant
ω	Oscillation frequency
A	Area
A_{R}	Richardson constant
$A_1, \ldots, A_n$	General coefficient
A_{ν}	Coefficient for impact ionization for carrier $\nu = \mathrm{n,p}$
B	Direct radiative recombination parameter
B_0	Bulk modulus
BV_{CE0}	Open collector–emitter breakdown voltage

BV_{DS}	Drain–source breakdown voltage
$BV_{\mathrm{DS\ RF}}$	RF drain–source breakdown voltage
BV_{GD}	Gate–drain diode breakdown voltage
BV_{GS}	Gate–source diode breakdown voltage
C	Correlation parameter
C	Capacitance in the passive model
C_{ij}	Elastic constants index i, j
C_{th}	Thermal capacitance
C_{ds}	Drain–source capacitance
C_{g}	Gate capacitance
C_{gd}	Gate–drain capacitance
C_{gs}	Gate–source capacitance
C_{jc}	Collector junction capacitance
C_{je}	Emitter junction capacitance
C_{pds}, C_{pgs}, C_{pgd}	Parasitic drain–source, gate–source, and gate–drain capacitances
$C_{\mathrm{gs},0}$,$C_{\mathrm{gd},0}$	Large-signal charge modeling coefficients
C_0	Static capacitance
C_{ss}	Parasitic capacitance in the dispersion model
C_{th}	Thermal capacitance
D_{nB},D_{pE}	Diffusivity of electrons in the base, holes in the emitter
$\boldsymbol{E}$	Local electric field
E_{A}	Acceptor energy
E_{a}	Activation energy
E_{break}	Breakdown field
E_{crit}	Critical field
E_{C}	Conduction band energy
$E_{\mathrm{F}}(x)$	Fermi energy at position x
$\mathrm{E_{f1}}$	Fermi-level correction energy
E_{g}	Bandgap energy
$E_{\mathrm{g},\Gamma}$	Bandgap energy in the Γ-valley
$E_{\mathrm{g},\Gamma 1}$, $E_{\mathrm{g},\Gamma 3}$	Bandgap energy in the Γ_1, Γ_3-valley
$E_{\mathrm{g,G-A}}$	Bandgap energy in the G–A valley
$E_{\mathrm{g,L}}$	Bandgap energy in the L-valley
$E_{\mathrm{g,L-M}}$	Bandgap energy in the L–M valley
$E_{\mathrm{g,X}}$	Bandgap energy in the X-valley
$E_{\mathrm{g},0}$	Bandgap energy at $T_{\mathrm{L}}= 0\,\mathrm{K}$
E_{V}	Valence band energy
G	Conductance in the passive model
G_{ass}	Associated gain
G_{p}	Power gain
$G_{\mathrm{m},2}$, $G_{\mathrm{m,d}}$, $G_{\mathrm{m,2,d}}$	Current parameters in Volterra approach
H	Hardness

I_0	Current parameter in the diode equation
I_0	Current at time $t = 0$
I_B, I_C, I_E	Base, collector, and emitter currents
I_D	Drain current
I_{Dmax}, I_{Dmin}	Maximum, minimum drain current
$I_{Dmax,RF}$, $I_{Dmin,RF}$	Maximum/minimum RF-drain current
I_{DS}	Drain–source current
I_{Dpp}	Peak-to-peak drain current
I_{Dsat}	saturated drain current
I_{DSn}, I_{DSp}	Drain–source current in the LS-modeling
I_{Dq}	Quiescent drain current
I_{DSS}	Saturated drain current, typically at $V_{GS} = 0\,\mathrm{V}$
I_G	Gate current
II_3	Ratio of input intermodulation 3rd order
IM_3	Intermodulation distortion ratio 3rd order
IMD_3	Intermodulation distortion 3rd order
I_{ij}	Current at port with index i,j
I_{opt}	Drain current optimized for noise figure
I_{pk}, I_{pk0}	Peak current parameter in LS-model
K_{bg}	Dispersion parameter in the LS-model
K_{trg}	Soft-breakdown pinch-off parameter
K_C	Fracture toughness
L	Channel length
L	Line inductance in the passive model
L_e	Emitter length
L_D, L_G, L_S	Drain, gate, and source inductances
L_D	Length of the depletion zone
L_{sb}	Soft breakdown function
L_{sb0}	LS-parameter for the soft-breakdown
L_{sd1}	Auxiliary function
MAG	Maximum available gain
M_C	Number of equivalent minima at the conduction band
MSG	Maximum stable gain
MTBF	Median time before failure
MTTF	Median time to failure
N_A	Acceptor doping concentration
N_B	Base carrier concetrations
N_C	Effective density of states of the conduction band
N_D	Donor doping concentration
N_{DC}	Donor doping concentration in the collector

N_E	Emitter carrier concentration
N_F	Noise figure
$N_{F,min}$	Minimum noise figure at optimum impedance
N_T	Concentration of traps
N_V	Effective density of states of the valence band
OIP_3	Output intercept point third order
P_i	Function in the charge model
PAE	Power-added-efficiency
P_{-1dB}	Output power at 1 dB compression
P_{DC}	DC-power
P_{diss}	Dissipated power
P_{in}	Input power
P_{ij}	Large-signal parameter for the pinch-off voltage
P_{out}	Output power
P_{SP}	Spontaneous polarization
P_{PZ}	Piezoelectric polarization
P_{sat}	Saturated output power
R	Resistance
R_{Con}	Contact resistance
R_D	Drain resistance
$R_{D,semi}$	Semiconductor contribution to R_D
R_L	Load resistance/impedance
$R_{S,met}, R_{D,met}$	Metal contribution to R_S and R_D
R_S, R_G, R_D	Parasitic source, gate, and drain resistances
R_S	Series resistance of diode
R_{band}	Band edge contribution to the source resistance R_S
R_{bb}	Base resistance
R_{chan}	Channel resistance
R_{ds}	Drain–source resistance
R_{gap}	Contribution of the contact gap to R_{bb}
R_{gd}, R_{gs}	Gate–drain/–source resistance
R_i	Input resistance
R_n	Equivalent noise resistance
R_{on}	On-resistance
R_{opt}	Optimum impedance
R_{spread}	Spread contribution to R_{bb}
R_{ss}	Parasitic output resistance in the dispersion model
R_{th}	Thermal resistance
S_{ij}	Scattering (S-) parameter, $i, j = 1, 2$

$S_\nu(f)$	Spectral noise density
T_{chan}	Channel temperature
T_{Debye}	Debye-temperature
$T_{\text{Drain}}, T_{\text{Gate}}$	Drain and gate noise-temperature
$T(E_x)$	Tunneling probability
T_{L}	Local lattice temperature
T_{sub}	Substrate temperature
T_0	Backside temperature in the LS-model
U	Unilateral gain
V_1	Variable in the Curtice model
$V_{\text{BE}}, V_{\text{CE}}, V_{\text{BC}}$	Base–emitter, collector–emitter, and collector–base voltages
$V_{\text{breakdown}}$	Breakdown voltage
$V_{\text{DS}}, V_{\text{GD}}, V_{\text{GS}}$	Drain–source, gate–drain, and gate–source voltages
$V_{\text{DS0}}, V_{\text{GS0}}$	Quiescent drain–source/gate–source voltage
$V_{\text{D,max,RF}}, V_{\text{D,min,RF}}$	Maximum/minimum RF-V_{DS}-voltage
$V_{\text{GS},X}$	Dispersion corrected V_{GS}-voltages
$V_{\text{GS},3}^{psat}, V_{\text{GS},3}^{plin}$	Linearity parameters
V_{bgate}	Breakdown parameter in the LS-model
V_{dgt}	Gate–drain voltage function
V_{kl}	Voltage at device port k,l
V_{knee}	Knee voltage
V_{p}	Pinch-off voltage
V_{p2}	Doping correction to the threshold voltage
V_{thr}	Threshold voltage
W	Wafer bow
$W_{\text{e}}, W_{\text{eb}}$	Emitter-(base) width
W_{g}	Gate width
$X_{\text{EB}}, X_{\text{E}}$	Thickness of the emitter, base
Y_{ij}	Y-parameter for $i, j = 1, 2$
Z_{ds}	Complex output conductance
Z_{ij}	Z-parameter for $i, j = 1, 2$
Z_0	Characteristic impedance
a, b, c, d, n	General parameters
$a_{\text{gate}-\text{lag}}, a_{\text{drain}-\text{lag}}$	Dispersion parameters for gate and drain-lag
$a_{\text{gate,cw}}, a_{\text{drain,cw}}$	Dispersion parameters comparing pulsed and cw
a_0, c_0	Lattice parameters
c_{300}	Heat capacity at $300\,\text{K}$
$c_{\text{L,AB}}$	Nonlinear coefficient of the thermal conductivity of the ternary semiconductor

d	Thickness, length
d_{AlGaN}	Thickness of AlGaN
d_{eff}	Effective gate-to-channel separation
d_{doping}	Channel layer thickness
d_{sub}	Substrate diameter
e_{ij}	Piezoelectric coefficient
f_{T}	Current gain cut-off frequency
$f_{\text{T,ext}}$	Extrinsic current gain cut-off frequency
f	Frequency
f_{c}	Frequency for $k = 1$
f_{c}	Lattice mismatch
f_{max}	Maximum frequency of oscillation
$f_{10\,dB}$	Maximum frequency which leaves 10 dB of power gain
$f(t, \mu, \sigma)_{\text{lognorm}}$	Log-normal distribution function
$f(t, \mu, \sigma)_{\text{norm}}$	Normal distribution function
f_1, f_2	Distribution function
g_{ds}	Output conductance
g'_{ds}m g''_{ds}	Derivatives of output conductance
$g_{\text{ds,ext}}$	Extrinsic output conductance
$g_{\text{ds,ext}}(\text{CW}(\text{RF or Pulsed}))$	Extrinsic output conductance for CW, RF, or pulsed operation
g_{m}	(Complex) transconductance
$g'_{\text{m}}, g''_{\text{m}}$	Derivatives of transconductance
g_{mi}	Intrinsic transconductance
$g_{\text{m,max}}$	Maximum transconductance
$h, \hbar$	(Reduced) Planck constant
h_{21}	Current–gain
$h(t)$	Thermal response function in the time domain
i_{ds}	Intrinsic drain source current
i_{II}	Impact ionization current in the LS-model
$i_{\text{d}}, i_{\text{g}}$	Noise current at drain and gate
k_{B}	Boltzmann constant
k_{f}	Fukui factor
$k_n, k_{rel,n}$	Drain-lag dispersion model parameters
k	Stability factor
$l_{\text{fps}}, l_{\text{fpd}}$	Field-plate extension to the source and drain side
$l_{\text{gd}}, l_{\text{gs}}$	Gate-to-drain/-to-source separation
l_{gg}	Gate-to-gate pitch
l_{g}	Gate length
mb	Doping coefficient
m_e	Free electron mass

m_n	Effective electron mass
$m_n(\Gamma\text{-K})$, $m_n(\Gamma\text{-A})$,	Effective electron mass at the
$m_n(\Gamma\text{-M})$	Γ-K/A/M transition
$m_n(\mathrm{X})$	Effective electron mass in X-valley
$m_{\nu,\mathrm{AB}}$	Effective carrier mass of the
	semi-conductor $\mathrm{A}_x\mathrm{B}_{1-x}\mathrm{N}$
$m_{n,\mathrm{l}}$	Longitudinal electron mass
$m_{n,\mathrm{t}}$	Transversal electron mass
m_p	Effective hole mass
$m_{p,\mathrm{h}}$	Effective heavy hole mass
$m_{p,\mathrm{l}}$	Light hole mass
$m_{p,\mathrm{so}}$	Spin–orbit hole mass
m^*	Tunneling mass
$\boldsymbol{n}$	A normal vector
n	Electron concentration
n	Ideality factor
n	Refractive index
n_{channel}	Channel charge density
n_{i}	Intrinsic carrier concentration
n_{sheet}	Sheet carrier concentration
p	Hole concentration
q	Elementary charge
$q_{\mathrm{bulk,traps}}$	Charge of bulk traps
q_{channel}	Channel charge
$q_{\mathrm{diel,interface}}$	Interface charge
q_{doping}	Doping charge
q_{initial}	Dynamic charge
$q_{\mathrm{semi,interfaces}}$	Charge at the semiconductor interfaces
t	Time
t	Thickness of the current-supporting layer
t_{ad}	Thickness of the adhesive
t_{sem}	Semiconductor thickness
t_{sub}	Substrate thickness
t_{subm}	Submount thickness
$\bar{v}_{\mathrm{ds}}, \bar{v}_{\mathrm{gs}}$	Static voltage components in
	dispersion model
$v_{\mathrm{ds}}, v_{\mathrm{d}}$	Intrinsic drain (source) voltage
v_{eff}	Effective carrier velocity
$v_{\mathrm{gs}}, v_{\mathrm{g}}$	Intrinsic gate (source) voltage
v_{peak}	Peak carrier velocity
v_{sat}	Saturated carrier velocity
x, y	Material composition parameter
y	Distance

List of Acronyms

2DEG	Two-Dimensional Electron Gas
3G	3rd generation (of mobile communication)
4G	4th generation (of mobile communication)
4H, 6H	Polytypes of SiC
AC	Alternating Current
ACLR	Adjacent Channel Leakage Ratio
ACPR	Adjacent Channel Power Ratio
ADS	Advanced Design System
AFM	Atomic Force Microscopy
AlN	Aluminum nitride
AESA	Active Electronically Scanned Array
ASIC	Application Specific Integrated Circuit
BCB	Bencocyclobutene
BEEM	Ballistic Electron Emission Spectroscopy
BGA	Ball-Grid-Array
BJT	Bipolar Junction Transistor
BN	Boron Nitride
BS	Backside Cooling
BV	Breakdown Voltage
CAD	Computer Aided Design
CAFM	Conductive Atomic Force Microscopy
CAIBE	Chemically-Assisted Ion Beam Etching
CAT-CVD	Catalytic Chemical Vapor Deposition
CAVET	Current Aperture Vertical Electron Transistor
CDMA	Code Division Multiple Access
CMOS	Complementary Metal Oxide Semiconductor
CMP	Chemical Mechanical Polishing
CTE	Coefficient of Thermal Expansion
CVD	Chemical Vapor Deposition
CW	Continuous Wave

c	cubic
DBF	Digital Beam-Forming
DC	Direct Current
DD	Drift-Diffusion
DLT(F)S	Deep Level Transient (Fourier) Spectroscopy
DHBT	Double Heterojunction Bipolar Transistor
DHFET	Double Heterojunction Field Effect Transistor
DHHEMT	Double Heterojunction High Electron Mobility Transistor
DOD	U.S. Department of Defense
DPD	Digital Predistortion
DRA	Driver Amplifier
EBIC	Electron-Beam Induced Current
ECR	Electron Cyclotron Resonance
ECMP	Electrochemical-Mechanical Polishing
EDGE	Enhanced Data Rates for GSM Evolution
EDX	Electron Diffraction
EER	Envelope Elimination and Restoration
ELO, ELOG, LEO	Epitaxial Lateral Overgrowth
ESD	Electrostatic Discharge
ET	Envelope-Tracking
EUV	Extreme Ultraviolet
EVM	Error Vector Magnitude
FC	Flip-Chip
FET	Field Effect Transistor
FIB	Focussed Ion Beam
FIT	Failures in Time
FOM	Figure-Of-Merit
GaAs	Gallium Arsenide
GaN	Gallium Nitride
GCPW	Grounded Coplanar-Waveguide
GSM	Global System for Mobile Communications
HBT	Heterojunction Bipolar Transistor
HCI	Hot Carrier Injection
HEMT	High Electron Mobility Transistor
HFET	Heterostructure Field Effect Transistor
HPA	High-Power Amplifier
HPSI	High-Purity Semi-Insulating (substrates)
HR	High-Resistivity
HTCC	High-Temperature Cofired Ceramics
HTCVD	High-Temperature Chemical Vapor Deposition
HVPE	Hydride Vapor Phase Epitaxy
IC	Integrated Circuit
ICP	Inductively Coupled Plasma
IDLDMOS	Interdigitated-Drain LDMOS

IF	Intermediate Frequency
II	Impact Ionization
IMPATT	IMPact Avalanche Transit Time
InAlN	Indium Aluminum Nitride
InAs	Indium Arsenide
InGaN	Indium Gallium Nitride
InN	Indium Nitride
JFET	Junction Field-Effect Transistor
LDMOS	Laterally diffused MOS
LED	Light-emitting Diode
LEEN	Low-Energy Electron-Excited Nanoscale Luminescence Spectroscopy
LNA	Low-Noise Amplifier
LO	Local Oscillator
LPCVD	Low-Pressure Chemical Vapor Deposition
LT	Low-Temperature
LTCC	Low-Temperature Cofired Ceramics
MAG	Maximum Available Gain
MBE	Molecular Beam Epitaxy
MC	Monte Carlo
MCM	Multi-Chip Module
MEMS	Micro-Electro-Mechanical Systems
MERFS	Micro Electromagnetic Radio Frequency System
MESFET	MEtal Semiconductor Field Effect Transistor
MHEMT	Metamorphic HEMT
MIC	Microwave Integrated Circuit
MIM	Metal–Insulator–Metal
MISFET	Metal Insulator Field Effect Transistor
MISHEMT	Metal Insulator Semiconductor HEMT
MMIC	Monolithic Microwave Integrated Circuit
MOCVD	Metal Organic Chemical Vapor Deposition
MOD	Ministry of Defense
MODFET	Modulation-Doped FET
MOMBE	Metal Organic Molecular Beam Epitaxy
MOS	Metal–Oxide–Semiconductor
MOSFET	Metal–Oxide–Semiconductor FET
MOSDHFET	Metal–Oxide Double Heterostructure FET
MOVPE	Metal Organic Vapor Phase Epitaxy
MSG	Maximum Stable Gain
MSL	Microstrip Line
MTBF	Mean Time Before Failure
MTTF	Mean Time to Failure
NBTI	Negative Bias Temperature Instability
NID	Non-Intentionally Doped
NiCr	Nickel Chromium

OFDM	Orthogonal Frequency-Division Multiplexing
ONO	Oxide–Nitride–Oxide
PA	Power Amplifier
PAE	Power-Added Efficiency
PAMBE	Plasma-Assisted Molecular Beam Epitaxy
PAR	Peak-to-Average Ratio
PAS	Positron Annihilation Spectroscopy
PCB	Printed Circuit Board
PCDE	Peak Code Domain Error
PEC	Photo-enhanced Electrochemical Etching
PECVD	Plasma-enhanced Chemical Vapor Deposition
PHEMT	Pseudomorphic HEMT
PIC	Polarization-Induced Charge
PL	Photoluminesce
PLM	Polarized Light Microscopy
PVT	Physical Vapor Deposition
RF	Radio Frequency
RIE	Reactive Ion Etching
RMS	Root-Mean Square
RTA	Rapid Thermal Annealing
Rx	Receiver
SAR	Synthetic Aperture Radar
SAW	Surface Acoustic Wave
SEM	Scanning Electron Microscopy
SH	Self-Heating
SHBT	Single Heterojunction Bipolar Transistor
SHFET	Single Heterojunction Field Effect Transistor
SHHEMT	Single Heterojunction High Electron Mobility Transistor
s.i., SI	Semi-Insulating
SiC	Silicon Carbide
SiCOI	Silicon Carbide on Insulator
SiCoSiC	Silicon Carbide on polySiC
SIMS	Secondary Ion Mass Spectroscopy
SiN	Silicon Nitride
SIP	System-In-Package
SIT	Static Induction Transistor
SKPM	Scanning Kelvin Probe Microscopy
SMD	Surface Mount Device
SoC	System on a Chip
SOI	Silicon On Insulator
SopSiC	Silicon on poly-crystalline SiC
SPDT	Single Pole Double Throw
SRPES	Synchrotron Radiation Photoemission Spectroscopy

SRH	Shockley–Read–Hall
SSPA	Solid-State Power Amplifier
STEM	Scanning Tunneling Electron Microscopy
TaN	Tantalum Nitride
TDD	Threading-Dislocation Density
TE	Thermionic Emission
TEM	Transmission Electron Microscopy
TLM	Transmission-Line Model
TMGa, TMAl, TMIn	TrimethylGallium, TriMethylAluminum, TriMethylIndium
T/R,TRX	Transmit/Receive
TWA	Traveling Wave Amplifier
TWTA	Traveling Wave Tube Amplifier
UWB	Ultra-Wide Band
VCO	Voltage-Controlled Oscillator
WBG	Wide BandGap
WBS	Wide Bandgap Semiconductor
WCDMA	Wideband Code Division Multiple Access
WiMAX	Worldwide interoperability For Microwave Access
Wz	Wurtzite
XRD	X-Ray Diffraction
Zb	Zincblende

1

Introduction

This monograph is devoted to the development of III-N semiconductor-based electronics for high-power and high-speed RF-applications. Material properties of these polar materials, the state-of-the-art of substrates, epitaxial growth, device and processing technology, modeling, and circuit integration, and examples are discussed. A full chapter is devoted to the critical aspect of device reliability. The work concludes with integration and packaging aspects specific to the new properties of III-N-based-circuits and subsystems.

In the second chapter, general material and transport properties, advantages, and theoretical electrical and thermal limits are presented. Further, the state-of-the-art for nitride-based substrates, materials, electronic devices, and circuits are reviewed systematically.

For epitaxial growth, both the aluminum and indium-based binary and ternary compounds are described with emphasis on AlGaN/GaN and In–Ga–N-based heterostructure systems in Chapter 3. Epitaxial growth techniques such as molecular beam epitaxy (MBE) and metal organic chemical vapor deposition (MOCVD) are analyzed systematically. Nitride-specific material characterization, doping, and material quality issues are analyzed. Substrate properties are reviewed systematically with respect to electronic requirements.

Currently a major focus of development is devoted to high electron mobility transistors (HEMTs) with gate lengths down to 30 nm and cut-off frequencies up to 190 GHz. Thus, for this class of devices, specific field-effect transistor problems such as Schottky and ohmic contacts, and lithography of optically transparent materials are discussed. State-of-the art recess processes and passivation technologies are analyzed. Bipolar device technology issues are reviewed.

In the device modeling and characterization Chapter 5, DC, small-signal, and noise characterization and modeling are presented with respect to nitride-specific questions. As frequency dispersion is a major source of performance and device degradation, the characterization and reduction of dispersion involving pulsed-characterization and other advanced techniques are discussed. Large-signal characterization and modeling are discussed for nitride devices, including the modeling of contacts, diodes, dispersion, and thermal aspects.

Chapter 6 discusses circuit examples for high-power RF amplifiers with a focus on increased impedance, thermal management, and high RF-power management between 0.5 and 100 GHz. Low-noise amplifiers are presented and analyzed for high-dynamic range, robustness, and high linearity. The last section of the chapter treats other circuits functions such as mixers and oscillators. Again, nitride-specific advantages and challenges are investigated.

In Chapter 7, nitride-specific device and circuit reliability issues and device failure mechanisms are analyzed and described systematically. The last chapter describes integration and packaging considerations, thermal-mounting and thermal-packaging considerations, for state-of-the-art amplifiers, and subsystems.

III-N Materials, and the State-of-the-Art of Devices and Circuits

In this chapter, general material and transport properties, advantages, and theoretical electrical and thermal limits of nitride semiconductors are presented with respect to electronic applications. This chapter further provides an overview of substrate materials. The state-of-the-art for nitride-based materials, electronic devices, and circuits are reviewed systematically. The last section defines application-specific requirements on nitride semiconductor devices from a more system oriented point of view. The chapter concludes with a number of problems.

2.1 State-of-the-Art of Materials Research

A systematic overview of the material properties is given in the following. For a systematic introduction, Fig. 2.1 depicts the bandgap at a lattice temperature of 300 K for various semiconductor materials as a function of lattice constant. The selection of the specific material constant as lattice constant in noncentrosymmetric materials may be arbitrary, however, is useful for a systematic introduction. With the bandgap of InN recently suggested to be about 0.8 eV and the bandgap of AlN to be 6.2 eV at room temperature, the III-N material system covers a very broad range of energies and thus emission wave lengths from the infrared to the deep ultraviolet unchallenged by any other available material. Electronically, a very broad range of bandgap energies is found resulting in extremely high bulk material breakdown voltages, which can be traded in for low-effective mass and high mobility of GaN with $m_n = 0.2\,m_e$, or even of InN with $m_n = 0.11\,m_e$ (potentially even $m_n = 0.04\,m_e$ [2.122]) by changing the material compositions in the heterostructures. At the same time, a second trade-off is imminent and different from any other semiconductor material system. Due to the strong polar material properties, the modification of the material composition results in dramatic modifications of the polar crystal properties, and thus of available

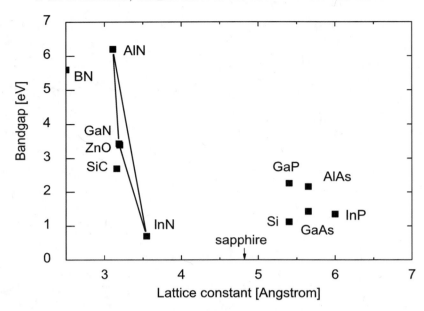

Fig. 2.1. Bandgap energy at $T_L = 300\,\mathrm{K}$ as a function of lattice constant of III-N semiconductors: Other III-V semiconductors are given for comparison

carrier concentrations obtained at the interfaces and thus available in the devices. Thus, and in some respect very similar to silicon, the huge success of III-N material is not mainly due to the intrinsic bulk material transport properties (especially the p-material properties do not compare well with n-properties in III-N materials), but due to the interface properties. In the case of silicon, the success is based on the formation of a native oxide which can be optimized and used tremendously, e.g., [2.411], and is recently replaced by other dielectrics, which also are well behaved on silicon, e.g., [2.136, 2.145].

In the case of III-N heterostructures, the interfaces allow for the formation of n-channels and intrinsically provide extremely high carrier concentrations $\geq 10^{13}\,\mathrm{cm}^{-2}$ through polarization engineering without further impurity doping. On the contrary, and unlike silicon, the semiconductor–dielectric interfaces remain a challenge in the III-N world, as suitable interfaces can be formed from a practical point of view, however, these interfaces and their long term behavior are not well understood mainly due to the high-sheet charge concentrations.

2.1.1 Binary Materials

The binaries of the elements Al–In–Ga–B–N are the basic materials for the semiconductor material class named Nitrides: gallium nitride (GaN), aluminum nitride (AlN), indium nitride (InN), and boron nitride (BN). Epitaxially grown silicon nitride (SiN), although not of crystalline quality, may be added from time to time. Boron nitride (BN) is still relatively immature as a semiconductor material [2.99]. Initial results on the material and device level are available [2.99], featuring the especially good thermal conductivity of BN, second only to diamond when semiconductor materials are compared. Good overview papers and collections are available, with a focus also on optoelectronic and general material properties, and, e.g., in [2.100, 2.101, 2.244, 2.315, 2.324, 2.369, 2.389]. Laser diodes are particularly addressed in [2.323]. The properties of SiN and SiO_2 are reviewed, e.g., in [2.440].

Crystal Structures for Electronic Applications

The III-N semiconductors can be found in three common crystal structures:

- Wurtzite
- Zincblende
- Rock salt

At room temperature GaN, AlN, and InN are found in the wurtzite structure, while BN prevails mostly in the cubic structure. The zincblende structure can also be found for GaN and InN for thin films, while for AlN no stable zincblende phase has yet been detected [2.99]. However, this work addresses also some mostly theoretical work on Zb–AlN. The rock salt phase is of no importance to electronic devices so far. In the wurtzite structure the growth is typically performed along the *c*-axis. Recently, growth along the m-plane has been reported [2.61, 2.342], as the resulting nonpolar material has a positive influence on diodes efficiency in optoelectronics.

Gallium Nitride (GaN)

Gallium Nitride (GaN) is the basic material of this material class which is typically used for all device layers requiring fast carrier transport with a high breakdown voltage. GaN is used as the channel material in various FETs and also as the base material in AlGaN/GaN HBTs, e.g., [2.311]. Most of the ohmic contact layers in any device incorporate binary n-doped and p-doped GaN. GaN can further be grown as a semi-insulating material with growth parameters close to those of the semiconducting layers.

Mechanical and Optical Properties

The crystal structure and the mechanical and thermal properties of GaN are discussed in a number of publications, e.g., in [2.4, 2.244, 2.315, 2.389]. The data of all III-N binaries are also compiled in table form in the Appendix. As an initial quantity, Table 2.1 compiles the data on the mass density. The Vickers hardness and fracture toughness of bulk GaN in comparison to other semiconductors are given in [2.92]. More recent results are presented in [2.524]. Table 2.2 compares the values of the hardness H and fracture toughness K_C.

Table 2.1. Mass density of III-N and other semiconductor materials

Material	GaN	AlN	InN	BN	Si	Ref.
Mass density (g cm^{-3})	6.1	3.23	6.81	3.48	2.33	[2.17, 2.99, 2.443]

Wz wurtzite, Zb zincblende, c cubic, d diamond

Table 2.2. Vickers hardness H and fracture toughness K_C of III-N and other semiconductor materials

Material	GaN	AlN	InN	BN	SiC	Si	Ref.
H (GPa)	12	14	11.2	55–65	33	9	[2.92, 2.451]
K_C (MPa m$^{1/2}$)	0.8	2.6	–	–	3.3	0.7	[2.92, 2.244]

Table 2.3. Elastic constants of wurtzite III-N semiconductors and other materials

	GaN (Wz)	AlN (Wz)	InN (Wz)	BN (c)	Ref.
C_{11} (GPa)	390	396	223	831	[2.99, 2.275, 2.372, 2.449, 2.485]
C_{33} (GPa)	398	373	200	–	[2.99, 2.275, 2.372, 2.449, 2.485]
C_{44} (GPa)	105	116	48	450	[2.372, 2.451, 2.485]
C_{12} (GPa)	145	137	115	190	[2.99, 2.275, 2.372, 2.449, 2.485]
C_{13} (GPa)	106	108	92	–	[2.275, 2.372, 2.449, 2.485]

Wz wurtzite, c cubic

Table 2.4. Elastic constants of zincblende III-N semiconductors

		GaN (Zb)	AlN (Zb)	InN (Zb)	BN (c)	Si	Ref.
C_{11}	(GPa)	293	304	187	820	165.8	[2.244, 2.371, 2.485]
C_{44}	(GPa)	155	193	86	480	79.6	[2.244, 2.371, 2.485]
C_{12}	(GPa)	159	160	125	190	63.9	[2.244, 2.371, 2.485]

Wz wurtzite, Zb zincblende, c cubic

Table 2.5. Coefficients of thermal expansion (CTE) of III-N semiconductors and other materials at room temperature

	GaN (Wz)	GaN (Sa)	AlN (p)	InN (p)	BN (c)	SiC	Sap.	Si	Ref.
α_a ($10^{-6} K^{-1}$)	3.1	3.8	2.9	3.6	1.15	3.2	4.3	2.6	[2.99, 2.222, 2.243]
α_c ($10^{-6} K^{-1}$)	2.8	2.9	3.4	2.6	–	3.2	3.2	2.6	[2.222, 2.243]

Wz wurtzite, Sa epitaxially on sapphire substrate, c cubic, p powder

c-BN is a particularly hard material, while GaN, AlN, and InN have nearly the same hardness. This hardness makes c-BN attractive in various ceramic materials. The elastic constants of wurtzite gallium nitride determined by Brillouin scattering are reported, for example, in [2.372] and are compiled in Table 2.3. A prediction of the high-temperature elastic constants of GaN, AlN, and InN is given in [2.381]. Deduced from this, a bulk modulus B_0 of GaN is found to be 210 GPa. For BN, a comparison of the different calculated and measured values is given in [2.99]. First principle calculations of the properties of zincblende AlN and GaN are given in [2.177]. The elastic constants of the zincblende phase are compiled in Table 2.4. The pressure dependence of the elastic constants of zincblende BN, AlN, GaN, and InN is analyzed in [2.178].

The linear coefficients of thermal expansion (CTE) at room temperature and at elevated temperatures are important for the growth. Those of wurtzite GaN have been measured between room temperature and $\geq$750 K for both bulk GaN and epitaxial layers grown on sapphire, e.g., in [2.243]. The data are compiled in Table 2.5. Table 2.5 further compares the CTE of III-N semiconductors to silicon carbide and other typical substrates materials. The differences in CTE and its temperature dependence have a similar impact to layer growth of heterostructures as the lattice constants, as discussed in Chapter 3. The thermal conductivity of GaN was initially reported in [2.423] and many of the references date back to this publication. More recent measurements, especially as function of dislocation density in thin layers, can be found in [2.120, 2.216, 2.264] and are discussed later with respect to the modeling and the importance of the crystalline quality. The data are compiled in Table 2.16.

Dielectric Constants

Compiling the basic dielectric properties, Table 2.6 gives both the static and the high-frequency dielectric constants. The dielectric constant of GaN is slightly lower than in silicon and GaAs (not shown). InN has the highest values of the three binary materials.

Overviews of further optical functions of GaN such as the refractive index are given in [2.3, 2.298]. The absorption spectrum of GaN at room temperature and the absorption coefficient are presented in [2.64].

Table 2.6. Dielectric constants of III-N and other semiconductor materials

Material	GaN (Wz)	AlN (Wz)	InN (Wz)	GaN (Zb)	BN (c)	Si	Ref.
ε_r	9.5 $\perp$, 10.4 $\parallel$	8.5	15.3	9.5	7.1	11.9	[2.3, 2.64, 2.492]
ε_r^{inf}	5.5	4.77	8.4	5.35	4.5	–	[2.3, 2.64, 2.492]

Wz wurtzite, Zb zincblende, and c cubic

Basic Transport Properties

Electronic transport in GaN is mostly understood, but a number of issues remain for further research. These include, e.g., the maximum carrier velocity v_{peak} in bulk material and at heterointerfaces, which are discussed later. General electrical properties of GaN, as well as AlN and $Al_x Ga_{1-x}N$, are compiled in [2.123]. An early, but accurate estimate of the carrier concentration dependence of the bulk mobility in GaN, InN, and AlN using the variational method is given in [2.64]. Based on more recent data, a mobility and carrier vs. doping concentration analysis for wurtzite MOCVD-grown bulk GaN is provided in [2.259]. Both donor and acceptor concentrations in the order of $10^{17} cm^{-3}$ are extracted. A specific mobility model for bulk GaN including the dependence on the free electron concentration is given in [2.308]; however, newer experimental results require an update of the actual parameters.

Table 2.7. Comparison of the low-field mobility values in various III-N bulk and 2DEG materials

Material	n/p	T_L (K)	N_D / N_A (cm^{-3})	μ ($cm^2 V^{-1} s^{-1}$)	Method	Ref.
GaN (Wz)	n	300	$N_D = 1 \times 10^{17}$	990	MC	[2.113]
GaN (Wz)	n	450	$N_D = 1 \times 10^{17}$	391	MC	[2.113]
GaN (Wz)	n	600	$N_D = 1 \times 10^{17}$	215	MC	[2.113]
GaN (Wz)	p	300	$N_A = 3.6 \times 10^{16}$	150	meas.	[2.123]
GaN (Wz)	n 2DEG	300	0	2,000	MC	[2.337]
GaN (Wz)	n	77	$N_D = 1 \times 10^{16}$	6,000	VP	[2.64]
GaN (Zb)	n	300	$N_D = 1 \times 10^{17}$	1,100	MC	[2.17]
GaN (Zb)	n 2DEG	300	0	2,100	MC	[2.337]
GaN (Zb)	p	300	$N_A = 1 \times 10^{13}$	350	meas.	[2.21]
AlN (Wz)	n	300	$N_D = 1 \times 10^{17}$	135	MC	[2.349]
AlN (Wz)	n	77	$N_D = 1 \times 10^{16}$	2,000	VP	[2.64]
AlN (Zb)	n	300	0	200	MC	[2.17]
AlN (Zb)	n	100	0	400	MC	[2.17]
InN (Wz)	n	300	$N_D = 1 \times 10^{17}$	3,000	MC	[2.35]
InN (Wz)	n	300	$N_D = 1.5 \times 10^{17}$	3,570	meas.	[2.114]
InN (Wz)	n	77	$N_D = 1 \times 10^{16}$	30,000	VP	[2.64]
InN (Zb)	n	100	$N_D = 1 \times 10^{17}$	9,000	MC	[2.17]
BN (c)	p	300	$N_A = 5 \times 10^{18}$	500	meas.	[2.253]

Wz wurtzite, *Zb* zincblende, *c* cubic, *VP* variational principle, *MC* Monte Carlo, *meas.* measured

Low-Field Mobility

Several predictions and measurements are available for obtaining estimates for the maximum low-field mobility in both bulk and 2DEG electron gas. Table 2.7 compiles the data of the low-field mobility with respect to temperature and impurity dependence of the mobility in III-N bulk materials. The maximum values from [2.64] are given for completely uncompensated material.

Recent advances in material growth show great improvements in channel mobility of optimized AlGaN/GaN 2DEG channel-layer material with Hall mobility values of up to $2{,}000\,\mathrm{cm^2\,V^{-1}\,s^{-1}}$ [2.435] at room temperature in agreement with theoretical predictions [2.337]. For the analysis of the mobility in bulk and 2DEG-GaN, the temperature dependence is plotted in Fig. 2.2 taken from [2.37] and data therein. The effects to influence mobility include:

– Phonon scattering by acoustic and optical phonons
– Ionized impurity scattering [2.37] at both background impurities and surface donors
– Threading dislocations [2.258, 2.329, 2.435]
– Alloy scattering [2.34, 2.414]

The maximum drift mobility of wurtzite bulk GaN is about $1{,}100\,\mathrm{cm^2\,V^{-1}\,s^{-1}}$. The hole mobility in bulk GaN is much lower with maximum values of $175\,\mathrm{cm^2\,V^{-1}\,s^{-1}}$. The reduced impurity scattering in 2DEG channels at

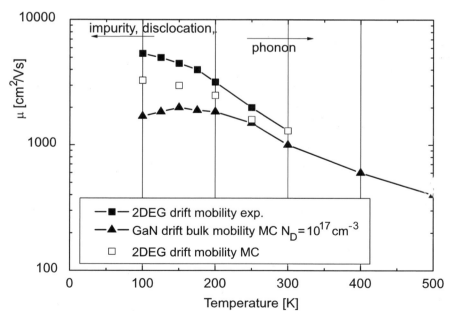

Fig. 2.2. Drift mobility as a function of lattice-temperature T_L for bulk GaN and 2DEG-GaN depicting the different limiting effects from [2.37, 2.329]

very low surface scattering levels yield 2DEG-mobility values of up to $2{,}000\,\mathrm{cm^2\,V^{-1}\,s^{-1}}$ for both wurtzite and zincblende material. A theory of charged dislocation scattering in GaN is given [2.258]. The fit of the temperature-dependent Hall data can be correlated with the dislocations observed by TEM. The impact of threading dislocations on the transverse mobility in GaN is further discussed in [2.493]. The model explains the impact of trap occupancy and related scattering on the mobility at different doping levels. Alloy disorder is a limiting factor to transport at GaN heterointerfaces, as discussed in [2.414]. Further analysis is provided in Chapter 3. AlN is an insulator due to the high activation energy of the donors and the large bandgap. The theoretical low-field mobility of AlN is discussed in [2.17]. It is found to be phonon-limited with the electron mobility values given in Table 2.7. Theoretical values for the drift mobility of $135\,\mathrm{cm^2\,V^{-1}\,s^{-1}}$ and $200\,\mathrm{cm^2\,V^{-1}\,s^{-1}}$ for fully uncompensated material have been found. Similar to InN, the hole transport properties of AlN are not understood very well. The predicted low-field drift mobility of InN at room temperature is found to be as high as $3{,}000\,\mathrm{cm^2\,V^{-1}\,s^{-1}}$ in [2.35] and $3{,}700\,\mathrm{cm^2\,V^{-1}\,s^{-1}}$ in [2.17]. At $100\,\mathrm{K}$, a maximum low-field mobility as high as $9{,}000\,\mathrm{cm^2\,V^{-1}\,s^{-1}}$ is predicted for zincblende InN material [2.17]. Variational principle calculations leads to predictions of $30{,}000\,\mathrm{cm^2\,V^{-1}\,s^{-1}}$ at $77\,\mathrm{K}$ for completely uncompensated material at low impurity concentrations of $10^{16}\,\mathrm{cm^{-3}}$. Again, hole transport in InN is not well understood due to the lack of proper p-doping [2.170].

High-Field Transport

Several MC calculations of the velocity-field characteristics of wurtzite GaN considering the high-field transport are available and given, e.g., in [2.37, 2.121]. There is a significant discussion on the extraction of delay times in HFETs and resulting carrier velocities, which disagree with the MC calculations, see also Sect. 2.2.7. Table 2.8 compiles the bulk saturation velocity and critical field parameters from various sources. The comparison reveals the increase of both the saturation velocity v_{sat} and the critical field E_{crit} compared to other high-speed materials such as silicon, GaAs, and SiC. Peak velocity values as high as $3 \times 10^7\,\mathrm{cm\,s^{-1}}$ are found for electrons in GaN. The differences between wurtzite and zincblende semiconductors are found to be insignificant. AlN has a very high critical field in agreement with the high bandgap energy, whereas the critical field of InN is lower than in GaN. The predicted electron peak velocity in InN can be as high as $4 \times 10^7\,\mathrm{cm\,s^{-1}}$.

Apart from bulk material, MC simulations in 2DEG AlGaN/GaN channels including hot-phonon and degeneracy effects are provided in [2.380] by a combined MC and Schrödinger–Poisson analysis. The wavefunctions for the confined electrons are calculated. The simulation suggests that the degeneracy in the 2DEG reduces the electron drift velocity, while hot phonons reduce the drift velocity and increase the electron energy relaxation time. The energy relaxation time at RT extract by MC simulation amounts to

Table 2.8. Comparison of the velocity-field characteristics in various bulk materials

Material	n/p	T_L (K)	E_{crit} (kV cm^{-1})	v_{peak} (10^7 cm s^{-1})	v_{sat} (10^7 cm s^{-1})	N_D/N_A (cm^{-3})	Ref.
GaN (Wz)	n	300	175	2.5	–	2×10^{16}	[2.500]
GaN (Wz)	n	300	150	2.7	2.5	1×10^{16}	[2.37]
GaN (Wz)	n	300	140	2.9	1.8	1×10^{17}	[2.121]
GaN (Wz)	n	77	150	3.2	2.7	1×10^{16}	[2.37]
GaN (Wz)	p	300	–	–	0.48	–	[2.286]
GaN (Zb)	n	300	145	2.6	1.34	1×10^{17}	[2.17]
GaN (Zb)	p	300	–	–	0.92	–	[2.286]
AlN (Wz)	n	300	450	1.7	1.4	1×10^{17}	[2.121, 2.349]
AlN (Wz)	n	300	447	2.3	2.16	1×10^{17}	[2.113]
AlN (Wz)	p	300	–	–	–	–	–
AlN (Zb)	n	300	550	1.8	1.59	1×10^{17}	[2.17]
AlN (Zb)	p	300	–	–	–	–	–
InN (Wz)	n	300	65	4.2	1.8	1×10^{17}	[2.121]
InN (Wz)	n	300	52	3.4	–	1×10^{17}	[2.113]
InN (Zb)	n	300	45	2.9	1.4	1×10^{17}	[2.17]
InN (Zb)	p	300	–	–	–	–	–

Wz wurtzite, *Zb* zincblende, *c* cubic

0.3 ps at $10\,\text{kV cm}^{-1}$ [2.380]. Experimental determination of the hot electron energy relaxation time in MBE-grown n-GaN is presented in [2.521]. The extracted value amounts to 0.2 ps at at a carrier concentration of $10^{18}\,\text{cm}^{-3}$. Similar field-dependent investigations in AlGaN/GaN heterostructures based on microwave noise are given in [2.281]. RT data yields energy relaxation times of 0.3 ps at $10\,\text{kV cm}^{-1}$ and 1 ps at $2\,\text{kV cm}^{-1}$. The transient electron transport based on MC simulations in wurtzite GaN, InN, and AlN is discussed in [2.121]. In the transient situation very high theoretical overshoot values are observed. However, the paper also depicts the trade-off between transient overshoot and saturated velocity in bulk FET-like transient transport.

Band Structure of GaN

The band structure of GaN has been analyzed in a number of publications, e.g., in [2.108, 2.127, 2.128, 2.238, 2.522]. However, it is not yet fully understood, especially with respect to the higher energy bands. A comprehensive overview of the band parameters of wurtzite and zincblende GaN, AlN, and InN is given in [2.485] in a continuation of the great III-V overview in [2.486]. A consistent set of parameters is presented for both the wurtzite and zincblende binary III-N materials. From an experimental point of view, absorption coefficient, energy gap, exciton binding energy, and recombination lifetime of Wz–GaN are measured and reported in [2.317]. From these both bandgap, exciton energies, and radiative constants can be derived.

Table 2.9 compiles available mass parameters from band structure calculations and measurements. Various predictions exist for the effective electron mass in GaN, e.g., in [2.128]. The conduction band electron effective mass in wurtzite GaN is measured to be $0.2\,m_e$, as reported in [2.91]. The effective mass parameters are further collected from the band structure calculations in [2.127, 2.128, 2.238, 2.486]. The heavy hole masses $m_{p,h}$ are further distinguished along the x- and z-direction. The hole effective mass in p-GaN and the influence of band splitting and band anisotropy on free hole statistics

Table 2.9. Comparison of the effective mass parameters in various III-N semiconductor materials

	m_n (m_e)	$m_n(\Gamma - K)$ (m_e)	$m_n(\Gamma - A)$ (m_e)	$m_n(\Gamma - M)$ (m_e)	Ref.
GaN (Wz)	0.2	0.36	0.27	0.33	[2.91]
AlN (Wz)	0.48	0.42	0.33	0.40	[2.349]
InN (Wz)	0.11 (0.05)	–	–	–	[2.122, 2.486]

	$m_n(\Gamma)$ (m_e)	$m_{n,l}(X)$ (m_e)	$m_{n,t}(X)$ (m_e)	Ref.
GaN (Zb)	0.15	0.5	0.30	[2.108, 2.485]
AlN (Zb)	0.25	0.53	0.31	[2.108, 2.485]
InN (Zb)	0.07–0.11	0.48	0.27	[2.485]

	$m_{p,h}$ (m_e)	$m_{p,l}$ (m_e)	$m_{p,so}$ (m_e)	Ref.
GaN (Wz)	1.4	0.3	0.6	[2.108, 2.244]
AlN (Wz)	3.52 (z)	3.53 (z)	0.25 (z)	[2.108, 2.244]
AlN (Wz)	10.42 (x)	0.24 (x)	3.81 (x)	[2.108, 2.244]
InN (Wz)	1.63	0.27	0.65	[2.108, 2.244]

		$m_{p,h}$ (m_e)	$m_{p,l}$ (m_e)	$m_{p,so}$ (m_e)	Ref.
GaN (Zb)	[100]	0.74	0.21	0.33	[2.108]
GaN (Zb)	[111]	1.82	0.18	0.33	[2.108]
GaN (Zb)	[110]	1.51	0.19	0.33	[2.108]
AlN (Zb)	[100]	1.02	0.35	0.51	[2.108]
AlN (Zb)	[111]	2.85	0.30	0.51	[2.108]
AlN (Zb)	[110]	2.16	0.31	0.51	[2.108]

Wz wurtzite, *Zb* zincblende

in wurtzite GaN are discussed in [2.400]. The degeneracy effects are strong and require detailed investigations to obtain the effective hole mass at the particular density-of-states.

Bandgap Energies

The bandgap energy parameters of different crystal structures in III-N materials are compared in Table 2.10. Further, the intervalley separation energies in k-space are of practical importance for the high-field transport of electrons and holes. They are compiled in Table 2.11 for both wurtzite and zincblende structures. The uncertainty of the band structure and energy gap prevails specifically for AlN and InN for the higher bands.

The data of wurtzite GaN and AlN are based on the first principle calculations in [2.441] and the erratum [2.442]. At heterointerfaces between two semiconductors, the energy band discontinuities and bandgap alignments are of high importance, as they determine the energy barriers which the carriers have to surmount. All III-N materials lead to so-called type-I transitions. However, in conventional heterostructures, the alignments are symmetrical with respect to the growth order starting from the substrate. In highly polar semiconductors, such as the III-N material system, this is not necessarily the case, due to the strain effects depending on the substrate (bottom), and the resulting strain and thus polarization charge in the thin top layer. The valence band discontinuities for GaN, AlN, and InN have been complied, e.g., in [2.275]. The valence band alignment energies are found for the three binary materials, as given in Table 2.12. The additional transitions toward ternary material are also discussed later. BN is not included in this comparison, as the cubic phase BN cannot be grown lattice-matched to the other materials.

Table 2.10. Comparison of the bandgap parameters in various bulk materials at 300 K

		(Wz)	Ref.			(Zb)	Ref.
GaN $E_{g,\Gamma1}$	(eV)	3.43	[2.121]	$E_{g,\Gamma}$	(eV)	3.38, 3.1, 3.2	[2.108]
GaN $E_{g,\Gamma3}$	(eV)	5.29	[2.121]	$E_{g,X}$	(eV)	4.57, 4.7	[2.108]
GaN $E_{g,L-M}$	(eV)	5.49	[2.121]	$E_{g,L}$	(eV)	5.64, 6.2	[2.108]
AlN $E_{g,\Gamma1}$	(eV)	6.2, 6.12	[2.121, 2.245]	$E_{g,\Gamma}$	(eV)	5.94, 6.0	[2.108]
AlN $E_{g,L-M}$	(eV)	6.9	[2.121]	$E_{g,X}$	(eV)	5.1, 4.9	[2.108]
AlN $E_{g,G-A}$	(eV)	7.2	[2.121]	$E_{g,L}$	(eV)	9.42, 9.3	[2.108]
InN $E_{g,\Gamma1}$	(eV)	0.77	[2.501]	$E_{g,\Gamma}$	(eV)	0.75	[2.486]
InN $E_{g,G-A}$	(eV)	4.09	[2.121]	$E_{g,X}$	(eV)	2.486	[2.486]
InN $E_{g,\Gamma3}$	(eV)	4.49	[2.121]	$E_{g,L}$	(eV)	5.79	[2.486]

Wz wurtzite, *Zb* zincblende

Table 2.11. Comparison of the intervalley separation energies in various bulk materials at 300 K

		(Wz)	Ref.			(Zb)	Ref.
GaN –	(eV)	–	–	Γ	(eV)	–	–
GaN $\Gamma - 3$	(eV)	1.9	[2.121]	$\Gamma - X$	(eV)	1.4	[2.17, 2.244]
GaN $L - M$	(eV)	2.1	[2.121]	$\Gamma - L$	(eV)	1.6–1.9	[2.244]
AlN Γ	(eV)	–	–	–	(eV)	–	
AlN $\Gamma - (L - M)$	(eV)	0.7	[2.121, 2.441]	–	(eV)	–	
AlN $\Gamma - K$	(eV)	1.0	[2.121, 2.441]	–	(eV)	–	
InN –	(eV)	–	–		(eV)	–	
InN $\Gamma - (M - L)$	(eV)	2.9–3.9	[2.244]		(eV)	–	
InN $\Gamma - A$	(eV)	0.7–2.7	[2.244]		(eV)	–	

Wz wurtzite, *Zb* zincblende

Table 2.12. Comparison of the bandgap alignment parameters in various bulk materials at 300 K

Top	Bottom (eV)	GaN (Wz)	AlN (Wz)	InN (Wz)	Ref.
GaN	(Wz)	0	0.7	1.05	[2.275]
AlN	(Wz)	0.7	0	1.81	[2.275]
InN	(Wz)	1.05	1.81	0	[2.275]

Top	Bottom (eV)	GaN (Zb)	AlN (Zb)	InN (Zb)	Ref.
GaN	(Zb)	0	0.85	0.51	[2.324]
AlN	(Zb)	0.85	0	1.09	[2.324]
InN	(Zb)	0.51	1.09	0	[2.324]

Wz wurtzite, *Zb* zincblende

Further empirical pseudopotential calculations of wurtzite GaN and InN are given in [2.522]. For the complete band structure, a nonlocal pseudopotential calculation of the III-nitride wurtzite phase materials system is performed in [2.128] for binary compounds such as GaN, AlN, and InN. Band structure calculations of wurtzite-type GaN and AlN are given in [2.298, 2.441]. Particularly the complicated valence band structure and the effective mass parameters of the wurtzite nitrides are obtained. The cubic approximation is found to be fairly successful in the analysis for the valence-band structures of the wurtzite-type nitrides.

Table 2.13. Comparison of the direct recombination parameters in various bulk materials at 300 K

	GaN (Wz)		AlN (Wz)		InN (Wz)	
B (cm^3 s^{-1})	1.1×10^{-8}	[2.317]	0.4×10^{-10}	[2.489]	2×10^{-10}	[2.533]

Wz wurtzite

The band structure in zincblende GaN, AlN, and AlGaN is described in [2.108]. Particularly, the energies of Γ, X, and L valley of Al$_x$Ga$_{1-x}$N as a function of mole fraction x of are calculated.

Table 2.13 gives the direct recombination parameters of the III-N semiconductors for the direct band structure. As can be derived, the direct recombination in GaN is very strong and about an order of magnitude higher than, e.g., in GaAs.

Aluminum Nitride (AlN)

Second to GaN, AlN is the most important binary material in the III-N material family for electronic applications and is mostly used as its ternary compound Al$_x$Ga$_{1-x}$N, e.g., in barriers heterostructures. It is characterized to be an insulator due to the high-bandgap energy and the high-activation energy of donors. Binary AlN is usually grown nucleation layer to start the growth on s.i. SiC or sapphire substrates, e.g., [2.96] and as an interlayer at the channel/ barrier interface, e.g., [2.66].

Mechanical and Optical Properties

Basic properties such as the crystal structure, mechanical properties, and thermal properties of AlN are compiled in [2.287]. The mass density of AlN is much smaller than in GaN or InN, as given in Table 2.1. As depicted in Tables 2.2 and 2.3, thermal expansion and Vickers hardness of AlN are relatively similar to those of GaN. Elastic constants of AlN are given for the wurtzite phase in Table 2.3 and for the zincblende phase in Table 2.4.

The intrinsic thermal conductivity of AlN is determined in [2.428]. Its high-thermal conductivity is better than that of any other semiconductor apart from BN, SiC, and diamond, as is compiled in Table 2.16. This makes AlN a potentially attractive substrate material. The optical functions, such as the refractive index n, of AlN are given in [2.287]. The optical functions are also compiled in [2.260], including the dielectric functions and the absorption functions.

Basic Carrier Transport Properties

Although not of primary importance to most devices except in very thin layers close to the channel, transport in AlN is relatively well investigated by MC simulations, also with respect to the understanding of the transport in $Al_xGa_{1-x}N$ for various material compositions x. An early compilation of the transport data of $Al_xGa_{1-x}N$ and AlN is given in [2.123], yielding the low mobility of wurtzite AlN for both n- and p-type material, as given in Table 2.8. The low-field electron mobility of AlN is calculated to be $135\,cm^2\,V^{-1}\,s^{-1}$ at room temperature and at a doping concentration of $10^{17}cm^{-3}$.

High-Field Transport

The electron high-field transport in wurtzite AlN is investigated by MC simulations in [2.349]. Further MC simulations on the transient transport in AlN are given in [2.121]. The characteristics are determined by the relatively high-effective electron mass of $0.48\,m_e$, the large bandgap of $\approx6.2\,eV$, and the very small intervalley separation. The resulting velocity-field characteristic yields a very high critical field of $450\,kV\,cm^{-1}$ [2.349], as shown in Table 2.7. The saturation velocity in AlN reaches $1.4\times10^7\,cm\,s^{-1}$ at room temperature, while the peak velocity in the bulk is calculated to be $1.7\times10^7\,cm\,s^{-1}$ at a doping concentration of $1\times10^{17}\,cm^{-3}$.

Band Structure of AlN

The high-bandgap energy of AlN of $6.2\,eV$ at room temperature allows the bandgap in $Al_xGa_{1-x}N$ to be modified in a broad range from the value of GaN to that of AlN. With the bandgap of InN found to be even smaller than that of GaN, an even wider range is available for the material $In_xAl_{1-x}N$. Early calculations of the band structure of wurtzite and zincblende AlN are collected in [2.236]. The band structure of wurtzite AlN is calculated in [2.128]. Electronic band structure properties of zincblende AlN are further calculated by the empirical pseudopotential method in [2.108]. From these calculations, effective electron and hole mass parameters can be obtained, as compiled in Table 2.9. The direct recombination parameter is depicted in Table 2.13, taken from [2.489] and used for the development of light-emitting diodes with very high bandgap and resulting photon energy.

Indium Nitride (InN)

InN and its compounds $In_xGa_{1-x}N$ and $In_xAl_{1-x}N$ so far are not yet widely used in electronic devices [2.230]. When used, both InAlN/GaN [2.143] as well as AlGaN/InGaN [2.425] heterostructures have been reported. The indium contents are low, e.g., in [2.85, 2.150], to achieve the lattice match to GaN buffer layers. The MOCVD growth of InN is complicated caused by the high-growth temperature and resulting defect background concentrations and the high amount of residual nitrogen vacancies due to the higher growth temperatures required relative to MBE growth, as described in Chapter 3. The MBE growth of InN is under development and allows improved material quality and thus the use of the full range of material composition in the material $In_xGa_{1-x}N$. High-quality InN has been recently grown by MBE, e.g., [2.114,2.501]. A full review of the epitaxial growth is given in [2.38]. A bulk electron mobility of $3{,}570\,cm^2\,V^{-1}\,s^{-1}$ at $300\,K$ is obtained. The mobility at $150\,K$ is as high as $5{,}100\,cm^2\,V^{-1}\,s^{-1}$. P-type InN has been recently reported in [2.170], which is essential for the realization of bipolar or optoelectronic pn-devices.

Mechanical and Optical Properties

An initial compilation of mechanical, optical, and thermal properties of InN is given in [2.449, 2.451]. Mechanical data are compiled in Tables 2.1 and 2.2; however, the data are relatively uncertain, due to the lack of real bulk InN material [2.449]. Elastic constants of InN are given in Table 2.3. The thermal expansion coefficients CTE in both wurtzite and zincblende structures are compiled in Table 2.5. The CTE and the lattice constants suggest the growth on sapphire substrates [2.54]. The thermal conductivity and the heat capacity of InN are still primarily based on estimates and extrapolations, as explained in [2.449], see also Table 2.16. The optical functions of InN are compiled in [2.299]. Due to the new research on samples with improved material quality, the bandgap and optical functions of InN are reconsidered, e.g., in [2.278, 2.279, 2.425]. This fact also had a drastic impact on the calculation of transport properties and the higher band, which have not yet been fully repeated based on the new band structure.

Basic Carrier Transport Properties

An early compilation of the transport properties in InN is given in [2.54, 2.453]. The early data are characterized by high background concentrations of 10^{19}–$10^{20}\,cm^{-3}$ and associated low mobilities. An early evaluation of the InN mobility as a function of temperature and compensation ratio using variational principle calculations is given in [2.64]. In theory, very high mobility values of $4{,}400\,cm^2\,V^{-1}\,s^{-1}$ are calculated, which, however, have not been fully reached experimentally. An electron mobility of $1{,}200\,cm^2\,V^{-1}\,s^{-1}$ and a sheet carrier concentration of $1.2 \times 10^{14}\,cm^{-2}$ have been obtained experimentally at the InN/AlN interface, as reported in [2.307]. Polarization engineering of n-InAlN/GaN HFETs and the effect on DC- and RF-performance are described in [2.186, 2.187]. The use of $In_xAl_{1-x}N/GaN$ is critical, as the polarization-induced charge (PIC) is a much stronger function of the material composition x than in the AlGaN/GaN material system, as explained later. P-channel $In_xGa_{1-x}N$ HFETs based on polarization doping are demonstrated in [2.450]. Hall measurements indicate a 2D-hole gas (2DHG) mobility of approximately $700\,cm^2\,V^{-1}\,s^{-1}$ at a low temperature $T_L = 66\,K$.

High-Field Transport

Mostly, the simulated results on transport in InN are still based on the assumption of a direct bandgap of InN of $1.89\,eV$ and an associated band structure.

This was recently corrected to the much smaller bandgap value of $\approx 0.77\,\mathrm{eV}$ at room temperature [2.501]. Most of the initial MC simulations thus have to be taken with a grain of salt. Electron transport in wurtzite–InN is calculated by Leary in [2.348]. Further ensemble MC calculations on the wurtzite material by Bellotti et al. are given in [2.35]. The intervalley separation energies are given in Table 2.11. As depicted in Table 2.8, the velocity-field characteristics yield a maximum carrier velocity of up to $4.2 \times 10^7\,\mathrm{cm\,s^{-1}}$ at a critical field of about 52–$65\,\mathrm{kV\,cm^{-1}}$ in the homogenous bulk case [2.113, 2.121]. These properties are promising, however, compared to GaAs or InGaAs material they are not really surprising when considering the low effective mass, the low bandgap, and maximum electron velocity. Transient-transport calculations in InN given in [2.121] report velocity overshoot values above $10^8\,\mathrm{cm\,s^{-1}}$. Quite outstanding theoretical HFET cut-off frequencies are derived from these simulations as upper bounds for a theoretical pure InN-channel-FET performance.

Band Structure

The data for the band structure of InN are taken, e.g., from the band structure calculations in [2.128], which do not yet account for the recent modification of InN bandgap energy. More experimental data on the electronic structure of MBE-grown InN are provided in [2.307]. The effective electron mass of InN has been found to be $0.11\,m_e$ [2.486], as compiled in Table 2.9. This effective mass value has to be considered high relative to the bandgap of $0.77\,\mathrm{eV}$, when we compare it to other semiconductors, such as InAs or InP. Initial values of the optical bandgap of InN were reported by Tansley in [2.450]. This leads to a value of $1.89\,\mathrm{eV}$ at room temperature. More recent reports on the optical properties of InN, resolving the discrepancies of several publications at different growth and doping conditions, are given in [2.278, 2.279]. The bandgap of InN is found to be $0.77\,\mathrm{eV}$ at room temperature [2.425], as given in Table 2.10. The difference is explained by the existence of oxy-nitrides, which have a much larger bandgap. This low bandgap means a really conventional electronic III-V material, e.g., with respect to the expected breakdown of devices. Optically it means that a very broad range of wave lengths is available in the III-N systems from deep ultraviolet to red. The intervalley separation energies of both wurtzite are compiled in Table 2.11, based on the bandgap of $1.89\,\mathrm{eV}$. Currently, there is little information on the intervalley energies of Zb–InN.

Boron Nitride (BN)

Boron nitride can be found in several crystallographic forms. The most important insulating or semiconducting form of BN is cubic material, which is metastable under normal conditions [2.53, 2.99, 2.253]. Ceramic BN is widely used for industrial tools as abrasive. The great advantage of cubic-BN is its Vickers hardness [2.253] and its good thermal conductivity, as compiled in Table 2.16. Even ultraviolet pn-diodes can be formed from c-BN, as given in [2.300]. Despite the material growth problems, ultraviolet LED can be formed [2.491], which operate up to very high temperatures of 530°C [2.300].

Mechanical and Optical Properties

The mass density of cubic c-BN is given in [2.99], as compiled in Table 2.1. The outstanding Vickers hardness of BN can be observed in Table 2.2 [2.99]. Elastic constants of cubic BN are compiled in Table 2.3. The thermal expansion for cubic BN is given in Table 2.5. The values for the cubic phase are very similar to those of Zb–GaN. Thermal conductivity measurements and heat capacity measurements of cubic BN will be given in Table 2.16. The thermal conductivity of BN amounts to values of up to $750\,\mathrm{W\,m^{-1}K^{-1}}$, which is higher than in any other semiconductor. The theoretical limit found is as high as $1{,}300\,\mathrm{W\,m^{-1}K^{-1}}$.

Basic Carrier Transport Properties

Basic mobility evaluation of cubic BN as a function of carrier concentration have been provided, e.g., in [2.253]. Cubic BN is typically p-doped [2.135]. A Hall mobility of $500\,\mathrm{cm^2\,V^{-1}\,s^{-1}}$ is found at a carrier concentration of $5 \times 10^{18}\,\mathrm{cm^{-3}}$ [2.253]. For n-type material, only few experimental data exist, e.g., [2.300, 2.496]. As c-BN is an insulator [2.496], the transport is characterized by high-activation energies of the carriers of ≥ 0.2 eV.

High-Field Transport

Band structure calculations are given in [2.71] with the result that cubic BN is an insulator with a bandgap of 7.1 eV. Thus, no MC transport calculations are available for c-BN.

Dielectric Breakdown Fields

Table 2.14 compiles measured dielectric electric breakdown fields for various III-N bulk materials. The dielectric breakdown field of GaN is reported to be about $3\,\mathrm{MV\,cm^{-1}}$. AlN has a much higher breakdown field due to the higher bandgap, which can be even higher than the value given in Table 2.14 [2.388]. The dielectric breakdown field of InN is not well investigated. The latter is due to the high trap concentrations N_T and the new findings for the fundamental bandgap. The breakdown fields of bulk c-BN reported in [2.53] vary from 2 to $6\,\mathrm{MV\,cm^{-1}}$.

Table 2.14. Comparison of the bulk breakdown fields in various bulk materials at 300 K

	GaN (Wz)	Ref.	AlN (Wz)	Ref.	InN (Wz)	Ref.	BN (c)	Ref.
E_break $(\mathrm{MV\,cm^{-1}})$	3.3	[2.61]	8.4	[2.388]	1.2	[2.244]	2–6	[2.253]

Wz wurtzite, *c* cubic

Band Structure of BN

Band structure calculations of cubic BN are available [2.496]. Similar calculations, but not of cubic phase BN, are provided in [2.237]. The bandgap of cubic BN is found to be similar to AlN and reported to be $E_g \approx 6.4\,\text{eV}$ at RT in [2.253]. Further calculations of cubic BN in pseudopotential local density formalism are provided in [2.497]. The calculated charge density of the BN is very similar to other III-V semiconductors. A defect analysis of Be, Mg, and Si in cubic BN is reported in [2.135, 2.448]. The substitution of B by Mg or Be typically leads to the p-type behavior of the grown material.

2.1.2 Material Limitations

The transport properties of III-N materials and the limitations of these materials have been discussed above. Further issues are discussed in the next paragraphs.

Recombination, Generation, and Breakdown

Recombination and generation are of fundamental importance for optoelectronic and bipolar electronic devices. However, due to the influence of electron trapping in FETs, these effects also critically determine their device performance. For high-field effects at increased driving forces, an experimental evaluation of impact ionization in GaN is discussed in [2.229]. The impact ionization parameters are extracted from a AlGaN/GaN HFET device with a gate length of $0.9\,\mu\text{m}$. For the high-field region of $\geq 10^6\,\text{V\,cm}^{-1}$ the data can be fitted to the classical relation:

$$\alpha_n = A \cdot \exp(-K/E), \tag{2.1}$$

$$\alpha_n = 2.9 \times 10^8 \exp(-3.4 \times 10^7/E). \tag{2.2}$$

For lower fields, this expression needs to be modified, as the field dependence is weaker due to the weakness of real carrier multiplication. The relations in (2.1)/(2.2) suggest that the critical field is a factor of 8 higher when compared to GaAs-data taken from a similar extraction procedure. MC simulations of electron impact ionization in both zincblende and wurtzite GaN are provided in [2.212]. At comparable fields, the electrons are cooler in the wurtzite structure, thus the ionization rates are lower than in the zincblende phase. A similar study on the electron initiated impact ionization in Zb–GaN can be found in [2.340]. The simulations suggest a very soft breakdown threshold similar to the device findings in [2.229]. MC simulations of hole-initiated impact ionization in both GaN-phases are discussed in [2.339]. The critical field for the hole-initiated II in Zb–GaN are found to be similar to those for the electron-initiated II and amount to $3\,\text{MV\,cm}^{-1}$. For the wurtzite structure, the breakdown is similar for high-electric fields, while for lower field, the hole II-rate appears to be greater. Electron-initiated impact ionization

Table 2.15. Breakdown fields, bandgap energies, and dielectric constants for various semiconductor materials

Material	Breakdown field $(MV\,cm^{-1})$	E_g (eV)	ε_r (−)	Ref.
Si	0.3	1.12	11.9	[2.463]
GaAs	0.4	1.43	12.5	[2.463]
InP	0.45	1.34	12.4	[2.463]
GaN (Wz)	3.3	3.43	9.5	[2.463]
AlN (Wz)	8.4	6.2	8.5	[2.388, 2.463]
InN (Wz)	1.2	0.7	15.3	[2.244, 2.451]
BN (c)	2–6	6.4	7.1	[2.253]
4H–SiC	3.5	3.2	10	[2.463]
6H–SiC	3.8	2.86	10	[2.463]
Diamond	5	5.6	5.5	[2.463]

in $Al_xGa_{1-x}N$ is evaluated by ensemble MC simulation in [2.56] for the full material composition range. The results obey the simple expression, as given in (2.1). As expected, the critical fields increase with the increase in the material composition and bandgap. Very low impact ionization rates are observed for $Al_xGa_{1-x}N$ in general. A study of RF-breakdown in bulk GaN and GaN MES-FETs is given in [2.112]. It is found that the RF-breakdown voltage increases with the frequency of the applied RF-large-signal excitation. The difference is explained by the time-response of the particle energy. The critical field for impact ionization at AlGaN/GaN HFETs is found to be around $3\,MV\,cm^{-1}$ at room temperature, as reported in [2.95]. This agrees well with the breakdown field in bulk GaN. Breakdown fields and related material properties are again compiled for comparison in Table 2.15 for various homogeneous materials. Generally, the increase of the breakdown fields with increasing bandgap is observed. The wide bandgap materials have breakdown fields, which are an order of magnitude higher than those of conventional semiconductors. The substrate material SiC and diamond have breakdown fields similar to III-N semiconductors.

2.1.3 Thermal Properties and Limitations

In addition to their electrical limits III-N semiconductor devices are subject to strong thermal (self-)heating. Table 2.16 compiles the thermal properties of several binary semiconductor materials. The temperature dependence of the thermal conductivity is modelled according to:

$$\kappa_L(T_L) = \kappa_{300\,K} \cdot \left(\frac{T_L}{300\,K}\right)^\alpha, \tag{2.3}$$

as also stated in Chapter 8.

Table 2.16. Thermal properties of III-nitride binary and ternary materials at 300 K

	κ_{300} (W K^{-1} m^{-1})	α (−)	c_{300} (J K^{-1} kg^{-1})	Ref.
Si	148	−1.35	711	[2.406]
GaAs	54	−1.25	322	[2.356]
InP	68	−1.4	410	[2.356]
GaN (Wz)	130	−0.43	491	[2.120, 2.423]
AlN (Wz)	285	−1.57	748	[2.428]
InN (Wz)	38.5, 45, 80,176	−	325	[2.223, 2.428, 2.449]
BN (c)	749	−	600	[2.99]
6H–SiC	390	−1.5	715	[2.57]
6H–SiC	490	−	690	[2.244, 2.331]
4H–SiC	330	−	690	[2.244, 2.416]
V-doped SiC	370	−	690	−
Sapphire	42	−	750	−
Diamond	2,000–2,500	−1.85	520	[2.405]

Thermal Conductivity and Heat Capacity

Silicon serves as a reference. The thermal conductivity of silicon is not reached by GaAs. On the contrary, GaN has a thermal conductivity similar to silicon. The thermal conductivity at 300 K κ_{300} of AlN is better than the κ_{300} of GaN, while InN has a very low value [2.223]. BN has the best value of all III-N materials. Initial determination of thermal properties of GaN is given in [2.423]. Further thermal data is compiled in [2.4]. The intrinsic thermal conductivity for AlN is determined in [2.428]. The thermal conductivity of GaN is often quoted dating back to the work of Sichel [2.423]. However, several investigations are available which especially focus on the effects of dislocations on the thermal conductivity. A good overview of the data are presented and the methods of measurements are compiled in [2.120]. The effect of dislocations on the thermal conductivity in GaN is investigated experimentally in [2.264]. The measurements show a dramatic increase of the thermal conductivity at reduced dislocation density of 10^8 cm^{-2}, especially at temperatures below 200 K. The theoretical predictions in [2.216] support a maximum value for the thermal conductivity in GaN of nearly 200 W K^{-1} m^{-1}. The experimental data for the thermal conductivity of InN ($\leq$45 W K^{-1} m^{-1}) is much lower than the theoretical predictions of 176 W K^{-1} m^{-1} [2.223].

The data for the heat capacity are also compiled in Table 2.16. The heat capacity is given for constant pressure. For SiC, the measured value of the polytype 6H–SiC is used for SiC in general [2.356]. The heat capacity of the binary semiconductors is lower than of the substrate material. More analysis with respect to the packaging materials and to the dynamic response is provided in Chapter 8.

2.1.4 Ternary and Quaternary III-N Materials

The existence of ternary and even quaternary materials in the III-N system is a fundamental advantage relative to other wide bandgap semiconductor materials, such as SiC. The possibility of growing $Al_xGa_{1-x}N$, $In_xGa_{1-x}N$, and $In_xAl_{1-x}N$ in heterostructures with the III-N binaries allows bandgap engineering. This has tremendous impact on the electronic and optoelectronic application of the materials. The material parameters of the quantity P are combined by quadratic interpolation in the following two approaches:

$$P_{A_xB_{1-x}N} = P_A \cdot x + P_B \cdot (1 - x) + C_{P,AB} \cdot x \cdot (1 - x). \qquad (2.4)$$

In the second approach, (2.4) can be written in another way, i.e.:

$$P_{A_xB_{1-x}N} = a + b \cdot x + c \cdot x^2 \qquad (2.5)$$

resulting in different coefficients, which can be directly correlated with the binary materials. Sometimes (2.5) is extended to a third-order polynomial:

$$P_{A_xB_{1-x}N} = a + b \cdot x + c \cdot x^2 + d \cdot x^3. \qquad (2.6)$$

All relevant quantities will be analyzed in the following sections using the aforementioned formulae in quadratic interpolation. If necessary, cubic or other interpolation schemes will be mentioned.

Aluminum Gallium Nitride ($Al_xGa_{1-x}N$)

$Al_xGa_{1-x}N$ is the most important ternary compound, as the lattice-mismatch relative to GaN can be effectively controlled for nearly all material compositions. A distinction may be required for wurtzite and zincblende AlGaN materials, as the zincblende material has a transition from a direct to an indirect semiconductor [2.235].

Mechanical and Optical Properties

For mass density, Vickers hardness, and dielectric constants, typically no bowing is applied, and the values can be interpolated linearly with high precis'on between the binary values of GaN and AlN. The heat capacity of the ternaries is interpolated linearly as given in (2.7):

$$c_{L,AB} = (1 - x) \cdot c_{L,A} + x \cdot c_{L,B}. \qquad (2.7)$$

The thermal conductivity of $Al_xGa_{1-x}N$ is interpolated as the following equation (2.8), taken from [2.357]. Data for the thermal conductivity of $Al_xGa_{1-x}N$ is given in [2.255]. The derived parameters are given in Table 2.17.

$$\kappa_{AB} = \left(\frac{(1 - x)}{\kappa_A} + \frac{x}{\kappa_B} + \frac{x \cdot (1 - x)}{C_{\kappa,AB}} \right)^{-1}. \qquad (2.8)$$

Table 2.17. Interpolation of the thermal conductivity κ in III-N ternary materials

	$C_{\kappa,AB}$ $(\mathrm{W\,m^{-1}\,K^{-1}})$	Ref.
$Al_xGa_{1-x}N$	6.95	[2.255]
$In_xGa_{1-x}N$	1.4	[2.356]
$In_xAl_{1-x}N$	3.3	[2.356]

Similarly, the temperature dependence α_{AB} is interpolated linearly.

$$\alpha_{AB} = (1 - x) \cdot \alpha_A + x \cdot \alpha_B. \tag{2.9}$$

The strong material composition dependence of the thermal conductivity is found for all ternary III-V semiconductors [2.356]. A thermal conductivity of $25\,\mathrm{W\,m^{-1}K^{-1}}$ is found for $Al_{0.4}Ga_{0.6}N$ [2.255].

Basic Carrier Transport Properties

An evaluation of temperature-dependent transport properties in zincblende $Al_xGa_{1-x}N$ and $In_xGa_{1-x}N$ semiconductors using MC simulations is given in [2.17]. The data for (2.6) are compiled in Table 2.18. The table further gives the interpolation of the low-field mobility according to the harmonic mean:

$$\frac{1}{\mu_{AB}} = \frac{x}{\mu_A} + \frac{1 - x}{\mu_B} + \frac{x \cdot (1 - x)}{C_{\mu,AB}}. \tag{2.10}$$

The database for the mobility of $In_xAl_{1-x}N$ and $In_xGa_{1-x}N$ is still relatively poor. However, initial values are provided based on the investigations in [2.65]. The alloy-scattering potential is found to be important for the bowing of the drift mobility of both $In_xGa_{1-x}N$ and $In_xAl_{1-x}N$, both for room temperature and at 77 K [2.65]. The alloy scattering is considered relatively less important for $Al_xGa_{1-x}N$ in [2.65]. However, the value extracted for the mobility of $Al_xGa_{1-x}N$ in [2.113] show also a strong dependence on the material composition. This is confirmed by the numerical study of the alloy scattering in AlGaN and InGaN [2.34].

The values for the bowing of the drift mobility are given in Table 2.18.

High-Field Transport

Table 2.19 compiles values for the bowing of the saturation velocity, which is calculated both according to the polynomial approach of (2.6) and according to quadratic bowing, as given in (2.14).

As can be found in [2.17], the critical field and the peak velocities typically do not vary linearly as a function of material composition for $Al_xGa_{1-x}N$ or $In_xGa_{1-x}N$ between the binary constituents. Again a polynomial approach is used, as given in Table 2.19. In another approach, quadratic bowing can be applied:

$$v_{\text{sat,AB}} = (1 - x) \cdot v_{\text{sat,A}} + x \cdot v_{\text{sat,B}} + x \cdot (1 - x) \cdot C_{v_{\text{sat,AB}}}, \qquad (2.11)$$

$$v_{\text{peak,AB}} = (1 - x) \cdot v_{\text{peak,A}} + x \cdot v_{\text{peak,B}} + x \cdot (1 - x) \cdot C_{v_{\text{peak}},\text{AB}}, \quad (2.12)$$

$$E_{\text{crit,AB}} = (1 - x) \cdot E_{\text{crit,A}} + x \cdot E_{\text{crit,B}} + x \cdot (1 - x) \cdot C_{E_{\text{crit}},\text{AB}}. \qquad (2.13)$$

Table 2.18. Interpolation of the mobility μ in III-N ternary compounds for different models at RT

$\mu(x)$	a $(\text{cm}^2\,(\text{Vs})^{-1})$	b $(\text{cm}^2\,(\text{Vs})^{-1})$	c $(\text{cm}^2\,(\text{Vs})^{-1})$	d $(\text{cm}^2\,(\text{Vs})^{-1})$	N_{D} (cm^{-3})	Ref.
$\text{Al}_x\text{Ga}_{1-x}\text{N}$ (Zb)	1,157	$-1,329$	-283	671	10^{17}	[2.17]
$\text{In}_x\text{Ga}_{1-x}\text{N}$ (Zb)	465	755	-61	676	10^{17}	[2.17]

$\mu(x)$	$C_{\mu,\text{n,AB}}$ $(\text{cm}^2\,(\text{Vs})^{-1})$	$C_{\mu,\text{p,AB}}$ $(\text{cm}^2\,(\text{Vs})^{-1})$	Ref.
$\text{Al}_x\text{Ga}_{1-x}\text{N}$ (Wz)	40	1e6	[2.113, 2.356]
$\text{In}_x\text{Ga}_{1-x}\text{N}$ (Wz)	97	1e6	[2.65]
$\text{In}_x\text{Al}_{1-x}\text{N}$ (Wz)	1e6	1e6	[2.65, 2.452]

Wz wurtzite, *Zb* zincblende

Table 2.19. Interpolation of the saturation velocity v_{sat}, the peak velocity v_{peak}, and E_{crit} in III-N ternary compounds

v_{sat}	a $(10^7\,\text{cm s}^{-1})$	b $(10^7\,\text{cm s}^{-1})$	c $(10^7\,\text{cm s}^{-1})$	N_{D} (cm^{-3})	Ref.
$\text{Al}_x\text{Ga}_{1-x}\text{N}$	1.3425	0.574	-0.3215	10^{17}	[2.17]
$\text{In}_x\text{Ga}_{1-x}\text{N}$	1.3286	0.3657	-0.2857	10^{17}	[2.17]

	$C_{v_{\text{sat}},\text{AB}}$ $(10^7\,\text{cm s}^{-1})$	$C_{v_{\text{peak}},\text{AB}}$ $(10^7\,\text{cm s}^{-1})$	$C_{E_{\text{crit}},\text{AB}}$ (kV cm^{-1})	N_{D} (cm^{-3})	Ref.
$\text{Al}_x\text{Ga}_{1-x}\text{N}$	-3.85	-0.1	50	10^{17}	[2.17, 2.356]
$\text{In}_x\text{Ga}_{1-x}\text{N}$	0.35	-1	-25	10^{17}	[2.17]
$\text{In}_x\text{Al}_{1-x}\text{N}$	–	–	–	–	–

		E_{crit} (kV cm^{-1})	N_{D} (cm^{-3})	Ref.
GaN	(Wz)	225	2×10^{16}	[2.211, 2.500]
GaN	(Wz)	140	10^{17}	[2.121]
AlN	(Wz)	450	10^{17}	[2.121]
InN	(Wz)	65	10^{17}	[2.121]
GaN	(Zb)	150	10^{17}	[2.17]
AlN	(Zb)	550	10^{17}	[2.17]
InN	(Zb)	45	10^{17}	[2.17]

The parameters are also given in Table 2.19. The values of the velocities of the binaries are given in Table 2.8, while the critical fields E_{crit} are repeated in Table 2.19 from Table 2.8.

Band Structure

Band structure properties of ternary $\text{Al}_x\text{Ga}_{1-x}\text{N}$ alloys are given in [2.127, 2.235]. The bowing of the effective masses m_ν and the bandgap E_g are described by quadratic interpolation:

$$m_{\nu,\text{AB}} = (1 - x) \cdot m_{\nu,\text{A}} + x \cdot m_{\nu,\text{B}} + x \cdot (1 - x) \cdot C_{m_\nu,\text{AB}}, \qquad (2.14)$$

$$E_{\text{g,AB}} = (1 - x) \cdot E_{\text{g,A}} + x \cdot E_{\text{g,B}} + x \cdot (1 - x) \cdot C_{E_\text{g},\text{AB}}. \qquad (2.15)$$

For $\text{Al}_x\text{Ga}_{1-x}\text{N}$, the dependence of the fundamental bandgap on alloy material composition and on pressure are discussed in [2.410]. A relatively strong bowing is observed for $\text{Al}_x\text{Ga}_{1-x}\text{N}$. The bowing parameters are compiled in Table 2.21. Further information on the dependence of the bandgap on the material composition of $\text{Al}_x\text{Ga}_{1-x}\text{N}$ for $(0 \leq x \leq 1)$ based on ultraviolet photodetectors grown on sapphire are given in [2.489]. The parameters for the bowing of the masses and bandgap are given in Table 2.20 and Table 2.21. As can be seen the bowing is mostly neglected. Band structure nonlocal pseudopotential calculations of the III-N wurtzite phase materials of the ternary alloys $\text{Al}_x\text{Ga}_{1-x}\text{N}$, $\text{In}_x\text{Ga}_{1-x}\text{N}$, and $\text{In}_x\text{Al}_{1-x}\text{N}$ are given in [2.127]. The data include a systematic collection of the effective mass bowing parameters at the relevant conduction band minima. The interpolation of the bandgap energies is given in Table 2.21. The recent changes for InN are included. Distinctions of the bowing are necessary in the zincblende structure, as direct-indirect transitions occur for $\text{In}_x\text{Al}_{1-x}\text{N}$ and $\text{Al}_x\text{Ga}_{1-x}\text{N}$. For $\text{Al}_x\text{Ga}_{1-x}\text{N}$, this occurs at $x = 0.69$ [2.254], while for $\text{In}_x\text{Al}_{1-x}\text{N}$, this occurs for $x = 0.187$ [2.256].

Table 2.20. Interpolation of the effective masses in III-N ternary materials

	$C_{m_\text{n},\text{AB}}$ (Wz)	$C_{m_\text{p},\text{AB}}$ (Wz)	Ref.
$\text{Al}_x\text{Ga}_{1-x}\text{N}$	0.0048	0	[2.127, 2.485]
$\text{In}_x\text{Ga}_{1-x}\text{N}$	0	0	[2.485]
$\text{In}_x\text{Al}_{1-x}\text{N}$	0	0	[2.485]

Wz wurtzite, *Zb* zincblende

Table 2.21. Interpolation of the bandgap energy in III-N ternary materials

	$C_{E_\text{gap},\text{AB}}$ (Wz)	$C_{E_\text{gap},\text{AB}}$ (Zb)	Ref.
$\text{Al}_x\text{Ga}_{1-x}\text{N}$	$-0.7, -1.33$	$\Gamma{:}-0.7$, $X{:}-0.61$	[2.410, 2.485]
$\text{In}_x\text{Ga}_{1-x}\text{N}$	-1.4	-1.4	[2.279, 2.485]
$\text{In}_x\text{Al}_{1-x}\text{N}$	-2.5	$\Gamma{:}-2.5$, $X{:}-0.61$	[2.485]

Indium Gallium Nitride ($In_xGa_{1-x}N$) and Indium Aluminum Nitride ($In_xAl_{1-x}N$)

The importance of InN and its ternary compounds is due to the smaller bandgap relative to GaN, allowing for a broader variety of layers for bandgap engineering also into the visible optical range for optoelectronic devices. High-quality $In_xGa_{1-x}N$ layers were recently grown by MBE, e.g., in [2.114, 2.501], mostly on sapphire substrates. $In_xAl_{1-x}N$ is lattice-matched to GaN for $x = 0.17$, which has recently drawn attention to this material for HEMT device applications [2.230, 2.485].

Mechanical and Optical Properties

Initial layer preparation and optical properties of $Ga_{1-x}In_xN$ thin films are demonstrated and compiled in [2.352]. The initial material is mostly poly-crystalline. The thermal conductivity and the heat capacity of $In_xGa_{1-x}N$ are interpolated from the binaries, as shown in (2.7) and (2.8) with parameters from Table 2.16 and Table 2.17.

Basic Carrier Transport Properties

Similar to InGaAs and InAlAs, initial expectations for InN and its compounds yield lower effective mass than GaN [2.128]. However, good quality films, especially for high In-contents, have not been realized with MOCVD growth, e.g., [2.181]. MBE growth has allowed InN to be grown with improved material quality, e.g., [2.86]. An initial compilation of the transport properties in $In_xGa_{1-x}N$ is given in [2.54, 2.453] and in the references therein.

High-Field Transport

MC analysis of the wurtzite $In_xGa_{1-x}N$ is given in [2.113]. The analysis is based on the wrong bandgap of InN. The analysis of peak velocities and mobility of the ternary material is strongly dependent on the alloy scattering in the resulting ternary material [2.34]. The parameters for the peak velocity v_{peak}, the saturation velocity v_{sat}, and the critical field E_{crit} are compiled in Table 2.19. MC simulations of the temperature-dependent transport properties of zincblende $In_xGa_{1-x}N$ are given in [2.17] and are also compiled in Table 2.19. The critical field E_{crit} decreases strongly with the indium content. Similarly the saturation velocity of $In_xGa_{1-x}N$ increases from GaN to InN. Very few data are available on the transport properties of $In_xAl_{1-x}N$. Al-rich $In_xAl_{1-x}N$ will certainly reveal insulating properties similar to AlN. $In_{0.17}Al_{0.83}N$ is the most important composition to understand. Calculations and initial data on the low-field mobility can be found in [2.452]. The prediction yields a low-field mobility of $\leq 200\,cm^2\,V^{-1}\,s^{-1}$ for $x \leq 0.5$. High-field transport in InAlN is so far not understood very well.

Band Structure

For the band structure, nonlocal pseudopotential calculations of $In_xGa_{1-x}N$ and $In_xAl_{1-x}N$ are given in [2.127] and are still based on a fundamental bandgap of 1.89 eV. The conduction band offset of InGaN/GaN is reported in [2.530]. The ratio of conduction to valence band offset is 58:42. With the improvements in material quality based on MBE growth, small bandgap bowing has been reported, e.g., in [2.501]. The bandgap bowing has been thoroughly investigated with respect to both bowing of InGaN and the bandgap value of InN in [2.86, 2.278]. InN is found to be a small bandgap semiconductor with a bandgap energy of 0.77 eV at RT [2.86]. The data are extracted from optical absorption, photoluminescence, and photomodulated reflectance. The higher bands are still under consideration, e.g., [2.163, 2.373, 2.523].

2.2 Polar Semiconductors for Electronics

Next to the basic knowledge on the bulk material properties, further substantial material analysis is required. III-N semiconductors provide the occurrence of significant electrical polarization effects that dominate especially the interfaces [2.499]. This fact deserves discussion in an extra section, since the effects are much stronger than in any of the semiconductors in use so far. A good overview is given in [2.499]. Precisely defining doping and doping profiles is one of the core competencies of silicon CMOS electronics, e.g., [2.32, 2.149]. However, for III-N devices, control of the polarization effects gains an importance similar or even greater than the impact of the impurity doping profiles in other semiconductor materials. Fig. 2.3 gives the crystal structure of wurtzite GaN, as described, e.g., in [2.11, 2.12]. The noncentrosymmetric GaN crystal leads to a strong ionicity and a residual electrical polarity of the semiconductor. Fig. 2.3 depicts the planar nature of the polarity. The analysis of the crystal structure further reveals the physical mechanisms of the charge control, the overall available carrier concentration, and their dependence from the layer structure.

2.2.1 Spontaneous Polarization

A basic understanding of the impact of material growth conditions by various methods in the AlN/GaN/InN material system on the polarization charge is achieved and reported in [2.11, 2.12, 2.90]. The findings are based on the crystal polarization constants of Bernardini et al. in [2.36]. The importance of both spontaneous and piezoelectric components is pointed out there, in contrast to earlier publications, e.g., [2.22]. The spontaneous polarization occurs along the c-axis of wurtzite structures and leads to strong electric fields of 3 MV/cm [2.12]. Results for the PIC of the model are compared in [2.12] to experimental results for HEMT channels achieved at the date of publication; in the meantime, additional progress has been achieved for the improvement

Ga   N

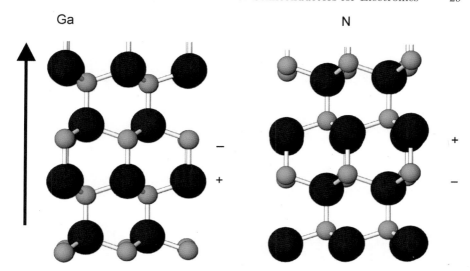

Fig. 2.3. Crystal structures of GaN for (**a**) (*left*) Ga-face, and (**b**) (*right*) N-face growth

in III-N material quality, i.e., for the reduction of defects. Basic findings for III-N layers grown by the two most important growth techniques, MOCVD and MBE, presented in these and other publications include:

- The absolute value of the piezoelectric constants of III-N semiconductors are ten times higher than in conventional III-V and II-VI semiconductors [2.36].
- Very similar mobility values, sheet carrier concentrations, and material quality are achieved for III-N semiconductors grown by MBE and MOCVD [2.96].
- Spontaneous polarization is more important than piezoelectric polarization for $Al_x Ga_{1-x}N/GaN$ heterostructure interfaces [2.12, 2.36].
- Spontaneous polarization can be as important as piezoelectric polarization or even more important for In(Al)N/GaN interfaces and still lead to very high sheet carrier densities [2.12, 2.36, 2.231].
- The polarity of the polarization can be changed and even be inversed as a function of strain and thus of material composition.
- A difference exists between MBE- and MOCVD-grown material with respect to the variety of the polarity at the charged interfaces [2.12, 2.90].
- Theoretical and experimental results agree to a certain extent [2.11].
- Uncertainties arise from the impact of surface conditions [2.11] and Fermi-level pinning at the passivation interfaces, dislocations [2.11], possible polarity intermixing; the latter, however, is unlikely in good structures.
- Surface roughness and nonabrupt interfaces are considered less influencing the sheet carrier concentration.

Table 2.22. Parameter sets for the spontaneous polarization of III-N binary semiconductors

Material	GaN ($C\,m^{-2}$)	AlN ($C\,m^{-2}$)	InN ($C\,m^{-2}$)	Ref.
P_{SP}	−0.029	−0.081	−0.032	[2.36]
P_{SP}	−0.034	−0.090	−0.042	[2.485]

GaN layers can either be grown Ga-face(-up) or N-face(-up), compare Fig. 2.3. For the orientation of the polarization in this book, the positive direction points from the metallic cation (Ga,In) to the nitride anion (N) along the c-axis. Thus, in the case of a Ga-face interface, the polarization points toward the substrate, while for a N-face interface, the polarization is directed toward the surface of the layer sequence. For further evaluation, the devices are either (a) considered to be grown on a thick buffer, relaxing the mismatch from the substrate to the lattice constant of the device layers, or (b) the strain is residual in the device material. For the polarization analysis Table 2.22 gives parameter sets for the spontaneous polarization for the binary III-N materials from various sources. For all materials, the effective spontaneous polarization is negative and the magnitude increases from GaN to InN to AlN.

Interfaces between Binary and Ternary Materials

The following interpolation schemes can be used for the spontaneous polarization in the GaN/AlN, the InN/GaN, and the InN/AlN material system:

$$P_{SP,Al_xGa_{1-x}N/GaN}(x) = \left(-0.052 \cdot x - 0.029 \right) (Cm^{-2}), \qquad (2.16)$$

$$P_{SP,In_xGa_{1-x}N/GaN}(x) = \left(-0.003 \cdot x - 0.029 \right) (Cm^{-2}), \qquad (2.17)$$

$$P_{SP,In_xAl_{1-x}N/GaN}(x) = \left(0.049 \cdot x - 0.081 \right) (Cm^{-2}). \qquad (2.18)$$

The methodology and the parameter values for the AlGaN/GaN interface are taken from [2.36]. A similar theoretical study of the macroscopic polarization and the impact on the GaN HEMT in a nonlinear model of the material composition is given in [2.527].

2.2.2 Piezoelectric Polarization

In addition to spontaneous polarization, piezoelectric effects were first considered to be responsible for the sheet carrier densities in GaN devices, e.g., in [2.525]. Early C–V measurements of the piezoelectrically induced charge (spontaneous polarization is not considered explicitly) in $Al_xGa_{1-x}N/GaN$ heterostructure FETs are given in [2.525] for both MBE- and MOCVD-grown material. The piezoelectric polarization is based on strain to the III-N crys-

Table 2.23. Piezoelectric constants e_{ij}, dielectric constants ϵ_{ij}, and lattice parameter a_i, c_i sets of III-N binary wurtzite semiconductors

Material	GaN	AlN	InN	BN	Ref.
e_{33} $(\mathrm{C\,m^{-2}})$	0.73	1.46	0.97	–	[2.36]
e_{31} $(\mathrm{C\,m^{-2}})$	−0.49	−0.6	−0.57	–	[2.36]
e_{33} $(\mathrm{C\,m^{-2}})$	–	1.55	–	–	[2.12]
e_{31} $(\mathrm{C\,m^{-2}})$	–	−0.58	–	–	[2.12]
e_{15} $(\mathrm{C\,m^{-2}})$	–	−0.48	–	–	[2.12]
e_{33} $(\mathrm{C\,m^{-2}})$	1	–	–	–	[2.338]
e_{31} $(\mathrm{C\,m^{-2}})$	−0.36	–	–	–	[2.338]
e_{15} $(\mathrm{C\,m^{-2}})$	−0.3	–	–	–	[2.338]
ϵ_{11} $(-)$	9.5	9.0	–	–	[2.29, 2.36]
ϵ_{33} $(-)$	10.4	10.7	–	–	[2.29, 2.36]
a_0 (Å)	3.189	3.112	3.54	2.534	[2.11]
c_0 (Å)	5.185	4.982	5.705	4.191	[2.11]

tal and a displacement of the anion-sublattice to the cation-sublattice, see Fig. 2.3. This can lead to very high electric fields of $2\,\mathrm{MV/cm}$.

Table 2.23 gives a summary of the piezoelectric constants of the different III-N binary materials [2.11] for comparison and further evaluation. The lattice parameters have been added where necessary. In various publications, e.g., [2.36], the piezoelectric component is considered to have the highest uncertainty relative to the other components. The piezoelectric polarization induced by strain along the c-axis in a strained layer of wurtzite-type III-N semiconductors is calculated, according to [2.11]:

$$P_{\mathrm{PZ}} = 2 \cdot \frac{a - a_0}{a_0} \left(e_{31} - e_{33} \cdot \frac{C_{13}}{C_{33}} \right). \tag{2.19}$$

In (2.19), C_{13} and C_{33} denote the elastic constants, which require special attention due to the residual strain in the semiconductor layers on top of the buffer. a and a_0 are the lengths along the hexagonal edge, which are similarly modified. For the lattice, elastic, and piezoelectric constants, the following linear interpolations are used for the $\mathrm{Al}_x\mathrm{Ga}_{1-x}\mathrm{N/GaN}$ material system [2.12]:

$$a_0(x) = (-0.077 \cdot x + 3.189) \cdot 10^{-10}\,(\mathrm{m}), \tag{2.20}$$

$$c_0(x) = (-0.203 \cdot x + 5.189) \cdot 10^{-10}\,(\mathrm{m}), \tag{2.21}$$

$$C_{13}(x) = (5 \cdot x + 103)\,(\mathrm{GPa}), \tag{2.22}$$

$$C_{33}(x) = (-32 \cdot x + 405)\,(\mathrm{GPa}), \tag{2.23}$$

$$e_{33}(x) = 0.73 \cdot x + 0.73\,(\mathrm{C\,m^{-2}}), \tag{2.24}$$

$$e_{31}(x) = -0.11 \cdot x - 0.49\,(\mathrm{C\,m^{-2}}). \tag{2.25}$$

For the $In_xAl_{1-x}N/GaN$ material system, where GaN forms the channel, the interpolation reads:

$$a_0(x) = (0.418 \cdot x + 3.112) \cdot 10^{-10} \, (\text{m}), \tag{2.26}$$

$$c_0(x) = (0.723 \cdot x + 4.982) \cdot 10^{-10} \, (\text{m}), \tag{2.27}$$

$$C_{13}(x) = (-16 \cdot x + 108) \, (\text{GPa}), \tag{2.28}$$

$$C_{33}(x) = (-149 \cdot x + 373) \, (\text{GPa}), \tag{2.29}$$

$$e_{33}(x) = -0.49 \cdot x + 1.46 \, (\text{C m}^{-2}), \tag{2.30}$$

$$e_{31}(x) = 0.03 \cdot x - 0.6 \, (\text{C m}^{-2}). \tag{2.31}$$

For the $GaN/In_xGa_{1-x}N$ material system, where GaN forms the barrier layer with the higher bandgap energy the interpolation reads growing the InGaN pseudomorphically on GaN:

$$a_0(x) = (0.351 \cdot x + 3.189) \cdot 10^{-10} \, (\text{m}), \tag{2.32}$$

$$c_0(x) = (0.52 \cdot x + 5.189) \cdot 10^{-10} \, (\text{m}), \tag{2.33}$$

$$C_{13}(x) = (-11 \cdot x + 103) \, (\text{GPa}), \tag{2.34}$$

$$C_{33}(x) = (-176 \cdot x + 405) \, (\text{GPa}), \tag{2.35}$$

$$e_{33}(x) = 0.24 \cdot x + 0.73 \, (\text{C m}^{-2}), \tag{2.36}$$

$$e_{31}(x) = -0.08 \cdot x - 0.49 \, (\text{C m}^{-2}). \tag{2.37}$$

The calculation of the PIC derived from the above polarization results is described later.

2.2.3 Device Design Using Polarization-Induced Charges

Deduced from the above considerations, the resulting charge can be calculated for the various interface combinations and growth conditions in several device configurations. In this context, mechanisms of the 2D-electron gas formation in AlGaN/GaN HEMTs are investigated by synchrotron radiation emission spectroscopy in [2.160]. The surface Fermi-level at the $Al_xGa_{1-x}N$ interface is found to be pinned at an energy of 1.6 eV below the conduction band minimum independent of the Al-content. Initial measurements of the piezoelectrically induced charge in AlGaN/GaN HFETs are given in [2.525]. For a heterointerface, the following formulae hold for the interface charge σ_B derived from the tensor of the wurtzite crystals mentioned above [2.11]:

$$\sigma_B = \mathbf{n} \cdot \mathbf{P}, \tag{2.38}$$

$$= P_{\text{bottom}} - P_{\text{top}}, \tag{2.39}$$

$$= P_{\text{SP,bottom}} - [P_{\text{SP,top}} + P_{\text{PZ,top}}], \tag{2.40}$$

$$= \sigma_{B,SP} + \sigma_{B,PZ}. \tag{2.41}$$

Deduced from (2.38)–(2.41), various interface combinations and devices can be constructed for their use in heterostructure FETs and HBTs. Some will be discussed later. The situations are displayed graphically in Fig. 2.4 with focus on electron devices. Hole-based devices are neglected due to the poor transport properties in the III-N system. For an electron channel, two conditions must be fulfilled:

- A positive interface charge must be present at the interface based on the overall polarization
- The bandgap in the semiconductor, to where the electrons are driven, must be smaller than of the other layer (as III-N semiconductor only form type I transitions).

I: $Al_xGa_{1-x}N/GaN$ Single Heterojunction HEMT

For an $Al_xGa_{1-x}N$ layer grown on top of a relaxed GaN layer, spontaneous polarization and piezoelectric polarization are parallel and under tensile strain. Depending on the growth conditions, the following situations occur:

1. In the case of Ga-face, i.e., a anion surface, the polarity of the parallel polarization components results in an electron GaN channel formation at the interface.
2. In the case of N-face or cation-type growth, however, the orientations of both the polarization components are changed and thus no electron channel formation occurs at the AlGaN/GaN interface.

In this context it is an important assumption for the charge analysis that no additional material or layers are grown on top of the AlGaN barrier layer. Additional surface states and the interaction with additional interfaces reduce the carrier concentration at the interface under consideration. This is also true for nucleation layer sequences used to compensate the lattice-mismatch directly at the substrate–buffer interface.

II: $GaN/In_xGa_{1-x}N$ Single and Double Heterojunction HEMT

For the interface of $In_xGa_{1-x}N$ with various concentrations grown pseudomorphically on top of a relaxed GaN buffer, the $In_xGa_{1-x}N$ layer is under compressive strain for all material compositions. Spontaneous and piezoelectric polarization are antiparallel and the piezoelectric polarization is stronger than the spontaneous polarization. The following situations can occur, depending on growth conditions:

1. In the case of Ga-face growth, the residual polarization points toward the GaN, thus the effective charge at the interface is negative, and no electron channel is formed.
2. In the case of N-face growth, the effective interface charge is positive and a 2DEG forms a channel in the $In_xGa_{1-x}N$.

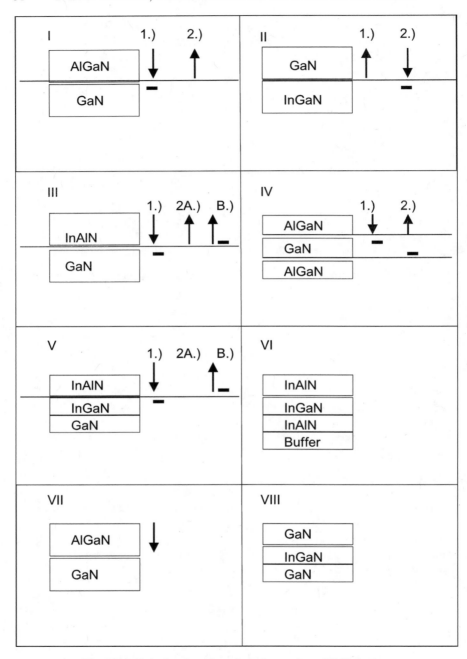

Fig. 2.4. Polarization situations for various III-N devices

III: $In_x Al_{1-x} N/GaN$ Single Heterojunction HEMT

For the interface $In_x Al_{1-x} N/GaN$, the two binaries InN and AlN forming $In_x Al_{1-x} N$ have a bandgap both smaller and larger than the bandgap of GaN, which makes the situation more complicated. The GaN buffer is grown relaxed on the substrate, thus the $In_x Al_{1-x} N$ layer on top is under different strain unless for $In_{0.17} Al_{0.83} N$, which is lattice-matched to crystalline GaN.

1. For Ga-face growth, both polarization components are parallel for $x < 0.17$, and an electron channel forms in the GaN layer for $x < 0.3$ due to the resulting overall polarization, e.g., [2.231].
2. For N-face growth the interesting situation arises for $x > 0.3$. Both polarization types are antiparallel in the $In_x Al_{1-x} N$ with the following distinction:
   A.) For $0.3 < x < 0.6$, no electron channel forms in the $In_x Al_{1-x} N$ as the bandgap of $In_x Al_{1-x} N$ is still bigger than of GaN.
   B.) For $0.6 \leq x \leq 1$ the bandgap of GaN is larger than of $In_x Al_{1-x} N$, thus a 2DEG and thus an electron channel forms in the $In_x Al_{1-x} N$.

IV: $Al_x Ga_{1-x} N/GaN/Al_y Ga_{1-y} N$ Double Heterojunction HEMT

The realization of double heterojunction devices is very desirable due to enhanced carrier confinement for both HEMT and HBT and improved backside isolation for FETs. The assumption of a relaxed GaN buffer on top of the substrate has to be dropped in this case. A good trap-free $Al_y Ga_{1-y} N$ material has to be grown relaxed on a substrate such as SiC or sapphire with a material composition grading of the $Al_y Ga_{1-y} N$ buffer and leading to an $Al_x Ga_{1-x} N/GaN/Al_y Ga_{1-y} N$ heterostructure. Double heterostructure FETs are realized, e.g., in [2.289]. A smaller composition y is chosen for the buffer layer to reduce the mismatch toward the substrate. When growing AlGaN on top of SiC, either N-face or Ga-face polarization can be chosen for the first interface:

– In the case of a Ga-face growth, a conventional double heterojunction FET forms with an electron channel at the upper heterointerface.
– In the case of an N-face growth, an inverted double heterojunction FET forms with an electron gas at the lower interface under the assumption, that the upper barrier does not modify the strain of the GaN channel layer.

Typically, a Ga-face approach will be chosen which improves the isolation of the channel toward the buffer layer and thus typically reduces the I_{Dmin} for FETs. Further, in [2.289], the transconductance g_m is increased as compared to the single heterojunction device.

V: $In_x Al_{1-x} N/In_y Ga_{1-y} N$ Double Heterojunction HEMT on GaN Buffer

In the $Al_x Ga_{1-x} N/GaN$ material system, the sheet carrier concentration density in the channel is increased by the increase of x and at the same time limited by the increase of defect concentration in the $Al_xGa_{1-x}N$ layer for rising x and the onset of relaxation. An $In_yGa_{1-y}N$ channel layer is a more general case of the situation III of the InAlN/GaN HFET. An $In_xAl_{1-x}N/In_yGa_{1-y}N/GaN$ double heterojunction FET can be formed in this case. The formation of electron or hole gases is dependent on both material compositions x and y. In [2.230], mole fractions of $0.08 \leq x \leq 0.27$ and ($0 \leq y \leq 0.18$) are suggested for the formation of HFETs, based on the analysis of the maximum strain values possible in the $In_xAl_{1-x}N$ barrier. $In_{0.17}Al_{0.83}N$ is lattice-matched to GaN. Strain analysis based on a relaxed GaN buffer yields the following distinction [2.230]:

- For $In_xAl_{1-x}N$ with $x \leq 0.17$ the $In_xAl_{1-x}N$ is under tensile strain and the piezoelectric field in the channel is in favor of the increase in electron concentration.
- For $In_xAl_{1-x}N$ with $x \geq 0.17$, the $In_xAl_{1-x}N$ is under compressive strain, and the strain reduces the electron concentration [2.231].
- Assuming an $In_{0.17}Al_{0.83}N$ barrier lattice-matched to GaN, the channel material $In_yGa_{1-y}N$ can be varied:
- The compressive strain in the $In_yGa_{1-y}N$ can be varied in the range $0.08 \leq x \leq 0.27$ in order to further enhance the charge carrier density and the upper $In_xAl_{1-x}N/In_yGa_{1-y}N$ interface. A value of $y = 0.1$ and $x = 0.17$ is suggested in [2.230] to double the drain current density, as compared to the $In_{0.17}Al_{0.83}N/GaN$ structure.

The situation V 2B.) in Fig. 2.4 becomes very unlikely in this case, as the bandgap of the InGaN in the channel reduces with increasing y. Very high currents are predicted for these HFETs in [2.230] with maximum current levels beyond $4\,A\,mm^{-1}$. Technical realizations yield maximum drain currents of $0.4\,A\,mm^{-1}$, e.g., in [2.232]. High-saturated drain current levels of $1.33\,A\,mm^{-1}$ have been reported in [2.84].

VI: $In_xAl_{1-x}N/In_yGa_{1-y}N/In_zAl_{1-z}N$ Double Heterojunction HEMT on Arbitrary Buffer

The assumption of the growth on relaxed GaN buffer must be further modified for the growth on arbitrary substrates. Double heterojunction devices can be formed in the $In_xAl_{1-x}N/In_yGa_{1-y}N$ material system. An $In_zAl_{1-z}N$ buffer will be grown on top of a substrate. The selection $z = 0.17$ yields a situation with the buffer lattice-matched to GaN. Arbitrary mole fractions yield the most complicated situation, i.e., a double $In_xAl_{1-x}N/In_yGa_{1-y}N/In_zAl_{1-z}N$ heterojunction HEMT with three material compositions x, y, and z. The situation is similar to that of the AlGaN/GaN/AlGaN heterostructure device; however, as the two polarization components are antiparallel, more distinctions are necessary. The distinction yields a similar analysis for the material composition as in the case of a single heterojunction using InAlN grown on a GaN layer. However, the transition to any substrate is critical, as a second channel can form at the nucleation layer/InAlN buffer interface.

VII: $Al_xGa_{1-x}N/GaN$ NPN-Single Heterojunction Bipolar Transistor

Polarization engineering in III-N HBTs is a secondary concern relative to the present severe issues like collector leakage due to the impact of dislocations, and processing issues, such as aggressive etch processes with subsequent surface damage increasing base ohmic contact resistance, e.g., [2.269, 2.283]. Material growth is, of course, of critical importance in the bipolar device; however, it is important, mostly with respect to mismatch reduction. Subsequent bandgap engineering requires the inclusion of the polarization effects [2.282], as their neglect would be misleading. A very simple AlGaN/GaN HBT can be grown using a thick GaN n-doped subcollector and a GaN n-doped collector on top of a substrate. A p-doped GaN base is added, as performed, e.g., in [2.383]. The emitter layer consists of $Al_xGa_{1-x}N$ and typically reveals material compositions $x = 0.1$ [2.513]. The AlGaN emitter is typically capped with a GaN layer which is highly Si-doped, and the transition is either graded or abrupt. The GaN-base p-doping, typically Mg, is degenerate with nominal doping levels of $p \geq 10^{19}$ cm^{-3}, although the activation is relatively low, i.e., $\leq 10^{18}$ cm^{-3}. This is due to the high activation energy [2.382] of the deep acceptor Mg with $E_A - E_V \approx 170$ meV. The growth at the base/emitter heterointerface will thus be Ga-face with a channel-like (donor-like) positive polarization charge at the emitter(AlGaN)/base(GaN) junction. This further results in a negative polarization (acceptor-like) charge at the graded AlGaN/GaN–Si interface near the top contact. Deduced from this, an additional electron barrier is created in the conduction band at the base/emitter junction and within the emitter grading. In [2.282], these additional peaks are flattened by smoothing the interface by material grading. This feature allows the prevention of hole injection into the emitter without hindering the electron transport. Several other issues will be discussed in the context of polarization engineering and its subsequent effects. The carrier lifetimes in the base at the GaN pn-junctions at different mobility levels are discussed in [2.141], yielding values between 24 and 240 ps. The impact of the polarization doping to the base layer is discussed in [2.23, 2.420]. Normal collector-down device/emitter up configuration see reduced p-doping for the Ga-face growth, while collector-up devices experience additional carrier concentration through polarization doping in the p-layers.

VIII: $GaN/In_xGa_{1-x}N/GaN$ Double Heterojunction Bipolar Transistor (DHBT)

Due to the limited doping activation of Mg in GaN and to several other issues in AlGaN/GaN HBTs, $GaN/In_xGa_{1-x}N/GaN$ npn-DHBTs have been developed and successfully demonstrated, e.g., in [2.271]. Again, polarization engineering is only a secondary concern. The advantage of $In_xGa_{1-x}N$ as a base material is caused by the increase of the p-doping activation, e.g., up to 7×10^{18} cm^{-3} as reported in [2.269]. This is based on the lower acceptor activation energies in p-$In_xGa_{1-x}N$. The base layers are grown with a material composition of $0.17 \leq x \leq 0.25$ [2.311], yielding a doping activation of $4 \times$

$10^{18}\,\mathrm{cm}^{-3}$. A double heterojunction is desirable due to the higher bandgap of GaN in the collector and the increased leakage current of a GaN/InGaN single heterojunction device, as mentioned in [2.269]. On the other hand, the conduction band spike at the collector/base interface has to be reduced, e.g., by material grading in the base towards the lower interface [2.269].

2.2.4 Analytical Calculation of Channel Charge Concentrations

For the calculation of the resulting charge at the heterointerface, the following interpolation formulae have been suggested in [2.22, 2.525], assuming no additional doping in the barrier and no surface effects apart from the interface under consideration:

$$n_{\mathrm{sheet}} = \frac{\sigma(x)}{e} - \left(\frac{\epsilon_0 \cdot \epsilon(x)}{d_{\mathrm{AlGaN}} \cdot e^2} \right) [e \cdot \phi_{\mathrm{B}}(x) + E_{\mathrm{F}}(x) - \Delta E_{\mathrm{C}}(x)]. \qquad (2.42)$$

Derived from (2.42), the maximum sheet carrier concentration of $6 \times 10^{13}\,\mathrm{cm}^{-2}$ at a GaN/AlN interface can be calculated [2.12]; however, it strongly depends on the degree of relaxation at the GaN/AlN interface. Recent HFET device examples with very high sheet carrier densities can be found in [2.68] using $\mathrm{Al_{0.34}Ga_{0.66}N/GaN}$ HFETs which yield sheet carrier densities of $1.45 \times 10^{13}\,\mathrm{cm}^{-2}$. Fig. 2.5 gives the dependence of the sheet resistance and Fig. 2.6 of the 2D-sheet carrier concentration from the material composition at an $\mathrm{GaN/Al_xGa_{1-x}N}$ heterostructure interface presented in recent publications. The impact of both barrier thickness and substrate is neglected in Fig. 2.5 and Fig. 2.6. Further the insertion of doping and of an AlN interlayer is depicted, for details see Chapter 3. Fig. 2.5 suggests a statistically proven linear variation of the sheet resistance from 500 to $200\,\Omega\,\mathrm{sq}^{-1}$. Fig. 2.6 shows a variation of the sheet carrier concentration of $7 \times 10^{12}\,\mathrm{cm}^{-2}$ for $x = 0.15$ to $1.6 \times 10^{13}\,\mathrm{cm}^{-2}$ for $x = 0.5$. At the same time, the influence of the additional n-doping in the barrier layers on the carrier concentration can be seen in Fig. 2.6, as discussed later. Both MBE and MOCVD provide similar charge concentrations as a function of the mole fraction.

2.2.5 Doping Issues

Independent of the impact of polarization as a source of carriers, bulk doping is of fundamental importance for the definition of material and device characteristics, as in any semiconductor. Review on the doping issues of III-N materials is supplied, e.g., in [2.87, 2.370, 2.437]. Some doping parameters are given in Table 2.25. III-N semiconductors are doped with impurities such as Si, Ge, Se, O, Mg, Be, and Zn. Typical unintentional impurities are C, H, O, and grown-in defects, such as vacancy and antisite point defects [2.417].

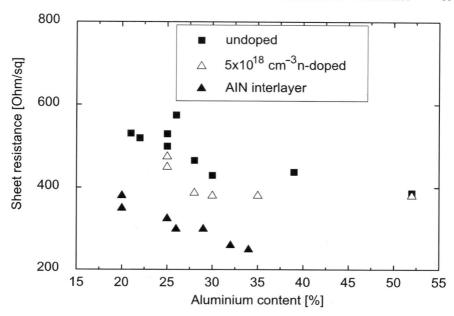

Fig. 2.5. Sheet resistance of $Al_xGa_{1-x}N/GaN$ heterostructures as a function of aluminum content x

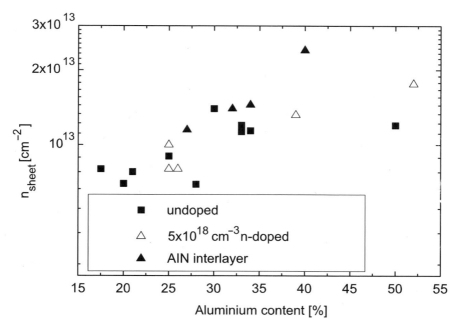

Fig. 2.6. Sheet carrier concentration in $Al_xGa_{1-x}N/GaN$ heterostructures as a function of aluminum content x

N-Doping

In electronic devices, n-doping is of critical importance for δ-doped high-voltage HFETs and the definition of contacts in FETs and HBTs. NID-GaN is found to be typically n-type. This is caused by either nitrogen vacancies, Ga interstitials or oxygen incorporation depending on the growth method [2.351].

Silicon doping is the typical choice for intended n-doping. The activation energy of silicon in GaN is 5–9 meV which allows effective doping [2.351]. Apart from the polarization doping, silicon doping can serve as additional carrier source in the channel. Given the high-operation voltages, the stability of such silicon doping in the barrier layer is crucial, as δ-doping is affected by the field peaks in any HEMT.

P-Doping

The reduced performance of p-doping of GaN is the most critical issue reducing optoelectronic device performance in III-N materials. This includes diodes [2.322, 2.325, 2.432], laser diodes [2.326], and bipolar transistors [2.383] on the electronic side. The issues involve:

- High-contact resistances and nonohmic behavior at the base [2.383] due to:
  - Reduced current gain at RT due to high-thermal activation energy of Mg of 120–200 meV as an acceptor [2.225]
  - Reduced activation of Mg due to the formation of Mg–H during MOCVD growth
- Mg MOCVD reactor memory leading to soft pn-junction doping profiles and reduced confinement of p-doping to the base layer [2.383] for both MBE- and MOCVD-growth.

Photoluminescence measurements in Mg-doped GaN grown by metal organic vapor-phase epitaxy (MOVPE) are described, e.g., in [2.248]. An ionization energy of the acceptor of 173 meV is concluded. The bandgap narrowing for high doping concentrations is found to be in the order of 10 meV. For MBE growth, the following modifications are necessary:

- No Mg-reactor memory occurs
- Sharper doping profiling is possible

The maximum doping activation of Mg in GaN is thus of the order of $\leq 10^{18}\,cm^{-3}$ at room temperature. More details with respect to the growth conditions will be given in Chapter 3.

2.2.6 Surfaces and Interfaces

Due to the aforementioned properties of III-N heterostructures, surfaces and interfaces require special attention in this book. The principal properties of the different interfaces shall be discussed here, while their technological realization is considered in Chaps. 3 and 4.

III-N/III-N Interfaces

Apart from the polarization effects already discussed in detail, several effects at a semiconductor/semiconductor interface occur, due to the change of the material or the material composition:

1. An energy alignment of the valence band and conduction band is observed.
2. Strain effects occur due to the modification of the lattice constants at the interface.
3. The interface is a preferred location for states, e.g., [2.160].
4. In III-N semiconductors, the differences in the polarization occur at the interface, leading to excess carrier concentrations.
5. Due to the quantization effects near the interface, transport occurs very close to the interface, which is especially true for the III-N materials due to the specific band structures.

The next paragraphs compile the enumerated issues.

Energy Alignment and Bandgap Energies

The energy alignment occurs in three different types of transitions in any semiconductor, e.g., [2.356]. The $GaN/Al_xGa_{1-x}N$ semiconductor/semiconductor interface is a type I interface. For both electrons and holes, energy barriers are formed in one direction. Thus channel or charge-accumulation areas can form, if the appropriate carriers are provided. The conduction band energy offset alignment of a ternary material $A_xB_{1-x}N/CN$ is described, e.g., for the $Al_xGa_{1-x}N/GaN$ interface:

$$\Delta E_C(x) = 0.7 \cdot [E_g(Al_xGa_{1-x}N) - E_g(GaN)]. \qquad (2.43)$$

The bandgap energy for $Al_xGa_{1-x}N$ as a function of material composition can be described as:

$$E_g(x) = x \cdot E_g(AlN) + (1-x) \cdot E_g(GaN) + (1-x) \cdot x \cdot C_{E_g,AlGaN}. \qquad (2.44)$$

The bowing factor $C_{E_{gap},AB}$ is already given in Table 2.21. The valence band offset energy ΔE_V is then determined accordingly from the following equation:

$$E_g(Al_xGa_{1-x}N) = \Delta E_C + \Delta E_V + E_g(GaN). \qquad (2.45)$$

The above relations can be generalized for several other material combinations $A_xB_{1-x}N/C_xD_{1-x}N$ in the III-N system. Table 2.24 summarizes the parameters for the energy alignment for the combinations in the III-N world. Overviews are given in [2.203, 2.429]. Table 2.24 gives the full summary of all parameters according to (2.43)–(2.45). The alignment of the ternary semiconductors depends critically on the bowing parameters of the bandgap, which are discussed in Table 2.21. The modifications of the bandgap energies due

Table 2.24. Parameters for energy alignment of the III-N binary and ternary interfaces at RT

Material	Interface type (–)	E_g (eV)	E_g (eV)	ΔE_C (eV)	ΔE_V (eV)	Ratio (–)	Ref.
AlN/GaN (Wz)	Type I	6.2	3.43	1.97	0.8	0.7/0.3	[2.11]
AlN/GaN (Zb)	Type I	5.1	3.38	0.92	0.8	0.53/0.47	[2.203]
InN/AlN (Wz)	Type I	0.77	6.2	3.62	1.81	0.66/0.34	[2.203]
InN/AlN (Zb)	Type I	0.75	5.1	3.31	1.04	0.76/0.24	[2.203]
InN/GaN (Wz)	Type I	0.77	3.43	1.61	1.05	0.6/0.4	[2.203]
InN/GaN (Zb)	Type I	0.75	3.38	1.58	1.05	0.6/0.4	[2.203]
GaN/Al$_{0.2}$Ga$_{0.8}$N	Type I	3.43	3.79	0.252	0.108	0.7/0.3	[2.230]
GaN/In$_{0.17}$Al$_{0.83}$N	Type I	3.43	4.7	0.66	0.61	0.52/0.48	[2.230]
In$_{0.1}$Ga$_{0.9}$N/ In$_{0.17}$Al$_{0.83}$N	Type I	3.03	4.7	0.868	0.801	0.52/0.48	[2.230]
GaN/In$_{0.2}$Ga$_{0.8}$N	Type I	3.43	3.58	0.09	0.06	0.6/0.4	[2.203]

to the different directions of the polarization at the interface and in quantum wells are discussed, e.g., in [2.498]. The energy levels in In$_x$Ga$_{1-x}$N/GaN quantum wells are analyzed for thickness between 23 and 130 Å. By variation of the composition the energy levels can be adjusted to the needs of the application.

Bulk and Interface Traps

Typical trap energies and other defect properties observed experimentally at the heterojunctions and in the bulk are given in Table 2.25. Good overview of defects in bulk materials are given in [2.89, 2.221, 2.315, 2.389, 2.390]. The occurrence of traps is of critical importance. It is partly caused by the heteroepitaxy, i.e., the growth on nonnative substrates. As the wide bandgap properties are modified by these traps with activation energies in the range of conventional semiconductors, such as Si or GaAs, the breakdown properties are critically affected, i.e., reduced. Table 2.25 compiles typical trap parameters. At the interfaces, traps lead to the pinning of the Fermi-level, as is described in [2.160]. The pinning of the Fermi-level in GaN is found to be about midgap, which puts GaN in a similar situation as GaAs. Other reports suggest a Fermilevel pinning at $E_C - 0.5\,\text{eV}$ [2.250] due to nitrogen vacancies or at $E_C - 0.8\,\text{eV}$ due to oxygen. Further details are given below.

Strain in III-N Semiconductors

As is discussed with the polarization effects, strain is a very important parameter in the design of substrate semiconductor buffer. In addition, the lattice-mismatch among different semiconductor layers is important for the resulting material quality and crystal stability. Table 2.26 compiles the lattice constants a_0 and the resulting lattice-mismatch along a_0 between the various III-N mate-

Table 2.25. Trap and doping parameters in III-nitride materials

	Doping	Energy (eV)	Donor	Ref.
GaN	n	$E_C - 0.44$	–	[2.490]
GaN	n	$E_C - 0.5$	–	[2.250]
GaN	n	$E_C - 0.005$	Si	[2.351]
GaN	p	$E_V + 0.22$	Mg	[2.350, 2.475]
GaN	p	$E_V + 0.23$	C	[2.350]
GaN	p	$E_V + 0.25$	Be	[2.350]
$Al_x Ga_{1-x} N$	n	$E_C - 0.017$ $(x = 0)$	Si	[2.351]
$Al_x Ga_{1-x} N$	n	$E_C - 0.054$ $(x = 0.18)$	Si	[2.351]
$Al_x Ga_{1-x} N$	n	$E_C - 0.090$ $(x = 0.6)$	Si	[2.351]
$Al_x Ga_{1-x} N$	p	$E_V + 0.4$	Mg	[2.475]

Table 2.26. Lattice constant a_0 (matrix diagonal) and mismatch for the III-N binary materials at 300 K

Material	GaN (Å)(%)	AlN (Å)(%)	InN (Å)(%)	SiC (Å)(%)	Sapphire (Å)(%)	Silicon (Å)(%)	Ref.
GaN (sub.)	3.189	2.3	10.6	3.3	14.8	–	[2.449]
AlN (sub.)	2.4	3.119	12	1.0	12.5	–	[2.323]
InN (sub.)	10.6	12	3.5446	14	25.4	–	[2.323]
SiC (sub.)	3.3	1.0	14	3.081	–	–	[2.323]
Sapphire (sub.)	14.8	12.5	25.4	11.5	2.777 (4.758)	–	[2.323]
Silicon (sub.)	16.9	18.7	7.6	–	–	3.84 (5.431)	[2.419]

rials and substrates. The lattice-mismatch is calculated according to

$$f_c = \frac{a - a_0}{a_0}. \qquad (2.46)$$

For SiC substrates, the lattice-mismatch is calculated from the lattice constant a. For silicon and sapphire the calculation needs some explanation. Silicon is a well investigated material and only the III-N-relevant parameters shall be given. GaN can be grown in the hexagonal phase on the 111-plane of the silicon. In this case the effective lattice constant of 3.84 Å can be used. For sapphire, the growth is along the 0001-direction. This leads to an effective constant of 2.77 Å. Applying the rules of Matthews and Blakeslee [2.280] or Fischer [2.11] for the pseudomorphic growth of III-N materials, the critical thicknesses given in Table 2.27 are obtained for the semiconductor/semiconductor transitions. Table 2.27 gives estimates of the critical thickness for typical material compositions. $Al_{0.2} Ga_{0.8} N$ can only be grown with a thickness of 20 nm on top of fully relaxed GaN. The less conservative calculation approach of Fischer [2.115] leads to a critical thickness of 40 nm for $Al_{0.2} Ga_{0.8} N$. Thus AlGaN relaxation can occur for the composition $0 \leq x \leq 0.4$ for thicknesses

Table 2.27. Critical thicknesses obtained by various methods

Material	Thickness (nm)	Method	Ref.
$Al_{0.2}N_{0.8}N$ on GaN (sub.)	20	calc. (MB)	[2.11]
$Al_{0.2}N_{0.8}N$ on GaN (sub.)	40	calc. (Fischer)	[2.11]
AlN on GaN (sub.)	2.9	measured	[2.199]
$In_{0.1}Ga_{0.9}N$ on GaN (sub)	10	calc. elastic	[2.146]
$In_{0.17}Al_{0.83}N$ on GaN (sub)	∞	–	[2.231]
$In_{0.27}Ga_{0.77}N$ on GaN (sub)	15	calc. (Fischer)	[2.231]

MB Matthews/Blakeslee

of up to 30 nm in AlGaN barriers grown on GaN. $In_xAl_{1-x}N$ can be grown lattice-matched on GaN for $x = 0.17$. The critical thickness of $In_{0.1}Ga_{0.9}N$ on GaN is 10 nm. This leaves ample opportunities for heterostructure design.

III-N/Dielectric Interfaces

The physical nature of the III-N semiconductor surfaces toward vacuum, gases, liquids, or any dielectric material determines device performance to a high extent. A key property of GaAs materials as compared to silicon-based device technology is the lack of an unpinned GaAs/oxide interface [2.117, 2.118]. Similar surfaces prevail for III-N devices with SiN. For GaN, a Fermi-level pinning close to a midgap energy level is described in [2.483]. Using SiO_2 as a dielectric, the interface GaN/SiO_2 is found to have a very low surface charge density, as reported in [2.276, 2.277]. For the $Al_{0.24}Ga_{0.76}N$ surface, experimental evidence for the Fermi-level pinning is found in [2.154, 2.160]. The underly-

Table 2.28. Fermi-level (surface potential) energies of III-N semiconductors at vacuum and dielectric interfaces

Material	Surface potential (eV)	Ref.
n-GaN/Vacuum	$E_C - 1.2$	[2.303]
n-GaN/Vacuum	$E_C - 2.8$	[2.477]
n-GaN(Si)/Vacuum	$E_C - 1.5, E_C - 0.2$	[2.205]
GaN/SiN	$E_C - 2.4$	[2.82]
GaN/Si_3N_4	$E_C - 0.27$	[2.25]
GaN/SiO_2	$E_C - 0.13$	[2.25]
$Al_{0.34}Ga_{0.66}N$/Vacuum	$E_C(Al_{0.34}Ga_{0.66}N) - 1.65\,eV$	[2.154]
$Al_{0.24}Ga_{0.76}N$/Vacuum	$E_C(Al_{0.24}Ga_{0.76}N) - 1.6\,eV$	[2.160]
$Al_{0.35}Ga_{0.65}N$/Vacuum	$E_C(Al_{0.35}Ga_{0.65}N) - (1.4\ to\ 1.7\,eV)$	[2.209]
$GaN/Al_{0.35}Ga_{0.65}N$/Vacuum	$E_C(Al_{0.35}Ga_{0.65}N) - (1.4\ to\ 0.6\,eV)$	[2.208]
$In_{0.3}Ga_{0.7}N$/Vacuum	$E_C(In_{0.3}Ga_{0.7}N)$	[2.247, 2.477]
InN/Vacuum	$E_C(InN) + 0.6\,eV$	[2.477]

ing data are obtained by synchrotron radiation photoemission spectroscopy (SRPES). Table 2.28 compiles available surface and interface data. Due to the more ionic bond between gallium and nitrogen relative to more covalent semiconductors such as GaAs and InP, the pinning or Fermi-level stabilization at the surface should be less pronounced, as argued in [2.429]. However, several reports exist on the stabilization of the Fermi-level for GaN, $Al_xGa_{1-x}N$, and InN [2.154, 2.160].

III-N/Metal Interfaces

The properties of III-N/metal interfaces are essential for the performance of nearly all device applications.

N-type Ohmic Contacts

The formation of ohmic contacts is a fundamental component in the fabrication of semiconductor devices. For n-contacts in III-N, intrinsic doping is the most important contender for reducing the contact resistance. Due to the wide bandgap properties, the typical contacts yield special formation properties such as high-annealing temperatures between 500°C [2.197] and 900°C [2.31]. Special geometric definitions, such as recessed ohmic contacts [2.198] or non-alloyed implanted contacts [2.526], have been suggested. At the same time, good ohmic contacts are also more easily achieved for III-N semiconductors, due to the high sheet carrier density at the interface, due to both the impact of polarization or due to impurity doping, especially using silicon.

P-type Ohmic Contacts

P-contacts are very well investigated due to their fundamental importance in the field of optoelectronics, e.g., in [2.325]. P-contacts issues prevail due to both the reduced doping activation and deep activation energies relative to the valence band. These properties of p-doping limit the performance of optoelectronic devices, which is one reason for the great number of publications on this topic. Detailed overview data in this book are given in Chapter 4.3.1.

Schottky Contacts

The formation of Schottky contacts on III-N semiconductors is a very critical capability for two reasons. First, wide bandgap FETs are required to support very high fields because of the high operation voltages required, while the semiconductor/metal transitions form Schottky interfaces with barrier heights of ≤ 1.5 eV. Second, for high-power applications, the Schottky contacts are subject to the very high channel lattice temperatures ≥ 300°C. Thus, Schottky contacts currently form a weakpoint within the semiconductor device architecture, especially with respect to reliability. Further overview data are given in Chapter 4.3.2.

2.2.7 Transport Properties in Polarized Semiconductors

The polarization effects at the interface favor very strong quantization effects. The transport properties in this highly quantized situation [2.421] will be discussed again in this section.

Monte Carlo Simulations

Long before the realization of advanced GaN HEMTs, which allow the extraction of high-speed properties, a large number of predictions have been made, especially by ab-initio calculations or by Monte Carlo simulations based on these calculations [2.492]. Initial electron mobilities of GaN, InN, and AlN are calculated by variational principle in [2.64]. Initial MC simulations of the velocity-field characteristics in bulk wurtzite-phase GaN are presented in [2.37]. They predict a maximum electron velocity of 3.1×10^7 cm s^{-1} and a saturation field of 150 kV cm^{-1} at room temperature. The analysis is performed between 77 and 1,000 K. Simulations of the transient transport in GaN and InN are given in [2.121]. Most of the initial simulations mentioned so far do not consider the interaction of the electrons with longitudinal and transversal optical phonons correctly [2.19, 2.517], and further, in the case of InN, use the bandgap of 1.89 eV at RT, e.g., [2.348]. Fig. 2.7 gives the comparison of simulated and measured velocity-field characteristics of GaN. The data in Fig. 2.7 are taken from [2.37, 2.121]. Additionally, Fig. 2.7 depicts experimental

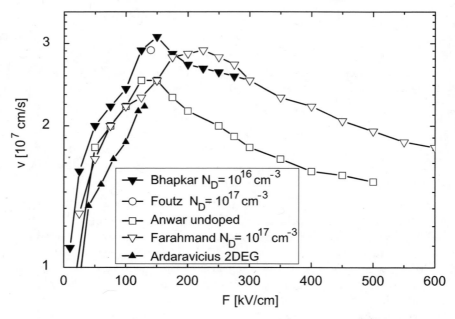

Fig. 2.7. Simulated velocity-field characteristics in bulk GaN by various authors

data of an AlGaN/GaN DEG which includes the interaction of the nonequilibrium phonon taken from [2.19]. Hot phonons and hot-electron penetration into AlGaN and GaN layers are reponsible for the lower than expected drift velocity. Ensemble Monte Carlo calculations within the density functional theory of GaN and ZnS are given in [2.119]. Ab-initio band structure calculations are used. These simulations propose both lower peak velocities and no occurrence of the Gunn effect in GaN. Device simulations for heterostructures including the quantization effects in the channel are given in [2.46]. The saturation velocity used is as low as $1.2 \times 10^7 \, \mathrm{cm \, s^{-1}}$. The electron relaxation time amounts to 0.1 ps for all materials. Very good agreement of measured velocities and MC simulations can be obtained [2.30], as also given in Fig. 2.7. Measured saturation velocities of $2.5 \times 10^7 \, \mathrm{cm \, s^{-1}}$ and a critical field at $180 \, \mathrm{kV \, cm^{-1}}$ are found for n-type bulk GaN. Peak velocities v_{peak} of $3.1 \times 10^7 \, \mathrm{cm \, s^{-1}}$ at an electric field of $140 \, \mathrm{kV \, cm^{-1}}$ are reported for AlGaN/GaN heterostructures. The obvious contradiction of several measurements and simulations will be discussed in the next section.

Velocity Measurements

Bulk and HFET MC simulations are not easily verified by experiment due to the sub-micron gate situation in FETs, the heterointerface situation in HFETs, and the strong field gradient near the gate [2.375]. Time-resolved electroabsorption measurements of the velocity-field characteristics in a bulk GaN pin-diode are presented in [2.500]. A time-of-flight technique is used. They predict a maximum electron velocity v_{peak} in bulk GaN of $1.9 \times 10^7 \, \mathrm{cm \, s^{-1}}$ and a saturation critical field of $E_{\mathrm{crit}} = 225 \, \mathrm{kV \, cm^{-1}}$. The measured peak velocity v_{peak} is still slightly lower than the prediction of the MC analysis. Other extraction techniques lead to stronger deviations. Derived from delay time analysis of HFETs, the maximum effective electron velocity in $Al_x Ga_{1-x} N$/GaN HEMTs, i.e., the quantized channel situation, with $l_{\mathrm{g}} = 1.3 \, \mathrm{\mu m}$ is found to be $1.2 \times 10^7 \, \mathrm{cm \, s^{-1}}$ in [2.5]. This is lower than the prediction of the MC simulations. Further extractions suggest that parasitic effects mask the real experimental effective velocity of $3.3 \times 10^7 \, \mathrm{cm \, s^{-1}}$ in a channel of AlGaN/GaN HFETs [2.42], which is found when considering the parasitic effects. Without the consideration of R_S and R_D, an effective velocity of $1.3 \times 10^7 \, \mathrm{cm \, s^{-1}}$ is extracted. The influence of charge control (charge nonconfinement or real-space transfer) at the heterostructure interface on carrier velocity is described in [2.462]. The source resistance R_S is found to be modulated in a nonlinear physical model. This modulation at high forward current I_D is correlated with the occurrence of space charge regions and carrier nonconfinement from the channel to the barrier under the source contact. Determination of quantum scattering lifetimes by Shubnikov–de Haas oscillations and parallel conduction are given in [2.409]. A ratio of the classical scattering time to the quantum scattering time of 6.1/1 is observed. This ratio indicates what proportion of the lifetime is represented by large angle scattering vs. all scattering events. The typical range found for the ratio for GaAs heterostructures is in the range 9–13 [2.409]. The results suggest that in addition to the optimization of the channel layers, also the parallel conduction in doped barrier layers has to be taken into account. With these considerations, all effects have been mentioned, which explain the inconsistency of measurements and simulations.

2.2.8 Polarization-Based Devices and Their Specific Properties

The strong polarization effects in III-N materials allow a number of special devices which are unique with respect to their properties. The first unique device is, of course, the classical III-N heterostructure FET, which yields extremely high channel carrier concentrations, even though there is no intentional impurity doping. For regular AlGaN/GaN HEMTs, the direct effects of spontaneous and piezoelectric polarization effects on the output characteristics of AlGaN/GaN HEMTs are discussed in [2.391]. The polarization critically influences the channel formation and thus also the output characteristics. The advantage of this paper is to show the direct correlation between device properties and polarization effects.

The high-dielectric constant of piezoelectric layers grown on top of a non-piezoelectric substrates can be effectively used in other electronic devices [2.338]. AlN [2.55] and GaN [2.242] surface-acoustic-wave (SAW) filters using the piezoelectric properties have been demonstrated. Epitaxially grown GaN piezoelectric thin-film SAW filters are grown on amorphous GaN buffers on sapphire. The low surface roughness of the material ensures good wave propagation while the good isolation between the electrodes ensure low electric losses. High-wave propagation velocities of $5,800\,\mathrm{m\,s^{-1}}$ and low insertion loss of $-7.7\,\mathrm{dB}$ are achieved [2.242]. The devices provide high fractional bandwidth and very low coefficients of frequency are obtained. Voltage controlled SAW filters on 2DEG AlGaN/GaN heterostructures are demonstrated in [2.129]. The external voltage applied to transducer is used to minimize the insertion losses in this case. This technology is suggested for cointegration with GaN MMICs. Further sensor devices are discussed with the applications at the end of this chapter.

2.3 Electrical and Thermal Limitations of Materials and Devices

Several predictions are being used to estimate the ultimate performance of nitride-based materials and devices. This section summarizes the most important figures-of-merit used in the analysis to predict the ultimate performance for fully developed device technologies.

2.3.1 Physical Modeling of Devices

Nearly all available simulation approaches have been used to model and predict the performance of III-N semiconductors and semiconductor devices. These include basic material research to obtain the band structure by density functional approaches [2.119], basic transport considerations to perform MC simulations [2.517], device drift-diffusion, and hydrodynamic multidimensional modeling approaches [2.46, 2.48], Monte Carlo approaches for both materials and devices [2.15, 2.111], as well as compact device modeling approaches [2.183, 2.392].

Material Modeling for Polar Devices

Material modeling from basic properties to devices is described, e.g., in [2.49, 2.111]. The complexities that arise in Monte Carlo-based modeling of noncubic symmetry semiconductors and their related devices are discussed in [2.50]. Owing to the increased size and number of atoms per unit cell, the band structure is far more complex in noncubic than in zincblende-phase semiconductors. This leads to a greater number of bands, smaller Brillouin zone, and an increase in the number of band intersections. Example articles are provided in [2.15, 2.49, 2.111, 2.492]. The detailed material properties have already been discussed in this chapter.

Device Simulation Including Polarization Effects

III-N device simulation is not very different from approaches in other semiconductor systems [2.406]. A quasi-2D-modeling of GaN-based heterostructure devices including quantum effects is given in [2.392]. An optimized effective mass-approach is coupled to a quasi-2D-model for the current flow. The PICs are included as surface charge in the Poisson equation which is solved self-consistently with the Schrödinger equation. The idealized model predicts very high drain currents of $2.5\,\mathrm{A\,mm^{-1}}$ for an $Al_{0.2}Ga_{0.8}N/GaN$ HEMT. On the HFET device level, full band Monte Carlo simulation of zincblende GaN MESFETs, including realistic impact ionization rates, are provided in [2.110]. The drain current in GaN MESFETs increases gradually with drain voltage V_{DS}. This is different from the impact ionization in GaAs MESFETs and it is related to the different impact ionization coefficients. Gate length scaling for $Al_{0.2}Ga_{0.8}N/GaN$ HFETs by 2D-full band MC simulation including polarization effect is analyzed in [2.15]. The polarization effects are included by constant interface charges at the barrier $(-\sigma)$ and channel interface $(+\sigma)$. From this, a cut-off frequency×gate length product $f_T \times l_g$ of $16\,\mathrm{GHz\,\mu m}$ can be deduced for a device with a gate length of $100\,\mathrm{nm}$. The maximum predicted output power P_{out} amounts to $46\,\mathrm{W\,mm^{-1}}$ for a gate length $l_g = 0.9\,\mathrm{\mu m}$, and $20\,\mathrm{W\,mm^{-1}}$ for a $100\,\mathrm{nm}$ device. The simulated on-state breakdown voltages are 300 and $60\,\mathrm{V}$, respectively. Several other approaches have been described to include strong polarization effects in device simulation. The carrier-concentration-dependent electron mobility in an AlGaN/GaN electron gas is analyzed by magnetoresistance and capacitance-conductance analysis in [2.185]. The effects of quantization are included in the mobility analysis. Scattering from interface states in AlGaN/GaN heterostructures is correlated with the high polarization field in the heterointerface. At temperatures higher than 200 K polar optical phonon scattering dominates the transport, yet both interface charge and roughness affect the mobility at the low and high sheet carrier density. For the compact models, a detailed physically based compact HFET model which includes thermal effects is given in [2.6]. The gradual channel approximation is used based on parameters on MC simulation. The thermal compression of a AlGaN/GaN device on sapphire substrate can be modeled very accurately. A theoretical study of GaN/AlGaN HEMTs by a compact model including a nonlinear formulation of the polarization is given in [2.527]. The cut-off frequency, transconductance, and current–voltage characteristics are computed in good agreement to measurements. The effect of the nonlinear polarization model on the sheet carrier density is also presented. An electrothermal Monte Carlo method is used for the investigation of submicrometer GaN HEMTs in [2.393]. The polarization effects are included according to (2.19). Additionally, fixed negative charges can be included to simulate surface traps. Peak velocities of 6×10^7 cm s^{-1} are reached at the drain side of the gate for a gate length $l_\mathrm{g} = 200$ nm. The maximum electron energies are as high as 2 eV.

Devices: Field Shaping and Breakdown Management

In addition to basic evaluation studies such as the comparison of zincblende-phase GaN, cubic-phase SiC, and GaAs MESFETs using a full band MC simulator [2.492], field shaping [2.183] and high-breakdown design are the principal applications of device simulations [2.358]. Such approaches allow the estimation and increase of maximum breakdown voltage BV_DS, and thus of the maximum output power P_out. Examples can be found [2.418, 2.465]. The impact of strain engineering on device level is stressed in [2.47]. Two findings are stressed: first, there are little changes in the DC- and transient-characteristics even for large stresses in overlayers. This is based on the stiffness of nitride semiconductors. A second finding focusses on the relaxation of the AlGaN. Piezoelectric polarization effects are more pronounced for devices with larger relaxation. Electric field shaping is presented in [2.358] for an AlGaN/GaN HEMT using field plates. The geometry of a field plate is optimized for reduced field peaking in the channel by 2D-simulation. A closed-form expression for the drain voltage dependence of the electric field using field plate is discussed in [2.183]. Direct analysis of the reverse gate leakage current I_G is provided by 2D-simulation in [2.182]. The mechanisms of direct tunneling, direct tunneling through a thin layer, and trap-assisted tunneling are distinguished by their field sensitivity. This allows to fit measured gate currents for various devices by 2D-simulation.

2.3.2 Devices: Figures-of-Merit

Apart from more specific simulation approaches, the prediction of the ultimate performance of semiconductor devices has a long-standing tradition. A great number of predictions are available for the promising material system of III-N semiconductors. Instead of repeating these various predictions, see, e.g., [2.301, 2.396, 2.462, 2.463], this work systemizes the figures-of-merits (FOMs) and calculates some principal dependencies. A good overview and comparison of different approaches is also given in [2.72].

Early initial predictions for wide bandgap compound semiconductors for superior high-voltage unipolar power devices are given in [2.415]. As given in (2.47), Johnson's FOM describes the power frequency product per unit width [2.167]:

$$\text{FOM}_{\text{Johnson}} = \frac{(v_{\text{sat}} \cdot E_{\text{crit}})^2}{2\pi}. \tag{2.47}$$

In (2.47), E_{crit} denotes the critical dielectric breakdown field and v_{sat} the effective saturation velocity of the material. A more microwave-oriented FOM is provided by Eastman in [2.132]. It combines microwave power, e.g., at 10 GHz, with the transit frequency f_{T} and the capacitive input reactance X_{c} [2.132], and the saturation velocity, as given in (2.48) [2.97]:

$$\text{FOM}_{\text{Eastman}} = P_{\text{out}}(\text{Class-A}) \cdot f_{10\,\text{dB}} \cdot R_{\text{L}} = 1.210^{23}\,\text{W}\,\text{Hz}^2\,\Omega. \tag{2.48}$$

$P_{\text{out}}(\text{Class-A})$ is the maximum output power in Class- A operation, $f_{10\,\text{dB}}$ is the maximum frequency yielding 10 dB of power gain, and R_{L} is the load impedance. This FOM is dedicated to the analysis of devices for the most important applications in high-power high-efficiency amplifiers. Wu [2.502] stated Baliga's FOM [2.26] as an efficiency measure for GaN HEMTs:

$$\text{FOM}_{\text{Baliga}} = \mu \cdot E_{\text{crit}}^2, \tag{2.49}$$

where again E_{crit} denotes the critical dielectric breakdown field and μ denotes the carrier mobility. High-efficiency devices have high-breakdown voltages (high E_{crit}) and low-access resistances (high mobility μ). The access resistances of III-N devices are typically higher, while the critical fields are significantly higher than in other fast semiconductors. This allows for efficient devices. Another FOM suggests maximizing the product of charge density n_{channel} and mobility μ to maximize gain and eventually, the current gain cut-off frequency f_{T}:

$$\text{FOM}_{\text{Gain}} = \mu \cdot n_{\text{channel}}. \tag{2.50}$$

Derived from this the channel charge density n_{channel} itself can also be maximized, as given in (2.51):

$$\text{FOM}_{n_{\text{channel}}} = n_{\text{channel}}. \tag{2.51}$$

Derived from (2.51), III-N HEMTs have a tremendous advantage in sheet carrier density as compared to other materials, while the absolute amount

Table 2.29. Properties relevant for low-noise applications for different HEMTs with a gate width $W_g = 2 \times 60\,\mu m$ and a gate length $l_g = 150\,nm$

Material (–)	$N_{F,min}$ at 10 GHz (dB)	G_{ass} at 10 GHz (dB)	BV_{DS} (V)	FOM_{LNA}	Ref.
GaAs PHEMT	0.5	14	6	168	[2.474]
InAlAs/InGaAs HEMT	0.3	16	4	213	[2.335]
AlGaN/GaN HFET	0.6	12	50	1,000	[2.220]

of carriers in the channel is not considered. Further, both (2.50) and (2.51) do not consider any breakdown issues, so an optimization only according to these FOM can be misleading, as the optimization may deteriorate breakdown, e.g., due to on-state channel breakdown. Fig. 2.8 illustrates Johnson's FOM, whereas Fig. 2.9 shows the channel Hall mobility versus charge carrier density as compared to values reported in the literature. As can be seen III-N HEMTs provide much higher sheet carrier densities for similar mobility values, as compared to GaAs HEMTs. A FOM from a thermal point of view is provided by Shenai et al. in [2.415]:

$$FOM_{Shenai} = \kappa_L \cdot \sigma_A E_{crit}. \tag{2.52}$$

In this case, κ_L is the thermal conductivity of the material in the heat-flow path (both semiconductor, substrate, and appropriate transition), σ_A is the electric drift-region-conductance per unit area in the semiconductor, and E_{crit} is the dielectric breakdown field. In terms of heat conductivity and conductance per unit area, III-N materials such as GaN and AlN are in the same order as Si, while the peak electric field at breakdown is significantly higher for III-N materials. For low-noise applications, the following FOM can be used, as defined by Nguyen in (2.53):

$$FOM_{LNA} = \frac{G_{ass} \times BV_{DS}}{N_{F,min}}. \tag{2.53}$$

Table 2.29 compares the minimum noise figure $N_{F,min}$, the associated gain G_{ass}, and the breakdown voltage BV_{DS} for a heterojunction FET with a gate width $W_g = 2 \times 60\,\mu m$ and a gate length $l_g = 150\,nm$. In this comparison, the linear impact of the breakdown voltage in (2.53) favors the AlGaN/GaN HEMT very much, while the actual low-noise performance is not outstanding. This FOM is thus suitable for applications which require both low-noise performance and extreme robustness.

2.3.3 III-N Devices: Frequency Dispersion

AlGaN/GaN HEMTs are found to be highly frequency-dispersive and did not yield the expected output power, especially in the beginning of device development, e.g., in [2.206]. Newer publications, e.g., [2.219], announced

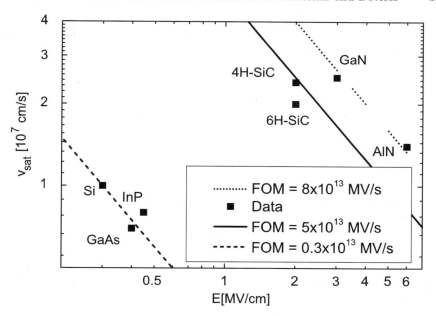

Fig. 2.8. Johnson's figure-of-merit in the E_{crit} vs. v_{sat} diagram

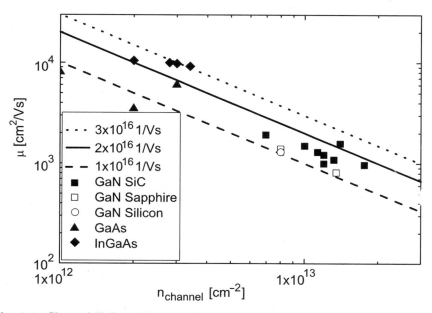

Fig. 2.9. Channel Hall mobility as a function of sheet charge carrier density n_{sheet} in GaN, GaAs, and InGaAs

dispersion-free devices and have to be taken with some caution; however, a lot of knowledge has been acquired for the reproducible control and reduction of dispersion. These results are discussed in more detail in Chapter 4. The aspect of performance degradation through dispersion, as described later, can be taken further. Related reliability issues are analyzed in Chapter 7.

Properties of GaN Semiconductors in Favor of Dispersion

One could ask why dispersion effects are so much correlated with III-N devices. In the GaAs and InGaAs material systems, frequency dispersion has been discussed in detail [2.51] but has not been found that crucial relative to the performance actually expected. In comparison to GaAs MESFETs, to (In)GaAs HEMTs, and other related devices, the physical origins of frequency dispersion in III-N HEMTs are more pronounced due to:

– Very high surface charge densities $n_{sheet} \geq 10^{13}\,cm^{-2}$
– Lattice-mismatch of semiconductor materials relative to the heterosubstrates
– The relatively high concentration of threading dislocations and other defects due to high epitaxial growth temperatures up to $1,100°C$ and high growth temperature gradients within the growth process
– The small absolute channel charge relative to the overall possible carrier charge in the whole device including buffer and nucleation layers

In the III-N material system vertical instead of lateral devices have also been realized, such a current aperture vertical electron transistors (CAVETs), e.g., in [2.516], which show reduced dispersion relative to the lateral devices, such as GaN MESFETs and HEMTs.

Observation of Dispersion

In a large number of publications, the following experimental findings are attributed to frequency dispersion; see, e.g., [2.206, 2.480, 2.535]. These include:

– Current collapse; i.e., the difference of continuous-wave (cw) currents measured from different quiescent bias, or a pronounced negative output conductance, which occurs even in devices with small gate width and is not primarily caused by thermal effects [2.336, 2.480]
– DC-current dispersion; i.e., the difference of cw-DC- or pulsed-DC from different quiescent bias [2.480, 2.483]
– Time dependence of the DC-device currents with time constants and related frequencies between the Hz and 100 MHz range [2.209, 2.483]
– RF-current slump; i.e., the difference of DC- and extracted RF-device current characteristics
– A (bias-dependent) shift of the threshold voltage
– RF-power slump; i.e., the reduction of the output power caused by RF-current slump [2.206]

- Premature power and gain compression [2.130]
- Gain reduction through a second virtual gate and thus possible gate length extension

The effects are certainly related, e.g., the RF-power slump in this enumeration is caused by the RF-current slump, i.e., reduction of $I_{DS,RF}$. The above-mentioned current reduction effects intermix with device degradation mechanisms. The suppression of dispersive effects is analyzed in Chapter 3 and Chapter 4. Further, the impact on performance issues and modeling of dispersion is discussed in Chapter 5 and the degradation-related issues are discussed in Chapter 7.

2.4 Substrates for Electronic Devices

Wide bandgap semiconductor substrates are components critical to the system electronics industry. The availability of affordable, freely available, large-diameter, high-quality substrates will heavily influence the final system introduction of wide bandgap devices and MMICs into systems. Good overview articles on the present status are available, e.g., in [2.9, 2.41].

2.4.1 Criteria for Substrate Choice

The following properties of substrates are evaluated in this book:

- Lattice-mismatch relative to the materials of the device layers
- Thermal conductivity κ_L and coefficient of thermal expansion (CTE)
- Maximum electrical isolation (at different temperatures)
- Price and price per area [2.41]
- Availability with respect to diameter (2–8 in.)
- Crystal quality and residual defect density [2.103]
- Surface properties and residual defects density [2.267]
- Wafer warp and wafer bowing [2.341]
- Mechanical and chemical properties with respect to thinning and viahole etching

Table 2.30 gives an overview of the most important physical properties of the substrate materials under consideration for III-N devices. These include lattice-mismatch, thermal conductivity, CTE, and isolation. Some aspects of Table 2.30 are discussed in detail. GaN is a very suitable substrate material choice; however, neither conducting nor semi-insulating GaN substrates are so far available in sufficient diameter and quality [2.106, 2.328]. Some promising results are mentioned in [2.384, 2.385]. GaN substrates can be either produced by growth on a foreign substrate or by a boule growth technique. 3-in. GaN grown by HVPE on sapphire substrates are reported in [2.80]. The situation is similar for AlN, which has also a very good thermal conductivity [2.404]. The thermal conductivity of SiC is outstanding. Due to growth conditions

Table 2.30. Basic properties of substrate and III-N semiconductor materials

Material	Lattice constant a (Å)	Mismatch to GaN (%)	κ_L (W m^{-1} K^{-1})	CTE (10^{-6} K^{-1})	Isolation (Ω cm)	Ref.
GaN	3.189	0	130	5.59	$\geq 10^9$	[2.341, 2.384]
6H SiC	3.08	3.4	490	4.2	$\geq 10^{11}$	[2.341, 2.384]
6H s.i. SiC	3.08	3.4	370	4.2	$\geq 10^{11}$	[2.341, 2.384]
Sapphire	4.758/$\sqrt{3}$	13	50	7.5	–	[2.123]
Silicon	5.4301	17	150	3.59	1–3 10^4	[2.341]
AlN	3.112	1	200	4.2	$\geq 10^{12}$	[2.341, 2.384]

and potential doping, the thermal conductivity of semi-insulating (s.i.) SiC can be reduced relative to pure crystalline SiC. The suitability of sapphire is strongly reduced due to its low thermal conductivity. All other substrate options except sapphire have a better thermal conductivity than GaN. Sapphire, even if thinned to 50 μm, reduces the reported power densities of devices in cw-operation to values of GaAs PHEMTs, e.g., in [2.16]. The difference in CTE of substrates and semiconductor materials affects both material growth and device reliability, because of the strain in the material. The maximum substrate isolation will further define the need to grow additional insulating semiconductor layers in the device. This will affect the attenuation of passive lines. For AlN, GaN, and SiC, the isolation has been developed to levels $\geq 10^9 \, \Omega$ cm at RT. Once the epitaxial layers are grown, one of the most critical consequences of the lattice-mismatch is the resulting bow of the wafer. Based on the data given in Table 2.30, an approximative scaling rule for the wafer bow is suggested in [2.341]:

$$W \approx t_{\text{sem}} \cdot d_{\text{sub}}^2 / t_{\text{sub}}^2. \tag{2.54}$$

Equation (2.54) formulates a relationship between the substrate diameter d_{sub}, the epitaxial layer thickness t_{sem}, and the substrate thickness t_{sub} based on a given bow, known, e.g., for given substrate and epitaxial thicknesses, and diameter. For a typical thickness of up to 4 μm for the epitaxial layers and a 2-in. wafer of a given thickness, a bow of 10 μm results. For a 4-in. wafer of the same thickness, this results in a bowing of 40 μm, which is close to the maximum tolerances of an optical stepper, as mentioned in [2.341]. This is mentioned here, as the bowing has to be accounted for during gate alignment and with respect to the mechanical stress during backside processing. The issues of properties, present availability, and crystal quality are discussed in the next sections for every type of material.

2.4.2 Silicon Carbide Substrates

Semi-insulating silicon carbide is one of the most attractive substrate materials for electronic applications, due to the favorable combination of lattice-mismatch, isolation, and thermal conductivity [2.9, 2.427]. It is available with increasing quality and diameters up to 4 in. [2.385]. Both polytypes 4H- and 6H–s.i. SiC are used. On the contrary, III-N optoelectronic applications use conductive SiC substrates [2.387] with similar diameters. Most of the outstanding electronic device results have been reported on semi-insulating SiC materials, e.g., [2.362].

Currently, the greatest amount semi-insulating SiC material is produced by Cree [2.427]. The growth method is based on physical vapor transport deposition (PVT). There is a number of other vendors for semi-insulating material, e.g., such as II-VI [2.103, 2.273] and Toyota in [2.320].

Norstel/Okmetic developed a high-temperature CVD (HTCVD) method for the growth of s.i. SiC substrates. II–VI produces undoped 2– and 3–in. 6H–SiC grown by PVT [2.14, 2.103, 2.273]. Detailed examples are given in [2.78, 2.159, 2.484]. Both growth methods PVT and HTCVD result in different substrate properties, e.g., with respect to residual mobility [2.246]. For conductive material, several vendors are available, e.g., [2.159, 2.359, 2.424].

A number of issues remain for the growth of SiC, so a comprehensive substrate analysis is still necessary. These issues include:

– Control and reduction of average micropipe defects density [2.427]
– General defect reduction, such as polytype inclusions and Si segregation [2.103, 2.274]
– Subsurface damage removal [2.273]
– Surface polishing to reduce average roughness [2.433]
– Availability of 2–6-in. wafers of increased and homogenous quality, e.g., [2.360]

Substrate isolation at RT is not considered a problem for SiC anymore, as levels beyond $10^9\,\Omega\,\mathrm{cm}$ at RT have been repeatedly reported [2.384]. This argument needs to be refined for high-temperature operation, as the substrate isolation mechanism is strongly temperature-dependent depending on the isolation method [2.529]. The availability of 4–6 in. diameters is desirable as most of the current GaAs fabrication lines are operated with these diameters.

SiC Substrate Characterization

Micropipes have been considered as the main obstacles for high quality SiC substrates and have attracted much attention for the definition of substrate quality, e.g., [2.103]. Micropipes are screw defects with a large Burger's vector which stretch through the complete crystals during bulk growth. Fig. 2.10 gives an optical image of a micropipe in a codoped semi-insulating SiC substrate. Polytype-control and background impurity control are of ultimate importance. Roughness is another criterion for substrates. Fig. 2.11 gives a microimage of a single micropipe for a surface with a residual roughness of 3 nm. Contrary to Fig. 2.11, Fig. 2.12 gives a microimage of a surface with a roughness RMS = 0.3 nm. The average micropipe density has been reduced

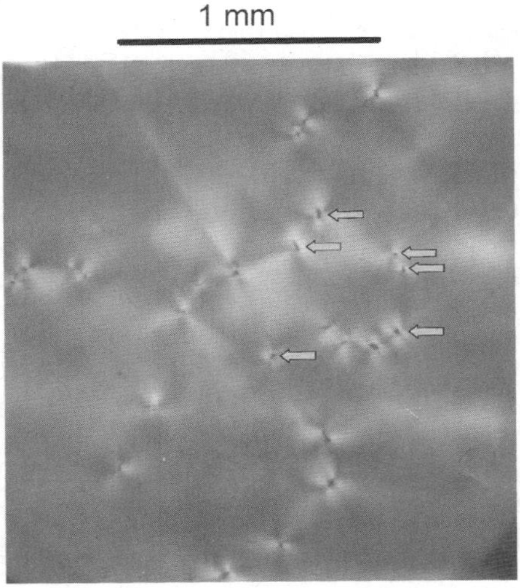

1 mm

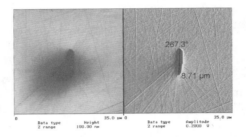

Fig. 2.10. Optical image of a micropipe cluster in s.i. SiC

Fig. 2.11. AFM image of a micropipe in s.i. SiC, RMS = 3 nm

Fig. 2.12. AFM image of a CMP-polished surface of a s.i. SiC, RMS = 0.3 nm

constantly during the substrate development, e.g., [2.384]. However, since micropipes define a killing defect for any active [2.28] and most passive devices (see Chapter 8), such as MIM capacitances, their average reduction is crucial, especially when increasing substrate diameter [2.433]. Filling micropipes by epitaxial growth is nearly impossible due to the typical geometry depicted in Fig. 2.11. Other dislocations than micropipes [2.103, 2.159] have an impact on material quality and on device performance. This is mainly due to the fact that they serve as nucleation centers for other defects in the epitaxial layers [2.274]. Basal-plane as well as screw dislocations have been found by high-resolution X-ray diffraction and synchrotron white beam topography in 4H–SiC [2.159]. The transformation mechanism of threading edge dislocations in the substrate into the epitaxial layer is discussed.

Subsurface damage [2.273] and grain boundaries, especially in the periphery of the crystals, are of significant importance for III-N device performance. Subsurface damage is not visible to some characterization methods. It becomes visible, e.g., after significant thermal or chemical treatment [2.103, 2.174]. The subsurface damage can be removed by the application of chemical–mechanical polishing, as reported in [2.174]. For the final substrate, several characterization methods are possible. Fig. 2.13 gives a comparison of two optical polarization images of 3-in. s.i. SiC substrates for the analysis of overall surface quality. The right image depicts a very good wafer, while the left wafer depicts a mediocre quality. Several other substrate characterization methods are applied. The polarized light microscopy (PLM) method is further explained in [2.267]. The procedure allows a fast method to analyze micropipe (clusters), screw dislocations, and domain boundaries. The wafers can further be analyzed with and without epitaxial layers in order to investigate the defect propagation into the epitaxial layers. Micro-Raman can be used for topography of SiC crystals

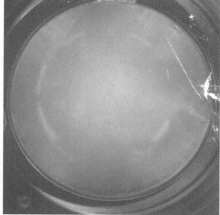

Fig. 2.13. Optical polarization images of two 3-in. s.i. SiC wafers

with a spatial resolution of $1\,\mu m$ [2.274]. Further, X-ray topography allows detailed analysis of the defect, as is performed, e.g., in [2.204]. The issues now are discussed with the respective substrate growth method.

Doped Conducting SiC Substrates in Electronic Applications

Doped conducting substrates are readily available by several industrial vendors for optoelectronic applications, e.g., [2.41]. N-type SiC substrates are ten times more cost-efficient than semi-insulating substrates and are presented, e.g., in [2.103, 2.316]. The conductivity is achieved by impurity doping with nitrogen. However, there are also significant demonstrations of high-power AlGaN/GaN HEMTs for electronic applications, e.g., [2.174] on n-doped SiC substrates using $\approx 10\,\mu m$-thick HVPE-based AlN interlayers. Very uniform device performance is given on this quasi-semi-insulating substrate approach. Further AlGaN/GaN HEMTs on p-type 6H–SiC substrates are given in [2.368]. Very high current levels $I_{Dmax} \geq 1.4\,A\,mm^{-1}$ and a cut-off frequency of 53 GHz are reported for devices with a gate length $l_g = 0.25\,\mu m$.

Semi-Insulating Codoped Substrates

For semi-insulating substrates with isolation levels of $\geq 10^6\,\Omega\,cm$ [2.83, 2.384], two basic mechanisms are used to obtain substrate isolation. In a first approach, SiC substrates are compensated by codoping, typically with vanadium [2.484]. The substrates suffer from the limited solulibility of vanadium in SiC [2.433]. The following substrate properties result from the codoping:

–   The deep traps activate at high-operation voltages and degenerate the performance of SiC MESFETs [2.484].
–   Vanadium codoping degrades bulk crystal quality due to increased local stress.
–   The micropipe density increases due to the codoping [2.535].
–   Codoping reduces lattice thermal conductivity by about 25% [2.83].

High-Purity Semi-Insulating (HPSI) Substrates

The negative effects of the codoping in s.i. SiC were discovered during SiC MESFET development, as discussed, e.g., in [2.433, 2.484]. SiC MESFETs on codoped substrates did not reach the expected power levels, which led to the development of undoped intrinsically compensated semi-insulating substrates. In this second approach, undoped high-resistivity or high-purity semi-insulating substrates (HPSI) [2.9, 2.294] with diameters up to 4 in. are under development with micropipe densities as low as $35\,cm^{-2}$. They promise the following properties [2.294]:

–   Increased uniformity suitable for MMIC processes
–   Increased resistivity and isolation

- Increased thermal conductivity [2.83]
- Reduced dispersive effects in the devices

However, in comparison to conducting SiC substrates, a smaller growth parameter window also has to be taken into account [2.535] for the growth of HPSI SiC substrates.

SiC Wafer-Size Scaling

Much of the III-N device results are so far achieved on 2-in. and 3-in. substrates. 3-in. s.i. SiC substrates are available with similar or improved quality relative to the 2-in. wafers [2.384]. Apart from the obvious cost improvement, wafer scaling is extremely desirable, as the history of GaAs has shown that uniformity of the devices eventually increases with increased substrate diameter, e.g., [2.446]. Further, equipment considerations are relevant, as most of the III-V processing facilities today are based on 4–6-in. equipment, which makes 2 and 3-in. processing a very cumbersome and expensive task. Four inch wafers with improving quality have been announced repeatedly [2.360, 2.385] and first production of GaN HEMT on 4-in. s.i. SiC substrate has been demonstrated.

2.4.3 Sapphire Substrates

Sapphire is a promising material due to the availability of cost-effective 2–4-in. diameter substrates [2.148]. It is widely used for development issues of electronic devices, and despite the lattice-mismatch, very impressive power densities of $\geq 12\,\mathrm{W\,mm^{-1}}$ and high-pulsed-output powers have been achieved [2.16, 2.66]. In addition to the adaption of the epitaxial parameters, the following substrate-related device issues have to be solved:

- General thermal management due to the reduced thermal conductivity
- Full wafer thinning:
    - Mechanical treatment of this hard material in the course of the thermal management, e.g., thinning from $500/250\,\mu\mathrm{m}$ to $50\,\mu\mathrm{m}$ as suggested in [2.16]
    - Impact of substrate quality during backside process
    - Wafer breakage due to the aforementioned issues
- Mismatch-induced stress and wafer bowing of thinned sapphire substrate due to the high lattice-mismatch to GaN
- Processing issues, e.g., in lithography due to the high substrate-related wafer bowing [2.16]

On the device level, results for AlGaN/GaN HEMTs on 4-in. sapphire substrates are given in [2.20]. The epitaxial growth gives a relatively high standard deviation of >4% for the Hall mobility. However, the high standard deviation can be well correlated with the standard deviation of the device parameters. The power performance of AlGaN/GaN HEMTs on sapphire substrate at 4 GHz is discussed in [2.478]. Power levels of $5\,\mathrm{W\,mm^{-1}}$ are reached at 2 GHz.

The epitaxial structure exhibits a thickness of $\geq 3\,\mu$m. Using an additional HVPE-grown buffer on sapphire substrates, the uniformity of DC- and RF-performance of MBE-grown AlGaN/GaN HEMTs is analyzed in [2.125]. The fast growth of a GaN buffer layer by HVPE is combined with MBE epitaxial layer growth on 2 in. wafers. The DC-parameters of 258 devices, such as threshold voltage and transconductance show very low standard deviation.

2.4.4 Silicon Substrates

Because of quantity of its use and its economic importance, silicon is the most important semiconductor and substrate material. Thus, the use of silicon as a substrate material is desirable, mostly for cost efficiency reasons. However, the use of silicon also introduces a number of critical issues and engineering challenges for III-N devices, as enumerated below:

- The impact of the lattice-mismatch relative to GaN with respect to material defects, quality, and device reliability [2.169].
- The difference in the thermal expansion coefficients, especially for high-power operation [2.328] (of similar gravity with respect to reliability, but not so often mentioned).
- The availability and cost of 2–6-in. insulating silicon substrates with isolation levels of 10–30 kΩ cm.
- The RF-transmission-line losses of passive lines for RF-frequencies [2.109].

The lattice-mismatch to III-N material in principle is a disadvantage; however, the substrate quality of silicon with respect to defects is outstanding compared to, e.g., SiC. Silicon substrates are virtually defect free, which helps a lot with device reliability [2.306]. To overcome the silicon issues, some very interesting epitaxial procedures have been developed, e.g., [2.168, 2.333], as will be described in Chapter 3. On the device level, AlGaN/GaN HEMTs on (111) silicon substrates are demonstrated, e.g., in [2.161, 2.162]. Very high power densities on silicon-substrates are reported by Triquint at 10 GHz [2.94, 2.109] and Nitronex at 2 GHz [2.52, 2.169]. Microwave and noise performance of 0.17 μm AlGaN/GaN HEMTs on (111) high-resistivity silicon substrates are reported. A minimum noise figure $N_{\mathrm{F,min}}$ of 1.1 dB is reported with an associated gain $G_{\mathrm{ass}} = 12$ dB at 10 GHz in [2.296]. The power density is 1.9 W mm^{-1} at 10 GHz [2.295, 2.297], while the PAE is only 18%. Promising reliability results for GaN HEMTs on silicon substrate have been given in [2.94] and [2.333] and are detailed further in Chapter 7.

2.4.5 GaN and AlN Substrates

Native GaN and AlN substrates have been only recently developed and used for electronic applications, although they are most important for optoelectronic applications. However, native substrates face increasing attention to the expected improvement in electronic device reliability, e.g., [2.73]. For native substrates, the lattice-mismatch is fully eliminated as compared to the heterosubstrates.

GaN Substrates

GaN substrates are only recently available in n-conducting and semi-insulating form with diameters suitable for electronic device fabrication, i.e., in 2-in. format. Free-standing GaN substrates are available based on two growth processes. First of all, GaN templates on sapphire can be used to reduce the defect density for optoelectronic devices, e.g., [2.33, 2.140]. The defect density is $8 \times 10^7 \, \text{cm}^{-2}$ for this kind of technique for a $10 \, \mu\text{m}$ layer grown by MOCVD. This kind of quasibulk approach can be used in a variety of ways, e.g., ELO or related approaches. In some approaches, the host substrate such as sapphire is removed leaving free-standing GaN. The second method is the growth of GaN boules, e.g., in [2.137, 2.328] by HVPE. Si-based n-doping is used for conductive substrates, while Fe doping is used to obtain s.i. substrates. Free-standing GaN substrates based on a similar procedure are provided by Sumitomo, e.g., in [2.140]. Impressive results of AlGaN/GaN HEMTs on free-standing GaN substrates have been demonstrated, e.g., in [2.73, 2.106]. The GaN substrates are fabricated by HVPE and are Fe-doped to achieve a resistivity of $10^6 \, \Omega \, \text{cm}$. The homoepitaxy changes the defect situation in the buffer completely, which makes these devices potentially more reliable. From an electronic device perspective, nitride devices are mostly unipolar devices. The bipolar devices face a number of substantial technological drawbacks, as described in Subsect. 2.5.6. Thus, the lattice-mismatch is relatively less important so far. In order to compare the impact of the crystal lattice constant and crystal quality, Fig. 2.14 gives the reported residual dislocation defect concentration as a function of the substrate lattice-mismatch relative to GaN for various substrate options. Fig. 2.14 shows, that only bulk GaN or ELO techniques (i.e. quasibulk GaN approaches) can provide the very low defect densities required especially for optoelectronic (laser) devices, whereas GaN electronic devices are typically grown with defect densities of 10^8–$10^9 \, \text{cm}^{-2}$ on various substrates.

AlN, ZnO, and Diamond Substrates

Native AlN substrates are also under discussion for optoelectronic applications [2.107, 2.403, 2.404, 2.407]. AlN itself is highly resistive and has a better thermal conductivity than GaN, actually the thermal conductivity is very similar to SiC. Further, compared to all other nonnative substrates, it has the lowest lattice-mismatch (Fig. 2.14). For the use of AlN for GaN-based lasers even the residual mismatch of the AlN relative to GaN lattice constant is too high to provide reliable GaN lasers without the use of additional concepts such as epitaxial regrowth techniques (ELO) to reduce the defect concentration [2.321]. This defect concentration of epitaxial layers grown on AlN substrates by migration enhanced MOCVD amounts to mid $10^6 \, \text{cm}^{-2}$. However,

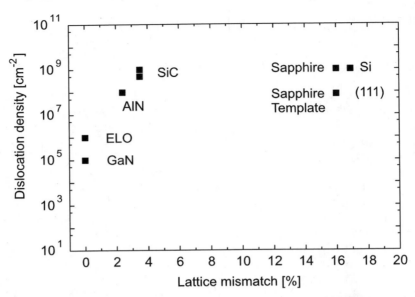

Fig. 2.14. Dislocation density as a function of lattice-mismatch relative to GaN

deep ultraviolet lasers are becoming feasible with the availability of a native substrate [2.404]. Free-standing AlN substrates with a thickness of 112 μm based on an Si(111) host substrate have recently been reported [2.224]. The boule growth of AlN by sublimation–recondensation is described as a growth method of the AlN substrates, e.g., in [2.404, 2.407]. High-purity AlN ceramics is sublimed in nitrogen atmosphere and recondensed on a seed substrate. GaN/AlGaN HEMTs on AlN substrate have been realized as reported, e.g., in [2.151]. The substrate diameter is so far limited; however, AlN substrates may enable more growth orientations and reduce the need of complex strain engineering. Electronic HEMTs on 30 μm-thick intermediate AlN layers grown by HVPE on SiC substrates are reported in [2.239]. The reduction of the gate-lag without surface passivation is attributed to the reduced amount of surface states related to dislocations and other defects. 10 μm-thick AlN layers are used in [2.174] to isolate conductive SiC substrates from the device. This approach is potentially a very low cost solution combining the advantageous thermal conductivity and isolation of AlN/SiC with a conductive low-cost substrate.

Further extending the material choices for substrates, wurtzitic zinc oxide (ZnO) materials and substrates are now available as large single crystals. For the optoelectronic applications, it is attractive due to a lattice-mismatch relative to GaN of only 2% and a large exciton binding energy of 60 meV. An overview of ZnO as a semiconductor material is described, e.g., in [2.257]. Next to the native substrate, also a developed wet-etch chemistry is available for ZnO as compared to the III-N. The advantages for electronic devices are the high breakdown fields and the high saturation velocity. CVD diamond provides the highest thermal conductivity of any potential substrate material. Recent reports in [2.165] provide AlGaN/GaN HEMTs on diamond substrate. The devices are initially grown on Si substrate and then transferred to the CVD diamond via an atomic attachment process. The devices are found to be relatively dispersive and the beneficial impact of the improved heat removal cannot be clearly demonstrated.

2.5 State-of-the-Art of Devices and Circuits

Numerous overviews on nitride devices and research issues are available due to its dynamic development, e.g., [2.353, 2.361, 2.401, 2.455, 2.464]. General overviews of GaN/AlGaN heterostructure devices for photodetectors and field-effect transistors are given, e.g., in [2.96, 2.301, 2.422, 2.465]. General properties of undoped AlGaN/GaN HEMTs for microwave power amplification are discussed in [2.96]. Early reports of AlGaN/GaN HEMTs can be found, e.g., in [2.190], while initial reports for GaN MESFETs can be found in [2.418]. Although it is impossible to provide a full picture of all the achievements, a systematic overview of the published key issues, to the best of the author's knowledge, is given in the following. Some of the results may be outdated quite soon; however, the story of the development is highlighted.

2.5.1 Nitride-Based Diodes

GaN Diodes

Nitride-based diodes are attractive due to their high breakdown voltages in combination with high-carrier velocities leading to higher conversion efficiencies [2.76] and improved temperature performance. Possibly GaN also provides the material property of negative feedback leading to Gunn oscillations, as discussed, e.g., in [2.408]. A theoretical discussion of GaN diodes with respect to negative differential resistance is provided in [2.8]. Simulation results for the operation of wurtzite and zincblende GaN Impact Avalanche Transit Time (IMPATT) diodes are given in [2.8, 2.363]. Further, the possibility of high-frequency operation at D-band (140–220 GHz) of GaN-based IMPATT diodes is discussed in [2.8].

SiC-Based Diodes

SiC diodes serve as a reference and target high-speed high-power density switching applications, e.g., [2.387]. They are applied for switch-mode power supplies, e.g., in computer servers [2.359]. Though the overall numbers of SiC Schottky diodes produced so far are very small relative to the mass market of silicon diodes, the high power density, the low R_{on} [2.494], and the high-temperature operation provide very interesting niche markets. Commercial

SiC diode process of Cree in [2.368, 2.427] and Infineon [2.434] are available with strong research efforts underway on circuit and application level, e.g., [2.366, 2.434]. The reduction of the reverse recovery current during conversion and the improvements in conversion efficiency relative to Si diodes are mentioned, especially at frequencies beyond 100 kHz. SiC diodes are further discussed for the replacement of converters in hybrid car power-conversion applications [2.359].

2.5.2 Power Electronics

III-N wide bandgap semiconductors are unique with respect to power electronics from low frequencies to the RF-domain. The motivation for the use of III-N-based diodes [2.76] and FETs [2.397] stems from their increased switching speed and thus their conversion capabilities and more efficient transmission and distribution of electric power at increased operation temperatures, e.g., [2.334]. The target is the development of a technology for fast and efficient GaN-based electrical switches at a wide range of power levels. General overviews of wide bandgap semiconductor power electronics are provided in [2.334, 2.469, 2.494]. Typical expectations and specifications are:

– High stand-off voltage, e.g., >1 kV [2.76, 2.438]
– High absolute current levels, e.g., up to 2 kA conducting current [2.76]
– A low forward voltage drop, e.g., ≤2% of the rated voltage
– Higher operation temperature (e.g., 225°C [2.334])
– High frequency switching operation, e.g., at 50 kHz and above

The use of GaN-diodes in low-frequency control applications is discussed in [2.59]. GaN is predicted to have insertion and isolation loss properties similar to other semiconductors. The trade-off between parasitic resistance, breakdown voltage, and off-state capacitance is discussed. GaN devices do not necessarily provide higher breakdown voltages, but GaN FETs provide lower switching energies for the same breakdown voltages. Principal design considerations for III-N high-power devices are given in [2.27]. Planar Schottky devices are fabricated to obtain principal device parameters. Based on that, AlGaN thyristor switches with breakdown voltages of 5 kV at current densities of $200\,A\,cm^{-2}$ with frequencies above 2 MHz are predicted based on the basic device parameters. Future military applications of wide bandgap power semiconductor devices are discussed in [2.105]. The high-power handling capabilities of the SiC and GaN devices in general allow more efficient power distribution at higher power levels and more compact device sizes. Estimates of the theoretical limit of lateral wide bandgap semiconductor devices for switching applications are given, e.g., in [2.396]. Lateral switching devices based on GaN and SiC are discussed for breakdown voltages of 1 kV and on-resistances of $1\,m\Omega\,cm^2$. The main advantage of the wide bandgap devices is the drastic reduction in gate width and chip area, and in power loss. At the same time the increase of the power density in the GaN FETs by a factor of

6 leads to an increase of the thermal dissipation and requires a significantly improved packaging technique.

RF-Switches

A high-power Tx/Rx RF-switch IC using AlGaN/GaN HFETs on sapphire substrate is described in [2.158]. The device is capable of handling a power of 46 dBm at 1 GHz with an insertion loss of 0.26 dB and an isolation of 27 dB. The high-breakdown voltage leads to a reduction of the number of FETs as compared to circuits realized with GaAs technology. A single pole double through (SPDT) switch implemented with GaN HFETs featuring a double recess technique is presented in [2.144]. A low insertion loss of 1.1 dB and an isolation of 21 dB can be realized. The double recess technique including an ohmic and a gate recess allows to optimize the access resistance and leakage current independently.

DC–DC-Converter

Fast and efficient DC–DC-converters are one application for AlGaN/GaN HEMTs. An overview of the suitability of wide bandgap semiconductors for this application is given in [2.460]. As an example, a 600 V fast DC–DC-converter AlGaN/GaN power HEMT is demonstrated in [2.397]. An optimized design including a source-terminated field plate delivers $R_{on} = 0.5 \, \text{m}\Omega \, \text{cm}^2$ and a maximum switching current density of $850 \, \text{A} \, \text{cm}^{-2}$. The power efficiency varies between 83% at 100 kHz and 75% at 500 kHz. A demonstration of a Class-E amplifier with a PAE of 89.6% at 27.1 MHz is given in [2.394, 2.395]. The applied drain voltage is 330 V. The devices are grown on n-SiC substrate and use a source-terminated field plate. Even higher breakdown voltages of 1.6 kV and low $R_{on} = 3.4 \, \text{m}\Omega \, \text{cm}^2$ are reported in [2.459]. With the application of conductivity modulation, even higher breakdown voltages are achieved, e.g., [2.471]. A typical FOM for power conversion and switching reads:

$$\text{FOM} = \frac{V_{\text{breakdown}}^2}{R_{\text{on}}} = q \cdot \mu \cdot n_{\text{sheet}} \cdot E_{\text{crit}}^2. \tag{2.55}$$

The limits of the material system used can be deduced with the approximation given in (2.55). A more advanced approximation including a surface limit is given in [2.459]. These limits are also depicted in Fig. 2.15. There are several examples for the trade-off between breakdown voltage and on-resistance for GaN HEMTs in comparison with Si and SiC devices, as given in Fig. 2.15. Data points are taken from [2.2, 2.63, 2.138, 2.213, 2.310, 2.398, 2.438, 2.445, 2.471, 2.472]. Silicon devices beyond the actual Si limit have been realized [2.310]. The third parameter, the switching time, is not explicitly mentioned in Fig. 2.15. However, turn-on delays of 7.2 ns can be reached with GaN FETs. This turn-on delay is a only a fraction of the Si MOSFET value. Further improvements are found for wide bandgap semiconductor with respect to heat dissipation and reduced chip area, as discussed in [2.396]. A recent general

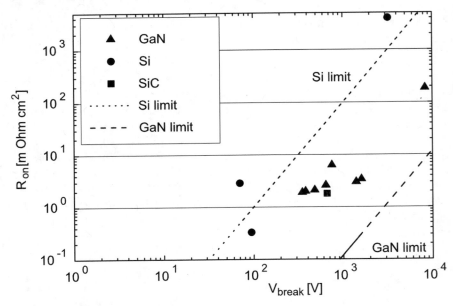

Fig. 2.15. On-resistance R_{on} as a function of breakdown voltage for switching applications

overview on the use of diamond for power applications is given in [2.207]. This material system promises very high breakdown voltages of 10 kV [2.396] and operation temperatures of 1,000°C, while this approach is still in its infancy.

2.5.3 RF-Metal Semiconductor Field-Effect Transistors (MESFETs)

MESFETs have been introduced as the first wide bandgap FETs into industrial production.

SiC MESFETs

A compilation of substrate and semiconductor material properties of SiC is given in [2.139]. Both static induction transistors (SIT) and SiC MESFETs are discussed. Pulsed-output power levels of 900 W of SiC SITs at 425 MHz with a drain efficiency of 80% are reported as well as output power levels of 80 W at 3.1 GHz with MESFETs. Substrate and epitaxial issues for SiC power devices are discussed in [2.433]. A classification of the residual defects is also given. A complete SiC MESFET technology, especially for S-band RF-applications, is provided by Cree [2.75]. The transition of this technology from discrete MESFETs to a high performance MMIC technology is described in [2.293].

The MMIC devices yield excellent wafer-to-wafer uniformity and a high power density of $4\,W\,mm^{-1}$ up to frequencies of 3.5 GHz. Silicon carbide MESFET performance and its application in broadcast power amplifiers between 0.5 and 0.9 GHz are reported in [2.454]. The SiC devices yield a much higher operation voltages of up to 55 V in comparison to Si LDMOS. This leads to increased output densities of $\geq 2\,W\,mm^{-1}$, but especially higher input and output impedances for SiC MESFETs. A comparison of SiC and GaN transistors for microwave power applications is given in [2.463]. The main distinction suggested is the power density of SiC MESFETs with $5–6\,W\,mm^{-1}$ up to X-band, while GaN is predicted to achieve up to $10\,W\,mm^{-1}$ up to 100 GHz on the device level.

GaN MESFETs

GaN MESFETs provide wideband properties similar to SiC MESFETs; however, both lack the potential advantages of heterojunctions. The advantage of GaN MESFETs relative to SiC MESFETs is the increased speed performance. This speed performance is, however, inferior to GaN heterostructure devices, e.g., as demonstrated in [2.62, 2.379]. A theoretical comparison by simulation of GaN MESFET-based on different GaN material phases wurtzite and zincblende GaN is given in [2.111]. The wurtzite structure yields higher breakdown voltages and lower transconductance than the zincblende structure. The simulated cut-off frequencies of wurtzite MESFETs are significantly lower, again based on the lower transconductance. Another systematic comparison of zincblende-phase GaN, cubic-phase SiC, and GaAs MESFETs using a full-band Monte Carlo simulator is given in [2.492]. A factor-of-four improvement of the breakdown voltage for GaN devices is found as compared to GaAs MESFETs for the same gate length of 100 nm. The experimental realization of doped-channel GaN heterojunction MESFETs are described in [2.62]. The devices with a gate length of 0.25 μm yield an output power density of $1.7\,W\,mm^{-1}$ at 8.4 GHz on sapphire substrates. The doped channel layer has a thickness of 50 nm and is n-doped with $n = 2 - 5 \times 10^{17}\,cm^{-3}$. The $Al_{0.15}Ga_{0.85}N$ layer has thickness of 18 nm. Epitaxially grown GaN junction FETs are given in [2.531]. The cut-off frequencies are $f_T = 6\,GHz$ and $f_{max} = 12\,GHz$ for a gate length of 0.8 μm. The channel mobility is $270\,cm^2\,V^{-1}\,s^{-1}$ for a free carrier concentration of $1.3 \times 10^{18}\,cm^{-3}$. The nominal p-doping of the pn-junction is $5 \times 10^{19}\,cm^{-3}$. The impact of p-doped overlayers on the dispersion in GaN junction FETs is described in [2.166]. The p-doped layer removes the low-frequency dispersion as it screens the channel from the surface states. This concept is promising, as the reproducibility of an epitaxial concept is higher than of a technological approach. However, the Mg trap depth is not sufficient to avoid carrier modulation, which leads to longer effective gate length and thus lower f_T and f_{max}. A GaN MESFET based purely on polarization doping through a graded AlGaN barrier layout without any impurity doping is compared to a conventional impurity n-doped GaN MESFET in [2.379]. The polarization-doped devices with a gate length of $0.7\,\mu m$ yield $f_T = 19\,GHz$ and $f_{max} = 46\,GHz$.

2.5.4 Metal Insulator Semiconductor Field-Effect Transistors (MISFETs)

Nitride MISFETs are very attractive candidates for improved performance of the gate diodes relative to nitride MESFETs and HEMTs with metal/semiconductor Schottky-gates. At the same time MISFETs require the mastery of the insulator/semiconductor interface. This interface affects threshold voltages and other device properties to a great extent. GaN-based MOSFETs (Metal Oxide Semiconductor FETs) have been demonstrated, e.g., in [2.191]. The comparison of the SiO_2/AlGaN/GaN MOSFETs with regular Schottky-gate HEMTs yields similar drain current densities, while the gate current density is reduced by six orders of magnitude for the MOSFET. The threshold voltage is shifted by $-3\,V$ for the MOSFET relative to the HEMT. A quarter micron SiN/AlGaN/GaN MISFET was developed as reported in [2.1]. Again the threshold voltage V_{thr} is shifted to negative values of $-7\,V$. This MISFET further yields a high power density of $6\,W\,mm^{-1}$ up to $26\,GHz$. The DC- and small-signal parameters of the MISFET are very similar to those of the HEMT. An overview of III-N insulating gate FETs is further provided in [2.193]. The low-leakage currents of the MISFETs found at RT are also reduced at temperatures of $300°C$. Further, the MISFET concept is applied to double heterojunction AlGaN/InGaN/GaN structures, which reduce the current collapse to a high extent [2.192, 2.426].

Power Performance of MISFETs

Several demonstrations of the power performance of III-N MISFETs are available [2.69, 2.74]. Reduced gate-leakage of MIS gate devices relative to MESFETs and consequently improved linearity at $4\,GHz$ are demonstrated in [2.69]. The third-order intermodulation ratio of $30\,dBc$ is reached with a power density of $1.8\,W\,mm^{-1}$ and an associated PAE$= 40\%$ at $2\,GHz$. SiO_2 is used as the gate dielectric. The microwave performance of undoped AlGaN/GaN MISFETs on sapphire substrates is given in [2.74]. A power density of $4.2\,W\,mm^{-1}$ at $4\,GHz$ is obtained at $V_{DS} = 28\,V$, again with a very negative threshold voltage $V_{thr} = -10\,V$. The extrinsic transconductance is $100\,mS\,mm^{-1}$. SiO_2/AlGaN/InGaN/GaN MOS double heterostructure FETs (MOSDHFETs) are demonstrated in [2.192, 2.426] with the introduction of the new channel material InGaN. The importance of the additional carrier confinement is to reduce the spillover into trapping states. A Cat-CVD Insulated AlN/GaN MISHFET is given in [2.142]. Very thin AlN layers with a thickness of $2.5\,nm$ are used as barrier layers. On top of this barrier a $3\,nm$ SiN layer is deposited by Cat-CVD. This layer combination allows a very small gate-to-channel separation, which is used combination with gate length of 60 nm to achieve a cut-off frequency $f_T = 107\,\text{GHz}$. The application of GaN MISFETs for low-loss switching is discussed in [2.217]. A 1 mm wide AlGaN/GaN MOSHFET yields switching powers of 43 W with 40 dB of isolation and 0.27 dB insertion loss at 100 MHz. The SiO_2 thickness is 7 nm in this case. The very negative threshold $V_{thr} = -9\,\text{V}$ is achieved for the device, which yields a maximum saturated pulsed-drain current of $2\,\text{A}\,\text{mm}^{-1}$.

2.5.5 High-Electron Mobility Transistors (HEMTs)

GaN HEMTs or heterojunction FETs (HFETs) are currently the most widespread and most advanced electronic nitride devices. They make full use of heterostructures and the advantageous breakdown and transport properties of undoped GaN. A huge number of compilations of GaN HEMT data is available, e.g., in [2.39, 2.96, 2.301, 2.302, 2.463]. Initial overviews are given in [2.40, 2.312, 2.422]. Initial microwave results of AlGaN/GaN HEMTs have been given in [2.40, 2.190, 2.506]. An overview of AlGaN/GaN HEMT technology is provided by Keller et al. and Mishra et al. from UCSB in [2.188, 2.301]. P_{sat} An overview of the application of GaN-based HFETs for advanced wireless communication is provided in [2.341]. Recent developments and trends in GaN HFETs in Europe are given in [2.455, 2.476]; overviews of the US activities are given, e.g., in [2.361, 2.401]. This section provides a systemized data collection which is presented with emphasis on the frequency aspect and the achievements relative to the theoretical predictions. To systemize device performance, Fig. 2.16 gives a compilation of the product cut-off frequency $f_T \times l_g$ as a function of gate length l_g. A product of 12 GHz µm is found for $l_g \geq 300\,\text{nm}$ for GaN HFETs. A GaAs reference value of 15 GHz µm is assumed. For gate lengths smaller than 300 nm, the short-channel effects become visible for GaN HFETs. Fig. 2.17 gives a compilation of the output power per unit gate width as a function of frequency. A distinction is made between pulsed- and cw-output power. The maximum power densities reach $12\,\text{W}\,\text{mm}^{-1}$ without additional measures at the gate, while output power densities $\geq 40\,\text{W}\,\text{mm}^{-1}$ have been reported using field plates [2.511]; however, reliability concerns prevail. Significant improvements of the output power up to 101 GHz have been reported as compared to all other semiconductor technologies [2.291, 2.378]. Similar to Fig. 2.17, Fig. 2.18 gives the absolute cw- and pulsed-output power as a function of frequency. The power levels on the transistor level reach 370 W under cw-operation conditions [2.488] at 2 GHz. Under pulsed-operation with low duty cycles, even higher output power values of up to 1 kW have been reported [2.268, 2.304, 2.503]. For low-noise operation, Fig. 2.19 gives the minimum noise figure $N_{F,min}$ as a function of frequency for both a single-recess GaAs PHEMT and a planar GaN HFET. The comparison yields promising minimum noise figure for the GaN HFET, which is slightly increased relative to the GaAs PHEMT [2.153, 2.262]. The next paragraphs systematically discuss the various frequency bands with respect to power and noise performance.

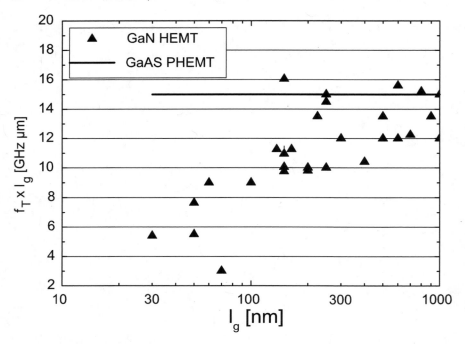

Fig. 2.16. Product cut-off frequencies $f_T \times l_g$ as a function of gate length l_g

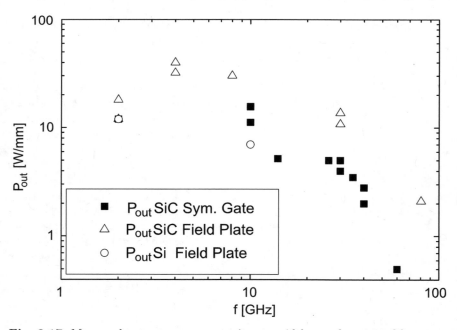

Fig. 2.17. Measured output power per unit gate width as a function of frequency

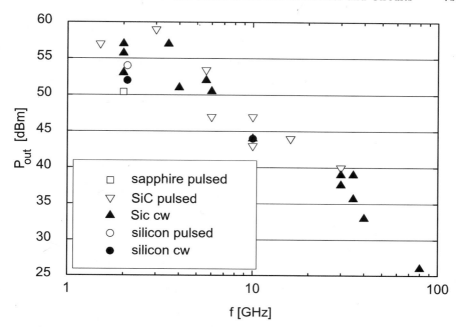

Fig. 2.18. Absolute cw- and pulsed-output power of AlGaN/GaN HFETs on various substrates as a function of frequency

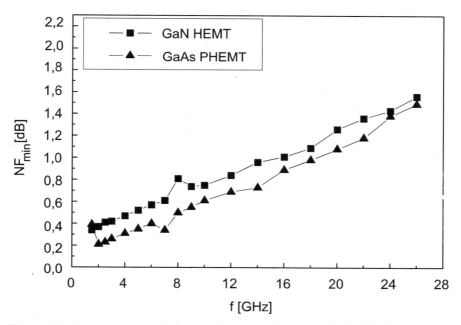

Fig. 2.19. The minimum noise figure $N_{F,min}$ for GaAs and GaN HEMTs as a function of frequency at a drain current $I_D = 100\,\mathrm{mA\,mm^{-1}}$ for $l_g = 150\,\mathrm{nm}$

L-Band and S-Band Frequencies (1–3 GHz)

GaN HEMTs are attractive for base-station applications, due to the increased operating voltages ≥ 28 V, higher impedance levels, and reduced thermal memory effects. Overviews and guidelines for future directions are given, e.g., in [2.309] and in [2.495].

Mobile Communication

Cree gave the first report of cw-power operation of GaN HEMTs on SiC with output power levels beyond 100 W with 108 W at 2 GHz in [2.362]. Higher power operation at 3.5 GHz has recently been demonstrated in [2.503] with output power levels of 550 W in pulsed-operation with low duty cycle at $V_{DS} = 55$ V in a single-stage design. On s.i. SiC substrate, an 80 W/96 W AlGaN/GaN HFET with a field-modulation plate is demonstrated by NEC in [2.343, 2.345]. The 24 mm device yields a peak PAE of 54% for an operation bias $V_{DS} = 32$ V. The 4 mm subcell yields a power density of 8.1 W mm^{-1} at $V_{DS} = 41$ V. An AlGaN/GaN field-plate-FET with a gate width of 32 mm using a recessed gate is demonstrated in [2.344]. The cw-power density is 4.7 W mm^{-1} and the peak PAE is 64% at 2 GHz and $V_{DS} = 47$ V. Similarly, a cw-output power of 179 W is then reported in [2.190] for a gate periphery of 48 mm. The linear gain is 9.3 dB at 2 GHz, the PAE is 64% at $V_{DS} = 46$ V. A similar transistor operated under one-carrier WCDMA conditions presents 280 W of single-ended-amplifier output power in [2.487] with the application of a gate width of 48 mm with a linear gain of 12.6 dB at $V_{DS} = 48$ V. The drain efficiency is 29% at 8 dB of backoff from the saturated power in WCDMA operation. The output power levels can be increased to 370 W with paralleled devices, again achieved under one-carrier WCDMA conditions at 2.14 GHz. In this case the linear gain is 11.2 dB at $V_{DS} = 45$ V. AlGaN/GaN HFETs on thinned sapphire substrates with 110 W of pulsed-output power at 2 GHz at $V_{DS} = 40$ V are presented by NEC in [2.16] with $W_g = 32$ mm device. The related cw-output power is much lower and reaches 22.6 W for a $W_g = 16$ mm device. The sapphire substrate is thinned to 50 µm in this case.

Fujitsu and Eudyna have demonstrated some of the most advanced device results with respect to base-station applications. Early results present a 36 W cw-AlGaN/GaN power HEMT using a surface-charge-controlled structure [2.198]. This structure combines a doped GaN cap to screen the polarization charges at the AlGaN/GaN interface with very well optimized material. Both the barrier thickness and the SiN passivation are optimized. An improved intermodulation distortion profile of such AlGaN/GaN HEMTs at a drain bias voltage $V_{DS} = 30$ V is reported in [2.318]. No sweet-spot behavior is found for two-tone measurements in deep class-A/B operation. Further technology development is described in [2.195, 2.197]. CW-operation of a 24 mm device at $V_{DS} = 50$ V is achieved with a peak PAE of 60% and an output power of 48 W at 2 GHz. Output power levels of 150 W in cw- and 174 W at 2.1 GHz in WCDMA-operation at $V_{DS} = 63$ V are reported in [2.172, 2.196] for a single packaged device. The PAE is 54% in cw-operation. For four-carrier WCDMA signals including the application of predistortion system, a peak efficiency of 40% is reached in Class-A/B operation while the WCDMA ACLR specification of $-50\,$dBc at $10\,$MHz offset is preserved. The gate width is $36\,$mm with a unit finger length of $400\,\mu$m. Based on these devices, a GaN HEMT amplifier in dual-device push–pull configuration is demonstrated in [2.171, 2.195]. The devices are again biased at $V_{DS} = 50\,$V. The cw-output power of the amplifier at $2.1\,$GHz is $136\,$W with an associated PAE of 53% and a linear gain of $12.9\,$dB. The amplifier is further characterized in one-carrier WCDMA operation. With the application of a DPD system, a peak output power of $250\,$W is reached. The associated drain efficiency is 37% at an ACLR of $-55\,$dBc. The average output power amounts to $46.5\,$dBm. The technology transfer and very promising reliability results are reported at $V_{DS} = 60\,$V [2.197, 2.330]. The devices can be operated at a compression level of P_{-3dB} at $V_{DS} = 60\,$V for $1,000\,$h at room temperature. The resistance of the ohmic contacts is found to stable within 6% at high current density of $2 \times 10^6\,$A$\,$cm^{-2} at $250°$C in nitrogen ambient. High-efficiency Class-E amplifiers in this technology are given in [2.473]. The amplifiers reach 82% of drain efficiency at $2.1\,$GHz, an output power of $10\,$W, and a linear gain of $19.5\,$dB. 45% of drain efficiency are reached for two-carrier WCDMA operation at ACLR of $-50\,$dBc. The technology has further been exploited for air traffic radar. Pulsed-output-power levels of $500\,$W at $1.5\,$GHz for radar operation are reported in [2.268]. The amplifier consists of four chips with a gate width of $36\,$mm each with an overall gate width of $144\,$mm in a dual push-pull configuration. Even higher output power levels are obtained at S-band in [2.304]. A maximum pulsed-output power of $59.3\,$dBm is obtained between 2.9–$3.3\,$GHz with a linear gain of $14\,$dB at $V_{DS} = 65\,$V and 10% duty cycle. The pulse width is $200\,\mu$s and the associated PAE amounts to 50%. Again a four-chip configuration is used with a single finger gate width of $375\,\mu$m and a total gate width of $4 \times 36\,$mm $= 144\,$mm.

RF Microdevices announced the commercial application of GaN FETs on $4\,$in. sapphire [2.412] and later on $3\,$in. s.i. SiC substrates in [2.134, 2.482]. The technology comprises very high power cw-densities of $\geq 22\,$W$\,$mm^{-1} at $2.14\,$GHz and $V_{DS} = 80\,$V. A $20\,$mm devices delivers more than $100\,$W of output power with an associated PAE of 55% at $V_{DS} = 48\,$V. Dual-gate devices are used for driver applications in [2.481]. Different dual-gate technologies are compared to single gate technologies. As an outcome a ground-terminated field plate on top of the SiN yields higher gain per stage, higher PAE, and better reliability under RF-drive.

Freescale provides AlGaN/GaN HEMTs on s.i. SiC for an industrial process [2.131]. Very high cw-power densities of 10–$11\,$W$\,$mm^{-1} are reached with PAE levels of 67% at $2.14\,$GHz. Devices with $12.6\,$mm periphery yield output power levels of $74\,$W or nearly $6\,$W$\,$mm^{-1} in cw-operation. The SiC substrates are thinned to $100\,\mu$m.

GaN devices on silicon substrates show much economic potential due to the low substrate cost. Conducting and high-resistivity silicon substrates are suitable candidates for operation of up to 10 GHz [2.109]. The performance of AlGaN/GaN HFETs fabricated on 100 mm silicon substrates for wireless base-station applications is further evaluated in [2.52]. Despite the high lattice-mismatch, very good reliability data are reported predicting 20 years of life-time based on a junction temperature of 200°C and 10% drift of the parameters. Cw-power densities of up to 12 W mm^{-1} at 2.14 GHz and $V_{DS} = 50$ V are reached with this AlGaN/GaN HEMT technology [2.169]. The gate width in this case is 100 μm, the maximum PAE 52%, and the power gain 15 dB. Recent reports announce cw-power levels of 150 W at 2.14 GHz with a drain efficiency of 65% and a linear gain of 16 dB, when operated at $V_{DS} = 28$ V [2.319]. The overall gate periphery is 72 mm in a single package with a single finger geometry of 0.7, μm × 200 μm. Promising reliability results have been reported in [2.168, 2.367]. Devices with a periphery of 36 mm are tested under RF-drive with an extrapolated drift of P_{sat} of ≤1 dB in 20 years at 28 V. On ampli-fier level, very promising PAE vs. ACLR results are reported in [2.201, 2.202] with PAE levels of 50% based on Nitronex GaN HEMTs. Envelope-tracking is used to obtain high PAE values of 50% for a nonconstant envelope input. An average output power of 37 W is reached at a peak-to-average ratio (PAR) of 7.67 dB. The linear gain in this operation is 10 dB. With the application of a digital predistortion the error vector magnitude (EVM) of 1.74% is reached with an ACLR of −51 dBc at 5 MHz offset. Specific WiMAX applications based on GaN on silicon are presented in [2.456]. Eight millimeter and 36 mm devices are tested under orthogonal frequency-division multiplexing (OFDM)-modulation between 3.3 GHz and 3.8 GHz. At 3.5 GHz, the 8 mm device deliv-ers an average output power of 1.5 W at 3.5 GHz, while the 36 mm device delivers 7 W. EVM is 2% and the drain efficiency (DE) amounts to ≥27%. The devices are fully packaged and qualified between −40°C and 85°C.

Several other processes are available. The performance of AlGaN/GaN HEMTs for 2.8 GHz and 10 GHz power amplifier applications is evaluated in [2.528] for military applications. Pulsed-output power levels of 41 W at 2.8 GHz with power densities up to 9.2 W mm^{-1} are obtained early in the development.

C-Band Frequencies (4–8 GHz)

C-band frequencies are used for both communication and radar applications and require high-power operation and high-temperature robustness in com-bination with high PAE. This frequency range typically cannot be reached by either LDMOS or SiC MESFETs. GaN HEMTs on s.i. SiC substrate using a field-plate gate concept are demonstrated at operating voltages of ≥90 V, resulting in cw-output power densities $P_{out} \geq 30$ W mm^{-1} at 4 GHz and 8 GHz [2.512]. Although the scaling of this performance to higher gate widths is questionable with respect to the thermal management, reliability, and efficiency, the fundamental high-speed power performance of GaN is visi-ble. At C-Band, the University of California reported a GaN HEMT on s.i. SiC with an output power of 2.8 W with a density of 18.8 W mm^{-1} and a PAE of 43%. A maximum PAE of 71% is reported with a power density of 6 W mm^{-1} at 4 GHz [2.67]. A 60 W and 100 W internally matched C-band GaN HEMT with $W_g = 24$ mm is given in [2.346, 2.347]. The cw-output power is 100 W at 5 GHz with 31% PAE and a linear gain of 12.9 dB at $V_{DS} = 56$ V. Applying pulsed-operation with 1% duty cycle, an output power of 156 W is found at 4 and 5 GHz [2.346]. Record cw-output power values of 140 W at 5.6 GHz from a single chip are reached with the application of $W_g = 50.4$ mm [2.173, 2.519]. A cell division concept is applied to suppress odd-mode loop oscillations while maintaining good gain. Record output power levels of 220 W at C-band are reported in [2.520] in a two-chip amplifier by improvements of device performance and reduction of the thermal resistance. 167 W are obtained from a single chip at 7 W mm^{-1}. Similarly 60 W of output power are obtained from a 16 mm chip at 45% PAE. These examples demonstrate the combination of high-gain and high-breakdown in GaN HFETs which have demonstrated much higher power densities, output power levels, and operation bias, while maintaining efficiency.

X-Band Frequencies: (8.2–12.4 GHz)

Power operation of AlGaN/GaN HEMTs at X-band is particularly attractive, due to the increase of operating voltages relative to GaAs while maintaining high-speed operation and thus guaranteeing high-gain and high-efficiency operation. The increase in operating voltage from typically 5–8 V for GaAs PHEMTs to ≥40 V will reduce the need to transport high currents on system level. Applications in the X-frequency band include transmit-receive (TRX) modules for naval and airborne phased array radar applications. Power-added efficiency is a major concern, both on the device and system level. Typical device targets are mentioned in [2.384]. PAE values of 60% are required at $V_{DS} = 40$ V with a power density of 6.4 W mm^{-1} for a device with 1.28 mm. Further the long term reliability of 10^6 h is required. Early reports of the power operation at X-band by the UCSB and Rockwell are given in [2.439, 2.507]. The output power density was 1.7 W mm^{-1} and 2.3 W mm^{-1} at that time for small devices. An early report of very high-power density of AlGaN/GaN HEMTs on s.i. SiC with 9.8 W mm^{-1} at 8 GHz is demonstrated by the UCSB in [2.504]. Basic predictions of the power operation limits are provided by Trew et al., focusing on the impact of the wide bandgap-material-properties on RF-power performance with respect to dispersion, e.g., in [2.462, 2.463, 2.465]. Output power densities of 10–12 W mm^{-1} are predicted at X-band. The experimental power-frequency limits of AlGaN/GaN HEMTs for X-band are evaluated by Eastman and coworkers, e.g., in [2.96, 2.98, 2.132]. Small periphery devices with $l_g = 300$ nm yield power densities of 11.7 W mm^{-1} at 10 GHz with an associated PAE of only 43% at $V_{DS} = 40$ V. At low V_{DS} bias, the PAE is as high as 55%. Geometrical considerations for the operation at X-band, e.g., the barrier-layer thickness, are evaluated in [2.457]. Triquint announced a power

MMIC process for radar applications at X-band [2.467]. Output power densities of $12\,W\,mm^{-1}$ at $10\,GHz$ have been reported with an associated PAE of 50%. An outstanding performance is also announced on Si substrates at X-band in [2.94]. The power densities for a small periphery device are as high as $7\,W\,mm^{-1}$ at $10\,GHz$ with an associated PAE of 37% at $V_{DS} = 40\,V$. Again, the PAE is limiting is this case.

In cooperation with General Electric, a power density of $9.2\,W\,mm^{-1}$ with $13.8\,W$ cw-output power at $10\,GHz$ and $55\,V$ drain bias using AlGaN/GaN HEMTs on SiC substrate are demonstrated by Lockheed in [2.18]. The associated PAE is 35%. At lower bias $V_{DS} = 30\,V$, the power density is $6.7\,W\,mm^{-1}$ with an associated PAE of 55%. Very promising hybrid single-stage X-band amplifiers are demonstrated by Hughes in [2.292] yielding output power values of $\geq 22.9\,W$ with a cw-PAE as high as 43%. More recent GaN MMICs in microstrip transmission-line technology are given in [2.313]. The devices with a gate length of $150\,nm$ use a field plate. Output power levels of $20\,W$ with 43% PAE are reached in pulsed-operation. The III-N activities of Raytheon for high-power and low-noise X-band applications are discussed in [2.210]. PAE, uniformity, and reliability are considered the main challenges for the application of GaN HEMTs in TRX modules. The thermal management considerations in [2.210] further predict an order of magnitude increase in heat removal within a TR module with the insertion of GaN technology. With the application of innovative power combining techniques, multichip solid-state power amplifiers with overall $150\,W$ output power levels are announced for X-band [2.79]. The PA is based on 16 chips similar to those reported in [2.81]. On hybrid packaged transistor level, Toshiba developed a $11.52\,mm$-gate-periphery GaN HEMT with a maximum output power of $81\,W$ in cw-operation. The PAE amounts to 34% at $9.5\,GHz$ and a gain compression level of $3\,dB$ [2.444]. The unit gate width is $160\,\mu m$ and no field plate technology is used for a gate length $l_g = 0.7\,\mu m$. A similar result is obtained with a gate periphery of $12\,mm$ and a gate length $l_g = 0.5\,\mu m$ with a hybrid AlGaN/GaN HEMT. The two chip device yields an output power of $\geq 60\,W$ and a maximum PAE of 35%. The linear gain is $10\,dB$ at a duty cycle of 10% and $V_{DS} = 40\,V$ in Class-B operation [2.518].

Double Heterostructure Devices

Double heterostructure devices overcome the disadvantage of the single heterostructure, which provides only limited electron confinement in the channel. The second heterojunction reduces the poor pinch-off characteristics and high-output conductance, which both degrade power performance and PAE at RF-frequencies for single heterojunction devices [2.289]. Initial reports on AlGaN/InGaN/AlGaN double heterojunction devices (DHFETs, DHHEMTs) based on the AlGaN/InGaN material system are given in [2.425], as also mentioned in [2.210]. The AlGaN/InGaN/GaN provided an initial output power of $6.3\,W\,mm^{-1}$ at $2\,GHz$ and $V_{DS} = 30\,V$. Very promising AlGaN/GaN/AlGaN

DHFETs on s.i. SiC for microwave and millimeter-wave power applications are given in [2.289], grown by MBE at HRL. The DHFETs yield higher transconductance and lower output conductance than SHHEMTs. PAE improvements of 10% are found at 10 GHz for a process with a gate length of 150 nm at $V_{DS} = 30$ V, while the power density is maintained.

Ku-Band (12.4–18 GHz) and K-Band (18–26 GHz)

Ku-band amplifiers find their applications in radio communication systems, such as satellite and mobile phone systems for point-to-point communication [2.314] and for military applications. Traveling-wave-tube amplifiers with very high efficiency levels of 70% and output power levels of 150 W with low weight are given in [2.288]. GaAs-based solid-state reference amplifiers for Ku-band operation can be found, e.g., in [2.314, 2.468], with a 40 W Ku-band high-efficiency hybrid solid-state power amplifier (SSPA). A good comparison of GaAs MMICs is given in [2.249]. Typical MMICs yield maximum output power levels of 5–10 W and PAE levels of $\geq$30%. References for conventional HPAs in efficiency-critical applications at K-band can be found in [2.24] based on InP HBTs. Output power levels of 870 mW are found at 20 GHz and 37% PAE. Initial reports of K-band power operation of III-N devices are given in [2.479, 2.505] with an output power density 3 W mm^{-1} at 18 GHz and 20 GHz for AlGaN/GaN HFETs on s.i. SiC substrate. AlGaN/GaN HEMTs operating at 20 GHz with a cw-power density $\geq$6 W mm^{-1} are reported in [2.312]. For applications requiring high efficiency, such as satellite communication, a power density of 3.2 W mm^{-1} with an associated PAE of 71% using AlGaN/GaN HEMTs at 20 GHz is reported in [2.399], again for a small gate width device. A power density of 6.7 W mm^{-1} at 18 GHz is reported in [2.227] with an associated PAE of 26%. Even higher power densities at 18 GHz have been achieved using field-plated gates, however, with a gain of only 5.8 dB in [2.226] for a small gate width of 100 μm. An output power density of 5.1 W mm^{-1} is reported for AlGaN/GaN HEMTs on silicon substrate in [2.93]. The devices have a periphery of 100 μm. Residual challenges for Ku-band devices include the increase of power-added efficiency for large periphery devices in applications requiring high efficiency, as GaAs PHEMTs and InP HEMTs have set high standards with respect to this requirement, e.g., [2.24, 2.466].

MM-Wave Frequencies

MM-wave operation of GaN FETs beyond 30 GHz has been demonstrated with improved material homogeneity and an increasing number of electron-beam T-gate processes being available on s.i. SiC, e.g., in [2.179].

Ka-Band (26.5–40 GHz)

GaN HEMTs are very attractive devices for Ka-Band MMIC applications due to their combination of high-speed and high-power performance [2.354].

State-of-the-art GaAs-PHEMT-based reference devices and MMICs can be found, e.g., in [2.214, 2.266] and the references therein. A record compact size MMIC with an area of only $9\,\text{mm}^2$ and $4\,\text{W}$ of output power level at $30\,\text{GHz}$ is given in [2.215]. Power levels of $10\,\text{W}$ at Ka-band can be reached with the grid amplifier concept [2.88] or up to $6\,\text{W}$ on MMIC level, see [2.266]. Early reports on short-gate-length AlGaN/GaN devices with a gate length $l_\text{g} = 0.12\,\mu\text{m}$ with high contact resistance R_C demonstrate current gain cut-off frequencies $f_\text{T} = 46\,\text{GHz}$ and $f_\text{max} \geq 100\,\text{GHz}$ in [2.58]. More recent devices with the same gate nominal gate length yield a cut-off frequency f_T of $120\,\text{GHz}$ [2.228]. State-of-the-art cw-power densities achieved at $26\,\text{GHz}$ with AlGaN/GaN HEMTs on s.i. SiC substrate are reported in [2.241] yielding a power density of $5\,\text{W}\,\text{mm}^{-1}$ and a linear gain of $7\,\text{dB}$ with an associated PAE of 30.1% at $V_\text{DS} = 25\,\text{V}$. Former TRW, now NGST, demonstrated power operation with $1.6\,\text{W}\,\text{mm}^{-1}$ power density at $29\,\text{GHz}$ and an associated PAE of 26% in [2.430, 2.431] at an early stage. Ka-band MMICs with pulsed-output power levels of $11\,\text{W}$ at $34\,\text{GHz}$ are mentioned [2.436]. Strong contributions to AlGaN/GaN HEMT power operation in Ka-band are provided by NEC. They demonstrated power densities of $7.9\,\text{W}\,\text{mm}^{-1}$ at $30\,\text{GHz}$ for a $2 \times 60\,\mu\text{m}$ device with a gate length of $250\,\text{nm}$ in [2.233]. The device with a gate length of $90\,\text{nm}$ yields an MSG of $8.3\,\text{dB}$ at $60\,\text{GHz}$. Similarly, a cw-output power of $2.3\,\text{W}$ at $30\,\text{GHz}$ is demonstrated in [2.184] yielding a power-added efficiency of up to 50%. Further, a saturated output power of $5.8\,\text{W}$ at $30\,\text{GHz}$ for a GaN with $W_\text{g} = 1\,\text{mm}$, a peak PAE of 43.2%, and a linear gain of $9.2\,\text{dB}$ are demonstrated in [2.156, 2.157] using a gate length of $0.25\,\mu\text{m}$ and small contact spacings of $0.7\,\mu\text{m}$. For frequencies targeting, e.g., missile radar applications at $35\,\text{GHz}$, Triquint demonstrated saturated power densities of $\geq 4\,\text{W}\,\text{mm}^{-1}$ with an associated PAE of 23% at $35\,\text{GHz}$ in [2.240]. The device has a linear gain of $7.5\,\text{dB}$ for a gate width $W_\text{g} = 200\,\mu\text{m}$. Even higher peak PAE values of 53% have recently been reported [2.179]. An output power of $3.5\,\text{W}$ at $35\,\text{GHz}$ is demonstrated in [2.510] using AlGaN/GaN HEMTs on s.i. SiC substrate with a gate width of $1.05\,\text{mm}$ and a gate length of $0.18\,\mu\text{m}$. The compression level for the peak power is $\text{P}_{-2\,\text{dB}}$. The prematched devices yield a linear gain of $\geq 8\,\text{dB}$ at $35\,\text{GHz}$. Further results of field-plate devices evaluated at $30\,\text{GHz}$ and $35\,\text{GHz}$ are given in [2.508]. Output power levels of $8\,\text{W}$ with an associated PAE of 31% at $31\,\text{GHz}$ are reached in cw-operation. With the inclusion of an InGaN back barrier layer, a power density of $13.7\,\text{W}\,\text{mm}^{-1}$ in cw-operation is obtained at $30\,\text{GHz}$ and $V_\text{DS} = 60\,\text{V}$. The associated PAE is 40% [2.509]. At even higher frequencies, $40\,\text{GHz}$ power operation is investigated in [2.194, 2.376] yielding power densities of $\geq 1.2\,\text{W}\,\text{mm}^{-1}$ and absolute power levels of $0.5\,\text{W}$. More recently, high performance AlGaN/GaN HEMTs with a power density of $2.8\,\text{W}\,\text{mm}^{-1}$ at $40\,\text{GHz}$ and a PAE of 10% are demonstrated in [2.44]. The gate length is $l_\text{g} = 0.18\,\mu\text{m}$, and the gate width is $W_\text{g} = 0.1\,\text{mm}$. Improved devices deliver $4.5\,\text{W,mm}^{-1}$ at $40\,\text{GHz}$ [2.45]. MMICs in this technology with output power levels of $5\,\text{W}$ at $30\,\text{GHz}$ are given in [2.43]. The dual-stage design gives a linear gain of $13\,\text{dB}$ and a PAE of 20%. The output periphery is 1.6 mm. Even higher power densities have been achieved at 40 GHz. The power density amounts to 10.5 W mm^{-1} and 34% PAE at $V_{DS} = 30$ V for a $W_g = 2 \times 75$ μm device [2.354]. The gate length is $l_g = 160$ nm.

V-Band (50–75 GHz) and W-Band (75–110 GHz)

Solid-state-based power generation and amplification between 50 and 120 GHz is typically limited to power levels of 25 dBm, see, e.g., [2.155, 2.447]. Operation both at high-power levels as well as higher operation voltages is therefore desirable. The impact of III-N devices can thus be substantial. AlGaN/GaN dual-gate HEMTs on s.i. SiC substrate are used to demonstrate V-band power operation at 60 and 65 GHz in [2.377]. Power gain of 8 dB is reached at 60 GHz, the maximum power density is 0.5 W mm^{-1} at 60 GHz. For a gate length $l_g = 140$ nm, a cut-off frequency $f_T = 91$ GHz is achieved in [2.164]. The f_{max} is a high as 122 GHz. Several results exist for the scaling of the gate length. Vertically scaled T-gated AlGaN/GaN HEMTs are reported in [2.45]. An extrinsic transconductance of 500 mS mm^{-1} is reported for devices with $l_g = 100$ nm. The cut-off frequencies are $f_T = 125$ GHz and $f_{max} = 174$ GHz. The first GaN W-band MMICs in microstrip technology are demonstrated by HRL [2.291]. The devices deliver output power levels of 316 mW at 80.5 GHz which have been obtained in cw-operation. The power density is 2.1 W mm^{-1}, the associated power gain amounts to 17.5 dB for three stages. On sapphire substrates, AlGaN/GaN HFETs with cut-off frequencies $f_T = 152$ GHz at $V_{DS} = 5$ V are reported for $l_g = 60$ nm. Even higher cut-off frequencies $f_T = 181$ GHz and $f_{max} = 186$ GHz are reported in [2.143]. When changing the barrier material to InAlN, $f_T = 172$ GHz is achieved with a gate length of 60 nm [2.143]. These devices are not fully passivated. Passivated AlGaN/GaN HEMTs on s.i. SiC yield $f_T = 153$ GHz for $l_g = 90$ nm [2.355] with the application of a Ge-spacer technology. More of such high-speed specific process modules will be required; however, the potential of III-N HEMTs for mm-wave operation is visible.

RF-Noise Performance

Low-noise operation of III-N HEMTs is attractive due to the combination of high-carrier velocities, good carrier confinement in the channel, and increased robustness with respect to high-input power levels relative to other high-speed devices such as InAlAs/InGaAs HEMTs, AlGaAs/InGaAs PHEMTs, or Si/SiGe HBTs. Various overviews [2.353] and reports are available on the low-noise performance of GaN FETs. A matched minimum noise figure $N_{F,min}$ as low as 0.72 dB at 8 GHz is reported in [2.262, 2.364] for a device with $l_g = 120$ nm and $W_g = 100$ μm. The associated gain is 12 dB at 8 GHz. The maximum G_{ass} is found for a drain current $I_D = 150$ mA mm^{-1}, while the minimum noise figure is found for $I_D = 100$ mA mm^{-1}. Minimum noise figures $N_{F,min}$ of AlGaN/GaN HEMTs on s.i. SiC of 0.75 dB at 10 GHz and 1.5 dB at $20\,\mathrm{GHz}$ are reported, respectively, in [2.220] for a $2 \times 60\,\mu\mathrm{m}$ device with $l_\mathrm{g} = 150\,\mathrm{nm}$. Fig. 2.19 gives the dependence of the noise figure $N_\mathrm{F,min}$ as a function of frequency for $I_\mathrm{DS} = 100\,\mathrm{mA\,mm^{-1}}$ in comparison to a GaAs PHEMT of the same gate length $l_\mathrm{g} = 150\,\mathrm{nm}$. The low-noise performance of GaN HEMTs on sapphire substrate as a function of aluminum concentration in the barrier layer on the device level is discussed in [2.261]. The intrinsic minimum noise figure $N_\mathrm{F,min}$ at a given frequency decreases nearly linearly with increasing Al-content between 20% ($1.7\,\mathrm{dB}$) and 35% ($0.92\,\mathrm{dB}$), measured at $12\,\mathrm{GHz}$. GaN HFETs on s.i. SiC substrate with excellent low-noise performance at low DC-power levels and $V_\mathrm{DS} = 1\,\mathrm{V}$ are demonstrated in [2.153]. Devices with a thin AlGaN Schottky barrier layers of $15\,\mathrm{nm}$ are compared to devices with $30\,\mathrm{nm}$ barrier thickness. The $N_\mathrm{F,min}$ is below $0.8\,\mathrm{dB}$ across the 2–12 GHz frequency-band with associated gains of better than $12.5\,\mathrm{dB}$ for the device with the thinner barrier. The associated gain G_ass increases by $2\,\mathrm{dB}$ and the minimum noise figure $N_\mathrm{F,min}$ is reduced by $0.2\,\mathrm{dB}$ for the devices with the thinner barrier. On silicon substrates, very promising noise figures $N_\mathrm{F,min}$ at $10\,\mathrm{GHz}$ are reported for $0.17\,\mu\mathrm{m}$ gate length devices in [2.295, 2.296]. The $N_\mathrm{F,min}$ amounts to $1.1\,\mathrm{dB}$ at $10\,\mathrm{GHz}$ with an associated gain of $12\,\mathrm{dB}$, measured at $V_\mathrm{DS} = 10\,\mathrm{V}$. The $N_\mathrm{F,min}$ at $18\,\mathrm{GHz}$ amounts to $1.75\,\mathrm{dB}$. Low-power linearity of GaN HFETs on s.i. SiC is discussed in [2.364]. Overdrive testing and studies for the use of GaN LNA in synthetic aperture radar applications are presented, e.g., in [2.31]. MMIC-based LNA circuit examples on s.i. SiC substrates are given, e.g., in [2.31, 2.102]. These are further discussed later and in Chapter 6.

Indium-Based (H)FETs

The use of indium is also investigated in III-N heterostructure devices. Improvements in speed and effective mass similar to the (In)GaAs material system are targeted when indium is used in the channel. Further, the use of indium is discussed in order to reduce some of the high surface-charge density and strain issues causing dispersion, e.g., in [2.85, 2.284, 2.327]. The piezoelectric effects can be omitted for certain material compositions in InAlN/GaN HFETs. Further, the conduction band off-set is increased as compared to typical $\mathrm{Al_{0.3}Ga_{0.7}N/GaN}$ HFETs [2.231]. HFETs can be realized in the (Al)GaN/InGaN, InAlN/GaN, and InAlN/InGaN systems. In addition, indium can further be used as a dopant in AlGaN/GaN HFETs to reduce interface roughness, e.g., [2.228, 2.240].

InAlN/GaN Heterostructure FETs

$\mathrm{In_{0.17}Al_{0.83}N/GaN}$ HFETs provide strong spontaneous polarization with very high sheet carrier concentrations $\geq 2 \times 10^{13}\,\mathrm{cm^{-2}}$ while the piezoelectric contribution is suppressed as the strain is minimized, as repeatedly stated [2.126, 2.186]. InAlN/GaN HFETs promise larger bandgap energies in the InAlN barrier, a larger conduction band off-set at the channel interface, and thinner barrier layers [2.284]. The latter may be useful for the realization of mm-wave devices with high aspect ratio [2.126]. Very high drain current densities of 4.5 A mm^{-1} are expected for the InAlN/GaN HFETs [2.230], based on the high-sheet carrier concentration already demonstrated [2.84]. In$_x$Al$_{1-x}$N/GaN single heterostructure HFETs grown by MOCVD on c-sapphire substrate are reported in [2.187]. With In contents x = 0.04 to x = 0.15 sheet carrier concentrations of 4×10^{13} cm^{-2} have been reached. For gate lengths $l_g = 0.7$ μm, cut-off frequencies $f_T = 11$ GHz and $f_{max} = 13$ GHz are achieved. The performance of InAlN/GaN HFET with 60 nm gate length and f_T values of 170 GHz is already discussed above [2.143]. Unstrained In$_{0.17}$Al$_{0.83}$N/GaN HEMTs on s.i. SiC are reported in [2.126]. With a gate length $l_g = 250$ nm a cut-off frequency $f_T = 43$ GHz is reached with a DC-transconductance of 355 mS mm^{-1} and a barrier thickness of 7.5 nm. The power densities at 10 GHz reach 2 W mm^{-1} with an associated PAE of 29%. The pinch-off voltage is -1.7 V.

InGaN/GaN Heterostructure FETs

On the device level, GaN/InGaN/GaN double heterostructure FETs are demonstrated in [2.85]. These DHHEMT structures are considered to be more stable with respect to device dispersion, as mentioned in [2.327]. Kelvin probe microscopy shows no change in the surface potential of InGaN DHHEMTs contrary to AlGaN/GaN HFETs. Device characteristics of GaN/In$_{0.15}$Ga$_{0.85}$N/GaN doped-channel double heterostructure HFETs are shown in [2.150]. The devices yield a maximum transconductance $g_{m,max}$ of 65 mS mm^{-1}, a maximum drain current of 272 mA mm^{-1}, and current gain cut-off frequencies $f_T = 8$ GHz and $f_{max} = 20$ GHz for a gate length of 1 μm. The output power at 1.9 GHz is 26 dBm for a gate width of 1 mm, i.e., 0.4 W mm^{-1}. P-channel InGaN HFET structures based on polarization doping without external acceptor doping are discussed in [2.534]. The channel sheet carrier concentration is 10^{11} cm^{-2} with a hole mobility of 700 cm^2 V^{-1} s^{-1} at $T_L = 66$ K. The maximum drain currents measured at room temperature (RT) are extremely low. MOSFET InGaN structures are also reported. AlGaN/InGaN/GaN DHFETs using SiO$_2$ as a passivation layer and gate dielectric layer yield a saturated output power density of 6.1 W mm^{-1} at 2 GHz and $V_{DS} = 30$ V [2.426]. The maximum transconductance $g_{m,max}$ amounts to 110 mS mm^{-1} for the MOSDHFET structure, while the simple DHFET structure yields 170 mS mm^{-1}. Pure InN/AlN MISFETs on Si(111) substrates are reported in [2.251]. A 26 nm-thick InN layer is grown by RF-plasma-assisted MBE (PAMBE) on AlN. A current density of ≥ 500 mA mm^{-1} is reached with a very low breakdown voltage of 3 V and a threshold voltage of -7 V. These examples show the promising principle operation of indium-based FETs so far. The main emphasis of In-based development focusses on material growth, while the device development requires the suppression of leakage currents and the stabilization of the InAlN interface toward the dielectric.

2.5.6 Heterojunction Bipolar Transistors (HBTs)

The family of III-N-based field-effect transistors has been heavily exploited, as discussed in the previous sections. The use of bipolar transistors has also been considered; however, substantial hurdles have delayed the development of III-N-based bipolar devices. At the same time, very few reports exist on bipolar devices in the SiC material system, e.g. [2.532].

AlGaN/GaN Heterojunction Bipolar Transistors

Complications with the p-doping in GaN [2.218,2.225] in epitaxial design [2.23] and issues with the processing [2.283] known from the optoelectronic devices have interfered with the successful development of bipolar AlGaN/GaN transistors.

Performance Predictions of PNP- and NPN-Devices

Various theoretical predictions have been made about the possible performance of AlGaN/GaN heterojunction bipolar transistors. Performance predictions for npn-$Al_xGa_{1-x}N$/GaN HBTs are given in [2.374]. Spontaneous and piezoelectric polarization are shown to have only minor impact on the device performance. Incomplete ionization and reduced activation of the base dopant are shown to be the main issues. Limiting factors and optimum design of npn- and pnp-AlGaN/GaN HBTs have been addressed and performances have been investigated by simulation in [2.311]. Again, the high-base resistance induced by the deep acceptor level is found to be one of the causes of limited current gain values for npn-bipolar devices. The analysis for pnp-devices indicates limited RF-performance caused by the reduced minority hole transport across the N-GaN base. Simulations of the characteristics of AlGaN/GaN heterojunction bipolar transistors are given in [2.234]. The 2D-simulations show the great potential of AlGaN/GaN HBT with high calculated β, high breakdown voltage BV_{CE0} of 55 V, and a maximum cut-off frequency $f_T = 18$ GHz. Similar optimistic predictions are given in [2.7]. Generally, AlGaN/GaN HBTs are discussed also for high operation temperatures, e.g., 300°C in [2.382], since the activation of the acceptor Mg-doping and the performance increase for higher operation temperatures. The measured current gain amounts to 10 for a $90\,\mu m^2$ device at this temperature due to the reduction of the base resistance.

AlGaN/GaN Heterojunction Bipolar Transistors Fabrication

GaN/AlGaN HBT fabrication issues are discussed in [2.383]. A standard III-V HBT process is applied to material grown by both MBE and MOCVD, while the annealing temperature of the ohmic contacts is increased to 700–800°C. Current gain levels of 10 are reached at elevated temperatures of 300°C for both material types. At the same time, the junction leakage is significantly increased at high temperature. However, remaining issues include these leakage currents and the reduced base conductivity. The main performance problem for HBTs is thus the reduced current amplification [2.283, 2.514]. Both the lifetime of 25 ps and diffusivity of electrons in Mg-doped GaN are found to be low compared to practical base width of, e.g., 100 nm. The collector leakage is correlated with the threading dislocation density. The maximum operation bias is 70 V. Selective area growth of AlGaN/GaN HBTs is used in [2.413] to analyze its potential advantages as compared to more traditional growth techniques by normal MOCVD. Both linear ohmic contacts and improved material quality and interfaces could be obtained by the use of this more complicated technique applying a two-step procedure for the base contact. Very high voltage operation $\geq$330 V with a high current gain of AlGaN/GaN HBTs using selectively regrown emitters on sapphire substrates is given in [2.513]. The high-breakdown voltage is attributed to the low-background doping level in the 8 μm-thick GaN collector and the suppression of leakage in the lateral base/collector junction due to etch damage in the 2–3 μm spacing. Further, the use of ELO growth, reducing the number of threading dislocations, significantly reduces the collector leakage [2.283]. Further analysis of the leakage currents is given in [2.514]. Anomalously high current gains at low current levels in AlGaN/GaN can be observed. The calculated extrinsic β of the devices can be misleading due to leakage and ohmic contacts parasitic contributions, while the intrinsic performance is low due to purely intrinsic effects. AlGaN/GaN HBTs thus have promising theoretical performance; however, this performance still cannot be exploited due to epitaxial and technological issues.

InGaN/GaN Heterojunction Bipolar Transistors

The problems of p-doping of binary GaN can be overcome by using InGaN as a base material. NPN-GaN/InGaN/GaN DHBTs on s.i. SiC substrates with current gains of 3,000 can be fabricated [2.270]. This is ascribed to the lower acceptor activation energies and higher overall doping activation, e.g., in $In_{0.14}Ga_{0.86}N$ [2.225, 2.269]. Active carrier concentrations of 10^{18} cm^{-3} can be achieved. The InGaN is regrown by MOVPE in this case.

Base transit time analysis in GaN/InGaN/GaN HBTs is performed in [2.70]. The contribution of the base transit time decreases with increasing temperature, while the collector transit time remains nearly constant. Similar to AlGaN/GaN devices, this behavior makes the devices more suitable for high-temperature operation [2.152]. Double heterostructure GaN/InGaN/GaN HBTs on s.i. SiC substrate are presented in [2.269, 2.271] and several of the doping and processing issues are resolved. The breakdown voltage of the DHBTs in [2.271] exceeds 50 V, while a low current gain is observed without the application of regrowth. Further improvements of the epitaxy by using base regrowth [2.270] and improved processing yield a current gain $\beta = 2,000$ in [2.271] for an emitter geometry of 50×30 μm^2. The maximum current is as high as 80 mA. Based of these developments, the further development of the GaN/InGaN (D)HBT is a promising candidate to overcome some of the epitaxial and processing limitations.

2.5.7 MMIC HEMT Technology

Apart from demonstrations on the device level, the demonstration of microwave integrated circuits (MICs) and monolithic microwave integrated circuits (MMICs) is a core competence for any RF-technology. The state-of-the-art is discussed in the following with focus on the III-N issues.

High-Power MMIC Technologies

Coplanar-Waveguide High-Power MMICs

Initial coplanar AlGaN/GaN HEMT MMICs on s.i. SiC substrates are reported in [2.133] for broadband applications. The transmission-lines have impedance levels between 30 and 80 Ω. MMICs with saturated output power levels beyond 37 dBm are realized between quasi-DC and 8 GHz. Several demonstrations of high-power coplanar MMICs on s.i. SiC are available for X-band. They are given in detail in Chapter 6. Coplanar K-band MMICs on sapphire substrate are reported in [2.332]. The devices exhibit a small-signal gain of 10 dB and a 3 dB bandwidth of more than 4 GHz between 20 and 24.5 GHz. A 200 μm-thick SiN is used for the MIM capacitances, while WSiN is used for the resistors. Further details on the passive components are given in Chapter 8.

Microstrip Transmission-Line High-Power MMICs

The first complete microstrip transmission-line high-power MMIC technology on s.i. SiC for SiC MESFETs and AlGaN/GaN HEMTs is presented by Cree in [2.362]. A 24 W high-power amplifier at 16 GHz is presented. Specific via etching techniques are required, which will be detailed in Chaps. 4 and 8. Triquint announces a full GaN HEMT MMIC technology on s.i. SiC with a saturated power density of 12 W mm^{-1} at 10 GHz and an associated PAE= 50% [2.467]. A similar microstrip MMIC process on silicon substrate is presented by Triquint in [2.109]. The silicon substrates are thinned to 100 μm in this case. The vias can be etched by standard Si processing equipment. The other passive components are not modified compared to the GaAs process. Further MMIC details are given in Chapter 6. At Ka-band, a comparison of both coplanar and microstrip line HEMT MMICs on s.i. SiC has been presented by HRL in [2.290]. The coplanar MMIC yields an output power of 1.6 W at 33 with 4 GHz bandwidth, while the MSL MMIC yields 2.2 W at 27 GHz with 8 dB gain between 24 and 32 GHz. When additional external tuning is used, a PAE of 27% is reached. The substrates are thinned to 50 μm. The via geometry is 30×30 μm^2. A recent report by NGST gives a dual-stage

MMIC power amplifier in microstrip technology with a pulsed-output power of 11 W at 34 GHz. The duty cycle is very low and unusual for the telecom application [2.436, 2.439].

Low-Noise MMIC Process Technologies

GaN HEMTs are attractive for LNA MMICs due to high sheet carrier densities n_{sheet} and high intervalley energy (≈ 1.5 eV). One of the first GaN LNA MMICs is presented in [2.102]. A coplanar-waveguide dual-stage broadband LNA yielding a minimum noise figure of 2.4 dB and a gain of 20 dB between 3 and 18 GHz is presented in [2.102]. The devices are biased at 2.5 V $\leq V_{DS} \leq 4$ V. The third-order intercept point is better than 34 dBm at 8 GHz, which demonstrates the good input linearity of the MMIC. The impact of linearity, high-input power levels, and a discussion of typical destruction levels in GaAs is given in [2.60]. Typical catastrophic destruction levels of GaAs PHEMT LNA MMICs amount to 20 dBm without additional circuitry [2.60]. GaN HEMTs with similar gate periphery can be driven to input power levels of 30 dBm without degradation of the output power. Further the exceptional linearity with an output intercept point of 43 dBm is proven. A full investigation of single- and dual-stage X-band LNAs, including linearity investigations, is presented in [2.220]. The trade-off between minimum noise figure at low $V_{DS} = 7$ V and good input and output linearity at high $V_{DS} = 25$ V is discussed. The noise figure of the full MMIC at X-band is mainly determined by the passive components. A coplanar C-band high-dynamic-range LNA MMIC has been presented in [2.515]. The noise figure is as low as 1.6 dB at 6 GHz and does not exceed 1.9 dB between 4 and 8 GHz. The input port of the circuit can endure up to 31 dBm of input power. The MMIC shows noise figures comparable to other semiconductor technologies, while the dynamic range and the survivability are strongly enhanced. A microstrip transmission-line technology on s.i. SiC substrate is used to demonstrate wideband dual-gate LNAs between 0.1 GHz and 5 GHz in [2.60]. A linear gain of 15 dB is found with a saturated output power of 33 dBm at a compression level at 13 dB. Further MMIC examples and design consideration are discussed in Chapter 6.

2.6 Applications Issues

A large number of possible applications has been suggested for III-N devices. Without taking into account all possibilities, some key applications are systemized in this section. A large number of overview papers have been provided on this issue in the literature. The impact of wide-bandgap microwave devices on defense and naval systems is discussed in [2.105, 2.189]. Specific overview articles on radar applications are given, e.g., in [2.104, 2.147, 2.263, 2.272]. The application in power systems is discussed in [2.470, 2.494]. Future directions and technology requirements of wireless communication systems are discussed, e.g., in [2.116, 2.309].

2.6.1 Broadband Communication

The most important nonmilitary application of III-N devices is the partial replacement of Si LDMOS transistors in base-station applications for mobile communication. For such broadband/multiband communication applications the following advantages for wide bandgap semiconductors are typically named:

- The increase in relative bandwidth for a given power level
- The higher output impedance level to match, i.e., higher impedance and reduced C_{ds} per unit gate width for load matching
- Increased efficiency through new circuit and system concepts
- Improved linearity for the same output power
- Thus increased linear output power
- A Reduction of memory effects
- Potentially, less effort for the thermal management based on high-temperature operation

Examples for the development of device and amplifiers in L-band and S-band are given in Chaps. 2 and 6.

2.6.2 Radar Components

The particular application of III-N solid-state high-power amplifiers and low-noise amplifiers is discussed for radars for several reasons. The particular applications include:

- Active electronically scanned arrays (AESA)
    - For airborne radars [2.104, 2.272],
    - For ground-based air defense radars [2.272],
    - For naval radars [2.147]
- For airborne and spaceborne reconnaissance systems, e.g., [2.263]

Radar applications are the main drivers in the development of the III-N electronic materials. The fundamental improvements through the use of wide bandgap amplifiers involve on the system level:

- Better power×aperture gain [2.535]
- The possibility of high-power ultra-wideband (UWB) systems [2.147,2.272]
- And higher reliability and survivability

The resulting improvements on the device and module level for defense applications include:

- Increasing power-added efficiency on system level by the increase of operation bias
- Elevated temperature operation allowing for reduced cooling [2.147]
- Reduced cooling through higher junction temperatures [2.124]

- Reduced limiter usage in the receiver path [2.272]
- Increased absolute output power levels [2.147]
- Improved reliability and increased system availability due to graceful degradation [2.263]
- Higher robustness with respect to supply voltage variations [2.263]
- Possibly the use of modulation schemes for increased functionality [2.147]

The frequency ranges involve S-band, C-band, X-band [2.104], and potentially Q-band [2.384]. Ultra broadband electronic warfare applications will further involve the range from 6–18 GHz [2.272]. In addition, Ka-band missile applications at 35 GHz are being discussed [2.179].

2.6.3 Electronics in Harsh Environments

As for other wide bandgap materials, nitride semiconductors have been suggested for their application in harsh environments, such as:

- Airborne operation [2.147]
- Space applications with high radiation levels, elevated temperatures, and strong temperature gradients [2.263]
- Space applications with requirements for reduced volume, reduced weight, and energy consumption [2.263]
- Car engines, i.e., microwave ignition [2.252] and energy-efficient DC–DC-conversion [2.470]

The particular functions include:

- RF-power sources and power generation [2.180]
- Highly linear RF-mixers [2.13]
- Robust RF-receivers [2.262, 2.364]
- DC–DC-conversion [2.470]
- Chemical sensors [2.365]
- Fluid sensors [2.285], biogen sensors [2.365], and (combustion) gas sensors [2.200, 2.365, 2.402, 2.458]
- Pressure sensors [2.175]
- Strain sensors [2.365]

A good overview of the sensor capabilities of GaN, especially with respect to its pyroelectric capabilities, is given in [2.10]. A sensor array using conventional semiconductor arrays for the operation regime between 200°C and 400°C is given in [2.461]. Various gaseous constituents can be discriminated. Gas sensors for hydrogen and propane are reported, based on MOSFETs in [2.176], and, based on n-type GaN and Pt-based Schottky contacts operating at 400°C, in [2.265]. Pressure sensors based on the pressure-induced conductivity changes in AlGaN/GaN HEMTs on Si substrates are reported in [2.175]. The pressure dependence can be used in pressure sensing applications; however, the pressure dependence poses also serious consequences on the variation of the wafer bow and overall strain when dicing and handling the device.

2.7 Problems

1. *Summarize the substrate situation for III-N electronic devices.*
2. *Given all the advantages named in the text, name at least three technical disadvantages of III-N semiconductors with respect to epitaxy and technology.*
3. *Name the advantages of double heterostructure devices relative to single heterostructure devices.*
4. *To this date, what are the shortcomings of AlGaN/GaN heterostructure bipolar transistors?*
5. *Do you expect a difference for of unipolar and bipolar devices with respect to the influence of the substrate? Discuss!*
6. *Calculate the cross-section of a copper wire to be able to transport 1 W of DC-power for an operation bias at 5 V and at 50 V.*

3

Epitaxy for III-N-Based Electronic Devices

Epitaxy is a core competency for the production of III-N devices. In this chapter, growth procedures for gallium-, aluminum-, and indium-based binary and ternary compounds are described with emphasis on AlGaN/GaN-, InGaN/GaN-, and (In)GaN/InAlN-based heterostructure systems. Nitride-specific material characterization, doping, and material quality issues are analyzed. Unlike Chapter 2, this chapter systematically reviews substrate properties with respect to the requirements of electronic device growth. This chapter mainly addresses the growth of both metal organic chemical vapor deposition (MOCVD) and molecular beam epitaxy (MBE). In general, MOCVD growth of GaN is performed at much higher temperatures (well above 1000°C) than MBE growth, which is typically performed at temperatures of about 700°C [3.205,3.206]. MBE growth enables the growth of very precise interfaces, which improve transport properties. The two methods can also be combined, e.g., to a methodology called migration-enhanced MOCVD, as described, e.g., in [3.44]. Other techniques such as hydride vapor phase epitaxy (HVPE) have also been mentioned, but do not play a dominant role for the fabrication of the active layers.

Surface Preparation

Specific surface preparation methods have been developed, which improve the surface roughness before epitaxial deposition and are used prior to the actual growth procedure. Further, specific procedures assure that the reactor spaces are not contaminated by extrinsic pollution. Standard surface-polishing techniques are based on plastic deformation by an abrasive, e.g., diamond. Electrochemical–mechanical polishing procedure (ECMP) improves the reproducibility of the epitaxial growth and is preferred to purely mechanical polishing (CMP), as the induced subsurface damage is reduced. A proprietary ECMP process of SiC applied by NovaSiC is mentioned in several publications, e.g., in [3.154,3.162,3.216]. A full description of a two-step ECMP procedure is given in [3.112]. The procedure is characterized by a balance of anodic oxidation and oxide removal [3.216] to obtain a defect-free surface of the substrate.

A high-temperature hydrogen etching process is used to further smooth the surface to atomic levels. On the device level, a correlation of the gate-leakage current and substrate defects in 2-in. SiC substrates is given in [3.126]. A clear correlation showing a reduced yield of the FETs for an increased defect density in a particular area is visible. The increased defect concentration in selected areas of the wafer leads to a statistical increase of the gate leakage current of about 100% measured at $V_{DS} = 20\,V$.

Substrate Surface Cleaning and Backside Metallization

Further wafer preparation includes the deposition of a Ti layer on the backside of the transparent substrate prior to growth to provide an effective heat transfer [3.162], typically for MBE growth only. A sputtered Mo thermal-contact layer is used for ammonia MBE in [3.213]. Various surface heating procedures have been reported for the in-situ preparation of the substrates within the reactor chambers or preparatory chambers. Prolonged heating of Si(111) for up to 10 h at 600°C with subsequent oxide removal by rapid thermal annealing is reported, e.g., in [3.180]. An ex-situ exposure to hydrogen at 1,600°C is reported in [3.205] prior to an in-situ surface cleaning by Ga deposition and desorption within the reactor chamber. The impact of reactor loading leakage and the impact of the contamination of sapphire substrates is discussed in [3.94]. Various cleaning conditions and sequences with prebake and flush for moisture removal prior to growth are discussed. A potential reaction is the composition of H_2O from hydrogen and Al_2O_3. Oxygen incorporation serving a shallow donor is considered a major obstacle for semi-insulating GaN. The investigation includes a prebake of the sapphire and a flush of the substrates by hydrogen between 10 min and 12 h. The optimized procedure includes both the prebake in-situ and no flush, and is critical in order to obtain semi-insulating GaN buffers.

3.1 The AlGaN/GaN Material System

The growth of AlGaN/GaN heterostructures is a basic technology for the production of high performance HEMT and HBT devices. General overview articles and comparisons between the various growth methods for the AlGaN/GaN system are given, e.g., in [3.36, 3.39, 3.138] . An overview of MBE growth of AlGaN/GaN heterostructures is given in [3.172].

3.1.1 Metal Organic Chemical Vapor Deposition (MOCVD)

Metal organic chemical vapor deposition (MOCVD) is the name of a growth technique that involves a dynamic flow in which gaseous reactants pass over a heated substrate and react chemically to form a semiconductor layer [3.36]. It is widely spread for the fast and precise growth of many III-V materials,

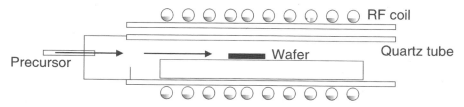

Fig. 3.1. Scheme of metal organic chemical vapor deposition (MOCVD) growth

e.g., for InGaP/GaAs [3.156] and $In_{0.53}Ga_{0.47}As/InP$ [3.61]. III-N MOCVD-based material growth is very attractive due to the large amount of experience and success available from the growth of optoelectronic devices, e.g., [3.141], and the increased growth rates relative to molecular beam epitaxy (MBE). Thus, MOCVD is widely used for the growth of heterostructures for electronic applications [3.94]. Fig. 3.1 gives a schematic view of MOCVD growth of III-N materials. The precursors TriMethyl-Gallium (TMGa), TriMethyl-Aluminum (TMAl), and TriMethyl-Indium (TMIn) react with ammonia (NH_3) on the hot substrate surface to form semiconductor layers. Additional sources such as nitrogen, silane (SiH_4), disilane (Si_2H_6) [3.48], and Bis(cyclopentadienyl) magnesium (Cp_2Mg) sources [3.105,3.121] are needed. The substrate is located on top of a graphite susceptor. Various reactor concepts have been developed [3.1,3.142] to accomplish and optimize the laminar flow of gases on larger area surfaces [3.7] and within multi-wafer concepts, e.g., in [3.48,3.192,3.221]. Both reactor concept and geometries, often kept proprietary, influence the growth conditions to a high extent. This fact also limits the comparability of parameters such as growth temperature, gas flows, and others parameters, which are adjusted externally. However, a principal understanding of the requirements for III-N growth can be obtained.

Growth Chemistry

The growth kinetics and chemistry of III-N semiconductors by MOCVD are described in various publications which shall be briefly mentioned here [3.138, 3.167, 3.169, 3.172]. The details of the complex reactions are not fully understood and especially the intermediate reactions are found to be very complex. The principal reaction to form GaN reads [3.142]:

$$Ga(CH_3)_3(v) + NH_3 \longrightarrow GaN(s) + 3CH_4(v)(l) \qquad (3.1)$$

A similar reaction occurs for AlN(s) replacing gallium by aluminum. The chemistry is governed by strong reactions of the precursors at room temperature. Further ammonia requires pyrolysis, which is not very efficient. Thus, high growth temperatures beyond 1000°C are used along with high V/III–ratios (>1000) and high gas velocities.

Optimized Conditions for MOCVD Growth

One of the key advantages for the AlGaN/GaN material system is based on the fact that the complete range of solid solutions from GaN to pure AlN is available using MOCVD growth. Typical growth rates for GaN are about $2\,\mu m\,h^{-1}$ [3.219]. Overview articles of the development of the growth conditions of AlGaN/GaN heterostructures are reported, e.g., in [3.34, 3.92, 3.168]. The aluminum content or mole fraction x in $Al_xGa_{1-x}N$/GaN heterostructures is varied between 0 and 0.6 in [3.92]. The defect density in the $Al_xGa_{1-x}N$ layer on GaN is found to increase significantly for $x \geq 0.3$ for typical barrier thicknesses of 18 nm for AlGaN/GaN HFETs. Further, the mobility in GaN channels along AlGaN/GaN hetero-interfaces decreases with decreasing barrier-layer thickness and with increasing Si-doping concentrations. Sheet carrier concentrations of $1.8 \times 10^{13}\,cm^{-2}$ are reached at $x = 0.44$. The AlGaN-spacer thickness between the channel and the Si-doped barrier layer is varied between 0 and 3 nm. A mobility of $1,650\,cm^2\,V^{-1}\,s^{-1}$ at RT is found for 3 nm thickness. The sheet carrier concentration decreases from 1.5 to $0.8 \times 10^{13}\,cm^{-2}$ with the increase of spacer thickness from 0 to 3 nm. Very good layer properties and reproducibilities have been reported for optimized growth conditions on two-inch and three-inch wafers, e.g., in [3.221]. The wafer-to-wafer variability of the sheet resistance mapping is $\leq 3\%$ on 3-in. s.i. (0001)SiC wafers. The critical growth parameters will now be discussed in detail.

Growth Temperature Schemes: Nucleation and Buffer for MOCVD

The growth temperatures critically determine the properties of the semiconductor layers. A typical temperature scheme for the growth of an AlGaN/GaN heterostructure sequence is given in [3.94]. On sapphire substrates, a 15–25 nm thick AlN nucleation layer is used in a growth temperature range of 500°C to 700°C. The consequent growth of the main GaN buffer layers is performed at $\geq 1,000$°C. For all temperatures, the changes of the growth temperature due to different thermal contacting at the backside of the substrate have to be considered. A GaN nucleation temperature of 550°C on 2-in. sapphire substrate is reported in [3.218]. A slightly misoriented grain morphology is obtained, which results in poor electrical properties. Growth at a temperature of 530°C for the nucleation layers is reported in [3.192] on 3-in. sapphire substrates. The growth of AlN as a nucleation layer on sapphire substrate with a temperature of 500 to 700°C is reported in [3.219]. Above the AlN nucleation layer, all layers are grown at much higher temperatures, especially the $Al_xGa_{1-x}N$ heterobarrier. AlGaN-based nucleation layers for growth on s.i. SiC substrates are reported in [3.185]. Defect-poor surfaces on top of GaN buffer layers with thicknesses up to 5 μm can be observed optimizing the alloy composition and thickness of the $Al_xGa_{1-x}N$ nucleation layers. The deposition temperatures of the $Al_{0.15}Ga_{0.85}N$ nucleation layer amount to 1,010°C on s.i. SiC and 1,040°C on sapphire for the same nucleation. The minimum Al-content is 6% for SiC nucleation and 15% for the nucleation on sapphire.

AlGaN nucleation on s.i. SiC substrates has also been reported in [3.102]. The thickness of the nucleation layer is 500 nm, followed by the GaN buffer. Growth of AlN nucleation layers on SiC is another standard procedure and has repeatedly been investigated, e.g., in [3.103]. The temperature is varied in [3.103] depending on the crystal structure of the n-type SiC substrate: A nucleation temperature of 1,080°C is used for the 6H–SiC substrates, while 980°C is used for 4H–SiC substrates. The thickness is 100 nm in both cases. The hotter growth for the 6H–SiC substrates shows smaller grains than for the growth on 4H–SiC, which means that the morphology of the AlN influences the quality of the subsequent layers. The specific nucleation of III-N materials on Si substrates is discussed below.

Growth Temperature Schemes: Channel and Barrier

The substrate temperatures reported for the growth of conducting GaN by MOCVD typically exceed 1,020°C [3.219]. The associated growth rate is $1.2 \,\mu m\, h^{-1}$. A growth temperature of 1,080°C for the growth of the main GaN layer is reported using hydrogen as a carrier gas in [3.93] on sapphire substrates. Growth temperatures of 1,180°C are reported for the GaN channel and the $Al_{0.26}Ga_{0.74}N$ heterobarrier in [3.174] on sapphire substrates. An analysis introducing additional layers in the buffer layer is given. The growth temperature is chosen accordingly for Al-, In-, and Ga-based binary and ternary semiconductors within the interlayer. A temperature of 1,125°C for the growth of $Al_{0.33}Ga_{0.67}N$ in the heterostructure is reported in [3.93]. The impact of the variation of the $Al_xGa_{1-x}N$ growth temperature between 1,060 and 1,125°C is reported in [3.92]. The Al-incorporation changes from 0.37 to 0.42 at otherwise constant conditions with the variation of the growth temperature from 1,060 to 1,125°C. Hot growth of the AlGaN layer is considered promising in [3.93] with respect to the output power, while a termination of the growth in the cool-down using nitrogen is found to be most suitable, again with respect to output power considerations of the $Al_{0.33}Ga_{0.67}N/GaN$ HEMTs realized on SiC. Highly-resistive thick $Al_{0.65}Ga_{0.35}N$ layers grown by MOCVD at 1,150°C are reported in [3.23]. When the growth temperatures of $Al_{0.65}Ga_{0.35}N$ are reduced to 920°C and silicon and indium codoping are applied, highly conductive $Al_{0.65}Ga_{0.35}N$ layers are reported. Mobility values of $22 \,cm^2\, V^{-1}\, s^{-1}$ and bulk carrier concentrations of $2.5 \times 10^{19}\, cm^{-3}$ are reached. The termination conditions of the semiconductor surfaces are discussed below with the specific methodology, e.g., cap layers [3.30, 3.93], or in-situ passivation layers, see [3.133].

MOCVD Gas Flow and Group V/III Ratio

Similar to the growth temperature discussed in the previous section, the element group-V/group-III ratio is of critical importance for the material properties. A nitrogen-rich growth is necessary in general because of the very different chemical bindings and, to a lesser extend, different atomic weights of the constituents N and group III-metals [3.142]. The absolute and relative gas flows and pressures are quantities that are strongly dependent on the susceptor and reactor geometries. Several reactor-specific investigations are available in the literature, e.g., in [3.1, 3.23, 3.92]. Optimization of the GaN nucleation layer on sapphire substrates as a function of nitridation conditions, group V/III ratio, growth temperature, and growth pressure is discussed in [3.218]. TMG flows of $32\,\mu$mol min^{-1} and NH$_3$ flows of $0.09\,$mol min^{-1}, i.e., N-rich conditions, are reported for GaN growth. The V/III ratio is thus 2,775. A group V/III ratio of 1,800 for GaN growth and 900–1,800 for AlGaN growth by MOCVD are reported in [3.39] for Al$_{0.35}$Ga$_{0.65}$N/GaN heterostructures on SiC and sapphire. A group V/III ratio of 2,000–2,600 is reported in [3.219] for the growth of GaN. In a comparison of two reactor concepts, a V/III ratio of 2,660 is reported for the growth of the GaN channel-layers on AlN nucleation layers in a so-called showerhead reactor in [3.219]. In a second quartz tube reactor, a group V/III ratio of 2,000 is used for the same TMG flow of $23\,\mu$mol min^{-1}. A V/III ratio of 2,100 is reported for GaN growth at a growth temperature of 1,030°C in [3.218]. The V/III ratio is increased to 2,775 for a growth temperature of 1,065°C to accommodate for the increased N-desorption.

Growth Pressure

The pressure in MOCVD is the most fundamental parameter next to temperature and V/III ratio. It basically determines the incorporation of the various constituents like the group-V and group-III materials, including group-IV dopants like Si and C. The influence of growth pressure on the morphology and carrier compensation in GaN and related alloys is described, e.g., in [3.219]. For very low pressures of 40 Torr, the GaN layers are highly resistive and yield very low carrier mobility nearly independent of the growth temperature. Very high growth pressures at 200 Torr and beyond lead to the loss of the resistive nature of GaN, e.g., at atmospheric pressure [3.6]. In an intermediate growth pressure range, both high mobility and high resistivity (i.e., low nonintentional doping) can be reached in GaN Layers. Optimized growth pressures of 130 and 65 Torr are reported for GaN layers [3.219].

Similar reports of a pressure of 76 Torr are given in [3.39] for Al$_{0.35}$Ga$_{0.65}$N/GaN heterostructures, and 100 mbar for Al$_{0.25}$Ga$_{0.75}$N/GaN HFET layers [3.48]. Further results at a growth pressure of 76 Torr are reported in [3.218] for the growth of the GaN nucleation layer on sapphire, while 150 Torr is used for the rest of the layers. This pressure, at which pressure prereactions occur, is strongly dependent on the reactor geometry. This critical pressure is found to be 150 Torr for GaN for the reactor used in [3.218]. For the growth of Al$_x$Ga$_{1-x}$N, the pressure dependence is modified depending on the Al mole-fraction [3.92]. For a given V/III ratio and growth temperature, the Al-mole fraction is particularly influenced by the partial pressures of Al and Ga, and their deviation from the equilibrium partial pressures, as further detailed in [3.92]. Thus, lower pressures than for optimized GaN layers are needed for good Al-incorporation in the $Al_xGa_{1-x}N$ layers, as reported in [3.219]. A medium growth pressure of 76 Torr is reported in [3.94, 3.185] for the growth of $Al_{0.15}Ga_{0.85}N/GaN$, due to the tendency of TMAl to form adducts with ammonia. At this pressure of 76 Torr, such prereactions are still not fully suppressed. Again, their occurrence is again strongly reactor-dependent. Relatively high pressure of 300 mbar is reported for the growth of $Al_{0.33}Ga_{0.67}N/GaN$ heterostructures in [3.93]. The high pressure growth also leads to increased incorporation of nonintentional doping such as C and Si.

Crystal Misorientation

The control of the misorientation of the substrate as compared to the III-N layers is of further importance for the layer quality. Growth of GaN on both on-axis and off-axis 6H–SiC substrates is reported in [3.103]. A 3.5° off-axis 6-H substrate is used for comparison. A strong impact on the surface conditions has been repeatedly reported, e.g., [3.103]. However, the impact of the misorientation on electronic device performance is not yet fully understood. This is due to the lack of SiC substrates that are oriented according to tight specifications with respect to the misorientation.

The slight misorientation from c-plane (0001)-sapphire substrates is well investigated for optoelectronic applications [3.58]. The emission from the photoluminescence of material on misoriented substrates is higher than for the on-axis material. The impact on GaN LED output power is further investigated in [3.228]. For the misorientation of 0.17° the morphology changed to small-step-type morphology. However, the optical output powers of the GaN LEDs are nearly identical despite the misorientation. The impact of the different optical surface properties has also to be considered.

For 4H–SiC and 6H–SiC substrates reports on the impact of the misorientation on layer and device performance are available [3.145, 3.171, 3.189, 3.195]. MOCVD-grown GaN epitaxial layers show a good morphology for on-axis substrates, while the crack density increases with the increase of the misorientation in [3.171].

This effect can be overcompensated by other growth parameters, such as the growth temperature, as shown for GaN diodes on 4H– and 6H–SiC in [3.145] and for heterostructures in [3.103]. An increase of the mobility and a carrier reduction in Si-doped GaN films is obtained with a slight misorientation of 3.5° compared to the on-axis layer. Positron annihilation spectroscopy shows that with increasing misorientation on 4H–SiC the concentration of nitrogen vacancies in GaN increases, while the number of shallow traps decreases [3.195]. This can potentially have an impact on device isolation in the buffer layer, if the misorientation is not controlled. With the application of MBE growth the importance of the misorientation rises, as described in [3.189]. The morphology of GaN grown on 4H–SiC(0001) with a very strong misorientation of 8° is strongly affected by undulations, while this not the case for a misorientation of ≤0.3°. Deduced from that, a controlled misorientation can be useful for the electronic device performance, if the growth parameters are properly matched.

Nucleation Layer

The conditions for the growth of the nucleation layer are critical for the defect concentrations, residual conduction, and the distribution of the defects and traps in all layers grown on top of it [3.103,3.185]. The nucleation layer is naturally highly dependent on the substrate choice. Several nucleation approaches have been suggested, both AlGaN (on SiC and sapphire), GaN (on sapphire), and AlN (on SiC and sapphire) nucleation have already been mentioned. The use of insulating AlGaN nucleation layers for the growth of AlGaN/GaN heterostructures on s.i. SiC and sapphire substrates is described in [3.185]. The nucleation layer thickness is 180 nm in this case. The TDD is about 5×10^9 cm^{-2}. The impact of the nitridation and nucleation conditions of the growth on the low-temperature GaN nucleation layer on the morphology on sapphire substrate is investigated in [3.218]. Dislocation densities estimated by XTEM of less than 10^8 cm^{-2} are reported for a 2 μm thick GaN layer. The grain orientation can be improved with increased nitridation temperatures. Further, the quality of the epitaxial GaN is improved by increased growth temperatures and pressures. In [3.219], AlN nucleation layers are grown on sapphire substrates. Thicknesses of 20 nm are fabricated at a growth temperature of 500–700°C. The TDD varied only slightly with values of $1-2 \times 10^9$ cm^{-2}. The nucleation on s.i. SiC can also be performed with AlN nucleation. The thickness of the nucleation layer is 120 nm in [3.183], which is higher than that for AlN nucleation on sapphire. The temperature dependence of the nucleation with AlN on SiC has already been discussed [3.30]. The AlN thickness is 100 nm in this case. The optimization of the nucleation layer is further monitored by the drain-lag type of pulsed-measurements for processed HFETs, as discussed in Chapter 5.

Buffer Isolation

The growth of the buffer layer in an AlGaN/GaN heterostructure is basically a trade-off between high isolation and surface morphology. For the buffer isolation, a minimum isolation requirement of 1×10^7 Ω cm–10^{11} Ω cm is suggested in [3.40,3.139]. Earlier examples include values of 10^5 Ω cm in [3.219] for high-resistivity GaN layers. The impact of the growth temperature of a GaN buffer between 1,070 and 1,150°C is discussed in [3.139]. Buffer isolation resistances between 10^4 and 10^{11} Ω cm are reached. Relatively higher growth temperature yield increased isolation in this regime. Further, it is mentioned that the isolation reduces by an order of magnitude when the test bias is switched from 2 to 40 V. The impact on surface morphology is also discussed in a trade-off, as hotter growth temperatures increase the surface roughness.

Buffer-Layer Thickness

The overall thickness of the epitaxial layers in the hetero-epitaxy is determined by the buffer layer, which varies between 350 nm [3.5], approximately 3 µm, e.g., in [3.139] and 5 µm in [3.132]. The impact of buffer layer thickness is discussed in more detail in [3.5]. A minimum thickness of 350 nm is used. The thickness is varied in the following trade-off. Thin buffer layers promise reduced thermal impedance and low growth time, while the growth of thicker layers reduces the effect of threading dislocation density TDD and the effect of the residual strain from the lattice-mismatch [3.132]. Further, the surface morphology improves for the buffer thickness between 2 and 3 µm on sapphire substrates, as mentioned in [3.40]. GaN buffer layers grown by MOCVD with thicknesses up to 5 µm are reported in [3.132]. Thicker buffer layers promise smoother growth fronts with improved interface roughness, reduced dislocation density, and further removal of the active region from the defective nucleation layers. The negative impact of thicker buffers on the thermal resistance to the substrate is further discussed in Chapter 8. The impact of various buffer materials and interlayers, such as InGaN, AlGaN, and GaN, on wafer bowing, grown in various ways by MOCVD, is discussed in [3.174]. In general, multilayer structures are found to enhance wafer bowing, see [3.5]. Recent reports demonstrate the introduction of thin (3 nm) InGaN layers in the buffer layer very close, i.e., at a distance of 10 nm, to the channel layer [3.154]. This is achieved in order to obtain a similar function as a real double heterojunction structure. A graded $Al_xGa_{1-x}N$-buffer approach grown by MOCVD is presented in [3.78]. The mole fraction x is varied from 0.3 at the nucleation layer to 0.05 near at the buffer channel interface. The thickness of the graded layer is 1 µm, the GaN channel thickness is 10 nm.

Codoping of the Buffer Layers

Unintentional codoping is a major problem for nonoptimized growth conditions [3.23, 3.94] in III-N semiconductors. Codoping introduces nonintentional background carrier concentration and thus persistent nonintentional conductivity. This fact has been the reason for a number of publications [3.6, 3.23, 3.94]. Nitrogen vacancies and oxygen serve as shallow donors [3.94] typically considered for GaN layers. Carbon prevails in the MOCVD reactor due to the methyl-precursors [3.28]. Mg and Fe are intentional dopants of interest, which, however, can be residual from reactor memory effects. In a number of the high-performance single-heterostructure devices, Fe or C [3.6, 3.63] are used to control the isolation in the buffer by pinning and thus stabilizing the Fermi levels through deep acceptors [3.28]. This codoping is one method used to control the Fermi-level in the buffer and to overcome the need for a second heterojunction. In [3.64], the use of Fe doping for applications in AlGaN/GaN HEMTs is described to improve the stability of the semi-insulating GaN layers and to reduce dispersion, especially the drain-lag. At the same time, the Fe doping does not modify the good structural quality of the layer. Thus the improved isolation is achieved without a high TDD necessary to obtain good isolation without codoping. The impact of high-temperature rapid thermal annealing (RTA) on the properties of Fe-doped semi-insulating GaN is investigated in [3.164] at temperatures between 750 and 1,050°C. The Fe distribution is found to be nonuniform and to increase close to the interfaces. The Fermi-level is pinned at $E_C - 0.5\,\mathrm{eV}$. The paper also states a strong residual gradient and diffusion of the Fe concentration in the buffer layer, partly based on a memory effect [3.15]. The reduction of dislocation density on both sapphire and s.i. SiC substrates through intentional Fe codoping is investigated in [3.15]. High-resistivity layers are obtained by both modulation doping and continuous bulk Fe-codoping. The Ferrocene flux is modulated to overcome the maximum doping level of $1 \times 10^{19}\,\mathrm{cm}^{-3}$. A more detailed study of the Fe doping at the regrowth interface of GaN on a GaN on sapphire template is given in [3.111]. The Fe doping of the interface is found to be crucial for the quality of the transistor layers on top of a 2-µm thick buffer. These thick Fe-doped layer leads to good mobility and increased carrier concentrations in the active transistor layers while no charge is found at the regrowth interface.

The impact of the excess incorporation of carbon on the trap states in MOCVD-grown material is investigated in [3.6]. Low-pressure growth at 76 Torr, compared to 760 Torr, leads to increased C incorporation in the GaN layers grown at 1,040°C. Carbon is considered to supply shallow but also deep acceptor levels in n-type GaN. This carbon contributes to the compensation of free carriers. C-doped samples are found to be highly resistive, while codoping with both Si and C leads to highly compensated layers. This finding is of critical importance, as O and C are typically present in any MOCVD reactor chamber. Oxygen incorporation during MOCVD growth is investigated in [3.23]. Higher growth temperatures at 1,150°C lead to reduced oxygen and carbon incorporation at levels of $10^{17}\,\mathrm{cm}^{-3}$, while the incorporation of carbon and oxygen can be strongly correlated. Growth temperatures of 920°C for $Al_{0.65}Ga_{0.35}N$ lead to oxygen concentrations of up to $1.5 \times 10^{19}\,\mathrm{cm}^{-3}$ and carbon concentrations of $1 \times 10^{18}\,\mathrm{cm}^{-3}$. In-codoping further suppresses the formation of native defects with acceptor character. Doping compensation is further of critical importance for the device behavior at high voltages, as explained in Chapter 5.

Dislocations in MOCVD Material

The threading dislocation density (TDD) was already discussed with the lattice mismatch in Chapter 2. The TDD is naturally dependent on the substrate type and quality. TDDs of $\approx 5 \times 10^8\,\mathrm{cm}^{-2}$ on s.i. SiC substrates and of low $10^9\,\mathrm{cm}^{-2}$ on sapphire substrates are reported for both MOCVD and MBE material in [3.39, 3.185]. These numbers are confirmed in a variety of publications, e.g., [3.36, 3.92]. MOCVD-grown material is analyzed by combined scanning capacitance microscopy and AFM measurements in [3.60]. Threading dislocations are found to be surrounded by a negative charge.

A TDD of $2 \times 10^9 \, \text{cm}^{-2}$ is mentioned for optimized MOCVD-based epitaxy of AlGaN/GaN on 4-in. silicon(111) substrate [3.204]. A good overview on the strain engineering on Si(111) is given in [3.34].

Epitaxial lateral overgrowth (ELOG) by MOVPE allows a reduction of the TDD below $\leq 10^9 \, \text{cm}^{-2}$ on sapphire substrates, as reported, e.g., in [3.86]. The leakage in AlGaN/GaN HBTs is reduced by an ELOG procedure with a concentration of threading dislocations of $10^8 \, \text{cm}^{-2}$ in the defect-reduced part. Areas with TDDs reduced to $\leq 10^6 \, \text{cm}^{-2}$ in the defect reduced part are reported in [3.86]. However, these stripes are typically 10µm wide only. The use of ELOG of GaN on conducting 6H–SiC substrates can yield dislocation densities as low as $2.2 \times 10^7 \, \text{cm}^{-2}$ in the wings (i.e., the defect-reduced area) and $2 \times 10^9 \, \text{cm}^{-2}$ in the windows (defect-rich area), as reported in [3.53]. GaN laser growth requires a TDD of $10^6 \, \text{cm}^{-2}$ obtained with an effective lattice mismatch of 0.01% to the heterosubstrate [3.34]. The TDD in bulk AlGaN grown by MOCVD on SiC substrates for solar blind detectors is found to be higher than in GaN layers. The overall TDD (mostly edge type) reach $2\text{–}5 \times 10^{10} \, \text{cm}^{-2}$ in the AlGaN, as shown in [3.157].

Partial dislocation annihilation by Si-δ-doping in GaN grown by MOCVD on silicon substrates is described in [3.31]. Screw dislocations are bent to neighboring dislocations, whereas edge dislocations are not affected by the burst of silicon during the growth. Fig. 3.2 depicts a GaN layer grown on sapphire(0001) (α-Al$_2$O$_3$) substrate to give a more qualitative image of the dislocation density. The GaN layer is etched by a photo-enhanced chemical (PEC) etching technique. The different etch rates in the area of the dislocations can be observed. Further, the extension of the dislocations is visible reaching 500 nm into the GaN layer. This fact requires a minimum buffer layer thickness to protect the actual device layers.

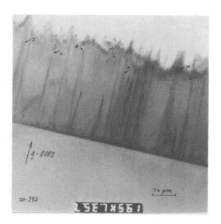

Fig. 3.2. SEM image of a PEC-etched GaN buffer on sapphire substrate

Impact of Growth on Carrier Mobility

Fig. 3.3 gives the temperature dependence of the drift mobility especially for temperatures lower than RT. Various mechanisms that influence the electron mobility can be separated. Data for both 2DEG AlGaN/GaN heterostructures and bulk GaN are given. The different contributions to the mobility lead to the overall mobility. The limiting effects dominating at different temperatures are the following:

− Impurity scattering dominating from below 70 K [3.144]
− Polar optical phonon scattering dominating above 70 K [3.21, 3.144]
− Acoustic phonon scattering [3.65, 3.101]
− Dislocation scattering
− Polar dipole scattering
− Interface scattering [3.4]

Impurity and polar optical scattering are the classical scattering mechanisms to reduce the mobility in the bulk, as described in [3.21], see Fig. 3.3. Acoustic phonon scattering is discussed in [3.65, 3.101] and found to be of significance. It can be distinguished from the other scattering mechanisms at temperatures below 20–50 K and plays a role in AlGaN/GaN heterostructures. Peak mobilities in 2DEGs of $80,000 \, \text{cm}^2 \, \text{V}^{-1} \, \text{s}^{-1}$ at 4 K are observed for sheet carrier concentrations of $0.4 \times 10^{12} \, \text{cm}^{-2} \leq n_{\text{sheet}} \leq 3 \times 10^{12} \, \text{cm}^{-2}$ [3.65].

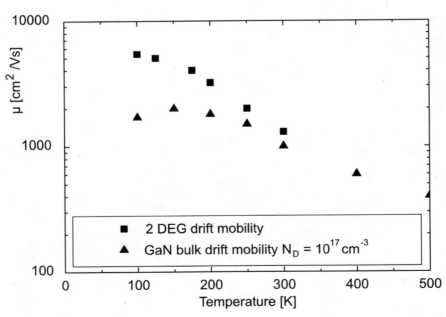

Fig. 3.3. Drift mobility as a function of lattice temperature in MOCVD material

Thus, for the III-N heterostructures, additional mechanisms prevail, which limit the mobility and which are discussed in [3.4, 3.80, 3.81]. Interface scattering is found to be relevant for carrier densities $\geq 7 \times 10^{12}\,\mathrm{cm}^{-2}$ [3.4]. For lower sheet carrier densities than $7 \times 10^{12}\,\mathrm{cm}^{-2}$, the mobility is limited by the interface charge and the related Coulomb scattering. Theoretical approaches to the additional dislocation scattering [3.80] and dipole scattering [3.81] are suggested. Dipole scattering of electrons in the heterostructures occurs due to the fluctuations from perfectly periodic dipole in highly polar semiconductors. Further, the high dislocation density of $10^8\,\mathrm{cm}^{-2}$ in AlGaN/GaN 2DEGs leads to an additional scattering mechanism [3.81]. However, mobilities of $51700\,\mathrm{cm}^2\,\mathrm{V}^{-1}\,\mathrm{s}^{-1}$ at $13\,\mathrm{K}$ are reached for carrier densities $\geq 2 \times 10^{12}\,\mathrm{cm}^{-2}$. The effect is different from the 3D-bulk impact of dislocations, as described in [3.120]. The drift mobility in heterostructure is further investigated, e.g., in [3.200]. The distinction of the carrier mobility between peak and the drift mobility in AlGaN/GaN heterostructures is made in [3.200]. Peak values of the mobility of $2,000\,\mathrm{cm}^2\,\mathrm{V}^{-1}\,\mathrm{s}^{-1}$ for low carrier densities of $0.4 \times 10^{13}\,\mathrm{cm}^{-2}$ are found on s.i. SiC substrates, while the actual drift mobility at high carrier densities of $1 \times 10^{13}\,\mathrm{cm}^{-2}$ amounts to $1,400\,\mathrm{cm}^2\,\mathrm{V}^{-1}\,\mathrm{s}^{-1}$ at room temperature. The peak mobility for layers on sapphire for comparable nominal barrier structures is only $1,300\,\mathrm{cm}^2\,\mathrm{V}^{-1}\,\mathrm{s}^{-1}$ at room temperature.

MOCVD-Grown Heterostructures

Nearly all electronic high-performance GaN FET and bipolar devices contain heterostructures of some kind. Thus, the specific growth of heterostructures and results are addressed in this section.

Heterostructure Transport in $Al_x\,Ga_{1-}N/GaN$

Fig. 3.4 depicts the extracted heterostructure Hall mobility data taken from a more statistical point of view from various publications, e.g., [3.13, 3.39, 3.102, 3.139]. The data on the AlGaN/GaN heterostructures are given for various substrates, including the channel resistivity contours. The Hall mobility values in general range between 500 and $2,000\,\mathrm{cm}^2\,\mathrm{V}^{-1}\,\mathrm{s}^{-1}$ depending on the substrate and material quality and some specific growth measures. The transport situation in heterostructures differs from a bulk approach, so that additional mechanisms influence the mobility, as is already discussed in chapter 2 and also further described in [3.82]:

- Alloy scattering mechanisms, especially at high sheet carrier densities
- Interface roughness
- Scattering for remote surface donors
- Scattering due to threading dislocations

Typical sheet resistances of 130–$500\,\Omega\,\mathrm{sq}^{-1}$ can be found for various compositions, as shown in Fig. 3.4. Numerous publications are available for the

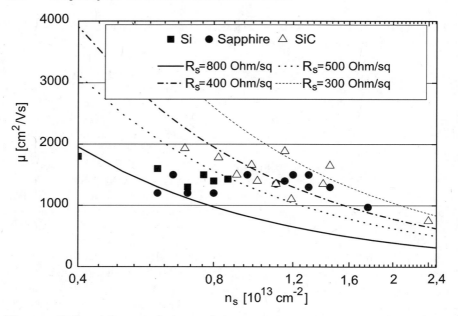

Fig. 3.4. Hall mobility as a function of sheet carrier concentration for AlGaN/GaN heterostructures grown on various substrates in MOCVD material

Al-content range between 10 and 35% [3.13, 3.39, 3.102, 3.139]. Even higher Al-contents ($x \geq 0.35$) in AlGaN/GaN MODFETs with Al-contents up to 52% are reported in [3.92, 3.136, 3.222]. The detailed influence of the MOCVD growth parameters will now be discussed.

Material Composition Dependence

The material dependence of the properties of the ternary materials is of great importance and determines several device parameters critically, as is already shown with the sheet resistance. Fig. 3.5 gives the Hall mobility data as a function of Al-content in the $Al_x Ga_{1-x}N$ barrier in MOCVD-grown heterostructures for completely undoped structures with a constant barrier thickness. The mobility is only weak function of the sheet carrier concentration up to a concentration of approximately 1×10^{13} cm^{-2}. Beyond that value, the Hall mobility decreases with increasing carrier concentration. Fig. 3.6 gives the Hall mobility as a function of carrier concentration in AlGaN/GaN heterostructures on sapphire for a given Al-mole fraction and barrier thickness with the Si δ-doping varied. The values for the sheet carrier concentration range from 4.5×10^{12} cm^{-2} to about 1.2×10^{13} cm^{-2}. The impact of doping of the $Al_{0.25}Ga_{0.75}N$ barrier can be observed in Fig. 3.6 for a given experiment at otherwise constant growth parameters. We see that the Si δ-doping in the barrier in the range 10^{17} cm^{-3} (nid) to 3×10^{19} cm^{-3} has a significant impact on the sheet carrier concentration, and that the doping has also a strong impact on

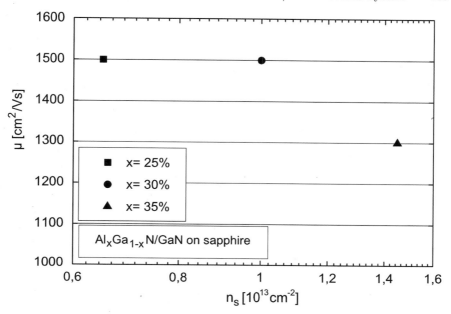

Fig. 3.5. Hall mobility as a function of sheet carrier concentration with the Al-content as a parameter in undoped MOCVD heterostructures on sapphire substrate

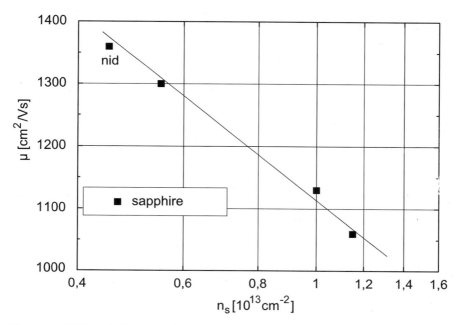

Fig. 3.6. Hall mobility as a function of sheet carrier concentration for different doping levels in the barrier for MOCVD material on sapphire and SiC substrates

the carrier mobility. Very similar data is found for growth on s.i. SiC in [3.48]. Additional effects arise from the introduction of AlN interlayers [3.139], as described further below. Depending on the growth parameters, sheet carrier densities of $2 \times 10^{13}\,\mathrm{cm}^{-2}$ can be reached [3.25, 3.92].

Uniformity

Epitaxial uniformity is a basic requirement for the reproducible device and MMIC production. The uniformity is currently given for 2-in. to 4-in. wafer sizes, e.g., [3.192]. For the average standard deviation of the sheet resistance, typically a value $\leq 1\%$ is specified for both 2-in. and 3-in. wafers on MOCVD material grown on sapphire and s.i. SiC substrates [3.192]. Fig. 3.7 shows the homogeneity of the sheet resistance within a multi-wafer growth reactor chamber for 2-in. s.i. SiC wafers, which achieves a standard deviation of 1%. The improvement in uniformity due to the transition from 2-in. to 3-in. s.i. SiC wafers is, e.g., investigated in [3.221]. The run-to-run reproducibility is significantly improved for the 3-in. growth, which is a prerequisite for reproducible device production. In a direct comparison, average standard deviations $\leq 3\%$ on 3-in. MBE-grown and MOCVD-grown material are reported in [3.221]. The substrate quality impact on the uniformity is strong, especially for SiC substrates with still evolving quality. Some general findings shall be listed here: With the increasing substrate diameter, improved substrate-related unifor-

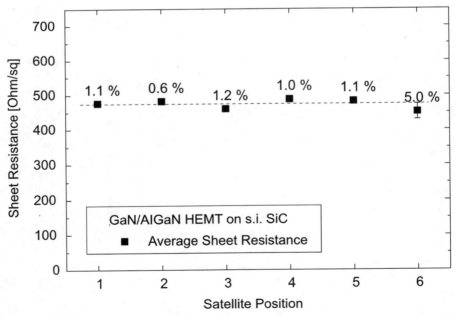

Fig. 3.7. Homogeneity of sheet resistance in MOCVD material within a $6 \times 2''$ reactor

mity and a wafer-to-wafer uniformity of AlGaN/GaN HEMT epitaxial wafers on sapphire and s.i. SiC substrates are reported [3.193]. The typical standard deviation of the sheet resistance is 2% on 4-in. sapphire substrates. Very reproducible AlGaN/GaN HEMTs are grown on n-type SiC with very low standard deviations of $\leq 1\%$ of the device parameters [3.96]. Typically, the substrate quality improves with the increase of the substrate diameter. This scaling benefit is similar to the findings of silicon technology and related substrate scaling. The contribution of the homogeneity of the material composition is investigated, e.g., in [3.161]. The material composition $x = 0.12$ of $Al_xGa_{1-x}N$ directly grown on s.i. SiC varied by $x \pm 0.005$ due to terrace steps on the growth surface.

Fig. 3.8 gives a topogram to demonstrate the on-wafer homogeneity of the sheet resistance on a 3-in. wafer. The homogeneity is very much influenced by the substrate quality, which is also visible in Fig. 3.8. The largest deviations of the sheet resistance can be correlated with substrate nonuniformities in the upper left edge shown in Fig. 3.8, resulting from substrate growth. Further particular growth issues to increase uniformity include the following:

– The uniform distribution of the defects such as micropipes and others in the substrates
– The uniformity of the temperature and of other growth parameters over the wafer
– The growth windows with respect to relative requirements and to relative deviations
– Such as the uniformity of the thermal contact of the substrate during growth

These issues are related to the particular reactor concepts, such as the wafer rotation during growth. These issues are now discussed in the following sections.

Material Analysis

Several state-of-the-art methods are used for the characterization of the III-N heterostructures. These advanced methods are necessary to optimize this complicated material system with regard to the great variety of effects. Atomic force microscopy (AFM) typically reveals surface roughness. The measurements are performed for areas with some $10 \, \mu m$ dimension, e.g., [3.204]. An example of an AFM image is given in Fig. 3.9 for the growth of GaN surface on sapphire substrate. A RMS roughness of 0.4 nm is observed. Structural topography by X-ray diffraction (XRD) and two-dimensional XRD maps reveal structural properties such as micropipes, dislocations, and grain boundaries of substrates and overgrown semiconductor layers [3.100]. The line widths in XRD measurements further allow the investigation of the lattice-mismatch and the Al-content in the barrier layer, as explained, e.g., in [3.204].

Photo-luminescence measurements at various temperatures reveal possible trap mechanisms, e.g., as stated in [3.219], between 2.25 and 3.0 eV. Fig. 3.10

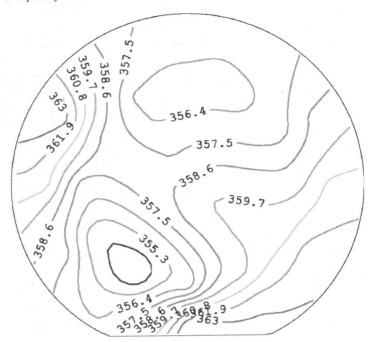

Fig. 3.8. On-wafer homogeneity of sheet-resistance of an AlGaN/GaN MOCVD-grown heterostructure on 3-in. s.i. SiC substrate, average $359\,\Omega\,\mathrm{sq}^{-1}\pm0.7\%$

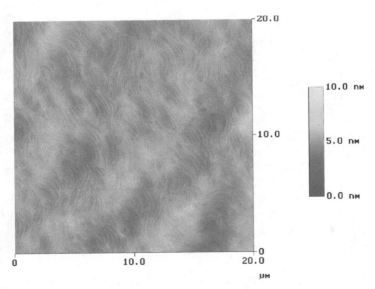

Fig. 3.9. Atomic force microscope image of a surface of a GaN layer MOCVD-grown on sapphire substrate

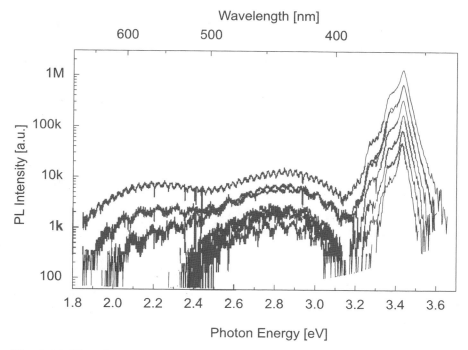

Fig. 3.10. Photoluminescence spectra of AlGaN/GaN heterostructures for different excitation energies

gives a spectrum of an AlGaN/GaN HEMT structure on s.i. SiC. There is yellow luminescence visible at a photon energy of ≈2.8 eV. The significance of this data is based on the electron recombination in high-resistivity films, potentially with edge-type threading dislocations [3.219] leading to high resistivity. An alternative explanation is the removal of a particular trap with changes of growth parameters.

MOCVD Growth on s.i. SiC(0001) Substrate

SiC(0001) requires the matching of the GaN layers to the 4H- or 6H-s.i. substrate. Direct growth of GaN on (0001) 6H-SiC leads to columnar grain with rough, faceted surfaces and high non-intentional net carrier concentrations [3.59]. Typical nucleation layers for growth on SiC are high-temperature AlN layers [3.215, 3.222]. Further details are given below. An example of multi-wafer epitaxy of AlGaN/GaN heterostructures on s.i. silicon carbide for power applications is given in [3.102]. Fig. 3.11 shows the sheet carrier concentration in $Al_xGa_{1-x}N$/GaN heterostructures as a function of Al-content on s.i. substrate for a mole fraction between 0 and 40%. A very linear dependence is visible for these samples from $x = 0.13$ to 0.35. As the Al-content defines the sheet carrier concentration at the interface, Fig. 3.12 gives the homogeneity of

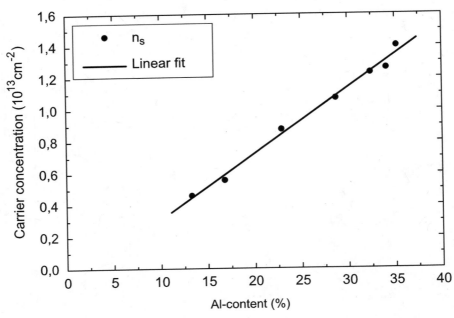

Fig. 3.11. Sheet carrier concentration as a function of Al-content for MOCVD-grown $Al_xGa_{1-x}N/GaN$ heterostructures on s.i. SiC substrate

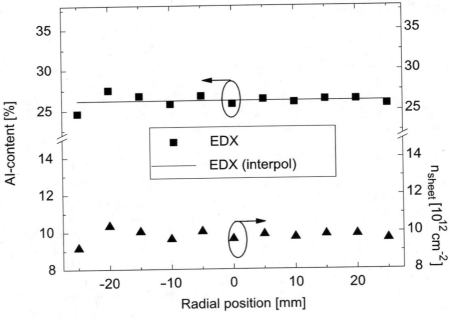

Fig. 3.12. Al-content and corresponding variation of the sheet carrier concentration as a function of radial position for a 2-in. s.i. SiC wafer

the Al-content determined by electron diffraction (EDX) and at the same time provides the calculated difference of the carrier concentration. Even a very small variation of $x \leq 0.2\%$ provides a significant variation in the sheet carrier concentration of $\approx 1 \times 10^{12}$ cm^{-2} over the 2-in. diameter. The n_{sheet} is calculated from the $n_{sheet}(x)$ extrapolated relation in Fig. 3.11.

MOCVD Growth on Sapphire(0001) Substrate

One of the crucial advantages of sapphire relative to SiC is the lack of micropipes and the commercial availability of larger substrate formats up to at least 6 in., as used for CMOS on sapphire with a different crystal orientation [3.95]. MOCVD growth on 4-in. sapphire substrates is reported in [3.7, 3.96]. MOCVD growth issues on sapphire substrates are the increased lattice-mismatch and the difference in thermal expansion between sapphire and GaN. The latter results is increased wafer bow [3.96] when the wafer is cooled to room temperature after growth. As an example, the wafer bowing can be reduced to 22 µm on top of 630 µm thick GaN buffer, as reported in [3.193]. Epiready 2–4 in. sapphire substrates have been used in [3.192]. The standard deviation of the sheet resistance is $\leq 1\%$ for all diameters. The Al-composition varies by 0.2% over the 3 in. wafer. The AlGaN thickness varies by 29.25 ± 0.75 nm. Fig. 3.9 depicts the AFM image of a GaN surface on sapphire substrate. The surface root mean square (RMS) roughness on sapphire amounts to only 0.4 nm for an area of 20×20 µm^2. The peak-to-peak roughness can be as high as 15 nm, which has several implications for the scaling of the devices to small gate lengths.

MOCVD Growth on Silicon(111) Substrate

Growth on silicon substrates is very attractive as a cost effective method and also relates the III-N material system to the most successful semiconductor system. First demonstrations of MOCVD-grown AlGaN/GaN HEMTs on silicon substrate have been claimed, e.g., in [3.29]. The results are obtained on p-type Si(111) substrates with a buffer layer of 500 nm thickness on top of a 30 nm AlN nucleation layer. An overview of the MOCVD growth on silicon substrate is given in [3.34]. The silicon material can either be conductive or highly resistive (30 kΩ cm). Highly resistive silicon is more expensive than conductive substrates and available only in smaller diameters. However, highly resistive silicon is advantageous to reduce the substrate RF-losses, especially at frequencies higher than 2 GHz [3.42]. Buffer growth is of great importance due to the increased lattice-mismatch between silicon and the III-N material system. A critical thickness of the buffer growth of 1 µm is mentioned in [3.34]. Furthermore, the difference in thermal expansion is pronounced, as explained in Table 2.5. Mismatch engineering is thus of critical importance. Patterned substrate techniques are described, e.g., in [3.85]. AlGaN/GaN HEMTs are fabricated on top of rectangular silicon ridges. The rectangular stripes have a maximum size of $168 \times 52 \, \mu m^2$, which limits the application of this techniques to small devices. The stress distribution in patterned GaN films grown by MOCVD on (111) silicon is investigated in [3.208]. The intentional cracking of the layers grown reduces the tensile strain in small area stripes. Low-temperature (LT) AlN-interlayers are used for strain engineering [3.34]. The LT-AlN layers are combined with high-temperature layer stacks. These methods can be used to reduce the threading dislocation density in the heterostructure layers [3.37]. Crack-free GaN material is reported in [3.79] on 2-in. p-type silicon grown by MOCVD. The buffer thickness is $2.5 \, \mu m$, which further reduces the dislocation density on top of specially designed layer sequences used for nucleation. The RMS surface roughness reported amounts to $0.64 \, nm$ determined by AFM. A commercial GaN HFET process on $100 \, mm$ diameter silicon substrates is given in [3.204], using GaN epitaxial layers up to $2 \, \mu m$ thickness in the SiGaNtic^{TM} process. The standard deviation of the $Al_{0.21}Ga_{0.79}N$ spectral-peak wavelength at $324 \, nm$ across a $100 \, mm$ wafer is as low as 0.6%. The Hall mobilities reach $1{,}430$–$1{,}500 \, cm^2 \, V^{-1} \, s^{-1}$ at a sheet charge carrier concentration of $\approx 8 \times 10^{12} \, cm^{-2}$. The difference in thermal expansion and lattice-mismatch have particular impact on the device reliability, which is further discussed in Chapter 7.

3.1.2 Molecular Beam Epitaxy (MBE)

MBE is the second attractive growth method for III-N semiconductors. The growth of semiconductors by MBE occurs via reactions between thermal-energy molecular, atomic, or ionized beams of the constituent elements on a heated substrate in an ultrahigh vacuum; see, e.g., in [3.36]. The advantage of MBE is a very precise definition of the interfaces and the increased flexibility of the polarity of the interfaces [3.87, 3.196]. These advantages are available at the expense of reduced growth rates relative to MOCVD growth. The growth rates of III-N MBE typically amount to 0.5–$1 \, \mu m \, h^{-1}$ [3.132]. Fig. 3.13 gives the schematic of a MBE growth chamber. The molecular beam of Ga and Al originate from effusion cell sources. An elemental group-V N-source is impossible due to the very high binding energy of the N_2. N radicals are thus generated either by a RF-plasma source [3.17, 3.132, 3.166] or by an ammonia source, as reported, e.g., in [3.194, 3.213]. The background pressure amounts to $\leq 10^{-11} \, mbar$ during MBE growth, as given, e.g., in [3.89]. The substrates are typically rotated during growth. Growth temperatures amount to $800°C$ for GaN growth. Growth conditions of MBE can be characterized by growth diagrams, given for III-V semiconductors, e.g., in [3.36, 3.88, 3.149].

State-of-the-Art of MBE Growth

Overviews on molecular beam epitaxial growth of III-N materials are given, e.g., in [3.50, 3.87, 3.172]. More general overviews of growth and doping of III-V nitrides including MBE growth are given in [3.39, 3.150]. AlN is typically used

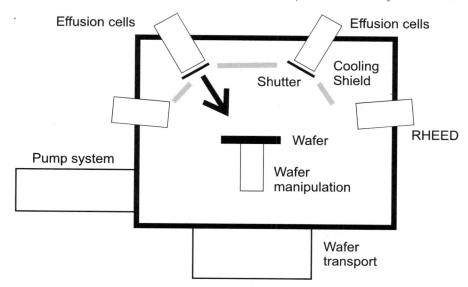

Fig. 3.13. Scheme of molecular beam epitaxial (MBE) growth

as a nucleation layer for MBE [3.90, 3.132, 3.216]. Issues in scaling nitride MBE systems to 4-in. single-wafer production are discussed in [3.188]. The scaling of the RF-plasma system is investigated, especially with respect to the small growth window in the Ga/N flux ratio. The second issue in wafer-size scaling is the growth temperature uniformity for increased wafer size. Compared to other III/V-MBE systems, the temperature control is more complicated due to the high growth temperatures of III-N materials, leading to high radiative losses. Further, the optical transparency of the substrates, such as SiC and sapphire, in the infrared complicates heat absorption. MBE-growth conditions are investigated in a number of publications, e.g., in [3.17, 3.41, 3.66, 3.67, 3.113, 3.165, 3.205, 3.213], and are detailed below. On the device level, high-power GaN/AlGaN HEMTs grown by plasma-assisted MBE operating up to 25 GHz are described in [3.124, 3.125, 3.132, 3.216]. An X-band power amplifier is given featuring a cw-output power of 22.9 W using four hybrid 1 mm cells. The amplifier yields a PAE of 37%. Further impressive device examples and a comparison of HEMTs grown either by plasma-assisted MBE or MOCVD are given in [3.39, 3.166]. Device examples of ammonia-based GaN MBE are given in [3.10]. AlGaN/GaN HEMTs with gate lengths of 130 nm feature very high drain current densities $\geq 1.3\,\mathrm{A\,mm^{-1}}$ and a cut-off frequency $f_\mathrm{T} = 103\,\mathrm{GHz}$.

Optimized Growth Conditions for MBE Growth

Similarly, as in MOCVD growth, optimized growth conditions for MBE are very system- or machine-specific. However, general trends will be outlined here.

Growth Temperatures for III-N MBE Growth

Before the actual growth procedure, the substrates are typically inserted into the growth chamber via a preparation chamber. The substrates are outgased in the preparation chamber. A typical substrate outgas temperature in the preparation chamber is reported to be 700°C in [3.90]. Further temperature steps for oxide removal reach 1,000°C. The reported nucleation temperature using AlN nucleation on sapphire substrate is 800°C, as reported in [3.150]. A nucleation temperature of 835°C on s.i. 6H–SiC substrates is used in [3.90]. The growth of the insulating GaN buffer layer on top of a nucleation layer is reported to be performed at 745°C in [3.90, 3.125] for Ga-rich conditions. The GaN channel layers are typically grown at 800°C using atomic nitrogen, as reported in [3.40]. The backside of the substrate is metalized by tungsten in this case. Thick bulk GaN layers are grown by plasma-assisted MBE between 600 and 800°C for different Ga/N flux ratios on SiC(0001), as reported in [3.205]. The threading dislocation density can be reduced from $10^{11}\,\mathrm{cm}^{-2}$ for a growth temperature of 600°C and Ga-lean growth to $10^{10}\,\mathrm{cm}^{-2}$ for a growth temperature of 800°C and Ga-rich growth. The nominal growth temperatures for GaN using ammonia-based MBE growth is 910–920°C and 880°C for AlN [3.10, 3.212]. A growth temperature for a bulk GaN layer of 860°C is reported in [3.194] for ammonia-based MBE on sapphire(0001). Magnetron sputter epitaxy is used in this case for the initial deposition of a AlN nucleation layer. The growth of quaternary $Al_xGa_{1-x-y}In_yN$ materials by plasma-assisted MBE on GaN templates is described in [3.137]. Substrate growth temperatures range from 590 to 650°C. The use of the MBE technique increases the material composition range, which is ($0 \leq x \leq 0.5$, $0 \leq y \leq 0.2$) in this case.

N-Group-III Ratio for MBE Growth

Similar to MOCVD growth, the V/III ratio is a critical parameter for the MBE growth, however, for a very different material transport. A distinction is made for RF-plasma-assisted MBE (PAMBE) and ammonia-based MBE. The V/III-ratio determines the quality of the epitaxial layers. The impact of the Ga/N flux ratio on trap concentrations and morphology in GaN grown by plasma-assisted MBE is discussed in [3.67, 3.68]. The distinction of the Ga-rich, intermediate, and N-stable regime is made. N-stable-grown GaN films are semi-insulating. GaN layers grown with ratios in the intermediate regime yield fewer pits with an atomically flat surfaces. At high Ga/N-ratio Ga droplet formation occurs. However, the mobility decreases significantly. The best material is thus grown within the intermediate regime. The best combination of surface morphology and material quality is achieved by a modulated technique varying alternately Ga-rich and Ga-lean growth, see, e.g., [3.33, 3.117]. The growth diagram for GaN, grown homoepitaxially by plasma-assisted MBE, is given in [3.67] for a growth temperature of 720°C. The surface morphology and GaN mobility are optimized for the various Ga-flux conditions. Similar surface structure diagrams between 550 and 700 K are given in [3.66].

The growth conditions such as V/III ratio and temperature for AlGaN are investigated for plasma-assisted MBE in [3.41]. The $Al_{0.45}Ga_{0.55}N$ layers are grown at temperatures between 650 and 750°C. Particularly, the O_2 incorporation is investigated. It is found that the incorporation is minimized for the higher growth temperatures.

For ammonia-MBE, carbon-doped insulating GaN MBE layers are grown on s.i. SiC substrates at a temperature of 910°C [3.211, 3.212, 3.214]. For the AlN nucleation in [3.194], argon and ammonia fluxes are adjusted leading to growth rates of $3.4\,nm\,min^{-1} \equiv 204\,nm/h$ at a very high deposition pressure of 1.4 mTorr. GaN-layer deposition pressures of 3×10^{-6} Torr are reached during the growth. The growth rates reach $1-2\,\mu m\,h^{-1}$ by the adjustment of the ammonia flow to 50 sccm.

Nucleation Layer Growth by MBE

The nucleation layer is of very high importance on various substrates, as this region with its polycrystalline inclusions contributes significantly to the quality of the MBE-grown active semiconductor layers and thus to device performance. The growth of an AlN nucleation layer by plasma-assisted MBE on s.i. 4H–SiC substrate is reported in [3.89]. The layer thickness is 100 nm, grown at a temperature of 835°C. The N_2 pressure is 2×10^{-5} Torr for the nucleation layer. A 60 nm thick AlN nucleation layer is used in [3.124] grown by PAMBE on s.i. SiC substrate under Al-rich growth conditions at 745°C. Similar thicknesses and conditions are reported in [3.125]. Hundred nanometer thick AlN nucleation layers are grown by RF-PAMBE at 835°C [3.90]. An AlN nucleation layer on sapphire with a thickness of 70 Å is used in [3.39] for RF-plasma-assisted MBE on sapphire substrate. Al-rich surfaces to obtain Ga-polarity are used. The GaN growth rate is $0.5\,\mu m\,h^{-1}$ at a growth temperature of 800°C. On silicon substrate, a 50–60 nm thick AlN nucleation layer is used, grown by reactive MBE in combination with a 500 nm special layer seque ice and a 1.5 µm thick GaN buffer [3.135]. The TDD amounts to $7-9 \times 10^9\,cm^{-2}$.

Growth and Isolation of the Buffer Grown by MBE

A large variety of buffer layers grown by MBE are reported for heteroepitaxy. The growth rate of GaN achieved by MBE in general is smaller than in MOCVD growth, e.g. 0.5 µm in [3.132]. This lower growth rate favors reduced layer thicknesses for MBE growth, especially for the buffer layer.

As an example for a optimized buffer sequence, after the growth of a 45 nm thick AlN nucleation layer on 4H-s.i. SiC, a 400 nm thick GaN two-layer sequence is used in [3.162, 3.166, 3.206] to improve the isolation, grown with a flux variation stepping from a low Ga-flux regime to a high Ga-flux regime. Further, the AlN and the first 100 nm are C-doped with CBr$_4$. This sequence is followed by a 1 µm thick undoped GaN buffer/channel-layer. Two micrometer GaN thick buffer layers grown by PAMBE on s.i. 4H–SiC substrate are reported in [3.132]. The growth rate is 0.5 µm h^{-1}.

Grown by ammonia-MBE on s.i. 4H–SiC substrate, a 2 µm thick highly insulating C-doped GaN buffer layer is reported in [3.213]. The layer is followed by a 200 nm undoped GaN channel layer grown at a nominal temperature of 920°C. Beryllium-doped GaN buffer layers grown by RF-plasma-assisted MBE on n-type 6H–SiC are reported in [3.90]. The leakage through a 1 µm thick buffer is reduced by the Be-doping at concentrations between 10^{18} and 10^{19} cm^{-3}. Even the lowest doping concentration leads to significant reduction of the leakage by three orders of magnitude.

On high-resistivity silicon substrate, a 1.5 µm-thick GaN buffer is reported in [3.135], grown by MBE on a 50 nm-thick AlN nucleation layer. Crack-free GaN bulk buffer layers with thicknesses up to 3 µm grown on silicon substrate are reported in [3.180]. Ammonia is used as a nitrogen source for the MBE. Buffer layers with thicknesses between 1 and 2.5 µm are investigated for AlGaN/GaN heterostructures. Hall mobility values up to 1,600 cm^2 V^{-1} s^{-1} are achieved at room temperatures. The TDD range between 5×10^9 and 10^{10} cm^{-2}.

The interface the substrate is especially susceptible to isolation issues due to the high concentration of defects. Indications for conduction parallel to the GaN/sapphire interface in PAMBE-grown material are investigated in [3.3]. The buffer material is etched to different thicknesses. The results show that the apparent carrier concentration increases near the interface, while the mobility decreases. Both interface conduction and impurity band conduction near the interface are considered to explain this behavior.

Codoping in MBE Material

Similar to MOCVD-grown material, intentional codoping is of great importance for the isolation of the nucleation and buffer layers. During MBE growth, this codoping is achieved mostly by carbon and (non-intentional) oxygen, as Fe sources are not available. Carbon introduces a midgap state which helps to increase the isolation in buffer layers. Codoping by carbon for MBE-grown GaN buffer material in the buffer is reported, e.g., in [3.162, 3.163, 3.166]. A 45 nm-thick C-doped nucleation layer is grown by PAMBE on s.i. SiC. This layer is followed by a sequence of GaN buffer layers. By adjustments of the CBr$_4$ pressure linearly from 15 and 5 mTorr, the background C-doping concentration is varied linearly from 1.2×10^{18}, 8×10^{17}, and 4×10^{17} cm^{-3}. A systematic investigation of carbon codoping via CBr$_4$ in PAMBE-grown material is reported in [3.162]. Doping levels of 2–6×10^{17} cm^{-3} significantly reduce the buffer leakage in AlGaN/GaN single heterostructures on s.i. SiC substrate. Unintentional oxygen incorporation into GaN layers is a major problem in

MBE growth. Several sources such as the background vacuum pressure, the nitrogen source, and the metal source are potential candidates. A systematic investigation is given in [3.41] for AlGaN layers grown by PAMBE on sapphire substrates. Ga-rich growth of AlGaN reduces the oxygen incorporation by a factor of three. Temperature reduction from 750 to 650°C significantly increases the oxygen incorporation. The nitrogen source is considered as a main source of oxygen in the MBE growth system, which is operated at a background pressure of 10^{-11} Torr. Intentional oxygen incorporation into GaN grown by PAMBE has been investigated in [3.165]. The doping is found to be controllable and reproducible up to doping level of 10^{18} cm^{-3}. Oxygen is an effective donor with low compensation and the incorporation is found to be strongly dependent on the growth polarity. PAMBE-grown GaN layers with N-polarity incorporate about 50 times as much oxygen than Ga-face material. Oxygen doping levels of 10^{22} cm^{-3} can be achieved.

Channel and Barrier Layers Grown by MBE

The growth conditions for the GaN channel layers is of primary importance for the material quality. Several detailed suggestions are available. GaN channel layers are grown by PAMBE at 735°C at a growth rate of 0.37 μm h^{-1} in Ga-rich conditions [3.89]. The Al$_{0.25}$Ga$_{0.75}$N heterobarrier layers are grown at the same temperature and the same nitrogen flow as GaN in this case. The GaN channel layers in [3.124] are grown by PAMBE on s.i. 6H–SiC under Ga-rich conditions at 745°C near the border of Ga-droplet formation. The upper half of Al$_{0.34}$Ga$_{0.66}$N layers are doped by Si to 1×10^{18} cm^{-3}. Similar growth temperatures for PAMBE on sapphire are used in [3.75, 3.77] with a growth temperature of 730°C for GaN, and 760°C for the subsequent AlN/Al$_{0.4}$Ga$_{0.6}$N growth. This temperature is significantly lower than the growth temperature for similar AlGaN layers grown by MOCVD. For ammonia(NH$_3$)-based MBE growth on s.i. SiC, the GaN channel layers are typically grown at higher temperatures, i.e., 920°C, as reported in [3.213]. A similar temperature of 910°C for GaN is reported in [3.10, 3.214] on 4H–SiC(0001). A growth temperature of 790°C is mentioned in [3.180] for growth on conducting p-type silicon(111) substrates. The resulting mobility data and sheet carrier densities are detailed below.

Dislocations in MBE Material

The TDD in the MBE-grown layers can be reduced to densities similar to those achieved for MOCVD-grown material. On s.i. SiC substrate, the dislocation density can be reduced to $\leq 10^9$ cm^{-2}, as observed by plan-view TEM in [3.17]. Plasma-assisted MBE (PAMBE) is used in this case.

Leakage current mechanisms in Schottky diodes are investigated in [3.133, 3.134] in plasma-assisted MBE material grown on GaN templates, which again are grown by MOCVD on sapphire substrates. A leakage mechanism is isolated during the investigation based on threading dislocations. The mechanism has an exponential temperature dependence, and so either a trap-assisted tunneling or a one-dimensional hopping mechanism is considered for the physical origin. A similar investigation of the impact of different dislocation types such as edge-, screw-, and mixed -type in GaN grown by PAMBE on MOCVD templates on sapphire substrates is given in [3.184]. A combination of scanning Kelvin probe microscopy (SKPM) and conductive atomic force microscopy (C-AFM) is used. Edge- and mixed-type dislocation are found to be negatively charged with limited contribution to the leakage, while pure screw dislocations, on the contrary, provide a conductive leakage path. The TDD obtained for the GaN templates on sapphire is found to be maintained through the subsequent MBE growth and is determined by electron-beam current microscopy (EBIC) to be $8 \times 10^8 \, \mathrm{cm}^{-2}$.

On resistive silicon substrate, the threading dislocation density amounts to $5\text{--}7 \times 10^9 \, \mathrm{cm}^{-2}$ reported in [3.32, 3.135] for MBE-grown material and observed by TEM. The overall layer thickness is $2.6 \, \mu\mathrm{m}$. Further investigations of p-type silicon(111) substrate in [3.180] demonstrate an increase of the threading dislocation density of GaN on silicon substrate relative to s.i. SiC or sapphire substrates by a factor of two in the same investigation.

MBE-Grown Heterostructures

The basic properties such as mobility and sheet carrier densities grown by MBE are very similar to those obtained by MOCVD growth. Fig. 3.14 gives the mobility as a function of Al-content in MBE-grown material. Similarly, Fig. 3.15 gives the sheet carrier concentration. The maximum sheet carrier concentrations reach $2 \times 10^{13} \, \mathrm{cm}^{-2}$. Similar tendencies for the MBE-grown material are observed as for MOCVD, compare Fig. 3.6.

Fig. 3.16 depicts the mobility as function of sheet carrier concentration for various substrates in MBE-grown material at 300 K. Again, very strong variations are observed for a given Al-content depending on the substrate type and the growth conditions, compare Fig. 3.7. The maximum Hall mobility is close to $2000 \, \mathrm{cm}^2 \, \mathrm{V}^{-1} \, \mathrm{s}^{-1}$ at RT. The results presented in Fig. 3.16 are now analyzed in detail for the different substrate types.

MBE Growth on s.i. SiC

High-performance heterostructures are achieved using both RF-plasma-source MBE and ammonia MBE [3.132, 3.214, 3.217]. Low-field mobility values of GaN/AlGaN heterostructures on 6H–SiC as high as $75,000 \, \mathrm{cm}^2 \, \mathrm{V}^{-1} \, \mathrm{s}^{-1}$ measured at $4.2 \, \mathrm{K}$ grown by PAMBE have been reported, while the RT mobility values amount to $1,400 \, \mathrm{cm}^2 \, \mathrm{V}^{-1} \, \mathrm{s}^{-1}$ [3.216]. An additional ECMP surface polish is reported to be crucial to the reproducibility of the heterostructure device results on the SiC substrate.

Sheet carrier densities of up to $2.1 \times 10^{13} \, \mathrm{cm}^{-2}$ have been obtained on s.i. SiC substrates using RF-PAMBE. The associated room temperature mobility is $1,000 \, \mathrm{cm}^2 \, \mathrm{V}^{-1} \, \mathrm{s}^{-1}$ [3.16]. Very thin barrier AlGaN layers

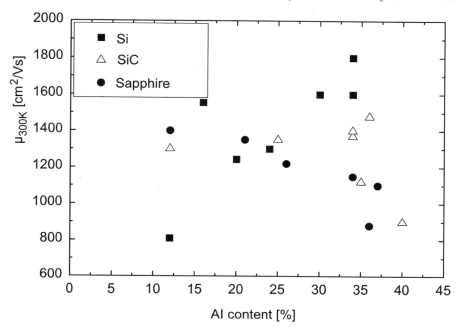

Fig. 3.14. Hall mobility as a function of Al-content in MBE-grown $Al_xGa_{1-x}N/GaN$ heterostructures on various substrates

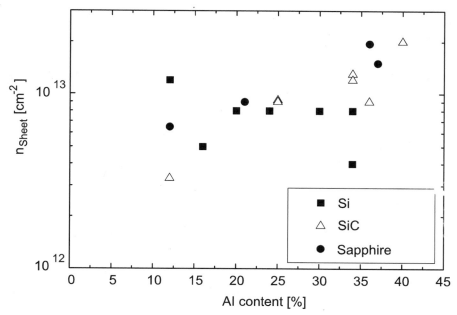

Fig. 3.15. Sheet carrier concentration as a function of Al-content in MBE-grown $Al_xGa_{1-x}N/GaN$ heterostructures on various substrates

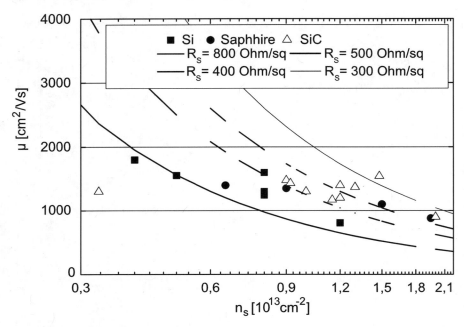

Fig. 3.16. Mobility in $Al_xGa_{1-x}N/GaN$ heterostructures on various substrates as a function of sheet carrier concentration

of 8 nm thickness lead to reduced sheet carrier densities of 8×10^{12} cm^{-2} due to the reduced polarization effect [3.75]. Very high cut-off frequencies ≥ 100 GHz of AlGaN/GaN HFETs on s.i. SiC substrates grown by ammonia MBE are reported in [3.10]. The AlGaN/GaN heterostructure mobility is 993 cm^2 V^{-1} s^{-1}, while a high sheet carrier density of 1.6×10^{13} cm^{-2} is measured. Inverted n-face AlGaN/GaN/AlGaN HFETs on s.i. SiC substrates show a room-temperature mobility of 860 cm^2 V^{-1} s^{-1}. The sheet carrier density amounts to 7.8×10^{12} cm^{-2} [3.26]. The latter result demonstrates the increased flexibility by MBE growth, as inverted polarization with N-face growth can be obtained.

MBE Growth on Sapphire

Most of the initial bulk material investigations and heterostructure results by MBE have been performed on sapphire substrates [3.39] or on GaN-templates grown on c-sapphire by MOCVD, e.g., [3.66–3.68, 3.133, 3.134, 3.137]. RT mobility values of 1,211 cm^2 V^{-1} s^{-1} are reported in [3.212] for $Al_xGa_{1-x}N/GaN$ heterostructures grown by ammonia-based MBE. The mobility is only a weak function of the Al-mole fraction at RT. The mobility at 77 K amounts to 5,660 cm^2 V^{-1} s^{-1} for $x = 10\%$, while it is 3,200 cm^2 V^{-1} s^{-1} for $x = 20\%$, and 2,600 cm^2 V^{-1} s^{-1} for $x = 30\%$. Further room temperature results for growth on sapphire are compiled in [3.214]. Using metal organic sources, III-N MBE can be performed leading to metal organic molecular beam epitaxy (MOMBE) [3.113]. Metal organic species are used for the group-III elements, while the group-N elements are generated by plasma or ammonia.

MBE Growth on Silicon(111)

Several reports exist for the MBE growth on silicon(111) substrate. MBE growth of AlGaN/GaN HEMTs on resistive Si(111) substrates using ammonia MBE are reported in [3.32,3.179,3.180]. The range for the resistivity in the substrate amounts to 4–20 kΩ cm for high resistivity [3.135] and 0.002 Ω cm [3.180] for p-doped substrates. The carrier densities reach 8×10^{12} cm^{-2} with room-temperature mobilities of 1,600 cm^2 V^{-1} s^{-1}, respectively [3.135]. The TDD is 5–7$\times10^9$ cm^{-3}. Mobilities of up to 2,100 cm^2 V^{-1} s^{-1} on Si(111) have been reported in [3.16] with Al-contents up to 50%. The report further claims no change in the surface morphology for such high Al-contents. The use of silicon substrates with substrate resistivities of 20 kΩ cm leads to very good noise performance using an AlGaN/GaN HEMT structure on silicon with ≥ 2 μm thickness. Good mobility values of 1,300 cm^2 V^{-1} s^{-1} are reported in [3.213] with an associated sheet carrier density of 3.3×10^{12} cm^{-2} grown by ammonia-based MBE. The mobility at 77 K amounts to 11,000 cm^2 V^{-1} s^{-1}. Picogiga reports MBE growth with an ammonia source on high resistivity silicon substrates with up to 4-in. diameter in [3.16]. The typical RMS roughness is 3–5 nm and the TDD is in the range of 10^9 cm^{-2}. The typical isolation of the buffer amounts to 10^6 Ω cm. Further, pseudomorphic growth of a AlGaN/GaN/AlGaN double heterostructure is reported to improve the buffer leakage.

3.1.3 MOCVD and MBE Growth on Alternative Substrates

Apart from the direct homo- and heteroepitaxial growth on a crystalline substrate, advanced alternative substrate methodologies have also been developed, e.g., [3.129]. A hybrid technique is used to fabricate AlGaN/GaN HEMTs on diamond substrate [3.84, 3.229]. Al$_{0.27}$Ga$_{0.73}$N/GaN heterostructures are conventionally grown on 2-in. silicon(111) substrates. The wafers are then mechanically frontside-mounted on a mechanical substrate. The silicon substrate is then removed. A 25 μm thick CVD grown polycrystalline diamond substrate is then atomically bonded to the GaN buffer. The procedure potentially combines the very good thermal conductivity of diamond with III-N semiconductor devices, as detailed in [3.229]. Remaining issues include the additional need of another mechanical substrate to enable the processing of the wafers. SiC on insulator (SiC on SOI or SiCOI) composite substrates are another substrate alternative, combining the advantages of crystalline and poly-SiC substrates and facilitating the substrate thinning possibilities [3.46]. With the application of the Smart Cut$^©$ and cleaving technology, a 270 nm SiC layer is bonded onto the oxide of a silicon(100) handle substrate [3.16].

The MOCVD growth of GaN on SiC is facilitated, as the lattice-constant of the SiC material is decisive for the HEMT layers, as reported in [3.217]. A growth recipe for GaN HEMT on SiC can be used. AlGaN/GaN HEMTs fabricated based on this procedure yield similar DC-characteristics and breakdown voltages as devices fabricated on silicon(111) substrates. The devices grown on the hybrid substrates show stronger thermal compression of the output characteristics. This is attributed to the increased thermal resistance R_{th} of the SiO$_2$ layer. The advantage of silicon and SiC are combined in the silicon on polycrystalline SiC (SopSiC) technology [3.108, 3.129]. A high resistivity silicon wafer is bonded on top of a polycrystalline SiC substrate. Smart-Cut splitting is used to remove some of the HR-silicon substrate, so that a crystalline substrate surface is maintained. The composite substrate has the good thermo-mechanical properties of polycrystalline SiC. This allows to increase the substrate diameter beyond the diameter of available s.i. SiC substrates. A MOCVD-based wafer technology on metallic nickel substrates is described in [3.209]. The metallic substrate is based on nickel electroplating and laser lift-off. Prior to lift-off, the Schottky diode epitaxial layers are grown by MOCVD on sapphire substrates. MBE growth of GaN on LiGaO$_2$ is described in [3.38]. Lithium gallate has limited thermal conductivity, while it has a very low lattice-mismatch relative to all other hetero-substrates available for III-N devices. As a true insulator, the RF-losses can be greatly reduced compared to compensated material. The material grown by MBE has a TDD of only 4–$5 \times 10^8 \, \mathrm{cm}^{-2}$. AlGaN/GaN HEMT on AlN templates on sapphire are reported in [3.8]. The TDD can be reduced to $1.5 \times 10^8 \, \mathrm{cm}^{-2}$ in this case.

3.1.4 Epitaxial Lateral Overgrowth (ELO)

Epitaxial lateral overgrowth (LEO) or (ELO(G)) is most important in the reduction of the dislocation density especially for III-N optoelectronic devices such as lasers [3.141]. A good overview of ELO is given in [3.51]. Threading dislocation densities as low as $2.2 \times 10^7 \, \mathrm{cm}^{-2}$ are reported in [3.53]. Several examples exist for ELO growth for electronic devices. The effect of threading dislocations on AlGaN/GaN HBT performance is reported in [3.127]. Device results obtained on conventional substrates and ELO substrates are compared. The reduction of the threading dislocation is found crucial for the reduction of the emitter–collector leakage, which is reduced by four orders of magnitude. A comparison of AlGaN/GaN HEMTs on sapphire substrates with and without grooves is provided in [3.45]. The devices with local grooves and increased crystalline quality yield an increase of drain current, transconductance, and output power. Further, the leakage currents are reduced.

3.1.5 Hydride Vapor Phase Epitaxy (HVPE)

HVPE is a very attractive option as a high-throughput and low-cost methodology for the growth of GaN quasi-substrates [3.159] due to its high growth rate, which enables very thick III-N layers [3.198]. The growth rates for HVPE of device layers are typically adjusted to 0.05–$0.3\,\mu m\,min^{-1}$, as reported in [3.109]. A good overview of the growth of GaN by HVPE is given in [3.198]. The principle of hydride vapor phase epitaxy for III-V element semiconductor is based on the flow of group-V hydride precursors and hydrogen chlorides in combination with nitrogen through a multizone furnace, typically consisting of quartz. The furnace has three functions and areas: a source zone, a mixing zone, and a deposition zone.

A quasi-substrate approach for epitaxial layer growth is presented in [3.159]. The HVPE layers are grown on GaN MOCVD templates on sapphire substrate. A detailed strain analysis is performed, which explains the strong wafer bending after removal of the sapphire substrate. Thicker sapphire substrates are suggested to reduce the wafer bending. On the device level, the fabrication and characterization of heterojunction GaN diodes is described in [3.35]. The active GaN layers are grown by HVPE on 4H–SiC substrates. Layers with thicknesses between 10 and $200\,\mu m$ grown by HVPE are investigated. GaN/AlGaN HEMTs grown by HVPE on AlN/SiC substrates are described in [3.109]. A $30\,\mu m$-thick insulating AlN layer is deposited on undoped conducting SiC substrates. The actual HEMT layer structure consists of a $3\,\mu m$-thick GaN layer and a $40\,nm$-thick $Al_{0.28}Ga_{0.72}N$ layer grown by MOCVD. The device yields relatively low low-frequency dispersion, which is attributed to the reduction of the lattice-mismatch and the consequent dislocation reduction in the very thick AlN buffer layer. The use of HVPE-grown buffer layers on sapphire substrate is described in [3.52]. MBE is used to grow the epitaxial layers. The uniformity of the device DC-parameters despite the more complicated approach. Complete growth of AlGaN/GaN HEMTs by HVPE on sapphire substrates is reported in [3.123]. Both buffer layers and epitaxial layers are grown by HVPE and yield an RMS surface roughness of $\leq 0.2\,nm$. AlGaN/GaN HEMTs with a gate length $l_g = 1\,\mu m$ are defined with a transconductance in excess of $110\,mS\,mm^{-1}$ and $I_{Dmax} = 0.6\,A\,mm^{-1}$.

3.2 Indium-Based Compounds and Heterostructures

The growth of (Al)In(Ga)N and related ternary and quaternary materials for electronic applications has not yet been as exploited as the growth of (Al)GaN [3.2]. The InGaN/GaN heterostructures are essentially developed for optoelectronic applications, e.g., [3.176]. The promise of very low effective masses of InN down to $0.05\,m_e$, see [3.49], and the remarkable transport properties [3.177] motivate the development of growth for electronics, e.g., [3.116, 3.130, 3.151, 3.175, 3.224]. The growth kinetics of InN and InGaN are reviewed, e.g., in [3.14, 3.225]. In general, the decomposition temperature of InN is relatively low, typically $\leq 600°C$, as mentioned in [3.73]. The growth of InAlN is described, e.g., in [3.74, 3.91]. Thus, the growth temperatures of InN by MOCVD typically amount to 300–$500°C$. This temperature range is typically too low for the growth of low-defect material, when MOCVD growth or ammonia-based MBE are being used. This is due to the decomposition rate of the ammonia gas in the reactor [3.73], as the temperature must satisfy the temperature conditions of NH_3 pyrolysis, while the dissociation of InN must be avoided [3.146]. The realization of double heterojunction FETs is very desirable, since the enhancement of the confinement effect in double hetero-junction is favorable for device performance [3.115]. In-inclusion may help to increase the flexibility of the barrier definition at the frontside and the back-side of the channel [3.116]. Further, increased sheet carrier densities can be achieved in InAlN/GaN heterostructures, as discussed below [3.91, 3.122].

3.2.1 MOCVD Growth of Indium-Based Layers

Early reports on MOCVD growth of InN on various substrates, such as Al_2O_3 or even GaAs, can be found in [3.226]. Growth is performed at atmo-spheric pressure on α-Al_2O_3, and good conductive properties are reached with intrinsic concentrations $n = 5 \times 10^{19}\,cm^{-3}$ and Hall mobility values of $300\,cm^2\,V^{-1}\,s^{-1}$. In more recent reports, low-temperature MOCVD growth of InN is analyzed, e.g., in [3.173]. Thin InN films of 100 nm thickness are grown at $530°C$ on low-temperature (grown at $400°C$) InN nucleation layers and on high-temperature GaN buffer layers. A near band-edge PL emission of 0.75 eV is observed already in this publication, which points to the revi-sion of the bandgap energy of InN. On the device level a comparison of $Al_{0.15}Ga_{0.85}N/GaN/Al_{0.15}Ga_{0.85}N$ and $Al_{0.15}Ga_{0.85}N/InGaN/Al_{0.15}Ga_{0.85}N$ double heterostructure HEMTs (DHHEMTs) grown by MOCVD on s.i. SiC substrates is given in [3.122]. The comparison of the devices at 77 K proves the superiority of the double heterostructure concept with respect to 2DEG mobil-ity and sheet carrier density. AlGaN/GaN/AlGaN DHHEMTs show enhanced mobility relative to the AlGaN/GaN single heterostructures. The mobility amounts to $>8,000\,cm^2\,V^{-1}\,s^{-1}$ for the DHHEMT at 77 K. The inclusion of a $In_xGa_{1-x}N$ layer with $x = 6\%$ leads to an increase of the mobility for a given sheet carrier density.

Growth Conditions: Material Transport and Temperature

Growth conditions of InN by MOCVD are characterized by low growth-temperatures and high growth-pressures [3.14]. It is the most difficult among the III-nitrides because of the low dissociation temperature of InN and the high equilibrium N_2 pressure. A good overview on the growth parameters and the substrate situation is given there [3.14]. InN growth on sapphire substrate by MOCVD reported in [3.226] yields InN with carrier densities of $5 \times 10^{19}\,cm^{-3}$ and a room-temperature Hall mobility of $300\,cm^2\,V^{-1}\,s^{-1}$. A very high V/III ratio of 10^5 is used for MOCVD in this case. TMI and ammonia are used as precursors at growth temperatures of 450–$550°C$ and reactor pressures of 0.1–$1\,bar$. The increase of the growth rate and the use of nitrogen increase the In incorporation at a given temperature, as reported in [3.36]. The impact of reactant gas on the growth of InN by MOVPE is reported in [3.227]. Growth rates of InN as high as $2\,\mu\mathrm{m}\,\mathrm{h}^{-1}$ are reached. A $0.5\,\mu\mathrm{m}$ thick InN layer yields intrinsic carrier concentrations of 10^{19}–$10^{20}\,\mathrm{cm}^{-3}$. The room-temperature Hall mobility range is 260–$120\,\mathrm{cm}^2\,\mathrm{V}^{-1}\,\mathrm{s}^{-1}$. Growth temperatures of 400–600°C have been used. A $1\,\mathrm{nm}$ thick $\mathrm{In}_{0.1}\mathrm{Ga}_{0.9}\mathrm{N}$ notch can be introduced into the GaN/AlGaN SHHEMT layer sequence to form a double heterostructure composed of $\mathrm{GaN/In}_{0.1}\mathrm{Ga}_{0.9}\mathrm{N/AlGaN}$ [3.154, 3.224]. The GaN buffer growth is performed at 550°C. The temperature is then increased to 810°C for the growth of the InGaN. A $11\,\mathrm{nm}$-thick GaN channel is grown at 920°C, followed by an $\mathrm{Al}_{0.3}\mathrm{Ga}_{0.7}\mathrm{N}$ barrier grown at 1,020°C [3.153]. The devices yield excellent cut-off frequencies of $\geq 150\,\mathrm{GHz}$ for a gate length of $100\,\mathrm{nm}$.

3.2.2 MBE Growth of Indium-Based Layers

Molecular beam epitaxy, especially plasma-assisted MBE, is advantageous for the growth of indium-based compounds, as the growth temperature can be modified significantly as compared to MOCVD growth. Nitrogen radicals can be generated in the plasma source independent from the growth temperature, which does not have to satisfy the requirements of NH_3 pyrolysis. This principal effect especially increases the material quality. A review of MBE growth of InN and InGaN is given in [3.146].

Growth Conditions: Temperature and Material Transport

Reports on the growth of InN films and InAlN/InN heterostructures similar to the InGaAs/InAlAs material system can be found in various publications, e.g., in [3.74, 3.76, 3.146, 3.199]. The bandgap energy of InN has been under constant investigation during the development of the material system. A good review of the recent material development and bandgap investigations of InN and InGaN is given in [3.147]. Recent estimates of the bandgap energy of InN derived from photoluminescence properties of undoped and Si-doped InN films grown by PAMBE amount to 0.76 eV at RT [3.76]. The impact of the interface and crystalline effects is further detailed. In general, the material quaity can be improved significantly by PAMBE growth, as compared to ammonia-based MBE and MOCVD. Indium incorporation and surface segregation during $\mathrm{In}_x\mathrm{Ga}_{1-x}\mathrm{N}$ growth by PAMBE are discussed in [3.24]. The temperature for the growth of InGaN on sapphire substrates is chosen between 580 and 620°C. A low-temperature GaN buffer is grown at 550°C. The growth rates for both GaN and InGaN amount to $200\,\mathrm{nm}\,\mathrm{h}^{-1}$. The material fluxes are adjusted to match $\mathrm{In}_x\mathrm{Ga}_{1-x}\mathrm{N}$ with an indium fractions of 0–0.5. High-quality InN films based on plasma-assisted MBE are reported in [3.73]. The growth temperature for the low-temperature InN buffer layer amounts to 240–420°C, while the actual InN layers are grown at 380–520°C. Both an annealing step at growth temperature and a rapid cooling at 200°C after the actual growth are applied. The mobility increases as a function of growth temperature between 380 and 490°C, whereas the intrinsic carrier concentration drops. The maximum Hall mobility at room temperature reaches 1,420 cm^2 V^{-1} s^{-1} at a background electron concentration of 1.4×10^{18} cm^{-3}. The corresponding growth temperature is 490°C. Earlier reports in [3.71] show Hall mobilities of 1,420 cm^2 V^{-1} s^{-1} at electron concentrations of 1.6×10^{18} cm^{-3} for 350 nm thick films. PAMBE growth, characterization, and properties of InN and In$_x$Ga$_{1-x}$N on sapphire substrates are further reported in [3.146]. The importance of the V/III ratio is pointed out in a trade-off between In-droplet formation (In-rich growth) and high N-partial pressure (to avoid dissociation). Considering both limitations the V/III ratio at the surface should be controlled to stoichiometry maintaining the the surface slightly N-rich [3.146]. The growth temperature is chosen between 460 and 550°C. Further, the importance of the nitridation procedure on the sapphire and the two-step growth with the low-temperature buffer are highlighted. The effects of atomic hydrogen irradiation on indium incorporation and ordering in In$_x$Ga$_{1-x}$N grown by RF-MBE at 640–700°C are discussed in [3.152]. The indium incorporation and ordering are enhanced by the presence of H$_2$. InAlN is grown by MBE at a temperature below 460°C for the incorporation of In due to the wek In-N bond [3.91]. The next paragraph addresses heterostructure growth.

3.2.3 Indium-Based Heterostructure Growth

Both (In)GaN/InAlN and InGaN/GaN heterostructures can be formed once indium sources are available in the growth system. Al-rich In$_y$Al$_{1-y}$N can be used effectively as barrier material for heterostructures with high sheet carrier densities [3.91]. Growth temperatures of 400 and 480°C are reported in [3.72] for the growth of In$_{0.15}$Al$_{0.85}$N barriers on GaN by PAMBE on sapphire substrates. Sheet carrier densities of 1.7×10^{13} cm^{-2} and associated Hall mobility values of 654 cm^2 V^{-1} s^{-1} at room temperature are reported using an additional AlN interlayer. After the growth of an AlN-nucleation layer at 900°C, the GaN buffer layer is grown at 730°C by PAMBE. The AlN interlayer with a nominal thickness of 1.3 nm is grown at 760°C. For the In$_{0.15}$Al$_{0.85}$N barrier layer, the growth is performed at 400 and 480°C. Better mobility and surface quality are achieved for the higher growth temperature. The subsequent processing of InAlN/GaN HFETs is also reported.

In$_{0.2}$Al$_{0.8}$N/GaN heterostructures grown by MOCVD on sapphire(0001) substrates are given in [3.107]. The layer sequence includes a low-temperature GaN nucleation layer, a 1.4 µm-thick GaN buffer layer, and a 10 nm In$_{0.2}$Al$_{0.8}$N barrier layer grown between 800 and 850°C. The Hall mobility and sheet carrier concentration amount to 2×10^{13} cm^{-2} and 260 cm^2 V^{-1} s^{-1}, respectively. InAlN/GaN heterostructures grown by MOCVD on silicon(111) substrate are reported in [3.210]. A 300 nm-thick (Al)GaN nucleation layer is deposited on high-resistivity 3-in. n-silicon substrates. A 1 µm-thick GaN buffer is capped by an $In_{0.16}Al_{0.84}N$ barrier layer. A 1 nm AlN layer can be included between channel and barrier. The barrier layer thickness is 20–25 nm. The Hall mobility amounts to $1{,}200\,cm^2\,V^{-1}\,s^{-1}$ without inclusion layer, and $1{,}800\,cm^2\,V^{-1}\,s^{-1}$ with AlN-inclusion layer. The related sheet carrier densities are 0.8×10^{13} and $1.5 \times 10^{13}\,cm^{-2}$. The findings in this section demonstrate the potential to increase the sheet carrier density in the heterostructures with the use of In and potentially to improve the associated mobility.

3.3 Doping and Defects

Doping and defects are fundamental for electronic device fabrication in any semiconductor system [3.202]. Defects in MOCVD- [3.143] and MBE-grown material [3.69] are found to be strongly growth-method dependent, as reported in various publications, e.g., [3.57, 3.118, 3.170]. Defects can be categorized in point, line, and areal defects [3.119, 3.170]. The important point defects are vacancies and interstitials, the line defects are threading dislocations, and areal defect are stacking faults [3.119]. These defects are thus discussed for each growth method. Good overview articles on defects in III-N materials are given in [3.118, 3.138, 3.160]. Early overview data for defect related donors, acceptors, and traps in GaN grown by MOCVD, MBE, and HVPE are given in [3.118]. The impact of surface stoichiometry and polarity on the defect structure for GaN is discussed in [3.178]. Point defects can be deliberately generated by various types of irradiation, such as electrons, protons, ions, γ-ray, and metal deposition by electron beam. The signatures obtained from the defect creation can then be compared to as-grown materials. It is stressed in [3.118] that shallow defects are dominating and that impurities such as Si and O are dominant in early as-grown GaN layers rather than defects. Threading dislocations in n-GaN are reported to have an acceptor-like behavior [3.118].

A very recent extensive review of the luminescence of defects in GaN, InN, and AlN is given in [3.170]. GaN is considered to break the long-standing paradigm that high defect densities precludes electronic device performance. Minimum nonintentional acceptor concentrations are found to be $10^{15}\,cm^{-3}$ even in the purest material. This also leads to a minimum of uncontrolled compensation in n-type material. Zn, Mg, C, Si, H, O, Be, Mn, and Cd are considered the most important impurities. The dominant role of extended defects such as treading dislocations and point defects such as nitrogen vacancies is also discussed in [3.160]. It has further been found in GaN and the related materials AlN, InN, InGaN, InAlN, AlGaN, that hydrogen is present in relatively high concentrations in as-grown samples, especially in p-type GaN [3.148]. It has been shown experimentally that the amount of oxygen present during material growth can have a strong influence on the background conductivity, indicating that it is a shallow donor in GaN. Carbon can also be a significant residual impurity in MOCVD-grown nitrides. Its source is typically the precursor. Typical characterization methods for defects include [3.118] the following:

- Temperature-dependent Hall measurements for the determination of the donor and acceptor concentrations
- X-ray diffraction (XRD)
- Transmission electron microscopy (TEM)
- Photoluminescence (PL)
- Atomic force microscopy (AFM)
- Positron annihilation spectroscopy (PAS)
- Magneto-resonance
- Deep-level transient spectroscopy (DLTS) [3.118]

The analysis of threading dislocations has already been mentioned. A more detailed analysis is presented in the following.

3.3.1 MOCVD Growth

Shallow- and deep-level defects in n-doped GaN grown by MOCVD are described in [3.54].

Doping in n-Doped and p-Doped MOCVD Material

Silicon introduces a shallow donor level in GaN at $E_C - 0.022\,eV$ [3.54, 3.56] and is found to be the dominant donor impurity in unintentionally (nid)-doped material. A deeper donor is also found at $E_C - 0.034\,eV$. Oxygen is another donor for GaN [3.160]. It is found to have a strong impact on the n-type background conductivity. It is prevailing due to the precursor NH_3 in MOCVD-grown material. A donor state at $E_C - 0.078\,eV$ is mentioned.

Mg and Zn are the most important acceptors in a number of p-dopants for III-N semiconductors, such as C-, Mg-, and Be-. They generally are not very efficient acceptors. The p-doping of GaN is complicated by the following effects, which have been named repeatedly [3.143]:

- Large thermal activation energy of 120–200 meV [3.54]
- Hydrogen passivation of MOCVD-grown GaN:Mg [3.143, 3.160] and
- Significant Mg reactor memory leading to broad Mg profiles, reported, e.g., in [3.62, 3.186]

Early overviews of the compensation mechanisms of p-doping are given by Nakamura in [3.143] and further, e.g., in [3.105, 3.121]. Mg has been described early as an acceptor in GaN at an electronic level of approx. $E_V + 0.2\,eV$, e.g., in [3.54]. It is the most important acceptor for GaN. A midgap trap level is isolated in p-type GaN with a photoionization energy of 1.8 eV. However, none of the deep levels is found to be of sufficient concentration to compensate the shallow acceptors. Mg accumulation effects in bulk samples at doping levels of 3×10^{19} to $9 \times 10^{19}\,cm^{-3}$ due to surface treatment and air exposure are reported in [3.62]. A strong surface band bending is observed of 1.2–1.6 eV for GaN:Mg. This is considered to be due to high-density surface states. Even strong wet etching by KOH and NH_4OH does not modify this high surface state density [3.62]. Hydrogen in general plays a dominant role for the passivation of acceptors [3.160]. Early reports of the activation of p-type GaN material by post-annealing steps at 800°C in nitrogen are given in [3.121, 3.143]. A more detailed overview on the impact of hydrogen is given in [3.160]. Carbon is a residual impurity in MOCVD-grown material. C is discussed to be both an acceptor and an amphoteric. Maximum hole concentrations of $3 \times 10^{17} \, cm^{-3}$ are reached at a mobility of $100 \, cm^2 \, V^{-1} \, s^{-1}$ [3.160]. P-doping of AlGaN is even more complicated than GaN, as reported, e.g., in [3.47]. The background is the even higher activation energy of Mg in AlGaN, the increased compensation due to Si in the precursors, and the interaction with Si, C, and O. The measured hole concentrations in AlGaN are even more strongly reduced than in GaN. Other p-dopants such as Ca, Zn, C, and Cd in the material have to be considered, as also reported in the overview on defects in GaN in [3.160].

Defects in MOCVD-Grown Material

Defect concentrations in the form of threading dislocations and other defects are of critical importance, as most of the electronic devices are grown by heteroepitaxy and a minimum mismatch relative to the semiconductor layers. A great number of reports exist on different defect types [3.138, 3.160, 3.170]. DLTS analysis of the barrier-layer material AlGaN is of importance due to the impact on dispersion in III-N FETs, similar to that of AlGaAs in GaAs PHEMTs. A donor-like deep-level defect in $Al_{0.12}Ga_{0.88}N$ grown by MOCVD on SiC substrate is found by DLTS at $E_C - 0.77 \, eV$, as reported in [3.55]. Deep levels are found by DLTS at threshold energies of 0.83 and 1.01 eV in n-doped GaN [3.54] grown by MOCVD. Further deep levels between $E_C - 0.87 \, eV$ and $E_C - 1.59 \, eV$ are reported. Luminescence, especially yellow luminescence, is of particular interest for electronic material analysis, e.g., by photoluminescence and surface photovoltage spectroscopy [3.181]. Yellow luminescence and related deep traps in undoped GaN grown by MOCVD are investigated in [3.22, 3.181]. For GaN layers the band energy of the luminescence is about 2.2 eV with a broad spectrum. It is present for different substrate types. The background of this emission are deep traps about 1 eV above the valence band which capture photoelectrons. Localized high current densities around screw dislocations in GaN films grown by MOCVD are investigated by ballistic electron emission spectroscopy (BEEM) in [3.18]. A correlation of the threading dislocations and charged traps is suggested, which leads to poor Schottky characteristics. Both donor- and acceptor-type behavior of threading dislocations are reported, as also stated in [3.118]. Apart from the detailed detection of single defects and traps, the correlation of defects with device performance is interesting. Deep traps in GaN MESFETs and their correlation with current collapse are analyzed in [3.55]. Two electron trap energies are isolated for this n-channel device. The defect energies of 1.8 and 2.85 eV below the conduction band have also been reported in other publications, e.g., in [3.22]. Variation of the edge dislocation concentration with the AlN nucleation layer design is directly correlated with the AlGaN/GaN reliability [3.57]. Traps can also be created by device operation and related degradation, as reported in [3.187]. Further analysis is provided in Chaps. 4 and 7.

3.3.2 MBE Growth

Surveys on doping and defects for MBE-grown materials are given, e.g., in [3.160, 3.170, 3.203].

Doping of n-doped and p-doped MBE Material

The most important n-dopant in MBE material is silicon. N-doped GaN material grown by MBE is characterized by a number of particular defects. Deep-level transient spectroscopy (DLTS) investigations in MBE-grown GaN material are given in [3.69]. The n-type GaN material is grown by MBE on MOCVD-grown templates on sapphire substrate. The characteristic capture-time constant obtained amounts to $8.6\,\mu s$ for a line-type deep trap at $E_C - 0.91\,eV$. A defect found at $E_C - 0.59\,eV$ does not show the behavior of a line or point defect. A comparison of the MBE growth of thin wurtzite GaN on 6H–SiC, silicon(111), and sapphire substrates is given in [3.190]. Cross-sectional TEM measurements of the n-type GaN materials reveal the domination of threading defects at the substrate/buffer or buffer/film interfaces at residual concentrations of 10^8–$10^9\,cm^{-2}$.

N-doping is typically performed by standard silicon effusion cells, e.g., [3.68]. Silane can be used as a silicon source for ammonia-based MBE [3.194]. Next to silicon, oxygen is also considered as a donor for MBE grown material [3.160]. Residual water vapor in MBE chambers or oxygen impurities leaked from the quartz containment vessel often employed in N_2 plasma sources is considered as the source.

P-doping of MBE material is different from that of MOCVD material. Typical for MBE, the doping profiles of MBE III-N material can be very sharp. Mg p-doping of GaN by PAMBE on MOCVD-grown GaN templates on sapphire substrates is described in [3.186]. MBE growth can cure two further disadvantages of GaN:Mg growth by MOCVD. It avoids hydrogen passivation of the Mg, and no Mg reactor memory is observed during MBE growth. The GaN:Mg layers are grown by PAMBE at 650°C. Mobilities $\mu_p = 24\,cm^2\,V^{-1}\,s^{-1}$ at $p = 1.8 \times 10^{17}\,cm^{-3}$ and $\mu_p = 7.5\,cm^2\,V^{-1}\,s^{-1}$ at $p = 1.4 \times 10^{18}\,cm^{-3}$ are reached. The incorporation of Mg is uniform and very sharp profiles can be realized due to the low growth temperature, as verified by SIMS measurements. The sharp profiles further allow very precise superlattice engineering and Mg profiling, as reported in [3.106]. Superlattice dimensions are varied between 2 and 14 nm in order to enhance the active hole concentration in the superlattice beyond 10^{18} cm^{-3}, especially at low temperatures.

Defects in MBE Material

Deep level defects in n-type GaN material grown by MBE are described, e.g., in [3.207] found by DLTS. Five deep traps are reported between $E_C - 0.23$ eV and $E_C - 0.96$ eV. The capture cross-sections amount to 2×10^{-15} cm^2 (0.59 eV) and 3×10^{-14} cm^2 (0.91 eV). Similar point defect serving as electron traps are found at $E_C - 0.06$ eV and at $E_C - 0.9$ eV. Capture cross-sections of 5.0×10^{-15} and 7.4×10^{-17} cm^2 are determined for electron levels of 0.21 and 0.23 eV in PAMBE-grown AlGaN/GaN/AlGaN heterostructure material [3.158]. The 0.21 eV defect is related to a nitrogen vacancy, while 0.23 eV is associated with extended defects.

Defect microstructure of thin wurtzite GaN films grown by MBE on SiC substrates, silicon substrates, and sapphire substrates are analyzed by TEM in [3.190]. Threading dislocations are line defects along the growth direction. The defect microstructure is dominated by threading defects that originated at the buffer/substrate or buffer/epilayer interface. A defect density of 10^8–10^9 cm^{-2} is found to be prevailing depending on the substrate. This density is residual despite the reduction of defects by the use of a 1 μm-thick buffer layer. HVPE-grown material has also been investigated in several publications, e.g., in [3.170, 3.223]. Three traps are isolated in [3.223] at energies $E_C - 0.65$ eV, $E_C - 0.61$ eV, and $E_C - 0.27$ eV.

The previous section has demonstrated the impact of impurities and defects; however, the most important requirement for electronics is their avoidance and control during material growth, which is part of this chapter. Control of processed-induced defect creation will be addressed in Chapter 4.

3.4 Epitaxial Device Design

In the previous sections, a number of specific growth features and their characteristic methodologies have been discussed for the InGaN/AlGaN/InAlN/GaN material system. This section addresses particular features for device growth. As an example, Fig. 3.17 gives the layer sequence of advanced HFETs for b th the III-As and the III-N material system. Fig. 3.17 visualizes and compares the concept of a double heterojunction and the specific device features with respect to epitaxial growth, as also proposed, e.g., in [3.104]. These features are detailed in the following sections and compared, where necessary, to III-As devices.

3.4.1 Geometrical Considerations

Geometrical considerations are numerous in the design of III-N devices. Most data are available on the fabrication of HFETs.

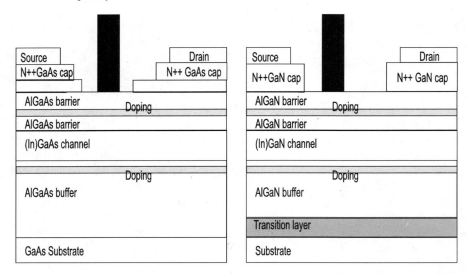

Fig. 3.17. Schemes of advanced double-heterojunction HFETs: (*left*) GaAs, (*right*) GaN

Barrier-Layer Design: Thickness

The barrier-layer thickness in HEMTs is a principal design feature. It is evaluated in [3.197]. Nonrecessed AlGaN/GaN HEMTs with barrier thicknesses of 10 and 20 nm are compared. As expected, the device with the thinner barrier layer shows reduced current due to the reduced polarization effects and higher transconductance. Further, a reduced saturated power density is obtained. RF-loadline measurements show the reduction of RF-full channel current as compared to DC full channel current and the increase in the RF-knee voltage for the devices with a smaller gate-to-channel separation. Thus there is a reduction in output power due to these surface-induced dispersive effects. The second effect on the output power at a given voltage is the expected reduced polarization effect, as also argued in [3.75]. The barrier thickness and mole fraction also strongly influence the vertical field distribution in the barrier, as measured by photoreflectance spectroscopy in [3.20]. The polarization effects lead to electrical fields $\geq 1\,\mathrm{MV\,cm^{-1}}$ [3.220]. A systematic investigation of the gate-to-channel-separation is performed in [3.75] for AlGaN/GaN HEMTs with a gate-length of 1 µm grown by PAMBE on sapphire substrates. The gate-to-channel separation is varied between 8 and 25 nm. The reduction of the gate-to-channel separation leads to a significant increase of the transconductance from 185 to 360 mS mm^{-1}. At the same time, the drain current at $V_{\mathrm{GS}} = 1\,\mathrm{V}$ is reduced from 1.1 to 0.68 A mm^{-1}. The positive threshold voltage shift is about 0.23 V per 1 nm change of the layer thickness. For very-high operation frequencies, a gate-to-channel separation of 11.3 nm is used in [3.77] for an HFET with a gate length of 60 nm. Very thin $Al_{0.2}Ga_{0.8}N$ layers in MOSHFETs are reported in [3.70]. A 7 nm-thick AlGaN layer leads to a threshold voltage of -0.3 V. The situation is close to a normally-off device. Al_2O_3 gate layers are used. The MOSHFET allows the overdrive of the device up to gate voltages of $+4$ V. Theoretical estimates for the impact of the barrier-layer thickness and the Al-content on the power performance and corresponding gain compression are discussed in [3.43]. An increase in output power is predicted with increasing barrier thickness and Al-content. Further geometrical issues for the design of (H)FETs include the following:

– The distance of δ-doping layers to the channel [3.48]
– The position of the (backside) InGaN layer to the channel [3.26, 3.153, 3.154]
– The thickness of doped GaN layers in recess processes, e.g., in [3.98]

The distance of the δ-doping of AlGaN/GaN HEMTs is optimized in a similar procedure as for GaAs PHEMTs, as reported in [3.48]. The considerations include scattering of the carrier in the channel and also doping efficiency. The optimum value amounts to 3–6 nm. The thickness of GaN cap layers needs to be optimized with respect to the surface depletion, as performed in [3.98]. In recess structures, the layer depth needs to be adjusted to the GaN etch process, as further described below and in Chapter 4.

Barrier-Layer Design: Mole Fraction

The Al-mole fraction in the barrier is a critical parameter in $Al_xGa_{1-x}N/GaN$ and $In_xAl_{1-x}N/GaN$ HFETs. Increased Al-contents promise higher sheet carrier concentrations in the channel at nearly constant mobilities. Devices with Al-concentrations of $\geq 35\%$ are reported, e.g., in [3.94,3.222]. The findings with an increase of the mole fraction in the barrier layer include the following:

– Eventually a saturation of the sheet carrier concentration due to relaxation of the AlGaN barrier layer, and grain formation, which both reduce the polarization effects
– Reduced doping efficiency of the Si doping
– Increased ohmic contact resistances

Typically, optimized single heterostructure FETs have Al-contents of 20–30% [3.204]. Al-contents between 26 and 52% are investigated on 4-in. sapphire substrates in [3.136]. The increase in Al-content leads to reduced surface quality and increased gate leakage, especially for Al-contents of 52%. The geometrical considerations have already been discussed above.

3.4.2 Growth of Cap Layers

Cap layers are introduced to moderate the transition at the semiconductor-passivation interface. In many publications, thin undoped GaN layers are reported for improved contact formation, e.g., in [3.12, 3.216]. The thicknesses are 1 nm in [3.32, 3.135], 3 nm in [3.12], and up to 5 nm in [3.166, 3.216]. The specific formation of electron and hole gases in cap layers can be investigated by photoreflectance spectroscopy [3.19, 3.191]. It becomes obvious, that small variation in the cap layers thickness in the nm-range have strong impact to the field distribution. Highly n-doped caps are generally used in any semiconductor system to facilitate ohmic contact formation in order to increase the density of states at the metal semiconductor interface. This is especially necessary for wide bandgap semiconductor devices, which require very high annealing temperatures to reach ohmic-type contacts. Highly-doped GaN caps have further repeatedly been suggested to reduce dispersive effects, e.g., in [3.12, 3.75, 3.97]. The resulting devices show reduced dispersion even for the unpassivated situation [3.182]. However, the high doping levels lead to prohibitive gate-leakage currents, which unfortunately reduce the operation voltage without the applications of further steps such as recessing to $V_{DS} = 15$ V only [3.182]. P-doped GaN caps with very high nominal Mg concentrations of 1×10^{20} cm^{-3} have been used to minimize RF- and DC-dispersion in AlGaN/GaN HEMTs, as reported in [3.12, 3.30]. The physical effect of high n- and p-type doping concentration is similar and is based on the compensation of traps and related charge at the semiconductor/passivation interface. The effect of MOCVD growth-termination conditions at the top cap layer above the barrier layer and the impact of wafer cool-down on the performance of the total GaN FETs is further investigated in [3.93]. A modest improvement in the device power density is observed for wafers, if, after deposition of the AlGaN layer under constant growth conditions, the samples are cooled down to room temperature in pure nitrogen. Insulating low-temperature-grown GaN-cap layers are used for the reduction of the dark currents in Schottky-barrier photodetectors [3.110]. The low-temperature GaN is grown at 560°C and reduces the dark current by four orders of magnitude. Superlattice cap layers have been recently proposed in order to reduce the source resistance of AlGaN/GaN HFETs [3.140]. Fifty nanometer thick superlattice-cap layers followed by a 20 nm n^{+}-doped GaN cap are proposed. The layer thicknesses of the AlGaN barrier and of the superlattice layers are optimized with the aim to reduce the potential barrier at the AlGaN/GaN channel interface below the ohmic contacts. This leads to on an overall reduction of the access resistances.

3.4.3 Doping

III-N devices are intrinsically doped by unintentional impurities and by polarization doping. In several reports, the intentional doping is completely based on the polarization mechanism, e.g., [3.39]. However, conventional impurity doping is used repeatedly in both HEMT [3.48] and HBT [3.128] structures for several reasons. On the device level, the impact of barrier doping on the performance of unpassivated AlGaN/GaN/SiC HEMTs grown by MOCVD is reported on in [3.13]. Conventionally doped unpassivated AlGaN/GaN heterostructures are found to be less sensitive to dispersive mechanisms due to shielding of the channel from the barrier/dielectric interface. The effects of the variation of δ-doping and geometry in AlGaN/GaN HEMTs on SiC substrate are described in [3.48]. Both the position and the nominal doping concentration of the δ-doping are varied for spacer thicknesses of 3–9 nm. The sheet charge is optimized for a spacer thickness of 6 nm, while the mobility increases with increased distance. A doped-channel AlGaN/GaN HFET with isoelectronic doping in the channel is described in [3.83]. The thickness of the doped channel layer is 70 nm in this case, which is much larger than the typical extension of an AlGaN/GaN channel layer, which is estimated to be $\leq$10 nm.

3.4.4 AlN Interlayer

The channel-barrier interface in HFETs has been optimized with respect to several aspects, e.g., interface roughness [3.139,3.155] and trap reduction [3.9, 3.82]. To that end, AlGaN/GaN high-power microwave HEMTs have been optimized by the introduction of a thin AlN interlayer with a thickness of a few monolayers, e.g., in [3.183]. The concept has repeatedly been published with different focuses [3.9,3.11,3.139]. The insertion of the thin AlN layer produces a larger effective conduction band discontinuity ΔE_C. This increase in effective ΔE_C is due to the polarization-induced dipole in the AlN [3.183]. This AlN interlayer concept implies the following advantages:

- Increased effective bandgap discontinuity [3.183]
- Decreased alloy scattering [3.9,3.183] and thus increased channel mobility
- Increased mobility for a given sheet carrier concentration [3.139]
- Increased sheet carrier density [3.139]
- Reduced interface trapping [3.9]

Additional issues to be considered are the following:

- The introduction of an additional tunneling barrier near the ohmic contacts
- Potentially a diode-like access resistance [3.99], similar to RTD

This kind of diode-like behavior is caused by the potentially strong band bending at various bias conditions. If additionally an AlN interlayer is introduced to serve as a barrier layer, a resonance is possible between the contact and the diode. Fig. 3.18 gives the sheet carrier concentration and Hall mobility of AlGaN/AlN/GaN heterostructures as a function of AlN interlayer thickness, as shown in [3.139]. An optimized mobility is found for the thickness of $\approx$1 nm with mobility values of 1,700 V s cm^{-2}. The sheet carrier concentration increases steadily from 0.9 to 1.4×10^{13} cm^{-2}. Fig. 3.19 further gives the sheet resistance as a function of AlN-interlayer thickness. A minimum is found at an interlayer thickness of about 1 nm. Further analysis of the drift mobility as a function of carrier concentration in [3.9] shows that the effect of the AlN layer specifically enhances the mobility at high carrier concentrations. In another

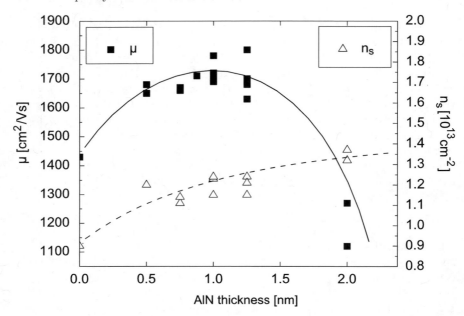

Fig. 3.18. Sheet-carrier concentration and mobility as a function of AlN interlayer thickness

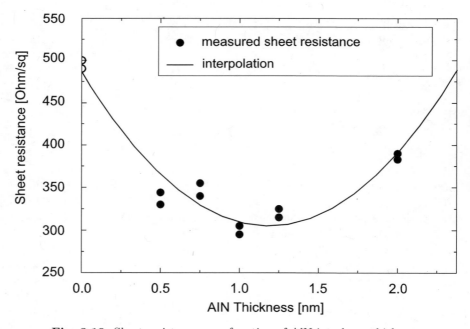

Fig. 3.19. Sheet resistance as a function of AlN interlayer thickness

device concept the AlGaN barrier layer can also be completely removed and the effect of the barrier is then performed entirely by the AlN-interlayer, as reported in [3.155].

3.4.5 Channel Concepts

Simple planar HFET structures use the extended buffer layers under the barrier as channel layers. Near the interface, the GaN is highly conductive. Nominally, the channel layer is grown at a similar temperature to that used for the buffer layer, e.g., [3.92]. More advanced composite-channel concepts with varying Al-mole fractions in the channel/barrier have been introduced in [3.114]. The first channel forms at the $Al_{0.3}Ga_{0.7}N/Al_{0.05}Ga_{0.95}N$ interface, while the second channel forms at the $Al_{0.05}Ga_{0.95}N/GaN$ interface. It has further been suggested that this graded-channel concept enhances linearity at 2 GHz; however, the improvements in linearity claimed are within the statistical scattering of the measurement data. Doped HFET channels with isoelectronic Al-doping in the channel have been described in [3.83]. The Al-doping is used to improve the crystalline quality of the channel layer by reducing the amount of threading dislocations serving as acceptors. This leads to an overall enhancement of the mobility. However, the absolute values of the Hall mobility reported can also be achieved by other means. Further channel concepts are based on double heterostructures. The layers for this concept can be grown by both MOCVD and MBE. $Al_{0.3}Ga_{0.7}N/GaN/Al_{0.05}Ga_{0.95}N$ based on MBE growth are described in [3.131]. The Al-fraction in the buffer has to be optimized with similar care as a GaN buffer with respect to unintentional doping, lattice mismatch, and defect mitigation. The Al-fraction thus cannot be chosen as freely as for a conventional AlGaAs/GaAs/AlGaAs HEMT layer structure.

3.4.6 Epitaxial In-Situ Device Passivation

Typically, the dielectric passivation in III-N devices is introduced by process-technological means, e.g., [3.136, 3.204]. This procedure is sometimes called ex-situ passivation. SiN can further be deposited in-situ, i.e., directly in the reactor growth chamber during MOCVD growth, as reported, e.g., in [3.201]. A thin SiN layer (4 nm) is deposited in the MOCVD chamber by flowing disilane and ammonia. The deposition is performed at a temperature of 980°C with a deposition rate of $0.1\,\text{Å}\,\text{s}^{-1}$ [3.27]. The growth temperature of the SiN is thus lower than that typically used for the AlGaN barrier. MBE layers can also be passivated in-situ; however, this is not done in the same MBE reactor chambers, but using subsequent MOCVD growth. Such a procedure thus includes an interruption of the growth process, as reported, e.g., in [3.166]. The more technologically based features of device design, such as recesses, contacts, implantation, and device passivation issues, are discussed in Chapter 4.

3.5 *Problems*

1. *Discuss the major differences of III-N growth by MOCVD, HVPE, and MBE with respect to relative advantages/drawbacks!*
2. *For which kind of devices does MBE growth have unique properties?*
3. *Discuss the possible advantages and drawbacks of indium-based FETs!*
4. *Discuss epitaxial measures for dispersion control!*
5. *What are typical growth issues for III-N HBT layers?*
6. *Which particular improvements/drawbacks can be found for GaN heteroepitaxy of HFET layers on GaN substrates?*

4

Device Processing Technology

A major focus of the development of III-N devices is devoted to process technology. III-N-based high electron mobility transistors (HEMTs) with gate lengths down to 30 nm and cut-off frequencies up to 180 GHz have been reported [4.91, 4.92]. Thus, for this class of devices, specific field-effect transistor problems, such as Schottky and ohmic contact definition, implantation, device isolation, lithography of transparent materials, and etching are discussed in this chapter. State-of-the-art recess processes and device passivation technologies are analyzed. Despite the limited number of publications, bipolar device technology issues are reviewed. As in every other semiconductor material system, the processing of nitride-based materials requires the development of a lot of specific technology, related mostly, but not exclusively, to the wide bandgap material properties. This specific development is addressed in this chapter.

4.1 Processing Issues

A large number of technological issues have to be solved to achieve III-N devices optimized for specific applications. A guideline to solve these issues, seen from a very general point of view, includes the following:

- The separation of process technology issues from epitaxial issues to achieve high uniformity as addressed, e.g., in [4.188]
- The improvement of the reproducibility of the processing steps and their results from wafer-to-wafer and run-to-run
- The improvement of the uniformity of process technology and yield
- The scaling of the process technology for different substrate diameters [4.23, 4.188]
- The increase and stabilization of long-term device reliability

With respect to improved FET and bipolar device performance, the processing issues include particularly the realization of the following:

- Devices with
    - High transconductance for high gain
    - High efficiency
    - High power at high-voltage operation
- Devices with high gain, low RF-noise, and low-phase noise

These requirements particularly include the need for the following:

- Good device isolation
- The realization of low-contact and access resistances
- FETs with
    - Improved gate barriers and reduced gate and drain leakage
    - Domain breakdown management by field shaping in the gate region to reduce leakage and to increase breakdown voltages
- Sufficiently high current levels in a trade-off with the necessary breakdown voltages
- Improved techniques for high current gain in bipolar devices
- Reduced low-frequency dispersion in all of its characteristics
- Backside processing of very hard and inert substrates materials with respect to thinning and viahole-etching

The measures to improve the devices are discussed with each processing step further below; however, these issues include the following:

- The introduction of new device concepts, such as
    - Single-field and multiple-field plate FET concepts with different electrical terminations
    - Polarization-engineered devices
- The understanding of critical processing steps, such as
    - Surface pre- and postprocessing
    - Implant and annealing processes
    - Vertical and lateral structuring
- The introduction of new materials, such as
    - III-N specific passivation layers
    - New contact metal sequences

These issues and requirements will now be discussed in detail.

State-of-the-Art

Details of device process technology are subject to intellectual property issues and thus to reduced publicity. Early reviews of III-N technology development are given, e.g., in [4.22, 4.263]. The overviews in general stress the requirement of specific elimination of trapping effects. Further overviews of III-N processing technology are given, e.g., in [4.55, 4.124, 4.188]. Typical process flows consist of

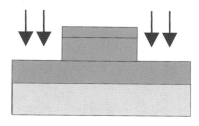

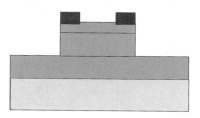

Fig. 4.1. Process flow of a GaN FET: (**a**) (*left*) mesa isolation, (**b**) (*right*) ohmic contact formation

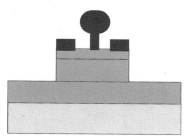

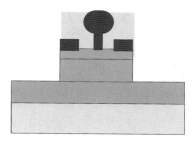

Fig. 4.2. Process flow of a GaN FET: (**c**) (*left*) gate definition, (**d**) (*right*) device passivation

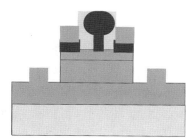

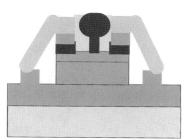

Fig. 4.3. Process flow of a GaN FET: (**e**) (*left*) second metal, (**f**) (*right*) galvanic metal

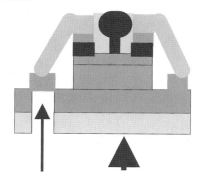

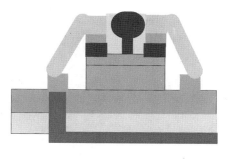

Fig. 4.4. Process flow of a GaN FET: (**g**) (*left*) thinning and etching, (**h**) (*right*) backside metallization

conventional III-V technology modules with selected III-N specific adaptions, such as rapid thermal annealing with very high temperatures. An advanced manufacturable GaN HEMT technology on 4-in. diameter high-resistivity Si substrates is described in [4.81]. It is fully based on optical stepper technology. Six mask levels are used for the formation of AlGaN/GaN HFETs. A similar high-voltage technology is proposed by Eudyna/Fujitsu in [4.128]. The processing is performed on 3-in. s.i. SiC substrates. The ohmic contacts are formed with a recess structure. A commercial GaN/AlGaN-based HFET device technology on s.i. SiC substrate is presented by RFMD in [4.221,4.222]. The processing can be performed on a GaAs production line, while some nitride specific steps are inserted. Typical descriptions of basic GaN HEMT process flows are given, e.g., in [4.97]. A typical process flow is depicted in Figs. 4.1–4.4. The processing of FETs thus includes the following steps, but not necessarily in this order:

1. Device isolation
2. Ohmic contact formation
3. Gate recess definition
4. Gate definition and first metal formation
5. Partial and full device passivation
6. Galvanic processing, including the passive components
7. Backside processing
8. Dicing

Additional steps, such as ohmic recess formation [4.127], ohmic implantation [4.289], (multiple) gate-recess, and field-plate formation [4.283] can be added. Typical processes following this processing order can be found, e.g., in [4.22, 4.124, 4.166]. Typical modifications to this process flow include the following:

– The exchange of the order of ohmic definition and device isolation, especially when applying implantation [4.22]
– Early device passivation to reduce surface pollution before and during processing, e.g., [4.68, 4.244]
– A different processing order when opening the passivation for the SiN-assisted gate [4.81]

The processing steps will now be discussed in detail.

4.2 Device Isolation

Device isolation is one of the primary process steps and increasingly important due to the high-voltage operation proposed for III-N FETs. The two principal technologies of mesa etching and implantation are described in the following.

4.2.1 Mesa Structures

Mesa isolation is the simplest technique for the structuring of devices and for their electrical isolation from their next neighbors. The active semiconductor layers between different devices are physically removed, typically by etching down to an insulating layer or to the substrate. In the case of III-N FETs, the etch is performed mostly to the insulating GaN buffer, as shown in Fig. 4.1. Typically, residual currents measured between devices in mesa structures have to be reduced to levels of $\leq 10^{-9}$ A at voltages of 50 V [4.188] and above in order to ensure reliable device isolation. This is equivalent to resistances of $10\,M\Omega\,cm$ and above. The resulting issues from the mesa process include the following:

- The control and reduction of surface currents
- The isolation of interconnect lines along conducting semiconductor layers at the mesa ridges
- The control of etched surface morphology

Several mesa processes have been described in the literature. A Cl_2/He-based ICP-RIE plasma dry etch is used in [4.30]. A Cl_2-based ECR process is proposed for device isolation in [4.55]. The typical etch depth is 200 nm. A medium RF-power Cl_2-based RIE etch process is described in [4.28]. The mesa etch depth is 120 nm. The etch rates for the mesa isolation reach $1\,nm\,s^{-1}$ at RF-power levels of 100 W. A BCl_3-based RIE process is described in [4.58]. The RF-power level is 10 W and the etch rate reaches $6\,nm\,min^{-1}$ in this case. The addition of CH_4 to BCl_3/H_2/Ar during ICP-based RIE improves the anisotropy of the etch and reduces the mask erosion [4.152]. The key disadvantage of such mesa processes is the generation of three-dimensional device structures on the wafer. This fact increases the complexity for the passive connection lines. Further mesa ridges are created that pose additional challenges for high-voltage isolation.

4.2.2 Ion Implantation for Isolation

Planar device processing is thus very desirable as it avoids the issues resulting from the mesa structures, such as interconnect lines above mesa ridges and isolation of horizontal and vertical surfaces. Ion bombardment represents a very attractive technique: Ion implantation processes allow isolation [4.19] and planar area selective doping [4.108] at the same time. Implantation for isolation is used in several state-of-the-art industrial processes, e.g., [4.81].

Isolation by Implantation

Overview articles on implantation in wide bandgap semiconductors with focus on GaN are provided, e.g., in [4.82, 4.147, 4.201, 4.214]. Ion mass, ion energy, and implantation temperature are the critical parameters. Damage build-up is to be reduced and post-implant annealing is to be reduced. H, He, and N implant isolation of n-type GaN is discussed in [4.19]. He-implantation leads to resistivities of $10^{10}\,\Omega\,\mathrm{cm}$ at room temperature. The temperature stability of the resistivity of the implanted structures up to $T_\mathrm{L} = 300°\mathrm{C}$ is also mentioned. The implanted layers have a resistivity of $10^4\,\Omega\,\mathrm{cm}$ at $300°\mathrm{C}$. Zn^+ implantation for device isolation is reported in [4.193]. Similarly, highly resistive GaN layers formed by ion implantation of Zn along the C-axis are described in [4.199]. Zn implantation is performed at an ion energy of $350\,\mathrm{keV}$ and a dose concentration of $1.9 \times 10^{14}\,\mathrm{cm}^{-2}$ and leads to a Zn concentration between 10^{17}–$10^{18}\,\mathrm{cm}^{-3}$. The isolation levels obtained in both GaN layers and AlGaN/GaN heterostructures reach $10^{11}\,\Omega\,\mathrm{sq}^{-1}$, while a thermal activation energy of $0.67\,\mathrm{eV}$ is found. The resistivity of the Zn layer even increases after RTA. A detailed analysis of the temperature-dependent implant damage during Ar^+ implantation is given in [4.254]. The implantation is performed at an energy of $150\,\mathrm{keV}$ at a dose of $3 \times 10^{15}\,\mathrm{cm}^{-2}$ at temperatures between 0 and $1,000°\mathrm{C}$. The implantation damage has typically two peaks, one near the surface and one in the bulk material in a depth range of 100–$200\,\mathrm{nm}$. The impact of implantation and post implant-annealing on the optical properties of GaN is described in many publications, e.g., [4.147] for Zn implant in [4.234]. The typical findings yield a curing of the optical damage after RTA, while the damage cannot be fully removed. Early annealing studies of MOCVD- and MBE-grown material for various implants, including Ar, Mg, Si, Be, C, and O, are given in [4.229]. An overview on disorder and thermal and ion-beam-assisted annealing is given in [4.147]. The associated annealing is performed under flowing NH_3 or N_2 gas. The recovery of the bulk material is very similar under both NH_3 and N_2, while the latter provides better surface quality. The use of ion implantation for ohmic contact formation is described in the next section.

4.3 Contact Formation

The formation of metal–semiconductor interfaces or contacts defines a second critical process step for any electronic device. Both ohmic and Schottky contacts are of critical importance with respect to device functionality, performance, and long-term reliability. Good overviews of ohmic contacts to GaN are given, e.g., in [4.82]. A great variety of metal-systems have been proposed for the formation of low-contact resistances and high thermal and electrical stability, which are discussed in the following.

4.3.1 Ohmic Contacts

The formation of ohmic contacts requires the solution of a number of issues for III-N semiconductor devices, which are now summarized.

Issues of Ohmic Contact Formation

The issue include the following:

– The increase of the density of states at the interface for wide bandgap semiconductors with a bandgap of $\geq 3.4\,\mathrm{eV}$
– The pretreat of the sensitive semiconductor surface prior to metal deposition
– The control of the metal morphology due to the high annealing temperatures of at least $600°\mathrm{C}$ [4.163] necessary due to the wide bandgap, including:

> The control of the vertical diffusion of metal by high temperature annealing [4.50]
> The control of the lateral diffusion of contact metals to avoid surface and gate leakage
> The control of the morphology for e-beam or optical pattern recognition

Fig. 4.5 summarizes a number of geometrical situations for the ohmic contacts. The different contact situations can be systemized as follows:

a. The metal is deposited on top of the barrier layer without annealing [4.289]
b. The metal is annealed into the barrier layer
c. The annealing is continued into the channel layer
d. The metal is deposited on a highly-doped cap
e. An ohmic recess is applied into the highly doped cap before annealing [4.128]
f. The recess is additionally applied to the barrier layer before annealing

In general, the formation of ohmic contacts at a semiconductor surface requires a high density of states near the interface to push the Fermi-level in the semiconductor close to the conduction band. The wide bandgap of III-N semiconductors renders this more difficult, while the polar properties can be used to engineer beneficial high densities-of-states at the interface. This is accomplished by superlattice engineering in the cap layers for ohmic contact formation in [4.191]. Several investigations of the current transport in various metal and semiconductor materials and interfaces are available, e.g., for GaN in [4.82] and for InGaN and InN in [4.257]. In the case of as-deposited metal contacts, thermionic emission is the dominating transport mechanism. For annealed contacts to InN and InGaN, field emission is dominating [4.154]. The annealing procedures are intended to provide strong vertical material diffusion, while lateral diffusion has to be minimized even at the high annealing temperatures required. A comprehensive study of the annealing characteristics of the contacts in AlGaN/GaN HEMTs on s.i. SiC substrate is provided in [4.165]. Variations of the annealing temperatures between 800 and 830°C and of the metal stack are investigated. The contact resistances vary between 0.5 and $2\,\Omega\,\mathrm{mm}$. The metal stacks contain Ti (10 nm), Al (50–100 nm), Ni

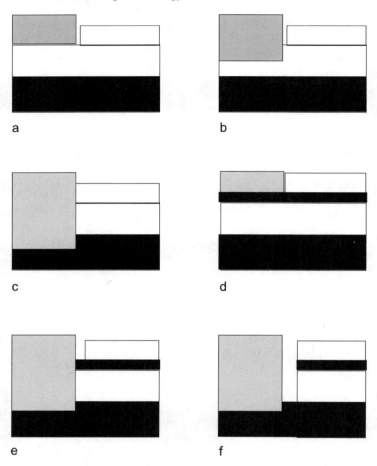

Fig. 4.5. Different ohmic contact situations

(20 nm), and Au (50 nm). The contacts covered by WSiN is found to be stable up to 400°C. For the analysis of the annealing procedure, Fig. 4.6 gives the measured contact resistance as a function of annealing temperature taken from [4.165] and others. Although the data given yields significant scattering, the general tendency of reduced contact resistances at higher temperatures is visible. At the same time, the morphology of the contacts deteriorates with increasing temperature. The annealing temperatures amount to up to 900°C, as reported, e.g., in [4.158, 4.252]. This results in very low contact resistances and leads potentially to problems with pattern recognition during e-beam lithography [4.274].

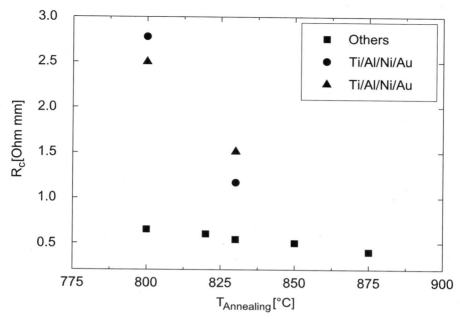

Fig. 4.6. Measured ohmic contact resistance as a function of published annealing temperatures from various publications

Ohmic Contact Material Systems

A large number of metal systems have been suggested for ohmic contact definition in the III-N semiconductor system. Table 4.1 gives an overview of metal systems and some details for both n- and p-type ohmic contact formation on GaN. The table further compiles annealing temperatures, applied impurity doping in the heterostructures, and resulting contact resistances. The table is based on annealing of metal sequences into the semiconductor layers without further measures, such as recesses, unless stated otherwise.

N-Type Ohmic Contacts

Early reports for ohmic contacts on n-type GaN are given in [4.162] using Ti/Al contacts alloyed at 900°C for 30 s. Improved ohmic contact stability is reported for Ti/Al/Pt/Au multilayer contacts to n-type GaN compared to Ti/Al/Au contacts [4.154]. The annealing temperature range is 750–950°C. Pt is used as a diffusion barrier for Au penetration. Ti/Al/Ni/Au contacts are suggested in [4.81] formed by rapid thermal annealing (RTA). They are used in the industrial processing of Nitronex and provide contact resistance of $0.75\,\Omega\,\mathrm{cm}$. A more systematic comparison of several metals and materials systems is given in [4.209] for GaN/AlGaN/GaN heterostructures. The comparison is given for the metal systems on top of GaN cap layers. The thermal stability of the contacts is compared for annealing temperatures up to 800°C and short-term annealing (typically 1 min). Afterwards, the devices are compared in long-term aging (200 h) at aging temperatures of up to 500°C. Ti/Al/Ni/Au multilayer ohmic contacts encapsulated by SiN are investigated in [4.137]. It is found that the composition of the SiN_x encapsulation layers critically affects the morphology of the contacts. This morphology, however, has great influence on edge definition, and thus on e-beam pattern mark recognition [4.137] during the subsequent lithography of the gates. The morphology of the contacts is especially important for processes with small margins, i.e., for small gate lengths and contact spacings [4.263]. The use of highly Al-doped-sputtered ZnO layers on top of AlGaN/GaN heterostructures provides a mediocre ohmic contact resistance of 2.7 Ω mm without any further contact annealing, as is reported in [4.197].

P-Type Ohmic Contacts

A very large number of publications exists for p-type ohmic contacts on GaN due to their importance for optoelectronic devices, e.g., [4.45, 4.226]. As the optoelectronic properties, such as optical transparency, do not a play any role for electronic devices, the different metal contact systems are compared with respect to their electric properties. The comparison is given in the lower part of Table 4.1. Because of the intrinsic properties of GaN, the p-type ohmic contact resistances are typically two orders of magnitude higher than that for

Table 4.1. Ohmic contact materials and properties of ohmic contacts on GaN

Material	Ann. temp. (°C)	Doping (cm^{-3})	Resistance $(\Omega \cdot cm^2)/(\Omega mm)$	Ref.
Au	500	n	$10^{-1} \, \Omega \cdot cm^2$	[4.162]
Al	500	n	$10^{-1} \, \Omega \cdot cm^2$	[4.192]
Ti/Al	900	n	$8 \times 10^{-6} \, \Omega \cdot cm^2$	[4.192]
Ti/Al	850	n	$1.6 \times 10^{-4} \, \Omega \cdot cm^2/3.4 \, \Omega \, mm$	[4.9]
Ti/Al/Au	750	n	$6 \times 10^{-6} \, \Omega \cdot cm^2$	[4.154]
Ti/Al/Pt/Au	850	n	$12 \times 10^{-6} \, \Omega \cdot cm^2$	[4.154]
Ti/Al/Ti/Au/WSiN	850/830	n	$0.5 \, \Omega \cdot mm$	[4.165, 4.166]
Ti/Al/Ti/Au	850	n	$0.6 \, \Omega \cdot mm$	[4.245]
Ti/Al/Ni/Au	830	n	$1.5 \, \Omega \cdot mm$	[4.165, 4.166]
Ti/Al/Ni/Au	n.a.	n	$0.75 \, \Omega \cdot mm$	[4.81]
Ni/Au	450	p	$1.7 \times 10^{-2} \, \Omega \cdot cm^2$	[4.226]
Ni/Au	600	p	$6.1 \times 10^{-4} \, \Omega \cdot cm^2$	[4.133]
Ni/Pd/Au	350–600	p	$1 \times 10^{-4} \, \Omega \cdot cm^2$	[4.45]
Ag (ZnO)	530	p	$5.5 \times 10^{-4} \, \Omega \cdot cm^2$	[4.11]
Pt/Re/Au	600	p	$1.4 \times 10^{-3} \, \Omega \cdot cm^2$	[4.210]

the best n-type contacts. Several overviews of the p-type ohmic contacts are available, due to their importance for laser and LED efficiency, e.g., in [4.189]. Ni/Au contacts on p-type GaN are proposed in [4.133, 4.226]. The annealing temperatures are 600 and 450°C. Contacts derived from this metal system using Ni/Pd/Au layer sequences are reported in [4.45] with low contact resistances. Contact resistances of $1 \times 10^{-4} \, \Omega \, cm^2$ are reported on Mg-doped GaN. Record low values of $4.5 \times 10^{-6} \, \Omega \, cm^2$ are claimed for heavily Be-doped GaN samples. Transparent Ag-based contacts are reported in [4.11] with contact resistances of $5 \times 10^{-4} \, \Omega \, cm^2$. Several other multilayer metal systems have been investigated featuring, e.g., the use of Re in Pt/Re/Au contacts in [4.210]. The main electronic application of p-contacts are metal-stacks used for base contacts in npn-GaN HBTs [4.173, 4.175, 4.211, 4.224, 4.281]. Ni/Pt/Au metal stacks are used for dry-etched p-type ohmic contacts [4.211]. As-deposited metal stacks are found to be rectifying. Annealing at progressively higher temperatures improve the ohmic behavior. However, even at annealing temperatures of 800°C, the p-type contacts are not purely ohmic. Pd/Au layer stacks are applied for regrown base contacts on GaN in [4.175, 4.281]. Ni/Au is used in [4.224] for the formation of regrown contacts on p-Mg-doped GaN. GaN/InGaN DHBTs promise improved p-contacts [4.173]. In this case, Pd/Au layer stacks have been used for selectively p-regrown contacts. Al/Au metal stacks are used for the n-type contacts.

Ion Implantation for Doping

Implanted contacts form an interesting alternative to reduce contact resistances in the access regions, as reported, e.g., in [4.289]. References are simple planar contacts, high-temperature annealed contacts, or diffusion-based contacts, as reported, e.g., in [4.289].

Doping Implantation for N-Type Ohmic Contact Formation

The implantation introduces additional mask steps into the process flow [4.289]; however, it provides an area-selective definition of conducting semiconductor layers. Doping implantation further allows for efficient channel definition and contacts without the need for epitaxial changes. At the same time, implanted contacts require annealing steps with very high temperatures. Early reports of high-temperature annealing of GaN, AlN, and InN at temperatures of $\geq 1,000$°C are given in [4.101]. A powder technique is used to allow for surface protection and to provide N_2 partial pressure during the annealing. Ion implantation of silicon for n-type and magnesium for p-type doping and subsequent high temperature annealing steps of GaN are described, e.g., in [4.297]. The implantation in bulk GaN is performed at an implantation dose of $5 \times 10^{14} \, cm^{-2}$ and energies of 180–250 keV using Si, Mg, or Mg^+. The annealing temperatures after implantation range from 700 to 1,100°C. The activation of ion-implanted silicon in AlN/GaN epitaxial films by annealing is discussed in [4.54]. The layers are implanted at an energy of 100 keV at a dose of 5×10^{14} cm^{-2}. The electrical activation of the silicon amounts to 19%. The annealing temperature is 1,150°C in this case. The isolated silicon donor energy is found to be 15 meV, deduced from Hall-effect measurements. On the device level, the fabrication of GaN PIN-diodes by Si$^+$ ion implantation is described in [4.108]. The electrical activation achieved amounts up to 25–30% at annealing temperatures between 1,150 and 1,200°C. The maximum annealing temperature for GaN layers grown by MBE is found to be approximately 800°C for GaN without impact on the structural properties, as reported in [4.296]. The yellow luminescence is significantly reduced after this annealing temperature. At even higher temperatures of up to 1,000°C, the structural decomposition of the material becomes visible. Thus, structural changes are to be expected for the material for the annealing temperatures proposed after implantation. A rapid thermal annealing procedure is described at very high annealing temperatures of up to 1,500°C for Si-doped GaN in [4.64]. The activation of Si implants is further discussed in [4.290]. The doses range between 5×10^{14} and 1.5×10^{16} cm^{-2} at energy levels of 100 keV. The resulting contact resistances of Ti/Al/Ni/Au contacts are as low as 0.02 Ω mm at doses of 5×10^{15} cm^{-2} and subsequent annealing. However, morphological changes have to be considered. The formation of n-type ohmic contacts in AlGaN/GaN HEMTs without any alloying is described in [4.289]. The ohmic contacts are formed by silicon implantation at 200°C with doses of 1.5×10^{15} cm^{-2} at energies of 30 and 60 keV. The Ni mask is removed after implantation and the sample surface is capped by AlN. The activation of the silicon is performed at 1,500°C with 100 bar overpressure for 1 min. Afterwards a conventional GaN HEMT process is applied. The resulting contact resistance amounts to 0.4 Ω mm measured in transmission-line measurement (TLM) structures. The morphology of the resulting contacts is greatly improved compared to conventional contacts with subsequent RTA. The formation of complete GaN junction FETs by ion implantation is described in [4.298]. N- and p-type doping is achieved by Ca- and Si-implantation followed by RTA at 1,150°C. No information is provided on the morphological changes.

Doping Implantation for Ohmic P-Type Contact Formation

P-doping of GaN is generally less efficient than n-doping. To resolve the low-effective activation of p-doping in GaN layers, several implantation studies have been performed. Mg-based doping with P-coimplantation for Mg is investigated in [4.297]. Electrical activation is achieved by RTA at temperature in excess of 1,000°C. The surface morphology is improved for both Ar and N$_2$ atmospheres. Effects of rapid thermal annealing (RTA) on beryllium implanted in-situ activated p-type GaN are described in [4.103]. The implantation is performed at an energy of 50 eV and a dose of 10^{14} cm^{-2}. Optimum RTA temperatures for the activation are found to be 1,100°C for 15 s. Multiple annealing steps are proposed to increase the effectiveness of Be-activation relative to single step annealing. No contact information is provided.

Regrown Ohmic Contacts

Apart from planar and implanted contacts, more complicated contact concepts and processes have been developed to address specific issues. Epitaxially regrown contacts can solve specific device issues, while they have the great disadvantage of loosing the natural sequence of epitaxial and subsequent technological process steps. Regrown contacts have particular advantages for the definition and protection of p-type contacts. A study to improve the current gain of AlGaN/GaN HBTs using a regrown emitter contact is given in [4.284]. The emitter contact is selectively regrown on the base layer, which is patterned by a dielectric mask. This technique is used to overcome the Mg-memory effect during MOCVD growth, which normally causes a displacement of the base–emitter junction. The technique is further useful to avoid etch damage of the base layer during processing. The etch damage is otherwise prohibitive, especially for the thin base layers necessary to increase the current gain in AlGaN/GaN HBTs [4.284]. A specific analysis of the p-n diodes in a comparison of regrown and conventional contacts is given in [4.282]. N-doped AlGaN layers are selectively regrown on the p-doped GaN surfaces. A Pd/Au metal sequence is deposited on the p-type GaN. Improved ideality factors and reverse leakage currents are obtained for the pn-diodes when growth pressure, growth temperature, and the semiconductor diode layer sequence are optimized compared to the as-grown p-type GaN. Regrown n-type contacts can also be used in FETs to reduce the access resistance. Self-aligned AlGaN/GaN HFETs with extrinsic transconductance values as high as $400\,\mathrm{mS\,mm^{-1}}$ are reported in [4.33]. They are based on a regrown ohmic contact concept for a gate length of $1.2\,\mu m$. Source and drain GaN layers are grown selectively by MOCVD. A 170 nm thick Si-doped GaN layer doped at $1\times10^{18}\,\mathrm{cm^{-3}}$ is grown on the etched surfaces. A layer sequence of Ti (20 nm)/Al (200 nm)/Ni (40 nm)/Au (40 nm) is deposited and annealed at 830°C for 30 s. The resulting source resistances are $0.95\,\Omega\,\mathrm{mm}$. AlN/GaN insulated gate heterostructure FETs with regrown n+ GaN ohmic contacts are reported in [4.122]. Four nanometer thick AlN is used as a gate insulator in this case. After selectively removing the AlN insulator, both source- and drain-contact n-type GaN layers are regrown on the GaN channel with a thickness of 150 nm. Ti/Al/Au contacts are deposited and annealed at 900°C at 30 s in N_2 atmosphere. The contact resistance achieved with the regrowth method is as low as $0.22\,\Omega\,\mathrm{mm}$.

4.3.2 Schottky Contacts

Metal–semiconductor Schottky contacts are the most critical elements in all those FETs that do not have a metal–oxide contact. Schottky-contact properties define charge control [4.238] and the hot spots of the electric fields in the gate region, as depicted in Fig. 4.7. This figure gives the simulated distribution of the electric field under the gate contact in an AlGaN/GaN HEMT, in this case with a gate length $l_g = 600\,\mathrm{nm}$. It can be seen that the field gradients

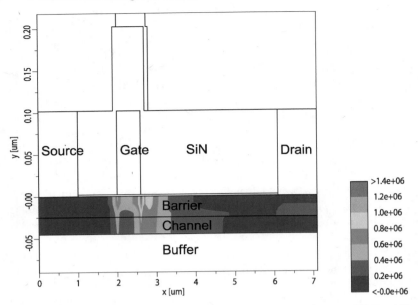

Fig. 4.7. Simulated field distribution in V/cm in an AlGaN/GaN HFET at $V_{DS} =$ 60 V

are located at the drain side end of the gate, and so a very precise gate definition is fundamental for FETs. Direct measurement of gate depletion regions
has been reported. High breakdown voltages of up to 405 V in AlGaN/GaN
HFETs with simple Schottky contacts have been measured, as reported, e.g.,
in [4.261]. A fourth terminal is applied to the device that can be individually
monitored. From this analysis the distribution of the depletion zone can be
directly measured. It is found that the extension of depletion zone strongly
varies with the voltage, which mitigates field peaking. It can be deduced from
this analysis and from Fig. 4.7 that the impact of the surface is critical to the
analysis. In a Schottky contact the forward current through a diode is considered ideal when $n = 1$, as given in (4.1) [4.78]. Nearly ideal Schottky contacts
with low ideality factors n, as defined in (4.1), avoid early gain compression
in FET power devices. The gate current is modeled:

$$I_G = I_0 \cdot \exp\left(\frac{q \cdot (V - I \cdot R_S)}{n \cdot k_B \cdot T_L}\right). \tag{4.1}$$

R_S denotes the series resistance of the diode and n the (non)-ideality. GaN
FETs initially suffer from relatively high ideality factors n [4.114] due to
various issues:

–   High surface roughness affecting the morphology of the semiconductor–
    metal interface

-   Isolation problems along the top semiconductor-passivation interface without the application of a gate or ohmic recess
-   High drain-current levels through high (polarization and impurity) doping
-   Nonohmic behavior of the access regions during the Schottky diode measurement

These issues are partly independent of the actual Schottky contact metal layer design and will be discussed further. The technological issues of the contact formation, however, include the following:

-   The reduction of leakage currents at high-voltage operation of up to 300 V
-   Stability with respect to high operation temperatures [4.110]
-   The effects due to the formation of a Schottky gate on a polar semiconductor (Ga-face, N-face polarity) [4.219]
-   Trapping behavior and dipole formation near the interface [4.219]
-   Adhesion issues of the gate contacts on relatively rough interfaces
-   The mechanical stability of the gates, e.g., with respect to the backside treatment of very hard substrate materials

Typically, the currents in a diode are modeled by a thermionic emission model:

$$I_\mathrm{G} = A \cdot T_\mathrm{L}^2 \cdot \exp\left(\frac{-q \cdot \phi_\mathrm{B}}{k_\mathrm{B} \cdot T_\mathrm{L}}\right) \exp\left(\frac{qV}{n \cdot k_\mathrm{B} \cdot T_\mathrm{L}}\right). \tag{4.2}$$

Tunneling introduces an additional mechanism, as described in various publications, e.g., in [4.83]. The tunneling probability $T(E)$ is modeled in one dimension according to

$$T(E_x) = \exp\left[-2\frac{\sqrt{2m^*}}{\hbar}\int_{x1}^{x2}\sqrt{\phi(x) - E_x}\,\mathrm{d}x\right]. \tag{4.3}$$

Detailed leakage mechanisms at GaN- and AlGaN-Schottky interfaces are described in [4.83]. Temperature-dependent measurements between 200 and 300 K allow a separation of tunneling and thermionic effects in (4.2) and (4.3). Effective barrier thinning through unintentional surface-defect donors is found to enhance the Schottky-barrier tunneling in AlGaN/GaN HFFTs and leads to increased leakage currents. The origins of leaky characteristics of Schottky diodes on p-doped GaN are discussed in [4.291]. The Mg concentration amounts to 5×10^{19} cm^{-3} near the interface. Au/Ni is used as metal stack. The defective surface region is considered critical for the formation of either Schottky or ohmic contacts for the same metal system. The physical background of the tunneling is the trap distribution near the interface, which enhances or suppresses the tunneling component. The influence of rapid thermal annealing on Ni/Pt/Au Schottky gates on nid-doped AlGaN is described in [4.183]. Pt diffusion after annealing at the temperature of 600°C leads to improved Schottky barrier height.

Ti/Ni/Pt/Au gates do not show any significant change after a similar annealing treatment, since interfacial Ti serves as a diffusion barrier. Ni is found to be crucial for the formation of a stable Schottky interface. Similarly, the impact of thermal annealing on AlGaN/Ni (3 nm)/Pt (30 nm)/Au (300 nm) Schottky diodes is investigated in [4.193]. The leakage current is significantly reduced after RTA. Two reasons are considered. One explanation is the Pt diffusion to the AlGaN layer, the second explanation is the reaction of Ni with the AlGaN surface, which reduces the surface trap density.

Very high operating temperatures are considered when Ir Schottky contacts are annealed at $500°C$ in O_2 ambient for 24 h [4.110]. The devices are found to be stable with respect to this temperature treatment. Thermal oxidation of diffused Ga is considered to improve the Schottky barrier ϕ_B in this case. A very high degradation temperature of $700°C$ is found for W and WSiN refractory contacts to n-GaN in [4.228]. Nb contacts degrade at a temperature of $300°C$. However, in general, the refractory gates to GaN are found to be more leaky than those to GaAs. Surface roughness of the semiconductor layers remains an issue, especially with regard to small-gate-length devices with a gate length below 300 nm. Roughness is a serious reliability issue as reported in [4.70]. An increase of the medium scale roughness is based on the cracking of the degraded AlGaN films. The local strain of AlGaN relaxes. Gate adhesion is further a major issue for the realization of the gates [4.9, 4.75]. No adhesion problems have been found for both Ni/Au- and Cu-gates, as reported in [4.9]. Pure Pt gates are considered to suffer from adhesion problems on GaN and AlGaN [4.75]. Other noble metals such as Au present similar problems [4.12].

The exposure of the gates to mechanical stress is another important issue. Additional metals on top of the gate heads introduces additional stress. Such nonuniform mechanical stress is evaluated by simulation in [4.25]. The III-N nitrides are found to be stiff, and the highest impact of stressed gate heads occurs for strongly strain-relaxed AlGaN. Further, during backside processing [4.255], the FETs devices are subject to the stress induced by substrate thinning and thermal exposure due to viahole etching adjacent to the gates. Experimental investigations regarding the mechanical stability are currently rare.

Metal Stacks for Schottky Contacts

Various metal systems have been proposed for use in Schottky contacts on III-N semiconductors. The pretreatment of the semiconductor is discussed below. Overview articles on metal stacks are found, e.g., in [4.163]. Table 4.2 compiles the most important characteristics of a large number of metals and metal systems. The Schottky barrier ϕ_B, the doping of the semiconductor, and the extracted ideality factors n are given. Out of the selection in Table 4.2, several options have been eliminated for state-of-the-art technologies. Ni/Au

Table 4.2. Schottky barriers and ideality factors on various semiconductor materials

Material	Schottky barrier $q \cdot \phi_B$ (eV)	Doping (cm^{-3})	Ideality (−)	Ref.
Au	1.32	4×10^{17}/n-GaN	1.24	[4.269]
Au	0.84–0.94	n-GaN	1.03	[4.78]
Cu	1.15	n-GaN	1.04	[4.9]
Ir	0.9–1.1	n-GaN	−	[4.110, 4.111, 4.113]
Mo	0.81–1.02	n-GaN	1.04–1.29	[4.208]
Nb	0.63	n-GaN	1.5	[4.228]
Ni	1.26	n-Al$_{0.15}$Ga$_{0.85}$N	−	[4.292]
Ni	0.71	n-Al$_{0.21}$Ga$_{0.79}$N	−	[4.113]
Ni/Au	0.97	n-GaN	1.05	[4.9]
Ni/Pt/Au	0.82	n-GaN	2.1	[4.193]
Pd	1.13	Al$_{0.21}$Ga$_{0.79}$N	−	[4.113]
Pt	1.1	2×10^{17}/n-GaN	−	[4.238]
Pt/Ti/Au	−	n-GaN	−	[4.166]
Re	0.82	1.5×10^{17}/n-GaN	1.1	[4.258]
Ru	1.1	−	−	[4.111]
Ti	0.58	4×10^{17}/n-GaN	−	[4.18]
Ti	1	n-Al$_{0.15}$Ga$_{0.85}$N	−	[4.292]
W	0.63	n-Al$_{0.21}$Ga$_{0.79}$N	−	[4.113, 4.228]
WSi	0.58	n-Al$_{0.21}$Ga$_{0.79}$N	−	[4.113]
WSiN	Leaky	1×10^{17}/n-GaN	−	[4.228]
Ni	2.0	p-GaN	−	[4.163]
Nb	Leaky	p-GaN	−	[4.228]
WSiN	0.8	p-GaN	−	[4.228]

is a favorite system in use. As a general trend, the Schottky barriers on AlGaN are higher than those on GaN. Nitronex reported the use of Ni/Au in [4.114]. Cree has also mentioned the use of Ni/Au in [4.275, 4.277]. Fujitsu/Eudyna and NEC [4.8] have reported on the use of Ni/Au in [4.118] for both HEMTs and MIS-HEMTs. RFMD also reported the use of Ni/Au in [4.221, 4.222]. Triquint reported the use of Pt/Au for high-power e-beam defined field-plate gates in [4.53, 4.153]. Other gate-stack options, such as Ir or W may be useful especially for high-temperature operation.

Cleaning and Pretreatment of the Gated Surface

Surface cleaning prior to metal deposition plays an important role for the yield and the homogeneity of the process, e.g., [4.114]. After opening the protecting SiN passivation layers [4.114] or resist layers prior to gate deposition, the remainders on the semiconductor surface have to be fully removed to increase process robustness. At the same time, etch-damage has to be avoided. Opening the SiN dielectric involves plasma dry etching, as proposed, e.g., in [4.245]. This procedure has to be optimized for low-semiconductor damage, as performed in [4.114]. ICP-RIE is used for this procedure, as plasma density and ion energy can be effectively decoupled and damage can be reduced. Thus, sufficient etch rates can be achieved without the use of high DC-substrate bias.

Resist and oxide removal is another critical step before gate formation. A wet NH_4OH treatment in 1:10 volume is applied in [4.183] to remove native oxides. A similar oxygen plasma followed by a 1:10 NH_4OH/H_2O treatment for 15 s is suggested in [4.238]. An O_2-plasma-ashing in combination with diluted $HCl:H_2O$ (ratio 1:1) treatment is used in [4.9] for the same purpose. The same HCl treatment is mentioned in [4.258]. Surface oxide removal by aqua regia ($HNO_3:HCl=1:3$) is suggested in [4.208]. The impact of a short HCl dip prior to gate–metal deposition is studied in [4.141]. A trade-off is found achieving either low-leakage and stronger current collapse or increased leakage and negligible current collapse. The trade-off is attributed to the existence of a thin interfacial oxide layer under the gate, which is partly or fully removed by the HCl etch. Residual oxide leads to reduced leakage but increased dispersion in the general trade-off between leakage and dispersion. The design of the gate head is discussed in the following section.

Device Annealing

Rapid thermal annealing is a crucial procedure for the formation of ohmic contacts, as has already been reported. Further, annealing can be instrumental in removing the etch damage, e.g., [4.204]. In addition, annealing can be effectively used to stabilize devices during or after processing with respect to long-term reliability [4.99]. A study of the post-annealing effects on the AlGaN/GaN HEMT performance after gate formation is given in [4.130]. The annealing is performed at 400°C for 10 min in N_2 atmosphere. The annealing is reported to reduce the number of surface defects with a time constant of $\leq 10\,\mu s$, resulting in a reduction of low-frequency dispersion observed by pulsed-DC-measurements with different time constants. Further annealing leads to the creation of traps with a time constant of $\geq 10\,\mu s$. Furthermore, the annealing of gate contacts has a large impact on the threshold voltage, as reported, e.g., in [4.59]. A device annealing is performed at 500°C and leads to a substantial change of the barrier height and the leakage currents. Similarly, gate annealing procedures have been developed to stabilize the gate process. Gate annealing in nitrogen atmosphere at 600°C for 5 min is reported in [4.193].

Gas Sensitivity

Any metal–semiconductor contact has a sensitivity to gases present at the metal–semiconductor interfaces. This sensitivity is based on the adsorption of the gases and the resulting modification of the space charge zone under the gate. This can possibly be exploited at extremely high operating temperatures $\geq 300°C$ due to the WBG properties of III-N semiconductors.

The hydrogen response of Pt-GaN Schottky diodes is investigated in [4.219] and in [4.132]. Hydrogen is found to dissociate on the Pt surface forming atomic hydrogen, which modifies the effective barrier height. Similar investigations of the hydrogen sensitivity of Pd/GaN diodes are reported in [4.271] and in [4.132]. The forward and reverse currents are significantly changed by the Schottky barrier reduction in the presence of hydrogen. The devices can be operated up to temperatures of 473 K. The changes of the effective Schottky barrier height in an atmosphere with 10% H_2 are typically in the order of 50 meV. Hydrogen-induced reversible changes in the drain current of $Sc_2O_3/AlGaN/GaN$ MOS-HEMTs are discussed in [4.119]. The gas-induced effect is based on the dissociation of the molecular hydrogen at the Pt contact, followed by the diffusion of the atomic hydrogen to the oxide/semiconductor ($Sc_2O_3/AlGaN$) interfaces, where the channel charge is modified, as already mentioned above.

4.4 Lithography

Lithography is a fundamental processing step for any semiconductor device. Optical lithography has been reported on for the formation of ohmic and gate contacts and further layers of III-N devices in [4.98, 4.166, 4.223]. Electron-beam gate lithography [4.263] has been used in the III-N material system for the gate formation of GaN FETs with gate lengths ≤ 300 nm. Minimum physical gate lengths $l_g = 30$ nm have been reported [4.93]. Even direct contact lithography is still used, mainly for contact definition, especially for 2-in. substrate diameter technologies, e.g., in combination with electron lithography [4.16, 4.126].

4.4.1 Optical Lithography

Optical stepper lithography provides parallel illumination of the wafer and thus provides a fast and thus cost effective lithography for mass production [4.61]. With the development in the silicon VLSI industry, optical gates down to physical lengths of 25 nm are being realized with UV [4.106, 4.187] and potentially with EUV optical stepper technology [4.232]. In the III-V semiconductor industry, the optical steppers used are typically i-line type and are based on 248 nm wavelength technology [4.61, 4.114]. Gate lengths down to at least 250 nm are typically realized, and the gate length can be decreased even to 150 nm, when phase shift masks are applied, e.g., [4.164]. The gate definition of GaAs and GaN FETs can be subdivided into the fundamental steps [4.61, 4.114]:

- Gate stem or trunk definition
- Opening of the dielectric [4.114]
- Resist removal
- Metal evaporation [4.114]
- Patterning of the gate head
- Gate head metal evaporation
- Gate head lift-off

The order of these steps can partly be changed and steps can be omitted based on the different processes, such as simple stem-gate, T-gate, or SiN-assisted gate.

For III-V semiconductors, Triquint announced an improved fabrication process-based on a dielectrically defined quarter micron optical T-gate for high-power GaAs PHEMTs, as reported in [4.61]. The improvement in power density in GaAs up to $2\,W\,mm^{-1}$ at $10\,GHz$ is attributed to reduced surface state density. A similar process has been reported for GaN FETs in [4.114]. This process enables early passivation of the semiconductor and thus protection of the semiconductor layers. The gate stem is defined by ICP etching of the SiN. A low-damage etch recipe is applied to provide accurate pattern transfer and critical dimension control. The absolute resolution of the i-line steppers typically used for III-V semiconductors amounts to at least $330\,nm$ [4.61, 4.164]. Phase-shift mask can be used to reduce the gate length for optically-defined devices to $0.15\,\mu m$, e.g., [4.239]. For GaN and related semiconductors, optical lithography on optically transparent substrates such as SiC or sapphire increases the complexity and raises several additional issues. These include the following:

- Reliable mark/pattern recognition of the optical stepper due to substrate transparency [4.98, 4.137]
- Increased surface roughness imposing problems with frontside/backside resolution [4.137]
- Radius-dependent angle illumination through wafer bowing

The increased accuracy of steppers relative to the direct contact lithography is necessary for gate and recess definition with positioning accuracies better than $100\,nm$. This is critical if the distances between the contacts are reduced to $\leq 500\,nm$ for high-speed operation.

4.4.2 Electron Beam Lithography

Electron beam processes provide increased resolution and positioning accuracy for the definition of gates. Devices with gate lengths down to $30\,nm$ have been reported for the III-N system [4.91, 4.93]. The absolute and relative accuracy of e-beam lithography amounts to a few nanometer, allowing for the gate definition of $14\,nm$ gate in VLSI [4.123]. At the same time, the big disadvantage of electron-beam technology is the linear illumination of the reticle, which is

Fig. 4.8. SEM image of an e-beam T-gate with a gate length $l_g = 150\,\text{nm}$

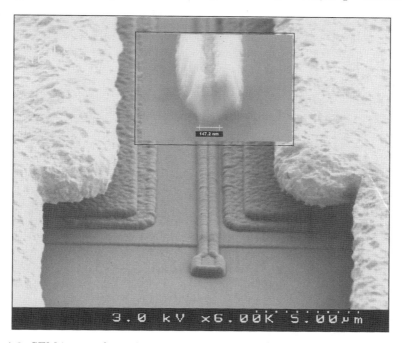

Fig. 4.9. SEM image of an e-beam T-gate with a gate length $l_g = 150\,\text{nm}$ including galvanic metal, smaller image: SEM image of an e-beam gate with a gate length $l_g = 150\,\text{nm}$ at higher resolution

time consuming. The double and triple-layer [4.113, 4.150] resist systems such as PMMA used for III-N devices are very similar to those developed for GaAs, e.g., [4.121]. A trilayer PMMA/P(MMAMAA)/PMMA resist is used for the definition of 120 nm gate in [4.150]. Mushroom-gate [4.126] and Γ-shaped gate structures [4.148] have been reported. Fig. 4.8 gives the SEM image of an e-beam gate of an AlGaN/GaN HFET with a gate length $l_g = 150$ nm. The good morphology and edge quality of the ohmic contacts is vital for the pattern recognition and positioning accuracy during the gate processing. Fig. 4.9 gives a similar image including the galvanic-metal layer. The contact spacing is asymmetric in this case. The T-shaped structure is visible on top of the gate stem in the enlargement in Fig. 4.9.

4.4.3 Field Plates and Gate Extensions

Conventional metal–semiconductor gate contacts provide a potential weakpoint for III-N devices. Devices with conventional gates do not exploit the full basic breakdown properties of wide bandgap semiconductors, as the highly uniform field distribution provides early breakdown. For this reason gate contact designs have been modified to enhance breakdown voltages by the use of gate extensions [4.237, 4.245], single field plates, [4.120, 4.294], and multiple field plates [4.283]. An example is given below in Fig. 4.13. Figures 4.10 and 4.11 show various concepts of the field plate and the naming convention of the geometry. The field plate can either be realized as an additional metal layer on top of the SiN layer covering the conventional gate stack, as depicted in Fig. 4.10 [4.279]. In the second concept the field plate can be directly integrated in the conventional gate stack, as shown in Fig. 4.11. In addition, a second source-terminated field plate or shield structure can be added [4.243, 4.245]. The most important parameter apart from the gate length l_g and the contact separations between source, gate, and drain are the length of the field plates, electrically connected either to the source, or the gate. The respective lengths, l_{fps} and l_{fpd}, are depicted in both Fig. 4.10 and Fig. 4.11. Further, the vertical height of the field plates above the semiconductor have a strong impact on the field distribution in the barrier and channel. Several simulation studies are available for the optimization of the geometry [4.120, 4.218], see also Fig. 4.13. The geometry of the field plate, i.e., the length l_{fpd} is varied and the resulting field distribution along the channel is visible. As a rule of thumb, the field plate length has to match the gate length, e.g., [4.280]. An exception is given for a very small gate length of 100 nm where the gate extension is longer, e.g., 360 nm, as reported in [4.237].

State-of-the-Art

The field-plate concept is widely used for silicon LDMOS [4.26] and GaAs HEMTs, e.g., for base station applications, as cited in [4.264]. The concept yields improvements in both power density and V_{DS} voltages [4.280]

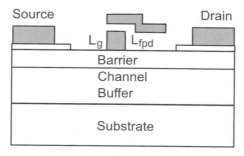

Fig. 4.10. Gate field plate and geometry (source- or gate-terminated) (type A)

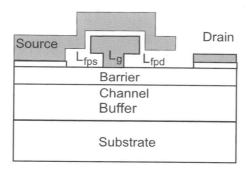

Fig. 4.11. Gate extension in combination with a source-terminated field plate (type B)

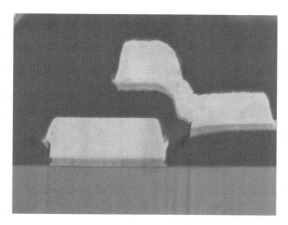

Fig. 4.12. SEM image of a gate with a gate length $l_g = 900\,\text{nm}$, including a field plate of type A

for all semiconductors. Further improvements discussed include device linearity [4.38] and device compression behavior, especially for GaN. Fig. 4.12 gives an SEM image of a gate structure with gate length $l_g = 900\,\text{nm}$, including

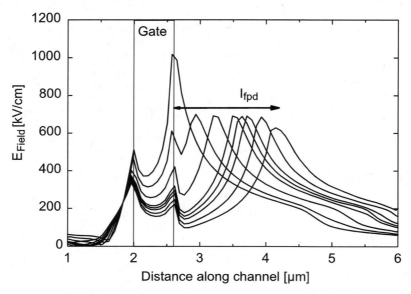

Fig. 4.13. Simulated field distribution along the channel of a conventional FET ($l_{\mathrm{g}} = 600\,\mathrm{nm}$, $l_{\mathrm{fps}} = 0$) as compared to a field-plated FET of various lengths l_{fpd} at $V_{\mathrm{DS}} = 60\,\mathrm{V}$

a field plate of type A of Fig. 4.10. Initial reports on the enhancement of the breakdown voltage through a field plate in AlGaN/GaN FETs are given in [4.294]. The overlapping gate structure is defined into a 150 nm thick SiN layer. The drain-side overlap of the gate l_{fpd} is 0.3 μm. Ando et al. reach output power densities $P_{\mathrm{sat}} = 12\,\mathrm{W\,mm^{-1}}$ at 2 GHz on s.i. SiC substrate using gate-recessed devices and field plates [4.5, 4.7]. The maximum V_{DS} operation voltage is 66 V. GaN HEMTs with a gate length $l_{\mathrm{g}} = 1\,\mu\mathrm{m}$ use gate extensions between 0.3–2 μm. The process is combined with a gate recess with an etch depth of 14 nm. The maximum two-terminal gate–drain breakdown voltage achieved is 200 V at $l_{\mathrm{fpd}} = 1\,\mu\mathrm{m}$. World records for RF-output power densities as high as 32 and $40\,\mathrm{W\,mm^{-1}}$ are reached with field-plated devices through optimization in [4.278, 4.279]. The length of the field late is varied between 0 and 1.1 μm. A typical increase of the gate capacitance of the field plate of 10–15% is considered as a guideline. The field plate consists of a layer stack of 10 nm Ti and 500 nm Au. Power and linearity characteristics of field-plated recessed-gate AlGaN/GaN HEMTs are reported in [4.38]. Devices with a gate length $l_{\mathrm{g}} = 0.6\,\mu\mathrm{m}$ use a gate-recess etch-depth of 12 nm and a 100 nm thick SiN nitride. Although limited by thermal issues, sapphire substrates can be used for the development of field plates to achieve output power densities $P_{\mathrm{sat}} = 12\,\mathrm{W\,mm^{-1}}$ at $V_{\mathrm{DS}} = 50\,\mathrm{V}$ in [4.37]. The SiN thickness is 70 nm in this case for devices with a gate length $l_{\mathrm{g}} = 0.7\,\mu\mathrm{m}$. The length of the field plate is $l_{\mathrm{fpd}} = 0.7\,\mu\mathrm{m}$. Conventional gate structures with multiple field plates of type A for power switches with high breakdown voltages are described in [4.283]. The additional degree of freedom with multiple field plates is used to further reduce the field peaking for gate-to-drain spacings of up to 24 μm for an AlGaN/GaN HFET with a gate length $l_g = 1.5$ μm. The nitride thickness is 180 nm. The length of the field plates is $l_{fpd} = 0.5$–0.7 μm for the first field plate and additional 0.5–0.7 μm for the second field plate. The maximum breakdown voltages claimed in [4.283] reach 900 V. The device with no field plate yields 250 V, the device with one field plate 650 V. The field plates in [4.283] are completely connected to the gate. However, this electrical connection or termination leaves an additional degree of freedom, as discussed later. Again for power-switching and power-supply electronics, field-plated AlGaN/GaN HFETs with breakdown voltages beyond 600 V have been reported, e.g., in [4.218]. In this case, the gate is a conventional metal stack and a source-terminated field plate is used, as depicted in the left part of Fig. 4.11. The gate length is $l_g = 1.5$ μm. The passivation is a combination of 360 nm SiN, followed by 600 nm of SiO_2. The study is continued with the application of the same overlapping structure from the drain contact. The additional drain field plate allows the reduction of the field peak at the drain edge of the ohmic contact without further increase of the GaN layer thickness. A very simple slant field plate for power applications is suggested in [4.51]. The gate can be realized in one lithography step with a self-aligned integrated field plate. Breakdown voltages $\geq 1,000$ V are obtained, which otherwise require multiple field plates with complicated multiple lithography steps.

The additional capacitance introduced with field plates needs consideration. Switching speed capabilities of field-plated AlGaN/GaN HEMTs up to 100 MHz are reported in [4.52]. In this case, three field plates are introduced. The gate length is 2 μm and the shift of the field plate extension is 0.6 μm for each field plate. The devices provide a breakdown voltage of 550 V. AlGaN/GaN MOS-HFETs including a field plate are reported in [4.2]. The gates with a gate length of 1.2 μm are deposited on a 16 nm thick SiO_2 layer deposited by PECVD. The overhang of the field plate (type A) is 1 μm. The devices are covered with a leaky SiN dielectric. Operation voltages of $V_{DS} = 55$ V are reported with this concept. Recent reports on innovative combinations of source and gate connected field plates are presented in [4.8] and [4.243]. The gate extension concept (type B) is combined with a source-terminated field plate. The devices are similar to those depicted in Fig. 4.11. The interlayer thickness of the SiN is 490 nm, which determines the separation of the two field plates. Output power densities up to 10 W mm^{-1} at 2.14 GHz for large-periphery devices are reached in pulsed-operation [4.38]. For operation frequencies of 10 GHz, gate extensions can also be used for smaller gate length of 0.35 μm, see [4.245]. A drain-side extension of 0.18 μm is used for an SiN thickness of 40 nm. This leads to output power density of 16.5 W mm^{-1} and 47% PAE at $V_{DS} = 60$ V for $W_g = 150$ μm. The improvements of the gate extension are especially visible at $V_{DS} \geq 25$ V.

Design Considerations and Trade-Offs

The major advantage of field-plated- and gate extension-HEMTs is the reduction of the electric field peaking, as depicted in Fig. 4.13 in comparison to a conventional FET without any field plate. The high gradients in carrier concentration near the gate lead to very strong field peaking. Further, the field plate leads to the depletion of the interface between semiconductor and the passivation layer. This is true for both the source and drain side of the gate contact. The application of the field plate thus leads to reduced dispersion, better pinch-off, improvement in device reliability, and to an increase of the effective gate length. The two typical situations for field plate designs in Fig. 4.10 and Fig. 4.11 shall be discussed in more detail. The first design (type A) is an additional metal on top of the gate contact separated from the gate by an additional dielectric layer. This requires additional processing steps in addition to the gate head formation. The potential of the field plate is either gate or source-connected (not visible). An independent fourth potential for the device is typically not considered. The approach requires the precise definition of the gate head, as the real geometry of the field plate is dependent in the shape of the gate head. The second design (type B) is given in Fig. 4.11. The approach combines a gate extension with a second field plate connected to the source contact. The gate-terminated field plate with the extension is easy to process and does not require additional process steps. Fig. 4.13 shows the enormous potential for the reduction of the maximum electrical field at $V_{DS} = 60$ V for a gate-terminated field plate of type A. The simulation shows a significant reduction of the electric field along the channel. Multiple field-plate designs of type B, as given in Fig. 4.11, can further reduce the field peaking. However, these advantages have to be carefully traded off against several technological issues and performance reductions, either related to gain, RF-bandwidth, and manufacturability. These include the following:

– The possible increase of the gate resistance R_G [4.168]
– The introduction of additional parasitic capacitances at source, gate, and drain [4.276]
– The introduction of additional processing steps and related processing uncertainty [4.37, 4.279]
– Additional reliability issues at the metal/insulator and semiconductor/insulator interfaces, e.g., due to the additional strain imposed by the metal layers

The possible increase of the gate resistance is an obvious effect. It is based on the effective decrease of the gate cross-section in some of the field-plate concepts, especially of type A, and the increase of the line resistance of the gate. A novel gate feed technology to reduce the gate resistance has been presented in [4.168]. Additional feeder structures are introduced for better connection of the gate finger, especially for very long gate fingers. Gate-connected field plates further lead to an additional contribution of gate-to-source capacitance

C_{gs} and gate-to-drain capacitance C_{gd} and thus lead to a reduction f_{max}, and thus of the power gain [4.280]. The effect of the C_{gd} increase is typically dominating. Source-connected field plates, however, increase the source-to-drain capacitance C_{ds} [4.276]. The trade-off for the optimization of the field plates includes reducing the field peaking to enhance breakdown, while preserving a good power gain and keeping the additional capacitive contributions low. These capacitive contributions can be understood from another point of view. The gate-terminated field plate has the effect of a partial increase of the gate length. As shown in Fig. 4.10, once the vertical heights are reduced to zero, the field plate is nothing else than an extension of the gate length l_g. For the source-terminated field plate, the effect on the capacitances is different. The gain parameters, represented, e.g., by the MAG/MSG or f_{max}, are not primarily affected. However, the source-to-drain capacitance C_{ds} is increased by a parasitic contribution [4.276], which has a strong impact on the output matching of the devices. As will be explained in Chapter 6, the capacitance C_{ds} has a limiting effect on both bandwidth and input/output matching. Thus, the parasitic contribution through the source-terminated field plate is critical, especially for broadband applications. The additional processing effort is obvious; however, it can be reduced, as the sole application of a gate extension does not involve additional lithography. Field plates as shown in Fig. 4.10 provide additional mask layers on the SiN, while the multiple field-plate approach involves further additional mask steps. For very small gate length of 100 nm [4.2], the definition of the field plate requires a second e-beam step for the definition of the gate head. The additional metallization layers further involve additional reliability issues. In the case of a source-terminated field plate, the full V_{DS} voltage drops between the field plate and the drain ohmic contact. As the optimized dielectric layers are typically very thin (100 nm), the maximum voltage of up to 200 V drops over this distance. However, the overall advantages dominate and successful realization of various field-plate concepts will make this approach a standard technology module for III-N FETs.

4.5 Etching and Recess Processes

The etching of semiconductors is a crucial step in semiconductor processing [4.172]. The generation of nonplanar structures is a basic need during the formation of semiconductor devices. Mesa etching was already described in Sect. 4.2.1. Controlled removal of material and etch-damage-control in critical device regions are important for the development of reproducible semiconductor process technologies. The tight crystalline bond strengths in GaN and other III-N semiconductors lead to the wide bandgap, but also make these semiconductors chemically inert and difficult to etch. This is especially true with respect to the polar nature of these semiconductors, where even very mild etching conditions can modify the surface charge densities significantly, e.g., [4.56]. The criteria for the evaluation of etching processes are enumerated and discussed in the following. They include the following:

- Etch rate, e.g., [4.203]
- Material selectivity [4.220]
- Etch anisotropy [4.204]
- Persisting etch damage [4.56]
- Surface morphology [4.56]
- Etch delay [4.29]

General overview articles on the etching of III-N semiconductors are provided, e.g., in [4.56, 4.172, 4.204].

4.5.1 Dry Etching

Dry plasma-based etching has become a dominant patterning technique for the etching of III-N semiconductors, e.g., [4.204]. This is especially true, because wet etching of III-N semiconductors is less effective than in other III-V semiconductor material systems [4.181, 4.204] and can hardly be called an industrially-suitable technology module. Dry etching of III-N materials was initially developed for highly anisotropic mesa etching, targeting high etch rates of $\geq 1\,\mu m\,min^{-1}$, smooth surfaces, and nonselective material removal [4.190]. Isotropic etching of III-N semiconductors leading smooth surfaces, low plasma-induced damage, and material-selective etching, has found increased interest, e.g., [4.204]. It is needed for the definition of recess and contact structures, however, this technique also suffers from severe obstacles.

Etch Methodologies

Several methodologies for plasma generation are being used for semiconductor dry etching. Dry etching is performed either by physical reaction or by chemical reaction or a combination of both. The most important basic techniques have been demonstrated for III-N semiconductors, e.g., [4.204]:

- Reactive ion etching (RIE) including reactive and sputtering material removal based on
    - Inductively coupled plasma (ICP-RIE) [4.43, 4.220]
    - Electron cyclotron resonance (ECR-RIE) [4.203]
- Ion beam etching methods such as chemically-assisted ion beam etching (CAIBE)

Pure RIE etching is based on reactive and physical material removal and intended for anisotropic etch profiles at relatively high etch rates. High-density plasma generation methods, such as ICP and ECR, use plasmas with increased ion densities. The advantages of ICP and ECR compared to pure RIE are based on the decoupling of the plasma density from the particle energy, so that anisotropy and damage in the semiconductor can be more effectively separated and thus controlled [4.43]. As reported below, the methodologies of ICP, ECR, and RIE can be combined [4.56, 4.69]. CAIBE is an ion beam etching method. It allows a decoupling of ion energy and ion density and thus a higher energy of the ions in the process. It produces high etch rates of $>100 \, \text{nm} \, \text{min}^{-1}$ through the introduction of a chemical component such as Cl_2 into the plasma. It is often used for mesa etching, e.g., in [4.126].

Plasma Chemistries

Plasma-based etching techniques use several plasma chemistries. A good overview article on the chemistry is given in [4.204]. Chlorine-chemistry etching is the substantial method for dry etching of the Al/Ga/In/N material system. Initial studies of the etching of GaN by RIE using BCl_3 and $SiCl_4$ are reported in [4.161]. Etch rates of $17.5 \, \text{Å} \, \text{s}^{-1}$ are reported using BCl_3 chemistry. Several studies are available with respect to the additive gases, RF-source power [4.43], acceleration bias [4.56], and discharge composition [4.203]. High-density-plasma-based reactive ion etching (ICP-RIE) of GaN, AlN, and InN is reported in [4.220]. Both Cl_2/SF_6 and Cl_2/Ar plasma chemistries are investigated. Etch rates of up to $680 \, \text{nm} \, \text{min}^{-1}$ are reached with the application of a high DC bias of $280 \, \text{V}$. At the same time, high GaN:AlN selectivities of 8:1 can be reached in a Cl_2/Ar plasma at $10 \, \text{mTorr}$, $500 \, \text{W}$ ICP-source power, and $130 \, \text{W}$ cathode RF-power. A study on inductively coupled plasma (ICP)-RIE Cl_2 etching of GaN, AlN, InN, InAlN, and InGaN is given in [4.43]. The DC-bias is again relatively high, i.e., $100 \, \text{V}$. The etch is strongly ion-assisted and thus anisotropic. Etch selectivities of InN of 6 over the other nitrides are achieved. An increase in etch rate with source power is found as a well as a decrease of the etch rate with pressure. The etch rates typically reach several $10 \, \text{nm} \, \text{min}^{-1}$, so that etch-depth control is a critical issue [4.43] considering device definition. A GaN:InN selectivity of 6.5:1 is reached at a pressure of $5 \, \text{mTorr}$.

The effects of added gases, such as H_2, Ar, and CH_4, on ICP-RIE Cl_2-dry etching of GaN and InGaN are reported in [4.156]. Pure Cl_2 plasma at a pressure of $10 \, \text{mTorr}$ produces strong anisotropic etching and very rough surface morphologies. The ICP power is 500–$2,000 \, \text{W}$, the RF-power amounts 100–$250 \, \text{W}$ at high DC-bias of up to $400 \, \text{V}$.

The significant etch delay observed during the RIE etching of GaN using Cl_2-chemistries is explained by persistent hard-to-etch surface oxides, e.g., in [4.29, 4.190]. This oxide can cause an etch delay of some $10 \, \text{s}$. This etch delay is equivalent to about 10–$15 \, \text{nm}$ of nominal etch depth, which is in the order of typical recess depths. This fact causes a big uncertainty for FET definition. A BCl_3 plasma pretreatment can remove the etch dead time, as reported in [4.29].

The etching of GaN by RIE and SF_6 chemistry is systematically investigated in [4.15] for DC-bias between -250 and $-400 \, \text{V}$. Etch rates of up to $167 \, \text{Å} \, \text{min}^{-1}$ are reached. The etch rate increases with increasing SF_6 flow. It is stressed that SF_6 is less corrosive than Cl_2 and Br_2 plasmas.

Electron cyclotron resonance-based reaction ion etching (ECR-RIE) of GaN using chemistries of H_2, He, Ar, and $CH_4/Ar/H_2$ are investigated in [4.56]. Consistent changes of plasma-induced damage are found for the increase of the DC-bias/ion energy. The damage is further correlated with the mass of the ions, e.g., for Ar.

ECR-based plasma etching of GaN, AlN, and InN using iodine or bromine plasma chemistries is reported in [4.203]. The etch rates for iodine chemistries are typically higher than those for Cl_2-chemistries, while bromine (HI)-based chemistries yield a reduction of the etch rate. The investigations are performed at a pressure of 1 mTorr for microwave powers between 200–1,000 W. InN yields higher etch rates for bromine chemistry than for chlorine chemistries.

Etch Damage and Contamination

Dry-etch-induced damage is of critical importance to the underlying semiconductor layers and their appropriate functioning in the device. Several investigations of etch damage are available, e.g., [4.190, 4.204, 4.212]. Etch damage modifies both the conductivity and the carrier mobility through the introduction of both defects and disorder [4.56]. This is also true for the optical properties of devices, where several investigations on etch damage and its recovery are available, e.g., as reported for ICP etching of LEDs in [4.157]. Several mechanisms have been proposed to cause etch damage in III-N semiconductors. Ion channeling of Ar^+ is suggested for low energy bombardment of GaN in [4.77]. It is found that GaN is more sensitive to ion channeling based on the decrease of the photoluminescence observed after ion bombardment. The second effect is atomic displacement of the lattice or amorphization. The effect of RIE and photo-assisted RIE on n- and p-doped GaN is investigated in [4.190]. The displacement damage is found to accumulate near the surface at low RF-powers of 50 W at medium −100 V DC-acceleration bias and short exposure times. Contamination is another critical problem in dry-etch processing. Oxygen contamination of GaN is reported in [4.161] after RIE etching using BCl_3. The samples that have undergone BCl_3 etching show increased content of oxygen in the top 10 nm of the GaN layer, determined by Auger spectroscopy. The oxygen is deduced to react from the surface with the damaged semiconductor layers. An impact on the first 10 nm of the device can be substantial. The impact of the presence of hydrogen in III-V semiconductor dry-etch processes is described in [4.63]. Hydrogen has two functions for the etching: its addition to the plasma considerably improves the surface morphology during the etch. At the same time hydrogen addition reduces the etch rate of mask materials such as Si_3N_4, which reduces mask erosion during etching. Residual hydrogen further passivates donors and acceptors in the semiconductors [4.63].

Several device issues can be traced to etch damage. The high access base resistance in III-N n-p-n bipolar transistors can be attributed to etch damage of the p-type contact during mesa formation, as reported, e.g., in [4.71]. P-type GaN is especially sensitive to etching, and a large number of publications is devoted to this problem. For the formation of contacts in optoelectronics, see [4.11, 4.133, 4.210, 4.226]. Annealing procedures can be used to reduce the ohmic contact resistances on p-type GaN in [4.210]. An RTA is performed at 600°C. Surface morphology is another issue after substantial material removal by dry etching in devices [4.156, 4.172]. The increase in the RMS through etching leads to microstructural changes that have a strong impact on devices. During mesa and etch-processes, several issues may occur, which, however, have not fully been detailed in the literature for III-N devices. They include the following:

- General adhesion problems of metals due to reduced surface quality after non-optimized etching [4.156]
- Mask erosion during processing [4.63, 4.266]
- Carrier passivation or compensation in the barrier layer of FETs [4.30]

Similar issues have been reported and solved for GaAs devices by optimization of the etch procedure. The impact of the recess etching on the barrier is discussed in more detail below.

4.5.2 Wet Etching

The ability to fabricate well-defined, reproducible, flat, and plane parallel interfaces is essential for the fabrication of III-N devices [4.73]. Wet etching is the second potential alternative for chemical etching of III-N semiconductors [4.172]. The lack of chemical reactivity of group III-nitrides to wet chemical etching has been repeatedly stated, e.g., in [4.73, 4.172]. As an advantage, wet etching generally avoids, or at least reduces, etch damage. As in any other semiconductor system, the reproducibility of wet etch is controlled mainly by the etchant transport at the semiconductor surface and can thus be very sensitive to the geometry of the actual etch situation [4.73].

Wet etching of GaN by hot H_3PO_4, NaOH, and KOH is described in [4.129] for a temperature range from RT to 250°C. KOH-based solutions etch AlN and InAlN, while the etching of GaN by KOH (30% mol) is found to be critical [4.233]. The etch rates reach $3.2 \, \mu m \, min^{-1}$. However, acids such as HCl, HNO_3, and H_2SO_4 do not reach reasonable etch rates. Generally, the etchants used have to be aggressive to reach reasonable etch rates. Further, typical etch temperatures are beyond 100°C, i.e., beyond the boiling point of pure water, so that other solvents, such as ethylene glycol, are suggested [4.233]. Highly anisotropic etch rates of up to $3.2 \, \mu m \, min^{-1}$ can be reached with H_3PO_4. KOH reaches similar etch rates in various solvents. Wet etching of polycrystalline AlN by KOH-containing photoresist remover is reported in [4.181]. As expected, the etch rates are found to depend strongly on the crystal quality of the polycrystalline AlN. This etching of AlN is further selective with respect to InN and GaN. Wet etching of InN by aggressive KOH and NaOH solutions is described in [4.76]. Acid solutions are found to be inappropriate, as no etch rate is observed. Alkaline solutions such as KOH and NaOH (33% weight) produce controllable surfaces. Etch rates of $10-100 \, \text{Å} \, \text{min}^{-1}$ are reached at $50°C$. Such solutions can be particularly useful for the removal of etch damage induced by dry etching.

Wet-Etch Enhancement Techniques

Given the low etch rates of wet etching, etch-rate-enhancing techniques have been investigated in several publications. Gate-recessing of GaN MESFETs using photo-accelerated electrochemical etching (PEC) is described in [4.205]. Hg-lamp-illuminated wet etching by KOH solution yields etch rates of $50 \, \text{nm} \, \text{min}^{-1}$ at a concentration of $0.02 \, \text{M}$ at an illumination of $40 \, \text{mW} \, \text{cm}^{-2}$. The resulting RMS roughness is $1.5 \, \text{nm}$ [4.288]. Simpler UV photo-enhanced methodologies have been suggested and are described in [4.13]. The light illumination occurs at $365 \, \text{nm}$ at $25 \, \text{mW} \, \text{cm}^{-2}$. The etch rate is $50 \, \text{nm} \, \text{min}^{-1}$. Photo-electrochemical wet etching of GaN using a KOH solution and broad area Hg-lamp illumination is further described in [4.287]. N^+-doped and nid-doped GaN could be etched with etch rates of $17-20 \, \text{nm} \, \text{min}^{-1}$, while p-doped GaN could not be etched at all by this technique. HeCd-laser-based photo-enhanced wet etching of GaN is given in [4.182]. Both dilute $HCl:H_2O$ (1:10) and 45% $KOH:H_2O$ (1:3) are used. This results in etch rates of a few hundred $\text{Å}/\text{min}$ for HCl and few thousand $\text{Å} \, \text{min}^{-1}$ for KOH. However, because of the aggressive chemistries, there is a lack of mask materials that can resist these chemistries [4.69]. A comparison of nitride Schottky diodes and AlGaN/GaN HFETs, etched by either ICP dry etching or KOH and H_3PO_4 PEC-wet etching, is reported in [4.235]. The PEC etch rates are found to be strongly dependent on the material composition of the $Al_xGa_{1-x}N$ ($0.17 \leq x \leq 0.44$). Etch rates as high as $400 \, \text{nm} \, \text{min}^{-1}$ can be found for H_3PO_4 and up to $2,000 \, \text{nm} \, \text{min}^{-1}$ for KOH, strongly dependent on the pH value. A selectivity of 12.6:1 is achieved for $Al_{0.17}Ga_{0.83}N/GaN$ at pH = 15. The comparison of the HFETs fabricated by PEC and dry etching yields lower leakage currents for the PEC devices.

4.5.3 Recess Processes

Technologically related to the etching of III-N semiconductors, the development of recess processes at both ohmic and gate contacts is very desirable. On the device level, this is due to the following effects:

-   A reduction of access resistances for the ohmic contacts [4.186] in FET or bipolar devices
-   An improvement of charge control in the gate region [4.58] of FETs
-   An increase in the gain×breakdown voltage product through decoupling of gain and breakdown [4.186, 4.266] in FETs

The issues resulting from this processing technique are discussed in the following sections.

Issues of Recess Processes

Fig. 4.14 gives typical sketches of single gate-recessed AlGaN/GaN HFET for both material-selective (left) and nonselective (right) etching. The critical issues of the recess definition involve the etch-process-control. The gate-recess requires the processing of the most sensitive device region. Both material selective [4.268] and nonselective [4.58, 4.153, 4.186] etch methodologies have been reported. For conventional dry etch processes, the initial selectivity of GaN to AlGaN amounts to 1:1–2 only, i.e., no selectivity, [4.266]. Thus, for nonoptimized conditions, selective plasma etching is not a reproducible process for III-N semiconductors. Further, because of the expected etch damage, an additional surface recovery process may be necessary, e.g., as mentioned for $Ar/Cl_2/CH_4$ etching in [4.266]. A RTA step is performed at a very high temperature of 700°C for recovery, which is a critical temperature. A second issue is the depth control for the uniform definition of small-gate-length FETs, suitable for highly reproducible processes, as suggested in [4.186]. The definition of etch-stop layers is critical despite of the low selectivity of most of the III-N materials. A third issue is plasma damage and plasma material introduction/contamination, as they are used for defining enhancement-mode AlGaN/GaN FETs without a recess in [4.30]. This nonintentional contamination is well-known in the processing of GaAs and requires special attention.

Examples of Gate-Recess Processes

Cl_2-based plasma chemistry is typically used for the processing of gate recesses in III-N FETs. A baseline low-damage $Cl_2/Ar/(CH_4)$-based gate-recess RIE etch processing is described in [4.266]. A RTA annealing step is used after the etching. The etch mixture is further developed from Ar/Cl_2-based plasma

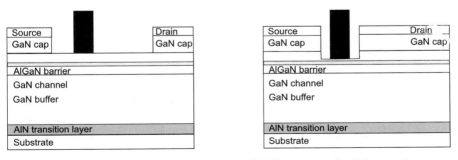

Fig. 4.14. Recessed AlGaN/GaN HEMT: (*left*) selective, (*right*) nonselective

involving an $Ar/Cl_2/CH_4/O_2$-based plasma rather than an $Ar/Cl_2/O_2$-gas-mixture. A selectivity of AlGaN to GaN of 16:1 is found with the improved plasma composition with optimized O_2-flow. A similar plasma mixture is used in [4.268]. However, a 5 nm thick GaN-doped cap layer is applied in the epitaxial layer sequence in this case. The recess is selectively etched through the GaN to the AlGaN layer. The contact resistances can be significantly reduced by the application of the recess process. Further, characterization of recess-gate AlGaN/GaN HEMTs on sapphire substrates is given in [4.149]. The gate-recess process involves nonselective Cl_2/Ar ICP-RIE etching using an etch DC-bias of $-50\,V$ and a pressure of $3\,mTorr$. This results in an etch rate of $12.5\,nm\,min^{-1}$. The gate-recess window has a geometrical width of $1\,\mu m$ and is centered between source and drain. Cut-off frequencies $f_T = 107\,GHz$ are reached with a gate length $l_g = 0.15\,\mu m$. The source-drain spacing is $2\,\mu m$ in this case. A systematic characterization of a Cl_2-based gate-recess process is given in [4.27]. The self-aligned gate recess involves a low-energy RIE process. The etching conditions are $15\,W$ of RF-power, a DC-voltage of $-10\,V$, and $10\,mTorr$ pressure. Devices based on three different etching times $(50–150\,s)$ are compared to unetched devices. To minimize the influence of surface conditions on etch depth, the etches are preceeded by an oxygen plasma and a $20\,s$ rinse with HCl:DI (1:10). The plasma treatment is used to remove residual photoresist from the exposed regions and the HCl dip is employed to remove any possible surface oxides. The etch depth is found to be nonlinear with the etching time. One problem of this approach, however, is based on the nonselective etch procedure, which makes depth control very critical, especially with regard to the nonlinearity of the etching time. Another systematic study of the Cl_2-based etching is provided in [4.28]. Again, a significant etch delay is observed and attributed to the formation of a thin surface oxide layer. This possible etch delay has a tremendous impact on the uniformity of the threshold-voltage for large area wafers, as it directly results in additional nonuniformities of the threshold voltage V_{thr}.

On the FET device level, the application of a gate-recess process for mm-wave applications and resulting improvements in gain and PAE are reported in [4.186]. The gate recess is etched non-selectively and the recess is fully filled with gate metal in this case. The gate-to-channel separation is reduced to $d_{gc} = 11\,nm$. Power and linearity characteristics of field-plated recessed-gate AlGaN/GaN HEMTs are given in [4.38] based on the recess process in [4.28]. In this case, the destructive three-terminal breakdown voltages of the recessed and nonrecessed devices are reported to be similar in the order of $110–120\,V$. Very high PAE values of 74% are achieved and very high operation voltages $V_{DS} = 90\,V$ on similar devices with field plate lead to a power density of $\geq 18\,W\,mm^{-1}$ [4.38].

Low gate-to-channel separations can also serve to produce enhancement-mode HEMTs. Further, etch damage during recess etching shifts the threshold voltage to more positive values, as reported, e.g., in [4.30]. In this case, a CF_4-treatment at an RF-power of $150\,W$ for $150\,s$ creates sufficient damage to create enhancement-mode AlGaN/GaN HEMTs. A comparison of conventional planar and gate-recessed AlGaN/GaN HEMTs with respect to the high temperature characteristics up to 200°C is given in [4.153]. In this case, the gate recess is formed by low-power RIE etching for a gate length $l_g = 0.25\,\mu m$ and is again nonselective. The main finding for the recessed HEMTs involves reduced device leakage at higher temperatures. The nonrecessed device show a dramatic increase of the gate currents with temperature. Further, the linear gain at 10 GHz is increased by more than 2 dB for all operation bias.

Enhancement-Mode HEMTs

The realization of normally-off transistors with III-N semiconductor is a delicate task, as also will be shown in Chapter 5. The combination of thin barrier layers, recess processes, and specific surface treatments [4.30] lead to the realization of enhancement-mode or normally-off transistors, which are particularly interesting for switching applications [4.217]. Gate recessing allows the definition of the threshold voltage independent of the layer growth. An enhancement-mode HFET with $V_{thr} = 0\,V$ is achieved for a gate-to-channel separation $d_{gc} = 7\,nm$ for an AlGaN/GaN heterostructure. The recess is non-selectively etched into the 30 nm-thick AlGaN-barrier. Enhancement-Mode Si_3N_4/AlGaN/GaN MISHFETs are reported in [4.265]. The layer structure consists of a 21 nm-thick AlGaN barrier layer and a 15 nm-thick SiN gate dielectric layer. This leads to a threshold voltage $V_{thr} = -4\,V$. An additional fluoride treatment with RIE in the gate area leads to a threshold voltage shift to $V_{thr} = 2\,V$. The exposure is performed by a powerful RIE CF_4-treatment with an RF-power of 150 W and a very long treatment time of 190 s. This procedure provides an enhancement-mode HEMT, however, based on a strong etch damage. Submicron AlGaN/GaN E-HEMTs are reported in [4.60, 4.185]. However, for gate lengths of 100–200 nm the product of gate length $l_g \times f_T$ is only 5–7 μm GHz and thus below the expectations, partly again due to etch damage. AlN/GaN HFETs with a small gate length $l_g = 100\,nm$ and a cut-off frequencies $f_T/f_{max} = 87/149\,GHz$ are reported in [4.95]. The AlN barrier is extremely thin and amounts to only 2.5 nm, which leads to a threshold $V_{thr} = +0.14$–$0.55\,V$. A special Cat-CVD passivation leads to the formation of the 2DEG in the channel, which is depleted without surface passivation.

A very positive threshold voltage $V_{thr} = 1\,V$ is achieved for AlGaN/GaN HEMTs through the application of 10 nm highly-doped p-GaN cap-layer on top of a 12 nm $Al_{0.22}Ga_{0.78}N$ barrier with a GaN channel [4.236]. At the ohmic contacts the p-doped cap layer is etched away.

Normally-off transistors can further be realized on p-GaN layers, with p-type channels, as shown in [4.60, 4.185]. The p-channel devices yield a very low maximum drain current I_{DS} of 10 mA mm^{-1}, which is about two orders of magnitude lower than that for n-channel (H)FETs.

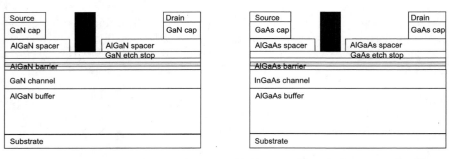

Fig. 4.15. Double Recessed HEMT: (*left*) GaN, (*right*) GaAs PHEMT

Advanced Recess Processes

As the etch methodologies still improve, more advanced etch techniques can be considered. These include ohmic-recess- and more advanced gate-recess-techniques such as double recesses [4.28]. Ohmic recesses have been proposed, e.g., in [4.127]. An etch is applied to reduce the access resistance and to control the lateral ohmic contact definition. Fig. 4.15 shows the proposal for a double gate-recess HFET device and compares it to a selectively-etched double recess GaAs PHEMT. The AlGaN/GaN material system involves the creation of channel layers through polarization charges contrary to the δ-doping of GaAs PHEMTs. The application of the double recess concept is rendered complicated due to the reduced selectivity of the AlGaN/GaN. These difficulties must be overcome to create GaN double-recessed HFETs [4.28]. A similar process has been proposed in [4.140], however, without further detailed results. The first demonstrations of AlInGaN/InGaN/GaN HFETs with a double gate-recess are provided in [4.1]. The AlGaN barrier for the double recess is etched with a BCl_3 and Cl_2 process without an etch-stop with the application of two etch bias conditions.

4.6 Surface Engineering and Device Passivation

This section addresses the technological means for engineering the ungated and the gated dielectric/semiconductor interfaces of III-N FETs in addition to the introduction in Chapter 2 and the analysis and modeling in Chapter 5.

4.6.1 Passivation of the Ungated Device Region

The material engineering of the semiconductor/dielectric interface of the ungated device region involves mostly device passivation. Physically, the deposited dielectric saturates dangling bonds at the ungated semiconductor interface to vacuum. The passivation layer modifies, i.e., mostly reduces the active trap concentration at the interface [4.105]. This passivation step is therefore very important, due to the high polarization-induced charges present in III-N devices. The strong occurrence of low-frequency and RF-dispersion in unpassivated III-N devices [4.246] is another hint to this importance (see also Chapter 2). Technologically, device passivation can be split into several process steps:

- Surface cleaning [4.171], i.e.,
  - Material removal [4.241]
  - Modification of the surface morphology [4.21]
- Pretreatment of the semiconductor surface, i.e.,
  - Chemical reaction at the surface (e.g., oxidation) [4.85],
  - Physical conditioning, (e.g., adsorption of gases) [4.206]
- Deposition of a dielectric, e.g., [4.89]
- Strain engineering by subsequent layers [4.241].

These enumerated steps cannot be fully separated in subsequent actual processing steps. For example, the deposition of a dielectric also modifies the surface morphology and the strain at the interface. Surface passivation has several functions for the device, which include the following:

- (Early) protection of surfaces open to the process flow, e.g., [4.81] from the influence of processing steps such as coatings, developer solutions, plasmas, and ambients [4.68, 4.114]
- Removal of surface material and intended damage [4.241]
- Modification, typically a reduction of surface states [4.139]
- Physical, mechanical, and chemical stabilization of the surface [4.262]
- Compensation of surface charge [4.171] by p-type [4.47, 4.48] and n-type device capping [4.128, 4.225]
- The modification of the surface electrical currents [4.242] by isolating or conductive dielectrics or conductive interlayers [4.139]

Technological solutions to the dispersion problem can be explained by mechanisms detailed in Fig. 4.16 and in the following.

A Model for Frequency Dispersion in an HFET

The understanding and avoidance of frequency dispersion is basically a simple exercise of dynamic charge conservation. Figures 4.16 and 4.17 show two basic situations for the interaction of dielectric–semiconductor interface with the device channel:

1. In the regular situation, I, the negative channel charge in an n-type FET is caused by a positive interface charge at the heterointerface if we assume polarization doping only. Intentional doping modifies the situation, however, does not provide additional information. The interface positive charge results in a negative surface charge at the top of the barrier (Fig. 4.16), which requires compensation at the surface. If the charge at the surface is not fully compensated in a static and dynamic sense,

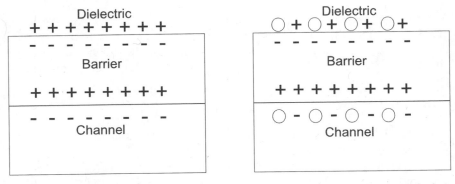

Fig. 4.16. Trapping effects at the interface dielectric–semiconductor and their impact on the FET channel

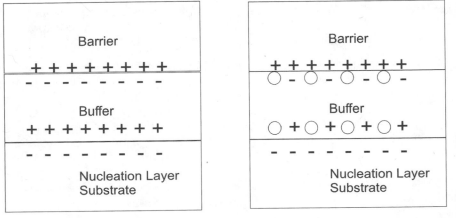

Fig. 4.17. Trapping effects in the buffer layer and their impact on the FET channel

frequency dispersive effects will occur. The modification of the surface charge during dynamic device operation will then directly interfere with the channel charge.

2. In the second situation, II, the empty charges represent the dynamic detrapping of the surface charges with different time constants. This results in a modification of the channel charge, again, meant dynamically. This dynamic trapping and detrapping is further fundamentally connected to the static and dynamic leakage currents at the semiconductor barrier/dielectric interface [4.142, 4.242]. The dynamics are influenced by various time constants involved.

These principle considerations are especially true for the ungated interface regions of the device; however, the situation is further strongly influenced by the effects at the backside of the channel [4.20, 4.216], as depicted in Fig. 4.17. These effects at the backside include the following:

– Bulk semiconductor trapping in the device buffer [4.136]
– Trapping/detrapping at the buffer–substrate interface [4.20]
– Leakage currents in the buffer layer [4.253]

Generation and recombination processes at buffer traps can modify the charge balance, as depicted in Fig. 4.17, similar to the dielectric/(Al,In)GaN interface. Further, because of the heteroepitaxy, the buffer region is highly disordered in the lower part and thus susceptible to carrier trapping. These trapping and detrapping effects at both interfaces and buffer are strongly modified by leakage currents [4.142, 4.242, 4.253], either through the buffer or at the interface. The leakage currents can occur for both DC-conduction or RF-currents. These principal effects are now detailed in the following sections.

4.6.2 Physical Trapping Mechanisms

A variety of trapping effects have been observed. General overview articles on the trapping effects in wide-bandgap microwave FETs are given, e.g., in [4.20, 4.21]. Summarizing the various findings on the mechanisms, Fig. 4.18 shows the location of traps in a typical AlGaN/GaN HFET. The device effects include the following:

1.) Semiconductor/dielectric surface traps [4.17, 4.139, 4.215, 4.260, 4.295]
2.) Barrier bulk traps [4.131]
3.) Interface traps at the channel/barrier interface
4.) Bulk traps in the buffer [4.20, 4.176]
5.) Interface traps at the substrate/semiconductor interface and in the nucleation layer [4.176, 4.231]
6.) Recombination/generation in depletion zones of the series resistances [4.144]

The underlying dynamic physical effects of both electrons and holes [4.259] are now summarized:

– Generation/recombination with surface traps at the passivation/barrier interface
– Generation/recombination with AlGaN barrier traps
– Generation/recombination with (In)GaN channel traps
– Generation/recombination with interface traps at the channel/barrier interface
– Generation/recombination with bulk traps in the buffer [4.20]
– Dynamic generation/recombination at the nucleation layer

The generation/recombination mechanisms involve the following:

– Shockley–Read–Hall generation/recombination,

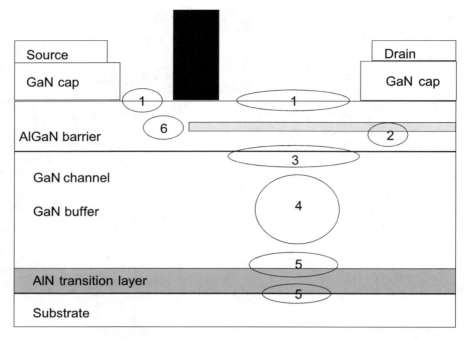

Fig. 4.18. Possible locations of traps in the GaN HFETs

– Trap-assisted band-to-band tunneling [4.215], especially
  – At the interfaces
  – With deep traps near midgap [4.198]
– Direct recombination
– Electroluminescence [4.195] (also hot carrier-induced) [4.176]

Additional (bulk and interface) traps can further be created by device degradation, e.g., [4.215].

4.6.3 Trap Characterization

Several investigations can be performed with regard to the impact of traps in III-N devices, e.g., as reported in [4.20, 4.21, 4.128]. Trapping effects on microwave performance, originating from the semiconductor/dielectric interface and in the buffer, have been investigated by a large number of methods, e.g., as detailed in [4.20, 4.135, 4.139, 4.159]. The nature of the traps is typically characterized by dynamic transient analysis. The related recombination time constants vary from seconds [4.260] to $1\,\mu s$ [4.102]. This includes also the thermal relaxation time of, e.g., $6\,\mu s$ [4.20, 4.176]. Even faster trapping has to be considered, as some devices show different dispersion when pulsed with time constants equivalent to frequencies faster than $1\,MHz$. Capture cross-sections have been determined in various publications, e.g., [4.21, 4.135]. Various methodologies are available to characterize traps and defects in III-N devices on the semiconductor level, e.g., in [4.260] and on device level in [4.262]. These methodologies include the following:

- Luminescence investigations [4.24]
- Photoionization spectroscopy [4.135]
- Surface potential analysis [4.139] and
- Deep-level transient spectroscopy (DLTS) [4.159]

Traps and defects prevail in both bulk material [4.24, 4.136] and at the various interfaces within the device [4.109]. A model detailed in [4.109] states the specific importance of the access region caused by the ungated surfaces. The investigations typically provide three types of information: the changes of the device terminal characteristics under pulsed-conditions [4.179], the energy characteristics and cross-sections of the traps [4.136], and their dynamic time constants [4.135]. Trap time constants between nanoseconds [4.179, 4.259] and several hundred seconds [4.139] can be found. Apart from the trap energies, cross sections, and time constants, additional information is extracted, i.e., the thermal relaxation time of the terminal currents of HFETs, e.g., $6\,\mu s$ mentioned in [4.20, 4.176]. The surface potential at the interface is additionally extracted and found to be modified, e.g., by UV illumination [4.135]. This demonstrates the impact of the surface-related trapping on the drain current [4.139].

Interface Trapping Effects

The interfaces in heterostructures are typical locations for the presence of traps, due to the abrupt modification of the crystal structure. Effects of interface traps on GaN (H)FETs are reported in [4.21, 4.249]. Some principal findings for the interface trapping are summarized:

- The effect of dispersion is often attributed to the surface effect, as device behavior differs significantly in the passivated and the unpassivated state [4.21, 4.216, 4.225]. This is, of course, true in general; however, the surface passivation has also secondary effects on the electric field distribution in the buffer.
- The investigation of surface-related drain-current dispersion effects in AlGaN/GaN HEMTs based on 2D-simulations is described in [4.177]. The impact of holes is stressed in this analysis. The extracted time constants amount to 10^{-4}–10^{-5} s. An activation energy of $0.3\,eV$ is extracted from the temperature dependence of the pulsed-DC-measurements. The application of a 2D-simulation framework makes surface-related effects responsible for both gate and drain lag effects.
- Drain-current transients and surface potentials can be analyzed by direct Kelvin probing in the ungated HFET region, as reported in [4.139]. The measured surface potential is found to be strongly modified by UV-laser-

illumination. It is found that surface states cause observed transients in the drain current of GaN HFETs when they trap electrons tunneling from the gate. The findings are supported by a theoretical model.

- Low-energy electron excited nanoscale-luminescence spectroscopy (LEEN) performed on AlGaN/GaN heterostructures yields yellow luminescence at energies of 2.18 and 2.34 eV [4.24]. The luminescence originates from near the 2DEG region. Strong strain is observed in the 2 nm-thick GaN cap layer on top of the $Al_{0.3}Ga_{0.7}N$ barrier, which yields strong optical emissions at energies ≤ 1.6 eV. The origins of the emissions are localized at the passivated interface region.
- Analysis of surface charge injection in passivated AlGaN/GaN FETs using an MOS test electrode is given in [4.194]. A major finding is that the charge center may not be located at the interface between the passivation and the semiconductor interface, but rather deep within the (bulk) dielectric overlayer. This charge may be trapped rather firmly in the passivation layer and not located in the interface states only.
- The mechanisms of the current collapse removal in field-plated HEMTs are also surface-related, as detailed in [4.143]. If the dielectric is highly insulating, the current collapse is found to be more pronounced than in the case of more conductive dielectrics under or near the gate. This confirms the direct relationship of leakage currents and dispersion removal at the interface [4.142].

Bulk Trapping Effects

Bulk material has to be considered for trapping effects, as typical electronic GaN devices require buffer layers with thicknesses of ≥ 500 nm to cover the mismatch of the heteroepitaxy, e.g., [4.114]. The effects found in buffer layers are summarized in the following.

- Studies performed by photoionization spectroscopy of traps in bulk GaN MESFETs are given in [4.135, 4.136]. A model is applied successfully that assumes a net transfer of charge from the conducting channel to the insulating buffer. The trapped carriers are reversely activated by photoionization. The procedure provides a dynamic model for drain current recovery in the MESFET using two trap levels. The trap levels isolated are consistent with persistent photoconductivity centers [4.21, 4.176]. A similar investigation is reported in [4.136] for AlGaN/GaN HEMTs grown on sapphire substrates by both MBE and MOCVD growth. Similar results in the photoionization spectra are obtained for the HEMTs grown by both methods. Similar phototransient and electroluminescence measurements are included for the diagnosis of trapping phenomena in GaN MESFETs in [4.176]. The analysis yields deep trap energies of 1.75, 2.32, 2.67, and 3.15 eV below the conduction band.
- Electroluminescence measurements on the same samples confirm the emission of light, especially at an increased operation bias $V_{DS} \geq 15$ V.

The deep traps are considered to be localized in the buffer or at the buffer/channel interface [4.176].

– Deep-level characterization via deep-level transient Fourier spectroscopy in Pt/GaN Schottky diodes on MBE RF-plasma-grown material is given in [4.159]. The existence of an additional intermediate buffer layer grown at temperatures of 690°C reduces the concentration of the deep trap levels at 0.45 eV below the conduction band by three orders of magnitude. Further, low-frequency excess noise measurements are used as a surface-sensitive measure for the analysis of traps, while the DLTFS measurements are used to analyze the buffer.

– The influence of threading dislocations caused by the heteroepitaxy of III-N materials has been analyzed in various ways. A comparison of photodetectors grown by heteroepitaxy for GaN on sapphire (0001) and by MOCVD homoepitaxy of SiC on 6H SiC is given in [4.247]. The comparison yields higher leakage currents in GaN on sapphire, due to increased threading dislocations. ELO growth has proven to mitigate that effect [4.145].

– High-injection hole-lifetimes of 15 ns are determined in bulk GaN Schottky rectifiers on freestanding GaN measuring the reverse recovery transient [4.115]. The substrates are initially grown by HVPE on sapphire substrates and the sapphire is subsequentially removed. This again hints at the impact of the mismatch to the material quality and related trapping.

These basic findings for the trapping are now further analyzed in the following sections.

Investigations of the Lattice Temperature Dependence of Trapping

Trapping and detrapping are processes that yield a strong lattice-temperature dependence. Thus, several investigations are available in order to analyze trap behavior in III-N devices in detail. The behavior of current collapse in AlGaN/GaN HFETs at cryogenic temperatures between 100 and 250 K are described in [4.198]. The main finding is a strong variation of the FET on-region in the output characteristics with temperature. Transient monitoring of the currents at various temperatures allows the isolation of two relaxation mechanisms when the injection time and the temperature of the device are individually varied. The time constants of the relaxation are in the range of seconds at 150 K in this case. The finding of two energy levels within the bandgap in [4.198] agrees with the results from [4.21, 4.176]. Further temperature-dependent investigations of the current collapse for AlGaN/GaN HFETs are given in [4.131]. Temperature-dependent DLTS measurements between 77 K and RT are reported. They reveal a small trap activation energy of 0.28 eV. However, the temperature dependence of the carrier concentration suggests the existence of DX centers in the AlGaN barrier. Further temperature-dependent investigations of the three terminal characteristics are given in [4.160]. The so-called kink effects in the cw-output characteristics of AlGaN/GaN HFETs are found to be more pronounced at cryogenic temperatures of 100 K than at RT, and the pulsed-DC-measurements provide similar characteristics. The paper further suggests the existence of a hole contribution to the gate current at low temperatures. We can deduce from these findings that trapping is persistent at both surface and in the bulk and that their impact on the three-terminal characteristics of AlGaN/GaN HFETs is consistent with findings in GaAs devices. However, the relative impact is more pronounced than is known in the GaAs world.

The Impact of the Initial State of the Device

When freshly processed devices are analyzed, the initial measurements often differ from a stable device status, if such status exists at all. A burn-in behavior is observed, which is partly, but not exclusively, attributed to the trapping effects, related current collapse, and reliability issues. Output characteristics, e.g., in [4.285], show a degradation when the device is measured repeatedly. As burn-in effect have also been observed for several other devices, e.g., InP HEMTs [4.35], the variation of the device properties is found to stop after a burn-in procedure [4.227] for optimized devices.

4.6.4 Technological Measures: Surface Preparation and Dielectrics

A minimum of surface treatment before and during passivation is necessary and unavoidable in any device processing. As a general finding, SiN- or other material-passivation greatly enhance the power performance of AlGaN/GaN devices, e.g., in [4.17, 4.20, 4.72, 4.117, 4.241, 4.246]. The effect on leakage current and dispersion, however, has to be treated in detail, e.g., [4.242].

Surface Preparation

During device fabrication, surface treatment, and cleaning conditions are very important and at the same time unavoidable, as discussed, e.g., in [4.57, 4.171]. Because of the polar nature of the III-N semiconductor materials, the status of the epitaxial surface is critical when the wafers are removed from the growth reactor. The wafer is thus exposed to air prior to further processing and device passivation. The exposure typical leads to an initial oxidation. Early passivation techniques within the process flow have been considered relevant in [4.114, 4.117], since the exposure of the semiconductor surface to air and any kind of processing medium is found to be critical, as also known from GaAs processes [4.61, 4.262]. Typical related process steps in the context of device passivation are ohmic contact formation, annealing, device isolation, and gate recess definition. As an alternative to early SiN ex-situ passivation, in-situ passivation in the growth reactor has also been suggested in [4.104] in order to avoid surface contamination by ambients. This agrees with findings in [4.10] that both the GaN and AlGaN surfaces are found to be highly corrosive when exposed to air or other ambients. This corrosion potentially requires the application of cleaning techniques.

Cleaning Procedures and Pretreatment

Given the sensitivity to any exposure, surface and interface cleaning procedures are vital for the reduction of surface states and dispersion, e.g., [4.178]. An improved fabrication process for obtaining high-power density AlGaN/GaN HEMTs has been suggested in [4.244]. The exposure of the critical ungated surface through harmful plasmas and chemical treatments is completely avoided. For other process variants various surface cleaning procedures have been proposed for the reduction or compensation of surface states, e.g., in [4.72, 4.178]. Before PECVD deposition, the surface in [4.72] is cleaned (in this order) by acetone, methanol, isopropanol, and water. The surface is then dipped in 30:1 buffered oxide-etch for 30 s. No significant changes in device characteristics are seen after these wet cleaning and etch steps. Several other treatments are reported. The surface treatment by a combined $HCl:HF:H_2O$ (1:1:2) and subsequent HF-buffered oxide-etch (14%) for 10 s in [4.49] is found to exert little influence on the carrier distribution and mobility. This and similar findings point to the existence of stable surface oxides, which can only be removed or modified by very aggressive techniques. The passivation in [4.49] is SiO_2. The effect of the pretreatment on pulsed-measurements is not reported. Subsequent deposition of the passivation, however, leads to a strong increase in channel-carrier concentration and reduced mobility. NH_3 low-power plasma pretreatment prior to SiN PECVD-based passivation is described in [4.57]. The NH_3 pretreatment is performed with the substrate held at 250°C. The gate-lag ratio observed during pulsed-operation and the output power degradation for long-term RF-operation can be significantly reduced. The increased device reliability is attributed to both the strengthening of the bonds at the interface and the incorporation of hydrogen, which passivates defects. A similar NH_3 treatment is reported in [4.117]. Dry plasma pretreatments are discussed in [4.178]. Plasmas based on SF_6, O_2, N_2O, NH_3, and N_2 are applied. Contrary to wet pretreatment NH_3-, N_2-, and combined NH_3/N_2-pretreatment delivers very strong changes and mitigates the drain-lag. The X-ray photoelectron spectroscopy (XPS) measurements reveal a correlation of the reduction of the carbon concentration on the AlGaN surface with the degree of the mitigation of the drain-lag. Carbon contamination of the surface in AlGaN/GaN HFETs can be mitigated by an air plasma descum and subsequent HCl dip prior to SiN deposition [4.14]. Auger spectroscopy reveals a correlation of the current collapse with the residual carbon contamination. The sensitivity of the breakdown voltage and related gate leakage and surface carrier concentrations to plasma damage in the AlGaN barrier is discussed in [4.216]. A surface-charge defect model is applied. Simulations show that a significant reduction of the surface charge to levels of 1×10^{12} cm^{-2} is necessary in order to reach suitable gate currents at voltages >100 V, even when a field-plate concept is applied.

Surface treatment of III-N semiconductors is especially important for p-type material. Hydrogen is typically incorporated strongly into the material during MOCVD growth, thus the layers are specifically sensitive to surface treatment. Hydrogen plasma treatment of p-type GaN is investigated using both hydrogen and deuterium to enable secondary ion mass spectroscopy (SIMS) investigations of the hydrogen effect [4.206]. Hydrogen plasma treatment leads to a strong hydrogen passivation of carriers in the p-type GaN. This is further confirmed by the analysis of the traps at 0.3 and 0.6 eV. Capacitance vs. frequency C(f)-measurements confirm the impact of the plasma treatment on the p-type doping-activation and the suppression of the trapping. At the same time, additional trapping at 0.4 eV is created. Based on these results, the deposition of dielectric material is now analyzed.

The Deposition of Dielectrics: Passivation

The deposition of dielectrics at the semiconductor/vacuum interface yields a large number of possibilities. With the application of optimized interface passivation, the interface neutralizes the net surface charge, as given in Fig. 4.16, in both static and dynamic sense. The net surface charge arises from a polarized GaN cap or AlGaN barrier and from the residual surface states resulting from dangling bonds, absorbed ions, or charge surface residual materials (e.g., oxides [4.10]).

Silicon Nitrides (Si_3N_4)

Silicon nitride is the most popular dielectric for the III-N material system, e.g., [4.17, 4.68, 4.114, 4.128, 4.139, 4.151, 4.246]. Typical effects of the passivation by SiN on III-N devices include the increase of the output power [4.246], reduction of frequency dispersive effects [4.72], and the modification of leakage currents [4.17, 4.242], often leading to a reduction of the breakdown voltages. Further long-term device reliability is positively affected [4.151]. Various methods and recipes exist for the deposition of SiN in its various modifications, e.g.,

- Plasma-enhanced chemical vapor deposition (PECVD) [4.17, 4.68, 4.241, 4.242, 4.246] for low damage deposition with low hydrogen content, including:
  - ECR-PECVD
  - ICP-PECVD [4.57]
- Catalytic chemical vapor deposition (CAT-CVD) [4.89, 4.90, 4.151]
- Others, such as epitaxial in-situ deposition [4.207].

The variations of PECVD deposition typically include the silane and ammonia flows [4.72], the deposition temperature [4.242], deposition pressure, deposition RF-power, and thickness [4.231]. The corresponding detailed engineering is discussed below.

Stress and Stress Engineering

Deposited SiN is an amorphous material that can be influenced in variety of ways. The actual deposition can be adjusted with respect to strain and stress on the underlying material. Stress ranges of the SiN_x on the semiconductor amount to 100–2,000 MPa, as reported, e.g., in [4.79, 4.293]. For SiN deposited by PECVD on AlGaN/GaN HEMTs, several growth conditions are analyzed in [4.246]. The deposition temperature range amounts to 150–300°C. The SiN, grown under NH_3-rich and higher temperatures conditions, gives the best power performance of the devices. The SiN is further characterized by the etch rate in buffered oxide etch. Nitrides with the lowest etch rate provide the best power densities in the III-N HFETs. A specific stress analysis of the SiN deposited by PECVD is reported in [4.241]. Dual frequency plasma deposition is used. Initial data suggests a strong impact of the induced stress. However, it is found that the surface damage through N-ion bombardment has a very strong influence on the surface, the carrier concentration, and the mobility. When this damage is avoided by a He precursor treatment and thus separated in the analysis, it is found that a uniform compressive stress on the surface has only a minimum impact on the polarization charges. The stress range amounts to -100 to 40 MPa meaning that the nitride is nearly stress-free. The mitigation of both leakage currents and current collapse by low-stress SiN_x with high refractive index is suggested in [4.116]. The in-situ CVD deposition of SiN in an MOCVD reactor is described in [4.39, 4.207]. The deposition occurs at 980°C using disilane and ammonia at a growth rate of $0.1 \, A \, s^{-1}$. The thickness of the deposited SiN amounts to 4 nm. Stress data is not provided although a very high strain of the initial SiN layers can be expected.

Temperature and Thickness of the Deposition

The SiN layer can be either composed of one or of several layers. The thickness of the initial and the subsequent passivation layers critically affects the stress and the permeability for ambients on the semiconductor layers. A PECVD-passivation of 350 nm thickness is reported in [4.72] based on silane (SiH_4) and ammonia (NH_3). The deposition is performed at 300°C after a dip in 30:1 buffered oxide etch for 30 s. The deposition leads to a significant increase in the RF-device output power. A more complete study of the various Si_3N_4-deposition parameters is given in [4.246]. Deposition SiN-thickness, RF-deposition power, and temperatures are investigated with respect to their impact on breakdown and RF-power. The thickness of the first SiN varies between 47 and 275 nm. An optimum for the RF-output power of the AlGaN/GaN HFET is achieved for a SiN thickness of 155 nm and a high deposition temperature of 300 °C, while the breakdown voltage is not critically affected by the SiN parameter variation. The increase in RF-output power is due to the increase in drain current.

The thickness of the PECVD-SiN determined by ellipsometry is 74–80 nm [4.57] deposited at a base plate temperature of 250°C. A SiH_4:NH_3:N_2 plasma recipe is used, which yields a refractive index of 2.03–2.09. The ICP RF-power is 35 W at 13.56 MHz. The crucial parameter is this investigation is not power performance but output power degradation and pulsed-DC-data after aging. The impact of hydrogen for the passivation of defects is discussed to lead to better stability.

The industrial process in [4.114] uses a 90 nm-thick SiN layer deposited by PECVD at 300°C and a pressure of 900 mTorr. After early passivation, this SiN layer is opened by an inductively coupled plasma (ICP)-etch for further processing. In combination with a low-damage ICP etch a very good device efficiency of 65% and output power density of 11 W mm^{-1} at 2 GHz are achieved. Further the thickness uniformity is of the order of 0.6%.

GaN HEMTs with SiN passivation deposited by catalytic chemical vapor deposition (Cat-CVD) are given in [4.89, 4.117, 4.151]. For the Cat-CVD deposited SiN in [4.151] the thickness is 50 nm after NH_3 pretreatment. The process was similarly developed for GaAs PHEMTs [4.86]. The Cat-CVD procedure with a NH_3 pretreatment provides the elimination of deposition plasma damage and thus reduces defect density at the interface. The output power degradation is 0.4 dB after 200 h of operation. For a similar Cat-CVD process, the deposition temperature is 250–300°C for a SiN thickness of 30 nm [4.90].

Based on these examples on the critical parameters strain, deposition-temperature, and -thickness the reliability aspect of SiN deposition is now discussed in more detail.

High-Field Effects in SiN

Similar to the semiconductor materials, the passivation materials and their interfaces are subject to high-field peaking and subsequent degradation effects. Further, the field distribution in the semiconductor and along the interfaces are critically affected by the SiN deposition so that the impact on reliability is strong. The typical degradation of AlGaN/GaN HEMTs with a legradation of I_{DSS} as a function of time can be compared to the degradation of the current in silicon CMOS devices, e.g., [4.34]. Shifts in the threshold voltage can be fully dominated by interface trap creation at increased bias. Thus, similar measures are required for the mitigation in III-N (H)FETs. The effect of the biasing to SiN-passivated AlGaN/GaN HEMT stressed at $V_{DS} = 17$ V and I_{DS} at $V_{GS} = 0$ V for only 1 h is strong. The PECVD deposition is performed at 300°C. The aging leads to a significant decrease in the transconductance g_m, a reduction of the saturated drain current I_{DS}, and a positive shift of the threshold voltage by at least 0.5 V [4.215]. The $1/f$-noise measurements after aging reveal a significant increase of the spectral noise density after aging. These results suggest both hot electron trapping and surface-state creation during device stress.

Possible effects of SiN passivation and high-electric field on short-term AlGaN/GaN HFET degradation are discussed in [4.131]. For a hot pinch-off stress ($V_{DS} = 20\,V$, $V_{GS} = -8\,V$) for 12 h, the unpassivated device shows a degradation of the saturated current of -37%. SiN passivation reduces the degradation; however, it cannot suppress it completely (-16%). The degradation is again attributed to hot electron trapping to surface states, which can be partially mitigated by the passivation. The modification of the gate lag before and after electric-field stress suggests a modification of the surface trap profiles. A similar investigation, again supported by optical response measurements, is reported in [4.184]. The current collapse can be suppressed by light illumination with energy smaller than the bandgap. The collapse is again attributed to the surface states localized in the gate-to-drain spacing rather than the gate-to-source spacing. In general, the deposition of the appropriate passivation is a key for the realization of reliable devices.

Oxide Passivation of the Ungated Interface

Oxide, e.g., SiO_2 passivation is suggested as an alternative to Si_3N_4 passivation for use in III-N HFETs. Various investigations exist to determine the different surface conditions modified by bias in MIS- and regular Schottky-gate structures, e.g., [4.10, 4.65, 4.85]. This paragraph concentrates on the passivation of the ungated regions while gate dielectrics are discussed below.

A comparison of SiO_2 and Si_3N_4 passivation on AlGaN/GaN HEMTs on silicon substrates is given in [4.17]. Both passivation layers are deposited by PECVD at 150°C (for SiO_2) and 300°C (for Si_3N_4). A higher surface trap density is considered to be responsible for the reduced power performance of the HEMTs with SiO_2 passivation. CV-investigations in MIS structures are reported in [4.10] for nitrided Ga_2O_3/SiO_2 films. The SiO_2 passivation is formed by remote O_2/He-based plasma using SiH_4 at 0.3 Torr. The SiO_2 films are deposited by remote PECVD. These improved oxide films provide a reduced density of interface states on n-GaN compared to Si_3N_4 on n-GaN. Effects of annealing on GaN–SiO_2-insulator MIM-interfaces is described in [4.174]. The deposition of the dielectric on n-type GaN is performed by low-pressure CVD (LPCVD) at 900°C using disilane and ammonia. The n-GaN is cleaned with H_2SO_4:H_2O_2 at 70°C. A very low surface charge of $3 \times 10^{11}\,cm^{-2}$ is found. This density increases after annealing at 1,100°C. The growth of MgO- and Sc-based oxides for their use in AlGaN/GaN HEMTs is described in [4.67]. MgO is found to produce lower interface state densities than Sc_2O_3. However, as a passivation material MgO is found to be degrading stronger than Sc_2O_3 in the annealing steps, which follow the actual passivation procedure. The application of oxides as gate dielectrics in MOSFETs is further discussed below.

Unconventional Dielectric Materials

A great number of more unconventional materials have been investigated for the passivation of III-N devices. They include polyimide (e.g., in [4.79]), low-k Bencocyclobutene (BCB) material passivation (e.g., in [4.267]), and insulators such as AlN [4.285]. A 4:1 polyimide-to-thinner mixture is spun on the surface of AlGaN/GaN HEMTs and cured at 300°C [4.79]. The effect of polyimide on the surface is attributed to the reduction in surface states and to the low stress of the polyimide on the surface and the nonoccurrence of plasma damage during deposition. The stress considered is lower than 70 MPa. This an order of magnitude lower than in typical SiN films. Low-k BCB passivation of AlGaN/GaN HEMTs is described in [4.267]. The thickness of the BCB amounts to 400 nm and is compared to a relatively thick Si_3N_4 layer of 300 nm. The inherent advantage of the BCB material is the reduced dielectric constant, the simple deposition, and the lack of deposition damage. The pulsed-DC-characteristics and the RF-output-power-results of BCB-passivated devices are similar to those with Si_3N_4 passivation. The physical nature of the stabilization of the semiconductor surface is not investigated in detail; however, it is found that the degradation of the transconductance and the drain current at 85°C is smaller than that for Si_3N_4.

The use of insulating materials of the III-N system is another option for surface deposition and, consequently, passivation. The growth of insulating AlN layers for the passivation of AlGaN/GaN HFETs by migration-enhanced epitaxy (MEE) is described in [4.104]. The growth temperature is 150°C with a growth rate of 800 Å h^{-1}. The total AlN thickness is 1,000 Å. AlN has the advantage of high thermal conductivity directly near the heat source. The physical nature of the AlN passivation is attributed to the hardness of the material, which reduces gate-bias-induced stress on the material, as explained in [4.230]. Surface passivation using AlN deposited by reactive magnetron sputtering on both MISFETs and HFETs is described in [4.285]. The thickness of the AlN passivation is 50 nm and it is deposited at a very high prcessing temperature of 880°C. The AlN passivation reduces both gate leakage and current collapse. Further, the repeatability of the DC measurements is improved for the AlN passivation.

The successful application of unconventional dielectric materials supports the idea that surface preparation prior to deposition has a similar impact on the device performance to that of the actual deposition of the dielectric.

4.6.5 Epitaxial Measures: Surface Preparation and Dielectrics

Being part of the growth process, epitaxial measures are very attractive for solving the issue of dispersion reproducibly already during the growth process. This is especially important for achieving reliability for high-voltage operation [4.127, 4.225]. Typical epitaxial measures involve the introduction of the following

- A highly or medium-n-doped cap [4.128] for high-power transistors without surface passivation [4.225]
- p-Capped GaN/AlGaN/GaN HEMTs [4.47, 4.48]
- AlN as a passivation material [4.104, 4.285]

– Superlattice-based contact approaches [4.196]
– In-situ deposition of SiN as a passivation material [4.104, 4.207]

The n-doped or p-doped GaN capping of AlGaN/GaN HFETs reduces the area of ungated AlGaN surface being exposed to technological treatment during the gate processing and thus the impact of trapping on the device performance. Whether the Schottky gate itself is placed on top of a GaN or an AlGaN surface is subject to the actual process design. Both options exist. Doping of the cap layers modifies the Fermi-level at the interface, as suggested and detailed, e.g., in [4.127]. The nominal doping of the p-doping in [4.47] amounts up to $10^{20}\,cm^{-3}$. N-doping levels of $5 \times 10^{18}\,cm^{-3}$ are reported in [4.225]. Superlattice-based capping approaches allow the precise definition of the Fermi-level at the interface to both the ohmic contact and to the ungated surface, as reported in [4.196]. The density of states near the ohmic contact is increased through the insertion of the highly-doped superlattice layer with 50 nm thickness. This is confirmed by an increase of the transconductance compared to a conventional n-type cap approach. An analysis of the power performance of AlGaN/GaN HEMTs grown by RF-plasma-assisted MBE including in-situ passivation by CVD is given in [4.207]. After MBE growth the samples are in-situ passivated with 4 nm SiN by high temperature CVD to form MISFETs and to reduce gate currents. The devices are then fully passivated by PECVD at 250°C. Similar approaches are used in [4.39].

4.7 Gate Dielectrics

Many publications for III-N FETs refer to the creation of AlGaN/GaN FETs with conventional metal–semiconductor Schottky gate contacts. Metal–insulator (MIS) or metal-oxide-based (MOS)FETs have been repeatedly suggested for the III-N material system, e.g., [4.3, 4.85, 4.94, 4.107, 4.118]. Several materials have been proposed for the insulating gate interfaces. These are discussed in the following sections. A distinction is necessary with respect to the impact of the dielectric to the gated and the ungated region: A dielectric can serve as a gate dielectric in the gated region and similarly in the ungated region, while it may also be useful to use a specific dielectric as a gate dielectric only and deposit a second dielectric for passivation of the ungated region.

Dielectric Materials

Several variations of gate dielectrics have been tested for III-N MISFETs, such as SiN and SiO$_2$ [4.3, 4.94]. Processing of gate dielectrics of GaN/AlGaN MOS-HFETs by thermal oxidation is reported in [4.107]. A 100 nm-thick Si layer is oxidized at 900°C in dry O$_2$ for 27–45 min. The resulting oxide thickness is 8–13 nm.

Surface passivation of GaN and GaN/AlGaN heterostructures and gate dielectric formation by dielectric Al_2O_3 and SiN films are reported in [4.85]. The critical defect creation during different plasma processing is discussed. ECR-H_2-treatment produced nitrogen-related vacancies in GaN and AlGaN, while ECR-N_2-treatment improved the surface quality. SiO_2 deposition created uncontrollable oxidation reactions. Further, the conduction band offset of SiN and $Al_{0.3}Ga_{0.7}N$ is considered to be 0.7 eV, which leads to strong gate leakage currents governed by Fowler–Nordheim tunneling.

High-quality oxide/nitride/oxide $(SiO_2/Si_3N_4/SiO_2)$ (ONO) insulator stacks for GaN-MIS structures with similarly low interface densities are presented in [4.65]. The stacks are deposited by jet vapor deposition. With respect to dielectric breakdown and leakage currents, these structures yield promising features. The interface trap density is extremely low and comparable to Si-MOS structures. However, III-N device performance is not provided.

Breakdown investigations of MgO/GaN MOSFETs by simulation are reported in [4.40] based on the assumption of an interface-state density of 3×10^{11} cm^{-2}. RF-plasma-deposited Sc_2O_3 is used as a passivating oxide and gate dielectric in [4.169, 4.170] with typical thicknesses of 40 nm. The influence of the gate-oxide thickness on Sc_2O_3/GaN MOSFETs is investigated in [4.41]. Thicknesses between 10 and 80 nm are investigated by simulation, based in this case on an interface-state density of 5×10^{11} cm^{-2}. The primary advantage found for this oxide is the mitigation of the virtual gate effect due to the low surface-state density and the reduction of the gate leakage.

Similar to silicon devices [4.250], high-k dielectrics are attractive materials to increase charge control in conventional GaN FETs. Several high-k materials have been tested for this purpose. An investigation of barium strontium titanate (BST) serving as a gate dielectric of AlGaN/GaN MOSFETs is reported in [4.80]. The $BaSrTiO_3$ is a solid-solution and in paraelectric state, which leads to very high dielectric constants (20–500). It is deposited by RF-magnetron sputtering as a gate dielectric with a thickness of 20 nm. This deposition also leads to a damage to the underlying GaN, which requires further investigations.

MISFETs

The principal advantage of metal–insulator–semiconductor (MIS) gate contacts is the reduction of the gate-leakage by orders of magnitude, e.g., [4.107]. This is a principle advantage relative to Schottky gates, especially in the forward direction of a diode. At the same time, the nature of the insulator–semiconductor interface has to be considered, especially also for the ungated region. It is still not fully resolved whether the Fermi-levels at the various dielectric/GaN interfaces are pinned to midgap for at least some of the dielectrics. High interface charge densities and related pinning of the Fermi-level near midgap make GaAs-based power MOSFETs nearly impossible.

Several studies are available for III-N semiconductors. Heterostructure AlGaN/GaN SiN-based MISFETs are reported in [4.46] based on (unstopped)

RIE etch definition of the gate insulator. A 27 nm gate dielectric passivation layer is deposited early in the process and then etched for gate foot definition. The thickness of the gate dielectric after etching amounts to 18 nm. The RF-output power density achieved with these devices amounts to 4.2 W mm^{-1} at 4 GHz for devices with a gate length $l_g = 600$ nm. Very thin SiN gate dielectric layers with a thickness of only 2 nm are reported in [4.94]. The layers are deposited by Cat-CVD using SiH_4 and NH_3 at a temperature of 300°C. Other dielectrics than SiN are used because the low conduction band discontinuity of the $SiN/Al_3Ga_{0.7}N$ interface is found to be 0.7 eV, while the discontinuity of the $Al_2O_3/Al_3Ga_{0.7}N$ interface is reported to be 2.1 eV [4.84]. Al_2O_3 oxide thicknesses as low as 3.5 nm are reported. Because of reduced Fowler–Nordheim tunneling, the leakage for the $SiN/Al_3Ga_{0.7}N$ interface can be reduced by replacing the SiN gate dielectric with Al_2O_3.

Sc_2O_3-based MOS-HEMTs have been reported in [4.170]. The gate leakage can be significantly reduced by 4–5 orders of magnitude depending on the annealing [4.41]. The 400Å of Sc_2O_3 are grown by RF-PAMBE at 100°C using Sc and O_2 from a plasma source. Good small-signal performance has been reported for Sc_2O_3-based MOSHEMTs in [4.170].

Very promising output power results are achieved based on GaN/AlGaN MOSHFETs with a 10 nm-thick SiO_2 for a gate length $l_g = 1.1\,\mu$m. The deposition is achieved by PECVD. The second passivation is silicon oxynitride (SiON) with a thickness of 75 nm [4.3].

4.8 Processing for High-Temperature Operation

The specific high-temperature operation of GaN devices is often considered to be an advantage, e.g., [4.138]. Many publications suggest the application of III-N materials for high-temperature device operation, e.g., [4.99, 4.138]. On the one hand, this operation refers to high ambient and substrate temperatures, e.g., in [4.153]; on the other hand, this operation mode is caused by the increased channel- and thus also contact-temperatures at ohmic and Schottky contacts, due to self-heating during high-power operation. At the same time, only relatively few reports exist for principal improvements of the processing to meet high temperature operation; see, e.g., [4.99, 4.155]. The specific technological requirements are discussed in this section. Reliable GaAs-type channel- and substrate-temperatures amount to ≤200°C. As is also stated in Sect. 7.5, channel temperatures $T_{chan} \geq 300$°C are investigated to meet and extrapolate the reliability FOMs for GaN devices.

WSiN-based ohmic contacts have been suggested in [4.99, 4.100]. WSiN serves as a diffusion barrier for the ohmic contacts to improve the morphology. Different sets of metallizations are fabricated employing Ti/Al/Au metal stacks. The Schottky-contact-stacks are also varied. WSiN/Au and Ir/Au contacts are found to be suitable for high-temperature operation after testing at 400°C for 120 h. The high-temperature stability of various metals and metal systems is further investigated in [4.209]. Re, Pt, Pd, Au, Ni, Ni/Au, Ni/Ga/Ni, Co, and Co/Au contacts are investigated for 200 h at a temperature of 500°C in N_2 atmosphere. The metals are placed on GaN-capped AlGaN/GaN heterostructures and the contacts are not encapsulated. Pt/Au stacks are found to provide very good thermal stability. Re and Ni/Ga/Ni metal stacks are found suitable for thermally stable Schottky contacts at high temperatures $\geq 400°C$. Post- and mid-processing full-wafer annealing is an additional processing step to improve and stabilize device performance, as reported, e.g., in [4.155]. The annealing can be performed at several stages of the processing. Annealing of the ohmic contacts is often performed at very high temperatures of $>800°C$ [4.165]. Thus the material has experienced a high temperature treatment in the early stages of the processing. A full wafer treatment can thus be applied either in the context of the gate processing or after the full process. The post-annealing temperature in [4.155] amounts to 400°C, which is consistent with the high channel temperature of 300°C expected for GaN.

4.9 Backside Processing

Backside or backend processes are required to enable area-efficient grounding of the devices and the use of microstrip transmission-lines. A backend process typically consists of the following process steps:

− Wafer thinning
− Viahole creation by etching or drilling
− Backside metallization and structuring

This sequence is well known and now discussed with respect to III-N substrates and materials.

4.9.1 Thinning Technologies

Typical substrate thicknesses used for epitaxy and frontside processing of III-N devices are thicker (350–650 μm) than the typical substrate thickness required for state-of-the-art devices and passive microstrip transmission-lines (50–150 μm). For a good thermal conductor such as SiC, thinning is not a requirement for thermal reasons, see Chapter 8. However, as the ratio of substrate thickness and transmission-line width defines the impedance for passive MMIC technologies, thinning is a requirement to reach reasonable line widths and overall MMIC sizes comparable to GaAs. For hybrid transistors, there is a benefit in gain, as thinning reduces the source inductance of the devices. This requires the development of a full wafer-scale thinning-technology in order to produce especially MMICs similar to GaAs, e.g., [4.4, 4.256] and references therein. Wafer thinning is typically performed by a combination of the following techniques: grinding, lapping, polishing, wet immersion, and spray etching [4.6].

Sapphire and SiC Substrates

Sapphire and s.i. SiC substrates have been used for both transistor and MMIC processes. Typical substrate thicknesses for through viahole and individual source viaholes amount to 50–100 μm, e.g., [4.125, 4.180, 4.255]. For a small number of processes, thinning of 2- and 3-in. s.i. SiC has been described. A thinning process to a thickness of 50 μm on 3-in. SiC wafers is demonstrated in [4.180]. This process can be used up to Ka-band frequencies, as the via inductance can be minimized to values similar to GaAs [4.240] or InP [4.74], where the minimum substrate thickness is also in the order of 50 μm. The reported inductances based on this thickness amount to 3.25 pH for three parallel round vias with 40 μm diameter. Thinning of full 2-in. wafers to 100 μm for X-band MMICs is reported in [4.255]. The inductance for a single via amounts to 14 pH. Thinning of sapphire substrates to 50 μm has been performed in [4.6]. This thickness is very necessary to reduce the thermal resistance of GaN HFETs on sapphire, as detailed in Chapter 8. The actual thinning process of sapphire is not detailed in [4.6]. However, from a purely mechanical point of view, the lattice mismatch of GaN layers on sapphire is critical, and wafer breakage is a critical issue. Silicon MOS devices on sapphire substrates have been thinned to 30 μm, as reported in [4.87].

Silicon Substrates

Thinning of silicon substrates is a standard technology, which is commonly used, e.g. [4.81]. Thinning of 2 GHz power bar GaN HFETs on silicon substrate to 150 μm (6 mil) thickness is described in [4.81]. A 4-in. full-wafer process is applied. Triquint provided an X-band GaN MMIC process with the thinning of silicon substrate to 125 μm (5 mil) [4.62].

4.9.2 Viahole Etching and Drilling Technologies

Via etching is typically accomplished with highly anisotropic ICP etching. The etching of GaAs substrates yields an etch rate of 2 μm min^{-1} for the definition of source vias [4.248]. A value for the etch rate can be found for the dry etching of InP with 0.6 μm min^{-1} [4.74].

Etching of SiC and Sapphire

Etching of SiC or sapphire is more difficult; however, it is reported in a number of publications [4.31, 4.32, 4.42, 4.66, 4.125]. Because of the inert properties of SiC and the required aggressive etch techniques, the masking of the vias is of critical importance. Aluminum is used as a mask material in [4.42] to etch SiC by SF_6/O_2 inductively coupled plasma (ICP) etching. The highly anisotropic etch is reached by a high bias voltage and a high source power of 500 W. Ninety seven micrometer-thick features and effective etch rates of 320 nm min^{-1} are reached.. The main part of the optimization of the SiC etch process is the understanding of the etch products, and the removal of the chemical byproducts, as performed by laser-induced fluorescence, e.g., [4.125]. The etch mechanism is either dominated by a more physical sputter process or by a more chemical reaction, thus a good trade-off needs to be found. The addition of O_2 or Cl_2 to the plasma determines the pathway of volatilizing additional C and thus increases the etch rate. Given the chemical inert material, very high etch rates of SiC are reported in [4.31] using an SF_6/O_2 plasma etch. The etch rates exceed $1.3 \, \mu m \, min^{-1}$ and allow viaholes into $330 \, \mu m$ thick substrates to be etched. The etch mask used in [4.31, 4.32] is Ni with a selectivity to SiC of about 50:1. The etch conditions are further investigated in [4.32]. Pressure, source power, substrate bias voltage, and the distance between the substrate holder and the source are investigated for an SF_6 helicon plasma for 4H–SiC. A maximum etch rate of $1.35 \, \mu m \, min^{-1}$ is achieved. A SiO_2 mask is used for the SF_6/O_2 etch of 3H–SiC substrate in [4.66]. The etch selectivity of SiC over SiO_2 is found to be as low as 2.6. A SF_6/chlorine/O_2 chemistry is used for the etching of 6H–SiC in [4.125]. Again etch rates are reported as a function of pressure, source power, and accelerating bias. Through-viaholes are reached in $140 \, \mu m$ thick substrates with an effective etch rate of $820 \, nm \, min^{-1}$. Very high coil powers $\geq 500 \, W$ and DC-bias $\geq 300 \, V$ are applied. The impact of Ar and Cl_2 on the etch is also investigated. Surface roughness after etching is found to be comparable to the SF_6 etch.

Sapphire is a very hard and inert material. Thus very few reports exist on the viahole etching of sapphire substrates. However, sapphire can be structured by aggressive etching techniques, e.g., by BCl_3-based ICP etching [4.112]. Cl_2, HCl, and HBr are added. The etch rate is $380 \, nm \, min^{-1}$. The etch selectivity to the photoresist used for the mask is very modest. Thus, sapphire is typically removed for optoelectronic LED applications rather than being thinned or structured, e.g., [4.213].

Etching of Silicon

Viahole etching through Si wafers is a standard VLSI technology, e.g., in [4.272]. However, the typical vias in VLSI technology are much smaller than that used for RF-devices. A source-ground-via structure process is described in [4.96] for a full 4-in. process. Conductive substrates are used to connect the frontside vias with the backside metal layer, thus no real through via is etched. The depth of the via is about $20 \, \mu m$, etched by a Cl_2-plasma-etch. Through-wafer RF-viaholes in silicon substrates with high aspect ratios of 8 and substrates thicknesses of $77 \, \mu m$ are reported in [4.273]. The related inductance amounts to $77 \, pH$ for a $4 \, \mu m$ via diameter. The technique is applied in a similar fashion to typical III-V substrate viaholes. The dry etching is performed by ICP-RIE at an etch rate of $2 \, \mu m \, min^{-1}$. The aspect ratio can be increased to 49:1.

Viahole Drilling Technologies

Apart from viahole dry-etching, laser drilling technologies have been developed [4.146, 4.202]. Such microprocessing is very attractive to structure very hard and inert materials especially for unthinned substrates. They further simplify the process by reducing the number of processing steps.

The process promises consecutive definition of high-aspect-ratio viaholes in SiC and sapphire substrates. A Q-switch Nd:YAG laser is used applying scanning optics in [4.146] at a repetition of 10 kHz. Investigations of the drilling parameters for CO_2 lasers are given in [4.134]. A Q-switched CO_2 laser is used for the drilling procedure. Using a typical repetition rate of 8 Hz, an etch rate between 230 and 870 μm min^{-1} is reported, with pulse energies of 60 mJ per pulse. The issues of the drilling include the removal of ionized debris, depth and geometry control, and a reduction of the residual surface contamination. Another full process is reported in [4.167]. A diode-pumped solid-state laser at 355 nm is used with a spot size of 15 μm. Typical pulse energies are 65 μJ per pulse at a pulse repetition rate of 20 kHz. The actual via diameter is 100 μm. The thickness of the gold plating is 5 μm. Drilling of viaholes into 100 μm thick sapphire substrates is reported in [4.251]. The concentric viaholes are filled with gold and serve as thermal vias to reduce the thermal resistance.

4.9.3 Viahole Metallization

Galvanic and metal processing of GaN RF-devices is a standard technology. Specific optimization is required with respect to the high-power and high-voltage operation. Several complete front- and backside high-voltage III-N processes have been demonstrated [4.114, 4.222]. A 5 μm-thick electroplated galvanic metal is used for the backside metal in a process devoted to power switching applications [4.96]. Devices with very high current levels of 150 A are interconnected. Similar metallization thicknesses have been used for III-N devices, e.g., a galvanic thickness of 4 μm in [4.81] for RF-power bar devices. Such galvanic thickness at the backside is similar to those typically performed in GaAs power processes [4.44, 4.88, 4.286]. Backside metal thicknesses of 4, 7, and 10 μm are reported even for 6-in. diameters and very thin GaAs wafers of 1 mil (25 μm) [4.44]. The associated stress of the wafers is found to be equal for all metal thicknesses, while the absolute wafer bow increases with the thickness for GaAs wafers. The differences in the CTE of galvanic metals and passivation layers is found to be a cause for reliability concerns of GaN HEMTs on 3-in. s.i. SiC substrates [4.270]. Aspects of all processing steps of frontside and backside technology have been discussed in this chapter. Further considerations are given in Chaps. 7 and 8 with respect to dicing [4.88], reliability, packaging, and integration.

4.10 *Problems*

1. Describe why GaN FETs are very sensitive to frequency dispersion!
2. Discuss the issues of the formation of Ohmic contacts to III-N semiconductors!
3. Describe the differences in etching methodologies between GaAs and GaN!
4. Discuss the critical processing steps of device passivation!
5. Summarize the most important trapping mechanisms in GaN FETs!
6. Describe the fundamental problems when processing an AlGaN/GaN npn-HBT!
7. Describe the issues of lithography for III-N devices!
8. Describe the alternatives for field-plate definition!
9. Clarify advantages and drawbacks of recess processes in GaN!

5

Device Characterization and Modeling

In this chapter, device modeling and characterization are presented with respect to nitride-specific issues. DC, RF-small-signal, and noise-characterization and modeling are discussed. Further, since frequency dispersion is a major source of performance- and device-degradation, the characterization and analysis of dispersion are discussed. This includes pulsed-characterization and other advanced techniques. Furthermore, large-signal characterization and modeling are discussed for nitride devices, such as the modeling of contacts, diodes, dispersion, and thermal aspects. Last, but not least, the modeling and characterization of linearity of devices are presented.

5.1 Device Characteristics

Device-terminal quantities, such as currents and voltages, and derived quantities, such as transconductance and conductances, are the most accessible device characteristics used for the evaluation of their performance. In the case of III-N devices, continuous-wave (cw) measurements can also be most misleading for performance evaluation, as frequency dispersion plays a dominant role in the functioning of III-N devices for power and RF-applications. Additional measurements and characterization are thus required. In the following, basic device properties will be evaluated as seen from a number of characterization and modeling perspectives.

5.1.1 Compact FET Analysis

Approximation of the DC-Quantities

Several device models are available from the analysis of silicon and GaAs FETs, which will be used in this work for the evaluation of GaN FETs. First of all, AlGaN/GaN HEMTs are dominated by their high sheet-carrier densities in the channel of $10^{13}\,\mathrm{cm}^{-2}$ and above. A calculation example of threshold

Table 5.1. Typical parameters for the calculation of the threshold voltage for an $Al_{0.3}Ga_{0.7}N/GaN$ HFET with $l_g = 300\,nm$

N_D (cm^{-3})	d_{doping} (nm)	ϕ_b (eV)	ΔE_C (eV)	E_{f1} (eV)	$\frac{q\sigma}{\epsilon}d_{gc}$ (V)	V_{p2} (V)	V_{thr} (V)
1e19	10	0.8	0.68 $(x = 0.3)$	0.026	4.98	1.84	−6.5

ε_r (−)	v_{eff} (cm s^{-1})	d_{eff} (nm)	C_0 (pF mm^{-1})	E_{crit} (kV cm^{-1})	n_c (cm^{-2})	n_{sheet} (cm^{-2})	$g_{m,max}$ (mS mm^{-1})
9.8	2.7e7	27	1	150	3e6	1e13	235

ΔV (V)	t (nm)	LD (nm)	I_{Dmax} (mA mm^{-1})
10	2.7	27	1,734

voltage V_{thr} in a classical model of a short-channel pulse-doped FET can be performed according to [5.103, 5.149, 5.151] including surface charge correction

$$V_{thr} = \frac{1}{q}\left(\phi_B - \Delta E_C - V_{p2} - \frac{q\sigma}{\epsilon}d_{gc} + E_{f1}\right). \tag{5.1}$$

ϕ_B denotes the intrinsic Schottky barrier height, ΔE_C is the conduction band discontinuity, σ the surface charge density, and E_{f1} denotes a small correction energy. The doping correction V_{p2} is calculated according to

$$V_{p2} = \frac{q \cdot N_D \cdot d_{doping}^2}{2 \cdot \epsilon}. \tag{5.2}$$

ϵ is the dielectric constant of the barrier layer, N_D originally is the donor concentration in the AlGaN layer [5.169], and d_{doping} originally is the channel thickness width. The model is based on a MESFET device with a homogenous doping of concentration N_D and a channel thickness d_{doping}. The model needs modification with respect to the polarization charge, as a significant amount of carriers is not generated by impurity doping. Further, the carriers in III-N HFETs are concentrated in a very small channel depth and not in the barrier. A typical example is given here. The idea of this calculation is based on the fact that the complete channel charge density n_c arises from barrier doping and surface charge at a gate-to-channel separation d_{gc} of 27 nm. The calculated threshold voltage is $V_{thr} = -6.5\,V$ with the influence of the polarization doping. It is visible in Table 5.1 that an enhancement-mode AlGaN/GaN HFET is difficult to realize. However, examples of the realization are given in [5.31, 5.94]. For the maximum transconductance $g_{m,max}$, a similar calculation example can be performed:

$$g_{m,max} = \frac{\partial I_{DS}}{\partial V_{GS}}\Big|_{V_{DS}=const.} \approx \frac{\epsilon \cdot v_{eff} \cdot W_g}{d_{eff}} \cdot \frac{1}{\sqrt{1 + \left(\frac{n_c}{n_{sheet}}\right)^2}}. \tag{5.3}$$

n_c is defined as

$$n_c = \frac{E_{crit} \cdot l_g \cdot C_0}{q}. \tag{5.4}$$

The critical field E_{crit} is the field for the onset of saturation and C_0 is the static capacitance. An approximation for the static capacitance C_0 is given below. Deduced from (5.3), a maximum transconductance of $235\,\mathrm{mS\,mm^{-1}}$ is found, which is in good agreement with typical AlGaN/GaN HFETs. Fig. 5.1 gives the measured transconductance g_m as a function of V_{GS} for a real device. A threshold voltage $V_{thr} = -4.5\,\mathrm{V}$, which increases for increasing V_{DS}, is observed by the linear extrapolation method. The maximum DC-transconductance amounts to $300\,\mathrm{mS\,mm^{-1}}$ and is relatively independent of the V_{DS} voltage. The maximum theoretical drain current I_{Dmax} can be calculated in a space-charge-zone limited model according to [5.88]

$$I_{Dmax} = \frac{2 \cdot \varepsilon_r \cdot \epsilon_0 \cdot v_{eff} \cdot \Delta V}{L_D^2} \cdot W_g \cdot t. \tag{5.5}$$

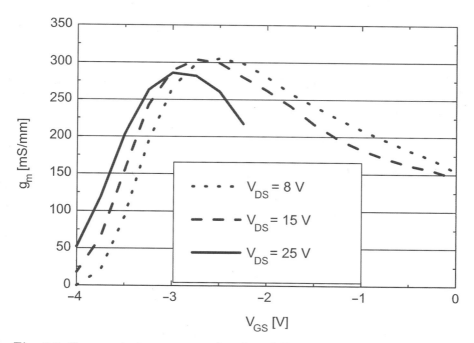

Fig. 5.1. Transconductance g_m as a function of V_{GS} with V_{DS} as a parameter for an AlGaN/GaN HEMT with $W_g = 2 \times 60\,\mu\mathrm{m}$ and $l_g = 150\,\mathrm{nm}$

Apart from the material constants, v_{eff} denotes the effective saturation velocity of the majority carriers, and ΔV the voltage drop over the depletion zone in the drain region. W_{g} denotes the gate width and t the effective thickness of the layer that supports the current. L_{D} is the effective depletion zone length. Deduced from this, a maximum drain current of $1{,}734\,\text{mA}\,\text{mm}^{-1}$ can be calculated, if the depletion zone L_{D} is assumed to be $27\,\text{nm}$ for a voltage drop $\Delta V = 10\,\text{V}$. The layer thickness t is assumed to be $2.7\,\text{nm}$ in agreement with Schrödinger–Poisson calculations, e.g., [5.158]. Fig. 5.2 gives the output characteristics of a GaN/AlGaN HEMT on s.i. SiC substrate measured in cw-mode up to $V_{\text{DS}} = 35\,\text{V}$ in comparison to a GaAs HEMT of the same gate width $W_{\text{g}} = 1\,\text{mm}$. The gate length is $l_{\text{g}} = 300\,\text{nm}$ in both cases similar to the example in Table 5.1. The GaN/AlGaN HEMT yields higher current, higher breakdown voltage, lower transconductance, reduced thermal effects for the same power density, and higher on-resistance and knee-voltage than the optimized GaAs device. The on-resistance R_{on} in a GaN HEMT can be rewritten as

$$R_{\text{on}} = R_{\text{S}} + R_{\text{D}} + R_{\text{chan}}, \tag{5.6}$$

where R_{S} and R_{D} denote source and drain resistances and R_{chan} the channel resistance. For III-N HEMTs, the DC-approximation of (5.6) may be critical, as the measured contribution of $R_{\text{S}} + R_{\text{D}}$ alone may leave no contribu-

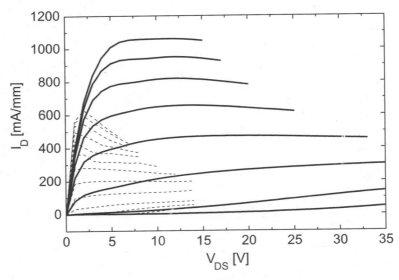

Fig. 5.2. CW-output characteristics of an AlGaN/GaN HEMT with V_{GS} ($2\,\text{V}$, step $-1\,\text{V}$) as a parameter compared to GaAs PHEMT output characteristics (step $0.2\,\text{V}$), gate width $W_{\text{g}} = 1\,\text{mm}$ and gate length $l_{\text{g}} = 300\,\text{nm}$

tion to R_{chan}. The DC-output conductance $g_{\text{ds,ext}}$, also depicted in Fig. 5.2, is defined as

$$g_{\text{ds,ext}} = \frac{\partial I_{\text{DS}}}{\partial V_{\text{DS}}}\Big|_{V_{\text{GS}}=\text{const.}} \tag{5.7}$$

A typical approximation of the open-channel contribution to g_{ds} is

$$g_{\text{ds}} = \frac{q \cdot \mu \cdot n_{\text{sheet}} \cdot W_{\text{g}}}{l_{\text{g}}}. \tag{5.8}$$

Deviations from the cw-values, calculated from (5.6) and (5.7), occur both during pulsed- and RF-operation. The impact of trapping is of significant importance, so that

$$g_{\text{ds,ext}}(\text{CW}) \neq g_{\text{ds,ext}}(\text{RF}) \neq g_{\text{ds,ext}}(\text{Pulsed}). \tag{5.9}$$

Equation (5.9) holds in similar fashion for the transconductance. Further below, the importance of additional derivatives will be considered.

Modeling the Resistances in an HFET

The source-resistances in HFETs can be modeled according to

$$R_{\text{S}} = R_{\text{S,semi}} + R_{\text{Con}} + R_{\text{S,met}} + R_{\text{band}}. \tag{5.10}$$

A similar model holds for the drain resistance R_{D}:

$$R_{\text{D}} = R_{\text{D,semi}} + R_{\text{Con}} + R_{\text{D,met}} + R_{\text{band}}. \tag{5.11}$$

The contributions of $R_{\text{D,semi}}$ and $R_{\text{D,met}}$ denote the ohmic contributions of the metal and the semiconductor. R_{Con} denotes the contact resistance of the semiconductor/metal transition. The principal difference between GaAs- and GaN-devices in this modeling approach are the band contribution R_{band} at the source, due to the increased bandgap discontinuity and the resulting additional resistive contributions. This is investigated particularly in [5.184]. The influence of the dynamic access-resistances in AlGaN/GaN FETs with regard to the linearity of the transconductance g_{m} and the cut-off frequency f_{T} has been discussed further in [5.131]. The increase of R_{S} is considered to occur due to the quasi-saturation of the electron velocity in the source region of the channel. The impact of the nonlinear source resistance is confirmed at low temperatures down to $120\,\text{K}$ [5.126]. Similar investigations with respect to the nonlinear source resistance can be found in [5.131]. The nonlinear source resistance is due to space-charge limited current conditions. A critical current density of about $40\,\text{MA}\,\text{cm}^{-2}$ is isolated. For current densities beyond that value, the resistance R_{S} increases critically.

Approximation of the Small-Signal RF-Quantities

Several definitions and approximations hold for the calculation of the cut-off frequencies and related quantities. f_T is defined and approximated as

$$f_T = f(h_{21}(f) = 0) \approx \frac{g_{mi}}{C_g}. \tag{5.12}$$

C_g denotes the total gate capacitance. Physical approximations of the intrinsic RF-transconductance are similar to those of the DC-transconductance. Dispersion effects are not considered in these first order approximations. The channel capacitance C_g for the short-channel MODFET can be approximated by several bias-dependent models, for example, as given in [5.151, 5.169]. A static approximation of the intrinsic gate capacitance yields [5.151]

$$C_0 = q \cdot K \cdot W_g \cdot l_g = \frac{\epsilon \cdot W_g \cdot l_g}{d_{gc}}. \tag{5.13}$$

Based on the values in Table 5.1, a value $C_0 = 0.964 \, \text{pF} \, \text{mm}^{-1}$ is found. In (5.13), d_{gc} denotes the separation of the physical gate to the channel layer. For GaN HFETs, this is a very well defined quantity as the vertical extension of the electron gas amounts to a few nanometer only. This value then needs modification with respect to the bias dependence.

Approximation of Large-Signal RF-Quantities

Many small-signal models exist for the approximation of the small-signal behavior of FETs. However, for large-signal operation, only a few basic approximation concepts exist. The output power P_{sat} of an RF-device can be approximated as

$$P_{sat} \approx \frac{\Delta I_D \cdot \Delta V_D}{8} = \frac{(I_{D,max,RF} - I_{D,min,RF}) \cdot (V_{D,max,RF} - V_{D,min,RF})}{8}, \tag{5.14}$$

where ΔI_D is the drain current during RF-swing and ΔV_D is the drain voltage RF-swing. This approximation does not consider nonlinear operation with the generation of harmonics. In first order the swing values are taken from pure DC-considerations. More elaborate investigations can be found in [5.116]. The dispersive effect of a knee walk-out is quantified by RF-swing and pulsed-DC-measurements. Deduced from the maximum drain current from $I_{D,max}$ in (5.5), the DC-knee voltage $V_{knee} = V_{D,min}$ is defined by the crossing of $I_{D,max}$ and R_{on} for $R_{chan} = 0$. There are no simple approximations for the definition of $V_{D,max}$ and $I_{D,min}$, once the devices are dispersive, i.e., once:

$$I_{D,max,DC} \neq I_{D,max,RF}, \tag{5.15}$$

$$V_{D,min,DC} \neq V_{D,min,RF}, \tag{5.16}$$

$$I_{D,min,DC} \neq I_{D,min,RF}, \tag{5.17}$$

$$V_{D,max,DC} \neq V_{D,max,RF}. \tag{5.18}$$

The first two relations (5.15) and (5.16) are typically considered as the knee walkout. A similar effect is found for high V_{DS} and low I_{DS} ((5.17)–(5.18)), typically considered poor RF-pinch-off. Fig. 5.3 gives an estimate of the possible saturated output power as a function of operation bias V_{DS}, assuming $I_{D,min} = 0$ and $V_{D,max} = 2 \times V_{D,op}$. It is clear from (5.14) that an increase of the knee voltage and a current reduction reduce the saturated output power. Fig. 5.3 gives the calculated values for operating voltages up to $V_{DS} = 150\,\mathrm{V}$ for various knee-voltage and drain current combinations. The output power densities are given without additional thermal and reliability considerations. The power-added efficiency (PAE) is defined as

$$\mathrm{PAE} = \frac{P_{out} - P_{in}}{P_{DC}} = \left(1 - \frac{1}{G_p}\right) \cdot \frac{P_{out}}{P_{DC}}. \tag{5.19}$$

The drain efficiency η_d is defined as

$$\eta_d = \frac{P_{out}}{P_{DC}}. \tag{5.20}$$

Approximations for the efficiency η_d can be derived from (5.14)

$$\eta_d = \frac{P_{out}}{P_{DC}} = \frac{\left(I_{D,max,RF} - I_{D,min,RF}\right) \cdot \left(V_{D,max,RF} - V_{D,min,RF}\right)}{8 \cdot I_{DC} \cdot V_{DS}}. \tag{5.21}$$

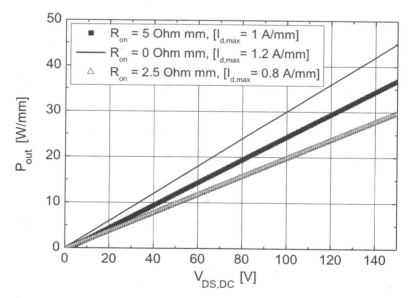

Fig. 5.3. RF output power estimate for different on-resistances and maximum drain current levels

The approximation of the efficiency depends strongly on the class of operation of the amplifier. However, if for class-A operation the relations $I_{DC} = (I_{D,max,RF} - I_{D,min,RF})/2$ and $V_{DS} = (V_{D,max,RF} - V_{D,min,RF})/2$ hold, (5.21) can be rewritten in a theoretical maximum:

$$\eta_d = \frac{1}{2}. \tag{5.22}$$

The impact of the minimum currents and of the knee voltage can be viewed if we assume $I_{D,max,RF} = 2I_{DC}$ and $V_{D,max,RF} = 2V_{DS}$:

$$\eta_d = \frac{1}{2} \cdot \left(1 - \frac{V_{D,min,RF}}{V_{DS}}\right) \cdot \left(1 - \frac{I_{D,min,RF}}{I_{DC}}\right). \tag{5.23}$$

The relative impact of the minimum voltages and currents on η_d becomes visible in (5.23). For III-N HEMTs, the high operation bias becomes favorable to mitigate the relative impact of the knee voltage in this equation. For other operation classes, the impact of the nonideal characteristics is even stronger, as the effect of nonideal pinch-off at high V_{DS} becomes more important, e.g., in class-B or class-F operation.

Characterization of Device Linearity

For the characterization of the linear operation of devices, no approximation for the third-order intermodulation product IM_3 can be given based on purely physical parameters. However, in a very simple model [5.136], some basic understanding can be obtained. The intrinsic drain–source current i_{ds} in a (MES)FET device can be rewritten as a function of the intrinsic bias

$$i_{ds} = \left[v_{gs} + \frac{g'_m}{2}v_{gs}^2 + \frac{g''_m}{6}v_{gs}^3\right] + \left[v_{ds} + \frac{g'_{ds}}{2}v_{ds}^2 + \frac{g''_{ds}}{6}v_{ds}^3\right]. \tag{5.24}$$

The derivative g'_m denotes $\partial g_m / \partial v_{gs}$ and g'_{ds} denotes $\partial g_{ds} / \partial v_{ds}$. Some very simple calculations show that the optimum load-resistance to cancel the second-order distortion amounts to

$$\frac{1}{R} = g_m \cdot \sqrt{\frac{-g'_{ds}}{g'_m}} - g_{ds}. \tag{5.25}$$

The third-order distortion terms cancel out for

$$\frac{1}{R} = g_m \cdot \sqrt[3]{\frac{-g''_{ds}}{g''_m}} - g_{ds}. \tag{5.26}$$

This simple calculation shows the importance of the derivatives of both conductances and of understanding them with respect to the matching of the device. As is seen in (5.7) and (5.27), frequency dispersive effects can be formulated in a very general fashion:

$$\frac{\partial^n I_{ij}}{\partial V_{kl}^n}(\text{DC}) \neq \frac{\partial^n I_{ij}}{\partial V_{kl}^n}(\text{RF}) \neq \frac{\partial^n I_{ij}}{\partial V_{kl}^n}(\text{Pulsed}).\qquad(5.27)$$

The indices i, j, k, l denote the various combinations of the ports of the three-terminal devices. The impact of dispersion on these higher order derivatives has not been exploited in detail so far. However, measurement approaches for the optimum determination of the RF-output loads up to 40 GHz with respect to linearity are available, e.g., for GaAs PHEMTs, GaN HEMTs, and Silicon LDMOS in [5.117, 5.119, 5.203].

Temperature Dependence of cw-DC-Characteristics

Temperature effects modify the device terminal characteristics. Fig. 5.4 gives the substrate–temperature-dependence of the output characteristics of an AlGaN/GaN HEMT on s.i. SiC substrate. The following thermal effects can be observed with rising substrate temperature T_{sub} and channel temperature T_{chan}:

- An increase of the on-resistance R_{on}
- A decrease of the output conductance g_{ds} for most of the ($V_{\text{DS}}, V_{\text{GS}}$) combinations, especially for high V_{GS}
- A reduction of the maximum current I_{Dmax}

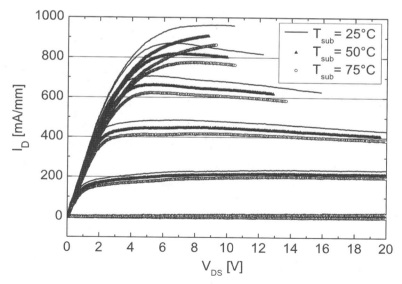

Fig. 5.4. CW-output characteristics with the substrate temperature T_{sub} as a parameter, $V_{\text{GS}} = 1\,\text{V}$, step $-1\,\text{V}$

Apart from the reduction of the maximum currents I_{Dmax}, the currents I_{D} are reduced also for lower V_{GS} branches, as shown in Fig. 5.4. This reduction in current for lower V_{GS} is not always the case, and is mostly dependent on the impact of trapping effects. The thermal impact on the maximum drain current I_{Dmax} in Fig. 5.4 can be approximated by a linear temperature dependence:

$$I_{\mathrm{Dmax}}(T) = I_{\mathrm{Dmax}}(T = 300\,\mathrm{K}) \cdot (1 - \alpha \cdot T). \tag{5.28}$$

Similar coefficients are found for the transconductance g_{m}. The coefficients α are typically of the order of $10^{-3}\,\mathrm{K}^{-1}$. Further details are given in Chapter 8.

HFET Input Characteristics

Fig. 5.5 gives the input characteristics, i.e., $I_{\mathrm{G}}(V_{\mathrm{GS}})$ of an AlGaN/GaN HEMT on s.i. SiC with a gate width $W_{\mathrm{g}} = 0.48\,\mathrm{mm}$ for various V_{DS} voltages. The input characteristics are determined by the diode-like behavior. Any occurrence of channel impact-ionization is not visible in Fig. 5.5; however, impact ionization has been reported for GaN HFETs [5.24], especially at cryogenic

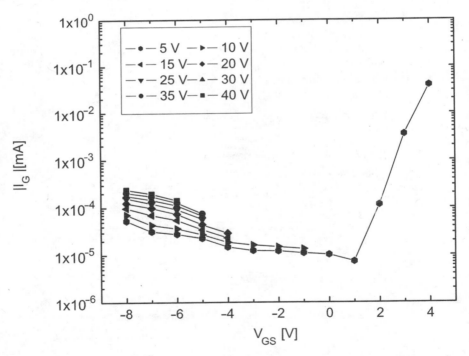

Fig. 5.5. Input characteristics of an AlGaN/GaN HEMT with $W_{\mathrm{g}} = 0.48\,\mathrm{mm}$ for various V_{DS} voltages

temperatures [5.175]. Additional ohmic contributions are visible in Fig. 5.5. The forward diode behavior is normally modeled according to

$$I_G(V_{GS}) = I_0 \cdot \left(\exp\left(\frac{n \cdot V_{GS}}{k_B T_L} \right) - 1 \right). \qquad (5.29)$$

n defines the ideality factor, which expresses the deviation from ideal exponential diode-type characteristics. The ideal diode case is $n = 1$. Because of the lack of maturity of the III-N FET diodes, very high ideality factors >2 have been reported initially [5.198], which were then optimized to levels of ≤1.5 [5.80].

FET Breakdown Voltage

Although III-N devices are considered for very high voltage operation, breakdown voltages and their mechanisms are still under discussion, e.g., [5.205]. Very high bias of operation are reported; however, the DC-two- and three-terminal breakdown voltages do not match the RF-operation. Very careful optimization of the DC-breakdown voltages for RF-GaN FETs beyond 300 V is reported in [5.127]. The procedure is completed by testing the breakdown voltages also at elevated temperatures [5.83].

Very high operation and breakdown voltages ≥900 V for power devices have been reported [5.84, 5.207]. This breakdown voltage is based on the drain current evaluation in dynamic three-terminal curve tracer measurements. The general mismatch of DC- and RF-breakdown is partly due to the high amount of traps in the epitaxial layers. The impact of the drain-to-source spacing of AlGaN/GaN transistors' breakdown characteristics in a trade-off with frequency response is discussed, e.g., in [5.193]. The current gain cut-off frequency increases with the reduction of the source-to-drain spacing, while the BV_{DS} breakdown voltage decreases. Different methods to measure breakdown voltages are available. Gate-to-source (BV_{GS}) and gate-to-drain (BV_{GD}) breakdown voltages are evaluated as

$$BV_{GD\ 2T} = V_{GD}|I_G = (1\,\text{mA mm}^{-1}), \qquad (5.30)$$

$$BV_{GS\ 2T} = V_{GS}|I_G = (1\,\text{mA mm}^{-1}). \qquad (5.31)$$

Typically, the two-terminal breakdown criterion for the gate current in GaN/AlGaN FETs is $I_G = 1\,\text{mA mm}^{-1}$, $0.1\,\text{mA mm}^{-1}$, or even lower. Fig. 5.6 gives the two-terminal breakdown voltages BV_{GS} and BV_{GD} for two types of AlGaN/GaN HFETs as a function of the gate-to-drain spacing l_{gd}. The difference in the two types of HEMTs is due to the barrier doping concentration and the gate length l_g. As a reference, the dependence of the gate–source diode breakdown on the gate–drain distance is also monitored. We see the

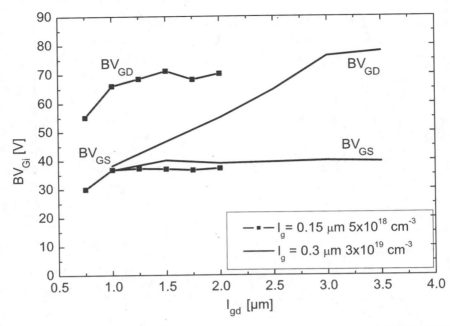

Fig. 5.6. Breakdown voltages BV_{GS} and BV_{GD} for two types of AlGaN/GaN HFETs as a function of gate–drain spacing l_{gd}

impact of both the gate length and of the barrier doping. A three-terminal version of the gate-to-drain breakdown voltage reads:

$$BV_{GD\ 3T} = V_{GD}|I_G = (1\,\mathrm{mA\,mm^{-1}})\ \&\ V_{GS}@\text{Pinch-off}. \qquad (5.32)$$

The three-terminal breakdown voltage BV_{DS} can be derived from

$$BV_{DS\ 3T} = V_{DS}\ \text{for}\ I_{DS} \le\ (1\,\mathrm{mA\,mm^{-1}})\ \&\ V_{GS}@\text{Pinch- off}. \qquad (5.33)$$

Further characterization methods involve dynamic curve tracer analysis [5.207], i.e., a dynamic or pulsed-type of analysis, typically of BV_{DS}, see (5.33). Beyond BV_{DS}, a sharp increase of the drain current I_D is used as a criterion [5.207]. For RF-large-signal operation, $BV_{DS\ RF}$ can be extracted from the direct measurement of the RF-output swing, as performed, e.g., in [5.116]. The dynamic limitation of the swing can be used directly to evaluate the maximum swing possible for the particular operation bias, frequency, and RF-loading of fundamental and harmonic frequencies [5.117]. Assuming the channel being the most sensitive part of the device, and assuming the breakdown field of GaN to be $3\,\mathrm{MV\,cm^{-1}}$, we obtain a maximum breakdown voltage of $300\,\mathrm{V}$ per $\mu\mathrm{m}$ contact separation. This upper limit assumes a homogenous field distribution between the contacts. This maximum value is naturally reduced by the highly inhomogeneous field distribution, traps, and the impact of interfaces and doping along the channel. In reality, values up to $80\,\mathrm{V\,\mu m^{-1}}$ have been reported [5.112]. In this context the stability of the breakdown measurement in a dynamic sense is of critical importance, as the device is biased in a critical situation. Thus the timing of the breakdown measurements needs to be chosen carefully to ensure repeatable measurement results [5.102].

5.1.2 Compact Bipolar Analysis

Calculation of the DC-Quantities

Principal DC-considerations are available for III-N bipolar devices. They resemble the well known relations of GaAs and InP HBTs, e.g., [5.151, 5.169]. The typical obstacles observed for the realization of low base-resistance GaN/AlGaN HBTs are compiled in [5.208]. The high extrinsic base resistance is a fundamental problem. The base resistance R_{bb} can be approximated by [5.150]

$$R_{\mathrm{bb}} = R_{\mathrm{Con}} + R_{\mathrm{spread}} + R_{\mathrm{gap}}, \tag{5.34}$$
$$R_{\mathrm{Con}} = \sqrt{\rho_{\mathrm{s}} \cdot \rho_{\mathrm{c}}}/2 \cdot L_{\mathrm{e}}, \tag{5.35}$$
$$R_{\mathrm{spread}} = \rho_{\mathrm{s}} W_{\mathrm{e}}/12 \cdot L_{\mathrm{e}}, \tag{5.36}$$
$$R_{\mathrm{gap}} = \rho_{\mathrm{s}} W_{\mathrm{eb}}/2 \cdot L_{\mathrm{e}}. \tag{5.37}$$

R_{Con} is the contact resistance, R_{spread} is the spreading resistance under the base, and R_{gap} is the base–emitter gap resistance. As seen in (5.35)–(5.37) all three components are high due to the high base resistivity ρ_{s} of p-doped GaN. ρ_{c} is the specific metal resistivity. For the current amplification β, the following calculations are often used. The common-emitter current gain

$$\beta = \frac{I_{\mathrm{C}}}{I_{\mathrm{B}}} \tag{5.38}$$

relates to the common base gain α as

$$\alpha = \frac{1}{\frac{1}{\beta} - 1}. \tag{5.39}$$

For the common base gain α, the following approximation is used for AlGaN/GaN HBTs [5.114]. α is rewritten to be composed as

$$\alpha = \gamma_{\mathrm{i}} \cdot \delta \cdot \alpha_{\mathrm{T}}. \tag{5.40}$$

Equation (5.40) includes the three factors: emitter injection efficiency γ_{i}, the recombination factor δ in the emitter base junction, and the base transport factor α_{T}. The recombination in Mg-doped GaN is high and the transport factor α_{T} low. A similar calculation of the respective quantities is given in [5.114], derived from [5.107]. The high base resistance in (5.34) leads to a strong reduction of the extrinsic current gain due to the relation

$$\beta_{\text{extr}} = \frac{I_{\text{C}}}{I_{\text{B,p}}} = \frac{D_{\text{nB}} X_{\text{E}} N_{\text{E}}}{D_{\text{pE}} X_{\text{B}} N_{\text{B}}} \cdot \exp\left(\frac{\Delta E_{\text{V}}}{k_{\text{B}} T}\right). \tag{5.41}$$

X_{E} and X_{B} denote the thicknesses of the base and emitter layer, N_{E} and N_{B} the carrier concentrations, and D_{nB} the diffusivities of electrons in base and D_{pE} of holes in the emitter. The impact of both the high Mg-concentration leading to reduced carrier lifetime and low diffusivity of electrons in a highly Mg-doped base is visible from (5.41). The open-collector breakdown voltage of a bipolar transistor can be approximated [5.177], as stated for AlGaN/GaN npn-HBTs in [5.97]:

$$\text{BV}_{\text{CEO}} = \frac{\varepsilon_{\text{r}} E_{\text{crit}}}{2 q N_{\text{DC}}} \cdot \left(1 - \alpha\right)^{1/mb}. \tag{5.42}$$

E_{crit} is the breakdown field of the GaN, N_{DC} is the donor concentration in the collector, and mb is a coefficient for the doping profiling in the collector. Dynamically measured collector–emitter breakdown voltages BV_{CEO} using curve-tracer measurements of 330 V in AlGaN/GaN HBTs have been reported [5.206].

Calculation of the Bipolar RF-Quantities

Only very few reports exist for the RF-analysis of III-N HBTs [5.36, 5.115]. For the current gain cut-off frequency f_{T} of HBTs a relation is used [5.96] similar to FETs:

$$f_{\text{T}} = \frac{1}{2\pi(\tau_{\text{e}} + \tau_{\text{b}} + \tau_{\text{c}})} = \frac{g_{\text{m}}}{2\pi \cdot (C_{\text{je}} + C_{\text{jc}})}. \tag{5.43}$$

For the maximum frequency of oscillation f_{max}, the following pproximation is given for bipolar devices based on f_{T}:

$$f_{\text{max}} = \sqrt{\frac{f_{\text{T}}}{8\pi \cdot R_{\text{bb}} \cdot C_{\text{jc}}}}. \tag{5.44}$$

This well-known relation again stresses the need to reduce the base sheet resistance in AlGaN/GaN HBTs. At the same time, the base resistance R_{bb} in III-N npn-transistors is comparably high. A distributed small-signal modeling of the base resistance is thus proposed in [5.115]. The model uses a distributed collector–base element-circuit, which is repeated for 25 times in order to model both capacitances and RF-amplification. Further, a transit-time-analysis in InGaN/GaN DHBTs and AlGaN/GaN HBTs is given in [5.36]. It is found that the emitter-transit-time τ_{e} dominates for collector current densities of $\leq 100\,\text{A}\,\text{cm}^{-2}$, while τ_{b} dominates beyond that density.

5.2 Frequency Dispersion

The continuous increase of the probing frequency from near DC to the GHz range reveals the most important effects for III-N semiconductor devices. The spectral effects observed between 1 Hz and a few gigahertz are referred to as frequency dispersion.

5.2.1 Dispersion Effects and Characterization

A large number of publications has stressed the impact of frequency dispersion on device performance [5.87, 5.120, 5.195], as is also stated in Chapter 4. The principal terminal effects of frequency dispersion have been discussed in Sect. 2.3.3 and Chapter 4.

Dispersive Effects

The characterization of frequency dispersion on the device-terminal level includes various advanced techniques, such as the following:

- Transient current measurements [5.87]
- cw-current measurements under different optical illumination conditions [5.122]
- Low-frequency excess noise measurements [5.157]
- Pulsed-IV measurements [5.120]
- Pulsed-RF S-parameter measurements [5.87]
- cw-power measurements [5.120]
- Pulsed-power and loadpull measurements [5.22]
- Power measurements under varying terminal load conditions [5.88]

The results of such measurements have been repeatedly reported.

- Strong transients in the current characteristics of GaN-based heterostructure FETs in the time-domain, as are given, e.g., in [5.87].
- The suppression of the DC-current collapse by light illumination with light energy smaller than bandgap energy is described in [5.122]. It shows the principal influence of defects and traps on the cw-device output characteristics.
- The influence of both the ungated device region and the surface states on $1/f$-noise is described in [5.194]. The main contribution of the excess noise is derived from the ungated region of the FET.
- In [5.46], a concise comparison of the drain excess noise in AlGaN/GaN HEMTs grown on silicon, SiC, and sapphire is given. The normalized low-frequency noise levels are strongly correlated with the leakage currents. The noise is found to stem from a source in the vicinity of the gate.
- Pulsed-DC-measurements reveal the effects of surface traps on the breakdown voltage and switching speed of GaN power switching HEMTs, as reported in [5.211]. The carrier trapping and detrapping is correlated with

the carrier injection time into the device. Trapping with deep energy levels is considered to be slower than trapping in shallow traps.

– Pulsed-IV and pulsed-RF S-parameter measurements are used for the extraction of large-signal models that assume substrate-related trapping effects [5.92, 5.116]. The V_{DS}-dependence of the drain-current dynamics is used to extract a more accurate prediction for the RF-power performance.

– Power measurements under load variation have been used to investigate the current collapse in [5.88]. The knee voltage is found to increase with the drain bias. RF-stress has an impact on dispersion and power characteristics of AlGaN/GaN HEMTs, as discussed in [5.67]. Surface engineering and passivation are of fundamental importance for RF-longterm reliability [5.62].

The effects for GaN FETs will now be enumerated and followed by more detailed analysis. They include the following:

1. Drain current collapse [5.122]
2. Drain and gate current dispersion, i.e., the difference of cw-, pulsed-, and extracted RF-current device characteristics
3. Time dependence of the device currents on the low-frequency (DC-MHz) scale
4. V_{DS}-dependent shift of the threshold voltage
5. RF-power-slump
6. Premature power/gain compression

These effects are now discussed in detail:

1. Current collapse (1) is the most common feature directly observed in the DC-measurement. It describes the change/collapse of the drain current I_D as a function of V_{DS}, which is not caused by thermal effects. From a different perspective, it describes the change of output characteristics when a different bias is applied before the output characteristic is measured.
2. Current dispersion (2) describes the difference of the DC-characteristics measured in cw-mode as compared to pulsed-measurements for different pulse widths and quiescent bias independent of the thermal effects. Fig. 5.7 gives the differences in the output characteristics for a pulse length of 1 μs, measured for different quiescent bias.
3. The time dependence (3) or transients in the time-domain of the DC-current I_D on the low-frequency time scale describes a device behavior similar to the current dispersion in (2).
4. A shift of the threshold voltage V_{thr} (4) is part of the general effects visible on the output characteristics in (1).
5. The effect of (5) denotes the lack of the RF-currents and voltages to reach areas in the (I_{DS}, V_{DS})-current–voltage plane, which are suggested to be the DC-current characteristics. This includes particularly the knee-voltage walk-out.

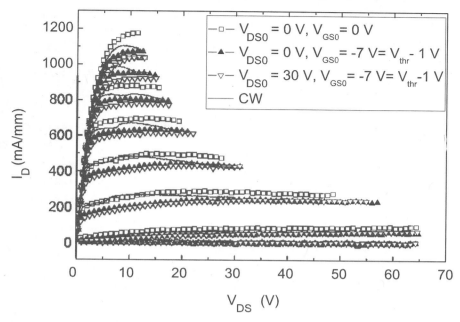

Fig. 5.7. CW- and pulsed-output characteristics that reveal the dispersion effects, $V_{GS} = -6$ to $1\,\text{V}$, step $1\,\text{V}$

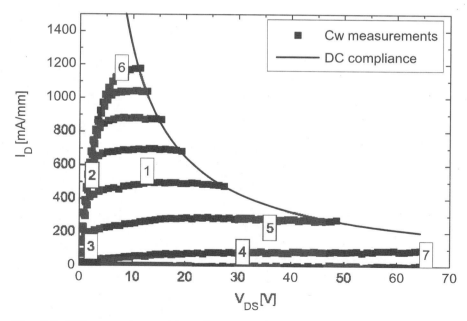

Fig. 5.8. Different quiescent biases for pulsed-measurements in the time-domain for the investigation of GaN HFETs

6. As a result of this compressive behavior, the P_{out} vs. P_{in} power characteristic yield a premature compression of gain or P_{out}, which is denoted as RF-power slump. The effects on the dynamic behavior are visualized below in Fig. 5.9.

Dispersion can be characterized by both time- and frequency-domain methods.

5.2.2 Dispersion Characterization and Analysis

Time-Domain Methods

Time-domain methods are used to separate dispersive effects. The following sequence of pulsed-DC-measurements is suggested, e.g., in [5.120, 5.182]. Fig. 5.8 clarifies two different characterization methods. The measurements include the following:

1. cw-measurements [5.101]
2. Pulsed-measurements at quiescent zero bias ($V_{DS0} = V_{GS0} = 0\,\text{V}$)
3. Pulsed-measurements at cold pinch-off V_{DS0} bias (e.g., $V_{DS0} = 0\,\text{V}$, $V_{GS0} = V_{thr}$)
4. Pulsed-measurements from hot pinch-off at operation bias V_{DS0} (e.g., $V_{DS0}=30\,\text{V}$, $V_{GS0} = V_{thr}-1\,\text{V}$). The timing of the measurements is of vital importance, as given in [5.182]. Additional measurements are useful for determining further effects:

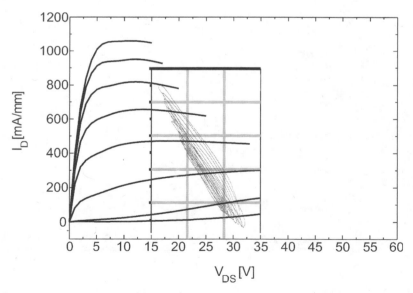

Fig. 5.9. RF-current-swing characteristics to reveal load and RF-dispersive effects

5. Pulsed-measurements from the full operation bias, e.g., at class-A bias at medium current bias ($V_{\mathrm{DS0}} = 40\,\mathrm{V}$, V_{GS0} for $I_{\mathrm{Dmax}}/2$)
6. Measurements in the full channel situation: on-state bias with V_{DS0} slightly above the knee voltage V_{knee} ($V_{\mathrm{DS}} = 10\,\mathrm{V}$, $V_{\mathrm{GS}} = 1\,\mathrm{V}$)
7. Pulsed-measurements near the breakdown voltage to reveal additional trapping effects

The procedure and its extensions allow the operation of the device at constant thermal conditions if the full-channel and class-A bias-situation are properly chosen, see Fig. 5.8. With the substrate temperature T_{sub} kept constant, self-heating (SH) due to constant dissipated DC-power on the dissipated power hyperbola is controlled. A low-duty-cycle (1% or lower) for the pulsed-measurements allows the minimization of the thermal influence of the pulses. Deduced from these measurements, the following figures-of-merit and dispersion parameters can be deduced for an operation bias $V_{\mathrm{DS}} = 40\,\mathrm{V}$:

$$a_{\mathrm{gate-lag}} = \frac{I_{\mathrm{DS,pulse},\,V_{\mathrm{GS0}}=V_{\mathrm{thr}}-2\,\mathrm{V},\,V_{\mathrm{DS0}}=0\,\mathrm{V}}}{I_{\mathrm{DS,pulse},\,V_{\mathrm{GS0}}=0\,\mathrm{V},\,V_{\mathrm{DS0}}=0\,\mathrm{V}}} \tag{5.45}$$

and

$$a_{\mathrm{drain-lag}} = \frac{I_{\mathrm{DS,pulse},\,V_{\mathrm{GS0}}=V_{\mathrm{thr}}-2\,\mathrm{V},\,V_{\mathrm{DS0}}=40\,\mathrm{V}}}{I_{\mathrm{DS,pulse},\,V_{\mathrm{GS0}}=V_{\mathrm{thr}}-2\,\mathrm{V},\,V_{\mathrm{DS0}}=0\,\mathrm{V}}}. \tag{5.46}$$

Further parameters are helpful to isolate the lagging:

$$a_{\mathrm{gate,cw}} = \frac{I_{\mathrm{DS,pulse},\,V_{\mathrm{GS0}}=V_{\mathrm{thr}}-2\,\mathrm{V},\,V_{\mathrm{DS0}}=0\,\mathrm{V}}}{I_{\mathrm{DS,cw}}} \tag{5.47}$$

and

$$a_{\mathrm{drain,cw}} = \frac{I_{\mathrm{DS,pulse},\,V_{\mathrm{GS}}=V_{\mathrm{thr}}-2\,\mathrm{V},\,V_{\mathrm{DS}}=40\,\mathrm{V}}}{I_{\mathrm{DS,cw}}}. \tag{5.48}$$

The ratios a_i can be evaluated at any pulsed-bias in the output characteristics. Typical bias to evaluate the dispersion to define a figure-of-merit for dispersion are the following:

– At the knee point, i.e., positive V_{GS} and $V_{\mathrm{DS}} = V_{\mathrm{knee}}$
– At a class-A bias, V_{DS} at operation and V_{GS} for $I_{\mathrm{Dmax}}/2$
– At high dynamic V_{DS} close to pinch-off, which also allows to estimate the isolation under pulsed-conditions

The evaluation allows to nominally separate the effect of gate-lag (5.45) and drain-lag (5.46). The measurements can be performed with several measurement configurations. More conventional pulsed-IV-measurements setups are proposed, e.g., in [5.149]. Recently, standardized DC-test equipment is often used [5.12, 5.188], which allows characterization of devices up to 10 A of current and 200 V of bias with pulse width $\leq 50\,\mathrm{ns}$. Such a system can be extended to pulsed-S-parameter measurements of devices with similar gate periphery and power levels, as reported in [5.197].

State-of-the-Art of Dispersion Analysis

Several approaches have been proposed for dispersion analysis. In a simplified model, the ratio of the pulsed-DC-measurements in (5.45) in the time-domain accounts for the influence of the gate bias, or gate-lag, while expression (5.46) accounts for the effect of the drain voltage or drain-lag. Fig. 5.8 depicts the various quiescent bias situations. In a first order approach, the gate-lag (comparison of quiescent situation (2) and (3) in Fig. 5.8) is caused by surface or near-gate-interface effects, while the drain-lag is caused by a backgating effect in the part below the channel, i.e., in the bottom substrate, nucleation layer, or buffer layer; see, e.g., [5.16]. The inclusion of measurements such as (5.47) and (5.48) allow the extraction of further information. Quiescent bias situation (5) is a typical class-A operation bias. Situation (6) further characterizes the maximum possible drain current I_{Dmax}. As with an initially saturated channel and barrier layer and low duty cycle, the maximum amount of carriers is available in a dispersive situation.

Temperature Dependence

The bias dependence of the mechanisms mentioned in previous sections are a good measure to isolate different dispersive effects. An accessible parameter is the substrate temperature. Typical findings for RF-dispersion down to cryogenic temperatures are reported, e.g., in [5.105, 5.128]. The findings include the following:

- Dispersion of the pulse output characteristics is more severe at a lower temperature $T_L = 150$ K, which implies that more electrons remain trapped and relax [5.128]. The output characteristics show more pronounced changes and gradients.
- Passivated HFET samples without dispersion at room temperature show increased dispersion at low temperatures of 150 K [5.105].
- The trap levels can be isolated more easily at lower temperatures, as the broadening of the levels is suppressed [5.121].
- Further, additional effects, such as the onset of impact ionization at cryogenic temperature, have been proposed [5.105].
- Investigations of frequency dispersion in AlGaN/GaN at elevated temperatures are given in [5.121]. The transition frequency of the dispersion shifts to higher frequencies with higher temperatures. The behavior is found to be consistent with the temperature behavior of the $1/f$-noise.

The general lattice-temperature dependence reduces the impact of trapping at temperatures $T_L \geq$ RT whereas at $T_L \leq$ RT the impact is enhanced. These findings support the characterization and analysis at room temperature.

Frequency-Domain Methods

Several frequency domain methods are available for the characterization of dispersion. As an example, Fig. 5.9 shows the measured output power swing of an AlGaN/GaN HEMT measured at 10 GHz in the RF-domain. The measurements are taken by a microwave transition analyzer (MTA) including four harmonics. Fig. 5.9 shows an RF-output swing at an operation voltage $V_{DS} = 30\,\text{V}$ with harmonics measured up to 50 GHz. Compared to the cw-DC and pulsed-DC output characteristics, deviations are visible especially at the knee voltage and in the near breakdown region. This powerful technique has been repeatedly applied, e.g., in [5.62, 5.64, 5.116]. It can be used for model verification [5.190] at the RF-frequency of interest. In the frequency-domain, the $1/f$-excess noise can be directly measured from 1 Hz up to at least 10 MHz. This allows the extraction of trap- and excess-noise information. As will be shown below in Fig. 5.42, trapping effects lead to additional contributions to the excess-noise spectrum and allow the isolation of trap-specific time constants.

5.2.3 Models for Frequency Dispersion in Devices

Several models exist that describe the physical mechanisms of frequency dispersion, either in parts or from a comprehensive point of view. This section further addresses the RF-modeling of these effects. Fig. 5.10 gives the RF-representations of dispersion effects, e.g., [5.87, 5.195].

The Concept of a Virtual Gate

One of the most critical effects of the surface charges in AlGaN/GaN FET is the effect of a second, so-called virtual gate. Increased surface-charge densities effectively lead to an increase of the area that serves as the gate metal and thus cause additional capacitance between surface and the channel charge. As a result, the channel-charge is modified on various time scales. In Fig. 5.10, this second virtual gate is represented by an additional capacitance and resistance (1). This gate has an effect similar to an extended gate length [5.87, 5.195].

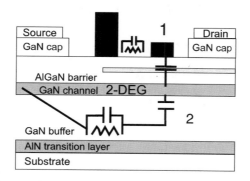

Fig. 5.10. RF-representations of dispersive effects

Backgating

A second, very prominent model for the impact of dispersion on the RF-properties is the backgating model, e.g., for AlGaN/GaN FET reported in [5.84]. In Fig. 5.10, a backside gate is depicted (2). This effect is very well known from silicon RF-MOSFETs [5.78] on conductive substrate, but it has also been described for GaAs MESFETs [5.169] and as a floating body effect for GaN/AlGaN HFETs due to the conductive nature of the AlN-interlayers on the heterosubstrate [5.128]. The interaction with a charge concentration in the buffer layer with the channel leads to an additional channel-buffer capacitance. The interaction is described for GaAs MESFETs by a junction capacitance model, as reported, e.g., in [5.169]. A dynamic current injection from the source side into the buffer interacts with the channel charge and leads to an additional RC constant.

A Charge-Control Model

Based on the previous analysis, a model will be developed to enhance the understanding of the various dispersive effects and, effectively, their suppression. As is pointed out in Sect. 2.2.3, understanding charge control is one of the most important needs within III-N FETs. The model is based on the analysis of the whereabouts of free carriers in the polar semiconductor materials, allowing exact elimination of the frequency-dependent and thus dispersive sources/drains of free carriers. The initial origin of the carriers in undoped III-N HFETs is manifold and is discussed widely for III-N FETs. The carriers originate from the following:

- Intentional and nonintentional (nid) doping of the bulk material
- δ-doping in the barrier and buffer layer
- Surface- and interface-charges due to polarization effects
- Generation/recombination with and without traps [5.157]
- The stress compensation and nonideal semiconductor layers due to the hetero-epitaxy, see Fig. 3.2

In the model, the effective free channel charge can be rewritten as the sum of the effective charges of the above mentioned mechanisms:

$$q_{\text{channel}} = q_{\text{diel,interface}} + q_{\text{semi,interfaces}} + q_{\text{bulk,traps}} + q_{\text{doping}} + q_{\text{initial}}. \quad (5.49)$$

Based on this analysis, the frequency dependence of the channel charge can be correlated with the frequency dependence of the charges q_i.

Absolute Charge Calculations

First of all, we need to calculate the absolute amount of charges in the channel relative to the source and drain at the surface to explain the sensitivity of the FETs to the changes. We assume a sheet-carrier density of $1 \times 10^{13}\,\text{cm}^{-2}$ for

Table 5.2. Parameters for charge and current analysis

l_g	Source drain (μm)	I_{Dmax}	n_{sheet} (cm^{-2})	Buffer thick. (μm)	Chan. thick.
0.25 μm	4 μm	1.2 A mm^{-1}	1×10^{13}	2	2.5 nm

Channel e$^-$	Charge	Buffer e$^-$	Buffer q	I_{DS}
4×10^8	64 pC	6×10^6	0.96 pC	4.8×10^7 A cm^{-2}

a 4 μm long HEMT channel with a gate length of 0.25 μm and a gate width of 1 mm. The channel thickness is assumed to be 2.5 nm, based on Schrödinger–Poisson solver analysis. This results in an overall channel number of 4×10^8 electrons or a charge of 64 pC, as detailed in Table 5.2. Relative to this amount, a buffer layer with an defect concentration of 10^{16} cm^{-3} and a thickness of 2 μm results in an overall charge of 6×10^6 electrons or 0.96 pC. Thus, an ideal buffer is not critical for the overall charge distribution. However, a charge contribution similar to the channel charge can be considered to be present at the following:

–   The dielectric/semiconductor interface
–   Substrate/buffer interface in the nucleation layer, depending on the nucleation growth

Based on calculations and the parameters in Table 5.2, the surfaces/interfaces can yield a charge contribution similar to the channel. This is one reason for the strong impact of dispersion in III-N devices. Further, the nucleation layer of ≥ 100 nm thickness can contribute a similar amount of electrons as the channel, if the growth is relatively immature.

Current Density Calculations

In addition to the charge analysis, the very high sheet-carrier concentrations at the channel interface lead to very high channel-current densities. Again based on the parameters from Table 5.2, channel-current densities of $\geq 10^7$ A cm^{-2} can be calculated. Such high current densities exceed even the highest current-densities in GaAs or InP HBTs by about two orders of magnitude [5.68]. This channel current density in GaN HFETs is further about three orders of magnitude higher than in Si power MOSFETs [5.55] and about one order of magnitude higher than in InGaAs HEMTs [5.37]. The latter is due to the lower band bending in the channel and the lower effective doping in (In)GaAs HFETs [5.37]. The high current densities lead to the following physical effects, which limit the current transport:

–   High-current crowding and related electromigration near the ohmic contacts [5.17]

- High current-density impact ionization for current densities $\geq 10^6 \, \mathrm{A \, cm^{-2}}$ [5.189]

These two effects eventually lead to the reduction of the densities during GaN FET device design in order to avoid high density effects, which are so far only known from bipolar and avalanche devices [5.184]. The impact of the current crowding in the source region is further analyzed in [5.185]. The current is space-charge limited [5.184] based on the following reasoning. Beyond a current density of $10^6 \, \mathrm{A \, cm^{-2}}$ space-charge effects occur due to the lack of a background impurity doping to compensate those. Thus, the effective source resistance R_S becomes a nonlinear function of the RF-drive power, which has also a strong impact on the power [5.184] and nonlinear performance of the device [5.108].

5.2.4 Suppression of Frequency Dispersion

Based on the modeling discussed in the previous section, the following means have been found useful for the reduction of the dispersion and high-current effects:

- The reduction of the defect density wherever possible within the device [5.194]
- The applications of appropriate technological surface/interface treatment [5.194] corresponding to the defect reduction
- The introduction of doped-cap layers that provide sufficient n-carriers to compensate the free surface states [5.85]
- A reduction of the channel current density, e.g., [5.84]

In addition to these insights for device design, dispersion can further be covered by the following:

- (Over)doping of the device with polarized and doping-based carriers [5.13]
- Device leakage of various kinds [5.81]
- The use of conducting passivation material [5.194]

These effects cover some of the frequency dispersive effects; however, they do not always help with the final reliable suppression of dispersion, as overdoping and device leakage lead to significant device reliability issues. These are discussed in Chapter 7.

5.3 Small-Signal Characterization, Analysis, and Modeling

5.3.1 RF-Characterization and Invariants

Several assumptions are typically made for the characterization of passive and active RF-devices. They are illustrated in Fig. 5.11. Multipole S-parameter

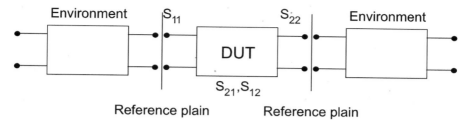

Fig. 5.11. Multipole characterization of the device-under-test (DUT)

measurements of RF-devices or device-under-test (DUT) are typically taken; see, e.g., [5.14, 5.47, 5.179, 5.201]. Derived from S-parameter measurements at a given impedance, e.g., $50\,\Omega$, all other high-frequency parameters, such as h-, Z-, Y-, and a-parameters, are available and can be converted to each other unambiguously. Several quantities can be derived from the S-parameters at the RF-reference planes given in Fig. 5.11. We derive from the S-parameters the maximum stable gain (MSG),

$$\text{MSG}(f) = \left| \frac{S_{21}}{S_{12}} \right|;$$ (5.50)

the maximum available gain (MAG),

$$\text{MAG}(f) = \left| \frac{S_{21}}{S_{12}} \right| \cdot \left(k \pm \sqrt{k^2 - 1} \right);$$ (5.51)

the unilateral gain U,

$$U(f) = \frac{\left| \dfrac{S_{21}}{S_{12}} - 1 \right|^2}{2 \cdot k \cdot \left| \dfrac{S_{21}}{S_{12}} \right| - 2 \cdot Re\left(\dfrac{S_{21}}{S_{12}} \right)};$$ (5.52)

and the device stability factor,

$$k(f) = \frac{1 - |S_{11}|^2 - |S_{22}|^2 + |S_{11}S_{22} - S_{12}S_{21}|^2}{2 \cdot |S_{12}||S_{21}|}.$$ (5.53)

Further invariants are usually derived. The current gain cut-off frequency is defined as

$$f_\text{T} = f(h_{21} = 0).$$ (5.54)

The maximum frequency of oscillation is derived based on the unilateral gain

$$f_\text{max} = f(U = 0);$$ (5.55)

and based on MAG/MSG

$$f_\text{max} = f(\text{MAG/MSG} = 0).$$ (5.56)

As a measure of the stability, the cut-off frequency for the stability f_c is defined:

$$f_\text{c} = f(k = 0).$$ (5.57)

These parameters are used for the characterization of III-N devices.

5.3.2 Common-Source HEMTs

The common-source FET configuration is mostly used for the application of FETs in power amplifiers. As an initial example, S-parameters of an AlGaN/GaN HEMT are measured between 0.5 and 20 GHz, as given in Fig. 5.12 for a gate width $W_g = 3.2$ mm and a gate length $l_g = 0.5$ μm. A large S_{21} is typical for GaN HFETs, further the output S-parameter S_{22} depicts a high impedance for a given gate width. Derived from these S-parameters, Fig. 5.13 gives the cut-off frequencies f_T and f_{max} calculated from the S-parameters measured at $V_{DS} = 7$ V as a function of gate length l_g for a planar device without gate recess. The results nicely follow a nearly constant product of $l_g \times f_T$ of 10 GHz μm down to a gate length of 150 nm. Furthermore, f_{max} values of 140 GHz are reached for devices with a periphery $W_g = 120$ μm. Fig. 5.14 gives the MAG/MSG at 40 GHz as a function of the gate width for a device with $l_g = 150$ nm. MAG/MSG values of 8 dB for devices with $W_g = 120$ μm gate width are reached. More than 6 dB of gain are found for an optimized

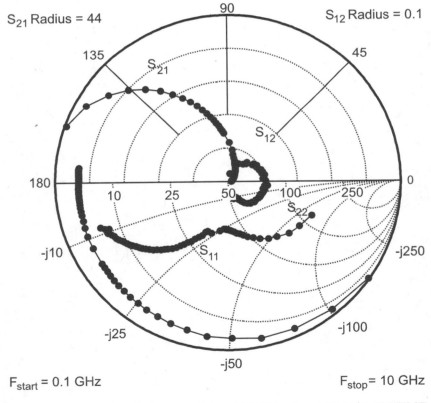

Fig. 5.12. S-parameters S_{ij} between 0.1 and 10 GHz of an AlGaN/GaN HEMT of gate length $l_g = 500$ nm measured at $V_{DS} = 30$ V for $W_g = 3.2$ mm

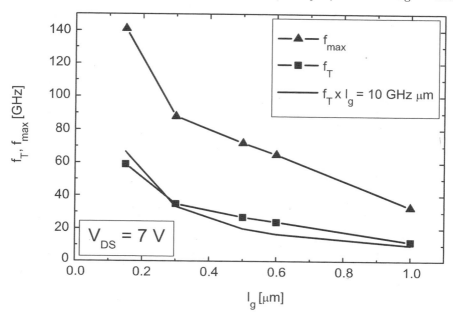

Fig. 5.13. f_T and f_{max} as a function of gate length l_g at $V_{DS} = 7\,V$ for an AlGaN/GaN HFET on s.i. SiC with $W_g = 2 \times 60\,\mu m$

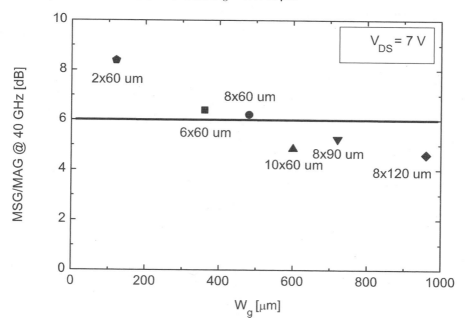

Fig. 5.14. MAG/MSG at $40\,GHz$ as a function of gate width W_g of an AlGaN/GaN HFET on s.i. SiC with $l_g = 150\,nm$

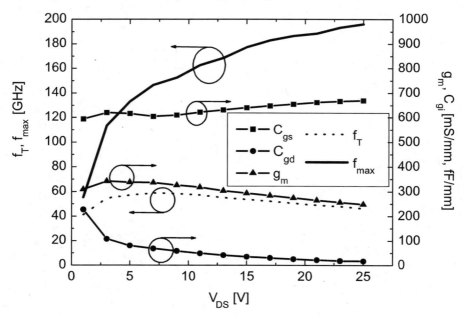

Fig. 5.15. f_T, f_{max}, g_m, C_{gs}, and C_{gd} as a function of V_{DS} for $l_g = 150\,\text{nm}$

device with a gate periphery $W_g = 8 \times 60\,\mu\text{m}$. Fig. 5.15 shows the cut-off frequencies f_T, f_{max}, and their most important constituting small-signal parameters, g_m, C_{gs}, and C_{gd} as a function of V_{DS} for a device with $l_g = 150\,\text{nm}$. The initial approaches for a typical small-signal model for GaN devices are similar or equal to those of GaAs PHEMTs, e.g., [5.14, 5.47, 5.179, 5.201]. Fig. 5.15 shows that the cut-off frequencies f_T are nearly constant as a function of V_{DS} at constant V_{GS}. The maximum frequency of oscillation rises to values up to 200 GHz at $V_{DS} = 25\,\text{V}$. This increase is based on the nearly constant transconductance g_m, the nearly constant C_{gs}, and the reduction of C_{gd} with rising V_{DS}. The physical background of the constant behavior of g_m and C_{gs} is based on the good carrier confinement of the electrons in the channel. An example of a comprehensive physics-based small-signal analysis of fully strained and partially relaxed high Al-content $\text{Al}_m\text{Ga}_{1-m}\text{N}/\text{GaN}$ HEMTs is given in [5.145]. The physical model allows close correlation of the current–gain cut-off frequency with device design parameter Al-content in the barrier. High Al-contents are found to be useful for increased transconductance. The advantages of high Al-content are traded for increased relaxation, increasing traps concentrations, and challenges with the ohmic contacting.

Temperature Dependence of the RF-parameters

The lattice temperature is another critical parameter, as GaN FETs are proposed for high-power and high-temperature operation. Fig. 5.16 gives the

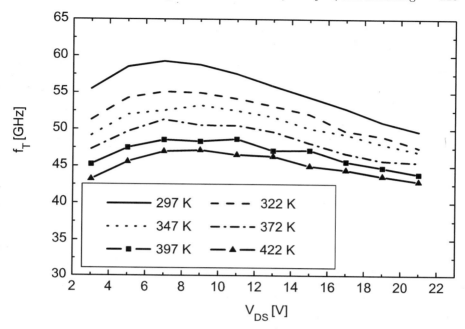

Fig. 5.16. Cut-off frequency f_T as a function of V_{DS} with the substrate temperature T_{sub} as a parameter

current gain cut-off frequency f_T as a function of V_{DS} with the substrate temperature T_{sub} as a parameter. The device has a gate length of 150 nm. The temperature at the backside on the unthinned SiC substrate is varied between room temperature and 422 K. A reduction of f_T of 25% for 100°C change in temperature can be derived from Fig. 5.16. Further, Fig. 5.17 gives the gain characteristics MAG/MSG at 40 GHz, f_T, and f_{max} as a function of V_{DS} with T_{sub} as a parameter for a device with $l_g = 150$ nm and a gate periphery $W_g = 8 \times 60 \,\mu$m. All three gain parameters in Fig. 5.17 show a similar gain reduction between 25 and 100°C, which amounts to about 25%.

Variations of AlGaN/GaN HEMT Structures

The variation and optimization of the layer structure of AlGaN/GaN HEMTs has a strong impact on the small-signal performance. As an example, Fig. 5.18 gives a comparison of the cut-off frequencies with different doping concentrations in two different HEMT structures at the same nominal gate lengths l_g. Fig. 5.18 shows that increased doping concentrations in the barrier lead to higher cut-off frequencies, while the breakdown voltages are reduced accordingly. This is depicted by the different V_{DS} ranges possible for the cw-S-parameter measurements.

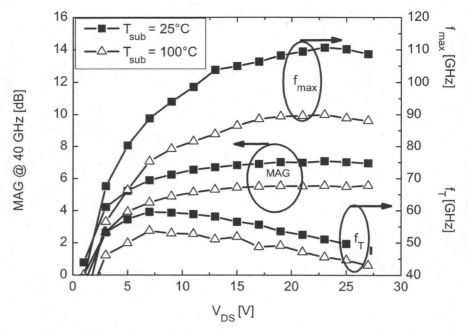

Fig. 5.17. MAG/MSG of an AlGaN/GaN HEMT on s.i. SiC at 40 GHz, f_T, and f_{max} as a function of V_{DS} with temperature T_{sub} as a parameter

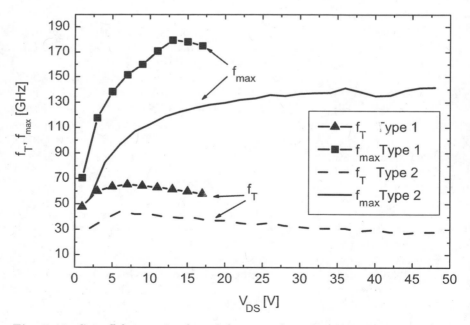

Fig. 5.18. Cut-off frequencies f_T and f_{max} as a function of V_{DS} for AlGaN/GaN HEMTs on s.i. SiC with two different doping concentrations

5.3.3 Dual-Gate HEMTs

Dual-gate and cascode AlGaN/GaN HEMTs are further attractive candidates for efficiently increasing the gain-per-stage in GaN RF-amplifiers. Two gates are placed in series within a typical HEMT. Fig. 5.19 gives an SEM image of an AlGaN/GaN dual-gate FET with a gate length $l_g = 150\,\mathrm{nm}$ on s.i. SiC substrate. Fig. 5.20 gives the comparison of the measured MAG/MSG gain for both a common-source and a dual-gate device as a function of frequency measured up to 110 GHz. The devices are compared for a gate width $W_g = 4 \times 45\,\mu\mathrm{m}$ and for a gate length $l_g = 150\,\mathrm{nm}$. The dual-gate device yields about double the available gain at a given frequency for a comparable V_{DS} bias (twice the V_{DS} of the common-source device). This allows operation up to the V-frequency band with gain levels of $\geq 10\,\mathrm{dB}$ at 60 GHz. Further examples of cascode- and dual-gate devices can be found, e.g., in [5.34]. The gate lengths of the first amount to 0.16 and 0.35 μm for the second allow to increase the breakdown voltage of $\geq 100\,\mathrm{V}$ while maintaining f_T values of 60 GHz.

5.3.4 Pulsed-DC- and RF-Characteristics

Pulsed-DC- and RF-measurements are of major importance for the under-standing of III-N device behavior. This is already stated in Sect. 5.2.1. A lot of measurement techniques have been developed for GaAs and Si-based devices [5.132, 5.133, 5.188, 5.197] using pulsed-techniques. As mentioned in Chapter 2, GaN MESFETs [5.183] and HEMTs [5.120] are characterized by standard-ized pulsed-measurements in order to separate dispersion and thermal effects.

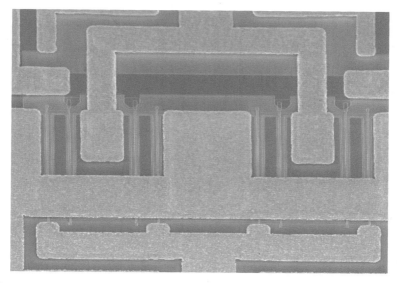

Fig. 5.19. SEM image of a dual-gate AlGaN/GaN HEMT with $l_g = 150\,\mathrm{nm}$

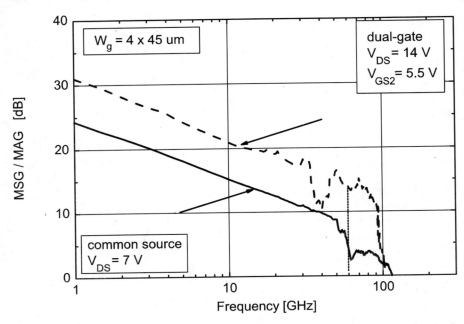

Fig. 5.20. MAG/MSG as a function of frequency for a common-source and a dual-gate AlGaN/GaN HEMT on s.i. SiC with $l_g = 150$ nm

Fig. 5.21 gives pulsed-DC-measurements of an AlGaN/GaN HEMT on s.i. SiC, which allow to separate thermal from the dispersive effects discussed previously. This can be achieved by comparing the pulsed-characteristics for low duty-cycle, typically below 1% [5.133], at different quiescent biases and different substrate temperatures between 25 and 100°C. Given the low duty-cycle, an isothermal-like device behavior can be assumed and isothermal effects with varying substrate temperature can be separated from self-heating effects [5.129]. As seen in Fig. 5.21, the influence of the substrate temperature between 25 and 100°C at low duty-cycle is relatively modest compared to the self-heating and the dispersive effects discussed previously. To separate the impact of thermal effects on small-signal RF-parameters, Fig. 5.22 compares the measured f_T for pulsed-conditions using an Agilent 85124 pulsed-DC- and RF-measurement-system. The measurements are performed for a gate width $W_g = 3.2$ mm and a gate length $l_g = 500$ nm for different quiescent bias. The duty cycle is chosen to be low, in this case 0.1%. The overall dissipated power at the maximum bias $V_{DS} = 45$ V is 22.5 W. The difference in the f_T especially for the hot pinch-off conditions is clearly visible and due to dispersive effects. The main focus of pulsed-measurements is not the pure evaluation of thermal effects, but the reduction of DC-power in the device and thus the reduction of self-heating to allow modeling at high bias. Very complex dispersion modes have been isolated in GaN HFETs, which require detailed

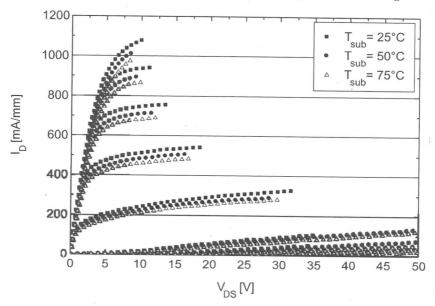

Fig. 5.21. Pulsed-output characteristics as a function of substrate temperature for a low duty-cycle of 0.1%, pulsed from ($V_{DS0} = 0\,V$, $V_{GS0} = -7\,V$)

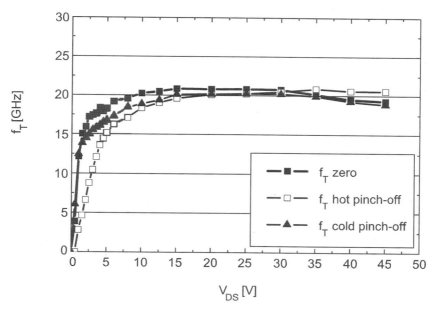

Fig. 5.22. Comparison of the cut-off frequency f_T of an AlGaN/GaN HEMT on s.i. SiC for pulsed-operation and different quiescent bias, zero: ($V_{GS0} = V_{DS0} = 0\,V$), cold pinch-off: ($V_{GS0} = -7\,V$, $V_{DS0} = 0\,V$), hot pinch-off: ($V_{GS0} = -7\,V$, $V_{DS0} = 30\,V$)

DC/RF-pulsed-analysis for modeling [5.197]. Thus, in the development and modeling of devices, biasing is possible at voltages that cannot be addressed in cw-mode, as shown in Fig. 5.22. A typical example of the application of DC/RF-pulsed-measurements is large-signal modeling for an operation bias V_{DS0}, where S-parameters at high V_{DS}-bias ($2 \times V_{DS0}$) and high current levels of $\geq I_{Dmax}/2$ are needed. However, they lead to prohibitive channel temperatures during the cw-measurements. Thus, pulsed-measurements are extremely useful.

5.3.5 Small-Signal Modeling

Small-Signal Analysis of FETs

The first approaches for the small-signal analysis of GaN FETs are similar to those of classical GaAs or InP HEMTs. This is due to the fact that GaN (H)FETs are very similar to GaAs PHEMTs or even to InP HEMTs, considering the specific shape of the device terminal characteristics. Examples for the modeling of GaAs are numerous and can be found, e.g., in [5.14, 5.47, 5.161, 5.179, 5.201]. These models have also been applied for the extraction of GaN FETs, e.g., in [5.200]. Fig. 5.23 gives the standard small-signal RF-topology for III-V FETs.

As with any FET, the device representation consists of a complex current source, a parallel output-capacitance C_{ds} and resistance R_{ds}, and an input capacitance C_{gs}. The parasitic shell is simple and similar to any microwave device. As such a model contains sufficient free parameters, the four complex S-parameters S_{ij} can be modeled well as a function of frequency, as long as the parasitic distributed effects do not play a dominant role. Thus, small-signal analysis is nearly always possible, even for very dispersive devices. The analysis of dispersion is possible also through small-signal measurements, as the virtual gate effect has significant impact on the elements extracted. Specific III-N device small-signal extraction procedures have been suggested, e.g., [5.76], which are derived from advanced GaAs methodologies [5.162]. This refers further to the modeling of the diodes and substrate effects. The characterization of forward AC behavior of GaN Schottky diodes yields a giant negative capacitance and nonlinear interfacial layer [5.198]. A similar finding is given in [5.42]. The small-signal approach of Schmale [5.162] is applied to GaN HFETs in [5.75, 5.76]. A 22-element model is applied for the extraction of a small-signal model, valid between 0 and 60 GHz for a GaN device with $l_g = 0.5\,\mu m$. The parasitic environment consists of two-shells in this case and is used beyond the cut-off frequency f_T of the device.

Substrate Effects in FETs

For AlGaN/GaN HEMTs on silicon substrates, additional effects have to be considered. The lossy and conductive substrates need to be modeled similar to

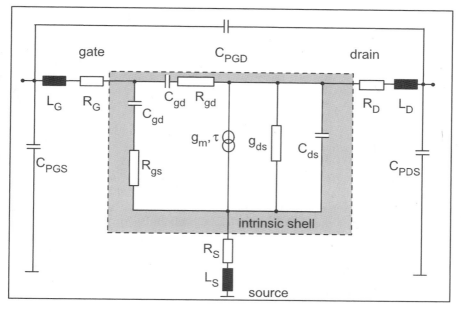

Fig. 5.23. Topology of a small-signal model of an AlGaN/GaN HFET

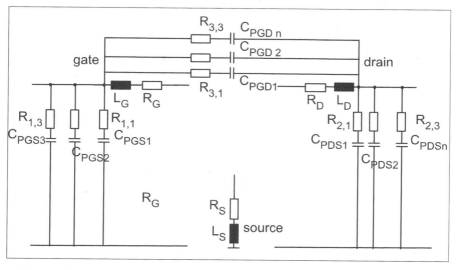

Fig. 5.24. Passive topology of a small-signal model of an AlGaN/GaN HFET

silicon RF-devices, e.g., [5.38]. The small-signal characteristics of microwave power GaN HEMTs on 4-in. Si wafers are described in [5.110]. The intrinsic shell of the HFET model is not modified. However, additional RC networks are included to model the residual substrate conductivity and the loading of the device. The conductive substrate leads to p-Si/GaN/metal structures, which are modeled as pin-diodes in series with the impedances reflecting the substrate conductivity and the device loading. The topology of the passive environment is depicted in Fig. 5.24. Multiple structures of the RC combinations are used.

Small-Signal HEMT Example on s.i. SiC Substrate

As an example of a complete set of small-signal parameters and the corresponding parasitic elements, Table 5.3 gives a full set of parameters for an AlGaN/GaN HEMT on s.i. SiC substrate with a periphery $W_g = 4 \times 60\,\mu m$ and a gate length $l_g = 150\,nm$. The elements are extracted by standard procedures, e.g., [5.179]. We see a low value of the capacitance C_{gd} at $V_{DS} = 25\,V$ and a reasonable ratio of the capacitances C_{gs}/C_{gd} of $\geq 6{:}1$. The R_{gd} is negligible in this extraction. The output output capacitance C_{ds} amounts to $0.27\,pF\,mm^{-1}$. Table 5.4 adds the parasitic elements in this small-signal approach with one parasitic shell. The source resistance is $0.8\,\Omega\,mm$, while the drain resistance R_D is higher due to a non-symmetric contact spacing. Both parasitic values are significant to the reduction of the gain at high RF-frequencies.

Bias and Geometry Dependence

The extracted elements feature several characteristics and dependencies. Several interesting effects can be observed. Fig. 5.25 gives the dependence of the intrinsic capacitances C_{gs} and C_{gd} on gate length. At gate lengths between 300 and 150 nm, the C_{gs}/C_{gd} ratio drops from 8 to 5 for this planar device. At the same time, the elements are bias-dependent. Fig. 5.26 gives the extracted bias-dependence of the RF-output-conductance g_{ds} and the output-capacitance C_{ds} as a function of drain–source bias V_{DS}. At constant gate bias V_{GS}, the variation of the extracted C_{ds} is about 15%, i.e., between 5 and 25 V.

Table 5.3. Small-signal parameters of an AlGaN/GaN HEMT on s.i. SiC with $W_g = 4 \times 60\,\mu m$ and $l_g = 150\,nm$ at $V_{DS} = 25\,V$

C_{gs} ($pF\,mm^{-1}$)	C_{gd} ($pF\,mm^{-1}$)	C_{ds} ($pF\,mm^{-1}$)	g_m ($mS\,mm^{-1}$)	τ (ps)	g_{ds} ($mS\,mm^{-1}$)	R_{gs} ($\Omega\,mm$)	R_{gd} ($\Omega\,mm$)
0.65	0.09	0.27	382	1.64	15	0.22	0

Table 5.4. Extracted parasitic elements for the coplanar device from Table 5.3

L_G (pH)	L_D (pH)	L_S (pH)	R_S ($\Omega\,mm$)	R_D ($\Omega\,mm$)	R_G (Ω)	C_{pgd} (fF)	C_{pgs} (fF)	C_{pds} (fF)
0	75	76	0.8	1.5	2	20	4	20

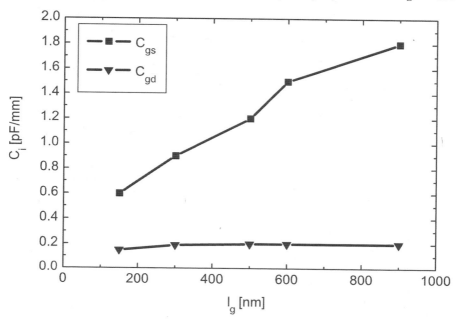

Fig. 5.25. Intrinsic capacitances C_{gs} and C_{gd} of an AlGaN/GaN HEMT on s.i. SiC as a function of gate length l_g at $V_{DS} = 7\,V$

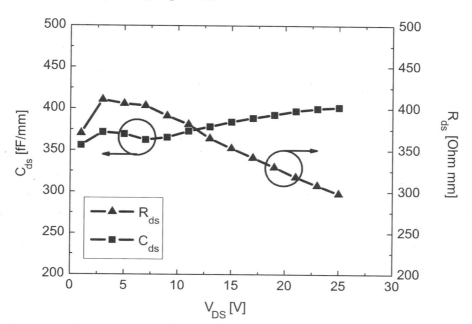

Fig. 5.26. RF-output-resistance R_{ds} and capacitance C_{ds} as a function of V_{DS} at constant V_{GS} for an AlGaN/GaN HFET with $l_g = 150\,nm$

This very moderate change, which is better than the change of C_{ds} in optimized LDMOS devices, can be exploited in advanced efficient circuit concepts. The extracted R_{ds} decreases by 25% with V_{DS} for the same device.

FET Transistor Delay Analysis

Based on the measurement of S-parameters and cut-off frequencies, FET delay times can be extracted. Various approaches have been published and been used for the delay time extraction for GaAs and InP devices, e.g., in [5.123, 5.168]. The typical finding for GaN HFETs is that the expected theoretical speed performance has not fully been exploited in real devices. Transistor delay analysis and extraction of the effective channel velocity for AlGaN/GaN HFETs are reported in [5.18, 5.130, 5.174, 5.176]. The approach according to Moll [5.123] assumes for the HFET device:

$$\frac{1}{2\pi \cdot f_T} = \tau_T = \tau_{RC} + \tau_D + \tau_{TR}. \tag{5.58}$$

In this case, τ_T is the total delay time, τ_{RC} is the channel-charge RC charging-delay, τ_D is the delay due to the extension of the depletion zone to the drain, and τ_{TR} is the transistor delay time. Parasitic de-embedding is crucial for the determination of realistic delays and velocities [5.180]. The extracted velocities for III-N devices from these approaches are typically lower than the simulated maximum velocities [5.54, 5.130]. Some issues in this discussion include the following:

- The parasitic effects of the source and drain region [5.131, 5.174, 5.180]
- The lack of the electron–phonon interaction in the MC simulations predicting the maximum velocities [5.210]

The stripping of masking parasitic effects, which can be accomplished in various ways [5.18, 5.180], is thereby of crucial importance. As is analyzed in [5.174], the parasitic effects can be eliminated, so that an intrinsic effective electron velocity of 1.86×10^7 cm s^{-1} in HFETs can be extract d. This latter value matches better some of the MC predictions for GaN bulk material [5.15]. At the same time, specific FET overshoot effects must be considered in the simulation for the comparison, e.g., [5.4, 5.144].

Example of Delay Time Analysis

As an example of the delay analysis, Table 5.5 gives the results of the extraction for an AlGaN/GaN device on s.i. SiC with gate lengths $l_g = 240$ and 120 nm. The delay time τ_{TR} in the channel scales with the gate length. The difference of the two examples is mainly based on the parasitic delay.

HBT Small-Signal Modeling

Al(In)GaN/GaN HBTs are so far only available with a limited gain performance. Npn-HBTs are characterized by high base resistances R_{bb} and high

Table 5.5. Extracted delay times for several AlGaN/GaN HEMTs

f_T (GHz)	l_g (nm)	τ_T (ps)	τ_{TR} (ps)	τ_{RC} (ps)	τ_D (ps)	Ref.
50	240	1.5	1.3	0	0.2	[5.50]
63	120	2	0.65	0.1	1.25	[5.174]

base contact resistances. This is reflected in the small-signal modeling. A distributed modeling is used for the description of the base transit time in AlGaN/GaN HBTs in [5.115]. A hybrid pi-small-signal-equivalent model of the active transistor is used, which is repeated 25 times to account for the base effect. InGaN/GaN HBTs have been proposed to solve the base resistance issues. The delay time analysis by simulation of npn-GaN/InGaN/GaN double heterojunction devices is provided in [5.36]. For a given base width, the GaN/InGaN/GaN DHBT is calculated to be superior in current gain cutoff frequency to the AlGaN/GaN/AlGaN DHBT. The simulated temperature dependence of the current gain cut-off frequency is found to be very weak between 300 and 500 K. Delay-time extraction examples of (In)GaN HBTs based on measurements are currently scarce.

5.4 Large-Signal Analysis and Modeling

Large-signal characterization and modeling are of great importance for III-N devices, which are genuine high-power devices intended for large-signal operation. A large number of large-signal characterization results exists for various types of substrates.

5.4.1 Large-Signal Characterization and Loadpull Results

Overview data for all substrates are already given in Chapter 2. Characterization examples of GaN HEMTs on s.i. SiC substrate can be found in [5.101, 5.116, 5.202]. They include fundamental, source, and harmonic load-pull analysis at various frequencies. A similar number of III-N devices has been investigated on sapphire substrates, especially for small gate widths of ≤200 μm, e.g., [5.101, 5.192]. Large-signal properties of AlGaN/GaN HEMTs on high-resistivity silicon substrates are reported, e.g., in [5.176, 5.181]. The characterization for III-N devices is in principle not different from that for GaAs or silicon power devices; however, it is more challenging due to the following issues:

- Higher power densities lead to more pronounced thermal effects
- Higher operation voltages require more safety measures
- More testing is required due to the impact of dispersion [5.74]

–   The high-voltage measurements require more safety measures in the case
    of device burn-out

These issues will be raised below wherever necessary.

Loadpull Measurement Results

As an initial example, Fig. 5.27 gives the measured output power, power-gain,
and power-added efficiency data of an AlGaN/GaN HEMT on s.i. SiC as
a function of input power at 10 GHz in cw-mode for a gate length $l_g = 250$ nm.
The tuning of the output load is performed in passive mode. The gate bias is
chosen for class-A/B operation with an quiescent current I_D with 10% of the
I_{Dmax}. The device in Fig. 5.27 yields power gain compression, although the
passive load is constant. Some of this behavior can still be tracked to effects of
dispersion and the onset of space charge effects [5.185]. However, a PAE= 50%
is reached at a low compression level of -2 dB at 10 GHz. The high PAE at low
compression is important in order to maintain high efficiency on system level.
The drain current I_D is depicted in Fig. 5.27, which increases with input power
due to the self-biasing. On the contrary, Fig. 5.28 gives output power, gain,
and power-added efficiency data as a function of input power at 40 GHz for an
AlGaN/GaN HEMT on s.i. SiC substrate with $l_g = 150$ nm, measured with
an active loadpull system. The active system is used in this case in order to

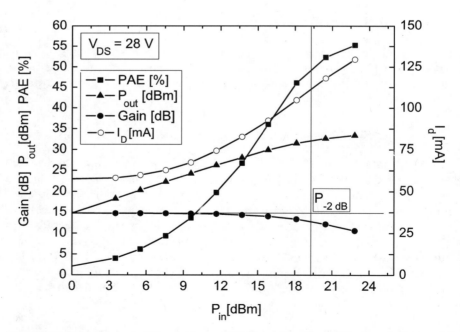

Fig. 5.27. Power sweep of P_{out}, gain, and PAE as a function of P_{in} for a $0.25\,\mu m \times 8 \times$
$60\,\mu m$ AlGaN/GaN HEMT at 10 GHz

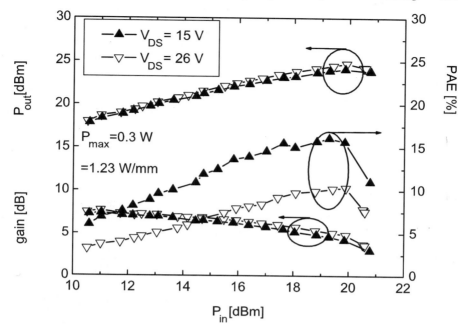

Fig. 5.28. Power sweep P_{out}, gain, and PAE vs. P_{in} of a $0.15\,\mu m \times 8 \times 30\,\mu m$ HEMT at 40 GHz

reach the required output load reflection coefficients for the $8 \times 30\,\mu m$ device at 40 GHz. Again a premature compression is visible; in this case, however, this is partly due to the change of the active load impedance. Pulsed-power measurements help to reduce the impact of the thermal effects, while the impact of the dispersion is dependent on the pulse width. A pulsed-DC- and RF-loadpull setup is presented in [5.22] for the evaluation of GaN HEMTs under pulsed-conditions including measurements of the intermodulation distortion.

Load Impedance Dependence

Fig. 5.29 gives loadpull contours of an AlGaN/GaN HEMT on s.i. SiC with a gate width $W_g = 8 \times 60\,\mu m$ measured at 10 GHz for an input power of 24 dBm. While in first order the optimum load impedance as a function of gate width at 2 GHz shows a nearly ohmic behavior in the Smith chart, for higher frequencies at 10 GHz an additional contribution is needed for the output capacitance C_{ds}. For a bias $V_{DS} = 25\,V$, an optimum output impedance reflection $\Gamma = (0.45/28°)$ for a $50\,\Omega$ system is found by loadpull measurements for the device in Fig. 5.29. As a major finding, III-N devices are always considered to be high impedance devices. However, a strong influence is also due to the parasitics device environment. This is shown for a passive load-pull measurement at 60 GHz in Fig. 5.30 for a device with $W_g = 2 \times 50\,\mu m$

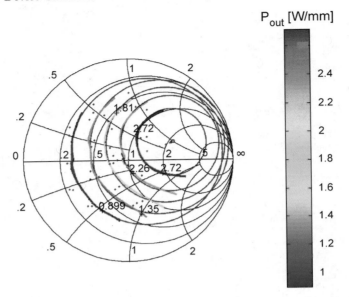

Fig. 5.29. P_{out} characteristics forming loadpull contours of an AlGaN/GaN HEMT with $W_{\text{g}} = 8 \times 60\,\mu\text{m}$ and $l_{\text{g}} = 300\,\text{nm}$ at 10 GHz at an input power of 24 dBm and $V_{\text{DS}} = 25\,\text{V}$, reference impedance 50 Ω

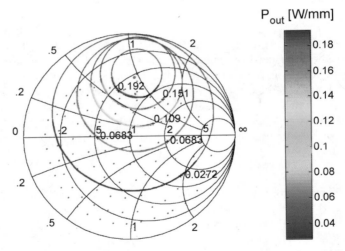

Fig. 5.30. P_{out} characteristics forming loadpull contours of an AlGaN/GaN HEMT with $W_{\text{g}} = 2 \times 50\,\mu\text{m}$ and $l_{\text{g}} = 100\,\text{nm}$ at 60 GHz at an input power of 7.1 dBm at $V_{\text{DS}} = 30\,\text{V}$, reference impedance 50 Ω

and a gate length $l_{\text{g}} = 100\,\text{nm}$. The intrinsic impedance level is transformed through the additional parasitic network representing the layout. This leads to strong deviations from the expected intrinsic high impedance at the device terminals, especially at frequencies higher than a few gigahertz. The optimum load has a strong imaginary part that renders the matching difficult despite the higher intrinsic impedance as compared to GaAs PHEMTs of the same gate length and width. Fig. 5.31 gives the dynamic loadline of an AlGaN/GaN HEMT similar to Fig. 5.29, measured at 10 GHz including higher harmonics at $V_{DS} = 30\,\text{V}$. From these measurements, the actual tuning of the output load can be controlled relative to the DC- and pulsed-output characteristics of the device. Further, the LS-prediction of the LS-modeling can be effectively verified. This analysis also allows the direct measurement of the impact of the knee walk-out and reduced RF-breakdown, see also [5.64, 5.186]. The thermal effect on the knee voltage can be suppressed by pulsed large-signal measurements [5.51]. As we see, the output swing in Fig. 5.31 is not fully symmetric to the DC-operation bias at $V_{DS0} = 30\,\text{V}$. The self-biasing with increasing input

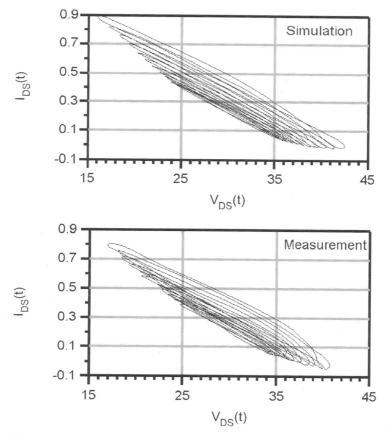

Fig. 5.31. Dynamic loadline of a GaN HEMT with $0.3\,\mu\text{m} \times 8 \times 125\,\mu\text{m}$ at 10 GHz and $V_{DS} = 30\,\text{V}$, simulation (*top*), measurement (*bottom*)

drive is also visible. Similar data is provided in [5.116]. A related approach is used in [5.192] to reconstruct the load impedance influence on the $I_D(V_{DS})$ characteristics of AlGaN/GaN HEMTs on sapphire at 4 GHz.

Bias Dependence of Large-Signal Measurement Results

Several experiments have been performed to investigate the fundamental frequency behavior [5.64]. The bias-dependent performance of high-power AlGaN/GaN HEMTs is described, e.g., in [5.202], featuring a maximum PAE of 60%. This bias dependence involves some important characteristics:

-   The maximum output power as a function of V_{DS}
-   The maximum PAE as a function of V_{DS}
-   The V_{GS} dependence, which defines the amplifier operation between class-A and C

Fig. 5.32 gives a comparison of the maximum in PAE as a function of V_{DS} bias for two different GaN/AlGaN HEMT technologies for the same gate length l_g. It can be seen that the maximum in PAE differs between the technologies investigated. The technology including a field plate yields PAE improvements, especially at higher V_{DS}-voltage. To estimate the potential for a given operation bias V_{DS}, Fig. 5.33 gives the calculated Cripps-load [5.41] and Cripps-output power up to an operation voltage of up to $V_{DS} = 150\,V$, in order to

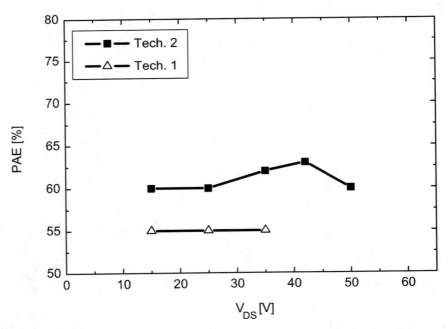

Fig. 5.32. PAE of an AlGaN/GaN HEMT on s.i. SiC as a function of V_{DS} with $W_g = 2 \times 400\,\mu m$ and $l_g = 500\,nm$ at 2 GHz

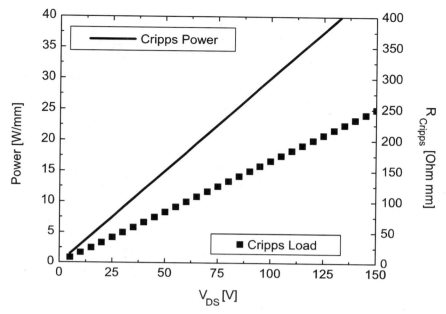

Fig. 5.33. Calculated Cripps-impedance and output power of an AlGaN/GaN HEMT with $I_{Dmax} = 1.2\,A\,mm^{-1}$ as a function of V_{DS}

allow a comparison to measured loadpull results. The assumed current I_{Dmax} is $1.2\,A\,mm^{-1}$. The results strongly suggest the higher impedance level for a GaN HEMT relative to GaAs or silicon-based devices. A load of $50\,\Omega\,mm$ is reached approximately at $V_{DS} = 25\,V$.

5.4.2 Large-Signal Modeling

The modeling of RF-devices is, in general, a highly complex problem with a great amount of degrees-of-freedom and dependencies [5.1]. A lot of research has already been performed into the inclusion of several effects into large-signal models, e.g., [5.7, 5.10, 5.43, 5.45, 5.111, 5.170]. At the same time, these models have to remain useful, i.e., they have to be simple and simulation-time-efficient while accurate for circuit simulation. As a simple example, spice models of GaN devices have been reported, e.g., in [5.70]. They allow for the analysis of the devices, especially in the time-domain.

What is Different About III-N Devices?

With the aforementioned arguments and the long tradition of semiconductor modeling, one has to ask, what is so special about the III-N devices and why spend extra effort on modeling them. Fig. 5.34 depicts the frequency spectrum, the associated physical effects, and some typical characterization methods for the large-signal modeling. III-N devices certainly do not differ in principle from any other semiconductor devices. However, some effects are more pronounced and require special attention. Further, some of the suggested new applications pose stringent requirements and open new aspects for characterization and modeling in general with regard to the frequency spectrum. First of all, on the device level GaN devices maintain the troublesome low-frequency-noise characteristics of any III-V FET device [5.194]. With regard to the domain of highly-linear power RF-amplifiers, they combine the handling of extremely high-power densities of $40\,\mathrm{W\,mm^{-1}}$ [5.204] at $4\,\mathrm{GHz}$, exceeding those that have been available so far. The high power levels also lead to higher linearity for a given output power level. At the same time, given the same power level as in conventional devices, reduced thermal effects are observed, especially on SiC substrate. This includes the reduction of critical memory effects that cannot fully be compensated by predistortion in silicon LDMOS devices, e.g., [5.21]. This reduction in memory effects allows to drive GaN devices at higher power levels and closer to the saturation applying memoryless predistortion. Higher frequencies in Fig. 5.34 and Fig. 5.35 become interesting, e.g., for efficiency improvements by means of harmonic termination. These findings are especially critical, because the system requirements drastically change. The new signal forms with nonconstant signal envelopes, especially of 3G/4G mobile-communication-systems, cover the full frequency range in Fig. 5.34 and Fig. 5.35, virtually from $0\,\mathrm{Hz}$ to the cut-off frequencies f_T and

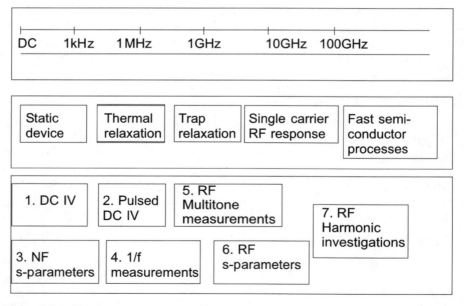

Fig. 5.34. Frequency spectrum and principal overview of RF-device effects and analysis tools

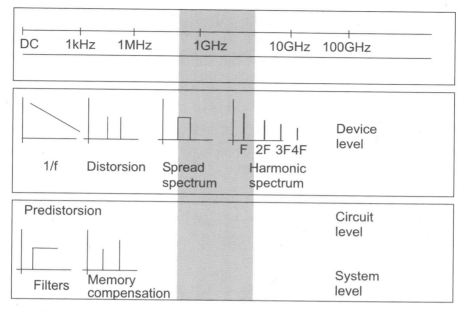

Fig. 5.35. Frequency spectra of importance for circuit and system design

f_{max} of the device. This complete frequency range is of primary importance, as the RF-carriers are modulated by extremely broad spectra, potentially $\geq 20\,MHz$. RF-frequency-dispersion thus remains a challenge for the modeling. All these facts relevant for the modeling are illustrated in Fig. 5.35 with respect to the resulting circuit and system design issues. On system level, mainly predistortion and filtering issues are important, as discussed further in [5.52].

LS-Analysis Based on Small-Signal Parameters

A large number of publications is available on large-signal models based on the extraction from small-signal S-parameters in the frequency-domain [5.10,5.43, 5.111]. These methodologies are well known from the GaAs FET world, where they are widely used [5.201]. The problem of large-signal model analysis and extraction can be broken down into two problems: the extraction of the charge parameters and the extraction of the current equations. First of all small-signal-based approaches have been adopted for GaN FETs, e.g., [5.7]. Derived from small-signal-based approaches, Fig. 5.36 gives the topology of a large-signal model suitable for III-N FETs. This topology in principal does not differ from any other III-V HFET device. However, the extraction procedures have to be modified. A description of the theoretical background of the principal modeling procedure in a state-space approach is given in [5.166].

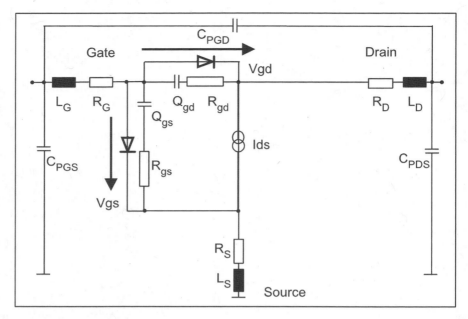

Fig. 5.36. Topology of a large-signal model of an AlGaN/GaN HFET

Analytical Approaches

Analytical current equations are widely used for the description of the DC-
and RF-currents, e.g., [5.10, 5.43, 5.190]. They combine simulation efficiency
with good modeling accuracy. Typical current equations for the description of
the transfer characteristics based on the cubic Curtice approach read [5.170]:

$$I_{DS} = (A_0 + A_1 \cdot V_1 + A_2 \cdot V_1 + A_2 \cdot V_1^2 + A_3 \cdot V_1^3) \tanh(\gamma V_{DS}), \quad (5.59)$$

where $V_1 = V_{GS} \cdot (1 + \beta \cdot (V_{DSO} - V_{DS})).$ \hfill (5.60)

Both the simple V_{GS} description and the simplified V_{DS} description do not
allow an accurate modeling of all the GaN HEMTs DC- and RF-characteristics.
However, the model and its derivations have been widely used as an initial
approach for modeling [5.44, 5.141, 5.142]. The model is simple and simulation-
time-efficient and can be tweaked efficiently even for advanced circuit simulation [5.141], however, only if applied correctly. The extraction and validation
of an analytical Curtice large-signal model for AlGaN/GaN HEMTs are provided in [5.63]. The analytical model uses the cubic Curtice-based current
equation approach [5.43, 5.45]. A model for a 0.3 μm×200 μm AlGaN/GaN
HEMT on sapphire substrate is extracted. It is verified between 0.5 and
26.5 GHz for the S-parameters and further by time-domain large-signal measurements at 4 GHz. A similar temperature-dependent nonlinear analytical
model for AlGaN/GaN HEMTs on s.i. SiC is given in [5.95, 5.199]. In this

case, the Curtice model of (5.59), (5.60) is slightly modified for the large-signal current equation and includes self-heating effects. The extraction procedure includes extensive pulsed-IV-measurements with pulse widths tuned to avoid the trapping time constants. These pulsed-measurements are also used to analyze the thermal effects and to extract the thermal coefficients and the thermal resistances. Cree reported the use of the modified Curtice current approach with four bias regions for the modeling of SiC MESFETs and GaN HEMTs [5.138, 5.141]. No additional information is given due to intellectual property constraints. The model is similar to the approach used in [5.56]. In general, the Curtice-based approaches need further modifications for the description of intermodulation and other linearity characteristics, as the higher order derivatives of the current equation are zero or trivial, due to the particular polynomial approach for the analytical current equation [5.181]. Other approaches, such as the Angelov and other analytical approaches, will be discussed below [5.181, 5.190].

Table-Based Models for GaN Devices

As an alternative to analytical approaches, table-based models have been developed, e.g., for GaAs devices in [5.8]. The table-based models yield higher modeling accuracy, while the typical numerical effort during simulation is higher, typically in the order of a factor of 1–2 in the simulation time compared to analytical approaches. The interpolation schemes have to be verified and analyzed critically in order to insure the accuracy and numerical efficiency [5.118] of the linearity modeling when using, e.g., cubic spline interpolation, see [5.154]. Table-based models for GaN HEMTs are given, e.g., in [5.118], based on GaAs approaches. The equivalent-model-elements are extracted from S-parameters and cw- and pulsed-DC-IV-measurements. Particularly, gridding and interpolation issues for dense and large gridding are discussed and optimized for numerical efficiency. The table-based data has been both extrapolated and interpolated from the measurements in order to enhance convergence speed and accuracy. Examples for further models are given below.

Aspects of Charge Modeling

The second issue of large-signal modeling is charge extraction and modeling. Both static and quasi-static approaches are used. Charge-storage is the major contributor to the linear behavior of a device, and, next to the current-source nonlinearities, a major source of nonlinearities [5.152]. A big challenge is to meet the requirement of charge conservation within the modeling approach or integrability [5.155] as this has also been an issue for the modeling of GaAs FETs. A general survey of nonlinear-charge modeling and its impact on the description of nonlinear device quantities is given in [5.152] for various device technologies such as Si, GaAs, and InP FETs. The modeling constraints of terminal charge conservation and energy conservation, and the necessary and sufficient conditions for the construction of unique device-specific nonlinear models from experimental device data are discussed. The focus of this paper is the accurate descrition of PAE, IP3, and ACPR. This generalized approach is further extended to GaN HFETs in [5.166]. Examples of GaN FETs modeling are given in [5.166] for the modeling in a state–space approach. This state-space approach is used as a general framework for various model descriptions of the dispersive features of the GaN HFETs.

Several charge modeling examples are available for FETs. The Curtice model uses a linearized model of C_{gs} and C_{gd} around the operating bias; for details see [5.170]. The Chalmers model provides the following equations for a GaAs PHEMT for the charges and capacitances:

$$C_{gs} = C_{gs,0}[1 + \tanh(\psi_1(V_{GS}))][1 + \tanh(\psi_2(V_{DS}))], \qquad (5.61)$$

$$C_{gd} = C_{gd,0}[1 + \tanh(\psi_3(V_{GS}))][1 - \tanh(\psi_4(V_{DS}))]. \qquad (5.62)$$

The functions ψ_i are polynomials of the voltages given in the argument. Charge modeling of GaN FETs is given in [5.7]. In this case the modeling is based on an extended charge approach derived from the GaAs PHEMT world in [5.6]. In some cases, intermodulation parameters and even ACPR can be reliably predicted, if charge-storage effects are included during the extraction of table-based LS-models, see [5.30, 5.118]. Charge conservation is traded off vs. simulation accuracy, and its violation poses the need for improved and consistent modeling. The issue of charge conservation is further discussed in [5.153]. The charge conservation issue is reduced to a local charge conservation approach. The dynamic charge relaxation can further be described by a relaxation time approach [5.59, 5.155]. The requirements of integrability or charge conservation are found crucial for the description of the dynamic relaxation due to thermal or trap relaxation effects.

Examples of Analytical Current Equations for GaN HFETs

The detailed description of the current relations is the first problem for the extraction of a large-signal model. The extraction of the RF-transconductance g_m and RF-output-conductance g_{ds} for the various bias and temperature situations remains a fundamental problem, e.g., [5.1]. Equations (5.63) and (5.64) give the basic current equations used for FETs according to Angelov [5.5]:

$$I_{DS} = I_{pk} \cdot (1 + \tanh(\psi)) \tanh(\alpha \cdot V_{DS})(1 + \lambda V_{DS}) \qquad (5.63)$$

$$\psi = P_1(V_{GS} - V_p) + P_2(V_{GS} - V_p)^2 + P_3(V_{GS} - V_p)^3 + ... \qquad (5.64)$$

The equations contain several fitting parameters, and further functions (P_i), which can be used to include dispersion, as detailed further. The typical LS-models of Angelov [5.5, 5.6, 5.9–5.11], Materka [5.111], and Curtice [5.43, 5.45, 5.170] are used for GaAs MESFETs and PHEMTs. The basic distinction between these models is not so much the accuracy, but the inclusion of thermal effects, the better description of the derivatives, and of the related linearity performance. The main difference between the models thus also includes the V_{DS}-dependence of the threshold voltage [5.30, 5.44] in order to model the particular shape of the transfer and output characteristics [5.5]. An extension of the Angelov approach [5.11] more suitable for GaN HEMTs is reported in [5.190] and reads

$$I_{DS}(V_{GS}, V_{DS}) = I_0(1 + \tanh(\psi_1)) + \frac{(1 + \lambda V_{DS})}{1 + \tanh(\psi_2)\psi_3} \cdot \tanh(\alpha V_{DS}), \quad (5.65)$$

$$V_p = V_{p0} - \delta \cdot V_{DS}, \quad (5.66)$$

$$\psi_1 = P_{11}(V_{GS} - V_p) + P_{21}(V_{GS} - V_p)^2 \\ + P_{31}(V_{GS} - V_p)^3, \quad (5.67)$$

$$\psi_2 = P_{12}(V_{GS} - V_p) + P_{22}(V_{GS} - V_p)^2 \\ + P_{32}(V_{GS} - V_p)^3, \quad (5.68)$$

$$\psi_3 = \alpha_3 \cdot (V_{DS} - V_{GS} - V_{DS0}) + \beta_3 \cdot (V_{DS} - V_{GS} - V_{DS0})^2 \\ + \gamma_3 \cdot (V_{DS} - V_{GS} - V_{DS0})^3. \quad (5.69)$$

This model has been modified again and applied directly to (dispersive) GaN HEMTs, as described in [5.7]. The modified current equations include

$$I_{DS} = 0.5 \cdot (I_{DSp} - I_{DSn}), \quad (5.70)$$

$$P_1 = g_{m,max}/I_{pk0}, \quad (5.71)$$

$$I_{DSp}(V_{GS}, V_{DS}) = I_{pk0} \cdot (1 + \tanh(\psi_p))(1 + \tanh(\alpha_p V_{DS})) \\ (1 + \lambda_p \cdot V_{DS} + L_{sb0} \exp(V_{GD} - V_{thr})), \quad (5.72)$$

$$I_{DSn}(V_{GS}, V_{DS}) = I_{pk0} \cdot (1 + \tanh(\psi_n))(1 - \tanh(\alpha_n V_{DS}))(1 - \lambda_n \cdot V_{DS}), \quad (5.73)$$

$$\psi_p = P_{1m}(V_{GS} - V_p) + P_{2m}(V_{GS} - V_p)^2 \\ + P_{3m}(V_{GS} - V_p)^3, \quad (5.74)$$

$$\psi_n = P_{1m}(V_{GS} - V_p) + P_{2m}(V_{GS} - V_p)^2 \\ + P_{3m}(V_{GS} - V_p)^3, \quad (5.75)$$

$$V_p(V_{DS}) = V_{pks} - \Delta V_{pks} + \Delta V_{pks} \tanh(\alpha_S V_{DS} + K_{bg} V_{bgate}), \quad (5.76)$$

$$P_{1m} = P_1(f(T)) \cdot [(1 + \Delta P_1)(1 + \tanh(\alpha_S V_{DS}))], \quad (5.77)$$

$$P_{2m} = P_2(f(T)) \cdot [(1 + \Delta P_2)(1 + \tanh(\alpha_S V_{DS}))], \quad (5.78)$$

$$P_{3m} = P_3(f(T)) \cdot [(1 + \Delta P_3)(1 + \tanh(\alpha_S V_{DS}))], \quad (5.79)$$

$$\alpha_p = \alpha_R + \alpha_S \cdot (1 + \tanh(\psi_p)), \quad (5.80)$$

$$\alpha_n = \alpha_R + \alpha_S \cdot (1 + \tanh(\psi_n)). \quad (5.81)$$

The modification mainly consists in increasing the flexibility to react to the effects of dispersion during the extraction, while keeping the number of parameters low. The introduction of K_{bg} controls the dispersion. The parameters ΔP_i allow the adjustment of the linearity. The temperature dependence of the carrier velocity is modeled in the functions $P_i(f(T))$. According to [5.7] these modifications allow an accurate modeling of the DC- and RF-S-parameters for GaN HFETs with small gate widths. The main issue quoted in [5.7] is the modeling of both increasing and decreasing transconductance with the drain voltage V_{DS}. A similar idea has been pursued in [5.190]. However, even if the small-signal parameters can be matched accurately for a wide bias- and frequency-range, the modeling of the large-signal and linearity characteristics is not certain and needs further consideration, for details, see [5.166].

Large-Signal Models: Dispersion Modeling

Modeling of modifications of the quasi-static behavior of devices is necessary due to the strong impact of both dispersion and thermal effects. This insight leads to the need of the separation of thermal effects from dispersion.

An Initial Approach for Dispersion Modeling

Initial modeling for the inclusion of dispersion of the output conductance g_{ds} and transconductance g_m [5.61] in an FET can be achieved by additional elements in the equivalent-circuit model. A very simple example of an equivalent circuit is given in Fig. 5.37. The modeling of the complex output conductance Z_{ds} and transconductance g_m reads

$$Z_{ds} = \frac{R_{ds} \cdot [1 + \omega\, C_{ss}\, R_{ss}]}{1 + i\,\omega\, C_{ss}[R_{ss} + R_{ds} + g_{m2} \cdot R_{ss} R_{ds}]} \qquad (5.82)$$

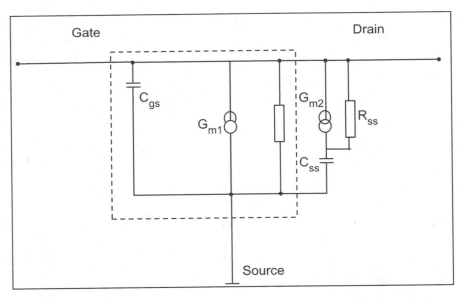

Fig. 5.37. Simplified dispersion equivalent circuit model for a GaAs FET

$$g_{\mathrm{m}}(t) = g_{\mathrm{m}1} - g_{\mathrm{m}2} \left[\frac{\mathrm{i}\, \omega\, C_{\mathrm{ss}}\, R_{\mathrm{ss}}}{1 + \mathrm{i}\, \omega\, C_{\mathrm{ss}}\, R_{\mathrm{ss}}} \right]. \tag{5.83}$$

In the DC-case ($\omega = 0$) and RF-case ($\omega = \infty$), the values of g_{m} and g_{ds} amount to

$$Z_{\mathrm{ds,DC}}(\omega = 0) = R_{\mathrm{ds}}, \tag{5.84}$$

$$g_{\mathrm{m}}(t)(\omega = 0) = g_{\mathrm{m}1}, \tag{5.85}$$

$$Z_{\mathrm{ds,RF}}(\omega = \infty) = \frac{R_{\mathrm{ds}}}{1 + R_{\mathrm{ds}}[1/R_{\mathrm{ss}} + g_{\mathrm{m}2}]}, \tag{5.86}$$

$$g_{\mathrm{m}}(t)(\omega = \infty) = g_{\mathrm{m}1} - g_{\mathrm{m}2}. \tag{5.87}$$

We see that the additional current generator $g_{\mathrm{m}2}$ and the additional RC combination modify both the transconductance g_{m} and output conductance g_{ds} in the requested manner. A physics-based large-signal model for HFETs including the substrate-induced drain-lag is reported in [5.92]. Similarly to Fig. 5.37, the equivalent circuit for a GaAs HFET is modified by additional elements. In the circuit model, electron generation/recombination in deep traps are expressed by a parallel circuit consisting of a diode and a resistor, which are deduced from SRH statistics. The model agrees well with two-dimensional simulation results and experimental current-transient data for large-signal voltage steps. An improved large-signal model more suitable for the simulation of integrated circuits is provided in [5.113]. It models the onset of the low-frequency dispersion in GaAs HFETs. The idea is very similar to that presented in Fig. 5.37. Fig. 5.38 gives the equivalent circuit of this model. In addition to the intrin-

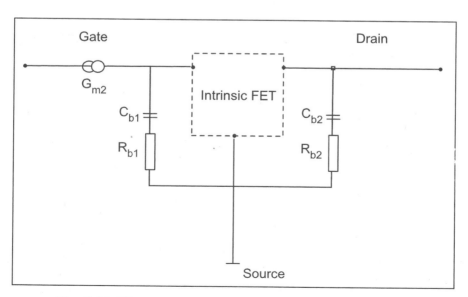

Fig. 5.38. Dispersion equivalent circuit model for a GaAs HFET

sic HFET model, an additional high-resistance network is implemented for the modeling of dispersion and current-lag. The output impedance and the pulse response of a GaAs HFET can be modeled precisely. All these initial approaches modify the relevant elements according to their behavior. A more extensive filter network is used for GaAs PHEMTs in [5.23] to model the pulsed-characteristics and even intermodulation distortion. Even the time-domain response of a 3GPP-WCDMA signal can be simulated in the baseband.

Modifications of the Current Equations

Modeling approaches for the inclusion of dispersive effects have already been presented with (5.65)–(5.69). Another approach is provided in (5.70)–(5.78). The dispersion is modeled according to the modification of the RF-drain-current-characteristics. The classical modification of the analytical current equation in the Angelov large-signal model is taken from [5.6]. The currents are initially modified as

$$I_{\mathrm{DS,RF}}\left[V_{\mathrm{GS}}(t),\,V_{\mathrm{DS}}(t)\right] = I_{\mathrm{DS,DC}}\left[V_{\mathrm{GS}}(t),\,V_{\mathrm{DS}}(t)\right] \tag{5.88}$$
$$+ \Delta I_{\mathrm{DS}}\left[V_{\mathrm{GS}}(t),\,V_{\mathrm{DS}}(t)\right] + \Delta I_{\mathrm{DS}}^{-}\left[V_{\mathrm{GS}}(t),\,V_{\mathrm{DS}}(t)\right].$$

The dispersion is included by the appropriate modification of the P-factors in (5.63) and (5.64). One example of this modification has been presented in [5.7] and (5.70)–(5.81).

Gate-Lag Modeling in Large-Signal Models for GaN HFETs

To this point, the initial modeling mainly modifies the bias-dependent small-signal equivalent circuit elements. A first approach to a dispersion correction in a real LS-model uses a frequency-dependent modification of the internal terminal bias, mostly of the V_{GS} bias, as reported in [5.79]. The gate voltage modification scheme is composed as

$$V_{\mathrm{GS},X} = \bar{v}_{\mathrm{gs}} + \alpha_1 \cdot (v_{\mathrm{ds}}(t) - \bar{v}_{\mathrm{ds}}) + \alpha_2 \cdot (v_{\mathrm{gs}}(t) - \bar{v}_{\mathrm{gs}}) \tag{5.89}$$

with $\bar{v}_{\mathrm{gs}}$ and $\bar{v}_{\mathrm{ds}}$ being the static voltage components. This new $V_{\mathrm{GS},X}$ is inserted into the current function:

$$I_{\mathrm{DS}} = f(V_{\mathrm{GS,X}}, v_{\mathrm{ds}}). \tag{5.90}$$

This methodology can be easily introduced into any large-signal model.

Another LS-model for the virtual gate effect in GaN HFETs is proposed in [5.39]. The model is based on the Agilent EEHEMT model and intends to model anomalous transients in the bias for high quiescent V_{DS0} based on surface states. A parasitic FET device is inserted in the drain path, including a voltage control circuit network, which controls the virtual gate and adjusts the gate voltage according to the V_{DS} bias. This virtual gate model can be very flexibly adjusted to the appropriate quiescent bias and drain pulse width situations. The verified pulse width range amounts to $100\,\text{ns} \leq t \leq 1\,\text{ms}$ for a GaN HFET with $l_g = 150\,\text{nm}$.

Drain-Lag Modeling in Large-Signal Models for GaN HFETs

A more elaborate description of the dispersion is truly two-dimensional with respect to V_{GS} and V_{DS}, as is suggested in [5.73, 5.74]. The idea of a virtual gate-lag model [5.39] can be extended to a drain-lag model based on the equivalence of drain-lag and self-backgating effects [5.73]. In first order, drain-lag effects are physically caused by trapping underneath the channel. The space charge created by the traps and changed by the drain bias adds a contribution to the (V_{DS}-dependent) pinch-off voltage V_{thr} and can be appropriately described. Depending on the time constants observed during the modeling of the feedback of this self-backgating, the virtual gate is dynamically adjusted once a drain voltage pulse is applied. The critical equation is [5.73]:

$$k_n = k_{rel\,n} \cdot g_m(V_{GS} - V_{thr}). \tag{5.91}$$

$k_{rel\,n}$ is the trap transient of the nth trap, which reacts to the drain current pulse, and translates into the dynamic adjustment of the actual control voltage at the gate terminal.

Large-Signal Models: Diodes and Ohmic Contacts

Additionally, the RF-modeling of diodes and ohmic contacts in FET has always been an important topic. Examples of the large-signal modeling of two-terminal GaAs RF-diodes are given, e.g., in [5.156]. The forward AC behavior of n-GaN Schottky diodes are analyzed in [5.198]. Negative capacitance and an additional interfacial layer are considered at the Schottky interface which leads to a nonlinear capacitive effect.

In the linear device region, ohmic contacts in wide-bandgap semiconductor devices yield a diode type contribution, as is pointed out, e.g., in [5.184]. The band structure near the ohmic contacts leads to nonlinear transport effects at both drain and source. The nonlinear source-resistance effects are described in [5.184, 5.185]. The source-resistor modulation and its bias dependence are discussed. As a result, a space-charge zone is extracted below the source which limits both RF-performance and amplifier linearity. The effect is primarily visible during the high current portion of the RF-cycle. However, a direct description of this nonlinear behavior in a lumped LS-model without physical device simulation has not been presented.

Soft-Breakdown and Impact Ionization

In the high-field-bias-domain, breakdown mechanisms in III-N devices can be included into the large-signal model, as reported in [5.186]. Additional reverse current generators to the gate current, similar to those already considered for dispersion modeling, are included parallel to the gate–source and source–drain diode. They can describe both tunneling-induced breakdown and channel breakdown. The extraction example in [5.186] gives a good agreement for the modeling of leaky GaN HFETs with a strong difference in the DC- and RF-breakdown voltages. Other typical modifications are made to the output conductance and can be incorporated due to soft breakdown or direct impact ionization, as performed for the Angelov-model in [5.6]. The modeling of the modification of the drain current due to impact ionization in [5.133] yields

$$i_{\mathrm{II}} = A \cdot I_{\mathrm{DS}} \exp\left(\frac{-B}{V_{\mathrm{DS}} - V_{\mathrm{DS0}}}\right). \qquad (5.92)$$

The inclusion of soft-breakdown into LS-modeling is further described in [5.6]. Soft-breakdown may not be a dominant issue in AlGaN/GaN devices; however, the particular behavior of small-gate-length devices ≤ 150 nm may require this feature. The drain current is modeled by the insertition of L_{sb} into the standard current equation according to

$$I_{\mathrm{DS}}(V_{\mathrm{GS}}, V_{\mathrm{DS}}) = I_0(1 + \tanh(\psi)) \cdot \tanh(\alpha V_{\mathrm{DS}}) \cdot$$
$$(1 + \lambda V_{\mathrm{DS}} + L_{\mathrm{sb}}), \qquad (5.93)$$
$$L_{\mathrm{sb}} = L_{\mathrm{sb0}} \cdot [\exp(L_{\mathrm{sd1}} V_{\mathrm{dgt}} + ...) - 1], \qquad (5.94)$$
$$V_{\mathrm{dgt}} = \frac{V_{\mathrm{DS}} - K_{\mathrm{trg}} V_{\mathrm{GS}}}{V_{\mathrm{tr}}}, \qquad (5.95)$$
$$L_{\mathrm{sd1}} = L_{\mathrm{d1}}(1 - L_{\mathrm{g1}} V_{\mathrm{GS}}). \qquad (5.96)$$

The dots in (5.94) indicate potential higher orders in the voltages. This modeling can be included to match the DC-gds, especially in the near-breakdown V_{DS} region.

Large-Signal Models: Thermal Analysis

Thermal and self-heating effects are very pronounced in any power semiconductor device, e.g., [5.167]. Several variants exists for the inclusion of thermal and especially self-heating effects in LS-models. The additional inclusion of the thermal effects in the Curtice model for GaN HFET is reported in [5.95]. The channel temperature T_{chan} to be used in a temperature-dependent version of (5.60) is modeled in a global approach in [5.71]

$$T_{\mathrm{chan}} = R_{\mathrm{th}} \cdot V_{\mathrm{DS}} \cdot I_{\mathrm{DS}} + T_0. \qquad (5.97)$$

R_{th} is the global thermal resistance including the temperature dependence of the thermal conductivity $\kappa(T)$ according to

$$\kappa(T) = \kappa(T_0) \cdot \left(\frac{T}{T_0}\right)^{-b}. \qquad (5.98)$$

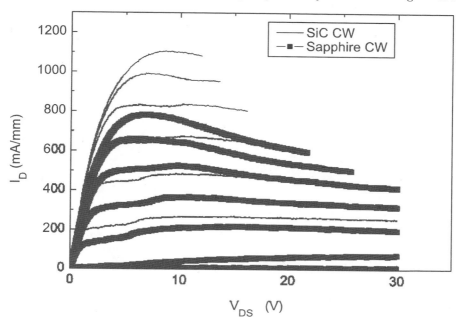

Fig. 5.39. CW-output characteristics to reveal the different thermal effects in SiC and sapphire, $V_{\mathrm{GS}} = -5$ to $1\,\mathrm{V}$, step $1\,\mathrm{V}$

The thermal conductivity for s.i. SiC at room temperature amounts to $3.3\,\mathrm{W}\,(\mathrm{cm\,K})^{-1}$. The temperature coefficient $b = -1.5$ is used. The particular R_{th} must include the mounting situation in the application, which may strongly differ from the (on-wafer) thermal situation, from which the model is extracted. Another global temperature concept is applied in [5.6]. It reads

$$T(T_0) = \frac{T_0}{\left[1 - \dfrac{\Theta(T_0) \cdot P_{\mathrm{diss}}}{4T_0}\right]^4}. \tag{5.99}$$

In this case, P_{diss} is the dissipated DC-power and $\Theta(T_0)$ is the temperature-dependent thermal resistance.

Extraction of the Thermal Resistance

A typical comparison of the different substrate properties for an AlGaN/GaN HEMT with a gate width $W_{\mathrm{g}} = 1\,\mathrm{mm}$ is given in Fig. 5.39. With otherwise nominally constant epitaxial and technological parameters, the strong difference in the drain current and derived DC-g_{ds} values due to thermal effects can be observed. The thickness of the substrates is $250\,\mu\mathrm{m}$ for sapphire and $370\,\mu\mathrm{m}$ for the s.i. SiC substrate at a constant backside temperature stabilized by a thermal chuck. The thermal impedance itself can be extracted from different methods:

- From the DC-characteristics [5.133]
- From pulsed-output characteristics [5.72, 5.88]
- From pulsed-RF-current characteristics
- From photocurrent measurements of the optical absorption spectra [5.146]
- From verified scaling rules [5.72]

The interaction and construction of thermal models from single gate fingers to complete power cells is given in [5.72]. Even an electrothermal model for complete output stages can be constructed. Further, the methodology allows the construction of a large-signal model from verified scaling rules for the parasitics. A large-periphery device is modeled from the thermal behavior of individual FET fingers, including their interaction. This construction is a very desirable procedure, as high power devices are not always accessible to direct modeling due the need to control high currents and high power levels [5.44]. The inclusion of multilayer and multifinger thermal effects into large-signal modeling of large-periphery GaAs FETs is described in [5.20]. A consistent method is used to derive the large-periphery behavior of the devices. The model incorporates one of the critical features often neglected in the LS-modeling procedure: the actual thermal mounting situation in the application. This includes the thermal-resistance and -capacitance after wafer thinning and after the actual mounting in modules. The impact of additional thermal viaholes and of nonuniform 3D-distributions are investigated in [5.19].

Dynamic Thermal Effects

The thermal behavior is a highly dynamic process, as described, e.g., in [5.95]. The timing and the transient behavior of static thermal effects in FETs is investigated in [5.132]. The work supplies a methodology for finding isothermal measuring conditions for GaAs devices using a limited amount of pulsed-DC- and pulsed-RF-measurements at pulse-widths between $2\,\mu s$ and $10\,ms$ for GaAs devices. Such near isothermal conditions can be used to investigate the substrate temperature dependence independently from the self-heating effects. Apart from the thermal resistance, in a first approach a thermal capacitance can be extracted to describe the dynamic thermal behavior with a thermal time-constant [5.95]:

$$\tau_{\text{thermal}} = R_{\text{th}} \cdot C_{\text{th}}. \qquad (5.100)$$

A thermal relaxation time of $4.2 \times 10^{-5}\,s$ is extracted for the Curtice approach in [5.95]. More thorough transient analysis is performed in [5.93, 5.209]. A very small relaxation time of $190\,ns$ is extracted by electrical transmission-line pulser measurements for a thermal resistance $R_{\text{th}} = 70\,K\,W^{-1}$. The simulated transient analysis in [5.209] for pulse width of 2–$5\,\mu s$ at 33% duty cycle shows that the maximum temperature is not reached within the pulse duration and is further strongly related to the substrate material. Thus, a very careful analysis is required for pulsed-operation with duty cycles between 10–100%. A more general procedure for the extraction of bias-dependent self-heating parameters is given in [5.2]. The procedure is based only on temperature-dependent electrical measurements. Both the thermal resistances and the thermal capacitances are extracted, which allows to describe the dynamic thermal behavior.

Behavioral Modeling Approaches

Behavioral modeling has recently been addressed for the modeling of RF-devices in order to enable a description of RF-devices for complex signal forms with nonconstant envelopes [5.1, 5.90]. For this very different modeling approach, a broad theoretical modeling base is available, e.g., in [5.187]. Direct extraction of the nonlinear HEMT model from vectorial LS-measurements is provided, e.g., in [5.163, 5.165]. The RF-IV characteristics of the device can be constructed purely from large-signal data. The general approach can be formulated as

$$I_1(t) = f_1(V_1(t), V_2(t), \dot{V}_1(t), \dot{V}_2(t), \ddot{V}_1(t), \ddot{V}_2(t), ..., \dot{I}_1(t), \dot{I}_2(t), ..), \quad (5.101)$$

$$I_2(t) = f_2(V_1(t), V_2(t), \dot{V}_1(t), \dot{V}_2(t), \ddot{V}_1(t), \ddot{V}_2(t), ..., \dot{I}_1(t), \dot{I}_2(t), ..). \quad (5.102)$$

The procedure has been used by a number of groups for the model extraction for AlGaN/GaN HEMTs. An artificial-neural-network (ANN) model is constructed from near-optimum-load large-signal measurements, as described in [5.164]. An X-band amplifier is constructed from an ANN model of a 1 mm device between 1–26 GHz. With this very general approach, measurement data from a four-channel vector large-signal measurement setup including the phase information of the signals is used for the modeling [5.163]. The specific adaption for GaN HEMTs includes the consideration of self-heating as an additional state variable apart from the six variables considered for a classical HEMT model.

5.5 Linearity Analysis and Modeling

One of the principal advantages of nitride devices is improved linearity compared to other device technologies at the same output power levels, e.g., [5.203]. Correspondingly, similar linearity levels for very much higher output power levels can be achieved [5.117]. The nature of these improvements can be attributed to the larger bandgap and thus increased operation voltages and output power capabilities [5.204]. A more detailed analysis, such as whether and to what extent GaN is really a more linear material, is not fully complete, e.g., [5.69, 5.71]. The full linearity is not yet exploited, especially due to the state of the technology with respect to dispersion. Charge trapping is known to be linked to electrical memory effects. Both the intermodulation levels and the side asymmetries of two-tone measurements are modified by charge trapping, as shown by Volterra-series analysis, e.g., in [5.27]. Class-AB operation is found to be very favorable for AlGaN/GaN HFETs with respect to gain, PAE, and linearity in [5.203]. Recent advances will be analyzed in the following section.

5.5.1 Basic Understanding

The inclusion of nonlinear effects into large-signal modeling is of fundamental importance. The inclusion of intermodulation distortion into large-signal modeling for GaAs and GaN devices has been investigated for some time, e.g., in [5.56, 5.69, 5.77, 5.143]. Typical verification procedures of the LS-models include a comparison of the LS-model with multitone intermodulation measurements, e.g., as performed in [5.77, 5.143]. ACPR LS-simulations and verification have only recently been performed, e.g., [5.23, 5.30, 5.152].

Modeling

Some of the conventional current modeling approaches, such as the Curtice model, are not fully suitable for linearity analysis and simulation. The inclusion of higher derivatives of the current and charge equations into the modeling, extraction, and verification is of fundamental importance for the LS-simulation and quantitative prediction of nonlinear effects. Thus, current modeling using polynomial current approaches, e.g., in [5.43, 5.45, 5.170], leads to trivial or even vanishing higher derivatives. Therefore, models with sufficient degrees of freedom are required for the efficient inclusion of nonlinear effects [5.30]. A broad number of publications is available for the modeling and understanding of the actual behavior of the intermodulation spectra, e.g., [5.90], and of the adjacent channel power ratio (ACPR) curve forms [5.49]. The basic nulling conditions of the intermodulation in GaAs MESFETs are discussed in [5.136], as already detailed earlier. The additional contribution of self-heating to intermodulation in GaAs FETs is described in [5.135] and reads

$$I_{\mathrm{D}} = I_{\mathrm{D},0}\left(T_{\mathrm{L}}\right) \cdot \left(1 - \delta \cdot P_{\mathrm{diss}}(t) \cdot h(t)\right). \tag{5.103}$$

$h(t)$ denotes the thermal impulse response function in the time-domain, when the current is sent through a low-pass filter. $I_{\mathrm{D},0}$ is the isothermal current. $h(t)$ is typically modeled with either a single-time-constant or few-time-constants response, e.g., [5.135]. An analysis of the relation of charge trapping and intermodulation in HEMT is further performed in [5.26]. A Volterra-series modeling approach is used for the drain current according to a Taylor-series expansion

$$I_{\mathrm{D}} = G_{\mathrm{m}} \cdot v_{\mathrm{g}} + G_{\mathrm{d}} \cdot v_{\mathrm{d}} + G_{\mathrm{m},2} \cdot v_{\mathrm{g}}^2 + G_{\mathrm{m},\mathrm{d}} \cdot v_{\mathrm{g}} \cdot v_{\mathrm{d}} + G_{\mathrm{d},2} \cdot v_{\mathrm{d}}^2$$
$$+ G_{\mathrm{m},3} \cdot v_{\mathrm{g}}^3 + G_{\mathrm{m},2,\mathrm{d}} \cdot v_{\mathrm{g}}^2 \cdot v_{\mathrm{d}} + G_{\mathrm{m},\mathrm{d},2} \cdot v_{\mathrm{g}} \cdot v_{\mathrm{d}}^2 + G_{\mathrm{d},3} \cdot v_{\mathrm{d}}^3. \tag{5.104}$$

This analysis yields the nonlinear transfer functions, as described, e.g., for GaAs FETs in [5.134].

Understanding of Intermodulation

More advanced understanding of the relevant factors of influence for analysis beyond RF-S-parameter and harmonic measurements has been achieved. The principal sources of nonlinearity in AlGaN/GaN devices are similar to those in any other FETs [5.137]. The analysis of the sources of nonlinearity yields the following:

- Quasi-static stationary analysis, e.g., [5.10, 5.43, 5.111], including
  Nonlinear current sources [5.136]
  Nonlinear charge sources [5.154]
  Diode analysis [5.156]
- Thermal effects [5.71, 5.95, 5.135], including thermal memory [5.21]
- The impact of dispersion and trapping on the linearity measures [5.27]
- The impact of the baseband impedance and matching networks [5.25, 5.29, 5.49]

Following this list, the principal effects can be separated. The nonlinear current source g_m is a natural source of nonlinear behavior, as the derivatives of the transconductance are typically nontrivial, compare (5.104). The same is true for the higher derivatives of the main capacitances C_{gs} and C_{gd}, which similarly are nontrivial. The principal impact of thermal effects on the linearity is due to the pulse response in (5.100), see [5.135]. The isolated frequency of the heating is a few kilohertz, which has an impact on the variation of intermodulation with frequency and thus memory effects. A typical time constant or frequency for the thermal dissipation is in the 1–10 kHz region; however, a second contribution is visible at higher frequencies up to the gigahertz regime, e.g. [5.209], which can be significant too. Thermal memory [5.21] is especially significant for signals with high peak-to-average ratio (PAR), as detailed further.

The temperature-dependent nonlinearities in GaN/AlGaN HEMTs are analyzed in [5.71] based on a physical device model. The model is calibrated to measurement data and predict the dependence of IM_3 on both temperature between 200 and 500 K and physical device parameters such as gate length. The relative change of f_{max} with temperature is found to be stronger for devices with $l_g = 100$ nm in comparison to devices with $l_g = 500$ nm. At the same time it is predicted that the GaN HFETs with shorter gates yield lower IM_3 and higher OIP$_3$ for a given gate width.

The bias dependence of the nonlinearities is of fundamental importance. A sweet-spot analysis can be performed, e.g., in [5.48], based on a Volterra-series modeling approach. This modeling approach allows a quantitative prediction of the sweet spots as a function of fundamental power. The model is further extended to random signals using a special harmonic balance machine. The intercept point behavior of GaAs FETs at Ka-band frequencies is analyzed in [5.119]. The impact of the gate biasing on the linearity is investigated, which is a typical procedure at all frequencies. The importance of the baseband

and higher-harmonics impedance with respect to intermodulation is further stressed in [5.25, 5.28]. The citations stress the impact of the parasitic bias networks and the resulting baseband impedance on the intermodulation and the resulting asymmetry of the intermodulation products. The variation of the distortion is as high as 4 dB with the baseband impedance. Further effects, such as the asymmetry of the intermodulation products, are more thoroughly investigated in [5.49]. The necessary condition for the asymmetry of the IMD generation is a significant reactive part of the baseband load impedance. However, these baseband and second harmonic effects must not be overridden by the real part of the IMD components. The memory-related contributions of intermodulation asymmetries will be discussed in the next section.

Memory Effects

Memory effects strongly influence the performance of transistors and power amplifiers in linear operation [5.178]. Their occurrence limits the system performance, as memory effects can only partially be compensated by typical linearization strategies [5.21] and thus effectively reduce the efficiency. This fact is very important for the application of GaN HFETs in base station applications, as systematic investigations have shown that GaN HFETs are better behaved with respect to thermal memory, as compared to all available semiconductor technologies, such as LDMOS and GaAs [5.21, 5.58, 5.178]. This includes the memory-related asymmetries of the intermodulation spectra, e.g., [5.90]. Memory effects can be separated into electrical [5.27] and thermal [5.21] relaxation effects within the semiconductor device. Electrical memory are caused by several nonlinear effects [5.49], e.g., by charge trapping, as discussed in [5.27]. The necessary condition for memory-like effects is a critical reactive load impedance in the baseband. Another factor that causes the electrical memory effect is the variation of terminal impedances over the input signal bandwidth around the carrier frequency, its harmonics, and the baseband frequencies [5.27], especially at high power levels, e.g., in a 90 W Si LDMOS device [5.21]. Thermal memory effects are caused by the gain variation through the temperature dependence of the electrical parameters are especially critical in statistical signals with high PAR and nonconstant envelope. Thermal memory, however, is especially critical as these variations can only partially be compensated [5.21]. Thermal compensators can be used in the predistortion to reduce the thermal memory by including a predictive function of the junction temperature. Another correction by a digital deterministic memory predistortion correction for LDMOS and GaN HEMTs is given in [5.52, 5.86]. The thermal compensation method proposed in [5.21] shows better compensation results than a purely memoryless compensation. Further, depending on the signal forms (e.g., EDGE vs. GSM vs. WCDMA signals vs. OFDM), and the amplifier concept [5.53, 5.57], the possibility to mitigate and correct the effects varies due to the nature of the signals [5.106]. Both baseband bandwidth and the number of carriers are of critical importance to the compensation, see [5.53]. The sensitivity of amplifiers to the different types of memory effects thus varies. A great amount of modeling of memory effects on the circuit level is described. Behavioral models based on several approaches have been proposed. A parallel Wiener model is proposed and validated to quantitatively describe memory effects in silicon bipolar base-station- and in GaAs-HFETs-handheld-PAs for different power levels [5.89]. CDMA signals are used for the extraction and validation of the model. Memory effects show a strong negative impact on the improvement of a memoryless predistortion algorithm applied for the 45 W power amplifier. A polynomial model for the IMD and spectral regrowth asymmetries for a 170 W LDMOS device is proposed in [5.90]. Both analog [5.56] and digital [5.21, 5.90] compensation methods for memory have been proposed to minimize the effect of asymmetric IMD and thus of memory. However, the promise of GaN devices is a reduction of memory in such a manner that the critical thermal memory effect can be reduced in order to allow the application of standard predistortion techniques without performance losses for complicated signal forms. This will be discussed with respect to GaN FETs in the next section.

5.5.2 Nitride-Specific Linearity Analysis

A number of linearity investigations have been performed for III-N devices [5.178]. Linearity and gain characteristics of AlGaN/GaN HEMTs are described in [5.203]. The impact of the band structure at the AlGaN/GaN heterojunction is mentioned. The reduction of f_T for increasing gate voltages V_{GS} as a function of rising V_{DS}-voltage is attributed to the scattering of electrons in the AlGaN. Subsequently, the electrons scatter in k-space into higher valleys and effectively loose speed, which results in a reduction of f_T [5.144]. Requirements to the physical device structure for low intermodulation-distortion in GaN/Al$_x$Ga$_{1-x}$N HEMTs are described in [5.104]. A case study is performed, which suggests an improvement in the distortion for high Al-content in the barrier and for devices with the inclusion of an AlN interlayer. Real-space transfer is reduced in this case, and the effective saturation velocity becomes independent of the gate voltage. Single-tone and two-tone time-domain large-signal characterization of GaN HFETs operated in class-A is given in [5.117]. The time-domain analysis yields nearly no phase shift for the fundamental and third harmonic output of the device well into compression for single-tone operation. This implies symmetric intermodulation products, which are experimentally confirmed by two-tone measurements. Fig. 5.40 gives the measured intermodulation product for a 1 mm device at 12 GHz at a two-tone measurement with 1 MHz tone-spacing at $V_{DS} = 20$ V. The measurement shows a nearly linear behavior in class-A/B for the IM_3 without sweet spot behavior. Reduced memory effects are a crucial advantage of GaN on s.i. SiC substrates due to improved thermal management of the actual semiconductor layer on top of a substrate with a very good thermal conductivity. Memory effects of GaN HFETs are analyzed in [5.58,5.178]. The ACLR spectra of GaN HEMTs on s.i. SiC are found to show a reduced amount of memory relative

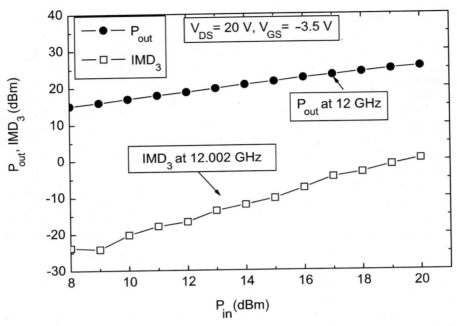

Fig. 5.40. Measurement of the output power P_{out} and intermodulation distortion IM_3 under two-tone conditions at 12 GHz with a two-tone spacing of 1 MHz

to Si, GaAs, and other semiconductor technologies. As an example, Fig. 5.41 gives the output spectrum of a one-carrier WCDMA signal for a GaN HEMT amplifier on s.i. SiC with a gate periphery $W_g = 32$ mm. The peak output power is 47 dBm at 1.95 GHz in Fig. 5.41. ACLR levels of -39 dBc at 5 MHz and ≤ -55 dBc at 10 MHz offset are found. The shoulder behavior of the signal at 5 and 10 MHz under one-carrier WCDMA operation and the spectral regrowth behavior show very little indication of memory effects and spectral regrowth [5.58].

Linearity Modeling and Simulation of GaN HFETs

Device modeling is performed and applied mostly for simple signal forms, i.e., simple two-tone measurements, despite the complicated waveform applied in today's communication systems. The conventional Chalmers model is used to model an AlGaN/GaN HFET with a gate periphery of 2 mm in [5.30, 5.181]. It is found in [5.32] for GaN FETs that the original Chalmers model cannot correctly predict the higher derivatives of the transconductance. Instead, the approach of Fager [5.30, 5.56] is used to match the higher derivatives. The measured output power and the measured intermodulation distortion in class-A, class-AB, and class-B with double-sweet spots can be precisely matched.

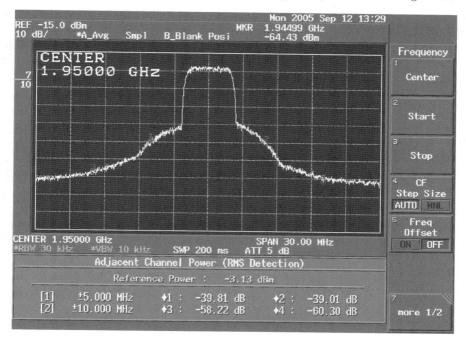

Fig. 5.41. One-carrier WCDMA spectrum at 1.95 GHz at a peak output power level of 47 dBm

The drain voltage dependence of the current is modeled according to

$$I_{DS} = \frac{\beta \cdot V_{GS,3}^2}{1 + \left(\frac{V_{GS,3}^{plin}}{V_L}\right)} \cdot (1 + \lambda V_{DS}) \tanh\left(\frac{\alpha \cdot V_{DS}}{V_{GS,3}^{psat}}\right), \qquad (5.105)$$

where the $V_{GS,3}$ is an effective V_{GS}-voltage modified according to [5.56]. $\alpha, \beta, \lambda, V_L, V_{GS,3}^{plin}$, and $V_{GS,3}^{psat}$ are parameters. For more complicated wave forms, especially for nonconstant signal envelopes, few investigations exist based on large-signal modeling of GaN FET [5.32, 5.48]. A detailed analysis is complicated, as the analysis in the frequency-domain only is not suitable, due to the nonconstant envelopes. Harmonic-balance simulations with an AM-modulated signal and a Gaussian-noise baseband modulation are provided in [5.48]. Examples of the more complicated signals are given in [5.32] combining harmonic balance and the artificial frequency mapping technique. OFDM signals can be simulated efficiently, as they use equally separated carriers.

GaN HFETs with Predistortion Correction

With the increasing interest in GaN devices, several GaN device technologies have been further subject to the LDMOS type of (software) predistortion linearization strategies with parameters matched to GaN. An analog dynamic gate bias technique for improved linearity of GaN FETs is described in [5.40]. A significant improvement of the third-order intermodulation of 10 dB is described, once the gain of the transistor is adjusted in accordance with the instantaneous envelope of the input signal to minimize AM-AM distortion. A large-signal model based on the Curtice approach is used to investigate the nonlinearities. The procedure also serves to improve the ACPR in WCDMA signals by 8 dB. Further, standard WCDMA procedures have been applied for the analog and especially digital predistortion of GaN FETs, e.g., [5.84,5.125]. GaN HEMTs on s.i. SiC substrates by Eudyna have been evaluated by four-carrier WCDMA signals with 5 MHz signal spacing. A significant improvement of the ACPR of $\geq$10 dB is achieved by the application of a digital predistortion [5.84]. A similar digital predistortion strategy has been applied in [5.125]. One-carrier WCDMA signals have been applied to a dual-stage amplifier with a gate periphery of 32 mm. The ACLR criterion is met for an average output power of 35 dBm between 1.8 and 2.7 GHz with the application of the DPD. Once the DPD is applied, the average output power can be increased to 40 dBm between 1.8 and 2.7 GHz meeting the 3GPP ACLR specifications. Error vector magnitude (EVM) and peak code domain error (PCDE) are also met after predistortion. Further examples are given in Chapter 6.

5.6 Noise Analysis

The section addresses the analysis and modeling of both low-frequency and RF-noise in III-N devices. The analysis is so far limited to FETs.

5.6.1 Low-Frequency Noise

Low-frequency noise is of great interest to process analysis [5.191] and to those device functions that are limited by phase noise, such as oscillators. It defines the minimum signal that can be processed by the device [5.65]. A strong correlation can be found of low-frequency noise, material quality, and device reliability [5.191]. Origins of low-frequency noise in FETs are well discussed in the literature, e.g., in [5.65, 5.66]. The sources of noise in the frequency range between 1 Hz and 10 MHz include the following:

- Fluctuations of the channel conductance and mobility [5.65, 5.66]
- Fluctuations of the sheet carrier density [5.65]
- Thermal activation of carriers from localized states in the bulk [5.173], at heterointerfaces [5.65], and at surfaces [5.194]
- Grain boundary motion [5.65]

As an example, Fig. 5.42 gives a low-noise spectrum of a GaN/AlGaN HEMT on s.i. SiC substrate with a gate width $W_g = 2 \times 30\,\mu m$ and a gate length $l_g = 150\,nm$ for various operation bias V_{DS}.

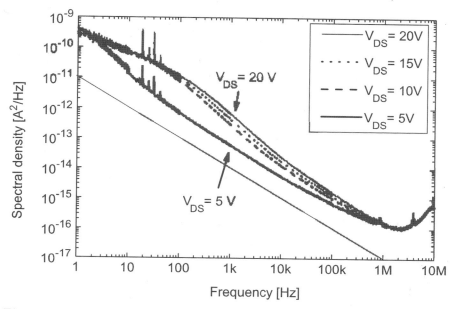

Fig. 5.42. Low-frequency noise spectrum of a GaN HFET between 1 Hz and 10 MHz as a function of drain voltage V_{DS} at a drain current $I_D = 100\,\mathrm{mA\,mm^{-1}}$

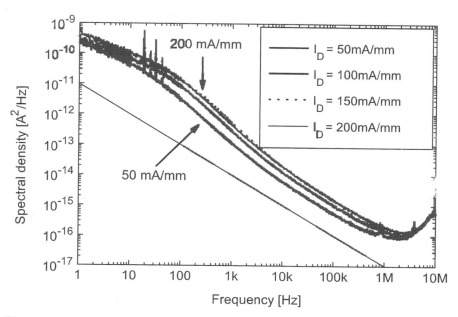

Fig. 5.43. Low-frequency noise spectrum of a GaN HFET between 1 Hz and 10 MHz as a function of drain current at $V_{DS} = 10\,\mathrm{V}$

The lowest $1/f$-noise level is found for the lowest V_{DS} voltage of 5 V. Fig. 5.43 gives the same spectrum as a function of drain current at a drain voltage $V_{DS} = 10$ V. An increase of the $1/f$-noise-floor is found with rising drain current I_{DS}. The spectrum between 1 Hz and 10 MHz is further characterized by a nonsteady behavior of the noise spectrum, which is affected by trapping, as also found for devices from other materials [5.65, 5.173]. The principal modeling for the low-frequency noise is typically [5.66]

$$\frac{S_\nu(f)}{I_{DS}^2} = \frac{\alpha_H}{N \cdot f}. \tag{5.106}$$

The Hooge-parameter α_H is dimensionless, f is the frequency, and N is the total number of carriers in the devices. Typical values of the Hooge-parameter amount to 10^{-4}, which change if the crystal is damaged [5.66]. In MOSFETs, the low-noise spectrum at low drain bias can be modeled according to

$$S_\nu(f) = \frac{e \cdot \mu \cdot \alpha_H}{L^2} \cdot \frac{I_{DS} \cdot V_{DS}}{f}. \tag{5.107}$$

Thus, the spectral noise density is proportional to both the source–drain current I_{DS} and the drain–source voltage V_{DS}. Deviations from the ideal frequency behavior of the current–noise spectral density are modeled according to

$$S_\nu(f) \sim \frac{1}{f^\gamma}. \tag{5.108}$$

A γ-value of 1.2 is found for the modeling of AlGaN/GaN HFETs on sapphire substrates in [5.91], extracted between 1 Hz and 100 kHz. Higher frequency measurements may be useful to separate the impact of fast traps, as can be seen in Fig. 5.42. The factors of influence on the $1/f$-noise are now discussed in detail. Low-frequency noise of doped-channel AlGaN/GaN HFETs on sapphire substrates is discussed in [5.91]. The noise-behavior observed gives $\gamma = 1$. The Hooge parameter is of the order of 10^{-2} and it is proportional to the channel width in this particular case. The noise is correlated with the high density of defects at the AlGaN/GaN interface, which leads to fluctuations of the channel charge.

A characterization of AlGaN/GaN MODFETs on sapphire substrates at low-drain bias of ≤ 1 V is given in [5.65]. Deviations from the purely $1/f$-behavior are reported in [5.65] as in various other publications. The lattice-temperature dependence of the low-frequency noise of MBE-grown samples shows a variation of the parameter γ between 0.85 and 1 for a temperature range between 130 K and RT. A linear scaling of the Hooge parameter with channel width indicates an interface-related effect [5.91, 5.139], possibly, since the AlGaN/GaN interface is the only interface in direct contact with the 2DEG. However, other explanations are also possible, such as a poor contact technology at the ohmic or Schottky contacts [5.139]. The dependence of the low-frequency noise on the silicon doping in the AlGaN barrier layer of the AlGaN/GaN HFET is analyzed in [5.173]. The findings include increased noise floor for the structure with barrier doping relative to the undoped structure and a deviation from the purely $1/f$-behavior especially for the doped structure. The increased low-frequency noise is attributed to the increased carrier trapping/detrapping with the impurity doping in the barrier. Effects of the surface passivation and barrier-layer defects on the low-frequency noise-spectral-density are described in [5.194]. The spectra between 1 Hz and 100 kHz are measured before and after passivation. The passivated device yields lower spectral low-noise density [5.194, 5.196]. The distinction between the gated and ungated region is made, as mentioned in [5.139]. The V_{GS}-dependence is used to distinguish the noise contributions of the series resistance and of the gated region with different noise exponents. The correlation of leakage current and low-noise density is used to explain the difference between passivated and unpassivated devices, as the number of carriers is increased due to the passivation. The substrate- and gate-current-dependence of the low-frequency noise is further investigated in [5.46]. The devices on sapphire and silicon substrate provide a low-noise spectral density similar to that of GaAs devices, while the AlGaN/GaN HEMTs on s.i. SiC substrate provide an increased $1/f$-noise level. However, for the devices on Si substrate, a correlation of the normalized leakage current and low-noise level is found, which indicates a processing issue rather than a substrate issue. Similar findings on the good correlation of the $1/f$-noise and the transients of DC-gate and DC-drain currents after illumination with light with $\lambda = 365$ nm in AlGaN/GaN HFETs are reported in [5.82]. The drain-current and gate-current noise are distinguished by measurements. Both contributions to the $1/f$-noise show a correlation with the transients of the DC-gate and drain currents. The correlation is not perfect, however; this suggests that the $1/f$-noise has the same physical origins as the dispersion.

5.6.2 RF-Noise Analysis and Characterization

III-N FETs are further attractive for RF-noise applications due to the combination of high-speed and high breakdown-voltage properties. In addition the high barrier discontinuity at the AlGaN/GaN interface suppresses potential noise sources. Fig. 5.44 gives the noise sources for the RF-noise in an AlGaN/GaN HFET. These sources of RF-noise include

- Ohmic (parasitic) resistances at source and drain
- Channel noise through velocity fluctuations [5.33]
- Interface contributions
- Surface contributions, such as leakage [5.147] and traps

This low-noise behavior is further analyzed in this subsection. Several investigations are available in the literature, e.g., [5.99, 5.100, 5.109]. A typical noise-equivalent circuit is depicted in Fig. 5.45. A standard small-signal equivalent circuit is used with two additional noise sources. In the standard small-signal

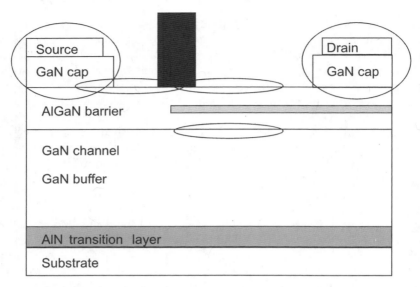

Fig. 5.44. RF-noise sources in AlGaN/GaN HEMTs

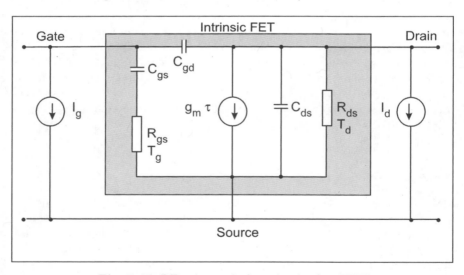

Fig. 5.45. RF-noise equivalent circuit of an FET

equivalent circuit, noise can be described by two noise-equivalent temperatures T_{Gate} and T_{Drain}. Several modifications of Fig. 5.45 are available for the intrinsic modeling of leaky HFETs, e.g., for InAlAs/InGaAs HFETs in [5.148,5.159]. The strong impact of the leakage on the noise in GaN HFETs is described in [5.159]. Additional considerations are necessary for the inclusion of impact ionization and other inductive effects [5.147]. The optimization of III-N FETs as in any other FET can be based on the Fukui-equation [5.60]:

$$N_{\mathrm{F,min}} = 10 \, \log \left(1 + \frac{k_{\mathrm{f}} \cdot f}{f_{\mathrm{T}}} \left[g_{\mathrm{mi}} (R_{\mathrm{G}} + R_{\mathrm{S}}) \right]^{0.5} \right). \qquad (5.109)$$

The requirements for low-noise operation are visible: high gain, high cut-off frequency, and low parasitic resistances are necessary to reduce the noise figures. The Fukui-factor k_{f} can be approximated by [5.109]

$$k_{\mathrm{f}} = 2 \cdot \left[\frac{I_{\mathrm{opt}}}{E_{\mathrm{C}} \cdot l_{\mathrm{g}} \cdot g_{\mathrm{mi}}} \right]. \qquad (5.110)$$

I_{opt} is the drain current level, which minimizes the noise figure. The data used in [5.109] for GaN HFETs did not fit the approximation in (5.110). However, the principal dependence can be used. For a given gate length l_{g}, the intrinsic g_{mi} is lower and the access resistances are higher than that for other semiconductors, whereas the critical field E_{crit} is significantly higher. A more detailed analysis is thus required. The dependence of RF-noise parameters on the Al-content in AlGaN/GaN HFETs is analyzed in [5.109, 5.160]. A comparison of the minimum noise figure and associated gain is given for different Al-contents in the barrier layer. The best RF-low-noise operation in [5.109] is found for the highest Al-content, as the gain and cut-off parameters are best for highest Al-content, while the parasitic elements are nearly independent of the Al-content. The results in [5.160] disagree with this finding, as the four noise parameters are fully independent of the Al-content. It is found that the quality of the channel material is decisive. Intrinsic noise-equivalent-sources and circuit-parameters for AlGaN/GaN HEMTs on s.i. SiC substrate are extracted in [5.98–5.100]. Three independent noise mechanisms are identified: velocity fluctuations, gate leakage [5.159], and traps, as given in Fig. 5.44. Fig. 5.45 gives the typical de-embedded noise-equivalent-circuit model for an HFET device. The noise theory of GaN HFETs is very similar to the modeling of RF-noise in GaAs, e.g., [5.140, 5.171], or InP (H)FETs, e.g., [5.3]. The differences for the modeling of GaN devices include the following:

－ Modified access resistances [5.131]
－ Better overdrive capabilities, i.e., robustness and linearity [5.35, 5.124]

Some examples are now given.

RF-Noise-Measurements and Modeling

RF-noise measurements are performed in a noise parameter measurement setup, e.g., based on a noise figure meter [5.160]. The measurements are performed in an input tuner systems typically extracting the full four noise parameters. Fig. 5.46 gives the measured noise figure N_{F} as a function frequency for a $W_{\mathrm{g}} = 2 \times 60 \, \mu\mathrm{m}$ AlGaN/GaN HEMT on SiC with $l_{\mathrm{g}} = 150 \, \mathrm{nm}$.

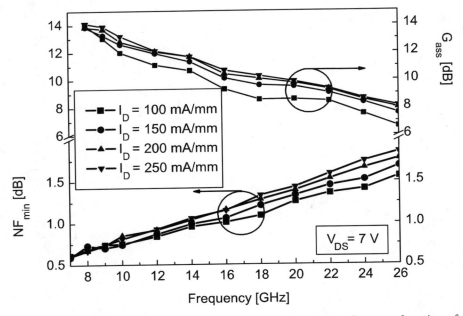

Fig. 5.46. Minimum noise figure $N_{F,min}$ and associated gain G_{ass} as a function of frequency for an AlGaN/GaN HEMT on s.i. SiC with $l_g = 150\,nm$

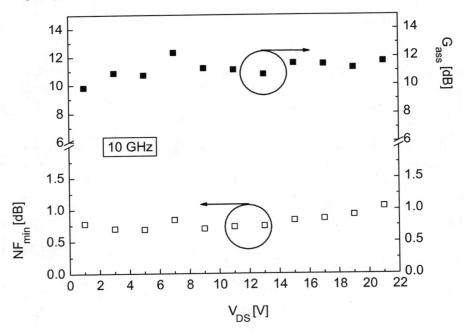

Fig. 5.47. Minimum noise figure $N_{F,min}$ and associated gain G_{ass} at 10 GHz vs. V_{DS} and I_D for an AlGaN/ GaN HEMT with $l_g = 150\,nm$

A minimum noise figure of 0.7 dB at 10 GHz for a drain current of 12 and 18 mA or 100 and 150 mA mm^{-1}, respectively, is observed. Fig. 5.47 gives the same minimum noise figure $N_{F,min}$ and the associated gain G_{ass} at 10 GHz as a function of V_{DS} bias and current level of 150 mA mm^{-1}. A very constant behavior of the noise figure $N_{F,min}$ and associated gain G_{ass} as a function of V_{DS} is observed for a gate length $l_g = 150$ nm. The Pospieszalski model can be extracted from the measurements in Fig. 5.46. The model reads for the gate and drain noise-sources [5.140, 5.172]:

$$I_{DS} = 4k_B \cdot \frac{\Delta f \cdot T_{Drain}}{R_{ds}}, \tag{5.111}$$

$$I_G = 4k_B \cdot \frac{\Delta f \cdot T_{Gate}}{R_i}. \tag{5.112}$$

A maximum drain temperature $T_{Drain} = 1{,}350$ K and a gate temperature $T_{Gate} = 300$ K are extracted at $V_{DS} = 7$ V and at drain current $I_{DS} = 150$ mA mm^{-1}. This result justifies the assumption of independent noise sources. The temperature $T_{Drain} = 1{,}350$ K is lower than the values found for a gate length of 700 nm in [5.160]. The extracted gate temperature T_{Gate} in [5.160] is in the same order as the room temperature. The Pucel or PRC model assumes for the noise sources [5.171, 5.172]

$$I_D = <i_d i_d^*> = 4k_B \cdot \Delta f \cdot T_L \cdot g_m \cdot P, \tag{5.113}$$

$$I_G = <i_g i_g^*> = 4k_B \cdot \Delta f \cdot T_L \cdot \frac{(\omega C_{gs})^2}{g_m} \cdot R. \tag{5.114}$$

The noise sources are modeled to be correlated $<i_d i_g^*>$ with the correlation parameter C. An extraction of the Pucel model for GaN HEMTs is provided in [5.160] in comparison to a Pospieszalski extraction. In general, the insertion of an AlN interlayer into the epitaxial structure improves the noise performance.

Input Linearity of III-N FETs

III-N FETs are attractive for applications that require a high degree of survivability under strong RF-illumination, e.g., [5.124]. A linearity analysis of low-noise AlGaN/GaN HEMTs on s.i. SiC substrate is given in [5.124]. The optimum biasing for the noise figure found is 15–20% of I_{Dmax} for a V_{DS} of 1–10 V. The ratio of the input distortion II_3 vs. carrier power is investigated. Apart from the promising minimum noise figure of 1 dB at 10 GHz, the linearity at high input power levels is particularly advantageous compared to GaAs HEMTs. Further results are given in Chapter 6.

5.7 *Problems*

1. Mention the specific differences of GaN FETs with respect to modeling relative to GaAs FETs.
2. What is the main difference between thermal and trap-related dispersive behavior?
3. What is particular to AlGaN/GaN HBT modeling relative to GaAs HBTs?
4. Describe the expected theoretical behavior of the IM_3-sweet-spots in a two-tone measurement of a GaN HFET!
5. Describe the differences of the WCDMA spectra of a GaN HFET relative to a spectrum of a silicon LDMOS!
6. What are the expected differences of GaN FETs as compared to state-of-the-art GaAs PHEMTs and InP HEMTs?

6

Circuit Considerations and III-N Examples

This chapter discusses circuit examples for III-N-based amplifiers with a focus on increased impedance, thermal management, and high RF-power management in the frequency range between a few MHz and at least 60 GHz. Low-noise amplifiers are presented and analyzed for high dynamic range, robustness, and high linearity. The last section treats highly linear power amplifiers for microwave and millimeter wave frequencies. Again, nitride-specific advantages and challenges, such as output power, linearity, and power-added efficiency (PAE), are investigated. The potential of III-N materials and devices are far from exploited in hybrid, MIC, and MMIC circuits. This chapter provides an overview of critical issues and circuit examples for nitride-based devices. As is pointed out in Sect. 2.5.7, there are relatively few reports on GaN-HEMT-based fully integrated MMICs in microstrip- or coplanar-waveguide passive technology so far, due to the tough requirements for a high-voltage passive technology.

6.1 Passive Circuit Modeling

Transmission-lines and transmission-line parameters are a critical issue for any integrated circuit device technology. Various transmission-line variants exist, e.g., coplanar-waveguide [6.145], grounded coplanar-waveguide [6.201], and microstrip transmission-lines [6.101].

6.1.1 Coplanar-Waveguide Transmission-Line Elements

Coplanar-waveguide transmission-lines or waveguides (CPW) are attractive, as they can be realized without the application of a backside process, since the ground component is guided on the frontside of the chip. A typical CPW line is given in Fig. 6.1. The position of the ground at the frontside is useful in terms of manufacturability; however, CPW is in general not as area-efficient as the microstrip line approach [6.6]. Further, CPW technology also requires

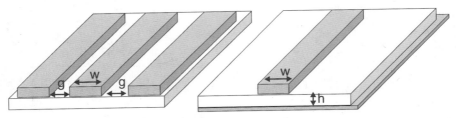

Fig. 6.1. Comparison of a coplanar-waveguide (*left*) and a microstrip (*right*) transmission-line

thinning of the wafer for thermal reasons or for mm-wave technology to suppress other propagation modes in the substrate, e.g, [6.138]. In general, the passive components in the III-N world are not very different from GaAs passive components, e.g., [6.169]. Microstrip-transmission-line models based on very few parameters are implemented in standard commercial simulators such as Agilent-ADS. III-N coplanar-waveguide models are similar to III-V passive models, however, require specific extraction. As native GaN substrates have not been used with hybrids and MMICs so far, hybrid integration of passive coplanar-waveguide components on AlN substrates for III-N devices is given in [6.66]. As AlN is a good insulator and easily available, transmission-lines, discontinuities, metal–insulator–metal (MIM) capacitors, and resistors are modeled for the realization of cost-effective hybrid AlN-substrates. Several MMIC processes are available on s.i. SiC substrate, partly also based on SiC MESFET MMICs [6.159]. Measured attenuation of coplanar-waveguides on 6H-, p-Type SiC, and high-purity semiinsulating 4H–SiC through 800 K is compared in [6.128]. The measured attenuation at RT for HPSI SiC at 10 GHz is 1 dB cm^{-1} and increases to 4.5 dB cm^{-1} at 773 K. Differences are found for different substrates. Conductive p-type 6H–SiC yield increased attenuation of coplanar-waveguides of 20 dB cm^{-1} at 1 GHz, which makes this substrate unsuitable for microwave operation. Further differences are found for the temperature dependence of the attenuation of p-type 6H–SiC and 4H-high-purity SiC. Due to the high value at RT, the increase in attenuation with temperature is not as pronounced as for HPSI SiC. Similar considerations apply for other conductive substrates for III-N devices, such as conductive or high-resistivity silicon. High-resistivity (HR) silicon with a resistivity of 1.6 MΩ cm with losses of 6.3 dB cm^{-1} at 20 GHz and RT can be fabricated, e.g., [6.190]. However, typical HR silicon has a lower resistivity of 10–30 kΩ cm [6.38] and higher RF-losses unless other measures are taken to minimize those, e.g., additional oxides [6.147]. The impact of the substrate conductivity on silicon transmission-lines is reviewed in [6.127].

Passive Modeling

Fig. 6.1 gives a comparison of a transmission-line in a coplanar and a microstrip technology. The signal line width w and the signal-to-ground spacing g, and

Table 6.1. Extracted electrical parameters of various passive elements on AlN, Alumina, s.i. SiC, and GaAs substrates

Element	Substrate	w (μm)	g (μm)	C_{exp} (pF m^{-1})	ε_r^{eff} (-)	Ref.
70 Ω CPW Line	AlN	45	75	99	–	[6.66]
50 Ω CPW Line	AlN	70	30	–	4.74	–
50 Ω CPW Line	Alumina	70	40	–	–	–
50 Ω CPW Line	GaN (250 μm)	70	30	–	5.20	–
50 Ω CPW Line	GaAs (625 μm)	70	45	–	6.93	–
50 Ω CPW Line	s.i. SiC (370 μm)	70	30	–	5.38	–
50 Ω CPW Line	s.i. SiC (370 μm)	25	13.3	–	5.40	–

the substrate thickness h are given. The modeling considerations are not different from other semiconductor devices. Modeling of coplanar-waveguide transmission-line on s.i. SiC is given in [6.141]. Models for a coplanar transmission-line library on GaN on s.i. SiC epitaxial layers are developed and are described, e.g., in [6.152]. The detailed modeling is derived from passive modeling on s.i. GaAs or other insulating dielectric substrates, e.g., [6.56,6.122]. In the model for a simple coplanar transmission-line the characteristic impedance is modeled according to [6.66]:

$$Z_0 = ((R + i\omega \cdot L)/(G + i\omega \cdot C))^{0.5}, \tag{6.1}$$

$$\gamma = ((R + i\omega \cdot L) \cdot (G + i\omega \cdot C))^{0.5}. \tag{6.2}$$

$R, L, G,$ and C denote the resistance, line inductance, conductance, and capacitance. Table 6.1 compiles parameters for transmission-line elements on AlN, Alumina, GaAs, and s.i. SiC substrates. w denotes the signal width and g the width of the signal-to-ground spacing. ε_r^{eff} denotes the calculated effective permittivity of a coplanar-waveguide. Table 6.1 shows the similarity of GaAs, AlN, and s.i. SiC. The typical modeling includes further structures, e.g., T-junctions, steps, cross-junctions, bends, and lumped elements, such as MIM capacitors of various types [6.159], and either NiCr- or other resistors, such as TaN- and TiN-resistors. The modeling is based on the well-known methodologies, e.g., [6.51].

6.1.2 Microstrip-Transmission-Line Elements

Microstrip-transmission-line GaN MMICs based on full passive libraries have been repeatedly reported, e.g., in [6.121,6.171]. The transmission-line parameters of GaAs technologies [6.39,6.51] can be adjusted as for any insulating dielectric substrate. All typical elements of a passive library can be described by conventional microstrip line models provided by commercial simulators. In addition to the modeling of the new substrate materials, conventional passive hybrid matching circuits are being used, especially for lower frequencies of

≤ 10 GHz, e.g., [6.195]. This is especially true, since the available generic substrates are either expensive (SiC), or (semi)-conductive (Si), or thermally critical (sapphire). The modeling of these hybrid substrates is not different from other conventional approaches, e.g., [6.101]. However, new circuit concepts, such as harmonic terminations, require improved passive modeling, which are discussed with the new circuit concepts.

6.2 High-Voltage High-Power Amplifiers

This section compiles the principle operation modes or classes for power amplifiers with emphasis on the classes suitable for high-voltage III-N devices as compared to other semiconductor technologies.

6.2.1 Basic Principles of High-Voltage High-Power Operation

The specific properties of III-N devices allow the application of more advanced circuit concepts. Good general overviews of the amplifier classes are given in [6.30, 6.34, 6.89, 6.135, 6.176].

Amplifier Classes

The amplifier classes in general differ in the mode of the quiescent DC-bias (class-A–class-C), the modulation of the DC-bias, the load fundamental and harmonic termination (class-D–class-G), the modulation of the load (Doherty), in phase/amplitude variation [6.62], and even the digital modulation of the input signal (class-S) [6.67]. Thus, the resulting waveforms differ significantly [6.30]. Fig. 6.2 illustrates the differences in bias and load conditions for some of the examples. The single classes are described for FETs in detail as follows:

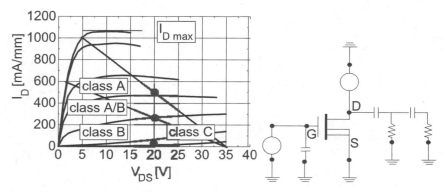

Fig. 6.2. Bias conditions and topology for different classes of operation: *left* (A–C), *right* (D–G)

–   Class-A: The device is biased at a quiescent current of $I_{\mathrm{Dmax}}/2$. This operation allows for maximum gain compared to the other classes; however, the maximum theoretical PAE is 50%. The linearity is good, as the bias is chosen to avoid the nonlinearities of the diodes [6.104].

–   Class-B: the device is statically biased at an operation V_{DS} with V_{GS} in a pinch-off condition with $V_{\mathrm{GS}} = V_{\mathrm{thr}}$. The maximum theoretical PAE amounts to 78.5% or $\pi/4\%$ [6.120, 6.191].

–   Class-A/B: the biasing used lies between the aforementioned conditions A and B mostly to increase efficiency and reduce quiescent power dissipation in a trade-off between the linearity and dissipated power at the different input power levels [6.180].

–   Class-C: the device is biased in extreme pinch-off with a quiescent $V_{\mathrm{GS}} \leq V_{\mathrm{thr}}$. The efficiency can be further increased at the expense of higher input power levels beyond 80% and reduced linearity. The theoretical efficiency limit for the class-C operation is 100%, while the associated output power of PAE = 100% is 0. This operation further requires increased breakdown voltages for the device, which favors III-N devices [6.30].

–   Class-D: The Class-D-type amplifier is a switch-mode amplifier. It consists of a switch device, typically realized with two transistors. The amplifier achieves high output powers and high efficiencies up to 100% through the phase separation of RF-current and voltage. A filter is required at the output which limits the bandwidth. The amplifier is further only suitable for phase modulation. As for all switch-mode amplifiers a high f_{T} is required [6.30] as compared to the operation frequency to reduce losses in the switching.

–   Class-E: This amplifier is again of switch-mode type, which favors transistors with high f_{T}. It consists of one transistor. The main characteristics of this amplifier class is an input driving similar to rectangular switching achieved [6.45, 6.46, 6.124]. This fact boost efficiency while the gain is reduced compared to a linear amplifier because of the need to overdrive. Again only phase modulation is possible. Harmonic termination is achieved through a relatively simple RCL circuit which taylors current and voltage swings. The output filters again limit the bandwidth while a high-breakdown voltage is required for the transistors [6.30]. Even when driven not fully into compression, Class-E amplifiers provide improvements in efficiency in more linear operation modes [6.95, 6.166].

–   Class-F: In this operation mode, the output waveform is shaped using an output harmonic termination technique leading to a square-type output voltage waveform with a half sine wave current wave forms [6.30, 6.63, 6.89, 6.146]. The challenge of Class-F amplifiers in general is the need to match a maximum number of harmonics [6.134] while this number of harmonics is limited for practical reasons in a hybrid environment [6.146] and at the same time correlated with the maximum achievable efficiency [6.134]. Several subclasses exist depending on the use of the harmonics [6.63, 6.136]. The inverted Class-F amplifier gives a half sine wave output voltage and

Table 6.2. Characteristics of amplifier classes from A to S on a scale from $-$ to $++$ for amplitude (AM) and phase modulation (PM)

Class	Quiescent DC-current	PAE (%)	Bandwidth rel.	Linearity	Comment
A	$I_{Dmax}/2$	50	$++$	$++$	
B	0	78.5	$++$	0	
A/B	$0 \leq I \leq I_{Dmax}/2$	78.5	$++$	$+$	
C	0	100	$-$	$-$	$P_{out} = 0$ at 100%
D	A/B	100	$-$ (filter)	$-$	Loss of amplitude
E	Switch	100	$-$ (filter)	$-$	Loss of amplitude
F	$-$	100	$-$ (filter)	$-$	
S	E,D	100		$++$	
Doherty	(A,B), F	78.5	$-$ (load)	$+$	
Chireix	AB,D,E,F	100	$-$	$++$	Switch AM, PM
ET	A/B	100	0	$+$	
EER	D,E	100	$-$	$++$	

a square type of output current. For all Class-F type amplifiers the linearity is reduced through the generation of harmonics at the input which requires input termination compensation [6.146].

More amplifier classes have been explicitly defined:

- Analog Class-S: both the DC-supply and the load are modulated, resulting in a maximum theoretical efficiency of 100% [6.30, 6.124]
- Digital Class-S in sigma–delta mode [6.64, 6.67]

Further amplifier types have been reported, such as:

- Doherty (load modulation) [6.21, 6.30]
- Chireix (outphasing) amplifiers [6.20]
- Envelope-tracking (ET) [6.182]
- Envelope elimination and restoration (EER) or Khan [6.77] amplifiers.

They are typical examples for efficient linear amplifiers. Table 6.2 summarizes the characteristics such as gain, bandwidth, linearity, and efficiency performance characteristics of different classes [6.30]. The original classes are extended, once the DC-bias and load are also modulated, leading to various forms of variants, e.g., in [6.30, 6.54]. Further, several variants exist for the envelope-tracking depending which time constants are used [6.182]. No optimum solution is available for all applications; however, the application of more advanced concepts is vital for efficient linear PAs. III-N devices favor several of these advanced concepts.

Amplifier Examples using III-N Devices

Many reports for the realization of the aforementioned amplifier classes and types using GaN HEMTs are available:

- For pure Class-A operation; see, e.g., [6.104]
- For pure Class-B operation; see, e.g., [6.92, 6.110, 6.120, 6.191]
- For Class-C operation; see, e.g., [6.189]

Simulations of Class-B operation of amplifiers based on GaN HBTs are provided in [6.76]. The simulations reveal an issue with a mismatch of the optimum load for optimum output power and optimum gain. Output harmonic termination techniques for AlGaN/GaN HEMT power amplifiers at 2 GHz using an active integrated antenna approach are discussed in [6.24]. The output termination of the second harmonic is short-circuited, while the third harmonic is terminated with an open structure. The real harmonic termination improves the PAE by 10–15% between 1.2 and 2.4 GHz. Class-C amplifiers are typically not considered for communication amplifiers as they reduce the amount of gain by at least 6 dB and strongly reduce linearity. Deep Class-C operation is used for III-N devices to enable the high-efficiency/high-power density operation at very high operation voltage in [6.189]. Further, Class-C is used in the load-modulation transistor in the Doherty concept [6.21, 6.96]. III-N Class-D-type RF-amplifiers have been considered in a number of publications [6.45], however, so far not been extensively used [6.52]. Class-D requires high f_T for the FETs, which are available in III-N devices. However, the loss of the amplitude information and the limited bandwidth make these amplifiers unsuitable for communication applications. III-N examples are given in [6.112] with 90% efficiency reached at 30 MHz. GaN HFETs enable the operation of this amplifier class, which typically used in the lower MHz regime, up to about 1 GHz [6.3, 6.135].

GaN Class-E and Class-F Amplifiers

GaN-based switch-mode Class-E hybrid amplifiers with 80–85% PAE at 2 GHz with associated output powers of 10 W are reported in [6.153]. The targeted bandwidth for the single-stage device amounts to 200 MHz from 1.9 to 2.1 GHz with 12 dB power gain. The matching is performed for a biasing to $I_{Dmax}/4$. The high f_T of the devices enables operation in Class-E up to 3 GHz with 75% associated PAE.

A two-stage class-E amplifier in [6.45] achieves 18.2 dB of gain at 2 GHz with an associated PAE= 50%. Similar results are reported by Eudyna in [6.166] with a linear gain of 19 dB at 2.1 GHz and 82% of maximum drain efficiency in a single-stage 10 W device operated at $V_{DS}= 50$ V. Again, the outstanding breakdown capabilities of GaN HEMTs are mentioned to overcome the high V_{DS}-voltage peaks of 135 V. A comparison between class-E power amplifiers employing LDMOS FETs at 1 GHz and SiC MESFETs at 2.14 GHz is given in [6.124]. The devices are oversized to achieve the appropriate efficiency, which is defined by the R_{on}. A proprietary LS-model based on the Curtice approach is used for simulation deep in class-B and class-C mode. Bandwidth information on the harmonic tuning for GaN HEMT class-E power amplifiers is given in [6.95]. The efficiency improvements can be obtained for a bandwidth of $\leq 200\,\mathrm{MHz}$ at $2.14\,\mathrm{GHz}$. Class-E amplifiers based on GaN HFETs have been subjected to DPD and one-carrier WCDMA signals in [6.80]. An EER/ET system is applied to obtain an average PAE of 48% and a PAR of 7.6 dB. 10 dB of gain are achieved with an average output power of 1.4 W. The predistortion reduces the EVM from critical 20 to 2.6% and thus below the critical specification of WCDMA communication. The efficiency of class-F and inverse class-F amplifiers is discussed for GaAs devices in [6.63]. Class-F and inverse class-F in theory do not differ in efficiency as RF-voltage and current are simply exchanged. In reality, inverse class-F and class-F amplifiers show differences in efficiency values for different quiescent currents I_{Dq} for both bipolar and FETs. The background of this behavior is the impact of the even-mode terminations at different current levels. Class-F amplifiers in GaN HFET field-plate MMIC technology have been demonstrated in [6.44, 6.47]. Very high output power densities of $\geq 6\,\mathrm{W\,mm^{-1}}$ and efficiency levels of up to 50% can be reached in this MMIC approach at 2 and 2.7 GHz. The current–gain cut-off frequency f_{T} amounts to 15–20 GHz, which is a factor of 6 higher than the frequency of operation. The absolute output power of the amplifier is 38 dBm. The simulated prediction of the PAE is 55–60%. With the use of more efficient GaN HFETs, hybrid class-F amplifiers with output power levels of 16 W and associated PAEs of 80% at $V_{\mathrm{DS}} = 42\,\mathrm{V}$ have been demonstrated recently [6.146]. The second to fourth harmonic are matched in this case. It is mentioned that the main efficiency increase is caused by the accurate matching of the first output harmonics similar to the findings in [6.134]. The input matching network provides a fundamental match and 2nd harmonic short. The gate periphery of the GaN HFETs is $W_{\mathrm{g}} = 3.6\,\mathrm{mm}$, the linear gain $\geq 15\,\mathrm{dB}$. Simulated class-F and inverse class-F amplifiers are again compared with the result, that the class-F amplifier leads to higher DE for a given output power in this comparison. This is based on the fact, that the Class-F amplifier has a lower average current for the same output swing. GaN-based class-F amplifiers have also been reported in [6.25] applying a second harmonic termination at 5.5 GHz in C-band. Even more advanced approaches based on GaN FETs, such as EER amplifiers [6.58, 6.79], have been proposed, as discussed later. The Chireix outphasing amplifier is realized based on Si LDMOS [6.62]. New Σ–Δ transmitter architectures are being proposed with extremely high requirements to the HPA transistors with regards bandwidth, speed, and robustness [6.22]. Class-S power amplifiers-based GaN HFETs have been mentioned in [6.124], so far without further realization. However, the potential of III-N for the advanced architectures is clearly visible.

6.2.2 General Design Considerations of III-N Amplifiers

III-N devices are attractive for amplifier design due to the higher intrinsic impedances on semiconductor level [6.100]. However, this fact alone does not necessarily imply that GaN HFET amplifiers are generally higher in impedance. A key consideration is the RF-transformation through the layout, especially in high gate width/multifinger devices or at higher frequencies, e.g., in Ka-band [6.31, 6.168]. At frequencies of 2 GHz, the intrinsic output matching is nearly ohmic, apart from the compensation of the output impedance C_{ds} [6.25] and of the parasitic contribution of layout and combining. The same is true for the input matching of the devices [6.180]. However, for large-periphery devices the situation is very different. The input impedance is very low, partly through the transformation of the parasitic access contributions to the device [6.181]. Apart from the parasitic layout contribution, the packaging contribution has to be considered, which adds significantly to the impedance. As an example, a dual push–pull packaging concept is used in [6.100] applying a dual impedance transformation, which modifies the actual device impedances with high parasitic contributions. Internal prematching concepts are also applied for large-periphery GaN HFETs, e.g., in [6.181] for class-AB, and in [6.166] for class-E operation.

6.2.3 Mobile Communication Amplifiers Between 500 MHz and 6 GHz

Benchmark: Silicon Laterally Diffused MOS (LDMOS) and GaAs Devices

Silicon LDMOS devices provide a sophisticated benchmark for the use of GaN HEMTs in third (3G) and fourth (4G) generation mobile communication base-station systems for high-bandwidth highly-linear applications. Good comparisons between different base-station device-technologies are given, e.g., in [6.99, 6.172, 6.174].

Linearity vs. Power-Added Efficiency for Silicon and GaAs

The fundamental trade-off between linearity and efficiency in highly linear operation is crucial for high-power communication applications. The efficiency at defined linearity constraints according to standards such as GSM or 3GPP standards is the key specification for any base-station system. The theoretical limits for this operation are explored in [6.172]. State-of-the-art performance for silicon LDMOS devices can be found in many publications, e.g., in [6.13, 6.42, 6.99, 6.172]. For a two-carrier W-CDMA signal the LDMOS HV6 device delivers an output power of 20 W at 2.1 GHz at an efficiency of 29% at an ACLR level of −37 dBc of 10 MHz offset [6.13]. The evolution of the LDMOS technology generations is discussed in [6.172]. Gate peripheries $W_g = 3 \times 50$ mm are needed to obtain output power levels of 100 W. Efficiency levels of 32% are reached for an ACLR of −37 dBc in two-carrier WCDMA operation with a linear gain of 18.5 dB. A gate length of 140 nm is used for the devices. A 200 W LDMOS-based Doherty amplifier is reported in [6.42]. A maximum PAE of 34% is found for WCDMA operation at 6 dB backoff and an APCR level of $-37\,$dBc. Similar results are provided in [6.99], which yield 29% PAE at $-37\,$dBc and an output power of 34 W and 2.1 GHz. The increase of PAE in highly linear amplifiers has become a major focus for LDMOS device development, which is well progressing. Even for higher operation voltages a 120 V interdigitated-drain LDMOS (IDLdMOS) on SOI substrate is discussed, e.g., in [6.196], breaking the LDMOS power limit.

GaAs-FET and -HFET high-voltage technologies have also been proposed, e.g., in [6.35, 6.109, 6.125]. The advantage of GaAs relative to LDMOS is the increased gain margin for a given gate length, which also more easily enables WiMAX applications at 3.5 GHz and beyond. A 45 W GaAs-power technology is demonstrated [6.125] with HFETs with a gate periphery of 32.4 mm and an operation voltage $V_{DS} = 26$ V for a gate length of 0.8 μm. The 6-in. wafers are thinned to 25 μm. A drain efficiency of 32% is achieved for an IM_3 of $-37\,$dBc with a two-carrier WCDMA signal for an average output power of 9.2 W at 2.14 GHz.

Properties of GaN HEMT Base Station Amplifiers

GaN HEMTs are a subject of active research for base-station applications. A general overview of the suitability of different device technologies for reconfigurable, multiband, multicarrier amplifiers is given in [6.36, 6.40]. Wide bandgap semiconductors provide the clear advantage of strongly reduced memory effects. Impressive GaN device results including DPD are provided, e.g., in [6.79, 6.109, 6.129, 6.180]. Possible advantages of GaN HEMTs in amplifiers relative to silicon LDMOS and GaAs HFETs are given in the following enumeration; see, e.g., [6.40, 6.174]:

- Higher impedance levels
- Increased bandwidth, i.e., multiband and multicarrier functionality [6.40] and increasing frequency agility at system level
- Increased efficiency in linear operation [6.174]
- Improved thermal properties, and thus reduced thermal memory effects [6.40]
- Increased gain levels for the same gate length

These advantages are now discussed in more detail.

Impedance Levels

Fig. 6.3 gives the optimum impedance of GaN HFETs at 2 GHz as a function of gate width derived for $V_{DS} = 28$ V operation, which is typical of base stations using LDMOS. For comparison with an improved technology, the optimum fundamental load impedances are given at an operation voltage of 48 V. The associated output power is also given. The output capacitance C_{ds} to compensate is nearly independent of the operation bias and can be directly obtained from the gate periphery in Fig. 6.3 and a constant small-signal value, e.g., $C_{ds} = 0.3\,\text{pF}\,\text{mm}^{-1}$.

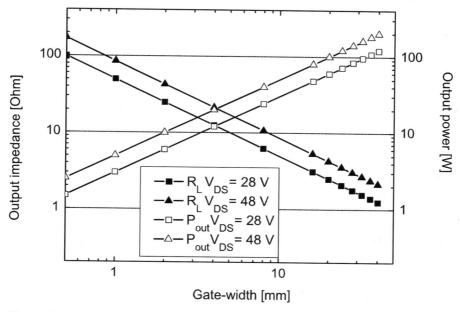

Fig. 6.3. Impedance levels and output power of a GaN power HEMT on s.i. SiC at 2 GHz as a function of gate width at $V_{DS} = 28$ V and $V_{DS} = 48$ V

Linearity vs. Power-Added Efficiency in GaN Devices

The efficiency of GaN amplifiers on circuit level for highly linear applications will be discussed starting from conventional class-AB concepts [6.73], push–pull concepts [6.100], Doherty amplifiers [6.21], and switch-mode amplifiers [6.47]. The considerations are finalized with reports on more advanced circuit concepts. General linearity considerations of GaN devices are given in a number of publications, e.g., [6.59, 6.110, 6.184]. High linearity of AlGaN/GaN HFETs on silicon substrates at 4 GHz is demonstrated in [6.175]. The intermodulation distortion is significantly better than in GaAs devices with the same output power density of 1 W mm^{-1}. High-power and linearity performance of GaN HEMTs on sapphire substrates is reported in [6.184]. A significant improvement of the IM_3 is found for the GaN FET at any output power level. Another study of large-signal linearity and efficiency of AlGaN/GaN MODFETs in [6.59] shows that the optimum impedances for maximum PAE and maximum P_{out} at 5 GHz are close (however, not identical) and well behaved. Further, the third-order intercept point is found to be insensitive to the gate bias. The properties of different Si, GaAs, and GaN FETs with respect to memory effects are discussed in [6.40]. The shape of the UMTS signal for different memory situations is discussed. For static nonlinearity, i.e., without memory effects, the spectral shaping has a convex structure on a logarithmic scale. For devices with memory, a concave behavior is typically observed. The real situation in devices is typically a combination of static and dynamic nonlinearity. The mixture of the two curvatures can then lead to an inflection point. In this investigation SiC MESFETs show no or minor memory effects relative to the strong effects observed in silicon LDMOS devices. The metric for memory effects and the application of a memoryless linearizer in [6.103] clearly suggest the strong improvements through memoryless compensation in GaN HFETs as compared to silicon LDMOS.

GaN HEMT Circuit Examples: Drivers

The need to reduce the number of stages in the base-station amplifier leads to a number of developments also for highly linear broadband driver amplifiers. A wide-band three-stage single-ended power module for multicarrier operation for drivers is given in [6.144]. A linear gain of 30 dB is reported for a bandwidth of 500 MHz between 1.8 and 2.3 GHz. The ACLR is lower than -50 dBc at an average output power of 19 dBm for a two-carrier WCDMA signal for an offset frequency $\Delta f = 10$ MHz. The use of dual-gate GaN HFETs for driver applications is discussed in [6.27, 6.177]. Linear gain levels of 30 dB per stage at 2.14 GHz can be reached at an operation bias of 48 and 60 V with similar efficiency levels. Further, RF-reliability is demonstrated. A broadband amplifier between 0.2 and 4 GHz delivers a WCDMA output power of 20 dBm at an ACLR of -45 dBc. A second example demonstrates a driver for WiMAX applications [6.27] between 3.2 and 3.8 GHz, again combining high gain (10 dB over the band) with broadband capability and linearity. This amplifier provides an output power of 27 dBm at an EVM of 2% in compliance with the WiMAX requirements.

GaN HEMT Circuit Examples: Final Stage

Several examples of power amplifiers based on GaN HEMTs exist.

GaN Class-A/B Amplifiers

A single-ended amplifier with a gate periphery $W_g = 48$ mm is reported applying a prematching concept within the device package. Output power levels of 370 W at 2.11 GHz are reached [6.111, 6.180]. The linearization of this amplifier is reported in [6.181]. For a two-carrier WCDMA signal, average output power levels of 60 W with an IM_3 of less than -50 dBc at 10 MHz offset are reached. The maximum drain efficiency of 24% is found at an average output power of 47.8 dBm and an ACLR of -35 dBc. This level of IM_3 is still in the range of modern linearization techniques to reach -45 dBc after linearization. Another single-ended GaN HFET amplifier with an output power level of 200 W is reported in [6.75]. The device is composed from two chips with a gate width of 36 mm each to reduce the thermal resistance. The bandwidth is 60 MHz at 2.11–2.17 GHz. A drain efficiency of 34% is reported at a peak-to-average ratio (PAR) of 8 dB or 45 dBm average under two-carrier WCDMA operation. The distortion of the GaN HFET is found to be worse than in a comparable silicon LDMOS. However, after the application of a DPD system, the ACLR of -50 dBc can be achieved at 8 dB backoff and an associated PAE of 34%. Fujitsu further reported a push–pull GaN HEMT amplifier consisting of two 36 mm devices with an overall peak output power level of 250 W [6.78]. For a four-carrier WCDMA signal, an ACLR of -50 dB is reached at an efficiency of 37% at 46 dBm average output power. The single 36 mm device reaches more than 150 W of cw-output power. A broadband push–pull amplifier is given in [6.73]. The device yields more than 50.1 dBm of peak output power at 1.95 GHz under 16-channel single-carrier WCDMA operation with an operation bias of 35 V and a gain of 12.9 dB. The 3 dB-bandwidth is 400 MHz. Broadband HPAs for WiMAX applications between 3.3 and 3.8 GHz are reported in [6.163, 6.164]. The single transistors yield output power levels of $\geq$38 dBm at PAE levels of $\geq$17% and EVM levels of 2% at $V_{DS} = 28$ V and 11 dB gain under OFDM operation.

Doherty Concepts

Good examples for highly efficient LDMOS power Doherty amplifiers for CDMA base stations are given in [6.15]. Efficiency levels of 40% are reached at 40 W of output power with an ACLR of -31 dBc at 2.1 GHz. The gain is $\geq$10 dB. GaN-based Doherty amplifiers are reported in [6.96]. The Doherty concept yields an efficiency of over 50% extending over 6 dB of dynamic range for output power levels of 1 W. A 40 W peak-power GaN Doherty power amplifier is reported in [6.21]. The power gain is 11 dB from 1.8 GHz to 2.5 GHz. Once linearized, a peak PAE of 65% and a linearity of -55 dBc at 5 MHz offset are obtained for a one-carrier WCDMA signal. More recently, very powerful GaN Doherty amplifiers are reported [6.165, 6.197]. A two-stage design with an average output power of 80 and 450 W peak power at 2.14 GHz is reported in [6.165]. 55% PAE are obtained at -6 dB backoff from the saturated output power. A careful design of the reference plane is applied transforming the input and output impedances of the main and the peak amplifier by means of the parasitics and the matching networks. Both main and peak amplifier consist of 2 × 36 mm devices. The linear gain amounts to 32 dB at 2.14 GHz. Based on similar devices, a Doherty with increased power range is proposed in [6.197]. Two devices with $W_g = 36$ mm each yield an average output power of 45 dBm at 50% at an ACLR of -38 dBc without the application of a DPD. The increased dynamic range is achieved through the asymmetrical bias of the peak and the main amplifier: the peak amplifier is biased at a lower V_{DS} of 40 V rather while the main amplifier is operated at $V_{DS} = 50$ V.

Envelope-Tracking and Envelope Elimination-Restoration Amplifiers

The envelope-tracking (ET) and the envelope elimination-restoration (EER) concept are attractive as they combine high-efficiency with high-linearity and high-bandwidth capability and offer further opportunities for predistortion. Wide bandwidth envelope-tracking amplifiers for orthogonal frequency-division multiplexing (OFDM) applications such as WiMAX are analyzed in [6.182]. A channel bandwidth of 16.25 MHz is used, the RF-bandwidth is 83.5 MHz at 2.4 GHz. The realization is based on GaAs MESFETs. A GaN HFET-based envelope-tracking amplifier with 50% PAE is demonstrated in [6.79] under single-carrier WCDMA operation. Digital predistortion is used on two levels. Memoryless DPD is used to compensate the gain variation over the envelope of the amplifier. Deterministic memory mitigation is used to further increase the linearity. The WCDMA signals have a reduced (clipped) PAR of 7.67 dB. An average output power of 37.2 W is measured for an ACLR-level of -52 dBc at 5 MHz offset from the carrier. Further, a class-E amplifier based on GaN HFETs is used in an envelope elimination restoration/envelope-tracking (EER/ET) system in [6.80]. With an unchanged PAE of 48%, improved linearity is achieved via predistortion of the class-E amplifier. The linearized EVM is 2.6% with predistortion, while only 20% is achieved without DPD. Recently, very high efficiency EER amplifiers have been reported based on 10 W GaN HEMTs [6.58]. The actual amplifiers are realized in class-F with class-C bias for the highly efficient PA and deliver efficiencies of up to 73% with harmonic control of the second and third harmonic at 2.14 GHz under OFDM operation.

GaN HFET Switch-Mode Amplifiers

Switch-mode amplifiers generally improve efficiency, however, suffer from reduced linearity and bandwidth due to the harmonic control, e.g., [6.34]. GaN-based switch-mode class-E hybrid amplifiers at 2, 2.7, and 3.5 GHz are reported in [6.153]. PAE values of 80–84% at 2 GHz, of 76–82% at 2.7 GHz, and of 72–78% at 3.5 GHz with associated output powers of 10 W are given. The linear gain is 11–12 dB at all frequencies. A device with a larger periphery yields 63 W with 75% associated PAE and $V_{DS} = 28$ V at 2 GHz. A GaN-HEMT-based class-E amplifier at 2.1 GHz is given in [6.166]. A drain efficiency of 45% at an ACLR of -50 dBc is reached with the application of a digital predistortion. A two-carrier WCDMA signal is transmitted with a PAR of 7.8 dB. The transmission of the amplitude information contradicts the original definition of a class-E amplifier as a switch-mode amplifier. The operation voltage is 50 V. Examples of class-F and class-D amplifiers based on GaN HEMTs have already been discussed above.

Final Stages: Linearization

Analog and digital predistortion concepts from the LDMOS world have been applied to GaN HFETs. A general finding for GaN HFETs is the very positive impact of linearization techniques [6.75, 6.103]. Possibly the strong effect in GaN HEMTs is due to the low P_{-1dB} relative to the saturated power and the soft power compression behavior of the device. Further, memory effects are strongly reduced [6.40]. The use of an analog prelinearization for GaN HFETs with an additional gate diode is described in [6.192]. The nonlinearity of the input capacitance is compensated by the application of the diode. High-efficiency feed-forward amplifiers using RF-predistortion linearizers for a modified silicon LDMOS Doherty amplifier are discussed in [6.117]. An average efficiency of 13.6% is reached at an output power of 45 W and an ACLR of -55 dBc. The linearized feed-forward amplifier has a 3.5% efficiency advantage over the nonlinearized version. Linearity and efficiency performance of GaN HEMTs on s.i. SiC substrate are given in [6.129] with the application of a PCM Sierra digital predistortion correction. The gate-width of the GaN HFET is 36 mm. Efficiency levels of 43% at an average output power of 11 W at 2.1 GHz are reached for two-carrier WCDMA signals at $V_{DS} = 28$ V. The DPD correction allows a 7 dB increase in linear output power and an increase of the PAE from 24 to 33%. The reduction of the quiescent bias to class-A/B leads to an increase of the ACLR of about 5 dB. An overview of linearization techniques and a comparison of GaN HFETs, GaAs HFETs, and silicon LDMOS devices is given in [6.91]. Future systems will require two-to four-carrier operation with up to 60 MHz bandwidth. Higher order IMD contributions are mentioned in [6.173] for GaN HFETs in comparison with GaAs FETs. The particular DPD implementation yields an improvement of the ACLR of about 20 dB for all three device technologies. System aspects, such as additional service costs at higher power densities, are also considered. Further the system aspects of a change of the operation voltage from $V_{DS} = 28$ V to 50 V are discussed [6.91, 6.173]. The need to adjust the distortion correction algorithms to the unique properties of each semiconductor technology is mentioned in [6.173]. The latter consideration gives additional potential to the GaN HFETs where the algorithms are not yet mature, i.e., not specific. It is further mentioned that the memory effects in the particular GaN PA are strong, however, can be very well suppressed by memory compensation. This concludes the base-station considerations.

6.2.4 C-Frequency Band High-Power Amplifiers

GaN HFETs are very attractive for the use in C-band (4–8 GHz), as the C-band frequencies cannot be reached by state-of-the-art SiC MESFETs and silicon LDMOS devices. Good references of packaged GaAs HEMTs for C-band applications with output power levels of 70 W and 50% PAE are given, e.g., in [6.179]. Initial reports of GaN-HEMT-based C-band high-power transistors and power amplifiers are given, e.g., in [6.25, 6.187]. The impact of second harmonic termination on the efficiency in C-band is described in [6.25]. PAE levels of 60% at 5.6 GHz are reached. Output power levels beyond 100 W at C-band are described in [6.69, 6.119]. CW-output-power levels of 60 W at 4 GHz and $V_{DS} = 61$ W are reached with an AlGaN/GaN HFET with $W_g = 24$ mm [6.155]. The device is internally matched. The linear gain is 10.2 dB with a maximum PAE of 42% in field-plate technology with a gate length $l_g = 0.5\,\mu$m. The output power is as high as 156 W at 4 GHz for pulsed-operation. The pulse width is 10 μs at a duty cycle of 1% at $V_{DS} = 50$ V. This leads to a power density of ≥ 6 W mm^{-1} for the same device with $W_g = 24$ mm. The passive elements of the packaged device are realized on alumina substrates in a package with 12.7×12.9 mm^2 [6.155]. This is approximately half of the area used for a GaAs FET. Power levels of 60 W, but drain efficiency levels of $\geq 50\%$ have been reported in [6.65]. The PAE is 45%. The gate width is $W_g = 16$ mm. Short and open stubs are used to realize a broadband matching of fundamental and second harmonic at the output. The relative bandwidth achieved is 15%. A new device passivation technique leads to output power levels $P_{out} = 100$ W in a single device without the application of a field plate [6.69]. The gate length is $l_g = 0.4\,\mu$m, the unit gate width amounts to 300 μm. The device yields a gate periphery $W_g = 50.4$ mm with a pulsed-output power density of 2.8 W mm^{-1}. The pulsed-output power is $P_{out} = 140$ W with an associated PAE of 25% at a bias $V_{DS} = 40$ V (duty cycle 2.5%, pulse width 20 μs). The linear gain is 10 dB. The large gate periphery used make the devices very susceptible to even- and odd-mode oscillations. A detailed analysis of a GaN FET multicell configuration is given in [6.199]. The probe insertion analysis of [6.162] is used. The conditions to be fulfilled for loop oscillations read in general

$$\text{loop phase } \angle\, \Gamma = 0 \tag{6.3}$$

$$\text{loop gain} |\Gamma| > 1 \ (0\,\text{dB}). \tag{6.4}$$

Odd-mode oscillations are suppressed by adding additional losses such as isolation resistors into the loop path and further by structuring the matching circuits around the semiconductors. A record output power of 170 W at 6 GHz has been reported combining four chips with $W_g = 11.52$ mm of gate periphery each in a package [6.160]. The chips are internally matched. Even higher output power levels are reported in [6.200]. A single chip amplifier yields an output power of 167 W with 7 W mm^{-1} power density at 5.6 GHz ($W_g = 24$ mm) with a strongly reduced duty cycle of 2.5% and 20 μs pulse width. In a two-chip configuration ($W_g = 48$ mm) 220 W of output power are obtained at 5.6 GHz with a 10% duty cycle and $V_{DS} = 60$ V. The increase of the gate-to-gate pitch to 30 μm is the key to reduce the thermal resistance to values acceptable for high power operation ($R_{th} = 9.6$ K W^{-1} mm^{-1}). The total package area is 8×8 mm^2. The integration of active antennas and the application of harmonic techniques at 7.25 GHz are discussed in [6.24]. A chip with $W_g = 1$ mm achieves an output power of 30 dBm and a peak PAE of 55% with a power gain of 14 dB for a gate length $l_g = 0.8\,\mu$m. A 30 dB suppression of the output-side harmonics is achieved. In summary, the output power levels and bandwidths for C-band HPAs can be dramatically increased as compared to the best GaAs HEMTs with further potential of area reduction and higher operation voltages.

6.2.5 X-Band High-Power Amplifiers

Efficient high-power amplifiers in the X-band (8.2–12.4 GHz) need to combine high-operation voltages and low-current levels with high-relative bandwidth and high efficiency at these frequencies. Such hybrid and MMIC amplifiers are key examples to be realized in GaN HEMT technology, e.g., [6.139].

X-Band Reference Amplifiers: GaAs PHEMT
and GaAs HBT-Based MMICs

GaAs MESFET, GaAs PHEMT, and InGaP/GaAs HBT solid-state amplifiers provide benchmarks for the comparison of GaN HEMT amplifiers for high-power high-efficiency operation in C-band [6.185], X-band, and Ku-band [6.14] applications. Not all relevant results may be published; however, some overview shall be given. General MMIC design strategies suitable for X-band have been reported, e.g., in [6.6,6.167]. CW-power levels of up to 2 kW at X-Band have been reported based on solid-state devices when several MMICs and MMIC-modules [6.4,6.41] based on solid-state power amplifiers are combined in power. Output power levels of up to 16 W at X-band and multiwatt (≥ 5 W) at Ka-band have been reported on GaAs chip level [6.84,6.178]. Area efficiency, bandwidth, and PAE are still being improved. Traveling wave tube (TWT) replacement by solid-state solutions is often considered, e.g., [6.32]; however, tubes are still strongly improving [6.97] and both the efficiency and the associated output power levels achieved with SSPAs are not yet fully appropriate for all applications, e.g., [6.178]. Solid-state multi-PA solutions in general feature the advantage of graceful degradation [6.4]. GaAs MMICs provide a stable, mature, and still improving reference for the evaluation of GaN HEMT MMICs. Following an impressive roadmap, GaAs device performance is still improving today and chip size is still continuously reduced, e.g., [6.49]. State-of-the-art GaAs microstrip PHEMT MMICs provide output power levels above 40 dBm [6.12] and maximum PAE levels above 50% at reduced output power levels, e.g., [6.183], while efficiency is traded for output power and bandwidth. Power densities of up to 1 W mm^{-1} can be reached on the GaAs MMIC level [6.19]. GaAs HBTs provide performance similar to GaAs HEMTs in X-band [6.28,6.43] and S-band, e.g., [6.126]. The typical chip size is 3.3×5.7 mm^2 [6.202] for a GaAs PHEMT PA with an output power of ≥ 10 W. A GaAs HBT 10 W amplifier has a chip size of 4.74×4.36 mm^2 in microstrip transmission-line technology, as reported in [6.28]. The linear gain is 18 dB with an efficiency of better 35% over 1 GHz of bandwidth. Further MMIC size reduction of X-band amplifiers is discussed, e.g., in [6.49]. Integration of several chips leads to power levels of 160 W [6.41]. Power levels as high as 2,000 W have been reported in solid-state-based systems [6.4]. Examples of GaAs PHEMT coplanar-waveguide HPAs are given, e.g., in [6.11]. The chip size is 4×4 mm^2. A maximum output power level of ≥ 5 W and power-added efficiencies of $\geq 50\%$ are reached.

X-Band GaN Hybrid Amplifiers

Several examples are available for hybrid high-power amplifiers based on GaN technology in X-band. A hybrid amplifier with a pulsed-output power of 40.7 W at 2.7 dB compression at 10 GHz is demonstrated with GaN HEMTs on s.i. SiC with $W_g = 12$ mm. The amplifier yields a PAE of 29% in [6.130]. The linear gain is 13.5 dB in class-A operation using source viaholes. A similar result is reported in [6.26,6.115,6.154]. A DC-pulse width of 20 μs is used, along with a pulse repetition of 2 kHz. The high-operation voltage is $V_{DS} = 55$ V and required to suppress the odd-mode oscillations. A single-stage hybrid GaN HFET X-band power amplifier with a saturated cw-output power of 22.9 W and 37% associated PAE at 9 GHz and $V_{DS} = 36$ V is given in [6.107]. Similarly, an output power of 21.9 W is reported with 42% of associated PAE. The saturated cw-power density yields 6.3 W mm^{-1} on the transistor level. The linear gain of the amplifier amounts to 9 dB with a measured bandwidth of 2 GHz. Further examples include a hybrid GaN HEMT push–pull amplifier in class-AB operation at 8 GHz, demonstrated in [6.92]. Output power levels of 36 dBm are reached at 5.2 GHz with a periphery $W_g = 2 \times 1.5$ mm. The amplifier yields a PAE of 42%. High-linearity and high-efficiency class-B amplifiers at 10 GHz, realized in GaN HEMT technology, are given in [6.120]. PAE levels of 54% are predicted for the push–pull configuration, however, are not verified by experiments. Power combining of hybrid GaN HFETs yields 60 W–81 W of output power, as reported in [6.161,6.198]. On hybrid packaged transistor level, Toshiba developed a GaN HEMT with a maximum output power of 81 W in cw-operation based on two 11.52 mm chips. The PAE amounts to 34% at 9.5 GHz and a gain compression level of 3 dB [6.161]. The backside is thinned to 150 μm to reduce the thermal resistance. The overall package size is 11×12.9 mm^2. The device is optimzed for an operation bias $V_{DS} = 20$ V and $I_{DS} = 4$ A. A similar result is obtained with a gate periphery of 2×12 mm and a gate length $l_g = 0.5$ μm with a hybrid AlGaN/GaN HEMT. The devices are thinned down to 100 μm. The two chip device yields an output power of ≥ 60 W and a maximum PAE of 35% at 10 GHz. The linear gain is 10 dB in Class-B operation [6.198]. The input and output impedance is 50 Ω. The bandwidth is 1 GHz with a PAE$\geq 28\%$. The overall package size is 21×12.9 mm^2.

X-Band GaN HEMT Circuits: Flip-Chip Technology

The flip-chip integration approach is favorable for power amplifiers, as low-cost passive carrier substrates with high-thermal conductivity can be combined with low-cost substrates with low-thermal conductivity for the active devices, e.g., [6.152]. A GaN-HEMT-based microwave power amplifier is demonstrated in [6.188] with 14 W of output power using GaN HEMTs grown on sapphire substrates. The active devices are flip-chip bonded on polycrystalline AlN substrates. The amplifier yields a -3 dB-bandwidth between 6–10 GHz with a peak gain of 9 dB. A peak output power of 14 W is reached with a power gain of 4 dB and an associated PAE of 25% at $V_{DS} = 25$ V. A power density of

$6.4\,\mathrm{W\,mm^{-1}}$ at $6\,\mathrm{GHz}$ at an operation voltage $V_{\mathrm{DS}} = 39\,\mathrm{V}$ in pulsed-operation is reached in a $50\,\mathrm{W}$ AlGaN/GaN HEMT power amplifier [6.187]. The output periphery is $W_{\mathrm{g}} = 8\,\mathrm{mm}$, while the output load reported amounts to $7\,\Omega$. Again, a flip-chip technology on AlN carrier substrate is applied. GaN HEMTs grown on sapphire substrates for microwave power amplification and integrated in flip-chip technology are demonstrated in [6.186]. Metal resistors are made by Ti with SiN dielectric. MIM capacitors are realized by Au with SiN dielectric. Maximum output power levels $P_{\mathrm{out}} = 7.6\,\mathrm{W}$ at $4\,\mathrm{GHz}$ are reached with a periphery $W_{\mathrm{g}} = 6\,\mathrm{mm}$. A broadband amplifier reaches output power levels of 1.6–$3.2\,\mathrm{W}$ between 3 and $9\,\mathrm{GHz}$ with a gate width $W_{\mathrm{g}} = 1\,\mathrm{mm}$.

X-Band GaN Monolithic Integrated Circuits

The monolithic integration of GaN HFETs at X-band frequencies (8.2–$12.4\,\mathrm{GHz}$) is very desirable in order to avoid losses in circuit performance with respect to efficiency and bandwidth. GaN MMICs with PAE= 40% are expected [6.140]. Losses occur when long bond wires or bumps are used, as mentioned, e.g., in [6.148]. Further, long bond wires introduce additional uncertainty for the realization of advanced circuit concepts, such as harmonic termination.

Microstrip Transmission-Line X-Band MMICs

A microstrip line passive environment is the technology of choice for the integration of GaN HFETs, due to reduced losses and smaller chip size relative to coplanar-waveguide approaches. While the process complexity is increased with the application of a backend process, standard design, and layout optimization flows can be used, based on microstrip transmission-lines [6.12, 6.28, 6.88, 6.167]. Chip-area reduction is the ultimate requirement for cost reduction also for GaN technology [6.49]. This can be most effectively performed in microstrip-line technology. As one of the first GaN MMICs in microstrip transmission-line technology on s.i. SiC, a dual-stage GaN HEMT MMIC is produced yielding pulsed-output power levels $P_{\mathrm{out}} \geq 20\,\mathrm{W}$ at $9\,\mathrm{GHz}$, as given in [6.26]. The associated power gain and PAE amount to $14\,\mathrm{dB}$ and 20%, respectively.

Recent progress of GaN MMIC technology development in X-Band is reported, e.g., in [6.139]. As a general tendency, single-minded performance targets have been modified to also meet reliability and reproducibility targets, e.g., with respect to the choice of power densities. Further, the efficient operation of GaN HEMTs at Q-band (33–$50\,\mathrm{GHz}$) is enforced [6.139], which is also beneficial for the efficiency targets at lower frequencies. At the same time, GaN HEMT solid-state devices with a bandwidth of over one decade between 2 and $20\,\mathrm{GHz}$, output power levels of $\geq 100\,\mathrm{W}$, and efficiencies of $\geq 30\%$ are proposed. Further reports of GaN microstrip transmission-line MMICs are given, e.g., in [6.99, 6.148, 6.170, 6.171]. The dual-stage designs in [6.148, 6.170]

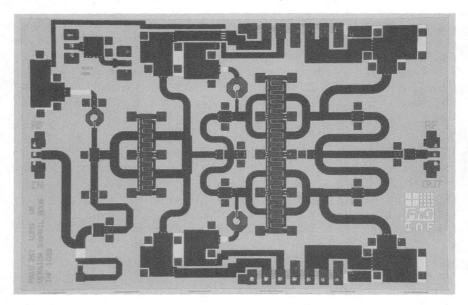

Fig. 6.4. Chip image of a dual-stage high-power amplifier in microstrip transmission-line technology, chip size $4.5\,\text{mm} \times 3\,\text{mm}$

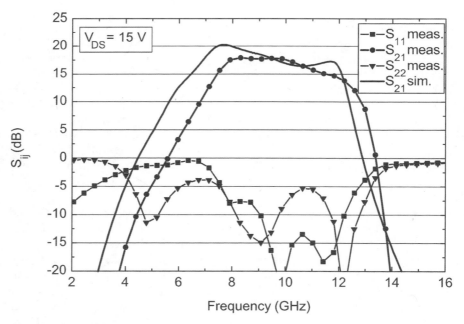

Fig. 6.5. S-parameters of the AlGaN/GaN HEMT MMIC amplifier in microstrip transmission-line technology

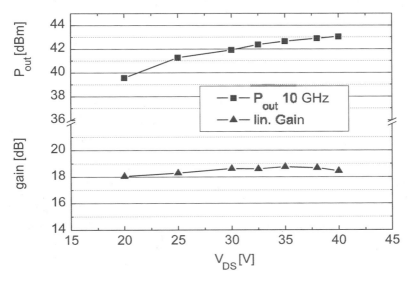

Fig. 6.6. Pulsed-output power and linear gain at 10 GHz as a function of V_{DS}

deliver output power levels of 20 W with a linear gain of 18 dB at $V_{DS} = 40$ V. For a second MMIC with 6 mm gate periphery, an output power at -1 dB compression beyond 14 W is achieved. The saturated output power is 22.4 W. The chip size is 4.5×3 mm^2 for the dual-stage, and 2.75×2.5 mm^2 for the single-stage device. Module integration of such X-band MMIC transmitter chains including an X-band driver amplifier is described, e.g., in [6.149]. Output power levels of 30 W at X-band are reached with the combination of two chips in the output stage. The thermal management and integration of such chips is of ultimate importance, see Chapter 8.

An X-Band Amplifier Example

The example is derived from [6.170, 6.171]. Fig. 6.4 gives the chip image of an X-band MMIC on s.i. SiC substrate. The chip size is 4.5×3 mm^2. The device technology is based on a gate length $l_g = 300$ nm. The design is tweaked to achieve the desired gain level at the upper edge of the frequency band, in this case 12 GHz. The forced matching in the input network of the first stage ensures gain flatness over the band. Interstage matching is performed with particular emphasis to the maximum input drive needs of the second amplifier stage over the full frequency band. Fig. 6.5 gives the S-parameters measured in cw-mode of the microstrip MMIC at $V_{DS} = 15$ V. Both a broad matching and a broad amplification S_{21} can be observed. This is a typical property of GaN X-band MMICs, sometimes even independent of the design methodology. The S-parameters are well reproduced by the simulation, performed with an optimized LS-model. The gain at the higher band edge increases with the increase of $V_{DS} > 15\,V$. Fig. 6.6 gives the measured pulsed-output power as a function of operation bias V_{DS}. The figure shows the increase of the power gain and output power with V_{DS} up to $40\,V$.

Coplanar-Waveguide Transmission-Line MMICs

From a processing perspective, coplanar-waveguide MMIC amplifiers can be realized more simply, since no backside process is required as compared to the microstrip approach, e.g., [6.9, 6.10, 6.169]. A coplanar-waveguide X-band linear power amplifier GaN HEMT MMIC is reported in [6.191]. The chip size is $6 \times 1.5\,mm^2$. An output power of $36\,dBm$ at $8\,GHz$ is reached and yields an intermodulation distortion IM_3 of $-35\,dBc$ at $1\,MHz$ offset and 34% of associated PAE. Single-stage coplanar-waveguide power amplifier MMICs yielding output power levels of $39\,dBm$ with an associated PAE of 33% and a linear gain of $10\,dB$ at $10\,GHz$ are presented in [6.9]. The chip size is increased for this coplanar chips, partly due to the need to connect the ground plane on the front side of the wafer. The device does not use the full set of passive lumped components for matching. A chip size of $4 \times 6.2\,mm^2$ is reported in [6.9]. A very compact coplanar MMIC design of a power amplifier reported in [6.81] delivers up to $16\,W$ of output power at $8\,GHz$ at an operation bias $V_{DS} = 28\,V$. The chip size is $2.2 \times 3.3\,mm^2$ in this case. The efficiency for this dual-stage MMIC is as high as 30%. The input and interstage matching are performed on small-signal level, whereas the output stage is matched to load-pull measurements.

Coplanar MMIC Example

The MMIC example is based on the considerations in [6.10, 6.169]. Fig. 6.7 gives the chip image of a dual-stage high-power amplifier in coplanar transmission-line technology. The chip size is $4.5 \times 3\,mm^2$. During the design phase a multistep procedure is used. Load-pull simulations of the power cells are performed with respect to gain, efficiency, and output power at a compression level given by the application. With respect to stability the load reflection is selected to have sufficient margin to the instable region where the small-signal gain would be larger than the maximum MSG/MAG. In a further design step, the matching networks of output, input, and interstage are designed. The power combining structures are included in this analysis and are part of the networks. The interstage network transforms the output load of the first stage into the input network of the second stage. Sufficient gain flatness is achieved by designing the interstage network for best match near the upper edge of the operation frequency. Stability analysis is performed according to [6.118]. Further, 2.5D- and 3D-EM-simulations need to be performed for more complicated structures and their application in the layout even for coplanar structures [6.37, 6.68, 6.168]. Fig. 6.8 gives the measured S-parameters of the coplanar amplifier in cw-mode. In general, the very broadband performance of the X-band amplifiers can be observed for the

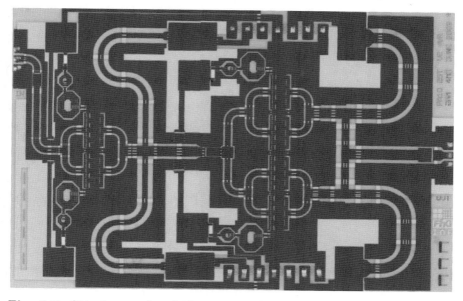

Fig. 6.7. Chip image of a dual-stage coplanar high-power amplifier, chip size $4.5 \times 3\,\mathrm{mm}^2$

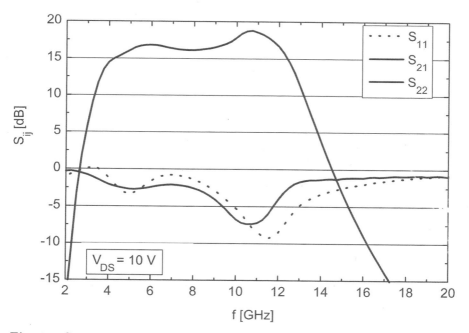

Fig. 6.8. S-parameters of a coplanar AlGaN/GaN HEMT MMIC X-band amplifier at $V_{DS} = 10\,\mathrm{V}$

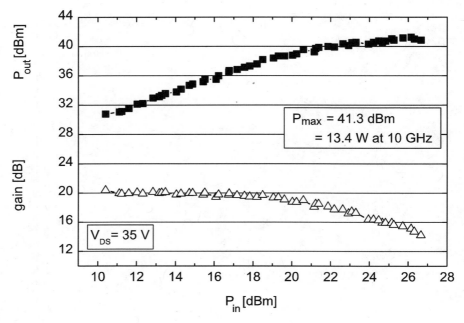

Fig. 6.9. Pulsed-output power and gain at 10 GHz and $V_{DS} = 35$ V

S_{21}. Fig. 6.9 gives the measured pulsed-output power and gain at 10 GHz. The MMIC delivers 13 W of saturated output power, which is equivalent to a power density of 3.3 W mm^{-1} at 10 GHz to a load of 50 Ω. Given the maturity of the technology, which is inferior to GaAs, the longterm advantages of GaN MMICs become visible. Further design considerations will be analyzed in Sect. 6.2.6.

6.2.6 Design, Impedance Levels, and Matching Networks

Matching and biasing of GaN HEMTs can be based on well-established techniques used for matching and power combining of GaAs devices, e.g., [6.6]. The realization of typical MMIC functions in GaN circuits, such as biasing and RF-matching, is very similar to that for GaAs. MIM capacitor values and resistors are very similar; however, as will be discussed in Chapter 8, the lumped elements have to be matched to the increased voltage and power levels. This also includes the design rules, which have to matched to increased power levels, increased electric fields, and higher thermal dissipation. RF-stability considerations of MMICs are discussed, e.g., in [6.118, 6.199]. Nyquist plots of the transfer functions can be used to completely characterize stability of an RF-amplifier. This is important as the (in-)stability circles of the GaN HEMTs are very close to the optimum matching loads. For an example of this property, see, e.g., [6.170].

GaN HEMT Circuits: Design Techniques

The intrinsic properties of wide bandgap GaN FETs are promising and have been repeatedly mentioned [6.121]. However, the use of appropriate design techniques is of critical importance to make use of these intrinsic material properties in circuits. Several examples for the extraction of optimum output load impedances of GaN HEMTs have been reported, e.g., [6.195]. For a GaN/AlGaN HFET power cell with a gate periphery of 1 mm on sapphire substrate, an optimum load of 32 Ω and a output capacitance $C_{ds} = 0.32\,\mathrm{pF\,mm^{-1}}$ are mentioned for an operation bias $V_{DS} = 27\,\mathrm{V}$. A more exact analysis of a 1 mm cell in X-band is given in [6.170]. The frequency-dependent distinction is made between the optimum load for output power, gain, and efficiency. The optimum impedance at 10 GHz of a 1.2 mm cell is $R_{opt} = 35\,\Omega$ for maximum power and $R_{opt} = 78\,\Omega$ for optimum PAE. A more systemized approach based on LS-models was already detailed in Chapter 5. A simple resistive load can be approximated by the Cripps-approach [6.29]:

$$R_{opt} = \frac{V_{DS}}{0.5 \cdot I_{pp}}. \tag{6.5}$$

V_{DS} and I_{Dpp} are the quiescent operation bias and the maximum drain current-swing (peak-to-peak) values in linear operation. Based on the operation voltage and input power, the modified Cripps-load can be calculated from a set of equations, which includes also the compensation of the output capacitance, as detailed in [6.29]. The load derived from this modified Cripps-approach can be used as a principle approximation for the verification of simulation results obtained by a large-signal model for frequencies up to about 12 GHz [6.170]. Above this frequency, the impact of the layout and associated impedance transformation becomes dominating.

Advanced Circuit Design Techniques

Circuit design of GaN circuits is similar to that of any other material system, such as GaAs. The tasks for device layout and circuit design include:

- Impedance transformation
- Selection of input matching, output matching, and relative bandwidth [6.88]
- Interstage matching
- Harmonic matching [6.25]
- Area reduction

A first pass design strategy for the use in GaAs X-band MMICs is described, e.g., in [6.167]. Such a strategy includes transistor size selection, characterization of the transistors at high power-level conditions, large-signal parameter extraction, broadband matching techniques, and layout; see [6.6]. Power design and power combining techniques have been described, e.g., in [6.6].

Matching techniques for narrowband and broadband designs require detailed analysis with respect to relative bandwidth, PAE, gain, linearity, and output power. Table 6.3 compares several properties (see also [6.6]) and examples of the application of matching techniques for GaN FETs. No amplifier concept or matching technique is found ideal for all applications, especially for broadband. The most profound trade-offs are among bandwidth, efficiency, and linearity, as seen in Table 6.3. The small-signal conjugate matching is widely used for input and output. However, it is not ideal for power matching due to the difference of R_{opt} for linear gain and output power, especially at high power levels due to the impact of harmonics [6.158]. Quarter-wavelength matching-approaches are attractive in general, as transmission-line elements can be easily used to emulate lumped components [6.195], especially for higher frequencies, while they yield additional advantages, such as half-wavelength transparency [6.30]. The decision, whether lumped-RCL elements or transmission-line elements are used, is based on the frequency and area available to place and fold extended transmission-line elements, especially at lower frequencies. Balanced (0–180°) amplifiers are typically used for simple high-bandwidth designs with increased output power, easy and reliable matching, and intrinsic redundancy. Examples for GaN-based PA designs are found in [6.113], for GaN HEMT low-noise designs, see [6.114]. Push–pull configurations can be used to increase efficiencies while maintaining good bandwidth, as performed with GaN HEMTs, e.g., in [6.53, 6.73, 6.100]. Feedback is used in various forms essentially to improve bandwidth at the upper band edge and the linearity of the design. A DC to 5 GHz design is presented based on feedback inductances [6.23].

Harmonic tuning is typically used to increase the efficiency [6.24, 6.25, 6.30]. As shown in Table 6.3 the termination has a limiting effect on the bandwidth unless very specific techniques are used [6.53].

For overall verification of the passive component models for MMIC layout, 3D- or 2.5D-electromagnetic simulations are typically used to verify the combination of the passive and active elements and their models, e.g., [6.148]. This allows a very precise prediction of the transmission-line elements, as reported in [6.33]. This is especially useful for nonlinear transmission-lines [6.105]. Sensitivity analysis towards process variations completes the design strat-

Table 6.3. Reported matching techniques for GaN FETs and different criteria

Matching	PAE	Bandwidth	Linearity	P_{out}	Gain	Ref.
Small–signal	+	0	++	0	++	[6.81]
Balanced	+	+		+	++	[6.113]
Feedback	− −	++		0	+	[6.113, 6.114]
Distributed	− −	++		+	+	[6.93]
LCR $\lambda/4$ transmission	−		++	+	0	[6.195]
Harmonic tuning	++	−	0	+	+	[6.24, 6.25]
Push–Pull	++	0	+	++	++	[6.53, 6.73]

egy [6.6, 6,89]. The sensitivity analysis is typically based on the variation of a variety of small-signal parameters and the inclusion of their intrinsic correlation into the analysis [6.89]. Newer methodologies for the design and integration approach for MMICs have been suggested, e.g., in [6.102]. The cointegration of MEMS for tunable impedance-matched networks is suggested. An X-band GaN HEMT power amplifier design using an artificial neural network modeling technique for the active GaN HFET is presented in [6.94]. A neural network is trained and used to model S-parameters of the device under specific operation conditions. This technique is considered quick, while the underlying device physics is omitted and no extraction of a model is necessary. The drawback is the lack of direct control of the extracted parameters. Transmitter design techniques have been described, e.g., in [6.88]. It is suggested that node architectures based on solid-state power amplifiers multiply output power while maintaining gain and efficiency. Traveling-wave-tube (TWT) replacement is considered as one application of multichip MMIC solutions [6.4]. It is discussed whether GaN MMICs significantly reduce the number of MMICs needed for this purpose and enable TWT replacement. As an example, pulsed-output power levels of 500 W and even 880 W have been realized with efficiencies of 50% in L-band and S-band [6.100, 6.108], based on GaN HEMT amplifiers. The device concepts employ four dies of devices integrated in one package in a push–pull concept. This example shows that GaN HEMTs simplify the architectures for a given output power.

6.2.7 Broadband GaN Highly Linear Amplifiers

GaN devices are extremely useful for broadband high-power high-efficiency amplification because of their high impedance levels and their high-gain device characteristics. Corresponding GaAs devices typically provide gate widths of up to $W_g = 16$ mm with power levels of 10 W [6.98]. Several GaN-based hybrid and MMIC-broadband amplifiers have been suggested in the literature, e.g., in [6.74, 6.82, 6.87, 6.93, 6.186]. LCR-matched power amplifiers are given in [6.195]. AlGaN/GaN HEMTs on sapphire substrates and flip-chip mounting on AlN passive substrates are used. A four-way Wilkinson combiner structure is used to match flip-chip devices with 4 mm of total gate periphery. The measured output power amounts to 8.5 W at 8 GHz when biased at $V_{DS} = 16$ V. Another amplifier delivers an output power of 35 dBm between 4 and 8 GHz, i.e., for a relative bandwidth of 50%. A cascode delay-matched power amplifier between 2 and 8 GHz is presented in [6.87], which yields 5 W of output power at 6 GHz, 1.35 W at 8 GHz, and a small-signal bandwidth between 1 and 9 GHz. A demonstration of a high-efficiency nonuniform monolithic GaN HEMT distributed amplifier is given in [6.93]. A four-stage dual-cascode concept is used for the gain cells in order to have a high gain-bandwidth product. The chip size is 2.5×1.4 mm^2. Power levels of ≥ 1 W are achieved between 3 and 12 GHz with a maximum of 3 W at 8 GHz. The maximum PAE is 25% at 3 GHz.

Ku-Frequency Band (14-18 GHz) MMIC Amplifiers

The Ku-frequency band is used for both civil and military applications requiring bandwidth, linearity, and efficiency [6.105]. Multioctave III-N device amplifiers with bandwidths between 2 and 20 GHz and efficiencies ≥30% are proposed, see [6.139]. Broadband reference MMICs based on GaAs HFETs are reported, e.g., for the frequency band 6–18 GHz, in [6.7, 6.85, 6.105]. The output power levels amount to ≥5 W in saturation and efficiencies beyond 22% with 12 dB of gain. The chip size is $5.8 \times 5.8\,\mathrm{mm}^2$. A GaN K-band linear amplifier in coplanar transmission-line technology on sapphire substrate is reported in [6.116]. A 3 dB bandwidth of 4.5 GHz is obtained with more than 10 dB gain between 20 and 24.5 GHz. A Ku-band dual-stage MMIC power amplifier with a 6 mm output periphery and a 3 mm wide driver stage yields a pulsed-output power of 24 W at 16 GHz [6.130]. The associated gain is 12.8 dB with a PAE of 22%. The supply voltage is 31 V. A packaged Ku-band MMIC amplifier in microstrip line technology on s.i. SiC is given in [6.150], which demonstrates the very good linearity of these amplifiers. The dual-stage microstrip transmission-line design operates between 9 and 19 GHz with a small-signal gain of 13 dB. The IMD3 is better than −30 dBc up to associated output power levels of 26 dBm. A 6–16 GHz linearized amplifier in microstrip-transmission-line technology on s.i. SiC substrate is described in [6.74] realized in quarter-micron GaN gate technology. A push–pull concept is used for the amplifier. Active FET nonlinear generators are used to achieve an improvement in carrier-over-intermodulation C/I-ratio of 8 dB. The output power level is 10 W. More recently, across the multioctave bandwidth 2–15 GHz, average cw-output power levels and associated PAE of 5.5 W and 25% are obtained [6.48]. Again a nonuniform distributed topology with five transistors is used in microstrip transmission-line technology. Maximum output power levels reach 6.9 W with 32% PAE at 7 GHz. The chip size is $4.45 \times 2.26\,\mathrm{mm}^2$. Similar results between 4 and 18 GHz are obtained in [6.105] in coplanar technology. However, the output power results are obtained in pulsed-operation with a duty cycle of only 5% with a gate width $W_\mathrm{g} = 2\,\mathrm{mm}$. As another example, Fig. 6.10 gives the chip image of a single-stage broadband linear power amplifier in coplanar technology on s.i. SiC substrate. Five GaN HEMT cascode GaN HEMTs with $l_\mathrm{g} = 150\,\mathrm{nm}$ are combined in a traveling-wave structure. Fig. 6.11 gives the measured S-parameters S_{ij} of the MMIC amplifier. The amplifier provides a small-signal gain of ≥8 dB between 1 and 20 GHz in a cascode device concept. The gate biases are $V_\mathrm{GS} = -3.5\,\mathrm{V}$ for the first gate and $V_\mathrm{GS} = 2\,\mathrm{V}$ for the second gate. This and other recent examples [6.48, 6.105] demonstrate the outstanding gain bandwidth and power capabilities of GaN HFETs for multioctave broadband applications.

6.2.8 GaN Mm-wave Power Amplifiers

GaN HEMTs are also very attractive for mm-wave operation from 30 GHz to at least 100 GHz [6.133] due to the material properties discussed in Chap-

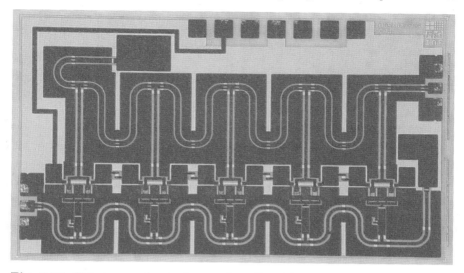

Fig. 6.10. Chip image of a single-stage broadband linear amplifier using cascodes

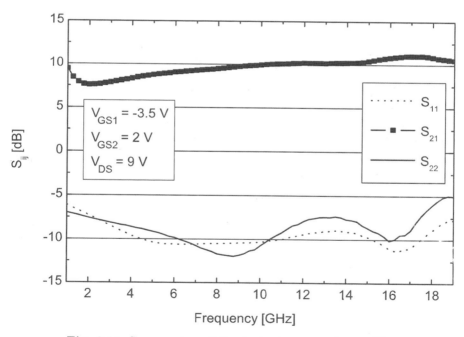

Fig. 6.11. S-parameters of the single-stage cascode amplifier

ter 2. For high frequency operation, e.g., at 30 GHz [6.168], 40 GHz [6.132], or even 80 GHz [6.106], the impedance situation changes dramatically. For a GaN

HEMT with small gate width of 100–300 μm with several hundred Ohms of intrinsic output impedance at 2 GHz, the layout transformation leads to a real output impedance of $\leq 50\,\Omega$ with a high inductive contribution, as is shown in Fig. 5.30 in Chapter 5. The complex impedance of a device with 8×60 μm gate width at 30 GHz yields $6.6 + i\,20.3\,\Omega$ [6.168]. Double heterojunction devices are suggested for improved pinch-off of short channel devices [6.106]. The intrinsic impedance is 110 Ω. Power gain levels of 7–8 dB per stage are targeted for mm-wave operation [6.140]. Interstage matching at mm-wave frequencies is thus a critical requirement to obtain sufficient power gain to 50 Ω impedance [6.168] at mm-wave frequencies. Very different impedance levels from the output of the first amplifier stage to the input of the second stage have to be matched similar to GaAs PHEMTs. The main distinction for the passive mm-wave design for GaN are relaxed design rules for high-power high-voltage operation, which at the same time reduce losses and minimize chip area. On the passive side mm-wave libraries of coplanar transmission-line elements have to be extracted similar to those for GaAs PHEMT, e.g., including the verification with 3D-EM simulation [6.55, 6.168]. Design using microstrip transmission-lines at mm-wave frequencies yields reduced source inductances [6.31], reduced area consumption [6.70], while the passive technology requires a higher amount of EM-simulations [6.70]. The use of microstrip transmission technology helps to suppress higher-order modes within the substrates [6.137] for mm-wave operation.

6.3 Robust GaN Low-Noise Amplifiers

In addition to the outstanding power performance, GaN HEMTs are promising candidates for robust low-noise applications. The low-noise performance of GaN HFETs is currently slightly inferior to that of GaAs PHEMTs, GaAs metamorphic, or InP HEMTs [6.57] for the same gate length [6.60, 6.132]. However, GaN HEMTs provide enormous advantages in terms of linearity [6.123], bandwidth [6.5], and robustness [6.123, 6.142] up to input power levels of $\geq 1\,$W.

6.3.1 State-of-the-Art of GaN Low-Noise Amplifiers

A number of examples for GaN LNA MMIC and hybrid LNAs are available. On the GaN HFET device level, minimum noise figures $N_{\mathrm{F,min}} = 0.4\,$dB at 6 GHz and 0.7 dB at 10 GHz are reported with associated gain levels $G_{\mathrm{ass}} = 14\,$dB and 11 dB, respectively [6.60]. Similar noise figures of 0.5 dB in the same frequency range (7–10 GHz) are reached at the full MMIC level using InP HEMT technologies, as reported in [6.57]. A combined low-noise high-power GaN HEMT technology for hybrid circuit approaches is given in [6.131]. Both low-noise operation with 1.7 dB of noise figure and 8.3 dB of associated gain at 8 GHz and power densities of 4.6 W mm^{-1} at $V_{\mathrm{DS}} = 25\,$V with 40% PAE at 10 GHz are reported. Wideband AlGaN/GaN HEMT low-noise amplifiers for highly survivable receiver electronics are presented in [6.16] using coplanar MMIC technology. A single-stage MMIC device on s.i. SiC substrate yields a noise figure of about 2 dB and a linear gain of 15 dB up to 3 GHz. The devices are subject to RF-stress tests with input power levels of 30 dBm. The output power P_{-1dB} is ≥ 1 W at 2 GHz. The associated third-order output intercept point (OIP3) is as high as 43 dBm at 2 GHz. C-band GaN HFET LNAs are reported in [6.142] with noise figures of ≤ 2.3 dB between 3 and 7 GHz. The dual-stage MMICs provide an associated gain of 20 dB for the same frequency range. The extrapolated output intercept point (OIP) is 26 dBm at 5 GHz. The devices are subject to input power levels of 36 dBm. A multioctave LNA MMIC from 0.2–8 GHz with ≤ 0.5 dB of noise figure at $V_{DS} = 12$ V and an output power $P_{-1dB} = 2$ W is given in [6.83]. The amplifier obtains high linearity with an OIP3 of 43.2–46.5 dBm and P_{-1dB} of 32.8–33.2 dBm (2 W) between 2 and 6 GHz. The associated PAE at P_{-1dB} is 28–31%. Further wideband LNAs based on dual-gate GaN technology are presented in [6.5]. A single-stage MMIC delivers 12.5–18 dB of gain and 1.3–2.5 dB noise figure between 1 and 12 GHz. Resistive feedback is used to allow wideband operation. The gate periphery is $W_g = 320$ µm with a gate length $l_g = 0.2$ µm. The output power of the MMIC is 25 dBm at $V_{DS} = 10$ V. The device can be driven to a compression level of P_{-25dB} with a destructive input power level of 38 dBm at 2 GHz. The temperature dependence of a GaN-based HEMT monolithic X-Band LNA between $-43°$C and $150°$C is investigated [6.151]. The LNA noise figure of 3.5 dB at room temperature increases to 5 dB at 150°C. At the same time, the associated gain decreases from 8.5 dB at RT to 5 dB at 150°C and 8 GHz. Investigations of the survivability of AlGaN/GaN HFETs for high-input drive are given in [6.18]. Two catastrophic failure mechanisms are identified. At low drain–source voltages (≤ 10 V), the forward turn-on of the gate diode can exceed the burnout limit, resulting in a sudden failure. Increasing the quiescent V_{DS} increases the peak drain-gate voltage and changes the failure mechanism to gate-drain reverse breakdown.

6.3.2 Examples of GaN MMIC LNAs

Robust GaN HEMT LNA MMICs for X-Band applications have been reported in [6.86]. Some findings in the course of the design and characterization are discussed later.

Single-Stage MMIC Small-Signal and Noise Performance

Fig. 6.12 gives the chip image of the single-stage coplanar design. The chip size is 2×1 mm^2. Fig. 6.13 compares the simulated and measured S-parameters of the amplifier at $V_{DS} = 7$ V and a drain current $I_D = 150$ mA mm^{-1}. The input and output match is better than 10 dB at 10 GHz and a maximum S$_{21}$ of

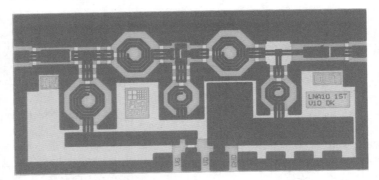

Fig. 6.12. Chip image of a single-stage GaN HEMT low-noise amplifier

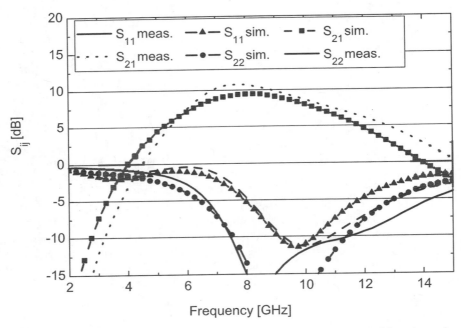

Fig. 6.13. Simulated and measured S-parameters S_{ij} as a function of frequency for a single-stage X-band LNA

10 dB is measured at 8 GHz. Fig. 6.14 provides the measured noise figure N_F of the single-stage design. The measured noise figure at 50 Ω is significantly increased relative to the $N_{F,min}$ of the single FET. This is due to the passive losses in the matching networks.

Small-Signal and Noise Performance of the Dual-Stage MMIC

Multistage MMICs are typical for LNAs, thus Fig. 6.15 compares the simulated and measured S-parameters of a dual-stage amplifier at $V_{DS} = 7$ V for

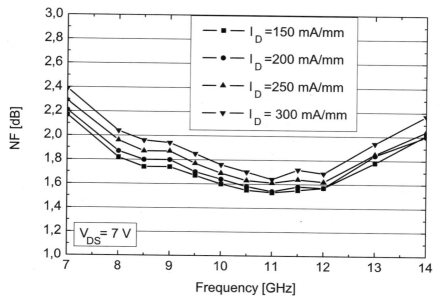

Fig. 6.14. Noise figure N_{F} of the X-band single-stage LNA as a function of frequency

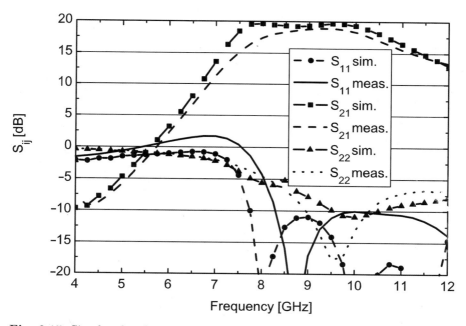

Fig. 6.15. Simulated and measured S-parameters S_{ij} as a function of frequency for the dual-stage X-band LNA

a drain current $I_D = 150\,\text{mA}\,\text{mm}^{-1}$. All drain current levels given are identical for the first and the second amplifier stage. A measured S_{21} of $\geq 18\,\text{dB}$ and a measured input and output match of better than $10\,\text{dB}$ are obtained at $10\,\text{GHz}$. The input and output of the devices are matched to the small-signal parameters.

Fig. 6.16 gives the measured noise figure N_F at $10\,\text{GHz}$ of the dual stage-design as a function of drain current I_{DS} and operation voltages V_{DS} between 1 and $20\,\text{V}$. The noise figure is nearly constant with V_{DS} while a stronger increase is visible for the increase of the current I_{DS}.

Fig. 6.17 gives the measured noise figure N_F at $V_{DS} = 7\,\text{V}$ as a function of frequency for the dual-stage LNA with the drain current as a parameter. A minimum noise figure $N_F = 1.6\,\text{dB}$ is observed for a drain current $I_D = 150\,\text{mA}\,\text{mm}^{-1}$ at $11\,\text{GHz}$. At $10\,\text{GHz}$, the device yields a noise figure of $1.8\,\text{dB}$. The drain current $I_D = 100\,\text{mA}\,\text{mm}^{-1}$ is not considered in Fig. 6.17, as the device did not provide sufficient gain for this bias. Similar operation bias are given in [6.83] for a $2\text{–}8\,\text{GHz}$ MMIC with $V_{DS} = 12\,\text{V}/200\,\text{mA}$ for the low power and $V_{DS} = 15\,\text{V}/400\,\text{mA}$ for the high-power setting for a $1.2\,\text{mm}$ device.

Large-Signal and Intermodulation Measurements

The cw output power measurement of the single-stage LNA for $V_{DS} = 25\,\text{V}$ at $50\,\Omega$ load at $10\,\text{GHz}$ gives a net linear gain of $10\,\text{dB}$, while a maximum cw-PAE of 24% is found. A maximum saturated output power of $26\,\text{dBm}$ is measured. The maximum net input power amounts to $23\,\text{dBm}$ or $1.66\,\text{W}\,\text{mm}^{-1}$ at a compression level of $P_{-7\text{dB}}$. Both the output power and the high-compression level demonstrate the robustness of the GaN LNAs. Intermodulation measurements at both input and output at $10\,\text{GHz}$ are given in Fig. 6.18. The output intercept point for two-tone excitation amounts to $27\,\text{dBm}$.

The input intermodulation IIP3-measurement of the single-stage MMIC as a function of V_{GS}- and V_{DS}-bias is given in Fig. 6.19. The V_{GS}-range in Fig. 6.19 covers similar drain current levels, as given in Fig. 6.16. The trade-off between linearity and noise figure can be observed by comparing Fig. 6.19 and Fig. 6.16. Lower noise figures N_F are achieved for reduced operation bias and low currents I_{DS} (Fig. 6.16), whereas the input linearity improves for increased operation bias V_{DS} and for higher current I_{DS}, as shown in Fig. 6.19. Fig. 6.19 clearly shows the combination of high-speed and WBG properties in GaN MMICs, i.e., high-linearity and low-noise figure.

6.4 Oscillators, Mixers, and Attenuators

In addition to their application in amplifiers, the outstanding material properties of III-N devices can be also used for other circuit applications and functions.

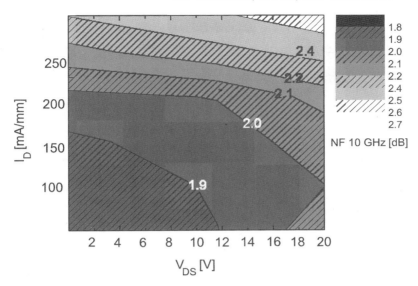

Fig. 6.16. Measured noise figure N_F in dB as a function of I_D and V_{DS} for the dual-stage X-band LNA at 10 GHz

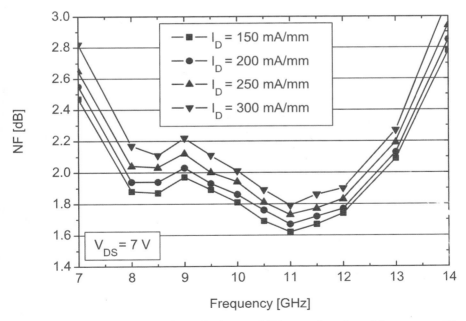

Fig. 6.17. Noise figure N_F of the dual-stage LNA as a function of frequency with the drain current as a parameter

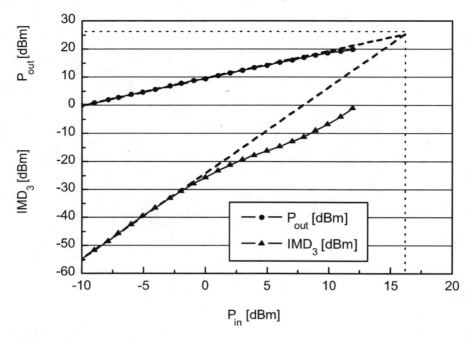

Fig. 6.18. Measured intermodulation distortion IM_3 as a function of input power for the single-stage X-band LNA at 10 GHz

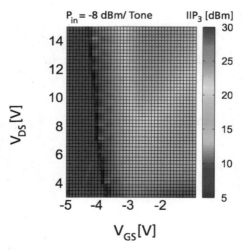

Fig. 6.19. Input intermodulation point IIP3 as a function of V_{DS} and V_{GS} at 10 GHz

6.4.1 Oscillators

GaN HEMTs are attractive candidates for robust high-power frequency sources, such as voltage controlled oscillators (VCOs), for applications in harsh environments, e.g., [6.50]. The high output-power densities allow MMIC design without the use of an additional output buffer amplifiers, e.g., in [6.72]. An example of a VCO with an output power of 35 dBm between 8.5 and 9.5 GHz is given in [6.72]. Output power densities of 2.1 W mm^{-1} are reached at $V_{DS} = 30$ V. The technology uses GaN HEMTs with a gate length $l_g = 150$ nm. A common-gate architecture is used as it maximizes device instability. The phase noise is estimated from the spectrum analyzer signal only, which is a critical procedure and thus shall not be detailed here. High-power single-ended Colpitts-type GaN oscillators using field-plated HEMTs are described in [6.194]. Both output power and operation bias can be significantly increased to 1.9 W and $V_{DS} = 40$ V with an output power density of 3.8 W mm^{-1} at 5 GHz. The DC-to-RF-conversion efficiency amounts to 21%, while the measured phase noise at 1 MHz offset from the carrier amounts to -132 dBc. A GaN HFET with $W_g = 4 \times 125$ μm is used. Similar low-phase noise AlGaN/GaN HEMT oscillators integrated with Ba$_x$Sr$_{1-x}$TiO$_3$ thin films are given in [6.193]. The high-k material Ba$_x$Sr$_{1-x}$TiO$_3$ is integrated for high-Q capacitances at the source and drain side of the devices. Phase-noise levels of -105 dBc at 100 kHz offset from the carrier with output power levels of 20 dBm are given. The DC- to RF-conversion efficiency is 12%. At higher frequencies of 39 GHz, a Q-band oscillator MMIC is presented in [6.90]. An output power of 25 dBm at $V_{DS} = 25$ V is reported at a record low-phase noise level of -120 dBc at 1 MHz offset from the carrier; however, the phase-noise measurement is again performed with a spectrum analyzer, which is very sensitive to locking. Differential GaN oscillators are reported in [6.143]. At an output power level of 23 dBm and 4.16 GHz a second-harmonic-suppression of -45 dB and a third-harmonic-suppression of -70 dB are reached. Initial results for ring oscillators in GaN digital circuits are given in [6.61]. The devices can be operated up to base plate temperatures of 265°C, which shows the suitability for operation of GaN VCOs in harsh environments.

6.4.2 GaN HEMT Mixer Circuits

Any communication system also requires the function of a mixer for frequency upconversion and downconversion. FET mixers are very attractive, as they can be used in active and passive mode, e.g., [6.8]. This allows a choice between conversion gain and conversion loss. The conversion gain is typically traded for the linearity of the mixer. GaN HEMT mixers are more advantageous than GaAs FET mixers, as they promise a higher dynamic range and better mixer linearity [6.2]. GaN high-power mixers with high-conversion gain are attractive candidates for communication systems because they enable the desired output power using a power amplifier with a small number of stages. Resistive mixer concepts using GaN HFETs have been reported repeatedly, e.g., in [6.2, 6.71]. A single-ended X-band mixer is presented with a chip size of 1.6 × 1.3 mm^2 [6.71]. The GaN HFET has a geometry of 0.15 μm ×1,000 μm. For an LO-signal of 10 GHz and an RF-signal of 12.4 GHz, the device has a conversion loss of −17 dB. The third-order input intercept point amounts to 40 dBm. This demonstrates the high-output power and linearity capabilities of GaN HFET mixers. C-band linear resistive GaN HFET mixers are described in [6.2]. A GaN HFET with a 10 × 50 μm periphery and a gate length $l_g = 0.7$ μm are used. The conversion loss of the GaN mixer is 7.3 dB with an local oscillator frequency (LO) of 250 MHz and an intermediate frequency of 4,925 MHz. The linearity is limited by the LO, which suggests that the robustness of the devices further determines mixer linearity. The third-order input intercept point is as high as 36 dBm. Dual-gate mixers provide the advantage of increased gain per chip area. Dual-gate mixers on s.i. SiC substrates are described in [6.156, 6.157]. A 0.7 μm × 300 μm device geometry is used for the realization of a mixer with a maximum output power of 19.6 dBm. The conversion gain is as high as 11 dB at 2 GHz. At 5 GHz, the conversion gain is 5 dB with an associated output power of 13 dBm. Dual-gate mixers for UWB (ultra-wideband) applications with a 2 GHz bandwidth at 24 GHz based on GaN FET on s.i. SiC substrates are presented in [6.156]. A 0.15 μm × 200 μm GaN HEMT is used. The maximum output power amounts to 8.9 dBm. The LO-signal at 24 GHz is modulated by a 0.4 ns pulse at the second gate. The output power from the mixer is found to be sufficient for direct antenna drive in the desired application. Again linearity, bandwidth, and robustness demonstrate the potential of GaN mixers.

6.4.3 Attenuators and Switches

GaN-HFET-based attenuators promise improved power handling while preserving high frequency operation and a wide dynamic range with good linearity. Broadband attenuators based on AlGaN/GaN HEMTs with high dynamic range are presented in [6.1]. The MMIC consists of four FETs in π-configuration, two connected in series, two in shunt-configuration. The MMIC provides an insertion loss of 4 dB, broadband operation from 0 up to 18 GHz, a high dynamic range of 30 dB, and a power handling capability of 15 W mm^{-1}. A simple optically-defined gate $l_g = 1$ μm gate length is used. The higher operation bias and breakdown capability lead to switches with very good isolation and better large-signal robustness. Based on these properties, also the electrostatic discharge (ESD) capabilities improves. As an example, a typical isolation for GaAs HFET switches is found to be about 30 dB. A GaN FET switch in single-pole-double throw (SPDT) configuration [6.71] yields an isolation of better than 40 dB from near DC to 3 GHz with an insertion loss of better than 1 dB. The compression power P_{-1dB} of 1 W is achieved with a 400 μm device. The third-order intercept point is 46 dBm and correlates well with the expectations.

6.5 *Problems*

1. *What is the estimated impedance advantage of an X-band power cell of 1 mm using a GaN HFET operated at 48 V relative to one using a GaAs PHEMT operated at 8 V?*

2. *Discuss the area advantage of a GaN MMIC relative to GaAs PHEMT MMIC for the same output power level of 10 W!*

3. *Calculate the Cripps-load at 2 GHz for an operation voltage of 50 V and 100 V for an output power of 5 W with an I_{Dmax} of 700 mA mm^{-1}.*

4. *Make a guess for the realization of a 100 W power amplifier at 4 GHz with respect to the power cell size. What are your constraints?*

5. *Which properties make GaN FETs suitable for advanced amplifier concepts?*

6. *Which device properties make GaN attractive for use in mixers?*

7. *Do you expect MMICs based on GaN HFETs to provide significantly better phase noise than GaAs PHEMTs? Give arguments!*

7

Reliability Aspects
and High-Temperature Operation

In this chapter, III-N-device-specific reliability issues and device failure mechanisms are discussed and analyzed systematically.

7.1 An Overview of Device Testing and of Failure Mechanisms

With the rapid innovation circles of the semiconductor industry, e.g., according to Moore's law [7.92], typical innovation times of 18 months have been maintained for decades now. This fact requires the application of accelerated testing procedures. Typically, extrapolated lifetimes of an electronic device amount to $\geq 10^6$ hours or 114 years at junction temperatures of 125°C.

7.1.1 Description of Device Degradation

The degradation of active electronic devices during operation is typically characterized by two modes. In the first mode, a gradual or parametrical change [7.132] occurs as a function of time of important figures-of-merit, such as drain and gate current I_D and I_G, transconductance g_m, current amplification β, or output power. A second mode is catastrophic failure or sudden total burnout. The latter is included in this testing procedure, as a total failure will also trigger the degradation criterion of choice. Several other application-specific temperature and degradation specifications exist, e.g., for base station applications [7.121] and further communication-, military- [7.107], and satellite-applications [7.6]. Accelerated testing procedures are required in order to timely ensure extrapolated lifetimes of ≥ 20 years at realistic temperatures of operation. Thus, in order to investigate the devices in realistic times scales, temperature accelerated testing is performed at elevated junction temperatures up to 300°C in the GaAs world for both HBT and (H)FETs, e.g., [7.82]. Typical assumptions of this procedure include degradation processes that are influenced by mechanisms such as diffusion of metal at the contacts, trap processes, or other processes that yield one-defined activation energy. The degradation is thus accumulated in one quantity. The activation energy is extracted according to

$$E_a = \frac{k}{\log_{10}(e)} \cdot \frac{d\log_{10}(MTTF)}{d(1/T)}. \tag{7.1}$$

From the gradual change of the FOMs, activation energies and median times to failure (MTTFs) can be extracted at a given temperature. The extraction is illustrated in Fig. 7.1. The testing is performed independently for various junction temperatures and the lifetimes extracted are plotted vs. temperature. From the qualified extrapolation, lifetimes can be derived for lower operation temperatures. This extraction procedure needs to be verified, as some degradation mechanisms do not obey accelerated degradation schemes, see, e.g., [7.72]. Other mechanisms are specific to low temperatures [7.86]. Further, some degradation mechanisms may be hidden in the high-temperature acceleration, such as, e.g., low-energy traps, see [7.57]. The criteria for failure applied to the particular data are typically a degradation of the drain current I_D of -5 or -10% [7.121], transconductance, β [7.2], or an output power degradation of $\leq 1\,dB$ [7.79, 7.121], typically even $0.2\,dB$. From a system point-of-view, even this measure is relatively high, as some system specifications only allow a component degradation of a few percent for the same temperature.

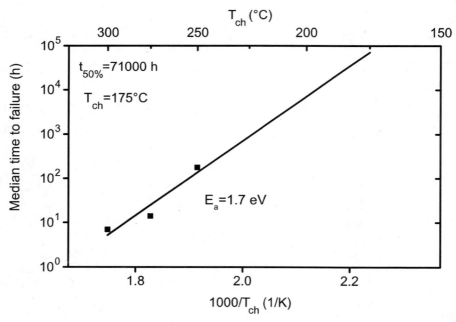

Fig. 7.1. Extraction of the $MTTF$ and activation energy of a GaN HFET [7.33]

The $MTTF$ at a given temperature T_L is calculated from the $MTTF$ at an elevated junction temperature T [7.110]:

$$MTTF\ (T_L) = MTTF\ (T) \cdot \exp\left(\frac{E_a}{k} \cdot \left(\frac{1}{T_L} - \frac{1}{T}\right)\right). \qquad (7.2)$$

Fig. 7.1 shows the extraction of the $MTTF$ and of the activation energy E_a. Some extraction procedures distinguish between more than one degradation mechanism, e.g., [7.25]. The bandgap energy of the main layers in the semiconductor device is an initial relative measure for interpreting the global quantity activation energy E_a. For GaN with a bandgap energy of 3.4 eV, activation energies of 1.7 eV [7.121] to 1.9 eV [7.102] have been reported. Very low activation energies of GaN devices, such as 0.38 eV, have initially been reported [7.47], which were too low to be of use. In general, based on the higher bandgap of the semiconductors similar or even improved reliability is eventually expected for GaN and other wide bandgap devices as compared to Si [7.137] and to GaAs [7.146] devices. However, this argument of the increase of the bandgap leading to better reliability has to be taken with a grain of salt, as even small bandgap devices, such as InP HEMTs with high In-content in the channel, yield activation energies beyond 1 eV and thus beyond the bandgap energy of the channel, as compiled and stated in [7.23]. Thus, the wide bandgap is beneficial, if

– The main degradation mechanism is related to the absolute bandgap(s)
– The same status of overall process development is achieved
– No additional low-energy mechanism prevail for III-N semiconductors, such as carrier-(de-)trapping or polarization effects

None of these conditions is fully secured for III-N devices. The development of III-N devices for reliability thus focusses on improved dielectrics and interfaces, better contacts (see Chapter 4), and trap-reduced semiconductor materials [7.91], see (Chapter 3). Similarly, a reduced sensitivity of III-N semiconductors toward bandgap-related degradation mechanisms, e.g., radiation damage [7.4] or electrostatic discharge (ESD) [7.121], is expected and also found to be reduced for bulk materials. However, III-N devices strongly rely on interfaces and heteroepitaxy, so that the semiconductor bulk property wide bandgap cannot be fully exploited for the reliability of the devices toward ESD and radiation.

Another way to express reliability is with the help of FIT rates [7.129]. They express the number of failures for an operation time of 10^9 h according to [7.59]

$$1\,FIT = 1\,failure/1 \times 10^9 \text{ device hours}. \qquad (7.3)$$

Both $MTTF$ and FIT rates are naturally related. Some applications may require a different FOM, such as the $MTBF$, the median-time-before-failure [7.57]. The $MTBF$ is based on the idea that even the first failure can fully destroy system performance. The extraction of the $MTBF$ requires a higher statistical effort than the extraction of the $MTTF$ [7.59] and is thus reported less often.

Device Failure Statistics

Device failure is a statistical process. This process is typically described by a number of statistical functions dependent on the failure mechanism [7.93]. The actual regression and the fitting error distinguishes between different functions, such as the log-normal, the normal function, and the Weibull regression function [7.44]. The log-normal distribution can typically be fitted to the failure distribution of semiconductor devices. The density probability function with the parameters μ and σ reads [7.64] for the log-normal distribution function

$$f(t, \mu, \sigma)_{\text{lognorm}} = \frac{1}{t\sigma\sqrt{2\pi}} \cdot \exp\left(-\frac{1}{2} \cdot \left(\frac{\ln(t) - \mu}{\sigma} \right)^2 \right), \qquad (7.4)$$

while it reads for the normal function

$$f(t, \mu, \sigma)_{\text{norm}} = \frac{1}{\sigma\sqrt{2\pi}} \cdot \exp\left(-\frac{(t - \mu)^2}{2\sigma^2} \right). \qquad (7.5)$$

The reproducibility of lifetime-test results is a key requirement for the system insertion of electronic devices. This is achieved through the parallel testing of nominally identical devices. The generation of accumulated device hours in excess of 10^5 h is a key requirement [7.44, 7.121] for the verification of the standard deviation of the results achieved. Multiple failure modes have also been analyzed and separated statistically [7.25]. Such a procedure allows the separation of effects and concentration on the most important failure modes, while control of the less demanding failure mode is maintained. In this procedure the device is operated under elevated temperature conditions to analyze early degradation, which is typical of the initial stages of GaN process development [7.108]. The distribution functions f_1 and f_2 of the two failure modes are composed as

$$f = (1 - x) \cdot f_1 + x \cdot f_2. \qquad (7.6)$$

Such a procedure can thus be used to separate the infant mortality of devices by burn-in procedures [7.44]. The actual mechanisms of the degradation will now be discussed in the following sections.

7.1.2 Degradation Mechanisms in III-N FETs

III-V HFET degradation is a well-investigated topic [7.24, 7.38, 7.70]. The effects of degradation of GaAs-based (P)HEMTs and of HEMTs based on the InAlAs/InGaAs material system [7.32] are well-known and discussed systematically in the literature. Overview articles can be found, e.g., in [7.38].

Device-performance degradation in general can be classified into two groups, one of which is reversible, including, e.g., simple thermal heating. Most important, however, are nonreversible effects, modifying or destroying device performance irreversibly for the remaining lifetime of the device. Some of these mechanisms are partially reversible, e.g., they can be recovered by light illumination or thermal heating. However, the problems resulting from this behavior make the device quasi-nonreversible for the application. The effects of HFET degradation typically include:

- The shift of the threshold voltage V_{thr} [7.32] and associated thermal runaway [7.1]
- Decrease of maximum drain current I_{Dsat} [7.121]
- Reduction of the DC- and RF-transconductance $g_{m,max}$ [7.23, 7.24] and similarly of all RF-gain parameters [7.38]
- Increase of the knee voltage under cw- or pulsed-DC-operation [7.29, 7.77, 7.88, 7.91]
- Change of the ideality factor and barrier height of the Schottky contacts [7.24]
- Change, i.e., mostly, but not always, increase of the gate currents and thus reduction of the breakdown voltage [7.22, 7.24, 7.133]

Physically, several nonreversible effects occur during device operation with an impact on reliability. The following effects are discussed for the bulk semiconductor or dielectrics:

- Occupation of traps in the semiconductor caused, e.g., by lattice mismatch of semiconductors and substrates [7.43] such as:
  - Buffer layer trap charging and detrapping [7.51, 7.97]
  - The propagation of substrate-related misfits and other dislocations from the nucleation layer to the surface [7.24]
  - Active semiconductor degradation by hot electrons [7.11, 7.40, 7.87, 7.123] and impact ionization [7.86]
- Relaxation of the material strain [7.61, 7.111]
- Thermal instability of the semiconductor at $\geq 600°C$ [7.31, 7.36]

At the device contacts the following effects occur:

- Gate contact degradation through:
  - Metal diffusion [7.143]
  - Gate sinking-induced degradation [7.24]
- Spiking of ohmic metals [7.24, 7.42]
- Ohmic contact degradation by:
  - Metal diffusion [7.19, 7.38]
  - Alloy creation [7.8] during operation

Transmission electron microscopy (TEM) cross-section images reveal no significant metal interdiffusion effects at gate and ohmic contacts in AlGaN/GaN HEMTs on s.i. SiC in [7.24]. In general the high annealing temperatures for

III-N devices lead to undefined metal–semiconductor interfaces, as reported in [7.42]. This effect is a reliability concern. Further, the high-annealing temperatures at or beyond the Debye temperatures of III-N semiconductors lead to structural modifications of the semiconductor layers. Most important we find the following degradation effects at the noncontacted or ungated surfaces and interfaces:

- Surface corrosion effects as proposed for GaAs PHEMTs in [7.41, 7.55, 7.138]
- Trap generation and occupation at the interface [7.40, 7.123]
- Degradation of passivation or of the dielectric due to trap charging and detrapping [7.89, 7.102]
- Surface breakdown [7.127]; and related degradation of isolation at mesa and implanted surfaces [7.85]

Further, environmental effects due to ambient gases and fluids have been investigated and include:

- Surface modification by fluorine [7.128] and hydrogen [7.9, 7.21, 7.90]
- Gas absorption at the polar surfaces [7.132]
- Donor or acceptor passivation of dopants by fluorine [7.128]

Further influences arise from radiation damage of various kinds

- Radiation-induced trap creation leading, however, to comparatively small changes in the device DC- and RF-characteristics [7.4].
- Semiconductor material degradation effects for γ-irradiation, while the metal–semiconductor interfaces, such as the Schottky diodes, degrade significantly [7.134].

For III-N devices additional degradation effects have to be considered:

- Change of mechanical stress from the on-wafer bowing of the semiconductor layers relative to the diced, MMIC, or packaging situation [7.62]. This includes:
  - Mechanical effects during the dicing process
  - Influence of the backside process on the frontside performance, e.g., during wafer thinning
- Change of the mechanical stress during die-attach

These effects have a strong impact on the final system application, as any device needs to be diced and mounted.

7.1.3 III-V HBT Device Degradation

III-V HBT reliability has been investigated in a large number of publications for GaAs-HBTs [7.2, 7.14, 7.75, 7.82] and InP-HBTs [7.103]. The most pronounced doping, interface, and contact issues have been solved [7.39] to make the GaAs HBT the most produced III-V electronic RF-device [7.46, 7.48]. The reliability of GaN-based HBTs is not well-investigated, due to the lack of devices with sufficient gain and overall performance. High-temperature operation of GaN HBTs at 400°C is investigated in a number of publications, e.g., [7.65, 7.147], as high-temperature operation increases the current gain in the existing devices. For III-N HBTs, a similar enumeration holds for the degradation effects in the overall device performance based on GaAs-HBT degradation [7.82]:

- Increase of the base-collector leakage [7.144]
- Instability of the p-doping and its activation in the base layers for npn-devices
- Contact issues due to the p-doping issues at the ohmic contacts

Physically, for III-N devices additional III-N-specific degradation mechanisms are to be expected especially due to the immaturity of the III-N bipolar approach:

- Ohmic contact degradation due to the nonideal ohmic properties for the p-contacts [7.147]
- A modified leakage path due to etch damage [7.144]

At the surfaces and interfaces the following effects are expected based on the remaining device issues:

- Device-passivation degradation
- Degradation of isolation at mesa surfaces [7.144]

With the development of (In)GaN HBTs with better performance, some of these issues will be resolved [7.65], also because of the rapid progress in the GaN HEMT [7.121] and III-N optoelectronic world [7.94].

7.2 Analysis of Nitride-Specific Degradation Mechanisms

Given the large number of publications on III-N electronic devices reviewed in Chapter 2, relatively little reliable and fully consistent data are available for the reliability of GaN devices. This is especially true with respect to increased operation voltages and to the new characteristics such as piezoelectric and spontaneous polarization effects. The few exceptions include the institutions involved in the base station market and military-oriented development, e.g., [7.119, 7.120]. Early reliability data of Cree is discussed in [7.108]. As a general finding, the reproducibility of the results is found insufficient in the early stages of development. Such results have been improved and the progress has continuously been reported, e.g., in [7.102]. This leads to the availability of initial GaN HEMT products [7.115, 7.141]. RFMD reported advances in reliability data in [7.17, 7.114] for their applications for the base station market. A significant stabilization of the device degradation results is found. The consistent evaluation of device reliability of GaN HEMTs by

Nitronex in [7.105, 7.119, 7.121] gives rise to the assumption, that the reliability issues of GaN HFETs can eventually be fully solved. This work is exclusively performed on silicon substrate. Some reliability data devoted to military applications with gate lengths $l_g = 250$ nm is given in [7.24]. Focussed ion-beam (FIB) images and SEM investigation show no contact issues after testing. The work specifically stresses the impact of the epitaxial layer structure on the device degradation. For higher frequencies, RF-degradation of small-gate-length GaN devices is discussed, e.g., in [7.13]. As a main finding for the reliability, the current degradation under DC-operation can be correlated with the RF-current degradation under RF-swing. Both current reductions in I_{DS} are found to be similar. In a series of publications [7.43, 7.77, 7.79], researchers at Triquint achieve a number of significant testing results. Effects of the AlGaN/GaN HEMT structure on RF-reliability are discussed in [7.79]. Three issues are stressed: first, the thickness of the AlGaN barrier layer; and second, the application of a GaN cap layer; and, third, the positive impact of a field plate on reliability.

7.2.1 DC-Degradation

DC-degradation experiments at elevated temperatures are the most natural testing routines of the device. Early reliability evaluation of AlGaN/GaN HEMTs grown on s.i. SiC substrate is given, e.g., in [7.78]. The findings include

- Reduction of the transconductance g_m
- Reduction of drain current I_{DSS}
- Change of the Schottky barrier height
- Increase of the on-resistance R_{on}

Electrical bias stress-related degradation of AlGaN/GaN HEMTs is reported in [7.71]. Typical findings for the device after DC-aging include:

- An increase of surface charge and increased transient behavior of the drain current verified by direct surface potential measurements
- A reduction of output power performance after stress [7.78]

The distinction between different stress-bias situations is further very important. Fig. 7.2 gives various bias conditions for testing. High-current I_{DSS} testing, low-current I_{Dq} testing in Class-A/B, and hot pinch-off testing at very high operation bias are typically compared. Pinch-off testing is compared to Class-A biasing, e.g., in [7.123]. In this case, Class-A biasing, or on-state operation of a AlGaN/GaN HFETs with a gate length $l_g = 0.25$ and 0.15μm at $V_{DS} = 25$ V, $P_{DC} = 6$ W mm^{-1}, is found to result in stronger device degradation than off-state stress at $V_{DS} = 46$ V and V_{GS} for pinch-off. Other publications, such as [7.67, 7.88], stress the importance of the off-state testing and Class-B or deep Class-A/B, see Fig. 7.2. A sudden degradation of leakage currents within the first hours of operation is found in [7.67]. Pinch-off testing is performed at a channel temperature $T_{chan} = 150°$C for a bias $V_{DS} = 40$ V

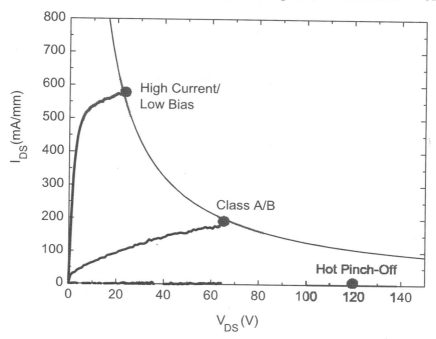

Fig. 7.2. Output characteristics of a GaN HFET with quiescent conditions for aging

and $I_{Dq} = 1.4\%$ of I_{Dmax} and devices with a gate length $l_g = 0.8\,\mu m$. The high-current test is typically found to be less critical, see [7.88]. All these findings are typical, though they do not always necessarily occur in parallel. Step-testing procedures can be applied increasing both bias and temperature with time. Degradation of AlGaN/GaN HEMTs on s.i. SiC under elevated-temperature step-lifetime-testing is further described in [7.24]. The testing starts at a channel temperature $T_{chan} = 150\,°C$ with a $15\,K$ step size. The aging is performed for $48\,h$ at each temperature up to $240\,°C$. A strong degradation of the GaN/AlGaN HEMTs is initiated at $195\,°C$. The actual degradation mechanism has not been fully isolated, however, the importance of the material quality on the propagation of the defects from the substrate is stressed. Short-time testing of AlGaN/GaN HEMTs on silicon is also described in [7.7]; however, even under such bias conditions, the devices do not show stable operation, contrary to the good reliability of GaN HEMTs on silicon in [7.121].

Testing Results

As a typical example of an early DC-testing result of an AlGaN/GaN HFET with $W_g = 50\,\mu m$, Fig. 7.3 gives the development of the DC-power density as a function of operation time at $V_{DS} = 20\,V$ at a backside temperature of $125\,°C$. A degradation of the DC-power is visible, due to the reduction of the current I_{DS} for a given bias V_{DS}–V_{GS} combination. The testing shows

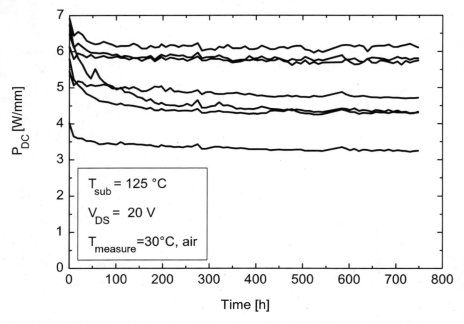

Fig. 7.3. DC-power as a function of time elapsed for GaN HEMTs with $l_g = 300\,\mathrm{nm}$ and a gate periphery $W_g = 50\,\mu\mathrm{m}$ [7.33]

a degradation in the first hours of the testing, which spreads the statistical device performance overtime. The device is cycled to room temperature for the DC-measurements. The reduction of the DC-current during aging of AlGaN/GaN HFETs is very similar to the degradation of advanced silicon MOSFETs, e.g., [7.20, 7.98]. This is especially true for the impact of electric fields to the dielectric interface. This can be modeled as an exponential function [7.28, 7.98]:

$$\frac{I(t)}{I_0} = b \cdot t^{-n}. \tag{7.7}$$

A similar expression is derived for the output power degradation in [7.28]:

$$\Delta P_{\mathrm{out}} = \frac{I(t)}{I_0} = (1 - b \cdot t^n)^2. \tag{7.8}$$

The data in [7.28] can be correlated to a negative bias temperature instability (NBTI) due to hot carrier injection (HCI), which is very similar to the degradation of Si MOSFETs. This description makes the degradation more predictable. The degradation mechanism in silicon MOSFETs is trap-assisted tunneling through the MOS interface. A similar mechanism at the SiN/(Al)GaN interface changes the performance for the GaN HFETs by dynamic trap occupation. The mitigation in [7.98] is performed by changes in the AlGaN/GaN process. To separate device yield from the actual degrada-

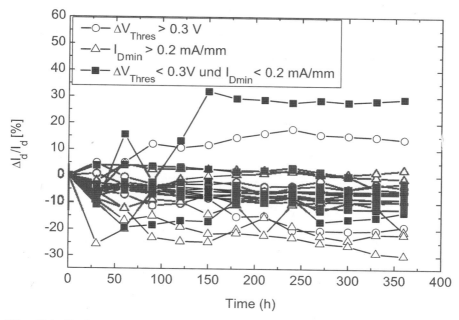

Fig. 7.4. Drain current I_D as a function of time elapsed for GaN HEMTs with $W_g = 50\,\mu m$ and $l_g = 300\,nm$ [7.33]

tion, Fig. 7.4 gives a typical degradation of the drain current at an elevated substrate temperature in an experiment similar to the one in Fig. 7.3 for a statistical selection of devices. In this set of devices, different modes of behavior are visible. Most of them, however, show a reduction of the drain current and respective power density as a function of time. The devices are sorted according to the derivation from the mean threshold value ΔV_{thr} and the initial value of I_{Dmin} prior to aging. For devices, which fulfill the moderate criteria given in the inset, the degradation is more homogenous. Devices with an initially high I_{Dmin} degrade stronger than the devices with low I_{Dmin}. The electrical background of the current degradation mode in Fig. 7.3 and Fig. 7.4 can be manifold:

- The aging translates into a positive shift of the device threshold voltage V_{thr}.
- This current degradation can be caused by an increase of the access resistances R_S and R_D.
- This behavior can be accompanied by a reduction of the transconductance g_m and the maximum current I_{Dsat}.

The most likely explanation in this case is a modification of the critical AlGaN/GaN and AlGaN/SiN interfaces due to carrier (de-)trapping, see [7.63]. This is confirmed by pulsed-DC-measurements prior to and after degradation, which suggest an increase of both the gate- and drain-lag. The latter

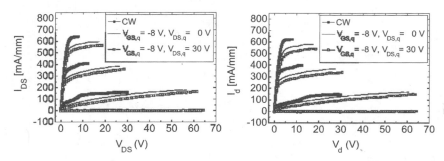

Fig. 7.5. Pulsed-testing of drain prior to (*left*) and after (*right*) aging at $V_{DS} = 120\,V$

finding also involves the buffer area of the device, as the drain-lag measurement (see (5.46)) can be influenced by both surface and buffer. As an example, Fig. 7.5 gives the pulsed-drain-lag measurement before and after testing the device at $V_{DS} = 120\,V$ and pinch-off. The pulse width is $1\,\mu s$ at a duty cycle of 0.1%. The cw-measurements are given for reference. Small changes of the output characteristics can be observed despite the high-bias testing. The aging thus leads to changes in the RF-effective IV-characteristics and can be correlated with the output power degradation in some cases, e.g., [7.77]. However, the time constants of the traps created or modified also have to be considered, as traps in III-N semiconductor are sometimes faster than the repetition rate and pulse width of the DC-pulse-testing.

7.2.2 RF-Degradation

RF-degradation testing is an advanced technique to analyze devices under real-operating conditions, typically high-power compression [7.60]. Failure modes in addition to those found during DC-testing can be analyzed, such as forward gate stress and dynamic (de-)trapping. Typical procedures to analyze the shift of the on-currents I_{DSS} and I_{Dq} as a function of time [7.12] while a certain power operation mode is maintained. The power, gain, and PAE changes are monitored in addition. This testing procedure, known from the silicon LDMOS world [7.16], has been applied to GaN HFETs, e.g., at $V_{DS} = 60\,V$ operation in [7.69]. The I_{Dq} degradation is the most critical failure mode for silicon LDMOS FET operation [7.16] and thus also considered for lifetime evaluation of GaN HFETs for the same application.

Several other testing procedures are available, e.g., [7.113]. For frequencies higher than $2\,GHz$, RF-reliability of AlGaN/GaN HEMTs grown by MBE on Si substrate is reported in [7.43]. Devices with a gate length $l_g = 300\,nm$ with a gate field plate are tested at $10\,GHz$ at operation voltages $V_{DS} = 20\text{--}40\,V$ and $I_{DS} = 200\,mA\,mm^{-1}$, i.e., in Class-A/B. The output power of the $4 \times 50\,\mu m$ GaN HEMT of $6.2\,W\,mm^{-1}$ at room temperature does not change after an initial small degradation at $V_{DS} = 40\,V$. In a similar experiment the effects of RF-stress on power characteristics and DC-pulsed-IV characteristics of AlGaN/GaN HEMTs with field plate gates with $l_g = 250\,\text{nm}$ are discussed in [7.77]. The device with the thinnest $Al_{0.32}Ga_{0.68}N$ barrier layer showed the slightest reduction in output power for thicknesses between 13 and 26 nm. Further, the positive impact of a GaN cap layer is mentioned. In addition, comparisons of pulsed-DC-measurements for various quiescent bias and at several stages of the RF-testing are a simple, though very sensitive, methodology for the analysis of the degradation of GaN HFETs. The impact of RF-stress on dispersion and power characteristics of AlGaN/GaN HEMTs is discussed in [7.56]. The saturated drain current I_{Dsat} is reduced after RF-tests at 5 GHz while the output power level is maintained. The dispersion of the transconductance measured between 50 Hz and 100 kHz is found to be nearly unchanged. UV-light-illumination leads to a drain current recovery. The analysis suggests the modification of traps in the gate–drain region as a result of the aging. The monitoring of low-frequency noise in correlation with the effects of RF- and DC-stress on AlGaN/GaN MODFETs is analyzed in [7.135, 7.136]. The principal finding for the impact of RF-stress is a significant shift in the threshold voltage V_{thr} of the GaN HEMT after the stress, while the $1/f$-noise floor is not strongly affected after stress. As an example of the output power degradation, Fig. 7.6 gives the development of the output power at 10 GHz over 500 h for four AlGaN/GaN HEMTs with $W_g = 0.48\,\text{mm}$ at a P_{-1dB} power compression operated at $V_{DS} = 30\,\text{V}$ and class-A/B. A change of output power with time is observed within margins of about 1.5 dB. The behavior can be separated in a decrease of the output power in the first 200 hours, which has been repeat-

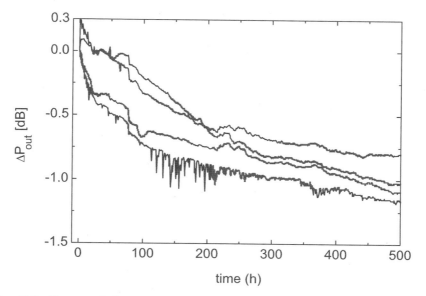

Fig. 7.6. Change of the output power of a GaN HEMT with $l_g = 300\,\text{nm}$ and $W_g = 480\,\mu\text{m}$ at 10 GHz vs. time [7.33]

edly presented [7.28] based on the negative bias temperature instability, as discussed with (7.8). It is further noted that some oscillations can occur in the test setup due to the changes of the device over time. The device degradation can be explained by the changes of several of the device characteristics with time:

- A change, i.e., mostly an increase of the knee voltage equivalent to a change of the access resistances
- A change, i.e., mostly a reduction of the maximum drain current I_{Dsat} [7.56]
- A change of the threshold voltage V_{thr}, for NBTI a positive shift, leading to a reduction in output power [7.28]. A negative shift of the threshold voltage leads to an increase of the drain current and output power over time
- A change, i.e., mostly an increase of the minimum or pinch-off current at high V_{DS} voltage [7.56, 7.145]
- A change, i.e., mostly a reduction of the breakdown voltage

The origin of actual changes of the output power and current can now be tracked.

7.3 Failure Analysis

Once the degradation of the terminal characteristics has been analyzed, physical sources for the degradation have to be isolated to mitigate the mechanisms through changes in epitaxial growth and process technology. Fig. 7.7 summarizes possible failure mechanisms in a GaN HFET in a device cross-section. Similar to any other III-V devices, the following issues have been discussed for device failure mechanisms of III-N FETs (for the numbering see Fig. 7.7):

- (3) Dislocation-induced degradation due to the lattice mismatch to the substrate [7.26]
- (3) Defect-induced changes of the minimum drain current I_{Dmin} [7.145]
- (1) Gate sinking [7.32, 7.38]
- (1) Other changes in the Schottky contacts [7.76]
- (5) Metal diffusion at ohmic contacts [7.73, 7.76, 7.84]
- (6) Crystal damage in the highly doped Si doping layer [7.38]
- (9) Cracking of the semiconductor to the stress and electric (polarization) fields [7.61]
- (2) Hot-carrier effects, mostly due to high-field gradient for nonoptimized devices [7.38, 7.109, 7.124]
- (4) (De-)trapping of the surface traps [7.38, 7.126]
- (7) Changes in the surface morphology [7.67]
- (8) Environmental effects through gas absorption [7.121]
- (11) Cracking issues due to the differences in CTEs of the materials in use [7.141]

The effects shown in Fig. 7.7 have already been grouped according to their physical nature.

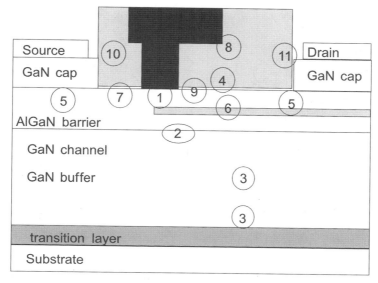

Fig. 7.7. Localized device failure mechanisms in GaN HEMTs on a heterosubstrate

7.3.1 Failure Mechanisms

Sometimes, but not always, the device failure mechanisms can be directly observed. Fig. 7.8 gives an example of a power bar. The breakdown path between two contacts and two gate fingers is optically visible, e.g., [7.76, 7.125]. However, such direct observation examples are not very numerous, as device failure may typically result in partial or total burnout due to high power densities, so that the device cannot be effectively analyzed, see Fig. 7.8. Several examples of failure-related analysis are available. The FIB/SEM/TEM cross-sections in [7.24] allow the analysis related to gate sinking, as is similarly performed for metamorphic devices, e.g., in [7.32, 7.121].

Apart from these classical effects enumerated, polar semiconductors face additional issues. A III-N FET, e.g., on s.i. SiC substrate, is fabricated on a heterosubstrate, resulting in an overall bowing of the wafer, as discussed in Chapter 2. This bow is modified several times during the processing:

– During epitaxy due to the lattice mismatch
– During processing due the metal deposition
– During wafer thinning
– During dicing and packaging

Such mechanical changes may have an impact on the III-N (H)FET performance, which can be dramatic in some cases. A more systematic study of the mechanical strain on the gate current degradation of AlGaN/GaN HFET is discussed in [7.62]. Uniaxial mechanical strain is applied to the device in a specific test jig. The gate current degradation is strongly affected due to

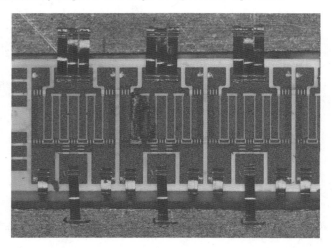

Fig. 7.8. Microscope image of a failed device

the change of the filling of traps in the AlGaN barrier at the drain side of the gate by the mechanical stress. The inverse piezoelectric effect has a similar mechanical effect during electrical stress and is thus related to purely mechanical stress. The related mechanical cracking of the AlGaN barrier layer due to the high polarization fields is described in [7.61]. A step-testing procedure is applied, which repeatedly provides a crack. This mechanical cracking is considered to be based on the piezoelectric properties and the resulting strong tensile strain once a voltage is applied. This principal finding hints to the need to evaluate the full consequences on the polarization in the devices.

Further studies involve impact of the source-gate-access region on reliability, as described in [7.41] similar to findings in other GaAs HFETs [7.122]. The electrostatic interaction of the source with the drain-end of the gate can have a major degrading impact on the effective breakdown voltage of GaN HEMTs.

A simple model for the impact of the gate length dependence of the reliability through thermal effects is given in [7.34]. Based purely on thermal consideration, the reliability drops as a function of gate length l_g. However, the model neglects the significant impact of the change in the hot carrier degradation due to the change of the gate length.

More advanced reliability issues are discussed in [7.141]. These include:

- Oxidation and intermetallic reaction of metals such as Ti, Al, Ni, Au in the ohmic contact
- Ohmic reactivity with the AlGaN
- Ion migration along dielectric surfaces
- Medium-/deep-level electron trap generation at high field
- Stress fracture of interconnect metals
- CTE mismatch of the various metals, dielectrics, and semiconductors

Some of these mechanisms have already been discussed. The ohmic contacts and the adjacent passivation are subject to ohmic oxidation, microcracking, and metal migration along the cracks. This leads to a reduction of the break-down voltages. Further, the reliability of MIM capacitances is investigated which is limited by trap-assisted band-to-band tunneling. This testing is very similar to the needs of passive elements on GaAs. However, due to the need for increased bias levels and parallel compact integration, the additional testing is needed. The degradation of the drain current of AlGaN/GaN HEMTs with implant isolation at 300°C is detailed in [7.85]. The off-state of the device is not affected, but with the flow of the carriers, a hot-carrier degradation mechanism is effective between gate and drain. In the case of a mesa struc-ture the effect is considered to be suppressed by the reduction of the side gate effect. This means, that two devices in the case of implantation can interact through the incomplete isolation, while the mesa process mitigates the inter-action. At the same time, a mesa process can also be critical, as discussed in Chapter 4. Parts of these limiting mechanisms are very process specific and some even trivial, such a metal breakage and CTE issues. However, they all require attention to ensure reliable operation of the full device in the system.

7.3.2 Reliability Case Studies

The effects resulting from the issues mentioned above can be grouped accord-ing to the most important failure mechanisms and are detailed here as case studies.

Degradation due to Unstable Contacts

High-contact annealing temperatures at or beyond the Debye temperature of GaN make the device susceptible to uncontrolled metal–semiconductor intermixing in alloys [7.141], uncontrolled chemical reactions [7.141], and uncontrolled interfaces and structures within the semiconductor [7.42]. FIB techniques and scanning tunneling microscope (STEM) images can help to analyze the contacts, as performed in [7.24]. The intermixing at the ohmic contacts leads to several possible reliability issues, such as [7.42]:

– Uncontrolled vertical or lateral diffusion of metal
– Spike structures with locally high current densities
– Reduction of nominally safe distances between contacts

Low-activation energies of only 0.9 eV are found during reliability testing for a more reliable and further developed SiC MESFET process [7.102]. This activation energy is correlated with the degradation of the ohmic contacts, which thus are improved to mitigate the degradation effect. Ohmic contacts are found to be degrading in AlGaN/GaN HFETs, e.g., in [7.141, 7.143]. On the contrary, no significant change of the ohmic metal in AlGaN/GaN HFETs prior to and after the stress test is found in [7.24]. The Ir-based Ohmic contact in [7.31] is optimized toward specific high-temperature operation at 350°C. Thus, several variants exist to built reliable ohmic contacts. However, also the secondary effects of ohmic contact formation and subsequent microcracking of the passivation have to be considered [7.141]. Low-annealing temperatures may be beneficial to suppress secondary effects, e.g., [7.69].

Schottky contacts have been investigated with regard to reliability in [7.1, 7.54, 7.125], especially at high temperatures [7.12]. The outdiffusion of Pt from the gate is found to be a major reliability concern in [7.60, 7.125]. The generation of an interfacial layer between the semiconductor and the gate contact are found to be the reason for an increase of the Schottky barrier height during stress in [7.121]. The analysis is confirmed by FIB images. Oxygen is found at the interface after aging in [7.60]. The oxide layer is found to be thinner near the gate edge. On the contrary, no metal interdiffusion at the gate contact is observed in [7.24] after DC-step-testing confirmed by STEM imaging. The failure mechanism observed in [7.29] can again be correlated with the oxidation of the AlGaN barrier near the gate after graceful degradation of the device RF-power performance at 10 GHz. Electron energy loss spectroscopy (EELS) maps of nitrogen and oxygen show oxygen at the critical gate–drain edge after aging. No electromigration or gate sinking is observed in the TEM images.

Ni/Au-based gate contacts are optimized and verified with respect to stable operation at channel temperatures of 300°C [7.1]. The optimization procedure is performed at elevated temperatures to guarantee the stability at reduced temperatures. Ni/Au contacts are tested for a gate length $l_g = 0.8\,\mu m$. Initial results provide sudden degradation of the gate contacts in the pinch-off test after some hours, as reported in [7.1, 7.58]. The degradation is correlated with the surface morphology and with the topological distribution of the gate currents. Hexagonal pits from the epitaxial growth are correlated with the increase of gate currents. A threshold level of $20\,\mu A\,mm^{-1}$ is defined for the reverse leakage at $V_{DS} = 40\,V$. The change of both reverse and forward gate currents after screening is insignificant at operation bias $V_{DS} = 50\,V$ and also as a function of temperature.

The forward gate current under high compression or switching is considered a reliability issue by the same group, e.g., [7.67, 7.68]. This contact-related issue is solved by the suggestion of a GaN MISHFET on a very similar device structure. However, no full reliability data is provided for this device type.

A trap analysis of these GaN–SiN-MISHEMTs is performed in [7.68]. A detailed pulsed-DC-analysis is performed from various quiescent bias, as detailed in Fig. 7.2 and Fig. 5.8. All quiescent bias are found to be stable. However, for the aggressive quiescent bias at I_{Dmax} at $V_{DS} = 50\,V$ with a dissipated power of 35 W/mm, a strong shift of the threshold voltage V_{thr} and transients of the output characteristics are observed. An improvement of the gate-insulating SiN dielectric is considered to enhance the reliability at high V_{DS} bias and forward-gate current conditions.

Further indirect effects of the gate metallization involve the impact of the gate on the underlying semiconductor through stress and strain. Such stress effects have been mentioned in [7.15].

Device Degradation Due to Changes in the Buffer Layer

The buffer in this study involves all the (insulating) semiconductor layers under the active device structure in the direction of the substrate. The buffer is a very large semiconductor volume relative to the actual active device. The buffer layer stack basically has three functions with physically contradicting demands:

- Compensation for lattice mismatch of various kinds
- Good thermal transport
- Very good electrical isolation at high bias

The simulated field distribution in the buffer is already analyzed in Fig. 4.7 in Chapter 4. We found a very inhomogeneous field distribution, especially at high V_{DS} bias. Thus the control of the trap concentrations and consequent charge control is most crucial for reliability, especially since the exact amount of carriers is stress-dependent. The optimization with respect to epitaxial defects in the buffer targets good electrical isolation [7.51] while maintaining a good surface morphology, i.e., to avoid hexagonal pits [7.67] and other structures. A detailed optimization procedure for reliability improvements is described in [7.51, 7.67]. The optimum defect engineering to be found leads to isolating defects which at the same time have no significant dynamic (de-)trapping behavior at all time range from hours to nanoseconds, even when the buffer is subject to high electric fields. Simulation studies on the impact of traps in AlGaN/GaN HFETs below the channel region are given in [7.45]. Iron buffer codoping has been suggested to enable good buffer isolation [7.53]; however, iron doping is subject to doping diffusion to the upper active device layers during growth and it is also subject to reactor memory which needs careful control. Even once the good isolation is obtained, the long-term stability of the material has to be shown, which has been proven, e.g., in [7.141]. The impact of the buffer engineering on the output power degradation is mentioned in [7.67]. After optimization of the growth conditions, which are not specified, the RF-reliability is enhanced with respect to the I_{Dq} degradation and the output power degradation [7.67]. Defect-induced changes of the minimum drain current on 3-in. s.i. SiC wafers are reported in [7.145]. With the application of a very fine wafer mapping the increase of the minimum current I_{Dmin} is correlated with micropipe-induced changes of the epitaxial layers. The abnormal growth on top of the micropipes is analyzed by SEM in cross-sections of the hexagonal pits.

Degradation Due to Unstable Interfaces

Several publications point to the importance of the dielectric/semiconductor and semiconductor/semiconductor interfaces, not only for the mitigation of interface-related dispersion, but, for very similar reasons, also for the gate leakage suppression and reliability enhancement [7.56, 7.67, 7.76]. The findings include:

– Charge modifications through surface-related techniques during and after device stress, e.g., [7.126]
– Sensitivity to light illumination in a study of the material composition in the barrier in [7.136]
– The impact of interface passivation, e.g., in [7.74, 7.117]
– Changes in the gate leakage currents for devices with an AlN interlayer at the AlGaN/GaN barrier interface [7.27]
– Indiffusion of hydrogen from the passivation in p-layers in optoelectronics devices [7.89]
– Strong changes of the HFETs in photoluminescence and lighting experiments [7.87, 7.91, 7.109]
– The beneficial impact of passivation early in the process on reliability [7.63]
– The positive impact of various field plates on device reliability [7.77]

The means to minimize these effects are mostly technology-based and strongly related to the passivation. The aim of a good and reliable device passivation is to:

– Stabilize the trap density at the dielectric/(Al)GaN interface [7.67]
– Reduce the possibility of creating additional traps [7.123] during operation
– Reduce the gate current (at high temperatures) [7.67]
– To achieve a lateral and vertical field reduction at the interface [7.43, 7.131]

The related hot-carrier degradation has been studied, e.g., in [7.109]. The investigations for GaN MESFETs show stronger degradation for unpassivated devices. Light illumination is used to discriminate the trapping effects at the surface from those in the buffer material. The experiments suggest a modification of the surface in the ungated device access region after hot-electron stress. The impact of the field plate for field engineering for AlGaN/GaN HFETs has been mentioned repeatedly, e.g., [7.77]. The effect of the field-peak reduction along the channel is especially beneficial at the sensitive semiconductor/passivation interface, as the field-related trap creation [7.133] and trap recharging can be reduced. This is especially true for multiple field-plate concepts [7.131]. Several studies are available for the optimization of the field-plate structures for various frequencies, e.g., [7.101]. Surface-related hot-carrier effects in AlGaN/GaN HFETs can be well distinguished from other effects, as the hot-carrier degradation has a negative temperature coefficient, as mentioned in [7.24]. That means this particular degradation reduces at higher ambient temperatures while other, such as diffusion, increase. Thus, despite the strong impact of the interface effects, other material-related effects can be separated [7.24, 7.79]. This concludes the discussion of devices under regular operation.

7.4 Radiation Effects

Wide bandgap semiconductor devices are promising candidates in radiation-harsh environments due to the relatively large bandgap and the reduced probability of current transport via trapping and other defects. A number of investigations have been published with respect to the impact of irradiation on III-N semiconductors. ^{60}Co γ-irradiation effects on n-GaN Schottky diodes are discussed in [7.4,7.134]. In general, a low susceptibility of the GaN semiconductor to radiation damage is found, whereas the metal/semiconductor interfaces may be the limiting factor because of the trap creation due to irradiation. Further effects of ^{60}Co γ-radiation on DC, RF, and pulsed-I–V characteristics of AlGaN/GaN HEMTs are discussed in [7.4]. The changes of the DC-characteristics are found to increase linearly with the dose of the γ-irradiation. The maximum pulsed-drain-currents I_{Dsat} increase with the irradiation level. In general, a high radiation hardness of GaN is found, while the radiation-induced changes are consistent with induced trap creation. Overviews of relevant traps are given, e.g., in [7.83]. Further γ-irradiation results are given in [7.116]. N-doped GaN layers are compared prior to and after ^{60}Co-γ-irradiation with a radiation doses of $10^{19}\,\mathrm{cm}^{-2}$. The slightly doped layer with $\mathrm{n} = 10^{17}\,\mathrm{cm}^{-3}$ shows a reduction of the carrier concentration, while a heavily doped layer with $10^{18}\,\mathrm{cm}^{-3}$ yields an increase after irradiation. Annealing at 550°C can recover the original values for mobility and carrier concentration. The annealing behavior of proton-irradiated $\mathrm{Al}_x\mathrm{Ga}_{1-x}\mathrm{N}/\mathrm{GaN}$ HEMTs on sapphire substrates is discussed in [7.18]. A high proton-irradiation dose $10^{14}\,\mathrm{cm}^{-2}$ at 1.8 MeV leads to a strong reduction of the transconductance and the saturated current by about 60%. Recovery of the I–V characteristics can be achieved by rapid thermal annealing at 800°C. Further results of proton irradiation on GaN layers are found in [7.50, 7.66]. The photoluminescence spectra after 2 MeV proton irradiation with fluences of up to $10^{16}\,\mathrm{cm}^{-2}$ show an increased hardness of GaN bulk material as compared to GaAs with respect to displacement damage. This makes the GaN bulk material suitable for space applications. The defects observed after proton irradiation are line defects. Additionally, new traps are introduced, which are not present after He-ion or electron bombardment. A remarkable hardness of GaN material with respect to ion damage is also found in [7.130]. For silicon ion doses of $10^{16}\,\mathrm{cm}^{-2}$ at energies of 90 keV, the interface is found to be the sink for damage build-up. Reversal of the disorder induced by the ion irradiation through annealing is found difficult. Neutron irradiation on GaAs double heterojunction HEMTs leads to trap creation [7.104]. Deep midgap traps are created, which affect both interfaces and semiconductor layer material. Both mobility and sheet carrier concentration are reduced after neutron irradiation at a dose of $10^{15}\,\mathrm{n\,cm}^{-2}$. Effects of neutron irradiation on Au/GaN Schottky diodes are given in [7.139]. Again a strong radiation hardness is found. Two deep electrical traps, one nitrogen vacancy related and one hydrogen related, are strongly influenced by neutron fluxes. In conclusion, systematic investigations show the predicted hardness of the wide bandgap semiconductor materials and devices, while device components such as contacts and dielectrics need further specific optimization.

7.5 High-Temperature Operation

The reduced thermal activation due to increased bandgap energies makes wide bandgap devices further promising for high-temperature operation beyond classical operation temperatures. This has been mentioned repeatedly for GaN devices, e.g., in [7.31, 7.36].

Comparison of Electronic Technologies

The maximum temperature for the operation of electronic devices for silicon and GaAs devices is limited. For low-power silicon on insulator (SOI) devices this temperature amounts to about 300°C [7.35]. For GaAs-HFETs, temperatures beyond 400°C are mentioned in [7.112, 7.142]. However, reliability constraints, e.g., of the contacts, lower this temperature for circuit operation to safer regions of operation, e.g., to 150°C, reported in [7.10]. For specific high-temperature operation, circuits allowing up to 250°C [7.49] and 300°C [7.37] have been presented. Techniques for the enhancement of RF-devices for high-temperature operation of GaAs devices are provided, e.g., in [7.95]. The GaAs devices are characterized by S-parameter measurements at elevated temperatures of up to 300°C, and the device changes relative to room temperature are used to modify the circuit design. The changes include increase in breakdown voltage, gain reduction, and shift of the threshold voltage. A typical principal criterion for the high-temperature stability of a semiconductor is the Debye temperature T_{Debye} [7.5]. Values of T_{Debye} are compiled in Table 7.1 for various semiconductors. Based on the comparison in Table 7.1,

Table 7.1. Debye temperatures (given at 300 K) for different semiconductors

Material	T_{Debye} (K)	Ref.
InAs	280	[7.80]
Si	625	[7.114]
GaAs	360	[7.80]
GaN	600	[7.36, 7.106]
AlN	1,150	[7.80]
InN	660	[7.80]
6H–SiC	1,200	[7.80]
4H–SiC	1,300	[7.36]
Diamond	1,860	[7.80]

SiC has been suggested as a genuine high-temperature semiconductor material [7.30]. Various SiC-device types such as MESFETs, MISFETs [7.140], junction FETs [7.30, 7.118, 7.118], and related circuits [7.81] have been suggested for operation temperatures beyond 500°C [7.52]. The maturing SiC-device- and MMIC-technology is a strong competitor to III-N devices, unless the nitride devices can make use of their higher speed performance. Principal high-temperature performance of III-N materials up to 800°C is discussed in [7.36]. AlGaN/GaN HEMTs grown by MOVPE on sapphire substrates are subject to temperature stress of up to 800°C, both in biased and unbiased conditions. Irreversible degradation of the intrinsic active heterostructures is found for operation above 600°C.

III-N Device High-Temperature Operation

The role for wide bandgap semiconductors in high-temperature electronics is discussed in [7.96]. The temperature dependence of device [7.3] and circuit performance as a function of backside temperature, packaging situation, and resulting junction temperature is of critical importance for system performance. On the device level, the importance of:

- The intrinsic carrier concentration
- The pn-junctions
- The thermionic leakage
- The reduction in carrier mobility
- Substrate leakage

are stressed for high-power high-temperature devices. High-temperature performance of AlGaN/GaN HEMTs on s.i. SiC substrates up to backside temperatures of 340°C has been discussed in [7.47]. The devices are found to operate in a stable short-term fashion up to backside temperatures of 250°C. The dissipated power density per area reached with SiC substrates is about $0.6\,\mathrm{MW\,cm^{-2}}$. The resistances R_S and R_D, the transconductance g_m, and the saturated current I_{DS} show a nearly linear temperature dependence between room temperature and 300°C. Further high-temperature effects for the RF-power characteristics for channel temperatures between RT and 300°C have been reported [7.1]. A linear gain of 12.3 dB is reported at $T_{chan} = 269°C$, which decreases nearly linearly by about 2 dB for a temperature change of 100°C at $V_{DS} = 50\,\mathrm{V}$. The saturated output power at 2.14 GHz decreases also nearly linearly by about 0.7 dB/100°C. The risk of thermal runaway in AlGaN/GaN HFETs is found to be low [7.1], based on a small shift in $V_{GS} \leq 0.01\,\mathrm{V}$ to maintain a stable quiescent current for a channel temperature range of 250°C. The gain parameter variation for a temperature range between $-25°C$ and 125°C is investigated for a variety of GaN technologies in [7.35]. Again, a linear reduction in the temperature of both the saturated drain current I_{DSS} and the transconductance g_m is found. This temperature behavior of GaN HEMTs is compared to GaAs PHEMTs. It is found that the temperature variation for the GaN devices is smaller than for the GaAs devices. The explanation given is that the saturated velocity v_{sat} in GaN channels is less modified by temperature than in (In)GaAs heterostructures. The high-bandgap energies and the high-heterostructure barriers in III-N devices, i.e., reduction of the thermionic leakage, may also serve as an explanation for this behavior.

Temperature Dependence of Physical Quantities

The intrinsic carrier concentration n_i within different semiconductors can be calculated from the Boltzmann statistics [7.96]:

$$n_i = \sqrt{N_C \cdot N_V} \cdot \exp\left(\frac{-E_g}{2k_B \, T_L}\right).$$
(7.9)

The temperature dependence of the bandgap E_g, of the effective densities of states at the conduction band N_C, and at the valence band N_V are calculated as

$$E_g = E_{g,0} \cdot \frac{\alpha - T_L^2}{\beta + T_L},$$
(7.10)

$$N_C = 2M_C \cdot \left(\frac{2 \cdot \pi \cdot k_B \cdot m_n \cdot T_L}{h^2}\right)^{3/2}, N_V = 2 \cdot \left(\frac{2 \cdot \pi \cdot k_B \cdot m_p \cdot T_L}{h^2}\right)^{3/2}.$$
(7.11)

Fig. 7.9 depicts the intrinsic carrier concentrations for various semiconductor materials calculated from (7.10) and (7.11). The differences between conventional materials, such as Si and GaAs, and wide bandgap semiconductors, such as GaN and SiC, are clearly visible. The exponential impact of the energy gap in (7.9) makes the fundamental difference. A requirement for a good high-temperature semiconductor is an intrinsic carrier concentration which is negligible at the operation temperature range relative o the relevant doping concentrations in the device. This reference concentration for III-N devices typically amounts to 10^{16} cm^{-3}. This value of the intrinsic concentration n_i is reached by classical materials such as Si and GaAs, whereas the wide bandgap materials do not reach this intrinsic value n_i even for temperatures of 1,000 K. A similar result can be obtained for the thermionic leakage at the heterointerfaces and contacts. The reverse leakage current at a junction can be calculated as

$$I \simeq A_R \cdot K \cdot T_L^2 \exp\left(\frac{-q \cdot \phi_B}{k_B \cdot T_L}\right).$$
(7.12)

Assuming a maximum Schottky barrier height of 90% of the individual bandgap performed by bandgap engineering as a theoretical limit, the leakage currents are compared in Fig. 7.10, normalized to the same leakage of Si at room temperature. Apart from the purely thermionic leakage, further

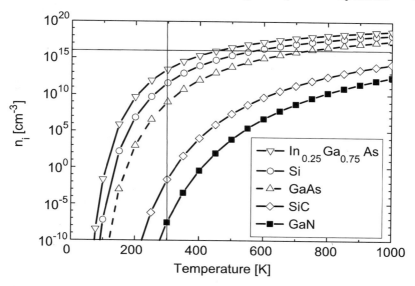

Fig. 7.9. Intrinsic carrier concentration n as a function of lattice temperature T_L for various semiconductors calculated from the Boltzmann statistics

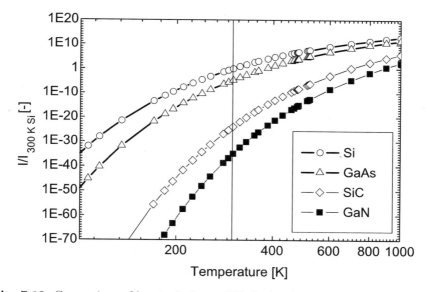

Fig. 7.10. Comparison of barrier leakage of Si, GaAs, GaN, and SiC as a function of lattice temperature T_L

leakage effects have to be considered. A second additional tunneling component is similarly influenced by increased temperature, as reported, e.g., in [7.99, 7.100, 7.148].

Chapter 8 will discuss the exact determination of the temperatures and integration aspects.

7.6 *Problems*

1. *Assume the proposed increase in reliability of a GaN FET relative to an $In_{0.25}Ga_{0.75}As$ channel FET is due only to the increase in bandgap (activation energy at midgap). What is the change in the lifetime at the channel temperature of $125°C$?*

2. *Based on what you know about GaN HFETs, is the assumption of question (1) reasonable? Give arguments!*

3. *Discuss the most pronounced failure mechanisms of GaN HFETs.*

4. *What is the most important difference between GaN HFETs and GaAs HEMTs with respect to substrate-related defects? Discuss!*

5. *Is a Schottky barrier of 90% of the bandgap a realistic assumption? What are the implications?*

6. *What are additional benefits of large bandgap devices apart from the intrinsic carrier concentrations? Think of traps and the related mechanisms.*

8

Integration, Thermal Management, and Packaging

This chapter describes device and MMIC integration, such as passive coplanar and microstrip circuit techniques, thermal management, and mounting and packaging considerations for state-of-the-art III-N devices and amplifiers.

8.1 Passive MMIC Technologies

So far there are only a few reports with a full description of integrated circuit technologies in microstrip transmission-line technology for GaN HFETs [8.49, 8.74, 8.101, 8.107], contrary to the well-developed GaAs technology [8.60]. Cree demonstrated a fully integrated MMIC process for GaN HEMTs and SiC MESFETs in [8.93, 8.107]. The substrates are thinned to $100\,\mu$m. Similar processes are available from NGST [8.39, 8.49] and HRL [8.74] with substrate thicknesses down to $50\,\mu$m. A microstrip transmission-line process has been demonstrated by Triquint on HR-silicon substrate. Insulating Si substrates with a resistivity of $30\,k\Omega\,cm$ are used [8.32]. The substrates are also thinned to $100\,\mu$m. MIM capacitances with a breakdown voltage of $150\,V$ are reported. Increased integration of GaN devices for base station applications, e.g., in driver amplifiers has been mentioned by RFMD in [8.20, 8.106], similar to the approach using silicon LDMOS for the same application; see [8.108]. Both spiral inductors and MIM- and MIS-capacitors are cointegrated.

8.1.1 Passive Element Technologies

The processing, characterization, and reliability of high-voltage high-power passive elements is of critical importance for GaN and SiC FET MMICs. Transmission-lines, airbridges, resistors, inductances, and MIM and other capacitances have to be hardened for the strict requirements of high-power, high-temperature, and high-voltage operation [8.124].

MIM Capacitances and Inductances

The availability of capacitive lumped elements is a key requirement for matching, DC-blocking, and other design functions on a MMIC. A number of recent publications have described the optimization of capacitances in the GaAs world for high-power operation [8.12,8.134]. Typically, SiN or SiO_2 are used in various forms [8.11, 8.80, 8.134]. For higher integration, Ta_2O_5 [8.30] can be used in the GaAs world to save area on the wafer. SiN thicknesses in the MIMs down to 25 nm have recently been found useful for GaAs devices [8.12]. A number of other high-k dielectrics have been developed for higher integration of capacitances in silicon IC processes [8.43,8.44,8.133]. The higher dielectric constants reduce the area consumption at high integration levels. Specific development is needed for high-voltage operation in MIMs for operation bias of ≥ 50 V. Cree reported the development of reliable MIM capacitances with a size similar to those for GaAs processes, however, with breakdown voltages increased by a factor of 3 [8.124]. The associated lifetime is 10^8 h. Typical breakdown voltages are at least 150 V. An SiO_2/SiN stack with thicknesses of 80/220 nm with an effective $\varepsilon_r = 6.7$ and a capacitance per unit area of $200\,pF\,mm^{-2}$ is given in [8.113].

Fig. 8.1 gives the measurement of the breakdown voltage of an optimized MIM capacitance integrated in a GaN HEMT process. It can be observed that these devices can be operated up to at least 150 V. Fig. 8.2 gives the

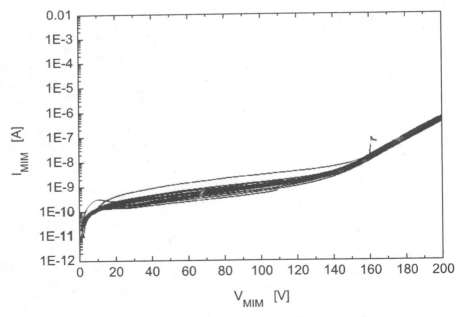

Fig. 8.1. Current–voltage measurement of various MIM capacitances at room temperature

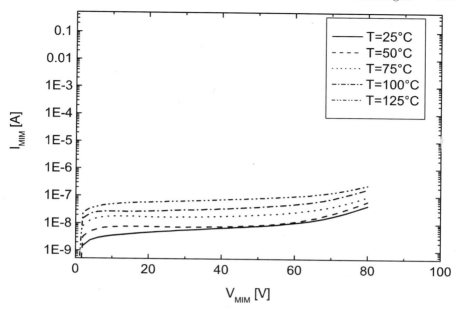

Fig. 8.2. Current–voltage characteristics of a MIM capacitance with the substrate temperature as a parameter

current voltage characteristics of an MIM capacitance for a different area as a function of temperature for a similar structure between 25 and 125°C. We see an increase in the leakage due to the temperature change. High-power operation may heat the passive devices on a MMIC, so an additional breakdown margin is needed, based on the measurements of Fig. 8.2. The substrate quality is important for the quality of the structures on top. Fig. 8.3 gives an example of the impact of a micropipe cluster on an MIM capacitance on s.i. SiC. This issue has been discussed, e.g., in [8.111]. However, the recent improvements in substrate quality have drastically reduced the occurrence of such effects. The high-power levels lead to the increase of the metallization thicknesses. Fig. 8.4 gives an image of a concentrated inductance with a thick galvanic metallization. The high-power operation requires the optimization of both metal thickness and the exact geometry of area-efficient lumped elements for high-voltage operation. This requires air gap adjustment for repeatable manufacturing, as the aspect ratio of thickness and air gap needs to be controllable during processing.

Passive Resistances

Lumped resistances can be integrated in various forms. First of all, ungated semiconductor layers can be used to obtain resistances for which the overall value does not have be controlled with great precision. This is typically true for high-resistance elements. For higher accuracy and lower resistance,

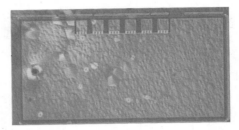

Fig. 8.3. Image of the impact of substrate quality on MIM metallization

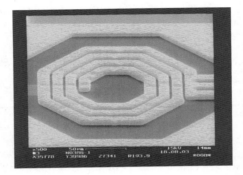

Fig. 8.4. Image of a lumped inductance with thick metallization layer

lumped ohmic resistances are typically integrated as TaN resistors with $25\,\Omega\,\mathrm{sq}^{-1}$ [8.50] or as NiCr resistors with a resistivity of $50\,\Omega\,\mathrm{sq}^{-1}$ [8.9] or $100\,\Omega\,\mathrm{sq}^{-1}$ [8.63]. TaN thin-film resistors for high resistance with $45\,\Omega\,\mathrm{sq}^{-1}$ and TiN with $45\,\Omega\,\mathrm{sq}^{-1}$ are reported on s.i. SiC in [8.113]. NiCr resistors with a sheet resistance of $20\,\Omega\,\mathrm{sq}^{-1}$ are reported by Cree in [8.93] for the integration of GaN HEMTs. Several other processes have been reported, some of which yield a sheet resistance of the NiCr of $50\,\Omega\,\mathrm{sq}^{-1}$, as is typical for the GaAs MMICs [8.103].

8.1.2 Microstrip Backend Technology

A microstrip transmission-line process on s.i. SiC substrate is rendered complicated due to the extreme material properties of SiC, such as hardness [8.68], chemical inertness, and overall substrate quality of s.i. SiC substrates. A microstrip backend process requires the solution of the following issues, as described in Chapter 4:

- Wafer mounting using an appropriate adhesive [8.76]
- Full wafer thinning [8.68]
- Via-etch
- Metallization
- Wafer removal

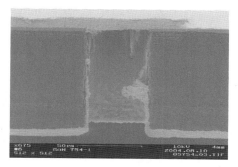

Fig. 8.5. Viaholes in a 100 μm thick s.i. SiC substrate

Fig. 8.6. Full power amplifier stage in microstrip transmission-line technology

Cree reported a process with 100 μm thick SiC substrates with 2- and 3-in. diameter and 75 μm diameter via holes in [8.93, 8.124]. Further references to a microstrip process on 3-in. s.i. SiC substrate for Ka-band are given in [8.74]. The thickness of the thinned s.i. SiC substrates is 50 μm in this case [8.74] with a via geometry of 30×30 μm. A thickness of 5 μm for the geometry of the galvanic is reported in [8.29]. The SiC MESFET process in [8.113] yields 100 μm substrate thickness. Fig. 8.5 and Fig. 8.6 give SEM images of viaholes in the output stage of a Gan HFET power amplifier on a 100 μm thick s.i. SiC substrate [8.120]. The viaholes are well covered by metal from the backside, which ensures good grounding of the frontside elements. Further the vertical shape of the viaholes is visible based on the anisotropic etch.

8.2 Integration Issues

Integration of III-N devices and MMICs is a key challenge for the application. [8.42, 8.51, 8.65, 8.67]. GaN devices pose additional problems for system engineers. These integration issues related to GaN-based hybrids and MMICs are:

- Increased power density per die and die area [8.51]
- Increased requirements for performance of conventional heat sinks/thermal contacts, as is discussed in Chapter 7

- DC- and RF-power transmission for increased power levels
- Transparent substrates [8.101]
- Substrates under epitaxial strain
- Increased operation temperatures [8.79].

The electrical issues are discussed in several publications, e.g., in [8.51, 8.101]. Further, assembly issues, such as the processing of optically transparent materials, pose new challenges for automatic packaging equipment, as mentioned in [8.101]. As bowing is a critical issue, the treatment of diced III-N devices under strain through heteroepitaxy is important to ensure reliable device operation. The challenge for increased operation temperatures with respect to packaging, especially in the 300–600°C range is mentioned in [8.79]. Currently operation temperatures similar to those for GaAs modules are being considered for TR-modules for the initial integration of III-N devices; see [8.51]. More specific high-temperature packaging techniques are discussed in [8.95, 8.125]. At these temperature levels, the packaging solutions are typically niche solutions.

Module Integration

Overview articles on transmit/receive (TRX) modules and related technologies for radar and other applications are given in [8.14, 8.42, 8.51, 8.52, 8.65, 8.67]. TRX modules have been under discussion for a long time [8.3, 8.51]; however, cost reduction [8.3, 8.62] and the automatic assembly aspects of a large number of hybrid components have so far restricted their application. TRX-module integration [8.3, 8.51] covers the different functions performed by a multifunction and multichip RF-module, as depicted in Fig. 8.7. This

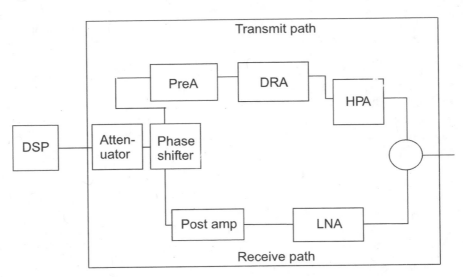

Fig. 8.7. Schematic of the RF-part of a transmit-receive module

includes phase shifting, attenuation, amplification, limiting, and module protection through circulators. The transmit path of a radar TR-module typically involves several driver (DRA) amplifiers and (balanced) high-power amplifiers (HPAs) in the transmit path. The receive path is composed of a limiting function, a low-noise amplifier (LNA), and possibly a second amplifier [8.42]. An overview of architectural aspects and requirements is given in [8.71]. Issues from TRX technology to system aspects and manufacturing aspects such as module calibration and statistics are covered. Trade-offs for the benefits of different module architectures are discussed further, e.g., in [8.3].

8.3 Thermal Management

Thermal management is of ultimate importance to III-N devices, as even very simple estimates can reveal, e.g., [8.36]. Very high channel temperatures are easily reached unless proper packaging and DC-limiting precautions are taken.

8.3.1 Thermal Analysis

Several overviews of thermal analysis of III-N devices are available, e.g., in [8.84, 8.109].

Basic Findings

Thermal analysis of AlGaN/GaN power HFETs is described, e.g., in [8.84]. DC-parameters, noise, and output power are analyzed between 380 and 540 K. The thermal coefficients of several quantities are extracted, as also detailed below. It is found in general that the temperature coefficients of the III-N device terminal quantities are lower that those of GaAs devices, e.g., [8.81]. This means that III-N devices are less sensitive to temperature changes. However, the dissipated power densities per unit area are much higher, which balances or even overcompensates this advantage. This situation is illustrated in Fig. 8.8 in the world map of thermal management, which has been presented in similar fashion is various publications. The increase of the RF-power densities and power levels approximately leads to the same increase in the dissipated power levels. The resulting dissipated power densities per area increase by a factor of 10 for the same application. This means that we change from power densities from $\leq 1\,\mathrm{W\,mm^{-2}}$ to the range of $1\text{–}10\,\mathrm{W\,mm^{-2}}$. The main information is hidden in Fig. 8.8. The change of the power density essentially leads passive-conduction cooling-system to their limits and requires solutions beyond, potentially, active cooling solutions.

Determination of Temperatures

Several techniques have been used for analysis of the actual temperature and temperature distributions in the materials and devices. Noninvasive infrared

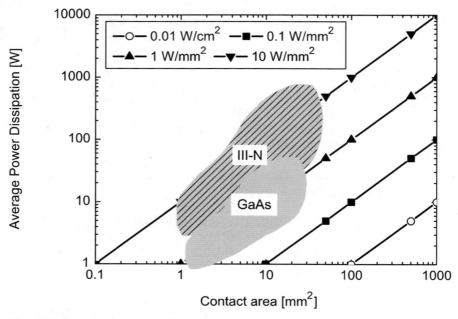

Fig. 8.8. Overview on thermally dissipated power vs. area for different technologies

camera measurements are reported in [8.30, 8.84, 8.109]. An infrared microscope is combined with an on-wafer loadpull setup. The variation of the temperature distribution as a function of input drive power can be monitored. This allows temperature control for realistic operation conditions. Additionally, Raman measurements of the lattice temperature in GaN devices with higher spatial resolution are given in [8.55]. The DUT can be investigated from the frontside and from the backside through the substrate, so that the heat generation can be directly monitored through the optically transparent substrate and the III-N layers. The spatial lateral resolution is submicron, while the depth resolution is ≈1 μm. Integrated Raman spectroscopy and IR thermography are used in [8.56, 8.96, 8.97]. The IR thermography is used for time-efficient of monitoring, e.g., for void control [8.66], with a diffraction-limited resolution of 2–5 μm, while the Raman spectroscopy allows better resolution of the peak temperatures near the gates. The Raman measurements and 3D-thermal simulations of the same structure have been brought to an agreement in the hot spot of the near gate region. Further techniques are presented with the results below. Fig. 8.9 gives the layout of a multifinger high-power GaN HFET. The most important thermal layout parameters, such as the gate-to-gate pitch l_{gg} and the gate width W_g are depicted. Other important parameters are the lateral and vertical boundary conditions.

Fig. 8.10 gives the principal thermal situations for a device. The device can be cooled by three approaches:

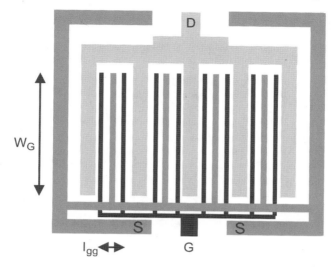

Fig. 8.9. Layout of an eight finger coplanar high-power GaN HFET

1. Purely from the backside
2. Purely from the frontside
3. In a combination of situation (1) and (2)

Based on this insight, Fig. 8.11 depicts a 3D-view of the thermally relevant parameters in a HPA for the most conventional situation (1) in Fig. 8.10. The parameters include:

- The overall extension of the chip ($X \times Y$)
- The thickness of the semiconductor t_{sem}
- The thickness of the substrate t_{sub}
- The thickness of the adhesive or solder t_{ad}
- The submount heat sink material and thickness t_{subm}
- The temperature T_{sub} and temperature distribution at the backside of the substrate
- The gate-to-gate pitch l_{gg}
- The gate width W_{g} and, with minor impact, the gate length l_{g}

The determination of the optimized parameters is discussed in Sect. 8.3.2.

8.3.2 Thermal Material Selection and Modeling

A detailed modeling of the relevant thermal parameters is provided in this section for proper thermal analysis. The temperature dependence of the thermal conductivity is modeled according to the exponential law in (2.3) in Chapter 2. A good compilation of models for the temperature of conventional semiconductor materials is given in [8.86]. Parameter values for the models for

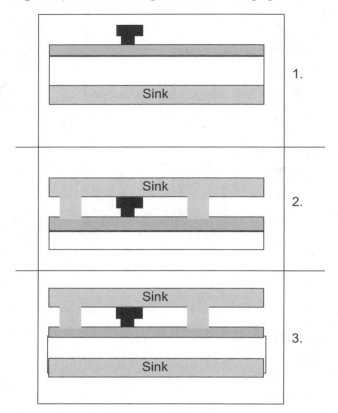

Fig. 8.10. Principal thermal situations

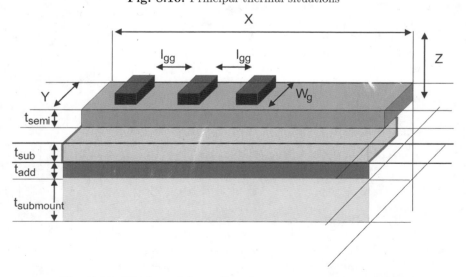

Fig. 8.11. 3D-view of thermally relevant parameters in an HPA

III-N semiconductors are again compiled in Table 8.1 with partial repetitions from Table 2.16. Examples of all relevant parameter values are mentioned, starting from the semiconductor materials, the substrate materials, the submounts, and the adhesives and solders. III-N semiconductors require the development of new heat spreader materials such as copper tungsten, copper diamond, and copper molybdenum with high thermal conductivity, which have a coefficient of thermal expansion (CTE) matched to the typical substrate materials.

Apart from the thermal conductivity the actual choice of the submount material critically depends on the matching of the CTE to the substrate of the semiconductor. Fig. 8.12 gives the world map of packaging materials. Except from natural and synthetic diamond, which has a strong mismatch to SiC, only the composites Cu/W, Cu/Mo, and Cu–diamond yield the proper CTE and a good thermal conductivity suitable to match that of SiC. Most of the pure metals are rather off in CTE. Crystalline AlN is another choice.

Table 8.1. Temperature-dependent thermal properties of III-N semiconductors and packaging materials

	κ_{300} $(\mathrm{W\,K^{-1}\,m^{-1}})$	α $(-)$	c_{300} $(\mathrm{J\,K^{-1}\,kg^{-1}})$	Ref.
Si	125	−1.65	711	[8.104]
GaAs	54	−1.25	322	[8.86]
GaN	160	−0.43	491	[8.36]
AlN crystal.	285	−1.57	748	[8.110]
InN	45	–	325	[8.54]
Si$_3$N$_4$	0.96	–	–	[8.36]
s.i. SiC (V-doped)	330–370		715	[8.28, 8.36]
6H–SiC	387	−1.49	715	[8.13]
High-purity s.i. SiC	490 ‖c	−1.61	715	[8.22]
Sapphire	28	−1	796	[8.91]
Diamond	1,800–2,500	−1.85	520	[8.70, 8.102]
Cu–W (25/75)	230	–	195	[8.25]
Co–Diamond	420, 650	–	–	[8.25, 8.135]
Mo–Cu (70/30)	170–210	–	276	
Epoxy	1	–	1000	[8.91]
Silver-filled epoxy	1.9	–	–	[8.25]
Au–Sn	68	–	–	[8.25]
AlN (ceramic)	150	−1.84	770	[8.91]
BeO	250	–	–	[8.70]
Cu	383	–	385	[8.25, 8.91]
Au	318	–	129	[8.36]

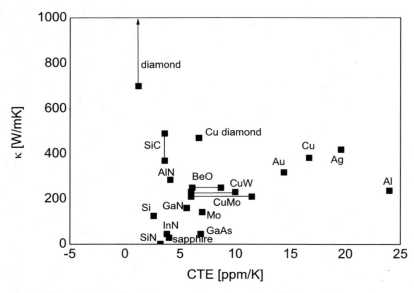

Fig. 8.12. Thermal conductivity as a function of coefficient of thermal expansion (CTE) at RT

Thermal Modeling of Semiconductors

Silicon is a reference for semiconductor materials [8.104] and also serves as a substrate material option for III-N devices [8.109]. The thermal properties of silicon are well investigated. Table 8.1 compares the thermal properties of the most important semiconductors with the properties of materials used in packaging and integration. GaN has a thermal conductivity very similar to that of silicon, thus, it is much better than the κ_{300} of GaAs. The low-temperature thermal conductivity of GaN between 4.2 and 300 K is given in [8.45]. AlN is available in various forms, either crystalline or ceramic. The high-thermal conductivity of the crystalline material is taken from [8.110]. Ceramic AlN is used in HTCC modules and various other substrate technologies, e.g., [8.75]. The thermal conductivity is lower than in crystalline AlN. However, the ceramic AlN still has a conductivity similar to silicon [8.91]. The properties of bulk InN are based on very few measurements only, e.g., [8.54]. InN is not available as a real bulk material, as the lattice mismatch to any substrate is significant. Ternary materials typically have much lower thermal conductivities than the binary semiconductors as measured for $Al_xGa_{1-x}N$ in [8.64]. The composition dependent modeling is given in (2.8) in Chapter 2.

Modeling of Substrate and Submount Materials

The doping- and polytype dependence of the thermal conductivity of s.i. SiC is critical for precise analysis of the semiconductor substrate, as is pointed out in [8.5]. High-purity SiC has a different thermal conductivity than (Va-) doped material, as mentioned in [8.5, 8.22, 8.94]. The thermal conductivity is further orientation dependent. Thermal and electrical conductivity and electrical properties of 6H–SiC are given in [8.5, 8.13]. The thermal conductivity is extracted between 300 and 500 K. Further, the thermal conductivity along the c-axis is found to be different from the conductivity along the a-axis [8.13, 8.36]. The mechanisms of the temperature dependence of the thermal conductivity of silicon carbide are discussed in [8.46] by theoretical separation of the limiting mechanisms. The impact of dislocations on the thermal conductivity at room temperature is mentioned. Dislocation densities of 10^8 cm^{-2} can significantly reduce the outstanding thermal conductivity of SiC to values of typical semiconductor material, such as GaN.

Diamond provides the best thermal conductivity of all materials with values of 1,800–2,500 W m^{-1} K^{-1}, which is an order of magnitude higher than the typical semiconductors and metals [8.102]. CVD diamond has been used for topside heat spreading for high-power InGaP/GaAs HBTs [8.35]. The coating of GaN HEMT with industrial CVD diamond thus is discussed in [8.102]. A low-temperature CVD deposition process has been developed, which requires deposition temperatures ≤500°C. The reduction of the thermal conductivity for thin layers has to be considered in this case.

Diamond and copper-tungsten micro-packages are described for the integration of GaN HFETs in [8.90]. Both flip-chip and backside cooling with the HFET soldered to diamond are considered. Copper-tungsten has a thermal conductivity similar to AlN and also the CTE can be matched accordingly. Copper-molybdenum is a light material and used for the integration GaAs power MMIC into modules, e.g., [8.69].

The adhesive and submount materials mentioned in Table 8.1 are further discussed below with the packaging examples.

8.3.3 Basic Thermal Findings, Heat Sources, and Thermal Resistances

Basic dependencies of the DC- and RF-device parameters on temperature are given in this section.

Temperature Coefficients

Fig. 8.13 gives the heat dissipation for a power amplifier at a given drain efficiency DE = 40% as a function of the operation voltage. The absolute amount of dissipated power is as high as 200 W at $V_{DS} = 75$ V, which means that such a device is thermally limited for a typical area of about 5–10 mm^2, see Fig. 8.8. The actual mounting situation will decide the channel temperatures.

Table 8.2 gives an overview of the temperature coefficients of device quantities on various substrates and mounting situations. The examples in Table 8.2 suggest that the temperature coefficients of the quantities in GaN FETs are

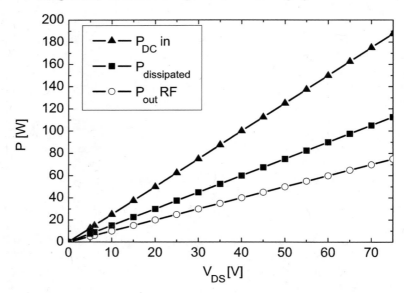

Fig. 8.13. DC-power, dissipated power, and RF-power as a function of V_{DS} operating voltage for an HPA at an assumed PAE of 40%

lower than for comparable GaAs PHEMTs [8.84]. However, as suggested in Fig. 8.8 and Fig. 8.13, the dissipated power levels and power densities per unit gate width and area are much higher than for the GaAs devices with a similar periphery, gate length, and area.

Table 8.2. Temperature coefficients of various HFET DC- and RF-quantities derived within the temperature ranges given

Substrate	Geometry $l_g \times W_g$ (µm)	Quantity	Coefficient	Range T_{sub} (K)	Ref.
GaN/SiC	0.35×250	I_{DSS}	$-14.8\,\text{mA mm}^{-1}/10\,\text{K}$	380–540	[8.84]
GaN/SiC	0.35×250	g_m	$-2\,\text{mS mm}^{-1}/10\,\text{K}$	380–540	[8.84]
GaN/SiC	0.35×250	f_T	$-0.24\,\text{GHz}/10\,\text{K}$	380–540	[8.84]
GaN/SiC	$0.35 \times 1,500$	P_{sat}	$-0.14\,\text{dB}/10\,\text{K}$	300–540	[8.84]
GaN/SiC	0.35×250	$N_{F,min}$	$0.07/10\,\text{K}$	380–540	[8.84]
GaN/SiC	0.35×250	R_n	$2.6\,\Omega/10\,\text{K}$	380–540	[8.84]
GaN/SiC	0.35×250	G_{ass}	$-0.114\,\text{dB}/10\,\text{K}$	380–540	[8.84]
GaN/SiC	$0.8 \times 2,580$	G_{ass}	$-0.23\,\text{dB}/10\,\text{K}$	316–573	[8.2]
GaN/SiC	$0.8 \times 2,580$	P_{sat}	$-0.06\,\text{dB}/10\,\text{K}$	316–573	[8.2]
GaN/SiC	0.3	g_m	$-5\,\text{mS mm}^{-1}/10\,\text{K}$	258–398	[8.2]
GaAs PHEMT	0.2	g_m	$-8.5\,\text{mS mm}^{-1}/10\,\text{K}$	233–423	[8.2]
GaAs PHEMT	1×150	g_m	$-12\,\text{mS mm}^{-1}/10\,\text{K}$	298–423	[8.18]
GaAs PHEMT	1×150	f_T	$-0.6\,\text{GHz}/10\,\text{K}$	298–423	[8.18]
GaAs PHEMT	1×150	P_{sat}	$-0.16\,\text{dB}/10\,\text{K}$	298–423	[8.18]

Thermal Resistances: Approximations

Thermal resistances express the global device temperatures once the dissipated power level is fixed. Several methodologies are available for the experimental and theoretical determination of the thermal resistances for a given mounting situation, e.g., [8.22, 8.36]. A closed-form determination of thermal resistance of GaN HFETs is presented in [8.22, 8.23]. The dependence on the layer thicknesses including host substrate, on the gate-to-gate pitch l_{gg}, on the gate length l_g, and on gate width W_g are considered and verified by finite element simulations. A simple analytical model is given in [8.36]. The thermal conductivities are calculated layer-wise in a column approach starting from the heat sources. This simplified procedure allows a construction of the thermal resistance within 10% of more elaborate simulations or measurements for the examples given. Input parameters are single-finger width W_g, gate-to-gate pitch l_{gg}, and number of gate fingers. This feature is useful if the heat spreading in the upper layers of the stack is sufficient to consider a 1D-problem. The particular influence of the thermal boundary resistance at the GaN/substrate interface to the performance degradation of GaN FETs is discussed in various publications, e.g., in [8.98, 8.118]. These findings are important, as this interface is typically optimized only with respect to electrical needs, such as isolation and dispersion requirements. A similar investigation of the effect of the thermal resistance at the GaN/SiC, GaN/sapphire, and GaN/AlN interface by simulation is provided in [8.33]. The diffuse mismatch approximation is used. The influence of the thermal boundary resistance is particularly important for the SiC substrate, which conducts the heat very well. For SiC, the relative contribution of the boundary to the overall thermal resistance for SiC is higher than for sapphire substrates. Experimental confirmation for the increase of the thermal resistance is provided in [8.98]. Conformal micro-raman thermography is used to determine the temperature at different depths within the device. Full 3D-models based in finite element method will be discussed below with the applications examples.

Self-Heating

Self-heating is the source of channel temperature increase, due to the enormous power levels per dye area in III-N devices. The limiting effects will now be discussed. 3D-thermal simulation of SiC MESFETs coupled with 2D-electrical simulations are reported in [8.10]. It is clearly shown in [8.10] that the heat transfer problem is a full 3D-problem. Thus, thermal effects are included in the simulation for comprehensive device optimization. Power handling limits and degradation of large area AlGaN/GaN RF-HEMTs on sapphire and SiC are discussed in [8.26] based on 3D-thermal simulations. The RF-output power density is correlated with the thermal impedance R_{th} of the device on SiC. The RF-power density reduces nearly linearly from 4 to 2 W mm^{-1} for a increase of R_{th} from 10.5 to 14.5 K mm W^{-1}. The dependence of the thermal impedance R_{th} on gate–gate pitch l_{gg} is discussed for various gate widths W_g

from 50 to 200 μm. For all gate widths, the thermal impedance reduces by 30% for the variation of the pitch from $l_{gg} = 20$ to 100 μm. Initial findings on dynamic properties of self-heating in high-power AlGaN/GaN HFETs on sapphire and SiC substrate are discussed in [8.38]. The DC-output characteristics are affected by the self-heating effects after less than 10^{-7} s. After 30 ns during the pulse period there is no reduction of the current with V_{DS} visible on s.i. SiC substrates. On sapphire substrates the thermal impedance is nearly an order of magnitude higher which leads to a higher thermal constants. It is further shown that the thermal impedance of the AlGaN/GaN HFET is mainly affected by the substrate material, as the active semiconductor layers are only about 1.5 μm thin.

Device Optimization Strategies

Device parameters are typically varied with regard to electrical performance. Based on the previous sections, potential thermal parameters for device optimization include:

- Variation of the single-finger length [8.22]
- Variation of the gate–gate spacings or pitch, typically between 20 and 100 μm [8.22, 8.83, 8.109]
- Optimization of the thermal interface from the substrate backside to the metal
- Increase of the chip area and optimization of the distribution of the devices on the chip
- (Microstrip) substrate thickness variation between 400 and 50 μm [8.109]

Minor thermal impact have:

- A variation of the gate lengths [8.22]
- A variation of the semiconductor layers thicknesses, e.g., the buffer layer between 0.5 and 3 μm

These thermal parameters similarly have a strong impact on the electrical performance, so trade-offs need to be evaluated comprehensively and thus examples will now be given.

8.3.4 Backside Cooling

Various mounting techniques have been proposed for semiconductor devices, and very innovative cooling concepts may be required for III-N devices. However, backside cooling is a very good reference for the understanding of the thermal situation. General comparisons of thermal impedances of GaN devices on sapphire, s.i. SiC, and silicon substrates for backside cooling are provided, e.g., in [8.28, 8.58]. GaN HEMTs on s.i. GaN substrates have not been analyzed extensively; however, the thermal situation is very close to the situation of GaN on silicon.

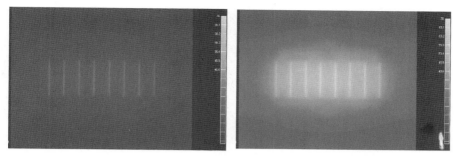

Fig. 8.14. Thermal image of a multifinger FET on SiC (*left*) and sapphire (*right*) substrate, unit $(K W^{-1})$

Silicon Carbide Substrates

SiC is the material with the highest heat conductivity relative to all other substrate choices except from diamond [8.102]. SiC has some unique properties. Its heat conductivity is approximately metal-like, e.g., compared to Cu or Au, see Table 8.1, [8.13, 8.22, 8.46]. Thus, the heat-spreading properties of a SiC substrate are very good and some thermal arguments known from GaAs or Si need to be revised. Contrary to GaAs and Si, thinning of SiC to 100 or 50 μm does not necessarily improve the overall thermal situation, as explained for diamond and SiC in [8.5, 8.136]. This is only the case if the thermal resistivity of the submount is better than of the SiC substrate. Further, thinning introduces a thermal barrier closer to the active device and may reduce the heat spreading, as the die-attach typically has thermal properties inferior to SiC. Thermal limitations on large-periphery SiC MESFETs, which have a thermal situation similar to GaN HFETs, are discussed in [8.123]. Maximum channel temperatures of 320°C are reached for a device with $W_g = 36$ mm gate periphery and 144 gate fingers with 250 μm single-finger gate width. Substrate thickness is 400 μm and the dissipated power $P_{diss} = 58$ W. The gate-to-gate pitch l_{gg} is not given.

Infrared and Raman temperature measurements on AlGaN/GaN HFETs on 4H–SiC substrate are given in [8.55, 8.56, 8.96]. The results show operation temperatures of 160°C near the gate finger at a power dissipation level of 20 W mm^{-1} in a 2×50 μm device on a 400 μm-thick SiC substrate. The gate-to-gate pitch is $l_{gg} = 50$ μm and the device is glued to a heat sink with epoxy with a contact area of 15 mm². Such measurements can be used effectively to calibrate the simulated temperature distribution to the real world, typically by means of the assumptions made for the heat source in the simulation. Fig. 8.14 depicts the temperature distribution in a multifinger GaN HFET for both SiC and sapphire substrates. The color code is given on the same scale for the $W_g = 1$ mm device with eight fingers. The maximum impedance is 20 K W^{-1} for the SiC and 69 K W^{-1} for SiC for the same substrate thickness. A crucial issue of the thermal simulation is to make sure that the substrate thickness t_{sub} is greater than the gate-to-gate pitch l_{gg}. If this is not the case, the thermal distribution will be very strongly affected by the boundary condition at the substrate/heat sink interface.

Sapphire Substrates

Careful thermal management for any electronic III-N device is of even greater importance on sapphire substrates, due to the limited thermal conductivity, which is even lower than that of GaAs (see Table 8.1 and Fig. 8.14). Despite the great Vickers hardness, substantial substrate thinning from typically 250–325 μm [8.130] is a requirement. The determination of the channel temperature in AlGaN/GaN HEMTs grown on sapphire and silicon substrates using a DC-characterization method is presented in [8.58]. A simple estimate of the channel temperature is given based on the threshold voltage shift and the change of the source resistance with temperature. The channel temperature T_{chan} is then derived from the reduction of the drain current with V_{DS}. For a device with $l_g = 0.45$ μm and $W_g = 50$ μm, a channel temperature $T_{chan} = 320°C$ is extracted for sapphire, while only $T_{chan} = 95°C$ are extracted for Si substrate. This ratio is a bit to high and reflects some error introduced by this purely electrical method. Thermal findings for GaN HFETs on sapphire substrate further include:

– Thinning of the sapphire substrate to 50 μm, as used for 100 W power operation in a 2 GHz power bar [8.4]
– Generally high-absolute thermal impedances that lead to very high operation temperatures

If we consider only backside cooling, sapphire limits the application of AlGaN/ GaN HFETs on sapphire to DC-dissipation power-densities approximately similar to those of GaAs for the same substrate thickness of 50 μm. For other mounting situations, such as flip-chip, sapphire can be a solution, as discussed, e.g., in [8.115].

Silicon Substrates

As a substrate material silicon has a thermal conductivity similar to that of the semiconductor GaN. Thus, it is well suited for power operation; however, it does not supply the same good heat spreading as SiC substrates. Further, the compensation of the high-lattice mismatch introduces an additional heat barrier near the substrate interface. GaN HEMTs on silicon substrates for wireless infrastructure applications are described in [8.109], featuring thermal design and related performance. A good comparison is given for the same nominal device, in this case a 10×200 μm device with a gate-to-gate pitch of 25 μm, but with different substrate materials. For the power density of 3 W mm^{-1}, the idealized ratio of the temperature increase is 160 to 75 to 50 K for the comparison sapphire vs. silicon vs. SiC substrate for a base plate temperature of 300 K. The comparison of the thermal impedance is given below. Direct on-wafer noninvasive thermal monitoring under microwave large-signal conditions are reported in [8.83] for similar GaN HFETs on silicon substrates. A comparison is performed between various ten finger devices with gate finger length of 100, 200, 300, and 400 μm. The gate-to-gate pitch is relaxed from 20 and 50 μm to obtain similar temperatures for the larger devices. The surface temperature observed for the GaN HFET with $W_g = 3$ mm by IR thermography varies between 120 and 200°C simply based on the change of the PAE during the power sweep. The silicon substrates are thinned to 150 μm is this case. This is thicker than for state-of-the-art silicon LDMOS substrates, which are typically thinned to thicknesses below 100 μm.

Comparison of the Substrates

The results for the thermal impedances of the backside cooling for III-N devices on various substrates are compiled in Table 8.3.

The comparison yields the best thermal resistance per unit gate width W_g for SiC, followed by silicon and sapphire. For the flip-chip integration, the thermal resistance is nearly independent of the substrate material. The variation of the results on sapphire substrate from [8.26] is due to a variation of the gate-to-gate pitch l_{gg} between 20 (highest impedance) and 100 μm (lowest impedance).

8.3.5 Flip-Chip Integration

Flip-chip integration is a technology for use in hybrid mounts in several semiconductor technologies, such as silicon [8.53] and GaAs [8.78]. Flip-chip integration of power HEMTs on GaN/SiC technology is discussed in [8.103]. Both

Table 8.3. Comparison of the substrate and mounting configurations

Substrate	Thickness (μm)	Gate width (mm)	Configuration	Impedance	Ref.
SiC	400	1	BS	17 K W^{-1}	[8.94]
SiC	300	0.2	BS	4–5 K mm W^{-1}	[8.26]
SiC	100	2	BS	22 K W^{-1}	[8.109]
Sapphire	–	1	BS	65 K W^{-1}	[8.94]
Sapphire	–	0.1	BS, diced	51.7 K mm W^{-1}	[8.115]
Sapphire	300	0.2	BS	50–90 K mm W^{-1}	[8.26]
Sapphire	100	2	BS	53 K W^{-1}	[8.26]
Silicon	100	18	BS	4.25–4.75 K W^{-1}	[8.109]
Silicon	100	2	BS	37 K W^{-1}	[8.109]
SiC	400	0.1	FC	4.9 K mm W^{-1}	[8.94]
Sapphire	–	1	FC	25 K W^{-1}	[8.94]

BS backside, *FC* flip-chip

electrical and thermal gold bumps are available. In this technology, thermal bumps are used directly within the active devices to remove the heat. Very initial large-periphery AlGaN/GaN MODFETs on sapphire mounted by flip-chip integration are reported in [8.117]. The effect of the thermal improvements by flip-chip bonding is proven by increases in the saturated currents I_{DSS}. Thermal management of AlGaN/GaN HFETs on sapphire substrate using flip-chip bonding with epoxy underfill is described in [8.115]. The passive integration is performed on AlN substrates using gold bumps. The impact of the sapphire substrate to the thermal resistance is found to be reduced, and the heat spreading within the GaN buffer is found to be significant. In this case, a reduction of the thermal impedance of the GaN buffer layer is suggested to optimize performance. Very high RF-power densities of $\geq 5\,\mathrm{W\,mm^{-1}}$ on sapphire can be reached with small device and by flip-chip integration of GaN devices with a gate width $W_{\mathrm{g}} = 2\,\mathrm{mm}$, as reported in [8.129]. A comparison of conventional and flip-chip bonding of AlGaN/GaN devices by photocurrent measurements is given in [8.94]. The technique allows an absolute temperature calibration of thermal simulations. The thermal resistance for backside-cooled GaN HFETs on sapphire amounts to $65\,\mathrm{K\,W^{-1}}$, for flip-chip bonding on sapphire to $25\,\mathrm{K\,W^{-1}}$, and to $17\,\mathrm{K\,W^{-1}}$ for backside-cooled SiC, as also shown in Table 8.2. The gate periphery is $W_{\mathrm{g}} = 10 \times 100\,\mathrm{\mu m} = 1\,\mathrm{mm}$. As an example,

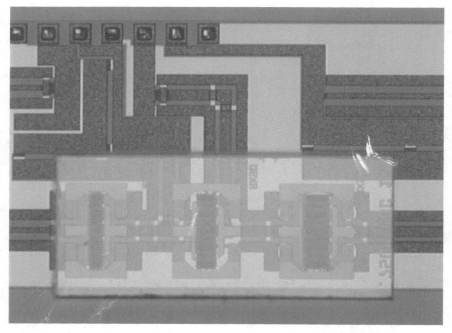

Fig. 8.15. Transparent GaN HEMTs flip-chip mounted on a passive coplanar carrier substrate

Fig. 8.15 gives the image of a GaN HEMTs on s.i. SiC substrate, which is integrated on a passive carrier substrate in coplanar transmission-line technology using gold bumps. The transparent SiC substrate is clearly visible and the bumps between the gate fingers can be seen from the top through the substrate.

Combined Integration Techniques

Even a thermal impedance of $17 \, K \, mm \, W^{-1}$ for ideally backside-cooled GaN HFETs on SiC may yield high channel temperatures for high-power operation. Thus, the combination of several techniques is discussed by various authors. Fig. 8.10 suggests a combination of frontside and backside cooling. This methodology is the most powerful option; however, at the same time it requires the highest integration effort, and a very careful adaption of the thermal coefficients of expansion is needed. A combination of flip-chip packaging and backside cooling has been suggested for GaN HFETs by RFMD in [8.105] with respect to thermal management for very high-power density operation. The thermal impedances in this case are strongly dependent on the actual thermal path to the heat sink, since isothermal heat sinks cannot easily be realized close to the device on both sides of the devices. A typical integration procedure for such an approach is a backside cooling combined with a flip-chip or ball grid array type of top substrate, as performed for silicon microprocessors [8.6]. Apart from combined flip-chip/backside integration, surface heat conductivity of dielectric material on top of the device can be used. CVD diamond has been proposed as a surface layer [8.102]. It is found that the thermal conductivity of a CVD diamond layer is higher than that of gold, even for very thin diamond layers with only $2 \, \mu m$ thickness.

8.3.6 Dynamic Thermal Effects

The dynamics of thermal processes are highly complex [8.59]. The main issue for thermal management is the determination of the appropriate relaxation mechanisms to describe nonconstant device excitation, compare (5.100) in Chapter 5. The dynamic behavior of device is characterized by complex geometries, temperature dependent material properties, and variations on various orders on spatial and temporal scales [8.127]. For large-scale thermal simulations this complex behavior of the devices is typically reduced to one or two time constants.

Values for the thermal relaxation time constant of the order of $10^{-7} \, s$ are mentioned for the thermal relaxation in [8.38]. Similar time constant of $190 \, ns$ are reported for AlGaN/GaN HEMTs on silicon [8.57]. These time constants have to be extracted along with the thermal resistance. For the long-term behavior of the thermal response a steady-state thermal behavior is assumed after $20 \, ms$ in simulations of silicon devices at low temperatures [8.59]. According to (5.100) the heat capacity becomes the second dominating material parameter, which is thus compiled in Table 8.1. It can be seen that the heat capacitances of the substrates of choice (SiC, sapphire, and silicon) are very similar. The heat capacitances of the actual semiconductor layers are not considered very important for the optimization, except for the optimization of the overall semiconductor thickness.

The input RF-drive-dependent temperature change of a GaN HFET is monitored by thermal imaging in [8.83]. The thermal distribution is found to be highly fluctuating with time, although steady input power levels are applied. Dynamic thermal effects under pulsed-DC-operation are further discussed in [8.82]. No time constants for the relaxation are given. The impact of the substrate choices sapphire and silicon for GaN HEMTs on the maximum dynamic temperature in pulsed operation is investigated by simulation in [8.131]. The typical analysis for pulsed operation is given for a very high power device in [8.73]. IR images are given for various pulsed widths. Further both power droop of $-0.3\,\mathrm{dB}$ and phase droop of $-3.9\,^\circ$ are measured for pulse widths of $200\,\mathrm{\mu s}$ for radar operation.

8.4 Device and MMIC Packaging

Packaging is a critical step for the integration of devices on the hybrid-package and module level. The integration typically includes:

– Wafer dicing
– Die-attach and related void control [8.66]
– Electrical connection including DC- and RF-substrate technology
– Cointegration [8.77]
– Oscillation control
– If needed, hermetic sealing

These steps will now be analyzed in detail for III-N devices.

8.4.1 Dicing

Semiconductor wafer dicing is typically performed by sawing, scribing, cleaving, or by the application of various laser techniques.

Sawing and Scribing

Sawing [8.16] or scribing of GaAs wafers are the typical cost-effective procedures for wafer dicing in industry [8.21, 8.122]. Typical issues of dicing are the reduction of ragged edges and gold burrs. Optimization strategies for sawing thus include proper blade selection leading to increased lifetime of the blade. The cutting speed and yield of sawing is not as good as with other dicing techniques. For SiC, sawing is the technique of choice [8.112], whereas sapphire is either diced by a combination of thinning, scribing, and subsequent cleaving [8.87], e.g., for optoelectronics, or the substrate is structured or removed by laser lift-off [8.24]. The optimization of scribing includes the stabilization of the tool quality for improved repeatability for high-volume production. For wafers thinner than 150 μm, dicing in general becomes more difficult, e.g., [8.88].

Laser Techniques

Laser dicing techniques are attractive due to the lack of direct physical contact of the tool components with the die. This results in reduced tool wear-out [8.40]. A water-jet-guided laser is suggested for optimized laser cutting of GaAs in [8.88]. In this case, a wavelength of 1,064 nm at 100 W of power is used. The water jet serves as a waveguide and helps to remove debris during the dicing process. UV-laser-assisted scribing of sapphire is discussed in [8.41]. This technique at a wavelength of 255 nm is used to scribe sapphire wafers of 100 μm thickness for III-N LEDs production. Very precise edges are possible with small cuts [8.87]. The technique is presented to overcome the lack of etch techniques for sapphire. A water-jet-guided laser is used for the damage-free dicing of SiC in [8.40]. A 1,064 nm at an average power of 56 W is used with a 40 μm thick water jet. The cutting speed for 380 μm thick substrates can be improved by about 40% relative to sawing. The technique can also be used to scribe SiC wafer with the application of a 532 nm laser and 23 μm thick water jet. Further, the edge quality is improved while the tool does not wear-out, due to the contact-free dicing procedure.

Impact of Dicing

In case of heteroepitaxy and of the lattice mismatch with the substrates, III-N layers have a defined bow after epitaxy, e.g., [8.85]. The pressure dependence of the conductivity in AlGaN/GaN HEMT is investigated in [8.47]. This is important, as the strain is modified on the chip level during dicing. The conductivity in the channel of an AlGaN/GaN HFET shows a linear change of $6.4 \times 10^{-2}\,\mathrm{mS\,bar^{-1}}$ through the application of compressive strain. A similar investigation gives a change of the capacitance $0.86\,\mathrm{pF\,bar^{-1}}$ for a membrane with a radius of 600 μm for both tensile and compressive strain. This is a strong dependence for an area which is similar to an FET, and the capacitance change is of the order of the total capacitance of an FET with $W_g = 1\,\mathrm{mm}$. This fact requires a strong control of the dicing changes in III-N devices and MMICs.

8.4.2 Die-Attach

The die-attach is the most important integration step of electronic devices, mostly for thermal reasons. It is composed of a positioning in all three directions (which shall not be discussed here) and an attach step. Several attach processes are available based on either solder or glue processes, e.g., [8.122].

An eutectic solder attach is used in [8.109] for the integration of GaN HEMTs on silicon substrates with an eutectic thickness of 25 μm. The die is soldered to a Cu/W package with a thermal conductivity $\kappa = 180\,\mathrm{W\,m^{-1}\,K^{-1}}$. This is a relatively thick eutectic layer, which strongly increases the thermal resistance of the stack. The optimization of the backside metal layer stack is also a subject of GaAs-based backside processes, e.g., [8.21, 8.122]. This includes:

- Precleaning wafer backside [8.21]
- Selection of layer sequences in the die-attach
- Selection of layer thicknesses
- Void reduction

Thermomechanical analysis of gold-based SiC die-attach assembly is described in [8.72]. Again, it is mentioned that the differences in CTE leads to strong mechanical stress to the mounted die, in this case a MEMS pressure sensor. The analysis of the die-attach allows proper compensation of the stress. Void control is typically achieved by removing the semiconductor and optically inspecting the eutectic. Void control of the die-attach can also be performed by thermal imaging, by acoustical microscopy, as performed and reported in [8.66], or by X-ray. Shear tests are performed to investigate the mechanical strength of the bonding. The impact of single voids on the associated temperature increase have been investigated for LDMOS devices, as reported, e.g., in [8.48]. The thermal resistance of the die-attach/solder layer is approximated in a 1D-approach as

$$R_{\mathrm{th}} = \frac{t_{\mathrm{ad}}}{\kappa A}. \tag{8.1}$$

Some thermal conductivities, typical thicknesses of the substrates, die-attach layers, and submount materials relevant to III-N devices are given in Table 8.4. Indium-based solder is typically not suitable for high-power operation, as the melting point of Indium is too low. AuSn solder is a typical die-attach for high-power semiconductors, e.g., [8.122]. Table 8.4 allows a relative comparison of the contributions to the thermal resistance per mm² assuming purely 1D-heat transport. It can be seen that the typical solder or glue layers have a critical impact on the overall thermal resistance of the stack. The optimization of the solder includes:

- Metal-material, temperature, and composition selection
- Metal thickness optimization
- Metal stack deposition [8.122] including:
    - Adjustment of the layer thickness to the leaching of Au
    - Solder stop integration to avoid via filling [8.122]
- Backside street integration

After optimization, the thermal conductivity of the eutectic can reach values in the same range as the semiconductor materials. Several effects have to be considered additionally. The so-called via blow-out, i.e., the filling of vias by the solder, is avoided by a stop layer. Such a layer directly develops an oxide,

Table 8.4. Comparison of thermal conductivities for typical materials for packaging

Material	Thickness (μm)	κ_L (W m^{-1} K^{-1})	Thermal impedance per mm^2 (K W^{-1})	Ref.
GaN	3	130	0.023	–
SiC	100	370	0.27	–
Silicon	100	150	0.66	–
Sapphire	50	50	1	–
AuSn	20	240	0.083	[8.36]
AuSn	25	–	–	[8.109]
AuGe	20	90	0.22	[8.109]
Epoxy	20	30–60	0.66–0.33	[8.36]
Silver epoxy	20	1.6–7.5	12.5–2.6	[8.36]
Cu–W	1,000	400	2.5	[8.36]
Cu–W	1,500	180	8.3	[8.109]

such as Ti, Ni, and Cr, as reported in [8.122]. Specifically for high-temperature operation, a die-attach scheme for silicon carbide for high-temperature operation at $T_{sub} = 500°C$ is reported in [8.17]. SiC diodes are attached to AlN substrates and packages by a specific metal attach scheme.

8.4.3 Package Technology Selection

The selection of the package has a strong impact on the device performance with respect to the system integration. Several optimization steps have been performed on the packages themselves, e.g., [8.70, 8.108]. General aspects of the packaging selection and optimization of hybrid transistors and MMICs include:

– Improved heat removal [8.66, 8.70]
– The need for sealing [8.126]
– The reduction of electrical losses, inductance, and capacitance [8.70]
– Cost reduction [8.62, 8.108]

Recent trends of integration with respect to III-N hybrid devices include:

– Cost reduction of conventional package technology in hybrid integration schemes [8.62]; this includes:
  – Introduction of new organic/plastic technologies replacing ceramic packages [8.1, 8.108]
  – Size reduction [8.128] and increase of packaging efficiency, i.e., reduction of the ratio of package area and semiconductor area [8.27]
– Integration of a number of chips per package [8.14, 8.108]

Initial GaN HFET examples for base station applications have mostly been realized in conventional ceramic packages, e.g., [8.77]. However, ceramic packages are being replaced by organic packages for the competitor LDMOS [8.108]. Recent trends show a similar integration schemes also for GaN HEMTs for base station applications in order to meet similar cost trends [8.116]. For higher frequencies and MMIC integration, [8.27] shows the historical evolution from hybrid transistor integration to multichip MMIC modules and micropackages. Conventional multichip-modules (MCM) concepts are currently applied, as also mentioned for space applications in [8.7]. Low-temperature-cofired-ceramics (LTCC) packaging is a conventional technique of choice, which is used in a variety of module applications, e.g., [8.62]. LTCC-technology is considered to be not very flexible and only cost effective at large production numbers. Further, larger line sizes and tolerances have to be taken into account [8.100]. Several approaches have been pursued to overcome these limitations, e.g., [8.62]. Cost reduction of the LTCC approach can also lead to the use of high-temperature cofired ceramics (HTCC) [8.128], in this case a single 12-layer package. An X-band multichip module including several PA and switch MMICs and ASIC are integrated. HTCC-based MCM packages are typically based on AlN dielectrics and can be used up to 60 GHz [8.75]. The packaging trends further pursued are based on the following technologies:

- Integrated multilayer ceramics modules such as
  - Low-temperature cofired ceramics (LTCC) for frequencies up to 76 GHz [8.7, 8.100, 8.137]
  - High-temperature co-fired ceramics (HTCC) [8.19, 8.128]
  - Photonic bandgap structures [8.137]
- (On the medium timescale) the introduction of
  - System-in-package (SIP) technology [8.7]
  - 3D-integration, as proposed in [8.89, 8.90, 8.132]

An example of the realization of a photonic bandgap structure is given in [8.137]. The LTCC and flip-chip integration of a GaAs MMIC approach is combined with the suppression of undesired modes between the chip ground-plane and the module ground through the use of periodic structures. The SIP approach integrates multiple components, such as GaAs HEMTs, on a high-density interconnect substrate, as performed in [8.121]. TaN and MIM capacitors are cointegrated with GaN HEMTs on AlN passive substrates.

A full 3D-integration of LTCC substrate including μ-ball grid-array (BGA) technology for the stacking on the motherboard is described in [8.90]. A full C-band transceiver including mixer, LNAs based on GaAs MESFET can be integrated by the application of this approach. The LTCC substrates are connected by μ-ball grid-array. Fully 3D-micro-electromagnetic systems (3D MERFS) are described for the integration in [8.31]. The idea of MERFS is based on the fact that there is no millimeterwave analog to the printed circuit board (PCB). Low cost-materials are used to decrease size and weight by an

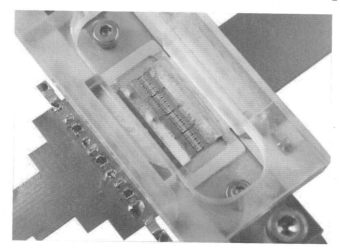

Fig. 8.16. Packaged GaN HEMT in standard ceramic housing

order of magnitude. About 1000 passive conductors and RF-components can be integrated in one substrate.

Thus parallel to the development of the active III-N devices high-performance passive module technologies are being developed increasing packaging density and isolation and line losses at the same time. These integration schemes will be used to integrate GaN RF-devices in future systems.

8.4.4 Thermal Management for Linear Applications

Thermal management is especially critical for linear RF-operation, as linear RF-operation is characterized by high back-off from the highly-efficient saturation point. Thus, the associated efficiency of the operation is much lower than in saturated operation, which leads to increased DC-dissipation and heat generation within the device, see [8.83]. The high-power densities per area of GaN devices aggravate the problem in linear operation, see Fig. 8.8.

Example of a Hybrid AlGaN/GaN HEMT on s.i. SiC

Stages of a base station amplifiers are typical examples for highly linear amplifier [8.77]. The integration task is very similar to the integration of silicon LDMOS or GaAs power FETs for the same application. The typical engineering tasks include:

– Thermal expansion coefficient matching [8.109]
– Bond wire engineering, as described, e.g., in [8.15, 8.70]
– Flange engineering for electrical connection [8.70]
– Cosubstrate integration [8.70]

The integration of a GaN HEMT with a periphery $W_g = 32\,\text{mm}$ in a ceramic package is shown in Fig. 8.16. The die is soldered with a AuSn eutectic to a Cu/W submount in a typical LDMOS package. The bonding approach is based on wire bonding in this example and includes no intentional pre-matching. This allows unrestricted broadband and multiband operation. Pre-matching of GaN FETs has also been suggested, e.g., in [8.77]. The internal matching provides input and output impedances of 3-j $4\,\Omega$ at $2.14\,\text{GHz}$ for two $W_g = 36\,\text{mm}$ GaN HFETs on silicon substrates, integrated in parallel into one ceramic package with single flanges. From a thermal point of view, both eutectic and gate-to-gate pitch l_{gg} have been optimized. The biasing of the transistor is still limited to deep class-A/B in order to reduce the quiescent current I_{Dq} and the associated dissipated power P_{diss}. Further, analog and digital predistortion enable the improvement of both peak and average output power and associated efficiency for given ACLR constraint.

GaN MMIC Integration

GaN MMIC integration includes the following issues, similar to power FET packaging:

- Heat spreading in a limited area
- Individual RF-connection
- Individual DC-connection
- Oscillation control within the cavity of the package [8.37]

The optimization starts on the device level. Specific MMIC-design procedures for the reduction of module costs are given, e.g., in [8.8]. This typically includes size reduction of the MMICs, higher integration with more elaborate passive components, and more compact and thus more risky transmission-line positioning. The latter requires very stable technologies and longstanding design experience on the specific process. MMIC packages are available in several forms as multichip hybrid modules [8.7]. This technology integrates microwave components, RF-feedthrough, and DC-bias/control junctions. The multichip modules can be realized in various packaging technologies. Low-cost surface mount device (SMD) packages based on organic materials are available, e.g., in [8.37, 8.119]. They can be used up to frequencies of $40\,\text{GHz}$ and even $60\,\text{GHz}$ for GaAs FET technologies. The integration requires less than $1\,\text{dB}$ of gain at the frequency of $40\,\text{GHz}$. The connection to a printed circuit board (PCB) is critical, and the losses can be minimized to $\leq 0.5\,\text{dB}$ at $60\,\text{GHz}$ [8.37]. Alumina-based ceramic MMIC packages have been introduced for frequencies up to $40\,\text{GHz}$ [8.114]. The package has an air cavity and is leadless. An example of the integration of a coplanar Ka-band driver GaN MMIC on s.i. SiC is depicted in Fig. 8.17. The module is realized in conventional split-block technology. The MMIC is integrated on a subcarrier which is CTE-matched to s.i. SiC. The eutectic solder requires specific optimization with respect to void control and thickness. The DC-supply and DC-oscillation

Fig. 8.17. Integrated GaN HEMT Ka-Band driver MMIC in a conventional split-block module

mitigation require new, area-efficient SMD components, which need to support the high-operation bias $V_{DS} = 30$ V. The module is thermally limited in cw-operation due to the backside cooling. The MMIC requires a quiescent current $I_{Dq} = 150$ mA mm^{-1} for suitable gain at Ka-band frequencies, which leads to a power dissipation $P_{diss} = 4.5$ W mm^{-1} at $V_{DS} = 30$ V for linear communication applications. Both the transition from the coplanar MMIC to the microstrip lines and the submount are visible in Fig. 8.17. Further, the integrated DC-control is visible.

8.4.5 Active Cooling

One basic finding from the previous sections is the thermal limitation of III-N hybrid and MMIC devices, see Fig. 8.8. The thermal limitation of electronics through thermal limitation is already mentioned in [8.138]. Thus, both advanced thermoelectrical cooling layers and embedded microchannel cooler concepts are suggested for the solution of this issue. SiC and silicon are relatively good heat spreaders, so active cooling techniques gain importance for III-N devices.

Thermal management of electronic devices using microchannels enabling local active cooling is described in [8.61]. 3D-microsystem structures are used to build 3D-microfluidic channels near the thermal hot spots of the electronic device. As another option, thermoelectric microcoolers for GaN FETs are described in [8.34]. The idea of microcoolers is to actively cool only the key power devices. Materials such as CVD diamond and AlN are used as substrates for the microcoolers. GaN is suggested to be the material to change the typical heat dissipation per area of an integrated GaAs device from currently $30\,\mathrm{W\,cm^{-2}}$ ($0.3\,\mathrm{W\,mm^{-2}}$, see Fig. 8.8) to several hundreds $\mathrm{W\,cm^{-2}}$ (1–$10\,\mathrm{W\,mm^{-2}}$) for GaN HEMTs with $20\,\mathrm{W}$ of output power. This range of power densities is also reached if we consider either the active area or the full chip size as a reference area. New concepts of internal device cooling are also considered. Internal thermoelectric cooling of laser diodes is described in [8.92]. The thermionic effect at the interfaces is used to cool the device. This principle is attractive, since this cooling occurs in or very near the hot spot active region. The procedure is suggested for electronics in [8.138].

8.5 *Problems*

1. *Compare the thermal properties of microstrip and coplanar transmission-line approaches of conventional GaAs FETs to GaN FETs on s.i. SiC.*
2. *Which mechanical material properties are important for the reliability of flip-chip packaging?*
3. *What is the principal difference between GaN on SiC substrates and other semiconductors for integration?*
4. *What is still the main drawback of 3D-integration? Think of both material and economic issues.*

9

Outlook

After about 16 years of outstanding progress in research and development, the challenging III-N material system is about to emerge into the open waters of system introduction. Some great achievements have been made, and others are yet to come: we will see higher operation frequencies, greater wafer formats, still higher output powers, and great results in linearity, efficiency, and bandwidth. However, as the experience of the GaAs devices has proven, a long way is still ahead until full use of material capabilities in subsystems and systems is achieved. Disappointments occur and will continue to arise due to reliability problems on the system level, especially since the proposed high-voltage operation has not been tested on the system-level so far. Further, new mechanisms of degradation at the proposed output power levels will be important, and extremely careful optimization will be necessary to sort out the details. Disappointments will also occur for industrialization, as the emerging markets will not be able to support all business plans. Schumpeter's law of the rise and fall of innovators and competitors will play a dominant role.

In the long run, great progress will be achieved especially for MOD systems in new radars, new electronic countermeasures, and there will be new means in communication and several other civil and military niche applications.

Civil applications in mobile communication will be seen; however, the everlasting silicon is driving down a roadmap of cost reduction at a pace never seen before, and though being technically inferior for some applications, it will keep its share. New applications for 3G and 4G communication systems will be technically possible; however, again the market, customer preferences, and cost issues will decide whether they will evolve. Some great lessons will be learned from the optoelectronic world, where the rise of III-N-based solid-state lighting will lead to a continuous development of materials, substrates, and technologies pushed by a strong consumer market. In an analogy, III-N optoelectronics will challenge the light bulbs, while III-N electronics will challenge the electronic equivalent, the tubes.

Finis in spe...

Appendix

Material Properties of Binary Compounds

Table 9.1 in a systematic way compiles the properties of the binary materials.

Table 9.1. Material properties of binary compounds at 300 K

Material	GaN	AlN	InN	SiC 6H
Crystal structure	Wz	Wz	Wz	Wz
Mass density ($kg\,m^{-3}$)	6,100	3,230	6,810	3,210
Melting temperature (K)	–	3,273	1,373	3,103
Lattice constant a (Å)	3.189	3.112	3.533	3.073
Lattice constant c (Å)	5.185	3.982	5.693	–
Thermal expansion ($10^{-6}\,K^{-1}$) $\Delta a/a$	5.59	4.2	3.548	4.3
Thermal expansion ($10^{-6}\,K^{-1}$) $\Delta c/c$	3.17	5.3	5.760	4.7
Static ε_r	9.5	8.5	15.3	9.66
RF-ε_r	5.47	4.77	8.4	6.70
Eff. e^--mass (m/m_0)	0.2	0.314	0.11	0.2
Bandgap ($E_{g,\Gamma}$) (eV) 300 K	3.434	6.22	0.8	3.0
Bandgap 2nd valley E_g (eV)	–	6.9	–	–
Bandgap 3rd valley E_g (eV)	–	–	–	–
Intervalley separation (eV)	1.5	2.5	1.1	–
Sound velocity ($m\,s^{-1}$)	5×10^3	1×10^4	5×10^3	–
Phonon energy $\hbar \cdot \omega_{LO}$ (meV)	91.2	99.2	89	–
Piezoelectric constant h_{pz} (cm^{-2})	0.5	0.55	0.3	–
Electron mobility 300 K ($cm^2\,V^{-1}\,s^{-1}$)	1,400	683	100	260
Hole mobility 300 K ($cm^2\,V^{-1}\,s^{-1}$)	170	14	–	100
v_{sat} electrons ($10^7\,cm\,s^{-1}$)	1.9	2.16	1.32	2
Thermal conductivity ($W\,K^{-1}\,cm^{-1}$)	1.3	2.85	0.38	3.7
Specific heat ($J\,K^{-1}\,kg^{-1}$)	491	748	325	670

References of Chapter 2

2.1. V. Adivarahan, M. Gaevski, W. Sun, H. Fatima, A. Kouymov, S. Saygi, G. Simin, J. Yang, M. Khan, A. Tarakji, M. Shur, R. Gaska, IEEE Electron Device Lett. **24**, 541 (2003)

2.2. A. Agarwal, M. Das, B. Hull, S. Krishnaswami, J. Palmour, J. Richmond, S. Ryu, J. Zhang, in *Device Research Conference*, State College, PA, 2006, pp. 155–158

2.3. I. Akasaki, H. Amano, in *Properties of Group III Nitrides*, No. 11 in EMIS Data reviews Series, ed. by J. Edgar (IEE INSPEC, London, 1994), Sect. 7.2, pp. 222–230

2.4. I. Akasaki, H. Amano, in *Properties of Group III Nitrides*, No. 11 in EMIS Data reviews Series, ed. by J. Edgar (IEE INSPEC, London, 1994), Sect. 1.4, pp. 30–34

2.5. M. Akita, S. Kishimoto, K. Maezawa, T. Mizutani, Electron. Lett. **36**, 1736 (2000)

2.6. J. Albrecht, P. Ruden, S. Binari, M. Ancona, IEEE Trans. Electron Devices **47**, 2031 (2000)

2.7. E. Alekseev, D. Pavlidis, Solid-State Electron. **44**, 245 (2000)

2.8. E. Alekseev, D. Pavlidis, Solid-State Electron. **44**, 941 (2000)

2.9. S. Allen, J. Milligan, Comp. Semicond. **9**, 25 (2003)

2.10. O. Ambacher, M. Eickhoff, A. Link, H. Hermann, M. Stutzmann, F. Bernardini, V. Fiorentini, Y. Smorchkova, J. Speck, U. Mishra, W. Schaff, V. Tilak, L. Eastman, Phys. Stat. Sol. C **0**, 1878 (2003)

2.11. O. Ambacher, B. Foutz, J. Smart, J. Shealy, N. Weimann, K. Chu, M. Murphy, A. Sierakowski, R. Dimitrov, A. Mitchell, M. Stutzmann, J. Appl. Phys. **87**, 334 (2000)

2.12. O. Ambacher, J. Smart, J. Shealy, N. Weimann, K. Chu, M. Murphy, R. Dimitrov, L. Wittmer, M. Stutzmann, W. Rieger, J. Hilsenbeck, J. Appl. Phys. **85**, 3222 (1999)

2.13. K. Andersson, V. Desmaris, J. Eriksson, N. Roersman, H. Zirath, in *IEEE International Microwave Symposium Digest*, Philadelphia, 2003, pp. 1303–1306

2.14. T. Anderson, D. Barrett, J. Chen, W.T. Elkington, E. Emorhokpor, A. Gupta, C. Johnson, R. Hopkins, C. Martin, T. Kerr, E. Semenas,

A. Souzis, C. Tanner, M. Yonanathan, I. Zwieback Material Science Forum, **457–460**, 75 (2004),

2.15. Y. Ando, W. Contrata, N. Samoto, H. Miyamoto, K. Matsunaga, M. Kuzuhara, K. Kunihiro, K. Kasahara, T. Nakayama, Y. Takahashi, N. Hayama, Y. Ohno, IEEE Trans. Electron Devices **47**, 1965 (2000)

2.16. Y. Ando, Y. Okamoto, H. Miyamoto, N. Hayama, T. Nakayama, K. Kasahara, M. Kuzuhura, in *IEDM Technical Digest*, Washington DC, 2001, pp. 381–384

2.17. A. Anwar, S. Wu, R. Webster, IEEE Trans. Electron Devices **48**, 567 (2001)

2.18. A.P. Zhang, L. Rowland, E. Kaminsky, J. Tucker, J. Kretchmer, A. Allen, J. Cook, B. Edward, Electron. Lett. **39**, 245 (2003)

2.19. L. Ardaravicius, A. Matulionis, J. Liberis, O. Kiprijanovic, M. Ramonas, L. Eastman, J. Shealy, A. Vertiatchikh, Appl. Phys. Lett. **83**, 4038 (2003)

2.20. S. Arulkumaran, M. Miyoschi, T. Egawa, H. Ishikawa, T. Jimbo, IEEE Electron Device Lett. **24**, 497 (2003)

2.21. D. As, D. Schikora, A. Greiner, M. Lübbers, J. Mimkes, K. Lischka, Phys. Rev. B **54**, R11118 (1996)

2.22. P. Asbeck, E. Yu, S. Lau, G. Sullivan, J.V. Hove, J. Redwing, Electron. Lett. **33**, 1230 (1997)

2.23. P. Asbeck, E. Yu, S. Lau, W. Sun, X. Dang, C. Shi, Solid-State Electron. **44**, 211 (2000)

2.24. M. Aust, A. Sharma, A. Chau, A. Gutierrez-Aitken, in *IEEE International Microwave Symposium Digest*, Honolulu, 2007, pp. 809–812

2.25. C. Bae, C. Krug, G. Lucovsky, A. Chakraborty, U. Mishra, J. Appl. Phys. **96**, 2674 (2003)

2.26. A. Baliga, IEEE Electron Device Lett. **10**, 455 (1989)

2.27. Z. Bandic, E. Piquette, P. Bridger, R. Beach, T. Kuech, T. McGill, Solid-State Electron. **42**, 2289 (1998)

2.28. H. Bang, T. Mitani, S. Nakashima, H. Sazawa, K. Hirata, M. Kosaki, H. Okumura, J. Appl. Phys. **100**, 114502 (2006)

2.29. A. Barker, M. Ilegems, Phys. Rev. B **7**, 743 (1973)

2.30. J. Barker, D. Ferry, D. Koleske, R. Shul, J. Appl. Phys. **97**, 063705 (2005)

2.31. A. Barnes, D. Hayes, M. Uren, T. Martin, R. Balmer, D. Wallis, K. Hilton, J. Powell, W. Phillips, A. Jimenez, E. Munoz, M. Kuball, S. Rajasingam, J. Pomeroy, N. Labat, N. Malbert, P. Rice, A. Wells, in *IMS Workshop Advances in GaN-based Device and Circuit Technology: Modeling and Applications*, Fort Worth, 2004

2.32. J. Beintner, Y. Li, A. Knorr, D. Chidambarrao, P. Voigt, R. Divakaruni, P. Pöchmüller, G. Bronner, IEEE Electron Device Lett. **25**, 259 (2004)

2.33. K. Bejtka, R. Martin, I. Watson, S. Ndiaje, M. Leroux, Appl. Phys. Lett. **89**, 191912 (2006)

2.34. E. Bellotti, F. Bertazzi, M. Goano, J. Appl. Phys. **101**, 3706 (2007)

2.35. E. Bellotti, B. Doshi, K. Brennan, J. Albrecht, P. Ruden, J. Appl. Phys. **85**, 916 (1999)

2.36. F. Bernardini, V. Fiorentini, D. Vanderbilt, Phys. Rev. B **56**, R10024 (1997)

2.37. U. Bhapkar, M. Shur, J. Appl. Phys. **82**, 1649 (1997)

2.38. A. Bhuiyan, A. Hashimoto, A. Yamamoto, J. Appl. Phys. **94**, 2779 (2003)

2.39. S. Binari, K. Ikossi-Anastasiou, J. Roussos, D. Parl, D. Koleske, A. Wickenden, R. Henry, in *Proceedings of the International Conference on GaAs Manufacturing Technology*, St. Louis, 2000, pp. 201–204

2.40. S. Binari, J. Redwing, G. Kelner, W. Kruppa, IEEE Electron Device Lett. **33**, 242 (1997)

2.41. J. Blevins, in *Proceedings of the International Conference on the GaAs Manufacturing Technology*, Miami, 2004, pp. 287–290

2.42. C. Bolognesi, A. Kwan, D. DiSanto, in *IEDM Technical Digest*, San Francisco, 2002, pp. 685–688

2.43. K. Boutros, B. Luo, Y. Ma, G. Nagy, J. Hacker, in *Compound Semiconductor IC Symposium Technical Digest*, San Antonio, 2006, pp. 93–95

2.44. K. Boutros, M. Regan, P. Rowell, D. Gotthold, R. Birkhahn, B. Brar, in *IEDM Technical Digest*, Washington DC, 2003, pp. 981–982

2.45. K. Boutros, W. Luo, K. Shinohara, in *Device Research Conference*, Santa Barbara, 2005, pp. 183–184

2.46. N. Braga, R. Mickevicius, R. Gaska, X. Hu, M. Shur, M. Khan, G. Simin, J. Yang, J. Appl. Phys. **95**, 6409 (2004)

2.47. N. Braga, R. Mickevicius, V. Rao, W. Fichtner, R. Gaska, M. Shur, in *Compound Semiconductor IC Symposium Technical Digest*, Palm Springs, 2005, pp. 149–152

2.48. N. Braga, R. Mickevicus, M. Shur, R. Gaska, M. Shur, M. Khan, G. Simin, in *IEDM Technical Digest*, San Francisco, 2004, pp. 815–818

2.49. K. Brennan, E. Bellotti, M. Farahmand, J. Haralson, P. Ruden, J. Albrecht, A. Sutandi, Solid-State Electron. **44**, 195 (2000)

2.50. K. Brennan, E. Bellotti, M. Farahmand, H. Nilsson, P. Ruden, Y. Zhang, IEEE Trans. Electron Devices **47**, 1882 (2000)

2.51. J. Brinkhoff, A. Parker, in *Workshop Applied Radio Science*, Hobart, 2004, pp. 1–8

2.52. J. Brown, W. Nagy, S. Singhal, S. Peters, A. Chaudhari, T. Li, R. Nichols, R. Borges, P. Rajagopal, J. Johnson, R. Therrien, A. Hanson, A. Vescan, in *IEEE International Microwave Symposium Digest*, Fort Worth, 2004, pp. 1347–1350

2.53. T. Brozek, J. Szmidt, A. Jabubowski, A. Olszyna, Diamond Relat. Mater. **3**, 720 (1994)

2.54. W. Bryden, T. Kistenmacher, in *Properties of Group III Nitrides*, No. 11 in EMIS Data reviews Series, ed. by J. Edgar (IEE INSPEC, London, 1994), Sect. 3.3, pp. 117–121

2.55. G. Bu AND D. Ciplys AND M. Shur AND L.J. Schowalter AND S. Schjman AND R. Gaska, IEEE Trans. Ultrason. Ferroelec. Frequency Control, **53**, 251 (2006)

2.56. C. Bulutay, Semicond. Sci. Technol. **17**, L59 (2002)

2.57. E. Burgemeister, W. von Muench, E. Pettenpaul, J. Appl. Phys. **50**, 5790 (1979)

2.58. J. Burm, K. Chu, W. Schaff, L. Eastman, M. Khan, Q. Chen, J. Yang, M. Shur, IEEE Electron Device Lett. **18**, 141 (1997)

2.59. R. Caverly, N. Drozdovski, C. Joye, M. Quinn, in *GaAs IC Symposium Technical Digest*, Monterey, 2002, pp. 131–134

2.60. S. Cha, Y. Chung, M. Wojtowicz, I. Smorchkova, B. Allen, J. Yang, R. Kagiwada, in *IEEE International Microwave Symposium Digest*, Fort Worth, 2004, pp. 829–832

2.61. A. Chakraborty, B. Haskell, S. Keller, J. Speck, S. Denbaars, S. Nakamura, U. Mishra, Jpn. J. Appl. Phys. **44**, L173 (2005)

2.62. Q. Chen, J. Yang, R. Gaska, M. Khan, M. Shur, G. Sullivan, A. Sailor, J. Higgins, A. Ping, I. Adesida, IEEE Electron Device Lett. **19**, 44 (1998)

2.63. L. Cheng, I. Sankin, N. Merret, J. Casady, W. Draper, W. King, V. Bondarenko, M. Mazzola, J. Casady, in *Device Research Conference*, State College, PA, 2006, pp. 159–160

2.64. V. Chin, T. Tansley, T. Osotchan, J. Appl. Phys. **75**, 7365 (1994)

2.65. V. Chin, B. Zou, T. Tansley, X. Li, J. Appl. Phys. **77**, 6064 (1995)

2.66. A. Chini, D. Buttari, R. Coffie, S. Heikmann, S. Keller, U. Mishra, Electron. Lett. **40**, 73 (2004)

2.67. A. Chini, D. Buttari, R. Coffie, L. Shen, S. Heikman, A. Chakraborty, S. Keller, U. Mishra, IEEE Electron Device Lett. **25**, 229 (2004)

2.68. A. Chini, R. Coffie, G. Meneghesso, E. Zanoni, D. Buttari, S. Heikman, S. Keller, U. Mishra, Electron. Lett. **39**, 625 (2003)

2.69. A. Chini, J. Wittich, S. Heikman, S. Keller, S. DenBaars, U. Mishra, IEEE Electron Device Lett. **25**, 229 (2004)

2.70. S. Chiu, A. Anwar, S. Wu, IEEE Trans. Electron Devices **47**, 662 (2000)

2.71. L. Chkhartishvili, J. Solid-State Chem. **177**, 395 (2004)

2.72. T. Chow, R. Tyagi, IEEE Trans. Electron Devices **41**, 1481 (1994)

2.73. K. Chu, P. Chao, M. Pizzella, R. Actis, D. Meharry, K. Nichols, R. Vaudo, X. Xu, J. Flynn, J. Dion, G. Brandes, IEEE Electron Device Lett. **25**, 596 (2004)

2.74. E. Chumbes, J. Smart, T. Prunty, J. Shealy, IEEE Trans. Electron Devices **48**, 416 (2001)

2.75. R. Clarke, J. Palmour, Proc. IEEE **90**, 987 (2002)

2.76. I. Cohen, T. Zhu, L. Liu, M. Murphy, M. Pophristic, M. Pabisz, M. Gottfried, B. Shelton, B. Peres, A. Ceruzzi, R. Stall, in *IEEE APEC*, Austin, 2005, pp. 311–314

2.77. P.T. Coleridge, R. Stoner, R. Flechter, Phys. Rev. B, **39**, 1120 (1989)

2.78. Compound Semiconductor, Epigress Licenses HTCVD Technology from Oktemic. Comp. Semicond. **6**, (2002)

2.79. Compound Semiconductor, Cap Wireless to use GaN Chips in X-Band Amplifiers. Comp. Semicond. **11**, (2005)

2.80. Compound Semiconductor, Hitachi Confirms 3-Inch GaN Substrate, Eyes 4-Inch. Comp. Semicond. **11**, (2007)

2.81. Compound Semiconductor, Cree Announces 40 W GaN Amplifier, First GaN MMIC. Comp. Semicond. **6**, 15 (2000)

2.82. T. Cook, C. Fulton, W. Meccouch, R. Davis, G. Lucovsky, R. Nemanich, J. Appl. Phys. **94**, 3949 (2003)

2.83. Cree, Silicon Carbide Substrates and Epitaxy: Product Specifications (2007), http://www.cree.com

2.84. A. Dadgar, M. Neuburger, F. Schulze, J. Bäsing, A. Krtschil, I. Daumiller, M. Kunze, K. Günther, H. Witte, A. Diez, E. Kohn, A. Krost, Phys. Stat. Sol. A **202**, 832 (2005)

2.85. I. Daumiller, M. Seyboth, C. Kirchner, M. Kamp, E. Kohn, in *Device Research Conference*, Denver, 2000, pp. 1–2

2.86. V. Davydov, A. Klochikhin, V. Emtsev, D. Kurdyukov, S. Ivanov, V. Verkshin, F. Bechstedt, J. Furthmüller, J. Aderhold, J. Graul, A. Mudryi, H. Harima, A. Hashimoto, A. Yamamoto, E. Haller, Phys. Stat. Sol. B **234**, 787 (2002)

2.87. C.V. de Walle, J. Neugebauer, C. Stampfl, in *Properties, Processing and Applications of GaN Nitride and Related Semiconductors*, No. 23 in EMIS Data reviews Series, ed. by J. Edgar et al. (IEE INSPEC, London, 1999), Sect. 8.1, pp. 275–280

2.88. M. DeLisio, B. Deckmann, C. Cheung, S. Martin, D. Nakhla, E. Hartmann, C. Rollison, J. Pacetti, J. Rosenberg, in *IEEE International Microwave Symposium Digest*, Forth Worth, 2004, pp. 83–86

2.89. G. Dimitrakopulos, P. Komninou, T. Karakostas, R. Pond, in *Nitride Semiconductors: Handbook on Materials and Device*, ed. by P. Ruterana, M. Albrecht, J. Neugebauer (Wiley-VCH, Weinheim, 2003), Chap. 7, pp. 321–378

2.90. R. Dimitrov, A. Mitchell, L. Wittmer, O. Ambacher, M. Stutzmann, J. Hilsenbeck, W. Rieger, Jpn. J. Appl. Phys. **38**, 4962 (1999)

2.91. M. Drechsler, D. Hofmann, B. Meyer, T. Detchprohm, H. Amano, I. Akasaki, Jpn. J. Appl. Phys. **34**, L1178 (1995)

2.92. M. Drory, J. Ager, T. Suski, I. Grzegory, S. Porowski, J. Appl. Phys. **69**, 4044 (1996)

2.93. D. Ducatteau, A. Minko, V. Hoel, E. Morvan, E. Delos, B. Grimbert, H. Lahreche, P. Bove, C. Gaquiere, J. DeJaeger, S. Delage, IEEE Electron Device Lett. **27**, 7 (2006)

2.94. D. Dumka, C. Lee, H. Tserng, P. Saunier, M. Kumar, Electron. Lett. **40**, 1023 (2003)

2.95. N. Dyakonova, A. Dickens, M. Shur, R. Gaska, Electron. Lett. **34**, 1699 (1998)

2.96. L. Eastman, V. Tilak, J. Smart, B. Green, E. Chumbes, R. Dimitrov, H. Kim, O. Ambacher, N. Weimann, T. Prunty, M. Murphy, W. Schaff, J. Shealy, IEEE Trans. Electron Devices **48**, 479 (2001)

2.97. L. Eastman. Personal Communication, 2002

2.98. L. Eastman, in *IEEE International Microwave Symposium Digest*, Seattle, 2002, pp. 2273–2275

2.99. J. Edgar, in *Properties of Group III Nitrides*, No. 11 in EMIS Data reviews Series, ed. by J. Edgar (IEE INSPEC, London, 1994), Sect. 1.2, pp. 7–21

2.100. J. Edgar (ed.), *Properties of Group III Nitrides*, No. 11 in EMIS Data reviews Series (IEE INSPEC, London, 1994)

2.101. J. Edgar, S. Strite, I. Akasaki, H. Amano, C. Wetzel (eds.), *Properties, Processing and Applications of GaN Nitride and Related Semiconductors*, No. 23 in EMIS Data reviews Series (IEE INSPEC, London, 1999)

2.102. G. Ellis, J. Moon, D. Wong, M. Micovic, A. Kurdoghlian, P. Hashimoto, M. Hu, in *IEEE International Microwave Symposium Digest*, Fort Worth, 2004, pp. 153–156

2.103. E. Emorhokpor, T. Kerr, I. Zwieback, W. Elkington, M. Dudley, T. Anderson, J. Chen, in *Proceedings of the International Conference on GaAs Manufacturing Technology*, Miami, 2004, pp. 139–142

2.104. J. Ender, H. Wilden, U. Nickel, R. Klemm, A. Brenner, T. Eibert, D. Nuessler, in *Proceedings of European Radar Conference*, Amsterdam, 2004, pp. 113–116

2.105. T. Ericsen, Proc. IEEE **90**, 1077 (2002)

2.106. K. Evans, Comp. Semicond. **10**, (2004)

2.107. S. Evans, N. Giles, L. Haliburton, G. Slack, S. Schujman, L. Schowalter, Appl. Phys. Lett. **88**, 062112 (2006)

2.108. W. Fan, M. Li, T. Chong, J. Xia, J. Appl. Phys. **79**, 188 (1996)

2.109. D. Fanning, L. Witkowski, C. Lee, D. Dumka, H. Tserng, P. Saunier, W. Gaiewski, E. Piner, K. Linthicum, J. Johnson, in *Proceedings of the International Conference on GaAs Manufacturing Technology*, New Orleans, 2005, p. 8.3

2.110. M. Farahmand, K. Brennan, IEEE Trans. Electron Devices **46**, 1319 (1999)

2.111. M. Farahmand, K. Brennan, IEEE Trans. Electron Devices **47**, 493 (2000)

2.112. M. Farahmand, K. Brennan, E. Gebara, D. Heo, Y. Suh, J. Laskar, IEEE Trans. Electron Devices **45**, 1844 (2001)

2.113. M. Farahmand, C. Garetto, E. Bellotti, K. Brennan, M. Goano, E. Ghillino, G. Ghione, J. Albrecht, P. Ruden, IEEE Trans. Electron Devices **48**, 535 (2001)

2.114. T. Fehlberg, G. Umana-Membreno, B. Nener, G. Parish, C. Gallinat, G. Koblmüller, S. Rajan, S. Bernardis, J. Speck, Jpn. J. Appl. Phys. **45**, L1090 (2006)

2.115. A. Fischer, H. Kühne, H. Richter, Phys. Rev. Lett. **73**, 2712 (1994)

2.116. G. Fischer, in *IEEE Wireless and Microwave Technology Conference*, Clearwater, 2004, p. FD-1

2.117. M. Fischetti, IEEE Trans. Electron Devices **38**, 634 (1991)

2.118. M. Fischetti, S. Laux, IEEE Trans. Electron Devices **38**, 650 (1991)

2.119. N. Fitzer, A. Kuligk, R. Redmer, M. Städele, S. Goodnick, W. Schattke, Semicond. Sci. Technol. **19**, S206 (2004)

2.120. D. Florescu, V. Asnin, F. Pollack, Comp. Semicond. **7**, 62 (2001)

2.121. B. Foutz, S. O'Leary, M. Shur, L. Eastman, J. Appl. Phys. **85**, 7727 (1999)

2.122. S. Fu, Y. Chen, Appl. Phys. Lett. **85** 1523 (2004)

2.123. D. Gaskill, L. Rowland, K. Doverspike, in *Properties of Group III Nitrides*, No. 11 in EMIS Data reviews Series, ed. by J. Edgar (IEE INSPEC, London, 1994), Sect. 3.2, pp. 101–116

2.124. G. Gauthier, Y. Mancuso, F. Murgadella, in *Proceedings of the European Gallium Arsenide Other Compound Semiconductors Application Symposium GAAS*, Paris, 2005, pp. 361–364

2.125. J. Gillespie, R. Fitch, N. Moser, T. Jenkins, J. Sewell, D. Via, A.C.A. Dabiran, P. Chow, A. Osinsky, M. Mastro, D. Tsvetkov, V. Soukhoveev, A. Usikov, V. Dmitriev, B. Luo, S. Pearton, F. Ren, Solid-State Electron. **47**, 1859 (2003)

2.126. J. Gillespie, G. Jessen, G. Via, A. Crespo, D. Langley, M. Aumer, H. Henry, D. Thomson, D. Partlow, in *Proceedings of the International Conference GaAs Manufacturing Technology*, Austin, 2007, pp. 73–76

2.127. M. Goano, E. Belloti, E. Ghillino, C. Garetto, G. Ghione, K. Brennan, J. Appl. Phys. **88**, 6476 (2000)

2.128. M. Goano, E. Belloti, E. Ghillino, G. Ghione, K. Brennan, J. Appl. Phys. **88**, 6467 (2000)

2.129. J. Grajal, F. Calle, J. Pedros, T. Palacios, in *IEEE International Microwave Symposium Digest*, Fort Worth, 2004, pp. 387–390

2.130. B. Green, K. Chu, E. Chumbes, J. Smart, J. Shealy, L. Eastman, IEEE Electron Device Lett. **21**, 268 (2000)

2.131. B. Green, H. Henry, J. Selbee, R. Lawrence, K. Moore, J. Abdou, M. Miller, in *IEEE International Microwave Symposium Digest*, San Francisco, 2006, pp. 706–709

2.132. B. Green, V. Tilak, V. Kaper, J. Smart, J. Shealy, L. Eastman, IEEE Trans. Microwave Theory Tech. **51**, 618 (2003)

2.133. B. Green, V. Tilak, S. Lee, H. Kim, J. Smart, K. Webb, J. Shealy, L. Eastman, IEEE Trans. Microwave Theory Tech. **49**, 2486 (2001)

2.134. D. Grider, J. Smart, R. Vetury, M. Young, J. Dick, B. Delayney, Y. Yang, T. Mercier, S. Gibb, C. Palmer, B. Hosse, K. Leverich, N. Zhang, J. Shealy, M. Poulton, B. Sousa, D. Schnaufer, in *Proceedings of the International Conference on GaAs Manufacturing Technology*, Miami, 2004, pp. 101–102

2.135. V. Gubaniv, E. Pentaleri, C. Fong, B. Klein, Phys. Rev. B **56**, 13077 (1997)

2.136. E. Gusev et al., in *IEDM Technical Digest*, San Francisco, 2004, pp. 79–82

2.137. A. Hanser, L. Liu, E. Preble, D. Tsetkov, M. Tutor, N. Williams, K. Evans, Y. Zhou, D. Wang, C. Ahyi, C. Tin, J. Williams, M. Park, D. Storm, D. Katzer, S. Binari, J. Roussos, J. Mittereder, in *Proceedings of the International Conference on GaAs Manufacturing Technology*, Vancouver, 2006, pp. 67–70

2.138. S. Harada, M. Kato, K. Suzuki, M. Okamoto, T. Yatsuo, K. Fukuda, K. Arai, in *IEDM Technology Digest*, San Francisco, 2006, pp. 1–4

2.139. G. Harris (ed.), in *Properties of Silicon Carbide*, No. 13 in EMIS Data reviews Series (IEE INSPEC, London, 1995)

2.140. J. Heffernan, M. Kauer, S. Hooper, V. Bousquet, K. Johnson, Phys. Stat. Sol. C **1**, 2668 (2004)

2.141. R. Hickman, J. van Hove, P. Chow, J. Klaassen, A. Wowchack, C. Polley, D. King, F. Ren, C. Abernathy, S. Pearton, K. Jung, H. Cho, J. LaRoche, Solid-State Electron. **44**, 377 (2000)

2.142. M. Higashiwaki, T. Mimura, T. Matsui, IEEE Electron Device Lett. **27**, 719 (2006)

2.143. M. Higashiwaki, T. Mimura, T. Matsui, Jpn. J. Appl. Phys. **45**, L843 (2006)

2.144. M. Hirose, Y. Takada, M. Kuraguchi, T. Sasaki, K. Tsuda, in *Compound Semiconductor IC Symposium Technical Digest*, Monterey, 2004, pp. 163–166

2.145. C. Hobbs, Leonardo, R. Fonseca, A. Knizhnik, V. Dhandapani, S. Samavedam, W. Taylor, J. Grant, L. Dip, D. Triyoso, R. Hegde, D. Gilmer, R. Garcia, D. Roan, M. Lovejoy, R. Rai, E. Hebert, H. Tseng, S. Anderson, B. White, P. Tobin, IEEE Trans. Electron Devices **51**, 971 (2004)

2.146. D. Holec, P.M.F.J. Costa, M. Kaspers, C. Humphreys, J. Cryst. Growth **303**, 314 (2007)

2.147. H. Hommel, H. Feldle, in *Proceedings of the European Radar Conference*, Amsterdam, 2004, pp. 121–124

2.148. Honeywell International, Honeywell Sapphire Products (2007) https://www51.honeywell.com/sm/em/products-applications/opto-electro-materials.html

2.149. A. Hori, H. Nakaoka, H. Umimoto, K. Yamashita, M. Takase, N. Shimizu, B. Mizuno, S. Odanaka, in *IEDM Technical Digest*, Washington DC, 1994, pp. 485–488

2.150. Y. Hsin, H. Hsu, C. Chuo, J. Chyi, IEEE Electron Device Lett. **22**, 501 (2001)

2.151. X. Hu, J. Deng, N. Pala, R. Gaska, M. Shur, C. Chen, J. Yang, G. Simin, M. Khan, J. Rojo, L. Schowalter, Appl. Phys. Lett. **82**, 1299 (2003)

2.152. J. Huang, M. Hattendorf, M. Feng, D. Lambert, B. Shelton, M. Wong, U. Chowdhury, T. Zhu, H. Kwon, R. Dupuis, IEEE Electron Device Lett. **22**, 157 (2001)

2.153. T. Hussain, A. Kurdoghlian, P. Hashimoto, W. Wong, M. Wetzel, J. Moon, L. McCray, M. Micovic, in *IEDM Technical Digest*, Washington DC, 2001, pp. 581–584

2.154. J. Ibbetson, P. Fini, K. Ness, S. DenBaars, J. Speck, U. Mishra, Appl. Phys. Lett. **77**, 250 (2000)

2.155. D. Ingram, Y. Chen, J. Kraus, B. Brunner, B. Allen, H. Yen, K. Lau, in *Radio Frequency Integrated Circuit Symposium Digest*, Anaheim, 1999, pp. 95–98

2.156. T. Inoue, Y. Ando, H. Miyamoto, T. Nakayama, Y. Okamoto, K. Hataya, M. Kuzuhara, in *IEEE International Microwave Symposium Digest*, Fort Worth, 2004, pp. 1649–1652

2.157. T. Inoue, Y. Ando, H. Miyamoto, T. Nakayama, Y. Okamoto, K. Hataya, M. Kuzuhara, IEEE Trans. Microwave Theory Tech. **53**, 74 (2004)

2.158. H. Ishida, Y. Hirose, T. Murata, A. Kanda, Y. Ikeda, T. Matsuno, K. Inoue, Y. Uemoto, T. Tanaka, T. Egawa, D. Ueda, in *IEDM Technical Digest*, Washington DC, 2003, pp. 583–586

2.159. H. Jacobsen, J. Birch, R. Yakimova, M. Syvajarvi, J. Bergman, A. Ellison, T. Tuomi, E. Janzen, J. Appl. Phys. **91**, 6354 (2002)

2.160. H. Jang, C. Jeon, K. Kim, J. Kim, S. Bae, J. Lee, J. Choi, J. Lee, Appl. Phys. Lett. **81**, 1249 (2002)

2.161. P. Javorka, A. Alam, N. Nastase, M. Marso, H. Hardtdegen, M. Heuken, H. Lüth, P. Kordos, Electron. Lett. **37**, 1364 (2001)

2.162. P. Javorka, A. Alam, M. Wolter, A. Fox, M. Marso, M. Heuken, H. Lüth, P. Kordos, Electron. Lett. **23**, 4 (2002)

2.163. D. Jenkins, J. Dow, Phys. Rev. B **39**, 3317 (1989)

2.164. G. Jessen, R. Fitch, J. Gillespie, G. Via, N. Moser, M. Yannuzzi, A. Crespo, J. Sewell, R. Dettmer, T. Jenkins, R. Davis, J. Yang, M. Khan, S. Binari, IEEE Electron Device Lett. **24**, 677 (2003)

2.165. G. Jessen, J. Gillespie, G. Via, A. Crespo, D. Langley, J. Wasserbauer, F. Faili, D. Francis, D. Babic, F. Ejeckam, S. Guo, I. Eliashevich, in *Compound Semiconductor IC Symposium Technical Digest*, San Antonio, 2006, pp. 271–274

2.166. A. Jimenez, D. Buttari, D. Jena, R. Coffie, S. Heikman, N. Zhang, L. Shen, E. Calleeja, E. Munoz, J. Speck, U. Mishra, IEEE Electron Device Lett. **23**, 306 (2002)

2.167. E. Johnson, RCA Rev. **26**, 163 (1965)

2.168. J. Johnson, J. Gao, K. Lucht, J.W.C. Strautin, J. Riddle, T. Therrien, P. Rajagopal, J. Roberts, A. Vescan, J. Brown, A. Hanson, S. Singhal, R. Borges, E. Piner, K. Linthicum, Proc. Elec. Soc. **7**, 405 (2004)

2.169. J. Johnson, E. Piner, A. Vescan, R. Therrien, Rajagopal, J. Roberts, J. Brown, J. Brown, K. Linthicum, IEEE Electron Device Lett. **25**, 459 (2004)

2.170. R. Jones, K. Yu, S. Li, W. Walukiewicz, J. Ager, E. Haller, H. Lu, W. Schaff, Phys. Rev. Lett. **96**, 125505–1 (2006)

2.171. K. Joshin, T. Kikkawa, in *Device Research Conference*, Long Beach, 2005, pp. 173–176

2.172. K. Joshin, T. Kikkawa, H. Hayashi, T. Maniwa, S. Yokokawa, M. Yokoyama, N. Adachi, M. Takikawa, in *IEDM Technical Digest*, Washington DC, 2003, pp. 983–985

2.173. Y. Kamo, T. Kunii, H. Takeuchi, Y. Yamamoto, M. Totsuka, T. Shiga, H. Minami, T. Kitano, S. Miyakuni, T. Oku, A. Inoue, T. Nanjo, H. Chiba,

M. Suita, T. Oishi, Y. Abe, Y. Tsuyama, R. Shirahana, H. Ohtsuka, K. Iyomasa, K. Yamanaka, M. Hieda, M. Nakayama, T. Ishikawa, T. Takagi, K. Marumoto, Y. Matsuda, in *IEEE International Microwave Symposium Digest*, Long Beach, 2005, pp. 495–498

2.174. M. Kanamura, T. Kikkawa, J. Joshin, in *IEDM Technical Digest*, San Francisco, 2004, pp. 799–802

2.175. B. Kang, S. Kim, F. Ren, J. Johnson, R. Therrien, P. Rajagopal, J. Roberts, E. Piner, K. Linthicum, S. Chu, K. Baik, B. Gila, C. Abernathy, S. Pearton, Appl. Phys. Lett. **85**, 2962 (2004)

2.176. B. Kang, F. Ren, B. Gilla, C. Abernathy, S. Pearton, Appl. Phys. Lett. **84**, 1123 (2004)

2.177. M. Kanoun, S. Goumri-Said, G. Merad, J. Cibert, H. Aourag, Semicond. Sci. Technol. **19**, 1220 (2004)

2.178. M. Kanoun, A. Merad, G. Merad, J. Cibert, H. Aourag, Solid-State Electron. **48**, 1601 (2004)

2.179. M. Kao, C. Lee, R. Hajji, P. Saunier, H. Tserng, in *IEEE International Microwave Symposium Digest*, Honolulu, 2007, pp. 627–630

2.180. V. Kaper, V. Tilak, H. Kim, R. Thompson, T. Prunty, J. Smart, L. Eastman, J. Shealy, in *GaAs IC Symposium Technical Digest*, Monterey, 2002, pp. 251–254

2.181. M. Kariya, S. Nitta, S. Yamaguchi, H. Amano, I. Akasaki, Jpn. J. Appl. Phys. **38**, L984 (1999)

2.182. S. Karmalkar, N. Satyan, D. Sathaiya, IEEE Electron Device Lett. **27**, 87 (2006)

2.183. S. Karmalkar, N. Soudabi, IEEE Trans. Electron Devices **53**, 2430 (2006)

2.184. K. Kasahara, H. Miyamoto, Y. Ando, Y. Okamoto, T. Nakayama, M. Kuzuhara, in *IEDM Technical Digest*, San Francisco, 2002, pp. 677–680

2.185. O. Katz, A. Horn, J. Salzmann, IEEE Trans. Electron Devices **50**, 2002 (2003)

2.186. O. Katz, D. Mistele, B. Meyler, G. Bahir, J. Salzman, in *IEDM Technical Digest*, San Francisco, 2004, pp. 1035–1038

2.187. O. Katz, D. Mistele, B. Meyler, G. Bahir, J. Salzmann, Electron. Lett. **40**, 1304 (2004)

2.188. S. Keller, Y. Wu, G. Parish, N. Ziang, J. Xu, B. Keller, S. DenBaars, U. Mishra, IEEE Trans. Electron Devices **48**, 552 (2001)

2.189. R. Kemerley, H. Wallace, M. Yoder, Proc. IEEE **90**, 1059 (2002)

2.190. M. Khan, A. Bhattarai, J. Kuznia, D. Olson, Appl. Phys. Lett. **63**, 1214 (1993)

2.191. M. Khan, X. Hu, G. Sumin, A. Lunev, J. Yang, R. Gaska, M. Shur, IEEE Electron Device Lett. **21**, 63 (2000)

2.192. M. Khan, M. Shur, G. Simin, Phys. Stat. Sol. A **200**, 155 (2003)

2.193. M. Khan, G. Simin, J. Yang, J. Zhang, A. Koudymov, M. Shur, R. Gaska, X. Hu, A. Tarakji, IEEE Trans. Microwave Theory Tech. **51**, 624 (2003)

2.194. R. Kiefer, R. Quay, S. Müller, K. Köhler, F. van Raay, B. Raynor, W. Pletschen, H. Massler, S. Ramberger, M. Mikulla, G. Weimann, in *Proceedings of the Lester Eastman Conference High Performance Devices*, Newark, 2002, pp. 502–504

2.195. T. Kikkawa, T. Maniwa, H. Hayashi, M. Kanamura, S. Yokokawa, M. Nishi, N. Adachi, M. Yokoyama, Y. Tateno, K. Joshin, in *IEEE International Microwave Symposium Digest*, Fort Worth, 2004, pp. 1347–1350

2.196. T. Kikkawa, E. Mitani, K. Joshin, S. Yokokawa, Y. Tateno, in *Proceedings of the International Conference on GaAs Manufacturing Technology*, Miami Beach, 2004, pp. 97–100

2.197. T. Kikkawa, M. Nagahara, N. Adachi, S. Yokokawa, S. Kato, M. Yokoyama, M. Kanamura, Y. Yamaguchi, N. Hara, K. Joshin, in *IEEE International Microwave Symposium Digest*, Philadelphia, 2003, pp. 167–170

2.198. T. Kikkawa, M. Nagahara, T. Kimura, S. Yokokawa, S. Kato, M. Yokoyama, Y. Tateno, K. Horino, K. Domen, Y. Yamaguchi, N. Hara, K. Joshin, in *IEEE International Microwave Symposium Digest*, Seattle, 2002, pp. 1815–1818

2.199. C. Kim, I. Robinson, J. Myoung, K. Shim, M. Yoo, K. Kim, Appl. Phys. Lett. **69**, 2358 (1996)

2.200. J. Kim, B. Gila, C. Abernathy, G. Chung, F. Ren, S. Pearton, Solid-State Electron. **47**, 1487 (2003)

2.201. D. Kimball, P. Draxler, J. Jeong, C. Hsia, S. Lanfranco, W. Nagy, K. Linthicum, L. Larson, P. Asbeck, in *Compound Semiconductor IC Symposium Technical Digest*, Palm Springs, 2005, pp. 89–92

2.202. D. Kimball, J. Jeong, P. Draxler, C. Hsia, P. Draxler, S. Lanfranco, W. Nagy, K. Linthicum, L. Larson, P. Asbeck, IEEE Trans. Microwave Theory Tech. **54**, 3848 (2006)

2.203. S. King, R. Nemanich, R. Davis, in *Properties, Processing and Applications of GaN Nitride and Related Semiconductors*, No. 23 in EMIS Data reviews Series, ed. by J. Edgar et al. (IEE INSPEC, London, 1999), Sect. C1.3, pp. 500–505

2.204. L. Kirste, S. Müller, R. Kiefer, R. Quay, K. Köhler, N. Herres, Mater. Sci. Eng. B **9**, 8 (2006)

2.205. K. Köhler, J. Wiegert, H.P. Menner, M. Maier, L. Kirste, J. Appl. Phys. **103**, 023706 (2008)

2.206. E. Kohn, I. Daumiller, P. Schmid, N. Nguyen, C. Nguyen, Electron. Lett. **35**, 1022 (1999)

2.207. E. Kohn, J. Kusterer, A. Denisenko, in *IEEE International Microwave Symposium Digest*, Long Beach, 2005, pp. 901–904

2.208. G. Koley, V. Tilak, L. Eastman, M. Spencer, J. Appl. Phys. **90**, 337 (2001)

2.209. G. Koley, V. Tilak, L. Eastman, M. Spencer, IEEE Trans. Electron Devices **50**, 886 (2003)

2.210. N. Kolias, T. Kazior, in *IMS Workshop Advances in GaN-based Device and Circuit Technology: Modeling and Applications*, Fort Worth, 2004

2.211. J. Kolnik, I.H. Oguzman, K.F. Brennan, R. Wang, P.P. Ruden, Y. Wang, J. Appl. Phys. **78**, 1033 (1995)

2.212. J. Kolnik, I. Oguzman, K. Brennan, R. Wang, P. Ruden, J. Appl. Phys. **81**, 726 (1997)

2.213. T. Komachi, T. Takayama, M. Imamura, in *IEDM Technical Digest*, San Francisco, 2006, pp. 915–918

2.214. J. Komiak, W. Kong, P. Chao, K. Nicols, in *IEEE International Microwave Symposium Digest*, Anaheim, 1999, pp. 947–950

2.215. K. Kong, B. Nguyen, S. Nayak, M. Kao, in *Compound Semiconductor IC Symposium Technical Digest*, Palm Springs, 2005, pp. 232–235

2.216. D. Kotchetkov, J. Zou, A. Balandin, D. Florescu, F. Pollak, J. Electron. Mater. J. Electron. Mater. **79**, 4316 (2001)

2.217. A. Koudymov, X. Hu, K. Simin, M. Ali, J. Yang, M. Khan, IEEE Electron Device Lett. **23**, 449 (2002)

2.218. P. Kozodoy, H. Xing, S. DenBaars, U. Mishra, A. Saxler, R. Perrin, S. Elhamri, W. Mitchel, J. Appl. Phys. **87**, 1832 (2000)

2.219. M. Krämer, R. Hoskesn, B. Jakobs, J. Kwaspen, E. Suijker, A. de Hek, F. Karouta, L. Kaufmann, in *Proceedings of the European GaAs and Related Compounds Application Symposium GAAS*, Amsterdam, 2004, pp. 75–78

2.220. D. Krausse, R. Quay, R. Kiefer, A. Tessmann, H. Massler, A. Leuther, T. Merkle, S. Müller, M. Mikulla, M. Schlechtweg, G. Weimann, in *Proceedings of the European Gallium Arsenide Other Compound Semiconductors Application Symposium GAAS*, Amsterdam, 2004, pp. 71–74

2.221. S. Kret, P. Ruterana, C. Delamarre, T. Benabbas, P. Dluzewski, in *Nitride Semiconductors: Handbook on Materials and Device*, ed. by P. Ruterana, M. Albrecht, J. Neugebauer (Wiley-VCH, Weinheim, 2003), Chap. 9, pp. 439–488

2.222. S. Krukowski, M. Leszczynski, S. Porowski, in *Properties, Processing and Applications of GaN Nitride and Related Semiconductors*, No. 23 in EMIS Data reviews Series, ed. by J. Edgar et al. (IEE INSPEC, London, 1999), Sect. 1.4, pp. 21–28

2.223. S. Krukowski, A. Witek, J. Adamczyk, J. Jun, M. Bockowski, I. Gregory, B. Lucznik, G. Nowak, M. Wroblewski, A. Presz, S. Gierlotka, S. Stelmach, B. Palosz, S. Porowski, P. Zinn, J. Phys. Chem. Solids, **59**, 289 (1998)

2.224. Y. Kumagai, T. Nagashima, A. Koukitu, Jpn. J. Appl. Phys. **46**, L389 (2007)

2.225. K. Kumakura, T. Makimoto, N. Kobayashi, Jpn. J. Appl. Phys. **39**, L337 (2000)

2.226. V. Kumar, G. Chen, S. Guo, B. Peres, I. Adesida, in *Device Research Conference*, Santa Barbara, 2005, pp. 61–62

2.227. V. Kumar, J. Lee, A. Kuliev, R. Schwindt, R. Birkhahn, D. Gotthold, S. Guo, B. Albert, I. Adesida, Electron. Lett. **39**, 1609 (2003)

2.228. V. Kumar, W. Lu, R. Schwindt, A. Kuliev, G. Simin, J. Yang, M. Khan, I. Adesida, IEEE Electron Device Lett. **23**, 455 (2002)

2.229. K. Kunihiro, K. Kasahara, Y. Takahashi, Y. Ohno, IEEE Electron Device Lett. **20**, 608 (1999)

2.230. J. Kuzmik, IEEE Electron Device Lett. **22**, 510 (2001)

2.231. J. Kuzmik, Semicond. Sci. Technol. **17**, 540 (2002)

2.232. J. Kuzmik, J. Carlin, T. Kostopoulos, G. Konstantinidis, A. Georgakilas, D. Pogany, in *Device Research Conference*, Santa Barbara, 2005, pp. 57–58

2.233. M. Kuzuhara, H. Miyamoto, Y. Ando, T. Inoue, Y. Okamoto, T. Nakayama, Phys. Stat. Sol. A **200**, 161 (2003)

2.234. D. Lambert, D. Lin, R. Dupuis, Solid-State Electron. **44**, 253 (2000)

2.235. W. Lambrecht, B. Segall, in *Properties of Group III Nitrides*, No. 11 in EMIS Data reviews Series, ed. by J. Edgar (IEE INSPEC, London, 1994), Sect. 5.2, pp. 163–166

2.236. W. Lambrecht, B. Segall, in *Properties of Group III Nitrides*, No. 11 in EMIS Data reviews Series, ed. by J. Edgar (IEE INSPEC, London, 1994), Sect. 4.3, pp. 135–140

2.237. W. Lambrecht, B. Segall, in *Properties of Group III Nitrides*, No. 11 in EMIS Data reviews Series, ed. by J. Edgar (IEE INSPEC, London, 1994), Sect. 4.2, pp. 129–134

2.238. W. Lambrecht, B. Segall, in *Properties of Group III Nitrides*, No. 11 in EMIS Data reviews Series, ed. by J. Edgar (IEE INSPEC, London, 1994), Sect. 4.4, pp. 141–150

2.239. J. LaRoche, B. Luo, F. Ren, K. Baik, D. Stodilka, B. Gila, C. Abernathy, S. Pearton, A. Usikov, D. Tsevtkov, V. Soukhoveev, G. Gainer, A. Rechnikov, V. Dimitriev, G. Chen, C. Pan, J. Chyi, Solid-State Electron. **48**, 193 (2004)

2.240. C. Lee, P. Saunier, J. Yang, M. Khan, IEEE Electron Device Lett. **24**, 616 (2003)

2.241. C. Lee, H. Wang, J. Yang, L. Witkowski, M. Muir, P. Saunier, M. Khan, Electron. Lett. **38**, 924 (2002)

2.242. S. Lee, H. Jeong, S. Bae, H. Choi, J. Lee, Y. Lee, IEEE Trans. Electron Devices **48**, 524 (2001)

2.243. M. Leszczynski, T. Suski, H. Teisseyre, P. Perlin, I. Grzegory, J. Jun, S. Porowski, T. Moustakas, J. Appl. Phys. **76**, 4909 (1994)

2.244. M. Levinshtein, S. Rumyantsev, M. Shur (eds.), *GaN, AlN, InN, BN, SiC, SiGe: Properties of Advanced Semiconductor Materials* (Wiley, New York, 2001)

2.245. J. Li, K. Nam, M. Nakarmi, J. Lin, H. Jiang, Appl. Phys. Lett. **83**, 5163 (2003)

2.246. Q. Li, A. Polyakov, M. Skowronski, M. Roth, M. Fanton, D. Snyder, J. Appl. Phys. **96**, 411 (2004)

2.247. S. Li, K. Yu, J. Wu, R. Jones, W. Walukiewicz, J. Ager, W. Shan, E. Haller, H. Lu, W. Schaff, Phys. Rev. B **71**, 161201 (2005)

2.248. P. Lim, B. Schineller, O. Schön, K. Heime, M. Heuken, J. Cryst. Growth **205**, 1 (1999)

2.249. C. Lin, H. Liu, C. Chu, H. Huang, Y. Wang, C. Liu, C. Chang, C. Wu, C. Chang, in *Compound Semiconductor IC Symposium Technical Digest*, San Antonio, 2006, pp. 165–168

2.250. Y. Lin, Q. Ker, C. Ho, H. Chang, F. Chien, J. Appl. Phys. **94**, 1819 (2003)

2.251. Y. Lin, S. Koa, C. Chan, S. Hsu, Appl. Phys. Lett. **90**, 142111 (2007)

2.252. K. Linkenheil, H. Ruoß, W. Heinrich, in *Proceedings of the European Microwave Conference*, Amsterdam, 2004, pp. 1561–1564

2.253. D. Litvinov, C. Taylor, R. Clarke, Diamond Relat. Mater. **7**, 360 (1998)

2.254. B.T. Liu, Appl. Phys. A, **86**, 539 (2007)

2.255. W. Liu, A. Balandin, Appl. Phys. Lett. **85**, 5230 (2004)

2.256. B.T. Liu, C.W. Liu, Optical Comm., **274**, 361 (2007)

2.257. D. Look, Mater. Sci. Eng. B **80**, 383 (2001)

2.258. D. Look, J. Sizelove, Phys. Rev. Lett. **82**, 1237 (1999)

2.259. D. Look, J. Sizelove, S. Keller, Y. Wu, U. Mishra, S. DenBaars, Solid-State Commun. **102**, 297 (1997)

2.260. S. Loughin, R. French, in *Properties of Group III Nitrides*, No. 11 in EMIS Data reviews Series, ed. by J. Edgar (IEE INSPEC, London, 1994), Sect. 6.2, pp. 175–189

2.261. W. Lu, V. Kumar, E. Piner, I. Adesida, IEEE Trans. Electron Devices **50** 1069 (2003)

2.262. W. Lu, J. Yang, M. Khan, I. Adesida, IEEE Trans. Electron Devices **48**, 581 (2001)

2.263. M. Ludwig, C. Buck, F. Coromina, M. Suess, in *IEEE International Microwave Symposium Digest*, Long Beach, 2005, pp. 1619–1622

2.264. C. Luo, D. Clarke, J. Dryden, J. Electron. Mater. **30**, 138 (2001)

2.265. B. Luther, S. Wolter, S. Mohney, Sensors Actuators B **56**, 164 (1999)

2.266. M. Lyons, C. Grondahl, S. Daoud, in *IEEE International Microwave Symposium Digest*, Fort Worth, 2004, pp. 1673–1676

2.267. X. Ma, T. Sudarshan, J. Electron. Mater. **33**, 450 (2004)

2.268. A. Maekawa, T. Yamamoto, E. Mitani, S. Sano, in *IEEE International Microwave Symposium Digest*, San Francisco, 2006, pp. 722–725

2.269. T. Makimoto, K. Kumakura, N. Kobayashi, in *Proceedings of the International Workshop on Nitride Semiconductors*, Nagoya, 2000, pp. 969–972

2.270. T. Makimoto, K. Kumakura, N. Kobayashi, in *International Conference on Solid-State Devices and Materials*, Tokyo, 2003, pp. 134–135

2.271. T. Makimoto, Y. Yamauchi, K. Kumakura, Appl. Phys. Lett. **84**, 1964 (2004)

2.272. Y. Mancuso, P. Gremillet, P. Lacomme, in *Proceedings of the European Microwave Conference*, Paris, 2005, pp. 817–820

2.273. C. Martin, T. Kerr, W. Stepko, T. Anderson, in *Proceedings of the International Conference on GaAs Manufacturing Technology*, Miami, 2004, pp. 291–294

2.274. E. Martin, J. Jimenez, M. Chafai, Solid-State Electron. **42**, 2309 (1998)

2.275. G. Martin, A. Botchkarev, A. Rockett, H. Morkoc, Appl. Phys. Lett. **68**, 2541 (1996)

2.276. K. Matocha, T. Chow, R. Gutmann, IEEE Trans. Electron Devices **52**, 6 (2005)

2.277. K. Matocha, R. Gutmann, T. Chow, IEEE Trans. Electron Devices **50**, 1200 (2003)

2.278. T. Matsuoka, H. Okamoto, M. Nakao, H. Harima, E. Kurimoto, Appl. Phys. Lett. **81**, 1246 (2002)

2.279. T. Matsuoka, H. Okamoto, M. Nakao, H. Harima, H. Takahata, H. Mitate, S. Mizuno, Y. Uchiyama, T. Makimoto, in *International Conference on the Solid-State Development and Materials*, Tokyo, 2003, pp. 132–133

2.280. J. Matthews, A. Blakeslee, J. Cryst. Growth **27**, 118 (1974)

2.281. A. Matulionis, J. Liberis, L. Ardaravicius, M. Ramonas, I. Matulioniene, J. Smart, Semicond. Sci. Technol. **17**, L9 (2002)

2.282. L.S. McCarthy, Dissertation, University of California Santa Barbara, Santa Barbara, 2001

2.283. L. McCarthy, I. Smorchkova, H. Xing, P. Kozodoy, P. Fini, J. Limb, D. Pulfrey, J. Speck, M. Rodwell, S. DenBaars, U. Mishra, IEEE Trans. Electron Devices **48**, 543 (2001)

2.284. F. Medjdoub, J. Carlin, M. Gonschorek, E. Feltin, M. Py, D. Ducatteau, C. Gaquiere, N. Grandjean, E. Kohn, in *IEDM Technical Digest*, San Francisco, 2006, p. 35.7

2.285. R. Mehandru, B. Luo, B. Kang, J. Kim, F. Ren, S. Pearton, C. Pan, G. Chen, J. Chyi, Solid-State Electron. **48**, 351 (2004)

2.286. C. Meng, G. Liao, J. Chen, in *IEEE International Microwave Symposium Digest*, Anaheim, 1999, pp. 1777–1780

2.287. W. Meng, in *Properties of Group III Nitrides*, No. 11 in EMIS Data reviews Series, ed. by J. Edgar (IEE INSPEC, London, 1994), Sect. 1.3, pp. 22–29

2.288. W. Menninger, R. Benton, M. Choi, J. Feicht, U. Hallsten, H. Limburg, W. McGeary, Z. Xiaoling, IEEE Trans. Electron Devices **52**, 673 (2005)

2.289. M. Micovic, P. Hashimoto, M. Hu, I. Milosavljevic, J. Duval, P. Willadsen, A. Kurdoghlian, P. Deelman, J. Moon, A. Schmitz, M. Delaney, in *IEDM Technical Digest*, San Francisco, 2004, pp. 807–810

2.290. M. Micovic, A. Kurdhoghlian, H. Moyer, P. Hashimoto, A. Schmitz, I. Milosavljevic, P. Willadsen, W. Wong, J. Duvall, M. Hu, M. Wetzel,

D. Chow, in *Compound Semiconductor IC Symposium Technical Digest*, Palm Springs, 2005, pp. 173–176

2.291. M. Micovic, A. Kurdhoglian, P. Hashimoto, M. Hu, M. Antcliffe, P. Willadsen, W. Wong, R. Bowen, I. Milosavljevic, A. Schmitz, M. Wetzel, D. Chow, in *IEDM Technical Digest*, San Francisco, 2006 pp. 425–428

2.292. M. Micovic, A. Kurdoghlian, A. Janke, P. Hashimoto, P. Wong, D. Moon, J. McCray, C. Nguyen, IEEE Trans. Electron Devices **48**, 591 (2001)

2.293. J. Milligan, J. Henning, S. Allen, A. Ward, P. Parikh, R. Smith, A. Saxler, Y. Wu, J. Palmour, in *Proceedings of the International Conference on GaAs Manufacturing Technology*, Miami, 2004, pp. 15–18

2.294. J. Milligan, J. Henning, S. Allen, A. Ward, J. Palmour, in *Proceedings of the International Conference on GaAs Manufacturing Technology*, New Orleans, 2005, p. 8.2

2.295. A. Minko, V. Hoel, G. Dambrine, C. Gaquiere, J. DeJaeger, Y. Cordier, F. Semond, F. Natali, J. Massies, H. Lahreche, L. Wedzikowski, R. Langer, P. Bove, in *Proceedings of the European Gallium Arsenide Other Compound Semiconductors Application Symposium GAAS*, Amsterdam, 2004, pp. 67–70

2.296. A. Minko, V. Hoel, S. Lepilliet, G. Dambrine, J. DeJaeger, Y. Cordier, F. Semond, F. Natali, J. Massies, IEEE Electron Device Lett. **25**, 167 (2004)

2.297. A. Minko, V. Hoel, E. Morvan, B. Grimbert, A. Soltani, E. Delos, D. Ducatteau, C. Gaquiére, D. Theron, J.C.D. Jaeger, H. Lahreche, L. Wedzikowski, R. Langer, P. Bove, IEEE Electron Device Lett. **25**, 453 (2004)

2.298. J. Miragliotta, in *Properties of Group III Nitrides*, No. 11 in EMIS Data reviews Series, ed. by J. Edgar (IEE INSPEC, London, 1994), Sect. 6.3, pp. 190–194

2.299. J. Miragliotta, in *Properties of Group III Nitrides*, No. 11 in EMIS Data reviews Series, ed. by J. Edgar (IEE INSPEC, London, 1994), Sect. 6.4, pp. 195–199

2.300. O. Mishima, K. Era, J. Tanaka, S. Yamaoka, Appl. Phys. Lett. **53**, 962 (1988)

2.301. U. Mishra, P. Parikh, Y. Wu, Proc. IEEE **90**, 1022 (2002)

2.302. U. Mishra, Y. Wu, B. Keller, S. Keller, S. Denbaars, IEEE Trans. Microwave Theory Tech. **46**, 756 (1998)

2.303. D. Mistele, O. Katz, A. Horn, G. Bahir, J. Salzmann, Semicond. Sci. Technol. **20**, 972 (2005)

2.304. E. Mitani, M. Aojima, A. Maekawa, S. Sano, in *Proceedings of the International Conference on GaAs Manufacturing Technology*, Austin, 2007, pp. 213–216

2.305. E. Mitani, M. Aojima, S. Sano, in *Proceedings of the European Microwave Integrated Circuits Conference*, Munich, 2007, pp. 176–179

2.306. M. Miyazaki, S. Miyazaki, Y. Yanase, T. Ochiai, T. Shigematu, Jpn. J. Appl. Phys. **34**, 6303 (1995)

2.307. K. Mkhoyan, J. Silcox, E. Alldredge, N. Ashcroft, H. Lu, W. Schaff, L. Eastman, Appl. Phys. Lett. **82**, 1407 (2003)

2.308. T. Mnatsakanov, M. Levinshtein, L. Pomortseva, S. Yurkov, G. Simin, M. Khan, Solid-State Electron. **47**, 111 (2003)

2.309. Y. Mochida, T. Takano, H. Gambe, in *IEDM Technical Digest*, Washington DC, 2001, pp. 14–21

2.310. P. Moens, F. Bauwens, J. Baele, K. Vershinin, E.D. Backer, E.S. Narayanan, M. Tack, in *IEDM Technical Digest*, San Francisco, 2006, pp. 919–922

2.311. C. Monier, F. Ren, J. Han, P. Chang, R. Shul, K. Lee, A. Zhang, A. Baca, S. Pearton, IEEE Trans. Electron Devices **48**, 427 (2001)

2.312. J. Moon, M. Micovic, P. Janke, P. Hashimoto, W. Wong, R. Widman, L. McCray, A. Kurdoghlian, C. Nguyen, Electron. Lett. **37**, 528 (2001)

2.313. J. Moon, D. Wong, M. Antcliffe, P. Hashimoto, M. Hu, P. Willadsen, M. Micovic, H. Moyer, A. Kurdhoglian, P. MacDonald, M. Wetzel, R. Bowen, in *IEDM Technical Digest*, San Francisco, 2006, pp. 423–424

2.314. K. Mori, Y. Sakai, S. Tsuji, H. Asao, K. Seino, H. Hirose, T. Takagi, in *IEEE International Microwave Symposium Digest*, Fort Worth, 2004, pp. 1661–1664

2.315. H. Morkoc, *Nitride Semiconductors and Devices*, No. 32 in Springer Series in Materials Science (Springer, Berlin Heidelerg New York, 1999)

2.316. S. Müller, J. Sumakeris, M. Brady, R. Glass, H. Hobgood, J. Jenny, R. Leonard, D. Malta, M. Paisley, A. Powell, V. Powell, V. Tsetkov, S. Allen, M. Das, J. Palmour, C. Carter, Eur. Phys. J. Appl. Phys. **27**, 29 (2004)

2.317. J. Muth, J. Lee, I. Shmagin, R. Kolbas, H. Casey, B. Keller, U. Mishra, S. DenBaars, Appl. Phys. Lett. **71**, 2572 (1997)

2.318. M. Nagahara, T. Kikkawa, N. Adachi, Y. Tateno, S. Kato, M. Yokoyama, S. Yokogama, T. Kimura, Y. Yamaguchi, N. Hara, K. Joshin, in *IEDM Technical Digest*, San Francisco, 2002, pp. 693–696

2.319. W. Nagy, S. Singhal, R. Borges, W. Johnson, J. Brown, R. Therrien, A. Chaudhari, A. Hanson, J. Riddle, S. Booth, P. Rajagopal, E. Piner, K. Linthicum, in *IEEE International Microwave Symposium Digest*, Long Beach, 2005, pp. 483–486

2.320. D. Nakamura, I. Gunjishima, S. Yamaguchi, T. Ito, A. Okamoto, H. Kondo, S. Onda, K. Takatori, Nature **430**, 1009 (2004)

2.321. N. Nakamura, MRS Bulletin **23**, 37 (1998)

2.322. S. Nakamura, Diamond Relat. Mater. **5**, 496 (1996)

2.323. S. Nakamura, S. Chichibu (eds.), *Introduction to Nitride Semiconductor Blue Lasers and Light Emitting Diodes* (Taylor & Francis, London New York, 2000)

2.324. S. Nakamura, G. Fasol, *The Blue Laser Diode* (Springer, Berlin Heidelberg New York, 1997)

2.325. S. Nakamura, M. Senoh, T. Mukai, Appl. Phys. Lett. **62**, 2390 (1993)

2.326. S. Nakamura, M. Senoh, S. Nagahama, N. Iwasa, T. Yamada, T. Matsushita, H. Kiyoku, Y. Sugimoto, Appl. Phys. Lett. **68**, 3269 (1996)

2.327. M. Neuburger, I. Daumiller, T. Zimmermann, M. Kunze, G. Koley, M. Spencer, A. Dadgar, A. Krtschil, A. Krost, E. Kohn, Electron. Lett. **39**, 1614 (2003)

2.328. J. Newey, Comp. Semicond. **8**, (2002)

2.329. H.M. Ng, D. Doppalapudi, T.D. Moustakas, N.G. Weimann, L.F. Eastman, Appl. Phys. Lett., **73**, 821, (1998)

2.330. J. Nikaido, T. Kikkawa, S. Yokokawa, Y. Tateno, in *Proceedings of the International Conference on GaAs Manufacturing Technology*, New Orleans, 2005, p. 8.1

2.331. O. Nilsson, H.H. Mehling, R. Horn, J. Fricke, R. Hofmann, S. Müller, R. Eckstein, D. Hofmann, High Temp. High Press. **29**, 73 (1997)

2.332. M. Nishijima, T. Murata, Y. Hirose, M. Hikita, N. Negoro, H. Sakai, Y. Uemoto, K. Inoue, T. Tanaka, D. Ueda, in *IEEE International Microwave Symposium Digest*, Long Beach, 2005, pp. 299–302

2.333. Nitronex, Nitronex Issued Patent, Demos 120 W Device (2003), http://compoundsemiconductor.net/articles/news/7/10/1/1

2.334. T. Nomura, H. Kambayashi, M. Masuda, S. Ishii, N. Ikeda, J. Lee, S. Yoshida, IEEE Trans. Electron Devices **53**, 2908 (2006)

2.335. Northrop Grumman Space Technology, Data sheet: ALH444 1–12 GHz Low Noise Amplifier (2005), http://www.velocium.com

2.336. S. Nuttinck, S. Pinel, E. Gebara, J. Laskar, M. Harris, in *Proceedings of the European Gallium Arsenide Other Compound Semiconductors Application Symposium GAAS*, Munich, 2003, pp. 213–215

2.337. R. Oberhuber, G. Zandler, P. Vogl, Appl. Phys. Lett. **73**, 818 (1998)

2.338. G. O'Clock, M. Duffy, Appl. Phys. Lett. **23**, 55 (1973)

2.339. I. Oguzman, E. Bellotti, K. Brennan, J. Kolnik, R. Wang, P. Ruden, J. Appl. Phys. **81**, 7827 (1997)

2.340. I. Oguzman, K. Brennan, J. Kolnik, R. Wang, P. Ruden, in *MRS Symposium*, vol. 395, 1st International Conference on Nitride Semiconductor, Boston, 1996, pp. 733–738

2.341. Y. Ohno, M. Kuzuhara, IEEE Trans. Electron Devices **48**, 517 (2001)

2.342. K. Okamoto, H. Ohta, S. Chichibu, J. Ichihara, H. Takasu, Jpn. J. Appl. Phys. **46**, L187 (2007)

2.343. Y. Okamoto, Y. Ando, K. Hataya, H. Miyamoto, T. Nakayama, T. Inoue, M. Kuzuhara, Electron. Lett. **39**, 1474 (2003)

2.344. Y. Okamoto, Y. Ando, K. Hataya, T. Nakayama, H. Miyamoto, T. Inoue, M. Senda, K. Hirata, M. Kosaki, N. Shibata, M. Kuzuhura, in *IEEE International Microwave Symposium Digest*, Fort Worth, 2004 pp. 1351–1354

2.345. Y. Okamoto, Y. Ando, H. Miyamoto, T. Nakayama, T. Inoue, M. Kuzuhara, in *IEEE International Microwave Symposium Digest*, Philadelphia, 2003, pp. 225–228

2.346. Y. Okamoto, A. Wakejima, Y. Ando, T. Nakayama, K. Matsunaga, H. Miyamoto, Electron. Lett. **42**, 283 (2006)

2.347. Y. Okamoto, A. Wakejima, K. Matsunaga, Y. Ando, T. Nakayama, K. Kasahara, K. Ota, Y. Murase, K. Yamanoguchi, T. Inoue, H. Miyamoto, in *IEEE International Microwave Symposium Digest*, Long Beach, 2005, pp. 491–494

2.348. S. O'Leary, B. Foutz, M. Shur, U. Bhapkar, L. Eastman, J. Appl. Phys. **83**, 826 (1998)

2.349. S. O'Leary, B. Foutz, M. Shur, U. Bhapkar, L. Eastman, Solid-State Commun. **105**, 621 (1998)

2.350. J. Orton, C. Foxon, in *Properties, Processing and Applications of GaN Nitride and Related Semiconductors*, No. 23 in EMIS Data reviews Series, ed. by J. Edgar et al. (IEE INSPEC, London, 1999), Sect. 8.5, pp. 300–305

2.351. J. Orton, C. Foxon, in *Properties, Processing and Applications of GaN Nitride and Related Semiconductors*, No. 23 in EMIS Data reviews Series, ed. by J. Edgar et al. (IEE INSPEC, London, 1999), Sect. 8.4, pp. 294–299

2.352. K. Osamura, S. Naka, Y. Murakami, J. Appl. Phys. **46**, 3432 (1975)

2.353. C. Oxley, Solid-State Electron. **48**, 1197 (2004)

2.354. T. Palacios, A. Chakraborty, S. Rajan, C. Poblenz, S. Keller, S. DenBaars, J. Speck, U. Mishra, IEEE Electron Device Lett. **26**, 781 (2005)

2.355. T. Palacios, N. Fichtenbaum, S. Keller, S. Denbaars, U. Mishra, in *Device Research Conference*, State College, PA, 2006, pp. 99–100

2.356. V. Palankovski, R. Quay, *Analysis and Simulation of Heterostructure Devices* (Springer, Wien New York, 2004)

2.357. V. Palankovski, S. Selberherr, in *Proceedings of the European Conference on High Temperature Electronics*, Berlin, 1999, pp. 25–28

2.358. V. Palankovski, S. Vitanov, R. Quay, in *Compound Semiconductor IC Symposium Technical Digest*, San Antonio, 2006, pp. 107–110

2.359. J. Palmour, in *Compound Semiconductor IC Symposium Technical Digest*, San Antonio, 2006, pp. 4–7

2.360. J. Palmour, A. Agarwal, S. Ryu, M. Das, J. Sumakeris, A. Powell, in *Proceedings of the International Conference on GaAs Manufacturing Technology*, New Orleans, 2005, p. 13.1

2.361. J. Palmour, J. Milligan, J. Henning, S. Allen, A. Ward, P. Parikh, R. Smith, A. Saxler, M. Moore, Y. Wu, in *Proceedings of the European Gallium Arsenide Other Compound Semiconductors Application Symposium GAAS*, Amsterdam, 2004, pp. 555–558

2.362. J. Palmour, S. Sheppard, R. Smith, S. Allen, W. Pribble, T. Smith, Z. Ring, J. Sumakeris, A. Saxler, J. Milligan, in *IEDM Technical Digest*, Washington DC, 2001, pp. 385–388

2.363. A. Panda, D. Pavlidis, E. Alekseev, IEEE Trans. Electron Devices **48**, 820 (2001)

2.364. P. Parikh, Y. Wu, M. Moore, P. Chavarkar, U. Mishra, B. Neidhard, L. Kehias, T. Jenkins, in *Lester Eastman Conference on Abstract Book*, Newark, 2002, pp. 56–57

2.365. S. Pearton, C. Abernathy, B. Gila, F. Ren, J. Zavada, Y. Park, Solid-State Electron. **48**, 1965 (2002)

2.366. R. Pierobon, S. Buso, M. Citron, G. Meneghesso, G. Spiazzi, E. Zanoni, in *11th Hetero Structure Technical Work*, Padova, 2001, pp. 57–58

2.367. E. Piner, S. Singhal, P. Rajagopal, R. Therrien, J. Roberts, T. Li, A. Hanson, J. Johnson, I. Kizilyalli, K. Linthicum, in *IEDM Technical Digest*, San Francisco, 2006, pp. 411–414

2.368. A. Ping, Q. Chen, J. Yang, M. Khan, I. Adesida, IEEE Electron Device Lett. **19**, 54 (1998)

2.369. J. Piprek (ed.), *Nitride Semiconductor Devices: Principles and Simulation* (Wiley-VCH, Weinheim, 2007)

2.370. K. Ploog, O. Brandt, J. Vac. Sci. Technol. A **16**, 1609 (1998)

2.371. A. Polian, in *Properties, Processing and Applications of GaN Nitride and Related Semiconductors*, No. 23 in EMIS Data reviews Series, ed. by J. Edgar et al. (IEE INSPEC, London, 1999), Sect. 1.3, pp. 11–20

2.372. A. Polian, M. Grimsditch, I. Grzegory, J. Appl. Phys. **79**, 3343 (1996)

2.373. S. Pugh, D. Dugdale, S. Brand, R. Abram, J. Appl. Phys. **86**, 3768 (1999)

2.374. D. Pulfrey, S. Fathpour, IEEE Trans. Electron Devices **48**, 597 (2001)

2.375. R. Quay, K. Hess, R. Reuter, M. Schlechtweg, T. Grave, V. Palankovski, S. Selberherr, IEEE Trans. Electron Devices **48**, 210 (2001)

2.376. R. Quay, R. Kiefer, F. van Raay, H. Massler, S. Ramberger, S. Müller, M. Dammann, M. Mikulla, M. Schlechtweg, G. Weimann, in *IEDM Technical Digest*, San Francisco, 2002, pp. 673–676

2.377. R. Quay, A. Tessmann, R. Kiefer, R. Weber, F. van Raay, M. Kuri, M. Riessle, H. Massler, S. Müller, M. Schlechtweg, G. Weimann, in *IEDM Technical Digest*, Washington DC, 2003, pp. 567–570

2.378. R. Quay, F. van Raay, A. Tessmann, R. Kiefer, M. Dammann, M. Mikulla, M. Schlechtweg, G. Weimann, in *Proceedings WOCSDICE*, Venice, 2007, pp. 349–352

2.379. S. Rajan, H. Xing, S. DenBaars, U. Mishra, D. Jena, Appl. Phys. Lett. **84** 1591 (2004)

2.380. M. Ramonas, A. Matulionis, L. Rota, Semicond. Sci. Technol. **18**, 219 (2003)

2.381. R. Reeber, K. Wang, MRS Internet J. Nitride Semicond. Res. **6**, 3 (2001)

2.382. F. Ren, C. Abernathy, J.V. Hove, P. Chow, R. Hickman, J. Klaasen, R. Kopf, H. Cho, K. Jung, J.L. Roche, R. Wilson, J. Han, R. Shul, A. Baca, S. Pearton, MRS Internet J. Nitride Semicond. Res. **3**, 41 (1998)

2.383. F. Ren, J. Han, R. Hickman, J.V. Hove, P. Chow, J. Klaasen, J. LaRoche, K. Jung, H. Cho, X. Cao, S. Donovan, R. Kopf, R. Wilson, A. Baca, R. Shul, L. Zhang, C. Willison, C. Abernathy, S. Pearton, Solid-State Electron. **44**, 239 (2000)

2.384. M. Rosker, in *Compound Semiconductor IC Symposium Technical Digest*, Palm Springs, 2005, pp. 13–16

2.385. M. Rosker, in *Proceedings of the International Conference on GaAs Manufacturing Technology*, New Orleans, 2005, p. 1.2

2.386. M. Rosker, in *IEEE RF IC Symposium Digest*, Honolulu, 2007, pp. 159–162

2.387. P. Roussel, Comp. Semicond. **9**, 20 (2003)

2.388. T. Ruemenapp, D. Peier, in *11th International Symposium on High-Voltage Engineering*, London, 1999, pp. 4.373–4.376

2.389. P. Ruterana, M. Albrecht, J. Neugebauer (eds.), *Nitride Semiconductors: Handbook on Materials and Device* (Wiley-VCH, Weinheim, 2003)

2.390. P. Ruterana, A. Sanchez, G. Nouet, in *Nitride Semiconductors: Handbook on Materials and Device*, ed. by P. Ruterana, M. Albrecht, J. Neugebauer (Wiley-VCH, Weinheim, 2003), Chap. 8, pp. 379–438

2.391. F. Sacconi, A.D. Carlo, H. Morkoc, IEEE Trans. Electron Devices **48**, 450 (2001)

2.392. F. Sacconi, A. Di Carlo, F. Della Sala, P. Lugli, in *Proceedings of the European GaAs Related Compounds on Application Symposium GAAS*, Paris, 2000, pp. 620–623

2.393. T. Sadi, R. Kelsall, N. Pilgrim, IEEE Trans. Electron Devices **53**, 2892 (2006)

2.394. W. Saito, M. Kuraguchi, Y. Takada, K. Tsuda, T. Domon, I. Omura, M. Yamaguchi, in *IEDM Technical Digest*, Washington DC, 2005, pp. 586–589

2.395. W. Saito, I. Omura, T. Domon, K. Tsuda, in *Compound Semiconductor IC Symposium Technical Digest*, San Antonio, 2006, pp. 253–256

2.396. W. Saito, I. Omura, T. Ogura, H. Ohashi, Solid-State Electron. **48**, 1555 (2004)

2.397. W. Saito, Y. Takada, M. Kuraguchi, K. Tsuda, I. Omura, T. Ogura, in *IEDM Technical Digest*, Washington DC, 2003, pp. 587–590

2.398. W. Saito, M. Kuraguchi, Y. Takada, K. Tsuda, Y. Saito, I. Omura, M. Yamaguchi, in *IEDM Technical Digest*, Washington DC, 2007, pp. 869–873

2.399. R. Sandhu, M. Wojtowicz, I. Smorchkova, M. Barsky, R. Tsai, J. Yang, H. Wang, M. Khan, in *Device Research Conference*, Santa Barbara, 2002, pp. 27–28

2.400. B. Santic, Semicond. Sci. Technol. **18**, 219 (2003)

2.401. P. Saunier, in *Proceedings of European Gallium Arsenide Other Compound Semiconductors Application Symposium GAAS*, Amsterdam, 2004, pp. 543–546

2.402. J. Schalwig, G. Müller, M. Eickhoff, O. Ambacher, M. Stutzmann, Sensors Actuators B **87**, 425 (2002)

2.403. L. Schowalter, S. Schujman, W. Liu, M. Goorsky, M. Wood, J. Granusky, F. Shahedipour-Sandvik, Phys. Stat. Sol. A **203**, 1667 (2006)

2.404. L. Schowalter, G. Slack, J. Whitlock, K. Morgan, S. Schujman, B. Raghothamachar, M. Dudley, K. Evans, Phys. Stat. Sol. C **0**, 1997 (2003)

2.405. M. Seelmann-Eggebert, P. Meisen, F. Schaudel, P. Koidl, A. Vescan, H. Leier, in *Proceedings Diamond*, Porto, 2000, pp. 744–749

2.406. S. Selberherr, *Analysis and Simulation of Semiconductor Devices* (Springer, Wien New York, 1984)

2.407. Sensor Electronic Technologies, SET Wins Contract for Device R&D on AlN. Comp. Semicond. **8**, (2004)

2.408. C. Sevik, C. Bulutay, Semicond. Sci. Technol. **19**, S188 (2004)

2.409. P. Shah, D. Smith, T. Griffin, K. Jones, S. Sheppard, IEEE Trans. Electron Devices **47**, 308 (2000)

2.410. W. Shan, J. Ager, K. Yu, W. Walukiewicz, E. Haller, M. Martin, W. McKinney, W. Yang, J. Appl. Phys. **85**, 8505 (1999)

2.411. A. Shanware, J. McPherson, M. Visokay, J. Chambers, A. Rotondaro, H. Bu, M. Bevan, R. Khamankar, L. Colombo, in *IEDM Technical Digest*, Washington DC, 2001, pp. 137–140

2.412. J. Shealy, J. Smart, M. Poulton, R. Sadler, D. Grider, S. Gibb, B. Hosse, B. Sousa, D. Halchin, V. Steel, P. Garber, P. Wilkerson, B. Zaroff, J. Dick, T. Mercier, J. Bonaker, M. Hamilton, C. Greer, M. Isenhour, in *GaAs IC Symposium Technical Digest*, Monterey, 2002, pp. 243–246

2.413. B. Shelton, D. Lambert, J. Huang, M. Wong, U. Chowdhury, T. Zhu, H. Kwon, Z. Weber, M. Benarama, M. Feng, R. Dupuis, IEEE Trans. Electron Devices **48**, 490 (2001)

2.414. L. Shen, S. Heikman, B. Moran, R. Coffie, N. Zhang, D. Buttari, I. Smorchkova, S. Keller, S. DenBaars, U. Mishra, IEEE Electron Device Lett. **22**, 457 (2001)

2.415. K. Shenai, R. Scott, B. Baliga, IEEE Trans. Electron Devices **36**, 1811 (1989)

2.416. S. Sheppard, K. Doverspike, W. Pribble, S. Allen, J. Palmour, L. Kehias, T. Jenkins, IEEE Electron Device Lett. **20**, 161 (1999)

2.417. J. Sheu et al., J. Phys. Condens. Mater. **14**, R657 (2002)

2.418. M. Shin, R. Trew, Electron. Lett. **31**, 498 (1995)

2.419. M. Shur, *GaAs Devices and Circuits* (Plenum, New York, 1987)

2.420. M. Shur, A. Bykhovski, R. Gaska, Solid-State Electron. **44**, 205 (2000)

2.421. M. Shur, R. Gaska, in *Compound Semiconductor IC Symposium Technical Digest*, Palm Springs, 2005, pp. 137–140

2.422. M. Shur, M. Khan, MRS Bulletin **22**, 44 (1997)

2.423. E. Sichel, J. Pankove, J. Phys. Chem. Solids **38**, 333 (1977)

2.424. SiCrystal AG, Silicon Carbide Product Specification (2004), http://www.sicrystal.com

2.425. G. Simin, X. Hu, A. Tarakji, J. Zhang, A. Koudymov, S. Saygi, J. Yang, A. Khan, M. Shur, R. Gaska, Appl. Phys. Lett. **40**, L1142 (2001)

2.426. G. Simin, A. Koudymov, H. Fatima, J. Zhang, J. Yang, M. Khan, X., A. Tarakji, R. Gaska, M. Shur, IEEE Electron Device Lett. **23**, 458 (2002)

2.427. R. Singh, J. Cooper, M. Melloch, T. Chow, J. Palmour, IEEE Trans. Electron Devices **49**, 665 (2002)

2.428. G. Slack, R. Tanzilli, R. Pohl, J. Vandersande, J. Phys. Chem. Solids **48**, 641 (1987)

2.429. L. Smith, R. Davis, in *Properties of Group III Nitrides*, No. 11 in EMIS Data reviews Series, ed. by J. Edgar (IEE INSPEC, London, 1994), Sect. 10.3, pp. 288–292

2.430. I. Smorchkova, M. Wojtowicz, R. Sandhu, R. Tsai, M. Barsky, C. Namba, P. Liu, R. Dia, M. Truong, D. Ko, J. Wang, A. Khan, IEEE Trans. Microwave Theory and Tech. **51**, 665 (2003)

2.431. I. Smorchkova, M. Wojtowicz, R. Sandhu, R. Tsai, M. Barsky, C. Namba, P. Liu, R. Dia, M. Truong, D. Ko, J. Wang, A. Khan, in *IEDM Technical Digest*, Washington DC, 2001, pp. 17.5.1–17.5.3

2.432. J. Song, D. Leem, S. Kim, J. Kwak, O. Nam, Y. Park, T. Seong, Solid-State Electron. **48**, 1597 (2004)

2.433. M. Spencer, J. Palmour, C. Carter, IEEE Trans. Electron Devices **49**, 940 (2002)

2.434. G. Spiazzi, S. Buso, M. Citron, M. Corradin, R. Pierobon, IEEE Trans. Power Electron. **18**, 1249 (2003)

2.435. D. Storm, D. Katzer, S. Binari, B. Shanabrook, X. Xu, D. McVey, R. Vaudo, G. Brandes, Electron. Lett. **40**, 1226 (2004)

2.436. D. Streit, A. Guitierrez-Aitken, M. Wojtowicz, R. Lai, in *Compound Semiconductor IC Symposium Technical Digest*, Palm Springs, 2005, pp. 5–8

2.437. S. Strite, in *Properties of Group III Nitrides*, No. 11 in EMIS Data reviews Series, ed. by J. Edgar (IEE INSPEC, London, 1994), Sect. 9.5, pp. 272–275

2.438. C. Suh, Y. Dora, N. Fichtenbaum, L. McCarthy, S. Keller, U. Mishra, in *IEDM Technical Digest*, San Francisco, 2006, pp. 911–913

2.439. G. Sullivan, M. Chen, J. Higgins, J. Yang, Q. Chen, R. Pierson, B. McDermott, IEEE Electron Device Lett. **19**, 198 (1998)

2.440. K. Sundaram, M. Deen, W. Brown, R. Sah, E. Poindexter, D. Misra (ed.), *Silicon Nitride and Silicon Dioxide Thin Insulating Films* (Electrochemical Society, 1999)

2.441. M. Suzuki, T. Uenoyama, A. Yanase, Phys. Rev. B **52**, 8132 (1995)

2.442. M. Suzuki, T. Uenoyama, A. Yanase, Phys. Rev. B **58**, 10064 (1998)

2.443. S. Sze, *Physics of Semiconductor Devices*, 2nd edn. (Wiley, New York, 1981)

2.444. K. Takagi, K. Masuda, Y. Kashiwabara, H. Sakurai, K. Matsushita, S. Takatsuka, H. Kawasaki, Y. Takada, K. Tsuda, in *Compound Semiconductor IC Symposium Technical Digest*, San Antonio, 2006, pp. 265–268

2.445. H. Takaya, K. Miyagi, K. Hamada, in *IEDM Technical Digest*, San Francisco, 2006, pp. 923–927

2.446. T. Tanaka, Y. Koji, T. Meguro, Y. Otoki, in *Proceedings of the International Conference on GaAs Manufacturing Technology*, Miami, 2004, pp. 295–298

2.447. O. Tang, K. Duh, S. Liu, P. Smith, W. Kopp, T. Rogers, D. Richard, in *GaAs IC Symposium Technical Digest*, Orlando, 1996, pp. 115–118

2.448. T. Taniguchi, J. Tanaka, O. Mishima, T. Ohsawa, S. Yamaoka, Appl. Phys. Lett. **62**, 576 (1993)

2.449. T. Tansley, in *Properties of Group III Nitrides*, No. 11 in EMIS Data reviews Series, ed. by J. Edgar (IEE INSPEC, London, 1994), Sect. 1.5, pp. 35–40

2.450. T. Tansley, C. Foley, J. Appl. Phys. **59**, 3241 (1986)

2.451. T. Tansley, E. Goldys, in *Properties, Processing and Applications of GaN Nitride and Related Semiconductors*, No. 23 in EMIS Data reviews Series, ed. by J. Edgar et al. (IEE INSPEC, London, 1999), pp. 123–128

2.452. T. Tansley, E. Goldys, in *Properties, Processing and Applications of GaN Nitride and Related Semiconductors*, No. 23 in EMIS Data reviews Series, ed. by J. Edgar et al. (IEE INSPEC, London, 1999), Sect. 4.5, pp. 135–136

2.453. T. Tansley, E. Goldys, in *Properties, Processing and Applications of GaN Nitride and Related Semiconductors*, No. 23 in EMIS Data reviews Series, ed. by J. Edgar et al. (IEE INSPEC, London, 1999), Sect. 4.4, pp. 129–134

2.454. F. Temcamani, P. Pouvil, O. Noblanc, C. Brylinski, P. Bannelier, B. Darges, J. Prigent, in *IEEE International Microwave Symposium Digest*, Phoenix, 2001, pp. 641–644

2.455. D. Theron, C. Gaquiere, J. de Jaeger, S. Delage, in *Proceedings of the European Gallium Arsenide Other Compound Semiconductors Application Symposium GAAS*, Amsterdam, 2004, pp. 547–550

2.456. R. Therrien, S. Singhal, W. Nagy, J. Marquart, A. Chaudhari, K. Linthicum, J. Johnson, A. Hanson, J. Riddle, P. Rajagopal, B. Preskenis, O. Zhitova, J. Williamson, I. Kizilyalli, in *IEEE International Microwave Symposium Digest*, San Francisco, 2006, pp. 710–713

2.457. V. Tilak, B. Green, V. Kaper, H. Kim, T. Prunty, J. Smart, J. Shealy, L. Eastman, IEEE Electron Device Lett. **22**, 504 (2001)

2.458. V. Tilak, K. Matocha, P. Sandvik, Phys. Stat. Sol. A **3**, 548 (2006)

2.459. N. Tipirneni, A. Koudymov, V. Adivaharan, J. Yang, G. Simin, M. Khan, IEEE Electron Device Lett. **27**, 716 (2006)

2.460. L. Tolbert, B. Ozpinecci, S. Islam, M. Chinthavali, in *IASTED International Conference on Power and Energy Systems*, Palm Springs, 2003, pp. 317–321

2.461. A. Tomchenko, G. Harmer, B. Marquis, J. Allen, Sensors Actuators B **93**, 126 (2003)

2.462. R. Trew, in *IEEE International Microwave Symposium Digest*, Seattle, 2002, pp. 1811–1814

2.463. R. Trew, Proc. IEEE **90**, 1032 (2002)

2.464. R. Trew, IEEE Trans. Electron Devices **52**, 638 (2005)

2.465. R. Trew, M. Shin, V. Gatto, Solid-State Electron. **41**, 1561 (1997)

2.466. Triquint Semiconductor, Advanced Product Information TGA2505 (2006), http://www.triquint.com

2.467. Triquint Semiconductors, Triquint Semiconductor and Lockheed Martin Announce Advanced Gallium Nitride Process with Improved Power, Efficiency, Stability (2003), http://www.triquint.com/investors/press/dspPressRelease.cfm?pressid=174

2.468. Triquint Semiconductors, 6.5 Watt Ku Band Power Amplifier (2004), http://www.triquint.com

2.469. D. Ueda, T. Murata, M. Hikita, S. Nakazawa, M. Kuroda, H. Ishida, M. Yanagihara, K. Inoue, T. Ueda, Y. Uemoto, T. Tanaka, T. Egawa, in *IEDM Technical Digest*, Washington DC, 2005, pp. 377–380

2.470. H. Ueda, M. Suimoto, T. Uesugi, T. Kachi, in *Proceedings of the International Conference on GaAs Manufacturing Technology*, Vancouver, 2006, pp. 37–41

2.471. Y. Uemoto, M. Hikita, H. Ueno, H. Matsuo, H. Ishida, M. Yanagihara, T. Ueda, T. Tanaka, D. Ueda, in *IEDM Technical Digest*, San Francisco, 2006, pp. 907–910

2.472. Y. Uemoto, D. Shibata, M. Yanagihara, H. Ishida, H. Matsuo, S. Nagai, N. Batta, M. Li, T. Ueda, T. Tanaka, D. Ueda, in *IEDM Technical Digest*, Washington DC, 2007, pp. 861–864

2.473. N. Ui, S. Sano, in *IEEE International Microwave Symposium Digest*, San Francisco, 2006, pp. 718–721

2.474. United Monolithic Semiconductors, Data sheet: 7–13 GHz Low Noise Amplifier (2000), http://www.ums-gaas.com

2.475. C. van de Walle, J. Neugebauer, C. Stampfl, in *Properties, Processing and Applications of GaN Nitride and Related Semiconductors*, No. 23 in EMIS Data reviews Series, ed. by J. Edgar et al. (IEE INSPEC, London, 1999), Sect. 8.1, pp. 275–280

2.476. F. van Raay, R. Quay, R. Kiefer, H. Walcher, O. Kappeler, M. Seelmann-Eggebert, S. Müller, M. Schlechtweg, G. Weimann, in *Proceedings of the European Gallium Arsenide Other Compound Semiconductors Application Symposium GAAS 2005*, Paris, pp 373 - 376

2.477. T. Veal, P. Jefferson, L. Piper, C. McConville, T. Joyce, P. Chalker, L. Considine, H. Lu, W. Schaff, Appl. Phys. Lett. **89**, 202110 (2006)

2.478. N. Vellas, C. Gaquiere, Y. Guhel, M. Werquin, D. Ducatteau, B. Boudart, J. Jaeger, Z. Bougrioua, M. Germain, M. Leys, S. Borghs, in *Proceedings of the European GaAs Related Compound Application Symposium on GAAS*, Milano, 2002, pp. 25–28

2.479. A. Vescan, R. Dietrich, A. Wieszt, A. Schurr, H. Leier, E. Piner, J. Redwing, Electron. Lett. **36**, 1234 (2000)

2.480. R. Vetury, PhD thesis, University of California Santa Barbara, Santa Barbara, 2000

2.481. R. Vetury, J. Shealy, D. Green, J. McKenna, J. Brown, S. Gibb, K. Leverich, P. Garber, M. Poulton, in *IEEE International Microwave Symposium Digest*, San Francisco, 2006, pp. 714–717

2.482. R. Vetury, Y. Wei, D. Green, S. Gibb, T. Mercier, K. Leverich, P. Garber, M. Poulton, J. Shealy, in *IEEE International Microwave Symposium Digest*, Long Beach, 2005, pp. 487–490

2.483. R. Vetury, N. Zhang, S. Keller, U. Mishra, IEEE Trans. Electron Devices **48**, 560 (2001)

2.484. F. Villard, J. Pringent, E. Morvan, C. Dua, C. Brylinski, F. Temcamani, P. Pouvil, IEEE Trans. Microwave Theory Tech. **51**, 1129 (2003)

2.485. I. Vurgaftman, J. Meyer, J. Appl. Phys. **94**, 3675 (2003)

2.486. I. Vurgaftman, J. Meyer, L. Ram-Mohan, J. Appl. Phys. **89**, 5815 (2001)

2.487. A. Wakejima, K. Matsunaga, Y. Okamoto, Y. Ando, T. Nakayama, T. Kasahara, H. Miyamoto, Electron. Lett. **41**, 1004 (2005)

2.488. A. Wakejima, K. Matsunaga, Y. Okamoto, Y. Ando, T. Nakayama, H. Miyamoto, Electron. Lett. **41**, 1371 (2005)

2.489. D. Walker, X. Zhang, A. Saxler, P. Kung, J. Xu, M. Razeghi, Appl. Phys. Lett. **70**, 949 (1997)

2.490. C. Wang, L. Yu, S. Lau, E. Yu, W. Kim, A. Botchkarev, H. Morkoc, Appl. Phys. Lett. **72**, 1211 (1998)

2.491. K. Watanabe, T. Taniguchi, H. Kanda, Phys. Stat. Sol. A **201**, 2561 (2004)

2.492. M. Weber, L. Tirino, K. Brennan, IEEE Trans. Electron Devices **50**, 2202 (2003)

2.493. N. Weimann, L. Eastman, D. Doppalapudi, H. Ng, T. Moustakas, J. Appl. Phys. **83**, 3656 (1998)

2.494. C. Weitzel, in *IEDM Technical Digest*, San Francisco, 1998, pp. 51–54

2.495. C. Weitzel, in *IEEE International Microwave Symposium Digest*, Seattle, 2002, pp. 285–288

2.496. R. Wentorf, J. Chem. Phys. **36**, 1990 (1994)

2.497. R. Wentzcovitch, K. Chang, M. Cohen, Phys. Rev. B **34**, 1071 (1986)

2.498. C. Wetzel, T. Takeuchi, H. Amago, I. Akasaki, Phys. Rev. B **61**, 2159 (2000)

2.499. C. Wood, D. Jena (eds.), *Polarization Effects in Semiconductors* (Springer, New York, 2008)

2.500. M. Wraback, H. Shen, J. Carrano, T. Li, J. Camphell, M. Schurman, I. Ferguson, Appl. Phys. Lett. **76**, 1155 (2000)

2.501. J. Wu, W. Walukiewicz, K. Yu, J. Ager, E. Haller, H. Lu, W. Schaff, Appl. Phys. Lett. **80**, 4741 (2002)

2.502. Y. Wu, Dissertation, University of California Santa Barbara, Santa Barbara, 1997

2.503. Y.-F. Wu, S. Wood, R. Smith, S. Sheppard, S.T. Allen, P. Parikh, J. Milligan, in *IEDM Technical Digest*, San Francisco, 2006, pp. 419–421

2.504. Y. Wu, D. Kapolnek, J. Ibbetson, P. Parikh, B. Keller, U. Mishra, IEEE Trans. Electron Devices **48**, 586 (2001)

2.505. Y. Wu, B. Keller, P. Fini, J. Pusl, M. Le, N. Nguyen, C. Nguyen, D. Widman, S. Keller, S. Denbaars, U. Mishra, Electron. Lett. **33**, 1742 (1997)

2.506. Y. Wu, B. Keller, S. Keller, D. Kapolnek, S. Denbaars, U. Mishra, IEEE Electron Device Lett. **17**, 455 (1996)

2.507. Y. Wu, B. Keller, S. Keller, N. Nguyen, M. Le, C. Nguyen, T. Jenkins, L. Kehias, S. Denbaars, U. Mishra, IEEE Electron Device Lett. **18**, 438 (1997)

2.508. Y. Wu, M. Moore, A. Saxler, T. Wisleder, U.K. Mishra, P. Parikh, in *IEDM Technical Digest*, Washington DC, 2005, pp. 583–585

2.509. Y. Wu, M. Moore, A. Abrahamsen, M. Jacob-Mitos, P. Parikh, S. Heikman, A. Burk, in *IEDM Technical Digest*, Washington DC, 2007, pp. 405–408

2.510. Y. Wu, M. Moore, A. Saxler, P. Smith, P. Chavarkar, P. Parikh, in *IEDM Technical Digest*, Washington DC, 2003, pp. 579–581

2.511. Y. Wu, M. Moore, A. Saxler, T. Wisleder, P. Parikh, in *Device Research Conference*, State College, PA, 2006, pp. 151–152

2.512. Y. Wu, A. Saxler, M. Moore, R. Smith, S. Sheppard, P. Chavarkar, T. Wisleder, U. Mishra, P. Parikh, IEEE Electron Device Lett. **25**, 117 (2004)

2.513. H. Xing, P. Chavarkar, S. Keller, S. DenBaars, U. Mishra, IEEE Electron Device Lett. **24**, 141 (2003)

2.514. H. Xing, D. Jena, M. Rodwell, U. Mishra, IEEE Electron Device Lett. **24**, 4 (2003)

2.515. H. Xu, C. Sanabria, A. Chini, S. Keller, U. Mishra, R. York, IEEE Microw. Wireless Compon. Lett. **14**, 262 (2004)

2.516. I. Yaacov, Y. Seck, U. Mishra, S. DenBaars, J. Appl. Phys. **95**, 2073 (2004)

2.517. S. Yamakawa, S. Aboud, M. Saraniti, S. Goodnick, Semicond. Sci. Technol. **19**, S475 (2004)

2.518. T. Yamamoto, E. Mitani, K. Inoue, M. Nishi, S. Sano, in *Proceedings of the European Microwave Integrated Circuits Conference* Munich, 2007, pp 173–175

2.519. K. Yamanaka, K. Iyomasa, H. Ohtsuka, M. Nakayama, Y. Tsuyama, T. Kunii, Y. Kano, T. Takagi, in *Compound Semiconductor IC Symposium Technical Digest*, San Antonio, 2006, pp. 241–244

2.520. K. Yamanaka, K. Mori, K. Iyomasa, H. Ohtsuka, M. Nakayama, Y. Kamo, Y. Isota, in *IEEE International Microwave Symposium Digest*, Honolulu, 2007, pp. 1251–1254

2.521. H. Ye, G. Wicks, P. Fauchet, Appl. Phys. Lett. **74**, 711 (1999)

2.522. Y. Yeo, T. Chong, M. Li, J. Appl. Phys. **83**, 1429 (1996)

2.523. Y. Yeo, T. Chong, M. Li, W. Fan, J. Appl. Phys. **84**, 1813 (1998)

2.524. I. Yonenaga, MRS Internet J. Nitride Semicond. Res. **7**, 6 (2002)

2.525. E. Yu, G. Sullivan, P. Asbeck, C. Wang, D. Qiao, S. Lau, Appl. Phys. Lett. **71**, 2794 (1997)

2.526. H. Yu, L. McCarthy, S. Rajan, S. Keller, S. DenBaars, J. Speck, U. Mishra, IEEE Electron Device Lett. **26**, 283 (2005)

2.527. T. Yu, K. Brennan, IEEE Trans. Electron Devices **50**, 315 (2003)

2.528. A. Zhang, L. Rowland, E. Kaminsky, J. Kretchmer, V. Tilak, A. Allen, B. Edward, in *IEEE International Microwave Symposium Digest*, Philadelphia, 2003, pp. 251–254

2.529. A. Zhang, L. Rowland, E. Kaminsky, J. Tucker, R. Beaupre, J. Kretchmer, J. Garrett, A. Vertiatchikh, G. Koley, H. Cha, A. Allen, J. Cook, J. Foppes, B. Edward, J. Electron. Mater.**32** 437 (2003)

2.530. H. Zhang, E. Miller, E. Yu, C. Poblenz, J. Speck, Appl. Phys. Lett. **84**, 4644 (2004)

2.531. L. Zhang, L. Lester, A. Baca, R. Shul, P. Chang, C. Wilson, U. Mishra, S. DenBaars, J. Zolper, IEEE Trans. Electron Devices **47**, 507 (2000)

2.532. F. Zhao, I. Perez-Wurfl, H. Chih-Fang, J. Torvik, B. Zeghbroek, in *IEEE International Microwave Symposium Digest*, Long Beach, 2005, pp. 2035–2038

2.533. B. Zhou, K. Butcher, X. Li, T. Tansley, Solid-State Electron. **41**, (1997)

2.534. T. Zimmermann, M. Neuburger, M. Kunze, I. Daumiller, A. Denisenko, A. Dadgar, A. Krost, E. Kohn, IEEE Electron Device Lett. **25**, 450 (2004)

2.535. J. Zolper, Wide Bandgap Semiconductor RF Electronics Technology, MTO Industry Briefing, Sept. 2001
http://www.compoundsemiconductor.net/cws/article/magazine/11332

References of Chapter 3

3.1. A. Alam, O. Schön, B. Schineller, M. Heuken, H. Jürgensen, Phys. Stat. Sol. A **180**, 109 (2000)

3.2. O. Ambacher J. Physics D **31**, 2653 (1998)

3.3. B. Ansell, L. Harrison, C. Foxon, J. Harris, T. Cheng, Electron. Lett. **36**, 1237 (2000)

3.4. J. Antoszewski, M. Gracey, J. Dell, L. Farone, T. Fisher, G. Parish, Y. Wu, U. Mishra, J. Appl. Phys. **87**, 3900 (2000)

3.5. J. Ao, T. Wang, D. Kikuta, Y. Liu, S. Sakai, Y. Ohno, Jpn. J. Appl. Phys. **42**, 1588 (2003)

3.6. A. Armstrong, A. Arehart, B. Moran, S. DenBaars, U. Mishra, J. Speck, S. Ringel, Appl. Phys. Lett. **84**, 374 (2004)

3.7. S. Arulkumaran, M. Miyoschi, T. Egawa, H. Ishikawa, T. Jimbo, IEEE Electron Device Lett. **24**, 497 (2003)

3.8. S. Arulkumaran, M. Sakai, T. Egawa, H. Ishikawa, T. Jimbo, T. Shibata, K. Asai, S. Sumiya, Y. Kuraoka, M. Tanaka, O. Oda, Appl. Phys. Lett. **81**, 1131 (2002)

3.9. R. Balmer, K. Hilton, K. Nash, M. Uren, D. Wallis, D. Lee, A. Wells, M. Missous, T. Martin, Semicond. Sci. Technol. **19**, L65 (2004)

3.10. J. Bardwell, Y. Liu, H. Tang, J. Webb, S. Rolfe, J. Lapointe, Electron. Lett. **39**, 564 (2003)

3.11. A. Barnes, D. Hayes, M. Uren, T. Martin, R. Balmer, D. Wallis, K. Hilton, J. Powell, W. Phillips, A. Jimenez, E. Munoz, M. Kuball, S. Rajasingam, J. Pomeroy, N. Labat, N. Malbert, P. Rice, A. Wells, in *IMS Workshop Advances in GaN-based Device and Circuit Technology: Modeling and Applications*, Fort Worth, 2004

3.12. J. Bernat, R. Pierobon, M. Marso, J. Flynn, G. Brandes, G. Meneghesso, E. Zanoni, P. Kordos, Phys. Stat. Sol. C **2**, 2676 (2005)

3.13. J. Bernat, M. Wolter, M. Marso, J. Flynn, G. Brandes, P. Kordos, Electron. Lett. **40**, 78 (2004)

3.14. A. Bhuiyan, A. Hashimoto, A. Yamamoto, J. Appl. Phys. **94**, 2779 (2003)

3.15. Z. Bougrioua, M. Azize, P. Lorenzini, M. Laügt, H. Haas, Phys. Stat. Sol. A **202**, 536 (2005)

3.16. P. Bove, H. Lahreche, J. Thuret, F. Letertre, B. Faure, in *Proceedings of the International Conference on GaAs Manufacturing Technology*, New Orleans, 2005, p. 4.3

3.17. O. Brandt, R. Muralidharan, P. Waltereit, A. Thamm, A. Trambert, H. von Kiedrowski, K. Ploog, Appl. Phys. Lett. **75**, 4019 (1999)

3.18. E. Brazel, M. Chain, V. Narayanamurti, Appl. Phys. Lett. **74**, 2367 (1999)

3.19. C. Buchheim, R. Goldhahn, G. Gobsch, K. Tonisch, V. Cimalla, F. Niebelschütz, O. Ambacher Appl. Phys. Lett. **92**, 013510 (2008)

3.20. C. Buchheim, A. Winzer, R. Goldhahn, G. Gobsch, O. Ambacher, A. Link, M .Eickhoff, M. Stutzmann, Thin Film Solids **450**, 155 (2004)

3.21. C. Bulutay, B. Ridley, N. Zakhleniuk, Phys. Rev. B **62**, 15754 (2000)

3.22. E. Calleja, F. Sanchez, D. Basak, M. Sanchez-Garcia, E. Munoz, I. Izpura, F. Calle, J. Tijero, J. Sanchez-Rojas, Phys. Rev. B **55**, 4689 (1997)

3.23. P. Cantu, S. Keller, U. Mishra, S. DenBaars, J. Appl. Phys. **82**, 3683 (2003)

3.24. H. Chen, R. Feenstra, J. Northrup, J. Neugebauer, D. Greve, MRS Internet J. Nitride Semicond. Res. **6**, 11 (2001)

3.25. A. Chini, R. Coffie, G. Meneghesso, E. Zanoni, D. Buttari, S. Heikman, S. Keller, U. Mishra, Electron. Lett. **39**, 625 (2003)

3.26. A. Chini, S. Rajan, M. Wong, Y. Fu, J. Speck, U. Mishra, in *Device Research Conference*, Santa Barbara, 2005, pp. 63–64

3.27. A. Chini, J. Wittich, S. Heikman, S. Keller, S. DenBaars, U. Mishra, IEEE Electron Device Lett. **25**, 55 (2004)

3.28. Y. Choi, M.P.H. Cha, B. Peres, M. Spencer, L. Eastman, IEEE Trans. Electron Devices **53**, 2926 (2006)

3.29. E. Chumbes, A. Schremer, J. Smart, Y. Wang, N. MacDonalds, D. Hogue, J. Komiak, S. Lichwalla, R. Leoni, J. Shealy, IEEE Trans. Electron Devices **48**, 420 (2001)

3.30. R. Coffie, D. Buttari, S. Heikmann, S. Keller, A. Chini, L. Shen, U. Mishra, IEEE Electron Device Lett. **23**, 588 (2002)

3.31. O. Contreras, F. Ponec, J. Christen, A. Dadgar, A. Krost, Appl. Phys. Lett. **81**, 4712 (2002)

3.32. Y. Cordier, F. Semond, P. Lorenzini, N. Grandjean, F. Natali, B. Damilano, J. Massies, V. Hoel, A. Minko, N. Vellas, C. Gaquiere, J. Jaeger, B. Dessertene, S. Cassette, M. Surrugue, D. Adam, J. Grattepain, R. Aubry, S. Delage, J. Cryst. Growth **251**, 811 (2003)

3.33. A. Corrion, C. Poblenz, P. Waltereit, T. Palacios, S. Rajan, U. Mishra, J. Speck, IEICE Trans.Electron. **E89**, 906 (2006)

3.34. A. Dadgar, A. Strittmatter, J. Bläsing, M. Poschenrieder, O. Contreras, P. Veit, T. Riemann, F. Bertram, A. Reiher, A. Krtschil, A. Diez, T. Hempel, T. Finger, A. Kasic, M. Schubert, D. Bimberg, F. Ponce, J. Christen, A. Krost, Phys. Stat. Sol. B **0**, 1583 (2003)

3.35. E. Danielsson, C. Zetterling, M. Östling, A. Nikolaev, I. Nikitina, V. Dmitriev, IEEE Trans. Electron Devices **48**, 444 (2001)

3.36. R. Davis, A. Roskowski, E. Preble, J. Speck, B. Heying, J. Freitas, E. Glaser, W. Carlos, Proc. IEEE **90**, 993 (2004)

3.37. Y. Dikme, M. Fieger, F. Jessen, A. Szymakowski, H. Kalisch, M. Woitok, P. van Gemmern, P. Javorka, M. Marso, R. Jansen, M. Heuken, Phys. Stat. Sol. C **0**, 2385 (2003)

3.38. W. Doolittle, S. Kang, A. Brown, Solid-State Electron. **44**, 229 (2000)

3.39. L. Eastman, V. Tilak, J. Smart, B. Green, E. Chumbes, R. Dimitrov, H. Kim, O. Ambacher, N. Weimann, T. Prunty, M. Murphy, W. Schaff, J. Shealy, IEEE Trans. Electron Devices **48**, 479 (2001)

3.40. L. Eastman, in *IEEE International Microwave Symposium Digest*, Seattle, 2002, pp. 2273–2275

3.41. C. Elsass, T. Mates, B. Heying, C. Poblenz, P. Fini, P. Petroff, S. DenBaars, J. Speck, Appl. Phys. Lett. **77**, 3167 (2000)

3.42. D. Fanning, L. Witkowski, C. Lee, D. Dumka, H. Tserng, P. Saunier, W. Gaiewski, E. Piner, K. Linthicum, J. Johnson, in *Proceedings of the International Conference on GaAs Manufacturing Technology*, New Orleans, 2005, p. 8.3

3.43. E. Faraclas, S. Islam, A. Anwar, Solid-State Electron. **48**, 1849 (2004)

3.44. Q. Fareed, R. Gaska, J. Mickevicius, G. Tamulaitis, M. Shur, M. Khan, in *Proceedings of the International Conference on GaAs Manufacturing Technology*, New Orleans, 2005, pp. 301–305

3.45. Z. Feng, S. Cai, K. Chen, K. Lau, IEEE Electron Device Lett. **26**, 870 (2006)

3.46. M. Fieger, Y. Dikme, F. Jessen, H. Kalisch, A. Noculak, A. Szymakowski, P.V. Gemmern, B. Faure, C. Richtarch, F. Letertre, M. Heuken, R. Jansen, Phys. Stat. Sol. C **2**, 2607 (2005)

3.47. J. Flynn, D. Keogh, F. Tamweber, M. Chriss, J. Redwing, in *WBG Semiconductors Workshop*, Copper Mountain, 2000, pp. n.a.

3.48. J. Flynn, H. Xin, J. Dion, E. Hutchins, H. Antunes, L. Corso, R.V. Egas, G. Brandes, Phys. Stat. Sol. C **0**, 2327 (2003)

3.49. S. Fu, Y. Chen, Appl. Phys. Lett. **85** 1523 (2004)

3.50. A. Georgakilas, H. Ng, P. Komninou, in *Nitride Semiconductors: Handbook on Materials and Device*, ed. by P. Ruterana, M. Albrecht, J. Neugebauer (Wiley-VCH, Weinheim, 2003), Chap. 3, pp. 107–192

3.51. P. Gibart, B. Beaumont, P. Vennegues, in *Nitride Semiconductors: Handbook on Materials and Device*, ed. by P. Ruterana, M. Albrecht, J. Neugebauer (Wiley-VCH, Weinheim, 2003), Chap. 2, pp. 45–106

3.52. J. Gillespie, R. Fitch, N. Moser, T. Jenkins, J. Sewell, D. Via, A.C.A. Dabiran, P. Chow, A. Osinsky, M. Mastro, D. Tsvetkov, V. Soukhoveev, A. Usikov, V. Dmitriev, B. Luo, S. Pearton, F. Ren, Solid-State Electron. **47**, 1859 (2003)

3.53. N. Gmeinwieser, K. Engl, P. Gottfriedsen, U. Schwarz, J. Zweck, W. Wegscheider, S. Miller, H. Lugauer, A. Leber, A. Weimar, A. Lell, V. Härle, J. Appl. Phys. **96**, 3666 (2004)

3.54. W. Götz, N. Johnson, D. Bour, C. Chen, H. Liu, C. Kuo, W. Imler, in *MRS Symposium*, First International Conference on Nitride Semiconductors, vol. 395, Boston, 1996, pp. 443–454

3.55. W. Götz, N. Johnson, M. Bremser, R. Davis, Appl. Phys. Lett. **69**, 2379 (1996)

3.56. W. Götz, N. Johnson, C. Chen, H. Liu, C. Kuo, W. Imler, J. Appl. Phys. **68**, 3144 (1996)

3.57. D. Green, , S. Gibb, B. Hosse, R. Vetury, D. Grider, J. Smart, J. Cryst. Growth **285**, 3144 (2004)

3.58. P. Grudowski, A. Holmes, C. Eiting, R. Dupuis, J. Appl. Phys. **69**, 3626 (1996)

3.59. A.D. Hanser, R.F. Davis in *Properties, Processing and Applications of GaN Nitride and Related Semiconductors*, No. 23 in EMIS Data reviews Series, ed. by J. Edgar et al. (IEE INSPEC, London, 1999), Sect. B2.2, pp. 386–395

3.60. P. Hansen, Y. Strausser, A. Erickson, E. Tarsa, P. Kozodoy, E. Brazel, J. Ibbetson, U. Mishra, V. Narayanamurti, S. DenBaars, J. Speck, Appl. Phys. Lett. **72**, 2247 (1998)

3.61. A. Hanson, S. Stockman, G. Stillman, IEEE Electron Device Lett. **13**, 504 (1992)

3.62. T. Hashizume, J. Appl. Phys. **94**, 431 (2003)

3.63. S. Heikman, S. Keller, S. DenBaars, U. Mishra, Appl. Phys. Lett. **81**, 439 (2002)

3.64. S. Heikman, PhD thesis, University of California Santa Barbara, Santa Barbara, 2002

3.65. E. Henriksen, S. Syed, Y. Ahmadian, M. Manfra, K. Baldwin, A. Sergent, R. Molnar, H. Stormer, Appl. Phys. Lett. **86**, 252108 (2005)

3.66. B. Heying, R. Averbeck, L. Chen, E. Haus, H. Riechert, J. Speck, J. Appl. Phys. **88**, 1855 (2000)

3.67. B. Heying, I. Smorchkova, C. Poblenz, C. Elsass, P. Fini, S. DenBaars, U. Mishra, J. Speck, Appl. Phys. Lett. **77**, 2885 (2000)

3.68. A. Hierro, A. Arehart, B. Heying, M. Hansen, U. Mishra, S. DenBaars, J. Speck, S. Ringel, Appl. Phys. Lett. **80**, 805 (2002)

3.69. A. Hierro, A. Arehart, B. Heying, M. Hansen, J. Speck, U. Mishra, S. Denbaars, S. Ringel, Phys. Stat. Sol. B **228**, 309 (2001)

3.70. M. Higashiwaki, S. Anantathanasarn, N. Negoro, E. Sano, H. Hasegawa, K. Kumakura, T. Makimoto, Jpn. J. Appl. Phys. **43**, L1147 (2004)

3.71. M. Higashiwaki, T. Matsui, Jpn. J. Appl. Phys. **41**, L540 (2002)

3.72. M. Higashiwaki, T. Matsui, Jpn. J. Appl. Phys. **43**, L768 (2004)

3.73. M. Higashiwaki, T. Matsui, J. Cryst. Growth **252**, 128 (2003)

3.74. M. Higashiwaki, T. Matsui, J. Cryst. Growth **251**, 494 (2003)

3.75. M. Higashiwaki, T. Matsui, Jpn. J. Appl. Phys. **43**, L1147 (2004)

3.76. M. Higashiwaki, T. Matsui, J. Cryst. Growth **269**, 162 (2004)

3.77. M. Higashiwaki, T. Matsui, Jpn. J. Appl. Phys. **44**, L475 (2005)

3.78. T. Inoue, T. Nakayama, Y. Ando , M. Kosaki, H. Miwa, K. Hirata , T. Uemura , H. Miyamoto, IEEE Trans. Electron Devices **55**, 483 (2008)

3.79. P. Javorka, A. Alam, M. Wolter, A. Fox, M. Marso, M. Heuken, H. Lüth, P. Kordos, Electron. Lett. **23**, 4 (2002)

3.80. D. Jena, A. Gossard, U. Mishra, J. Appl. Phys. **88**, 4734 (2000)

3.81. D. Jena, A. Gossard, U. Mishra, Appl. Phys. Lett. **76**, 1707 (2000)

3.82. D. Jena, I. Smorchkova, A. Gossard, U. Mishra, Phys. Stat. Sol. B **228**, 617 (2001)

3.83. C. Jeon, J. Lee, J. Lee, J. Lee, IEEE Electron Device Lett. **25**, 120 (2004)

3.84. G. Jessen, J. Gillespie, G. Via, A. Crespo, D. Langley, J. Wasserbauer, F. Faili, D. Francis, D. Babic, F. Ejeckam, S. Guo, I. Eliashevich, in *Compound Semiconductor IC Symposium Technical Digest*, San Antonio, 2006, pp. 271–274

3.85. S. Jia, Y. Dikme, D. Wang, K. Chen, K. Lau, M. Heuken, IEEE Electron Device Lett. **26**, 130 (2005)

3.86. M. Johnson, Z. Yu, J. Brown, N. El-Masry, J. Cook, J. Schetzina, J. Electron. Mater. **28**, 295 (1999)

3.87. M. Kamp, H. Riechert in *Properties, Processing and Applications of GaN Nitride and Related Semiconductors*, No. 23 in EMIS Data reviews Series, ed. by J. Edgar et al. (IEE INSPEC, London, 1999), Sect. B2.8, pp. 426–439

3.88. S. Karpow, Y. Kovalchuck, V. Myachin, Y. Pogorelskii, J. Cryst. Growth **129**, 563 (1993)

3.89. D. Katzer, S. Binari, D. Storm, J. Roussos, B. Shanabrook, E. Glaser, Electron. Lett. **38**, 1740 (2002)

3.90. D. Katzer, D. Storm, S. Binari, J. Roussos, B. Shanabrook, E. Glaser, J. Cryst. Growth **253**, 481 (2003)

3.91. D. Katzer, D. Storm, S. Binari, B. Shanabrook, A. Torabi, L. Zhou, D. Smith, J. Vac. Sci. Technol. B **23**, 1204 (2005)

3.92. S. Keller, G. Parish, P. Fini, S. Heikman, C. Chen, N. Zhang, S. DenBaars, U. Mishra, Y. Wu, J. Appl. Phys. **86**, 5850 (1999)

3.93. S. Keller, R. Vetury, G. Parish, S. DenBaars, U. Mishra, Appl. Phys. Lett. **78**, 3088 (2001)

3.94. S. Keller, Y. Wu, G. Parish, N. Ziang, J. Xu, B. Keller, S. DenBaars, U. Mishra, IEEE Trans. Electron Devices **48**, 552 (2001)

3.95. D. Kelly, C. Brindle, C. Kemerling, M. Stuber, in *Compound Semiconductor IC Symposium Technical Digest*, Palm Springs, 2005, pp. 200–203

3.96. T. Kikkawa, K. Imanishi, M. Kanamura, K. Joshin, in *Compound Semiconductor IC Symposium Technical Digest*, Palm Springs, 2005, pp. 77–80

3.97. T. Kikkawa, T. Maniwa, H. Hayashi, M. Kanamura, S. Yokokawa, M. Nishi, N. Adachi, M. Yokoyama, Y. Tateno, K. Joshin, in *IEEE International Microwave Symposium Digest*, Fort Worth, 2004, pp. 1347–1350

3.98. T. Kikkawa, M. Nagahara, T. Kimura, S. Yokokawa, S. Kato, M. Yokoyama, Y. Tateno, K. Horino, K. Domen, Y. Yamaguchi, N. Hara, K. Joshin, in *IEEE International Microwave Symposium Digest*, Seattle, 2002, pp. 1815–1818

3.99. A. Kikuchi, R. Bannai, K. Kishino, Phys. Stat. Sol. A **188**, 187 (2001)

3.100. L. Kirste, R. Moller, R. Kiefer, R. Quay, K. Köhler, N. Herres, Appl. Surf. Science **253**, 209 (2006)

3.101. W. Knap, E. Borovitskaya, M. Shur, L. Hsu, W. Walukiewicz, E. Frayssinet, P. Lorenzini, N. Grandjean, C. Skierbiszewski, P. Prystawko, M. Leszczynski, I. Grzegory, Appl. Phys. Lett. **80**, 1228 (2002)

3.102. K. Köhler, S. Müller, N. Rollbühler, R. Kiefer, R. Quay, G. Weimann, in *Proceedings of the International Symposium on Compound Semiconductors*, Lausanne, 2003, pp. 235–238, ed. by M. Ilegems

3.103. D. Koleske, R. Henry, M. Twigg, J. Culbertson, S. Binari, A. Wickenden, M. Fatemi, Appl. Phys. Lett. **80**, 4372 (2002)

3.104. N. Kolias, T. Kazior, in *IMS Workshop Advances in GaN-based Device and Circuit Technology: Modeling and Applications*, Fort Worth, 2004

3.105. P. Kozodoy, PhD thesis, University of California Santa Barbara, Santa Barbara, 1999

3.106. P. Kozodoy, Y. Smorchkova, M. Hansen, H. Xing, S. DenBaars, U. Mishra, A. Saxler, R. Perrin, W. Mitchel, Appl. Phys. Lett. **75**, 2444 (1999)

3.107. J. Kuzmik, A. Kostopoulos, G. Konstantinidis, J. Carlin, A. Georgakilas, D. Pogany, IEEE Trans. Electron Devices **53**, 422 (2006)

3.108. R. Langer, B. Faure, A. Boussagol, P. Bove, H. Lahreche, A. Wilk, J. Thuret, F. Letertre, in *Proceedings of the International Conference on GaAs Manufacturing Technology*, Vancouver, 2006, p. 4E

3.109. J. LaRoche, B. Luo, F. Ren, K. Baik, D.D. Stodilka, B. Gila, C. Abernathy, S. Pearton, A. Usikov, D. Tsevtkov, V. Soukhoveev, G. Gainer, A. Rechnikov, V. Dimitriev, G. Chen, C. Pan, J. Chyi, Solid-State Electron. **48**, 193 (2004)

3.110. M. Lee, J. Sheu, Y. Su, S. Chang, W. Lai, G. Chi, IEEE Electron Device Lett. **25**, 593 (2004)

3.111. W. Lee, N.J. Ryou, J. Limb, R. Dupuis, D. Hanser, E. Preble, N. Williams, K. Evans, Appl. Phys. Lett. **90**, 093509 (2007)

3.112. C. Li, I. Bhat, R. Wang, J. Seiler, J. Electron. Mater. **33**, 481 (2004)

3.113. T. Li, R. Campion, C. Foxon, S. Rushworth, L. Smith, J. Cryst. Growth **251**, 499 (2003)

3.114. J. Liu, Y. Zhou, R. Chu, Y. Cai, K. Chen, K. Lau, IEEE Electron Device Lett. **26**, 145 (2005)

3.115. J. Liu, Y. Zhou, J. Zhu, K. Lau, K. Chen, IEEE Electron Device Lett. **27**, 10 (2006)

3.116. J. Liu, Y. Zhou, J. Zhu, Y. Cai, K. Lau, K. Chen, IEEE Trans. Electron Devices **54**, (2007)

3.117. K. Liu, T. Tezukab, S. Sugitab, Y. Watarib, Y. Horikoshib, Y. Sua, S. Changa J. Cryst. Growth **263**, 400 (2004)

3.118. D. Look, Phys. Stat. Sol. B **228**, 293 (2001)

3.119. D. Look, Z. Fang, L. Polenta, MRS Internet J. Nitride Semicond. Res. **5**, W10.5 (2000)

3.120. D. Look, J. Sizelove, Phys. Rev. Lett. **82**, 1237 (1999)

3.121. H. Lu, I. Bhat, in *MRS Symposium*, First International Conference on Nitride Semiconductors, vol. 395, Boston, 1996, pp. 497–502

3.122. N. Maeda, T. Saitoh, K. Tsubaki, T. Nishida, K. Kobayashi, Phys. Stat. Sol. B **216**, 727 (1999)

3.123. M. Maestro, D. Tsvetkov, V. Soukhoveev, A. Usikov, V. Dmitriev, B. Luo, F. Ren, K. Baik, S. Pearton, Solid-State Electron. **47**, 1075 (2003)

3.124. M. Manfra, N. Weimann, Y. Bayens, P. Roux, D. Tennant, Electron. Lett. **39**, 694 (2003)

3.125. M. Manfra, N. Weimann, O. Mitrofanov, T. Wächtler, D. Tennant, Phys. Stat. Sol. A **200**, 175 (2003)

3.126. K. Matsushita, H. Sakurai, H. Kawasaki, Y. Takada, T. Sasaki, K. Tsuda, in *Proceedings of the International Conference on GaAs Manufacturing Technology*, New Orleans, 2005, p. 4.2

3.127. L. McCarthy, I. Smorchkova, H. Xing, P. Fini, S. Keller, J. Speck, S. DenBaars, M. Rodwell, U. Mishra, Appl. Phys. Lett. **78**, 2235 (2001)

3.128. L.S. McCarthy, PhD thesis, University of California Santa Barbara, Santa Barbara, 2001

3.129. G. Meneghesso, C. Ongaro, E. Zanoni, C. Brylinski, M. di Forte-Poisson, V. Hoel, J. de Jaeger, R. Langer, H. Lahreche, P. Bove, J. Thorpe, in *IEDM Technical Digest*, Washington DC, 2007, pp. 807–810

3.130. F. Medjdoub, N. Sarazin, M. Tordjman, M. Magis, M. di Forte-Poisson, M. Knez, E. Delos, C. Gaquiere, S. Delage, E. Kohn, Electronics Lett. **43** 691 (2007)

3.131. M. Micovic, P. Hashimoto, M. Hu, I. Milosavljevic, J. Duval, P. Willadsen, A. Kurdoghlian, P. Deelman, J. Moon, A. Schmitz, M. Delaney, in *IEDM Technical Digest*, San Francisco, 2004, pp. 807–810

3.132. M. Micovic, A. Kurdoghlian, P. Janke, P.H.D. Wong, J. Moon, L. McCray, C. Nguyen, IEEE Trans. Electron Devices **48**, 591 (2001)

3.133. E. Miller, D. Schaadt, E. Yu, X. Sun, L. Brillson, P. Waltereit, J. Speck, J. Appl. Phys. **94**, 7611 (2003)

3.134. E. Miller, E. Yu, P. Waltereit, J. Speck, Appl. Phys. Lett. **84**, 535 (2004)
3.135. A. Minko, V. Hoel, S. Lepilliet, G. Dambrine, J. DeJaeger, Y. Cordier, F. Semond, F. Natali, J. Massies, IEEE Electron Device Lett. **25**, 167 (2004)
3.136. M. Miyoshi, M. Sakai, S. Arulkumaran, H. Ishikawa, T. Egawa, M. Tanaka, O. Oda, Jpn. J. Appl. Phys. **43**, 7939 (2004)
3.137. E. Monroy, N. Gogneau, F. Enjalbert, F. Fossard, D. Jalabert, E. Bellet-Amalric, L. Dang, B. Daudin, J. Appl. Phys. **94**, 3121 (2003)
3.138. H. Morkoc, *Nitride Semiconductors and Devices*, Springer Series in Materials Science, vol. 32 (Springer, Berlin Heidelberg New York, 1999)
3.139. S. Müller, K. Köhler, R. Kiefer, R. Quay, M. Baeumler, L. Kirste, Phys. Stat. Sol. C **2**, 2639 (2005)
3.140. T. Murata, M. Hikita, Y. Hirose, Y. Uemoto, K. Inoue, T. Tanaka, D. Ueda, IEEE Trans. Electron Devices **52**, 1042 (2005)
3.141. N. Nakamura, in *MRS Bulletin*, Warrendale, 1998, pp. 1145–1156
3.142. S. Nakamura, S. Chichibu (eds.), *Introduction to Nitride Semiconductor Blue Lasers and Light Emitting Diodes* (Taylor & Francis, London New York, 2000)
3.143. S. Nakamura, N. Iwasa, M. Senoh, T. Mukai, Jpn. J. Appl. Phys. **228**, 309 (2001)
3.144. S. Nakamura, T. Mukai, M. Senoh, J. Appl. Phys. **71**, 5543 (1992)
3.145. Y. Nakano, J. Suda, T. Kimoto, Phys. Stat. Sol. C **2**, 2208 (2005)
3.146. Y. Nanishi, Y. Saito, T. Yamaguchi, M. Hori, F. Matsuda, T. Araki, A. Suzuki, T. Miyajima, Phys. Stat. Sol. A **200**, 202 (2003)
3.147. Y. Nasishi, Y. Saito, T. Yamaguchi, Jpn. J. Appl. Phys. **42**, 2549 (2003)
3.148. J. Neugebauer, C. Van de Walle Appl. Phys. Lett. 68, 1829 (1996)
3.149. S. Newstead, R. Kubiak, E. Parker, J. Cryst. Growth **81**, 49 (1987)
3.150. H. Ng, D. Doppalapudi, D. Korakakis, R. Singh, T. Moustakas, J. Cryst. Growth **189–190**, 349 (1998)
3.151. N. Okamoto, K. Hoshino, N. Hara, M. Takikawa, Y. Arakawa, J. Cryst. Growth **272**, 278 (2004)
3.152. Y. Okamoto, K. Takahashi, H. Nakamura, Y. Okada, M. Kawabe, Phys. Stat. Sol. A **180**, 59 (2000)
3.153. T. Palacios, A. Chakraborty, S. Heikmann, S. Keller, S. DenBaars, U. Mishra, IEEE Electron Device Lett. **27**, 13 (2006)
3.154. T. Palacios, A. Chakraborty, S. Keller, S. DenBaars, U. Mishra, in *Device Research Conference*, Santa Barbara, 2005, pp. 181–182
3.155. T. Palacios, L. Shen, S. Keller, A. Chakraborty, S. Heikman, D. Buttari, S. DenBaars, U. Mishra, Phys. Stat. Sol. A **22**, 837 (2005)
3.156. N. Pan, J. Elliot, M. Knowles, D. Vu, K. Kishimoto, J. Twynam, H. Sato, M. Fresina, G. Stillman, IEEE Electron Device Lett. **19**, 115 (1998)
3.157. G. Parish, PhD thesis, University of California Santa Barbara, Santa Barbara, 1999
3.158. C. Park, Y. Park, H. Lee, I. Yoon, T. Kang, H. Cho, J. Oh, K. Wang, Jpn. J. Appl. Phys. **44**, 1722 (2005)
3.159. T. Paskova, V. Darakchieva, E. Valcheva, P. Paskov, I. Ivanov, B. Monemar, T. Böttcher, C. Roder, D. Hommel, J. Electron. Mater. **33**, 389 (2004)
3.160. S. Pearton, J. Zolper, R. Shul, F. Ren, J. Appl. Phys. **86**, 1 (1999)
3.161. A. Petersson, A. Gustafsson, L. Samuelson, S. Tanaka, Y. Aoyagi, MRS Internet J. Nitride Semicond. Res. **7**, (2002)
3.162. C. Poblenz, P. Waltereit, S. Rajan, S. Heikman, U. Mishra, J. Speck, J. Vac. Sci. Technol. B **22**, 1145 (2004)

3.163. C. Poblenz, P. Waltereit, S. Rajan, U.K. Mishra, J.S. Speck, P. Chin, I. Smorchkova, B. Heying, J. Vac. Sci. Technol. B **23**, 1562 (2005)

3.164. A. Polyakov, N. Smirnov, A. Govorkov, A. Shlensky, S. Pearton, J. Appl. Phys. **95**, 5591 (2004)

3.165. A. Ptak, L. Holbert, L. Ting, C. Schwartz, M. Moldovan, N. Giles, T. Myers, P. van Lierde, C. Tian, R. Hockett, S. Mitha, A. Wickenden, D. Koleske, R. Henry, Appl. Phys. Lett. **79**, 2740 (2001)

3.166. S. Rajan, P. Waltereit, C. Poblenz, S. Heikman, D. Green, J. Speck, U. Mishra, IEEE Electron Device Lett. **25**, 247 (2004)

3.167. J. Redwing, T. Kuech in *Properties, Processing and Applications of GaN Nitride and Related Semiconductors*, No. 23 in EMIS Data reviews Series, ed. by J. Edgar et al. (IEE INSPEC, London, 1999), Sect. B2.7, pp. 416–425

3.168. J. Redwing, M. Tischler, J. Flynn, S. Elhamri, M. Ahoujja, R. Newrock, W. Mitchel, Appl. Phys. Lett. 69, 963 (1996).

3.169. F. Ren, J. Zolper *Wide Energy Bandgap Electronic Devices* (World Scientific, New Jersey, 2003)

3.170. M. Reshchikova, H. Morkoc, J. Appl. Phys. **97**, 061301–1 (2005)

3.171. M. Rudziski, P. Hageman, A. Grzegorczyk, L. Macht, T. Rödle, H. Jos, P. Larsen, Phys. Stat. Sol. C **2**, 2141 (2005)

3.172. P. Ruterana, M. Albrecht, J. Neugebauer (eds.), *Nitride Semiconductors: Handbook on Materials and Device* (Wiley-VCH, Weinheim, 2003)

3.173. P. Ruterana, M. Morales, F. Gourbilleau, P. Singh, M. Drago, T. Schmidtling, U. Pohl, W. Richter, Phys. Stat. Sol. A **202**, 781 (2005)

3.174. M. Sakai, T. Egawa, M. Hao, H. Ishikawa, Jpn. J. Appl. Phys. **43**, 8019 (2004)

3.175. W. Schaff, L. Hai, H. Jeonghyun, W. Hong, in *Proc. IEEE/Cornell Conf. High Perf. Devices*, Ithaca, 2000, pp. 225–231

3.176. F. Scholz, Progress in Crystal Growth and Characterization of Materials, **35**, 243 (1997))

3.177. F. Schwierz, V. Polyakov, in *Proc. Intl. Solid-State Integrated Circuit Technology ICSICT* Shanghai, 2006, pp. 845–848

3.178. D. Segev, C. Van de Walle, Europhys. Lett. **76** 305 (2006)

3.179. F. Semond, B. Damilano, P. Lorenzini, S. Vezian, N. Grandjean, M. Leroux, J. Massies, Appl. Phys. Lett. **75**, 82 (1999)

3.180. F. Semond, P. Lorenzini, N. Grandjean, J. Massies, Appl. Phys. Lett. **78**, 335 (2001)

3.181. I. Shalish, L. Kronik, G. Segal, Y. Shapira, Y. Rosenwaks, U. Tisch, J. Salzman, Phys. Rev. B., **59**, 9748 (1999)

3.182. L. Shen, R. Coffie, D. Buttari, S. Heikman, A. Chakraborty, A. Chini, S. Keller, S. DenBaars, U. Mishra, J. Electron. Mater. **33**, 422 (2004)

3.183. L. Shen, S. Heikman, B. Moran, R. Coffie, N. Zhang, D. Buttari, I. Smorchkova, S. Keller, S. DenBaars, U. Mishra, IEEE Electron Device Lett. **22**, 457 (2001)

3.184. B. Simpkins, E. Yu, P. Waltereit, J. Speck, J. Appl. Phys. **94**, 1448 (2003)

3.185. J. Smart, A. Schremer, N. Weimann, O. Ambacher, L. Eastman, J. Shealy, Appl. Phys. Lett. **75**, 388 (1999)

3.186. I. Smorchkova, E. Haus, B. Heying, P. Kozodoy, P. Fini, J. Ibbetson, S. Keller, S. DenBaars, J. Speck, U. Mishra, Appl. Phys. Lett. **76**, 718 (2000)

3.187. A. Sozza, C. Dua, E. Morvan, M. diForte Poisson, S. Delage, F. Rampazzo, A. Tazzoli, F. Danesin, G. Meneghesso, E. Zanoni, A. Curutchet, N. Malbert, N. Labat, B. Grimbert, J.D. Jaeger, in *IEDM Technical Digest*, Washington DC, 2005, pp. 590–593

3.188. M. Steen, M. Sheldon, R. Bresnahan, T. Bird, D. Gotthold, in *Proceedings of the International Conference on GaAs Manufacturing Technology*, New Orleans, 2005, p. 13.5

3.189. J. Suda, Y. Nakano, T. Kimoto, MRS Symposium **831**, 471 (1999)

3.190. B. Sverdlov, A. Botchkarev, G. Martin, A. Salvador, H. Morkoc, S. Tsen, D. Smith, in *MRS Symposium*, First International Conference on Nitride Semiconductors, vol. 395, Boston, 1996, pp. 175–180

3.191. H. Takeuchi, Y. Yamamoto, Y. Kamo, T. Oku, M. Nakayama, The European Physical Journal B **52**, 311 (2006)

3.192. T. Tanaka, Y. Koji, T. Meguro, Y. Otoki, in *Proceedings of the International Conference on GaAs Manufacturing Technology*, Miami, 2004, pp. 295–298

3.193. T. Tanaka, K. Takano, H. Fujikura, T. Mishima, Y. Kohji, H. Kamogawa, T. Meguro, Y. Otoki, in *Proceedings of the International Conference on GaAs Manufacturing Technology*, New Orleans, 2005, p. 4.1

3.194. H. Tang, J. Webb, Appl. Phys. Lett. **74**, 2373 (1999)

3.195. E. Tengborn, M. Rummukainen, F. Tuomisto, K. Saarinen, M. Rudzinski, P. Hageman, P. Larsen, A. Nordlund, Appl. Phys. Lett. **89**, 091905 (2006)

3.196. A. Thamm, O. Brandt, J. Hilsenbeck, R. Lossy, K.H. Ploog, in *Prooceedings of the International Symposium on Compound Semiconductors*, Monterey, 2000, pp 455–460

3.197. V. Tilak, B. Green, V. Kaper, H. Kim, T. Prunty, J. Smart, J. Shealy, L. Eastman, IEEE Electron Device Lett. **22**, 504 (2001)

3.198. A. Trassoudaine, R. Cadoret, E. Aujol, in *Nitride Semiconductors: Handbook on Materials and Device*, ed. by P. Ruterana, M. Albrecht, J. Neugebauer (Wiley-VCH, Weinheim, 2003), Chap. 4, pp. 193–240

3.199. R. Underwood, PhD thesis, University of California Santa Barbara, Santa Barbara, 1999

3.200. M. Uren, T. Martin, B. Hughes, K. Hilton, A. Wells, R. Balmer, D.H.A. Keir, D. Wallis, A. Pidduck, M. Missous, Phys. Stat. Sol. A **194**, 468 (2002)

3.201. R. Vandersmissen, J. Das, W. Ruythooren, J. Derluyn, M. Germain, D. Xiao, D. Schreurs, G. Borghs, in *Proceedings of the International Conference on GaAs Manufacturing Technology*, New Orleans, 2005, p. 13.3

3.202. C. Van de Walle, C. Stampfl, J. Neugebauer, J. Cryst. Growth **189/190**, 505 (1998)

3.203. C. Van de Walle, J. Neugebauer, C. Stampfl, in *Properties, Processing and Applications of GaN Nitride and Related Semiconductors*, No. 23 in EMIS Data reviews Series, ed. by J. Edgar et al. (IEE INSPEC, London, 1999), Sect. A8.2, pp. 275–280

3.204. A. Vescan, J. Brown, J. Johnson, R. Therrien, T. Gehrke, P. Rajagopal, J. Roberts, S. Singhal, W. Nagy, R. Borges, E. Piner, K. Linthicum, Phys. Stat. Sol. C **0**, 52 (2002)

3.205. P. Waltereit, S. Lim, M. McLaurin, J. Speck, Phys. Stat. Sol. A **194**, 524 (2002)

3.206. P. Waltereit, C. Poblenz, S. Rajan, F. Wu, U.K. Mishra, and J.S. Speck, Jpn. J. Appl. Phys. **43** L1520 (2004)

3.207. C. Wang, L. Yu, S. Lau, E. Yu, W. Kim, A. Botchkarev, H. Morkoc, Appl. Phys. Lett. **72**, 1211 (1998)

3.208. D. Wang, S. Jia, K. Chen, K. Lau, Y. Dikme, P. van Gemmern, Y. Lin, H. Kalisch, R. Jansen, M. Heuken, J. Appl. Phys. **97**, 56103 (2005)

3.209. S. Wang, S. Chang, K. Uang, B. Liou, in *Device Research Conference*, Santa Barbara, 2005, pp. 59–60

3.210. N. Watanabe, H. Yokoyama, M. Hiroki, Y. Oda, T. Kobayashi, T. Yagi, in *Compound Semiconductor IC Symposium Technical Digest*, San Antonio, 2006, pp. 257–260

3.211. J. Webb, H. Tang, J. Bardwell, S. Rolfe, Y. Liu, J. Lapointe, P. Marshall, T. MacElwee, Phys. Stat. Sol. A **188**, 271 (2001)

3.212. J. Webb, H. Tang, J. Bardwell, P. Coleridge, Phys. Stat. Sol. A **176**, 243 (1999)

3.213. J. Webb, H. Tang, J. Bardwell, P. Coleridge, Appl. Phys. Lett. **78**, 3845 (2001)

3.214. J. Webb, H. Tang, J. Bardwell, Y. Liu, J. Lapointe, T. MacElwee, Phys. Stat. Sol. A **194**, 439 (2002)

3.215. T. Weeks, M. Bremser, K. Ailey, E. Carlson, W. Perry, R. Davis Appl. Phys. Lett. 67, 401 (1995)

3.216. N. Weimann, M. Manfra, S. Chakraborty, D. Tennant, IEEE Electron Device Lett. **23**, 691 (2002)

3.217. N. Weimann, M. Manfra, T. Wächtler, IEEE Electron Device Lett. **24**, 57 (2003)

3.218. A. Wickenden, D. Koleske, R. Henry, R. Gorman, J. Culbertson, M. Twigg, J. Electron. Mater. **28**, 301 (1999)

3.219. A. Wickenden, D. Koleske, R. Henry, R. Gorman, M. Twigg, M. Fatemi, J. Freitas, W. Moore, J. Electron. Mater. **29**, 21 (2000)

3.220. A. Winzer, R. Goldhahn, G. Gobsch, A. Dadgar, A. Krost, O. Weidemann, M. Stutzmann, M. Eickhoff, Appl. Phys. Lett., **88**, 4101 (2006)

3.221. M. Wojtowicz, B. Heying, P.C.I. Smorchkova, R. Sandhu, T. Block, M. Aumer, D. Thomson, D. Partlow, in *Proceedings of the International Conference on GaAs Manufacturing Technology*, Miami, 2004, p. 299

3.222. Y. Wu, B. Keller, P. Fini, S. Keller, T. Jenkins, L. Kehias, S. Denbaars, U. Mishra, IEEE Electron Device Lett. **19**, 50 (1998)

3.223. L. Wu, W. Meyer, F. Auret, Phys. Stat. Sol. A **201**, 2277 (2004)

3.224. Y. Wu, M. Moore, A. Abrahamsen, M. Jacob-Mitos, P. Parikh, S. Heikman, A. Burk, in *IEDM Technical Digest*, Washington DC, 2007, pp. 405–408

3.225. F. Yam, Z. Hassan, Superlattices and Microstructures, **43**, 1 (2008)

3.226. A. Yamamoto, T. Shin-ya, T. Sugiura, A. Hashimoto, J. Cryst. Growth **189–190**, 461 (1998)

3.227. F. Yang, J. Hwang, Y. Yang, K. Chen, J. Wang, Jpn. J. Appl. Phys. **41**, L1321 (2002)

3.228. T. Yuasa, Y. Ueta, Y. Tsuda, A. Ogawa, M. Taneya, K. Takao, Jpn. J. Appl. Phys. **38**, L703 (1999)

3.229. J. Zimmer, G. Chandler, in *Proceedings of the International Conference on GaAs Manufacturing Technology*, Austin, 2007, pp. 129–132

References of Chapter 4

4.1. V. Adivarahan, M. Gaevski, M. Islam, B. Zhang, Y. Deng, M. Khan, IEEE Trans. Electron Devices **55**, 495 (2008)

4.2. V. Adivarahan, A. Koudymov, S. Rai, J. Yang, G. Simin, M. Khan, in *Device Research Conference*, Santa Barbara, 2005, pp. 177–178

4.3. V. Adivarahan, J. Yang, A. Koudymov, G. Simin, M. Khan, IEEE Electron Device Lett. **26**, 535 (2005)

4.4. V. Agarwal, D. Rawal, H. Vyas, J. Electrochem. Soc. **152**, G567 (2005)

4.5. Y. Ando, Y. Okamoto, K. Hataya, T. Nakayama, H. Miyamoto, T. Inoue, M. Kuzuhara, in *IEDM Technical Digest*, Washington DC, 2003, pp. 563–566

4.6. Y. Ando, Y. Okamoto, H. Miyamoto, N. Hayama, T. Nakayama, K. Kasahara, M. Kuzuhura, in *IEDM Technical Digest*, Washington DC, 2001, pp. 381–384

4.7. Y. Ando, Y. Okamoto, H. Miyamoto, T. Nakayama, T. Inoue, M. Kuzuhara, IEEE Electron Device Lett. **24**, 289 (2003)

4.8. Y. Ando, A. Wakejima, Y. Okamoto, T. Nakayama, K. Ota, K. Yamanoguchi, Y. Murase, K. Kasahara, K. Matsunaga, T. Inoue, H. Miyamoto, in *IEDM Technical Digest*, Washington DC, 2005, pp. 576–579

4.9. J. Ao, D. Kikuta, N. Kubota, Y. Naoi, Y. Ohno, IEEE Electron Device Lett. **24**, 500 (2003)

4.10. C. Bae, C. Krug, G. Lucovsky, A. Chakraborty, U. Mishra, J. Appl. Phys. **96**, 2674 (2004)

4.11. K. Ban, H. Hong, D. Noh, T. Seong, J. Song, D. Kim, Semicond. Sci. Technol. **20**, 921 (2005)

4.12. J. Bardwell, I. Foulds, B. Lamontagne, H. Tang, J. Webb, P. Marshall, S. Rolfe, J. Stapledon, J. Vac. Sci. Technol. A **18**, 750 (2000)

4.13. J. Bardwell, I. Foulds, J. Webb, H. Tang, J. Electron. Mater. **28**, 1071 (1999)

4.14. J. Bardwell, S. Haffouz, W. McKinnon, C. Storey, H. Tang, G. Sproule, D. Roth, R. Wang, Electrochem. Solid-State Lett. **10**, H46 (2007)

4.15. D. Basak, M. Verdu, M. Montojo, M. Sanchez-Garcia, F. Sanchez, E. Munoz, E. Calleja, Semicond. Sci. Technol. **12**, 1654 (1997)

4.16. F. Benkhelifa, R. Kiefer, S. Müller, F. van Raay, R. Quay, R. Sah, M. Mikulla, G. Weimann, in *Proceedings of the International Conference on GaAs Manufacturing Technology*, New Orleans, 2005, p. 8.4

4.17. J. Bernat, P. Javorka, A. Fox, M. Marso, H. Lüth, P. Kordos, Solid-State Electron. **47**, 2097 (2003)

4.18. S. Binari, H. Dietrich, G. Kelner, L. Rowland, K. Doverspike, D. Gaskill, Electron. Lett. **30**, 909 (1994)

4.19. S. Binari, H. Dietrich, G. Kelner, L. Rowland, K. Doverspike, D. Wickenden, J. Appl. Phys. **78**, 3008 (1995)

4.20. S. Binari, K. Ikossi, J. Roussos, W. Kruppa, D. Park, H. Dietrich, D. Koleske, A. Wickenden, R.L. Henry, IEEE Trans. Electron Devices **48**, 465 (2001)

4.21. S. Binari, P. Klein, T. Kazior, in *IEEE International Microwave Symposium Digest*, Seattle, 2002, pp. 1823–1826

4.22. S. Binari, W. Kruppa, H. Dietrich, G. Kelner, A. Wickenden, J. Freitas, Solid-State Electron. **41**, 1549 (1997)

4.23. K. Boutros, M. Regan, P. Rowell, D. Gotthold, B. Brar, in *Proceedings of the International Conference on GaAs Manufacturing Technology*, Miami, 2004, pp. 23–26

4.24. S. Bradley, A. Young, L. Brillson, M. Murphy, W. Schaff, L. Eastman, IEEE Trans. Electron Devices **48**, 412 (2001)

4.25. N. Braga, R. Mickevicius, V. Rao, W. Fichtner, R. Gaska, M. Shur, in *Compound Semiconductor IC Symposium Technical Digest*, Palm Springs, 2005, pp. 149–152

4.26. H. Brech, W. Brakensiek, D. Burdeaux, W. Burger, C. Dragon, G. Formicone, B. Pryor, D. Rice, in *IEDM Technical Digest*, Washington DC, 2003, pp. 359–362

4.27. D. Buttari, A. Chini, G. Meneghesso, E. Zanoni, P. Chavarkar, R. Coffie, N. Zhang, S. Heikman, L. Shen, H. Xing, C. Zheng, U. Mishra, IEEE Electron Device Lett. **23**, 118 (2002)

4.28. D. Buttari, A. Chini, G. Meneghesso, E. Zanoni, B. Moran, S. Heikman, N. Zhang, L. Shen, R. Coffie, S. DenBaars, U. Mishra, IEEE Electron Device Lett. **23**, 76 (2002)

4.29. D. Buttari, A. Chini, T. Palacios, R. Coffie, L. Shen, H. Xing, S. Heikman, L. McCarthy, A. Chakraborty, S. Keller, U. Mishra, Appl. Phys. Lett. **83**, 4779 (2003)

4.30. Y. Cai, Y. Zhou, K. Chen, K. Lau, IEEE Electron Device Lett. **26**, 435 (2005)

4.31. P. Chabert, J. Vac. Sci. Technol. B **19**, 212 (2001)

4.32. P. Chabert, N. Proust, J. Perrin, R. Boswell, Appl. Phys. Lett. **76**, 2310 (2000)

4.33. C. Chen, S. Keller, G. Parish, R. Vetury, P. Kozodoy, E. Hu, S. DenBaars, U. Mishra, Y. Wu, Appl. Phys. Lett. **73**, 3147 (1998)

4.34. K. Cheng, J. Lee, J. Lyding, Y. Kim, Y. Kim, K.P. Suh, IEEE Electron Device Lett. **22**, 188 (2001)

4.35. M. Chertouk, M. Dammann, K. Köhler, G. Weimann, IEEE Electron Device Lett. **21**, 97 (2000)

4.36. R. Cheung, S. Withanage, R. Reeves, S. Brown, I. Ben-Yaacov, C. Kirchner, M. Kamp, Appl. Phys. Lett. **74**, 3185 (1999)

4.37. A. Chini, D. Buttari, R. Coffie, S. Heikmann, S. Keller, U. Mishra, Electron. Lett. **40**, 73 (2004)

4.38. A. Chini, D. Buttari, R. Coffie, L. Shen, S. Heikman, A. Chakraborty, S. Keller, U. Mishra, IEEE Electron Device Lett. **25**, 229 (2004)

4.39. A. Chini, J. Wittich, S. Heikman, S. Keller, S. DenBaars, U. Mishra, IEEE Electron Device Lett. **25**, 55 (2004)

4.40. H. Cho, K. Lee, B. Gila, C. Abernathy, S. Pearton, F. Ren, Solid-State Electron. **47**, 1597 (2003)

4.41. H. Cho, K. Lee, B. Gila, C. Abernathy, S. Pearton, F. Ren, Solid-State Electron. **47**, 1757 (2003)

4.42. H. Cho, P. Leeungsnawarat, D. Hays, S. Pearton, S. Chu, R. Strong, C. Zetterling, M. Östling, Appl. Phys. Lett. **76**, 739 (2000)

4.43. H. Cho, C. Vartuli, C. Abernathy, S. Donovan, S. Pearton, R. Schul, J. Han, Solid-State Electron. **42**, 2277 (1998)

4.44. H. Chou, T. Lee, S. Huang, H. Weng, M. Tsai, J. Lee, M. Chertouk, D. Tu, P. Chao, C. Wu, in *Proceedings of the International Conference on GaAs Manufacturing Technology*, Scottsdale, 2003, p. 10.2

4.45. C. Chu, C. Yu, Y. Wang, J. Tsai, F. Lai, S. Wang, Appl. Phys. Lett. **77**, 3423 (2000)

4.46. E. Chumbes, J. Smart, T. Prunty, J. Shealy, in *IEDM Technical Digest*, San Francisco, 2000, pp. 385–388

4.47. R. Coffie, D. Buttari, S. Heikmann, S. Keller, A. Chini, L. Shen, U. Mishra, IEEE Electron Device Lett. **23**, 588 (2002)

4.48. R. Coffie, L. Shen, G. Parish, A. Chini, D. Buttari, S. Heikman, S. Keller, U. Mishra, Electron. Lett. **39**, 1419 (2003)

4.49. X. Dang, E. Yu, E. Piner, B. McDermott, J. Appl. Phys. **90**, 1357 (2001)

4.50. Y. Dora, A. Chakraborty, S. Heikman, L. McCarthy, S. Keller, S. DenBaars, U. Mishra, IEEE Electron Device Lett. **7**, 529 (2006)

4.51. Y. Dora, A. Chakraborty, L. McCarthy, S. Keller, S. DenBaars, U. Mishra, IEEE Electron Device Lett. **27**, 713 (2006)

4.52. Y. Dora, C. Suh, A. Chakraborty, S. Heikman, S. Chandrasekarana, V. Mehrotra, U. Mishra, in *Device Research Conference*, Santa Barbara, 2005, pp. 191–192

4.53. D. Dumka, C. Lee, H. Tserng, P. Saunier, M. Kumar, Electron. Lett. **40**, 1023 (2003)

4.54. R. Dupuis, C. Eiting, P. Grundowski, H. Hsia, Z. Tang, D. Becher, H. Kuo, G. Stillman, M. Feng, J. Electron. Mater. **28**, 319 (1999)

4.55. L. Eastman, V. Tilak, J. Smart, B. Green, E. Chumbes, R. Dimitrov, H. Kim, O. Ambacher, N. Weimann, T. Prunty, M. Murphy, W. Schaff, J. Shealy, IEEE Trans. Electron Devices **48**, 479 (2001)

4.56. C. Eddy, B. Molnar, J. Electron. Mater. **28**, 314 (1999)

4.57. A. Edwards, J. Mittereder, S. Binari, D. Katz, D. Storm, J. Roussos, IEEE Electron Device Lett. **26**, 225 (2005)

4.58. T. Egawa, G. Zhao, H. Ishikawa, M. Umeno, T. Jimbo, IEEE Trans. Electron Devices **48**, 603 (2001)

4.59. A. Endoh, Y. Yamashita, K. Ikeda, M. Higashiwaki, K. Hikosaka, T. Matsui, S. Hiyamizu, T. Mikura, Jpn. J. Appl. Phys. **43**, 2255 (2004)

4.60. A. Endoh, Y. Yamashita, K. Ikeda, M.H.A. Hikosaka, T. Matsui, S. Hiyamizu, T. Mimura, Jpn. J. Appl. Phys. **43**, 2255 (2004)

4.61. D. Fanning, L. Witkowski, J. Stidham, H. Tserng, M. Muir, P. Saunier, in *Proceedings of the International Conference on GaAs Manufacturing Technology*, San Diego, 2002, pp. 83–86

4.62. D. Fanning, L. Witkowski, C. Lee, D. Dumka, H. Tserng, P. Saunier, W. Gaiewski, E. Piner, K. Linthicum, J. Johnson, in *Proceedings of the International Conference on GaAs Manufacturing Technology*, New Orleans, 2005, p. 8.3

4.63. G. Franz, Phys. Stat. Sol. A **159**, 137 (1997)

4.64. M. Fu, V. Sarvepalli, R. Singh, C. Abernathy, X. Cao, S. Pearton, J. Sekhar, Solid-State Electron. **42**, 2335 (1998)

4.65. B. Gaffey, L. Guido, X. Wang, T. Ma, IEEE Trans. Electron Devices **48**, 458 (2001)

4.66. D. Gao, M. Wijesundara, C. Carraro, R. Howe, R. Maboudian, J. Vac. Sci. Technol. B **22**, 513 (2004)

4.67. B. Gila, J. Kim, B. Luo, A. Onstine, W. Johnson, F. Ren, C. Abernathy, S. Pearton, Solid-State Electron. **47**, 2139 (2003)

4.68. J. Gillespie, A. Crespo, R. Fitch, G. Jessen, D. Langley, N. Moser, D. Via, M. Williams, M. Yannazzi, in *Proceedings of the International Conference on GaAs Manufacturing Technology*, New Orleans, 2005, p. 13.2

4.69. S. Golka, W. Schrenk, G. Strasser, in *Proceedings of German Microwave Forum*, Vienna, 2005, pp. 189–192

4.70. D. Gotthold, S. Guo, R. Birkhahn, B. Albert, D. Florescu, B. Peres, J. Electron. Mater. **33**, 408 (2004)

4.71. J. Graff, E. Schubert, A. Osinsky, in *Proceedings of IEEE/Cornell Conference on High Performance Devices*, Ithaca, 2000, pp. 28–32

4.72. B. Green, K. Chu, E. Chumbes, J. Smart, J. Shealy, L. Eastman, IEEE Electron Device Lett. **21**, 268 (2000)

4.73. J. Grenko, C. Reynolds, R. Schlesser, K. Bachmann, Z. Rietmeier, R. Davis, Z. Sitar, MRS Internet J. Nitride Semicond. Res. **9**, (2004)

4.74. R. Grundbacher, R. Lai, M. Nishimoto, T. Chin, Y. Chen, M. Barsky, T. Block, D. Streit, IEEE Electron Device Lett. **20**, 517 (1999)

4.75. J. Guo, F. Pan, M. Feng, R. Guo, P. Chou, C. Chang, J. Appl. Phys. **80**, 1623 (1996)

4.76. Q. Guo, O. Kato, A. Yoshida, J. Electrochem. Soc. **139**, 2008 (1992)

4.77. E. Haberer, C.H. Chen, A. Abare, M. Hansen, S. DenBaars, L. Coldren, U. Mishra, E. Hu, Appl. Phys. Lett. **76**, 3941 (2000)

4.78. P. Hacke, T. Detchprohm, K. Hiramatsu, N. Sawaki, Appl. Phys. Lett. **63**, 2676 (1993)

4.79. M. Hampson, S. Shen, R. Schwindt, R. Price, U. Chowdhury, M. Wong, T. Zhu, D. Yoo, R. Dupuis, M. Feng, IEEE Electron Device Lett. **25**, 238 (2004)

4.80. P. Hansen, L. Shen, Y. Wu, A. Stonas, Y. Terao, S. Heikman, D. Buttari, T. Taylor, S. DenBaars, U. Mishra, R. York, J. Speck, J. Vac. Sci. Technol. B **22**, 2479 (2004)

4.81. W. Hanson, R. Borges, J. Brown, J. Cook, T. Gehrke, J. Johnson, K. Linthicum, S. Peters, E. Piner, P. Rajagopal, J. Robert, S. Singhal, R. Therrien, A. Vescan, in *Proceedings of the International Conference on GaAs Manufacturing Technology*, Miami, 2004, pp. 107–110

4.82. P. Hartlieb, R. Davies, R. Nemanich, in *Nitride Semiconductors: Handbook on Materials and Device*, ed. by P. Ruterana, M. Albrecht, J. Neugebauer (Wiley-VCH, Weinheim, 2003), Chap. 10, pp. 491–523

4.83. T. Hashizume, J. Kotani, H. Hasegawa, Appl. Phys. Lett. **84**, 4884 (2004)

4.84. T. Hashizume, S. Ootomo, H. Hasegawa, Appl. Phys. Lett. **83**, 2952 (2003)

4.85. T. Hashizume, S. Ootomo, T. Inakagi, H. Hasegawa, J. Vac. Sci. Technol. B **21**, 1828 (2003)

4.86. T. Hattori, G. Nakamura, S. Nomura, T. Ichise, A. Masuda, H. Matsumura, in *GaAs IC Symposium Technical Digest*, Anaheim, 1997, pp. 78–80

4.87. N. Hefyene, S. Cristoloveanu, G. Ghibaudo, P. Gentil, Y. Moriyasu, T. Morishita, M. Matsui, A. Yasujima, Solid-State Electron. **44**, 1711 (2000)

4.88. H. Hendriks, J. Crites, G. D'Urso, R. Fox, T. Lepkowski, B. Patel, in *Proceedings of the International Conference on GaAs Manufacturing Technology*, Las Vegas, 2001, pp. 181–184

4.89. M. Higashiwaki, N. Hirose, T. Matsui, IEEE Electron Device Lett. **26**, 139 (2005)

4.90. M. Higashiwaki, N. Hirose, T.M.A. Mimura, J. Appl. Phys. **100**, 033714 (2006)

4.91. M. Higashiwaki, T. Matsui, Jpn. J. Appl. Phys. **44**, L475 (2005)

4.92. M. Higashiwaki, T. Matsui, Jpn. J. Appl. Phys. **45**, L1111 (2006)

4.93. M. Higashiwaki, T. Matsui, T. Mimura, in *Device Research Conference*, State College PA, 2006, pp. 149–150

4.94. M. Higashiwaki, T. Matsui, T. Mimura, IEEE Electron Device Lett. **27**, 16 (2006)

4.95. M. Higashiwaki, T. Mimura, T. Matsui, IEEE Trans. Electron Devices **54**, 1566 (2007)

4.96. M. Hikita, M. Yanagihara, K. Nakazawa, H. Ueno, Y. Hirose, T. Ueda, Y. Uemoto, T. Tanaka, D. Ueda, T. Egawa, IEEE Trans. Electron Devices **52**, 1963 (2005)

4.97. J. Hilsenbeck, Dissertation, Technische Universität Karlsruhe, 2001

4.98. J. Hilsenbeck, F. Lenk, R. Lossy, J. Würfl, K. Köhler, H. Obloh, in *Proceedings of the International Symposium on Compound Semiconductors*, Monterey, 2000, pp. 351–356

4.99. J. Hilsenbeck, W. Rieger, E. Neubauer, W. John, G. Tränkle, J. Würfl, A. Ramakrishan, H. Obloh, Phys. Stat. Sol. A **176**, 183 (1999)

4.100. J. Hilsenbeck, W. Rieger, J. Würfl, R. Dimitrov, O. Ambacher, in *Proceedings of the International Symposium on Compound Semiconductors*, Berlin, 1999, pp. 507–510

4.101. J. Hong, J. Lee, C. Vartuli, J. Mackenzie, S. Donovan, C. Abernathy, R. Crockett, S. Pearton, J. Zolper, Solid-State Electron. **41**, 681 (1997)

4.102. S. Hsu, D. Pavlidis, in *GaAs IC Symposium Technical Digest*, San Diego, 2003, pp. 119–122

4.103. H. Huang, C. Kao, J. Tsai, C. Yu, C. Chu, J. Lee, S. Kuo, C. Lin, H. Kuo, S. Wang, Mater. Sci. Eng. B **107**, 237 (2004)

4.104. J. Hwang, W. Schaff, B. Green, H. Cha, L. Eastman, Solid-State Electron. **48**, 363 (2004)

4.105. J. Ibbetson, P. Fini, K. Ness, S. DenBaars, J. Speck, U. Mishra, Appl. Phys. Lett. **77**, 250 (2000)

4.106. S. Inaba, K. Okano, S. Matsuda, M. Fujiwara, A. Hokazono, K. Adachi, K. Ohuchi, H. Suto, H. Fukui, T. Shimizu, S. Mori, H. Oguma, A. Murakoshi, T. Itani, T. Iinuma, T. Kudo, H. Shibata, S. Taniguchi, T. Matsushita, S. Magoshi, Y. Watanabe, M. Takayanagi, A. Azuma, H. Oyamatsu, K. Suguro, Y. Katsumata, Y. Toyoshima, H. Ishiuchi, in *IEDM Technical Digest*, Washington DC, 2001, pp. 641–644

4.107. K. Inoue, Y. Ikeda, H. Masato, T. Matsuno, K. Nishii, in *IEDM Technical Digest*, Washington DC, 2001, pp. 577–580

4.108. Y. Irokawa, J. Kim, F. Ren, K. Baik, B. Gila, C. Abernathy, S. Pearton, C. Pan, G. Chen, J. Chyi, S. Park, Solid-State Electron. **48**, 827 (2004)

4.109. J. Izpura, Semicond. Sci. Technol. **17**, 1293 (2002)

4.110. C. Jeon, H. Jang, J. Lee, Appl. Phys. Lett. **82**, 391 (2003)

4.111. C. Jeon, J. Lee, J. Appl. Phys. **95**, 698 (2004)

4.112. C. Jeong, D. Kim, K. Kim, G. Yeom, Jpn. J. Appl. Phys. **41**, 6206 (2002)

4.113. G. Jessen, R. Fitch, J. Gillespie, G. Via, N. Moser, M. Yannuzzi, A. Crespo, R. Dettmer, T. Jenkins, in *GaAs IC Symposium Technical Digest*, San Diego, 2003, pp. 277–279

4.114. J. Johnson, J. Gao, K. Lucht, J. Williamson, C. Strautin, J. Riddle, R. Therrien, P. Rajagopal, J. Roberts, A. Vescan, J. Brown, A. Hanson, S. Singhal, R. Borges, E. Piner, K. Linthicum, Proc. Electrochem. Soc. **7**, 405 (2004)

4.115. J. Johnson, A. Zhang, W. Luo, F. Ren, S. Pearton, S. Park, Y. Park, J. Chyi, IEEE Trans. Electron Devices **49**, 32 (2002)

4.116. H. Kambayashi, T. Wada, N. Ikeda, S. Yoshida, in *GaN, AlN, InN and Related Materials*, ed. by M. Kuball, T. Myers, J. Redwing, T. Mukai. Materials Research Society Symposium Proceedings vol. 892, Warrendale PA, 2006, p. FF05-03

4.117. Y. Kamo, T. Kunii, H. Takeuchi, Y. Yamamoto, M. Totsuka, T. Shiga, H. Minami, T. Kitano, S. Miyakuni, T. Oku, A. Inoue, T. Nanjo, H. Chiba, M. Suita, T. Oishi, Y. Abe, Y. Tsuyama, R. Shirahana, H. Ohtsuka, K. Iyomasa, K. Yamanaka, M. Hieda, M. Nakayama, T. Ishikawa, T. Takagi, K. Marumoto, Y. Matsuda, in *IEEE International Microwave Symposium Digest*, Long Beach, 2005, pp. 495–498

4.118. M. Kanamura, T. Kikkawa, T. Iwai, K. Imanishi, T. Kubo, K. Joshin, in *IEDM Technical Digest*, Washington DC, 2005, pp. 572–575

4.119. B. Kang, R. Mehandru, S. Kim, F. Ren, R. Fitch, J. Gillespie, N. Moser, G. Jessen, T. Jenkins, R. Dettmer, D. Via, A. Crespo, B. Gila, R. Abernathy, S. Pearton, Appl. Phys. Lett. **84**, 4635 (2004)

4.120. S. Karmalkar, U. Mishra, IEEE Trans. Electron Devices **48**, 73 (2001)

4.121. T. Kato, K. Hayashi, Y. Sasaki, T. Kato, IEEE Trans. Electron Devices **34**, 753 (1987)

4.122. H. Kawai, M. Hara, F. Nakamura, S. Imanaga, Electron. Lett. **34**, 592 (1998)

4.123. H. Kawaura, T. Sakamoto, T. Baba, Y. Ochiai, J. Fujita, J. Sone, IEEE Trans. Electron Devices **47**, 856 (2000)

4.124. S. Keller, Y. Wu, G. Parish, N. Ziang, J. Xu, B. Keller, S. DenBaars, U. Mishra, IEEE Trans. Electron Devices **48**, 552 (2001)

4.125. F. Khan, B. Roof, I. Adesida, J. Electron. Mater. **3**, 212 (2001)

4.126. R. Kiefer, R. Quay, S. Müller, K. Köhler, F. van Raay, B. Raynor, W. Pletschen, H. Massler, S. Ramberger, M. Mikulla, G. Weimann, in *Proceedings of Lester Eastman Conference on High Performance Devices*, Newark, 2002, pp. 502–504

4.127. T. Kikkawa, M. Nagahara, T. Kimura, S. Yokokawa, S. Kato, M. Yokoyama, Y. Tateno, K. Horino, K. Domen, Y. Yamaguchi, N. Hara, K. Joshin, in *IEEE International Microwave Symposium Digest*, Seattle, 2002, pp. 1815–1818

4.128. T. Kikkawa, M. Nagahara, N. Okamoto, Y. Tateno, Y. Yamaguchi, N. Hara, K. Joshin, P. Asbeck, in *IEDM Technical Digest*, Washington DC, 2001, pp. 585–588

4.129. B. Kim, J. Lee, H. Park, Y. Park, T. Kim, J. Electron. Mater. **27**, L32 (1997)

4.130. H. Kim, J. Lee, W. Lu, Phys. Stat. Sol. A **202**, 841 (2005)

4.131. H. Kim, R. Thompson, V. Tilak, T. Prunty, J. Shealy, L. Eastman, IEEE Electron Device Lett. **24**, 421 (2003)

4.132. J. Kim, F. Ren, B. Gila, C. Abernathy, S. Pearton, Appl. Phys. Lett. **82**, 739 (2003)

4.133. J. Kim, J. Je, J. Lee, Y. Park, T. Kim, I. Jung, B. Lee, J. Lee, J. Electron. Mater. **30**, L8 (2001)

4.134. S. Kim, B. Bang, F. Ren, J. D'Entremont, J. Blumenfeld, T. Cordock, S. Pearton, J. Electron. Mater. **33**, 477 (2004)

4.135. P. Klein, S. Binari, J. Freitas, A. Wickenden, J. Appl. Phys. **88**, 2843 (2000)

4.136. P. Klein, S. Binari, K. Ikosso-Anastasiou, A. Wickenden, D. Koleske, R. Henry, D. Katzer, Electron. Lett. **37**, 661 (2001)

4.137. Y. Knafo, I. Toledo, I. Hallakoun, J. Kaplun, G. Bunin, T. Baksht, B. Hadad, Y. Shapira, in *Proceedings of the International Conference on GaAs Manufacturing Technology*, New Orleans, 2005, p. 13.4

4.138. E. Kohn, in *Proceedings of SODC*, Nanjing, 2000, pp. 13–16

4.139. G. Koley, V. Tilak, L. Eastman, M. Spencer, IEEE Trans. Electron Devices **50**, 886 (2003)

4.140. N. Kolias, T. Kazior, in *IMS Workshop Advances in GaN-based Device and Circuit Technology: Modeling and Applications*, Fort Worth, 2004

4.141. P. Kordos, J. Bernat, D. Gregusova, M. Marso, H. Lüth, Semicond. Sci. Technol. **21**, 67 (2006)

4.142. P. Kordos, J. Bernat, M. Marso, H. Lüth, F. Rampazzo, G. Tamiazzo, R. Pierobon, G. Meneghesso, Appl. Phys. Lett. **86**, 253511 (2005)

4.143. A. Koudymov, V. Adivarahan, J. Yang, G. Simin, M. Khan, IEEE Electron Device Lett. **26**, 704 (2005)

4.144. A. Koudymov, G. Simin, M. Khan, A. Tarakji, R. Gaska, M. Shur, IEEE Electron Device Lett. **24**, 680 (2003)

4.145. P. Kozodoy, J. Ibbetson, H. Mar, P. Fini, S. Keller, J. Speck, S. DenBaars, U. Mishra, Appl. Phys. Lett. **73**, 975 (1998)

4.146. O. Krüger, C. Scholz, R. Grundmüller, H. Wittrich, P. Wolter, J. Würfl, G. Tränkle, in *Euro-Med. Symposium on Laser Induced Breakdown Spectroscopy*, Crete, 2003, p. 22

4.147. S. Kucheyev, J. Williams, S. Pearton, Mat.Sci.Eng. R **33**, 51 (2001)

4.148. V. Kumar, J. Lee, A. Kuliev, R. Schwindt, R. Birkhahn, D. Gotthold, S. Guo, B. Albert, I. Adesida, Electron. Lett. **39**, 1609 (2003)

4.149. V. Kumar, W. Lu, F. Khan, R. Schwindt, A. Kuliev, J.Y.M. Khan, I. Adesida, in *IEDM Technical Digest*, Washington DC, 2001, pp. 573–576

4.150. V. Kumar, W. Lu, R. Schwindt, A. Kuliev, G. Simin, J. Yang, M. Khan, I. Adesida, IEEE Electron Device Lett. **23**, 455 (2002)

4.151. T. Kunii, M. Totsuka, Y. Kamo, Y. Yamamoto, H. Takeuchi, Y. Shimada, T. Shiga, H. Minami, T. Kitano, S. Miyakuni, S. Nakatsuka, A. Inoue, T. Oku, T. Nanjo, T. Oishi, T. Ishikawa, Y. Matsuda, in *Compound Semiconductor IC Symposium Technical Digest*, Monterey, 2004, pp. 197–200

4.152. B. Lee, S. Jung, J. Lee, Y. Park, M. Paek, K. Cho, Semicond. Sci. Technol. **16**, 471 (2001)

4.153. C. Lee, P. Saunier, H. Tserng, in *Compound Semiconductor IC Symposium Technical Digest*, Palm Springs, 2005, pp. 177–180

4.154. C. Lee, H. Kao, Appl. Phys. Lett. **76**, 2364 (2000)

4.155. J. Lee, D. Liu, H. Kim, W. Lu, Solid-State Electron. **48**, 1855 (2004)

4.156. J. Lee, K. Chang, I. Lee, S. Park, J. Electrochem. Soc. **147**, 1859 (2000)

4.157. J. Lee, C. Huh, D. Kim, S. Park, Semicond. Sci. Technol. **18**, 530 (2003)

4.158. H. Leier, A. Wieszt, R. Bethasch, H. Tobler, A. Vescan, R. Dietrich, A. Schurr, H. Sledzik, J. Birbeck, R. Balmer, T. Martin, in *Proceedings of European Gallium Arsenide Other Compound Semiconductors Application Symposium GAAS*, London, 2001, pp. 49–52

4.159. B. Leung, N. Chan, W. Fong, C. Zhu, S. Ng, H. Lui, K. Tong, C. Surya, L. Lu, W. Ge, IEEE Trans. Electron Devices **49**, 314 (2003)

4.160. C. Lin, W. Wang, P. Lin, C. Lin, Y. Chang, Y. Chan, IEEE Electron Device Lett. **26**, 710 (2005)

4.161. M. Lin, Z. Fan, Z. Ma, L. Allen, H. Morkoc, Appl. Phys. Lett. **64**, 887 (1994)

4.162. M. Lin, Z. Ma, F. Huang, Z. Fan, L. Allen, H. Morkoc, Appl. Phys. Lett. **64**, 1003 (1994)

4.163. Q. Liu, S. Lau, Solid-State Electron. **42**, 677 (1998)

4.164. T. Lodhi, J. McMonagle, R. Davis, D. Brookbanks, S. Combe, M. Clausen, M.F. O'Keefe, A. Collar, J. Atherton, in *Compound Semiconductor IC Symposium Technical Digest*, San Antonio, 2006, pp. 125–128

4.165. R. Lossy, P. Heymann, J. Würfl, N. Chaturvedi, S. Müller, K. Köhler, in *Proceedings of European Gallium Arsenide Other Compound Semiconductors Application Symposium GAAS*, Milano, 2002, pp. 23–27

4.166. R. Lossy, J. Hilsenbeck, J. Würfl, K. Köhler, H. Obloh, Phys. Stat. Sol. A **188**, 263 (2001)

4.167. R. Lossy, A. Liero, O. Krüger, J. Wüerl, G. Tränkle, Phys. Stat. Sol. C **3**, 482 (2005)

4.168. R. Lossy, A. Liero, J. Würfl, G. Tränkle, in *IEDM Technical Digest*, Washington DC, 2005, pp. 580–582

4.169. B. Luo, R. Mehandru, J. Kim, F. Ren, B. Gila, A. Onstine, C. Abernathy, S. Pearton, D. Gotthold, R. Birkhan, B. Peres, R. Fitch, N. Moser, J. Gillispie, G. Jessen, T. Jenkins, M. Yannuzi, G. Via, A. Crespo, Solid-State Electron. **47**, 1781 (2003)

4.170. B. Luo, R. Mehandru, J. Kim, F. Ren, B. Gila, A. Onstine, C. Abernathy, S. Pearton, D. Gotthold, R. Birkhan, B. Peres, R. Fitch, N. Moser, J. Gillispie, G. Jessen, T. Jenkins, M. Yannuzi, G. Via, A. Crespo, Solid-State Electron. **48**, 355 (2004)

4.171. B. Luo, R. Mehandru, J. Kim, F. Ren, B. Gila, A. Onstine, C. Abernathy, S. Pearton, R. Fitch, J. Gillespie, R. Dellmer, T. Jenkins, J. Sewell, D. Via, A. Crespo, Solid-State Electron. **46**, 2185 (2002)

4.172. L. Ma, K. Adeni, C. Zeng, Y. Jin, K. Dandu, Y. Saripalli, M. Johnson, D. Barlage, in *Proceedings of the International Conference on GaAs Manufacturing Technology*, Vancouver, 2006, pp. 105–109

4.173. T. Makimoto, Y. Yamauchi, K. Kumakura, Appl. Phys. Lett. **84**, 1964 (2004)

4.174. K. Matocha, R. Gutmann, T. Chow, IEEE Trans. Electron Devices **50**, 1200 (2003)

4.175. L. McCarthy, I. Smorchkova, H. Xing, P. Kozodoy, P. Fini, J. Limb, D. Pulfrey, J. Speck, M. Rodwell, S. DenBaars, U. Mishra, IEEE Trans. Electron Devices **48**, 543 (2001)

4.176. G. Meneghesso, A. Chini, E. Zanoni, M. Manfredi, M. Pavesi, B. Boudart, C. Gaquiere, in *IEDM Technical Digest*, San Francisco, 2000, pp. 389–392

4.177. G. Meneghesso, G. Verzellesi, R. Pierobon, F. Rampazzo, A. Chini, U. Mishra, C. Canali, E. Zanoni, IEEE Trans. Electron Devices **51**, 1554 (2004)

4.178. D. Meyer, J. Flemish, J. Redwing, in *Proceedings of the International Conference on GaAs Manufacturing Technology*, Austin, 2007, pp. 305–308

4.179. S.D. Meyer, C. Charbonniaud, R. Quere, M. Campovecchio, R. Lossy, J. Würfl, in *IEEE International Microwave Symposium Digest*, Philadelphia, 2003, pp. 455–458

4.180. M. Micovic, A. Kurdhoghlian, H. Moyer, P. Hashimoto, A. Schmitz, I. Milosavljevic, P. Willadsen, W. Wong, J. Duvall, M. Hu, M. Wetzel, D. Chow, in *Compound Semiconductor IC Symposium Technical Digest*, Palm Springs, 2005, pp. 173–176

4.181. J. Mileham, S. Pearton, C. Abernathy, J. MacKenzie, R. Shul, S. Kilcoyne, Appl. Phys. Lett. **67**, 1119 (1995)

4.182. M. Minsky, M. White, E. Hu, Appl. Phys. Lett. **68**, 1531 (1996)

4.183. N. Miura, T. Oishi, T. Nanjo, M. Suita, Y. Abe, T. Ozeki, H. Ishikawa, T. Egawa, IEEE Trans. Electron Devices **51**, 297 (2004)

4.184. T. Mizutani, Y. Ohno, M. Akita, S. Kishimoto, K. Maezawa, IEEE Trans. Electron Devices **50**, 2015 (2003)

4.185. J. Moon, D. Wong, T. Hussain, M. Micovic, P. Deelmann, H. Ming, M. Antcliffe, C. Ngo, P. Hashimoto, L. McCray, in *Device Research Conference*, Santa Barbara, 2002, pp. 23–24

4.186. J. Moon, S. Wu, D. Wong, I. Milosavljevic, A. Conway, P. Hashimoto, M. Hu, M. Antcliffe, M. Micovic, IEEE Electron Device Lett. **26**, 348 (2005)

4.187. T. Mori, M. Kase, K. Hashimoto, M. Kojima, T. Sugii, in *IEDM Technical Digest*, Washington DC, 2003, pp. 623–626

4.188. H. Morkoc, *Nitride Semiconductors and Devices.* Springer Series in Materials Science, vol. 32 (Springer, Berlin Heidelberg New York, 1999)

4.189. H. Morkoc, *Nitride Semiconductors and Devices.* Springer Series in Materials Science, vol. 32 (Springer, Berlin Heidelberg New York, 1999) Chap. 6

4.190. Z. Mouffak, N. Medelci-Djezzar, C. Boney, A. Bensaoula, L. Trombetta, MRS Internet J. Nitride Semicond. Res. **8**, 7 (2003)

4.191. T. Murata, M. Hikita, Y. Hirose, Y. Uemoto, K. Inoue, T. Tanaka, D. Ueda, IEEE Trans. Electron Devices **52**, 1042 (2005)

4.192. S. Nakamura, M. Senoh, T. Mukai, Appl. Phys. Lett. **62**, 2390 (1993)

4.193. T. Nanjo, N. Miura, T. Oishi, M. Suita, Y. Abe, T. Ozeki, S. Nakatsuka, A. Inoue, T. Ishikawa, Y. Matsuda, H. Ishikawa, T. Egawa, Jpn. J. Appl. Phys. **43**, 1925 (2004)

4.194. M. Neuburger, J. Allgaier, T. Zimmermann, I. Daumiller, M. Kunze, R. Birkhahn, D. Gotthold, E. Kohn, IEEE Electron Device Lett. **25**, 256 (2004)

4.195. G. Neumark, I. Kuskovsky, H. Jiang (eds.), *Wide Bandgap Light Emitting Materials and Devices* (Wiley-VCH, Weinheim, 2007)

4.196. M. Nishijima, T. Murata, Y. Hirose, M. Hikita, N. Negoro, H. Sakai, Y. Uemoto, K. Inoue, T. Tanaka, D. Ueda, in *IEEE International Microwave Symposium Digest*, Long Beach, 2005, pp. 299–302

4.197. K. Nishizono, M. Okada, M. Kamei, D. Kikuta, K. Tominaga, Y. Ohno, J. Ao, Appl. Phys. Lett. **84**, 3996 (2004)

4.198. S. Nuttinck, S. Pinel, E. Gebara, J. Laskar, M. Harris, in *Proceedings of European Gallium Arsenide Other Compound Semiconductors Application Symposium GAAS*, Munich, 2003, pp. 213–215

4.199. T. Oishi, N. Miura, M. Suita, T. Nanjo, Y. Abe, T. Ozeki, J. Appl. Phys. **94**, 1662 (2003)

4.200. T. Palacios, E. Snow, Y. Pei, A. Chakraborty, S. Keller, S. DenBaars, U.K. Mishra, in *IEDM Technical Digest*, Washington DC, 2005, pp. 787–789

4.201. S. Pearton, J. Zolper, R. Shul, and F. Ren, J. Appl. Phys. **86**, 1 (1999)

4.202. S. Pearton, C. Abernathy, B. Gila, F. Ren, J. Zavada, Y. Park, Solid-State Electron. **48**, 1965 (2002)

4.203. S. Pearton, C. Abernathy, C. Vartuli, Solid-State Electron. **42**, 2269 (1998)

4.204. S. Pearton, R. Shul, F. Ren, MRS Internet J. Nitride Semicond. Res. **5**, 11 (2000)

4.205. A. Ping, D. Selvanathan, C. Youtsey, E. Piner, J. Redwing, I. Adesida, Electron. Lett. **35**, 2141 (1999)

4.206. A. Polyakov, N. Smirnov, A. Govorkov, K. Baik, S. Pearton, B. Luo, F. Ren, J. Zavada, J. Appl. Phys. **94**, 3960 (2003)

4.207. S. Rajan, P. Waltereit, C. Poblenz, S. Heikman, D. Green, J. Speck, U. Mishra, IEEE Electron Device Lett. **25**, 247 (2004)

4.208. C. Ramesh, V. Reddy, C. Choi, Mater. Sci. Eng. B **112**, 30 (2004)

4.209. E. Readinger, J. Robinson, S. Mohney, R. Therrien, Semicond. Sci. Technol. **20**, 389 (2005)

4.210. V. Reddy, S. Kim, J. Song, T. Seong, Solid-State Electron. **48**, 1563 (2004)

4.211. F. Ren, J. Han, R. Hickman, J.V. Hove, P. Chow, J. Klaasen, J. LaRoche, K. Jung, H. Cho, X. Cao, S. Donovan, R. Kopf, R. Wilson, A. Baca, R. Shul, L. Zhang, C. Willison, C. Abernathy, S. Pearton, Solid-State Electron. **44**, 239 (2000)

4.212. F. Ren, J. Lothian, S. Pearton, C. Abernathy, C. Vartuli, J. Mackenzie, R. Wilson, R. Karlicek, J. Electron. Mater. **26**, 1287 (1997)

4.213. P. Roussell, in *Proceedings of the International Conference on GaAs Manufacturing Technology*, Vancouver, 2006, pp. 231–232

4.214. P. Ruterana, M. Albrecht, J. Neugebauer (eds.), *Nitride Semiconductors: Handbook on Materials and Device* (Wiley-VCH, Weinheim, 2003)

4.215. D. Sahoo, R. Lal, H. Kim, V. Tilak, L. Eastman, IEEE Trans. Electron Devices **50**, 1163 (2003)

4.216. W. Saito, M. Kuraguchi, Y. Takada, K. Tsuda, I. Otmura, T. Ogura, IEEE . Trans. Electron Devices **52**, 159 (2005)

4.217. W. Saito, Y. Takada, M. Kuraguchi, K. Tsuda, I. Omura, IEEE Trans. Electron Devices **53**, 356 (2006)

4.218. W. Saito, Y. Takada, M. Kuraguchi, K. Tsuda, I. Omura, I Omura, Jpn. J. Appl. Phys. **43**, 2239 (2004)

4.219. J. Schalwig, G. Müller, U. Karrer, M. Eickhoff, O. Ambacher, M. Stutzmann, L. Görgens, G. Dollinger, Appl. Phys. Lett. **80**, 1222 (2002)

4.220. R. Schul, C. Willison, M. Bridges, J. Han, J. Lee, S. Pearton, C. Abernathy, J. Mackenzie, S. Donovan, Solid-State Electron. **42**, 2269 (1998)

4.221. J. Shealy, in *GaAs-IC Symposium Short Course: Emerging Technologies from Defense to Commercial*, San Diego, 2003

4.222. J. Shealy, in *CSIC-IC Symposium Short Course: Emerging Technologies from Defense to Commercial*, Palm Springs, 2005

4.223. J. Sheats, B. Smith (eds.), *Microlithography: Science and Technology* (Marcel Dekker, New York, 1998)

4.224. B. Shelton, D. Lambert, J. Huang, M. Wong, U. Chowdhury, T. Zhu, H. Kwon, Z. Weber, M. Benarama, M. Feng, R. Dupuis, IEEE Trans. Electron Devices **48**, 490 (2001)

4.225. L. Shen, R. Coffie, D. Buttari, S. Heikman, A. Chakraborty, A. Chini, S. Keller, S. DenBaars, U. Mishra, IEEE Electron Device Lett. **25**, 7 (2004)

4.226. J. Sheu, Y. Su, G. Chi, P. Koh, M. Jou, C. Chang, C. Liu, W. Hung, Appl. Phys. Lett. **74**, 2340 (1999)

4.227. S. Sheu, J. Liou, C. Huang, IEEE Trans. Electron Devices **45**, 326 (1998)

4.228. K. Shiojima, D. McInturrf, J. Woodall, P. Grudowski, C. Eiting, R. Dupuis, J. Electron. Mater. **28**, 228 (1999)

4.229. E. Silkowski, Y. Yeo, R. Hengehold, M. Khan, T. Lei, K. Evans, C. Cerny, in *MRS Symposium*, First International Conference on Nitride Semiconductors, vol. 395, Boston, 1996, pp. 813–818.

4.230. G. Simin, A. Koudymov, A. Tarakji, X. Hu, J. Yang, M. Khan, M. Shur, R. Gaska, Appl. Phys. Lett. **79**, 2651 (2001)

4.231. D. Siriex, D. Barataud, R. Sommet, O. Noblanc, Z. Ouarch, C. Brylinski, J. Teyssier, R. Quere, in *IEEE International Microwave Symposium Digest*, Boston, 2000, pp. 765–768

4.232. S. Sivakumar, in *IEDM Technical Digest*, San Francisco, 2006, pp. 985–988

4.233. D. Stocker, E. Schubert, J. Redwing, Appl. Phys. Lett. **73**, 2654 (1998)

4.234. S. Strite, P. Epperlein, A. Dommann, A. Rockett, R. Broom, in *MRS Symposium*, First International Conference on Nitride Semiconductors, vol. 395, Boston, 1996, pp. 795–800.

4.235. Y. Su, S. Chang, T. Kuan, C. Ko, J. Webb, W. Lan, Y. Cherng, S. Chen, Mater. Sci. Eng. B **110**, 260 (2004)

4.236. C.S. Suh, A. Chini, Y. Fu, C. Poblenz, J.S. Speck, U.K. Mishra, in *Device Research Conference*, State College PA, 2006, pp. 163–166

4.237. Y. Sun, L. Eastman, IEEE Trans. Electron Devices **52**, 1689 (2005)

4.238. K. Suzue, S. Mohammad, Z. Fan, W. Kim, O. Aktas, A. Botchkarev, H. Morkoc, J. Appl. Phys. **80**, 4467 (1996)

4.239. H. Takenaka, D. Ueda, IEEE Trans. Electron Devices **43**, 238 (1996)

4.240. K. Tan, D. Streit, R. Dia, S. Wang, A. Han, P. Chow, T. Trinh, P. Liu, J. Velebir, H. Yen, IEEE Electron Device Lett. **12**, 213 (1991)

4.241. W. Tan, P. Houston, G. Hill, R. Airey, P. Parbook, J. Electron. Mater. **33**, 400 (2004)

4.242. W. Tan, M. Uren, P. Houston, R. Green, R. Balmer, T. Martin, IEEE Electron Device Lett. **27**, 1 (2006)

4.243. R. Therrien, S. Singhal, J. Johnson, W. Nagy, R. Borges, A. Chaudhari, A. Hanson, A. Edwards, J. Marquart, P. Rajagopal, C. Park, I. Kizilyalli, K. Linthicum, in *IEDM Technical Digest*, Washington DC, 2005, pp. 568–571

4.244. R. Thompson, V. Kaper, T. Prunty, J. Shealy, in *GaAs IC Symposium Technical Digest*, San Diego, 2003, pp. 298–300

4.245. R. Thompson, T. Prunty, V. Kaper, J. Shealy, IEEE Trans. Electron Devices **51**, 292 (2004)

4.246. V. Tilak, B. Green, H. Kim, R. Dimitrov, J. Smart, W. Schaff, J. Shealy, L. Eastman, in *Proceedings of the International Symposium Compound Semiconductors*, Monterey, 2000, pp. 357–363

4.247. J. Torvik, J. Pankove, B.V. Zeghbroeck, IEEE Trans. Electron Devices **46**, 1326 (1999)

4.248. D. Tossell, K. Powell, M. Bourke, Y. Song, in *Proceedings of the International Conference on GaAs Manufacturing Technology*, St. Louis, 2000, pp. 79–82

4.249. S. Trassaert, B. Boudart, C. Gaquiere, D. Theron, Y. Crosnier, F. Huet, M. Poisson, Electron. Lett. **35**, 1386 (1999)

4.250. H. Tseng, C. Capasso, J. Schaeffer, E. Hebert, P. Tobin, D.C. Gilmer, D. Triyoso, M.E. Ramon, S. Kalpat, E. Luckowski, W. Taylor, Y. Jeon, O. Adetutu, R. Hegde, R. Noble, M. Jahanbani, C.E. Chemali, B. White, in *IEDM Technical Digest*, San Francisco, 2004, pp. 821–824

4.251. Y. Uemoto, D. Shibata, M. Yanagihara, H. Ishida, H. Matsuo, S. Nagai, N. Batta, M. Li, T. Ueda, T. Tanaka, D. Ueda, in *IEDM Technical Digest*, Washington DC, 2007, pp. 861–864

4.252. M. Uren, T. Martin, M. Kuball, J. Hayes, B. Hughes, K. Hilton, R. Balmer, in *IMS Workshop Wide Bandgap Technologies*, Seattle, 2002

4.253. M. Uren, K. Nash, R. Balmer, T. Martin, E. Morvan, N. Caillas, S. Delage, D. Ducatteau, B. Grimbert, J. de Jaeger, IEEE Trans. Electron Devices **53**, 395 (2006)

4.254. I. Usov, N. Parikh, D. Thomson, R. Davis, MRS Internet J. Nitride Semicond. Res. **7**, 1 (2002)

4.255. F. van Raay, R. Quay, R. Kiefer, W. Fehrenbach, W. Bronner, M. Kuri, F. Benkhelifa, H. Massler, S. Müller, M. Mikulla, M. Schlechtweg, G. Weimann, in *Proceedings of European Gallium Arsenide Other Compound Semiconductors Application Symposium GAAS*, Paris, 2005, pp. 233–236

4.256. C. Varmazis, G. D'Urso, H. Hendricks, Semiconduct. Int. **23**, 87 (2000)

4.257. C. Vartuli, S. Pearton, C. Abernathy, J.M. Kenzie, M. Lovejoy, R. Shul, J. Zolper, A. Baca, M. Hagerott-Crawford, K. Jones, F. Ren, Solid-State Electron. **41**, 531 (1997)

4.258. H. Venugopalan, S. Mohney, Appl. Phys. Lett. **73**, 1242 (1998)

4.259. G. Verzellesi, R. Pierobon, F. Rampazzo, G. Meneghesso, A. Chini, U. Mishra, C. Canali, E. Zanoni, in *IEDM Technical Digest*, San Francisco, 2002, pp. 689–692

4.260. R. Vetury, PhD thesis, University of California Santa Barbara, Santa Barbara, 2000

4.261. R. Vetury, Y. Wu, P.T. Fini, G. Parish, S. Keller, S. DenBaars, U. Mishra, in *IEDM Technical Digest*, San Francisco, 1998, pp. 55–58

4.262. R. Vetury, N. Zhang, S. Keller, U. Mishra, IEEE Trans. Electron Devices **48**, 560 (2001)

4.263. G. Via, S. Binary, D. Judy, in *Proceedings of the International Conference on GaAs Manufacturing Technology*, Miami, 2004, pp. 19–22

4.264. A. Wakejima, K. Ota, K. Matsunaga, M. Kuzuhara, IEEE Trans. Electron Devices **50**, 1983 (2003)

4.265. R. Wang, Y. Cai, C. Tang, K. Lau, K. Chen, IEEE Electron Device Lett. **27**, 793 (2005)

4.266. W. Wang, Y. Li, C. Lin, Y. Chan, G. Chen, J. Chyi, IEEE Electron Device Lett. **25**, 52 (2004)

4.267. W. Wang, C. Lin, P. Lin, C. Lin, F. Huang, Y. Chan, G. Chen, J. Chyi, IEEE Electron Device Lett. **25**, 763 (2004)

4.268. W. Wang, P. Lin, C. Lin, C. Lin, Y. Chan, IEEE Electron Device Lett. **26**, 5 (2005)

4.269. X. Wang, L. He, J. Electron. Mater. **27**, 1272 (1998)

4.270. A. Ward, in *IMS Workshop WFI GaN Device and Circuit Reliability*, Honolulu, 2007

4.271. O. Weidemann, M. Hermann, G. Steinhoff, H. Wingbrant, A. Spetz, M. Stutz-mann, M. Eickhoff, Appl. Phys. Lett. **83**, 773 (2003)

4.272. J. Wu, J. del Alamo, K. Jenkins, in *IEDM Technical Digest*, San Francisco, 2000, pp. 477–481

4.273. J. Wu, J. Scholvin, J.D. Alamo, IEEE Trans. Electron Devices **48**, 2181 (2001)

4.274. Y. Wu, PhD thesis, University of California Santa Barbara, Santa Barbara, 1997

4.275. Y. Wu, B. Keller, P. Fini, S. Keller, T. Jenkins, L. Kehias, S. DenBaars, U. Mishra, IEEE Electron Device Lett. **19**, 50 (1998)

4.276. Y. Wu, M. Moore, T. Wisleder, P. Chavarkar, U. Mishra, P. Parikh, in *IEDM Technical Digest*, San Francisco, 2004, pp. 1078–1079

4.277. Y. Wu, S. Keller, P. Kozodoy, B. Keller, P. Parikh, D. Kapolnek, S. DenBaars, U. Mishra,

4.278. Y. Wu, M. Moore, A. Saxler, T. Wisleder, P. Parikh, in *Device Research Conference*, State College PA, 2006, pp. 151–152

4.279. Y. Wu, A. Saxler, M. Moore, R. Smith, S. Sheppard, P. Chavarkar, T. Wisleder, U. Mishra, P. Parikh, IEEE Electron Device Lett. **25**, 117 (2004)

4.280. Y. Wu, A. Saxler, M. Moore, T. Wisleder, U. Mishra, P. Parikh, in *Compound Semiconductor IC Symposium Technical Digest*, Palm Springs, 2005, pp. 170–172

4.281. H. Xing, P. Chavarkar, S. Keller, S. DenBaars, U. Mishra, IEEE Electron Device Lett. **24**, 141 (2003)

4.282. H. Xing, S. DenBaars, U. Mishra, J. Appl. Phys. **97**, 113703 (2005)

4.283. H. Xing, Y. Dora, A. Chini, S. Heikman, S. Keller, U. Mishra, IEEE Electron Device Lett. **25**, 161 (2004)

4.284. H. Xing, L. McCarthy, S. Keller, S. DenBaars, U. Mishra, in *Proceedings of the International Symposium Compound Semiconductors*, Monterey, 2000, pp. 365–369

4.285. Y. Liu, J. Bardwell, S. McAlister, S. Rolfe, H. Tang, J. Webb, J. Appl. Phys. **96**, 2674 (2004)

4.286. E. Young, H. Hendriks, G. Rojano, R. Baskaran, T. Ritzdorf, J. Klocke, in *Proceedings of the International Conference on GaAs Manufacturing Technology*, San Diego, 2002, pp. 180–183

4.287. C. Youtsey, I. Adesida, G. Bulman, Electron. Lett. **33**, 245 (1997)

4.288. C. Youtsey, I. Adesida, L. Romano, G. Bulman, Appl. Phys. Lett. **72**, 560 (1999)

4.289. H. Yu, L. McCarthy, S. Rajan, S. Keller, S. DenBaars, J. Speck, U. Mishra, IEEE Electron Device Lett. **26**, 283 (2005)

4.290. H. Yu, L. McCarthy, H. Xing, P. Waltereit, L. Shen, S. Keller, S. DenBaars, J. Speck, U. Mishra, Appl. Phys. Lett. **85**, 5254 (2004)

4.291. L. Yu, L. Jia, D. Qiao, S. Lau, J. Li, J. Lin, H. Jiang, IEEE Trans. Electron Devices **50**, 500 (2003)

4.292. L. Yu, D. Qiao, Q. Xing, S. Lau, K. Boutros, J. Redwing, Appl. Phys. Lett. **73**, 238 (1998)

4.293. H.V. Zeijl, L. Nanver, in *Solid-State Circuits Technical Conference*, Beijing, 1998, pp. 98–101

4.294. N. Zhang, S. Keller, F. Parish, S. Heikman, S. DenBaars, U. Mishra, IEEE Electron Device Lett. **21**, 373 (2000)

4.295. N. Zhang, B. Moran, S. DenBaars, U. Mishra, X. Wang, T. Ma, in *IEDM Technical Digest*, Washington DC, 2001, pp. 589–592

4.296. C. Zhu, W. Fong, B. Leung, C. Cheng, C. Surya, IEEE Trans. Electron Devices **48**, 1225 (2001)

4.297. J. Zolper, M. Crawford, A. Howard, S. Pearton, R. Abernathy, C. Vartuli, C. Yuan, R. Stall, J. Ramer, S. Hersee, R. Wilson, in *MRS Symposium*, First International Conference on Nitride Semiconductors, vol. 395, Boston, 1996, pp. 801–806.

4.298. J. Zolper, R. Shul, A. Baca, R. Wilson, S. Pearton, R. Stall, Appl. Phys. Lett. **68**, 2273 (1996)

References of Chapter 5

5.1. A. Alabadelah, T. Fernandez, A. Mediavilla, B. Nauwelaers, A. Santarelli, D. Schreurs, A. Tazon, P. Traverso, in *Proceedings of the European Gallium Arsenide Other Compound Semiconductors Application Symposium GAAS*, Amsterdam, 2004, pp. 191–195

5.2. K. Anderson, C. Fager, J. Pedro, in *IEEE International Microwave Symposium Digest*, Long Beach, 2005, pp. 1159–1162

5.3. Y. Ando, A. Cappy, K. Marubashi, K. Onda, H. Miyamoto, M. Kuzuhara, IEEE Trans. Electron Devices **44**, 1367 (1997)

5.4. Y. Ando, W. Contrata, N. Samoto, H. Miyamoto, K. Matsunaga, M. Kuzuhara, K. Kunihiro, K. Kasahara, T. Nakayama, Y. Takahashi, N. Hayama, Y. Ohno, IEEE Trans. Electron Devices **47**, 1965 (2000)

5.5. I. Angelov, L. Bengtsson, M. Garcia, in *IEEE International Microwave Symposium Digest*, Orlando, 1995, pp. 1515–1518

5.6. I. Angelov, L. Bengtsson, M. Garcia, IEEE Trans. Microw. Theory Tech. **44**, 1664 (1996)

5.7. I. Angelov, V. Desmaris, K. Dynefors, P. Nilsson, N. Rorsman, H. Zirath, in *Proceedings of the European Gallium Arsenide Other Compound Semiconductors Application Symposium GAAS*, Paris, 2005, pp. 309–312

5.8. I. Angelov, N. Rorsman, J. Stenarson, M. Garcia, H. Zirath, in *IEEE International Microwave Symposium Digest*, Anaheim, 1999, pp. 525–528

5.9. I. Angelov, H. Zirath, Electron. Lett. **28**, 129 (1992)

5.10. I. Angelov, H. Zirath, N. Rorsman, IEEE Trans. Microw. Theory Tech. **40**, 2258 (1992)

5.11. I. Angelov, H. Zirath, N. Rorsman, in *IEEE International Microwave Symposium Digest*, San Diego, 1994, pp. 1571–1574

5.12. F. Benkhelifa, R. Kiefer, S. Müller, F. van Raay, R. Quay, R. Sah, M. Mikulla, G. Weimann, in *Proceedings of the International Conference on GaAs Manufacturing Technology*, New Orleans, 2005, p. 8.4

5.13. J. Bernat, M. Wolter, M. Marso, J. Flynn, G. Brandes, P. Kordos, Electron. Lett. **40**, 78 (2004)

5.14. M. Berroth, R. Bosch, IEEE Trans. Microw. Theory Tech. **38**, 891 (1990)

5.15. U. Bhapkar, M. Shur, J. Appl. Phys. **82**, 1649 (1997)

5.16. S. Binari, P. Klein, T. Kazior, in *IEEE International Microwave Symposium Digest*, Seattle, 2002, pp. 1823–1826

5.17. J.R. Black, IEEE Trans. Electron Devices **16**, 338 (1969)

5.18. C. Bolognesi, A. Kwan, D. DiSanto, in *IEDM Technical Digest*, San Francisco, 2002, pp. 685–688

5.19. F. Bonani, G. Ghione, M. Pirola, C. Naldi, in *IEDM Technical Digest*, Washington DC, 1993, pp. 101–104

5.20. F. Bonani, G. Ghione, M. Pirola, C. Naldi, in *GaAs IC Symposium Technical Digest*, Philadelphia, 1994, pp. 141–144

5.21. S. Boumaiza, F. Ghannouchi, IEEE Trans. Microw. Theory Tech. **51**, 2427 (2003)

5.22. H. Bousbia, D. Barataud, G. Neveux, T. Gasseling, J. Nebus, J. Tessier, in *IEEE International Microwave Symposium Digest*, San Franscisco, 2006, pp. 1452–1455

5.23. R. Brady, G. Valdivia, T. Brazil, in *IEEE International Microwave Symposium Digest*, Honolulu, 2007, pp. 593–596

5.24. B. Brar, K. Boutros, R. deWarnes, V. Tilak, R. Shealy, L. Eastman, in *Proceedings of Lester Eastman Conference on High Performance Devices*, Newark, 2002, pp. 487–491

5.25. J. Brinkhoff, A. Parker, IEEE Trans. Microw. Theory Tech. **51**, 1045 (2003)

5.26. J. Brinkhoff, A. Parker, in *IEEE International Microwave Symposium Digest*, Fort Worth, 2004, pp. 799–802

5.27. J. Brinkhoff, A. Parker, in *Workshop Application Radio Science*, Hobart, 2004, pp. 1–8

5.28. J. Brinkhoff, A. Parker, M. Leung, IEEE Trans. Microw. Theory Tech. **51**, 2523 (2003)

5.29. B. Bunz, A. Ahmed, G. Kompa, in *Proceedings of the European Gallium Arsenide Other Compound Semiconductors Application Symposium GAAS* Paris, 2005, pp. 649–652

5.30. P. Cabral, J. Pedro, N. Carvalho, IEEE Trans. Microw. Theory Tech. **52**, 2585 (2004)

5.31. Y. Cai, Y. Zhou, K. Chen, K. Lau, in *Device Research Conference*, Santa Barbara, 2005, pp. 179–180

5.32. N. Carvalho, J. Pedro, W. Jang, M. Steer, in *IEEE International Microwave Symposium Digest*, Long Beach, 2005, pp. 801–805

5.33. C. Chen, M. Deen, Y. Cheng, M. Matloubian, IEEE Trans. E ectron Devices **48**, 2884 (2001)

5.34. C. Chen, R. Coffie, K. Krishnamurthy, S. Keller, M. Rodwell, U. Mishra, IEEE Electron Device Lett. **21**, 549 (2000)

5.35. Y. Chen, R. Coffie, W. Luo, M. Wojtowicz, I. Smorchkova, B. Heying, Y. Kim, M. Aust, A. Oki, in *IEEE International Microwave Symposium Digest*, Honolulu, 1998, pp. 137–140

5.36. S. Chiu, A. Anwar, S. Wu, IEEE Trans. Electron Devices **47**, 662 (2000)

5.37. Y. Chou, D. Leung, R. Lai, R. Grundbacher, M. Barsky, Q. Kan, R. Tsai, M. Wojtowicz, D. Eng, L. Tran, T. Block, P. Liu, M. Nishimoto, A. Oki, IEEE Electron Device Lett. **24**, 378 (2003)

5.38. . Colvin, S. Bhatia, K. O, IEEE J. Solid-State Circuits **34**, 1339 (1999)

5.39. A. Conway, P. Asbeck, in *IEEE International Microwave Symposium Digest*, Honolulu, 2007, pp. 605–608

5.40. A. Conway, Y. Zhao, P. Asbeck, M. Micovic, J. Moon, in *IEEE International Microwave Symposium Digest*, Long Beach, 2005, pp. 499–502

5.41. S. Cripps, in *IEEE International Microwave Symposium Digest*, Boston, 1983, pp. 221–223

5.42. G. Crupi, D. Xiao, D. Schreurs, E. Limiti, A. Caddemi, W. De Raedt, M. Germain, IEEE Trans. Microw. Theory Tech. **54**, 3616 (2006)

5.43. W. Curtice, IEEE Trans. Microw. Theory Tech. **28**, 448 (1980)

5.44. W. Curtice, in *IMS Workshop WMG: Solid-State Power Invades the Tube Realm*, Honolulu, 2007

5.45. W. Curtice, R. Camisa, IEEE Trans. Microw. Theory Tech. **32**, 1573 (1984)

5.46. A. Curutchet, N. Malbert, N. Labat, A. Touboul, C. Gaquiere, A. Minko, M. Uren, Microelectron. Reliab. **43**, 1713 (2003)

5.47. G. Dambrine, A. Cappy, F. Heliodore, E. Playez, IEEE Trans. Microw. Theory Tech. **36**, 1151 (1988)

5.48. N. de Carvalho, J. Pedro, IEEE Trans. Microw. Theory Tech. **47**, 2364 (1999)

5.49. N. de Carvalho, J. Pedro, IEEE Trans. Microw. Theory Tech. **50**, 2090 (2002)

5.50. D. DiSanto, C. Bolognesi, IEEE Trans. Electron Devices **53**, 2914 (2006)

5.51. S. Doo, P. Roblin, G. Jessen, R. Fitch, J. Gillespie, N. Moser, A. Crespo, G. Simpson, J. King, IEEE Microw. Wireless Compon. Lett. **16**, 681 (2006)

5.52. P. Draxler, J. Jeong, C. Hsia, S. Lanfranco, W. Nagy, K. Linthicum, L. Larson, P. Asbeck, in *IEEE International Microwave Symposium Digest*, Long Beach, 2005, pp. 1549–1552

5.53. P. Draxler, S. Lafranco, D. Kimball, C. Hsia, J. Jeong, J. van de Sluis, P. Asbeck, in *IEEE International Microwave Symposium Digest*, San Francisco, 2006, pp. 1534–1537

5.54. L. Eastman, V. Tilak, J. Smart, B. Green, E. Chumbes, R. Dimitrov, H. Kim, O. Ambacher, N. Weimann, T. Prunty, M. Murphy, W. Schaff, J. Shealy, IEEE Trans. Electron Devices **48**, 479 (2001)

5.55. J. Evans, G. Amaratunga, IEEE Trans. Electron Devices **44**, 1148 (1997)

5.56. C. Fager, J. Pedro, N. de Carvalho, H. Zirath, IEEE Trans. Microw. Theory Tech. **50**, 2834 (2002)

5.57. P. Fedorenko, J. Kenney, in *IEEE International Microwave Symposium Digest*, Honolulu, 2007, pp. 1453–1456

5.58. G. Fischer, in *IEEE Wireless and Microwave Technology Conference*, Clearwater, 2004, p. FD-1

5.59. M. Foisy, P. Jeroma, G. Martin, in *IEEE International Microwave Symposium Digest*, Albuquerque, 1992, pp. 251–254

5.60. H. Fukui, IEEE Trans. Electron Devices **26**, 1032 (1979)

5.61. J. Golio, M. Miller, G. Maracas, D. Johnson, IEEE Trans. Electron Devices **37**, 1217 (1990)

5.62. B. Green, K. Chu, E. Chumbes, J. Smart, J. Shealy, L. Eastman, IEEE Electron Device Lett. **21**, 268 (2000)

5.63. B. Green, H. Kim, K. Chu, H. Lin, V. Tilak, J. Shealy, J. Smart, L. Eastman, in *IEEE International Microwave Symposium Digest*, Boston, 2000, pp. 237–241

5.64. F. De Groote, O. Jardel, J. Verspecht, D. Barataud, J. Teyssier, R. Quere, in *Proc. ARFTG Microwave Measurement Symposium*, Washington DC, 2005

5.65. W. Ho, C. Surya, K. Tong, W. Kim, A. Botcharev, H. Morkoc, IEEE Trans. Electron Devices **46**, 1099 (1999)

5.66. F.N. Hooge, IEEE Trans. Electron Devices **41**, 1926 (1994)

5.67. S. Hsu, P. Valizadeh, D. Pavlidis, in *GaAs IC Symposium Technical Digest*, Monterey, 2002, pp. 85–88

5.68. M. Ida, K. Kurishima, N. Watanabe, T. Enoki, in *IEDM Technical Digest*, Washington DC, 2001, pp. 776–779

5.69. S. Islam, A. Anwar, in *IEEE International Microwave Symposium Digest*, Seattle, 2002, pp. 267–270

5.70. S. Islam, A. Anwar, IEEE Trans. Microw. Theory Tech. **14**, 853 (2004)

5.71. S. Islam, A. Anwar, IEEE Trans. Electron Devices **49**, 710 (2004)

5.72. R. Jansen, A. Smymakowski, M. Bahn, A. John, C. Rieckmann, A. Noculak, S. Chalermwisutkul, D. Klümper, in *IMS Workshop Advances in GaN-based Device and Circuit Technology: Modeling and Applications*, Forth Worth, 2004

5.73. O. Jardel, F. de Groote, C. Charbonniaud, T. Reveyrand, J. Teyssier, R. Quere, D. Floriot, in *IEEE International Microwave Symposium Digest*, Honolulu, 2007, pp. 601–604

5.74. O. Jardel, F. De Groote, T. Reveyrand, J.C. Jacquet, C. Charbonniaud, J. Teyssier, D. Floriot, R. Quere, IEEE Trans. Microw. Theory Tech. **55**, 2660 (2007)

5.75. A. Jarndal, G. Kompa, IEEE Trans. Microw. Theory Tech. **53**, 3440 (2005)

5.76. A. Jarndal, G. Kompa, in *IEEE International Microwave Symposium Digest*, Long Beach, 2005, pp. 1423–1426

5.77. A. Jarndal, G. Kompa, **54**, 2830 (2007)

5.78. M. Je, H. Shin, IEEE Electron Device Lett. **24**, 183 (2003)

5.79. K. Jeon, Y. Kwon, S. Hong, IEEE Microw. Guided Wave Lett. **7**, 78 (1997)

5.80. J. Johnson, J. Gao, K. Lucht, J.W.C. Strautin, J. Riddle, T. Therrien, P. Rajagopal, J. Roberts, A. Vescan, J. Brown, A. Hanson, S. Singhal, R. Borges, E. Piner, K. Linthicum, Proc. Electron. Soc. **7**, 405 (2004)

5.81. S. Karmalkar, N. Satyan, D. Sathaiya, IEEE Electron Device Lett. **27**, 87 (2006)

5.82. O. Katz, G. Bahir, J. Salzmann, IEEE Trans. Electron Devices **37**, 2250 (1990)

5.83. T. Kikkawa, K. Imanishi, M. Kanamura, K. Joshin, in *Proceedings International Conference GaAs Manufacturing Technology*, San Diego, 2002, pp. 171–174

5.84. T. Kikkawa, T. Maniwa, H. Hayashi, M. Kanamura, S. Yokokawa, M. Nishi, N. Adachi, M. Yokoyama, Y. Tateno, K. Joshin, in *IEEE International Microwave Symposium Digest*, Fort Worth, 2004, pp. 1347–1350

5.85. T. Kikkawa, M. Nagahara, T. Kimura, S. Yokokawa, S. Kato, M. Yokoyama, Y. Tateno, K. Horino, K. Domen, Y. Yamaguchi, N. Hara, K. Joshin, in *IEEE International Microwave Symposium Digest*, Seattle, 2002, pp. 1815–1818

5.86. D. Kimball, P. Draxler, J. Jeong, C. Hsia, S. Lanfranco, W. Nagy, K. Linthicum, L. Larson, P. Asbeck, in *Compound Semiconductor IC Symposium Technical Digest*, Palm Springs, 2005, pp. 89–92

5.87. E. Kohn, I. Daumiller, M. Kunze, M. Neuburger, M. Seyboth, T. Jenkins, J. Sewell, J. Norstand, Y. Smorchkova, U. Mishra, IEEE Trans. Microw. Theory Tech. **51**, 634 (2003)

5.88. A. Koudymov, G. Simin, M. Khan, A. Tarakji, R. Gaska, M. Shur, IEEE Electron Device Lett. **24**, 680 (2003)

5.89. H. Ku, M. McKinley, J. Kenney, IEEE Trans. Microw. Theory Tech. **50**, 2843 (2002)

5.90. K. Ku, J. Kenney, in *IEEE International Microwave Symposium Digest*, Phoenix, 2003, pp. 799–802

5.91. D. Kuksenkov, H. Temkin, R. Gaska, J. Yang, IEEE Electron Device Lett. **19**, 222 (1998)

5.92. K. Kunihiro, Y. Ohno, IEEE Trans. Electron Devices **43**, 1336 (1996)

5.93. J. Kuzmik, S. Bychikhin, M. Neuburger, A. Dadgar, A. Krost, E. Kohn, D. Pogany, IEEE Trans. Electron Devices **52**, 1698 (2005)

5.94. J. Lee, D. Liu, Z. Lin, W. Lu, J. Flynn, G. Brandes, Solid-State Electron. **47**, 2081 (2003)

5.95. J. Lee, K. Webb, IEEE Trans. Microw. Theory Tech. **52**, 2 (2004)

5.96. K. Lee, A. Dabiran, P. Chow, A. Osinsky, S. Pearton, F. Ren, Solid-State Electron. **48**, 37 (2004)

5.97. K. Lee, A. Dabiran, A. Osinsky, P. Chow, S. Pearton, F. Ren, Solid-State Electron. **47**, 1501 (2003)

5.98. S. Lee, K. Webb, in *IEEE International Microwave Symposium Digest*, Seattle, 2002, pp. 1415–1418

5.99. S. Lee, K. Webb, in *IEEE International Microwave Symposium Digest*, Fort Worth, 2004, pp. 1057–1060

5.100. S. Lee, K. Webb, V. Tilak, L. Eastman, IEEE Trans. Microw. Theory Tech. **51**, 1567 (2003)

5.101. H. Leier, A. Wieszt, R. Bethtasch, H. Tobler, A. Vescan, R. Dietrich, A. Schurr, H. Sledzik, J. Birbeck, R. Balmer, T. Martin, in *Proceedings of the European Gallium Arsenide Other Compound Semiconductors Application Symposium GAAS*, London, 2001, pp. 49–52

5.102. R. Leoni III, J. Bao, J. Bu, X. Du, M. Shirokov, J. Hwang, IEEE Trans. Electron Devices **47**, 498 (2000)

5.103. M. Li, Y. Wang, Electron Devices **55**, 261 (2008)

5.104. T. Li, R. Joshi, R. del Rosario, IEEE Trans. Electron Devices **49**, 1511 (2002)

5.105. C. Lin, W. Wang, P. Lin, C. Lin, Y. Chang, Y. Chan, IEEE Electron Device Lett. **26**, 710 (2005)

5.106. T. Liu, S. Boumaiza, F. Ghannouchi, IEEE Trans. Microw. Theory Tech. **53**, 3578 (2005)

5.107. W. Liu, *Handbook of III-V Heterojunction Bipolar Transistors* (Wiley, New York, 1998)

5.108. Y. Liu, R. Trew, G. Bilbro, in *IEEE International Microwave Symposium Digest*, Honolulu, 2007, pp. 597–600

5.109. W. Lu, V. Kumar, E. Piner, I. Adesida, IEEE Trans. Electron Devices **50**, 1069 (2003)

5.110. S. Manohar, A. Narayanan, A. Keerti, A. Pham, J. Brown, R. Borges, K. Linthicum, in *IEEE International Microwave Symposium Digest*, Seattle, 2002, pp. 449–452

5.111. A. Materka, T. Kacprzak, IEEE Trans. Microw. Theory Tech. **33**, 129 (1985)

5.112. K. Matocha, T. Chow, R. Gutmann, IEEE Trans. Electron Devices **52**, 6 (2005)

5.113. N. Matsunaga, M. Yamamoto, Y. Hatta, H. Masuda, IEEE Trans. Electron Devices **50**, 1194 (2003)

5.114. L. McCarthy, PhD thesis, University of California Santa Barbara, Santa Barbara, 2001

5.115. L. McCarthy, L. Smorchkova, P. Fini, M. Rodwell, J. Speck, S. DenBaars, U. Mishra, Electron. Lett. **38**, 144 (2002)

5.116. P. McGovern, J. Benedikt, P. Tasker, J. Powell, K. Hilton, J. Glasper, R. Balmer, T. Martin, M. Uren, in *IEEE International Microwave Symposium Digest*, Long Beach, 2005, pp. 503–506

5.117. P. McGovern, D. Williams, P. Tasker, J. Benedikt, J. Powell, K. Hilton, R. Balmer, T. Martin, M. Uren, in *IEEE International Microwave Symposium Digest*, Fort Worth, 2004, pp. 825–828

5.118. E. Mengistu, G. Kompa, in *Proceedings of the European Microwave Integrated Circuits Conference*, Manchester, 2006, pp. 292–295

5.119. T. Merkle, A. Tessmann, S. Ramberger, in *IEEE International Microwave Symposium Digest*, Seattle, 2002, pp. 453–456

5.120. S.D. Meyer, C. Charbonniaud, R. Quere, M. Campovecchio, R. Lossy, J. Würfl, in *IEEE International Microwave Symposium Digest*, Philadelphia, 2003, pp. 455–458

5.121. T. Mizutani, H. Makihara, M. Akita, Y. Ohno, S. Kishimoto, K. Maezawa, Jpn. J. Appl. Phys. **42**, 424 (2003)

5.122. T. Mizutani, Y. Ohno, M. Akita, S. Kishimoto, K. Maezawa, IEEE Trans. Electron Devices **50**, 2015 (2003)

5.123. N. Moll, M. Hueschen, A. Fischer-Colbrie, IEEE Trans. Electron Devices **35**, 879 (1988)

5.124. J. Moon, M. Micovic, A. Kurdoghlian, P. Janke, P.H.W. Wong, L. McCray, Electron. Lett. **38**, 1358 (2002)

5.125. T. Nass, D. Wiegner, U. Seyfried, W. Templ, S. Weber, S. Woerner, P. Klose, R. Quay, F. van Raay, H. Walcher, H. Massler, M. Seelmann-Eggebert, O. Kappeler, R. Kiefer, in *Joint Symposium on Opto- and Microelectronic Devices and Circuits*, Duisburg, 2006, pp. 133–136

5.126. A. Nidhi, T. Palacios, A. Chakraborty, S. Keller, U. Mishra, IEEE Electron Device Lett. **27**, 877 (2006)

5.127. J. Nikaido, T. Kikkawa, E. Mitani, S. Yokokawa, Y. Tateno, in *Proceedings of International Conference on GaAs Manufacturing Technology*, New Orleans, 2005, pp. 97–100

5.128. S. Nuttinck, S. Pinel, E. Gebara, J. Laskar, M. Harris, in *Proceedings of the European Gallium Arsenide Other Compound Semiconductors Application Symposium GAAS*, Munich, 2003, pp. 213–215

5.129. S. Nuttinck, E. Gebara, J. Laskar, H. Harris, IEEE Trans. Microw. Theory Tech. **49**, 2413 (2001)

5.130. C. Oxley, M. Uren, IEEE Trans. Electron Devices **52**, 165 (2005)

5.131. T. Palacios, S. Rajan, A. Chakraborty, S. Heikman, S. Keller, S. DenBaars, U. Mishra, IEEE Trans. Electron Devices **52**, 2117 (2005)

5.132. A. Parker, J. Scott, J. Rathmell, M. Sayed, in *IEEE International Microwave Symposium Digest*, San Francisco, 1996, pp. 1707–1710

5.133. A. Parker, J. Rathmell, IEEE Trans. Microw. Theory Tech. **49**, 2105 (2001)

5.134. A. Parker, J. Rathmell, IEEE Trans. Microw. Theory Tech. **51**, 588 (2003)

5.135. A. Parker, J. Rathmell, in *IEEE International Microwave Symposium Digest*, Fort Worth, 2004, pp. 803–806

5.136. A. Parker, J. Scott, Electron. Lett. **29**, 1961 (1993)

5.137. J. Pedro, N. Carvalho, *Intermodulation Distorsion in Microwave and Wireless Circuits* (Artech House, Boston, 2003)

5.138. R. Pengelly, in *Proceedings of Wireless and Microwave Technology Conference*, Clearwater, 2004, pp. RC-5

5.139. J. Peransin, P. Vignaud, D. Rigaud, L. Vandamme, IEEE Trans. Electron Devices **37**, 2250 (1990)

5.140. M. Pospieszalski, IEEE Trans. Microw. Theory Tech. **37**, 1340 (1989)

5.141. W. Pribble, in *IMS Workshop WMG: High Power Device Characterization and Modeling*, Honolulu, 2007

5.142. W. Pribble, S. Sheppard, R. Smith, S. Allen, J. Palmour, T. Smith, Z. Ring, J. Sumakeris, A. Saxler, J. Milligan, in *IMS Workshop Wide Bandgap Technologies*, Seattle, 2002

5.143. G. Qu, A. Parker, in *IEEE International Microwave Symposium Digest*, Baltimore, 1998, pp. 745–748,

5.144. R. Quay, K. Hess, R. Reuter, M. Schlechtweg, T. Grave, V. Palankovski, S. Selberherr, IEEE Trans. Electron Devices **48**, 210 (2001)

5.145. A. Rashmi, A. Kranti, S. Haldar, M. Gupta, R. Gupta, IEEE Trans. Microw. Theory Tech. **51**, 607 (2003)

5.146. P. Regoliosi, A. Reale, A.D. Carlo, P. Romanini, M. Peroni, C. Lanzieri, A. Angelini, M. Pirola, G. Ghione, IEEE Trans. Electron Devices **53**, 182 (2006)

5.147. R. Reuter, M. Agethen, U. Auer, S. van Waasen, D. Peters, W. Brockerhoff, F. Tegude, IEEE Trans. Microw. Theory Tech. **45**, 977 (1997)

5.148. R. Reuter, F. Tegude, in *IEEE International Microwave Symposium Digest*, Boston, 1998, pp. 137–140

5.149. P. Roblin, H. Rohdin, *High-Speed Heterostructure Devices* (Cambridge University Press, Cambridge, 2002)

5.150. M. Rodwell, M. Urteaga, T. Mathew, D. Scott, D. Mensa, Q. Lee, J. Guthrie, Y. Betser, S. Martin, R. Smith, S. Jaganathan, S. Krishnan, S. Long, R. Pullela, B. Agarwal, U. Bhattacharya, L. Samoska, M. Dahlstrom, IEEE Trans. Electron Devices **48**, 2606 (2001)

5.151. H. Rohdin, P. Roblin, IEEE Trans. Electron Devices **33**, 664 (1986)

5.152. D. Root, in *Proceedings of IEEE International Midwest Symposium Circuits and Systems*, Dayton, 2001, pp. 768–772

5.153. D. Root, S. Fan, in *IEEE International Microwave Symposium Digest*, Albuquerque, 1992, pp. 255–258

5.154. D. Root, M. Iwamoto, J. Wood, in *GaAs IC Symposium Technical Digest*, Monterey, 2002, pp. 279–282

5.155. D. Root, in *Proceedings of European Microwave Conference*, Cannes, 1994, pp. 854–859

5.156. D. Root, M. Pirola, S. Fan, W. Anklam, A. Cognata, IEEE Trans. Microw. Theory Tech. **41**, 2211 (1993)

5.157. S. Rumyantsev, N. Pala, M. Shur, E. Borovitskaya, A. Dimitriev, M. Levinshtein, R. Gaska, M. Khan, J. Yang, X. Hu, G. Simin, IEEE Trans. Electron Devices **48**, 530 (2001)

5.158. F. Sacconi, A. Di Carlo, F. Della Sala, P. Lugli, in *Proceedings of the European Gallium Arsenide Other Compound Semiconductors Application Symposium GAAS*, Paris, 2000, pp. 620–623

5.159. C. Sanabria, A. Chakraborty, H. Xu, M. Rodwell, U. Mishra, R. York, IEEE Electron Device Lett. **27**, 19 (2005)

5.160. C. Sanabria, X. Hongtao, T. Palacios, A. Chakraborty, S. Heikman, U. Mishra, R. York, IEEE Trans. Microw. Theory Tech. **53**, 762 (2005)

5.161. M. Schlechtweg, Dissertation, Universität Kassel, 1989

5.162. I. Schmale, G. Kompa, in *Proceedings of European Microwave Conference*, Munich, 1999, pp. 258–261

5.163. D. Schreurs, J. Verspecht, B. Nauwelaers, A.V. de Capelle, M. VanRossum, in *Proceedings of European Microwave Conference*, Tel Aviv, 1997, pp. 921–926

5.164. D. Schreurs, J. Verspecht, E. Vandamme, N. Vellas, C. Gaquiere, M. Germain, G. Borghs, in *IEEE International Microwave Symposium Digest*, Philadelphia, 2003, pp. 447–450

5.165. D. Schreurs, J. Wood, N. Tufillaro, D. Usikov, L. Barford, D. Root, in *IEDM Technical Digest*, San Francisco, 2000, 819–822

5.166. M. Seelmann-Eggebert, T. Merkle, F. van Raay, R. Quay, M. Schlechtweg, IEEE Trans. Microw. Theory Tech. **55**, 195 (2007)

5.167. K. Shenai, R. Scott, B. Baliga, IEEE Trans. Electron Devices **36**, 1811 (1989)

5.168. K. Shinohara, Y. Yamashita, A. Endoh, I. Watanabe, K. Hikosaka, T. Matsui, T. Mimura, S. Hiyamizu, IEEE Electron Device Lett. **25**, 241 (2004)

5.169. M. Shur, *GaAs Devices and Circuits* (Plenum, New York, 1987)

5.170. F. Sischka, *The Curtice Mesfet Model* (Agilent Technologies, Böblingen, Germany, 2001)

5.171. H. Statz, H. Haus, R. Pucel, IEEE Trans. Electron Devices **21**, 549 (1974)

5.172. J. Stenarson, M. Garcia, I. Angelov, H. Zirath, IEEE Trans. Microw. Theory Tech. **47**, 2358 (1999)

5.173. Y. Su, S. Wei, R. Wang, S. Chang, C. Ko, T. Kuan, IEEE Electron Device Lett. **24**, 622 (2003)

5.174. T. Suemitsu, K. Shiojima, T. Makimura, N. Shigekawa, Jpn. J. Appl. Phys. **44**, L211 (2005)

5.175. H. Sun, C. Bolognesi, Appl. Phys. Lett. **90**, 123505 (2007)

5.176. W. Sutton, D. Pavlidis, H. Lahreche, B. Damilano, R. Langer, in *Proceedings of the European Gallium Arsenide Other Compound Semiconductors Application Symposium GAAS*, Munich, pp. 209–212, 2003

5.177. S. Sze, *Physics of Semiconductor Devices*, 2nd edn. (Wiley, New York, 1981)

5.178. T. Takano, Y. Oishi, T. Maniwa, H. Hayashi, T. Kikkawa, K. Araki, Microw. Opt. Tech. Lett. **45**, 551 (2005)

5.179. P. Tasker, J. Braunstein, in *IEEE International Microwave Symposium Digest*, Orlando, 1995, pp. 611–614

5.180. P. Tasker, B. Hughes, IEEE Electron Device Lett. **10**, 291 (1989)

5.181. R. Therrien, S. Singhal, J. Johnson, W. Nagy, R. Borges, A. Chaudhari, A. Hanson, A. Edwards, J. Marquart, P. Rajagopal, C. Park, I. Kizilyalli, K. Linthicum, in *IEDM Technical Digest*, Washington DC, 2005, 568–571

5.182. J. Tirado, J. Sanchez-Rojas, J. Izpura, IEEE Trans. Electron Devices, **54**, 410 (2007)

5.183. S. Trassaert, B. Boudart, C. Gaquiere, D. Theron, Y. Crosnier, F. Huet, M. Poisson, Electron. Lett. **35**, 1386 (1999)

5.184. R. Trew, in *IEEE International Microwave Symposium Digest*, Seattle, 2002, pp. 1811–1814

5.185. R. Trew, Y. Liu, W. Kuang, R. Vetury, J. Shealy, IEEE Trans. Microw. Theory Tech. **54**, 2061 (2006)

5.186. R. Trew, Y. Liu, H. Yin, G. Bilbro, J. Shealy, R. Vetury, P. Garber, M. Poulton, in *IEEE International Microwave Symposium Digest*, San Franscicso, 2006, pp. 643–646

5.187. T. Turlington, *Behavioral Modeling of Nonlinear RF and Microwave Devices*, (Artech House, Boston and London, 2000)

5.188. N. Ui, Y. Tajima, in *IMS Workshop WMG: High Power Device Characterization and Modeling*, Honolulu, 2007

5.189. S. Vainshtein, V. Yuferev, J. Kostamovaara, IEEE Trans. Electron Devices **50**, 1988 (2003)

5.190. F. van Raay, R. Quay, R. Kiefer, H. Massler, M. Schlechtweg, G. Weimann, in *IEEE International Microwave Symposium Digest*, Philadelphia, 2003, pp. 451–454

5.191. L.K.J. Vandamme, IEEE Trans. Electron Devices **41**, 2176 (1997)

5.192. N. Vellas, C. Gaquiere, F. Bue, Y. Guhel, B. Boudart, J. de Jaeger, M. Poisson, IEEE Electron Device Lett. **23**, 246 (2002)

5.193. A. Vertiatchikh, L. Eastman, Electron. Lett. **39**, 876 (2003)

5.194. A. Vertiatchikh, L. Eastman, IEEE Electron Device Lett. **24**, 535 (2003)

5.195. R. Vetury, N. Zhang, S. Keller, U. Mishra, IEEE Trans. Electron Devices **48**, 560 (2001)

5.196. S. Vitusevich, M. Petrychuk, S. Danylyuk, A. Kurakin, N. Klein, A. Belyaev, Phys. Stat. Sol. A **202**, 816 (2005)

5.197. D. Wandrei, Y. Tajima, in *IMS Workshop WMG: High Power Device Characterization and Modeling*, Honolulu, 2007

5.198. C. Wang, C. Zhu, G. Zhang, L. Li, IEEE Trans. Electron Devices **50**, 1145 (2003)

5.199. J.W. Lee, S. Lee, K.J. Webb, in *IEEE International Microwave Symposium Digest*, Phoenix, 2001, pp. 679–682

5.200. A. Wieszt, R. Dietrich, J. Lee, A. Vescan, H. Leier, E. Piner, J. Redwing, H. Sledzik, in *Proceedings of the European Gallium Arsenide Other Compound Semiconductors Application Symposium GAAS*, Paris, 2000, pp. 260–263

5.201. J. Wood, D. Root, in *IEEE International Microwave Symposium Digest*, Boston, 2000, pp. 685–688

5.202. Y. Wu, P. Chavarkar, M. Moore, P. Parikh, U. Mishra, in *IEDM Technical Digest*, Washington DC, 2001, pp. 951–953

5.203. Y. Wu, P. Chavarkar, M. Moore, P. Parikh, U. Mishra, in *IEDM Technical Digest*, San Francisco, 2002, 697–699

5.204. Y. Wu, M. Moore, A. Saxler, T. Wisleder, P. Parikh, in *Device Research Conference*, State College, PA, 2006, pp. 151–152

5.205. Y. Wu, A. Saxler, M. Moore, R. Smith, S. Sheppard, P. Chavarkar, T. Wisleder, U. Mishra, P. Parikh, IEEE Electron Device Lett. **25**, 117 (2004)

5.206. H. Xing, P. Chavarkar, S. Keller, S. DenBaars, U. Mishra, IEEE Electron Device Lett. **24**, 141 (2003)

5.207. H. Xing, Y. Dora, A. Chini, S. Heikman, S. Keller, U. Mishra, IEEE Electron Device Lett. **25**, 161 (2004)

5.208. H. Xing, D. Green, L. McCarthy, I. Smorchkova, P. Chavarkar, T. Mates, S. Keller, S. DenBaars, J. Speck, U. Mishra, in *Bipolar Circuits Technology Meeting*, Minneapolis, 2001, pp. 125–130

5.209. J. Xu, W. Yin, J. Mao, IEEE Microw. Wireless Compon. Lett. **17**, 55 (2007)

5.210. S. Yamakawa, S. Aboud, M. Saraniti, S. Goodnick, Semicond. Sci. Technol. **19**, S475 (2004)

5.211. N. Zhang, B. Moran, S. DenBaars, U. Mishra, X. Wang, T. Ma, in *IEDM Technical Digest*, Washington DC, 2001, pp. 589–592

References of Chapter 6

6.1. E. Alekseev, S. Hsu, D. Pavlidis, in *Proceedings of European Microwave Conference*, Paris, 2000, pp. 1–4

6.2. K. Andersson, V. Desmaris, J. Eriksson, N. Roersman, H. Zirath, in *IEEE International Microwave Symposium Digest*, Philadelphia, 2003, pp. 1303–1306

6.3. P. Asbeck, J. Mink, T. Itoh, G. Haddad, Microw. J. **42**, 22 (1999)

6.4. H. Ashoka, J. Ness, A. Robinson, M. Gourlay, J. Logan, P. Woodhead, D. Reuther, in *IEEE International Microwave Symposium Digest*, Baltimore, 1998, pp. 1149–1153

6.5. M. Aust, A. Sharma, Y. Chen, M. Wojtowicz, in *Compound Semiconductor IC Symposium Technical Digest*, San Antonio, 2006, pp. 89–92

6.6. I. Bahl, in *International Workshop on Integrated Nonlinear Microwave and Millimeterwave Circuits*, Duisburg, 1994, pp. 71–91

6.7. A. Barnes, M. Moore, M. Allenson, in *IEEE International Microwave Symposium Digest*, Denver, 1997, pp. 1429–1432

6.8. W. Baumberger, IEEE J. Solid-State Circuits **29**, 1244 (1994)

6.9. A. Behtash, H. Tobler, F. Berlec, H. Leyer, V. Ziegler, B. Adelseck, T. Martin, R. Balmer, D. Pavlidis, R. Jansen, M. Neuburger, H. Schumacher, in *IEEE International Microwave Symposium Digest*, Fort Worth, 2004, pp. 1657–1660

6.10. F. Benkhelifa, R. Kiefer, S. Müller, F. van Raay, R. Quay, R. Sah, M. Mikulla, G. Weimann, in *Proceedings of International Conference on GaAs Manufacturing Technology*, New Orleans, 2005, p. 8.4

6.11. A. Bessemoulin, R. Quay, S. Ramberger, H. Massler, M. Schlechtweg, IEEE J. Solid-State Circuits **38**, 1433 (2003)

6.12. W. Bösch, J. Mayock, M. O'Keefe, J. McMonagle, in *IEEE International Radar Conference*, Arlington, 2005, pp. 22–26

6.13. H. Brech, W. Brakensiek, D. Burdeaux, W. Burger, C. Dragon, G. Formicone, B. Pryor, D. Rice, in *IEDM Technical Digest*, Washington DC, 2003, pp. 359–362

6.14. D. Bryant, K. Salzmann, R. Hudgens, in *IEEE International Microwave Symposium Digest*, Atlanta, 1993, pp. 1373–1376

6.15. J. Cha, J. Kim, B. Kim, J. Lee, S. Kim, in *IEEE International Microwave Symposium Digest*, Fort Worth, 2004, pp. 533–536

6.16. S. Cha, Y. Chung, M. Wojtowicz, I. Smorchkova, B. Allen, J. Yang, R. Kagiwada, in *IEEE International Microwave Symposium Digest*, Fort Worth, 2004, pp. 829–832

6.17. T. Chang, W. Wu, J. Lin, S. Jang, F. Ren, S. Pearton, R. Fitch, J. Gillespie, Microw. Opt. Tech. Lett. **49**, 1152 (2007)

6.18. Y. Chen, R. Coffie, W. Luo, M. Wojtowicz, I. Smorchkova, B. Heying, Y. Kim, M. Aust, A. Oki, in *IEEE International Microwave Symposium Digest*, Honolulu, 2007, pp. 307–310

6.19. Y. Chen, C. Wu, C. Pao, M. Cole, Z. Bardai, L. Hou, T. Midford, in *GaAs IC Symposium Technical Digest*, San Diego, 1995, pp. 281–284

6.20. H. Chireix, Proc. IRE **23**, 1370 (1935)

6.21. K. Cho, W. Kim, J. Kim, S. Stapleton, in *IEEE International Microwave Symposium Digest*, San Francisco, 2006, pp. 1895–1899

6.22. J. Choi, J. Yim, J. Yang, J. Kim, J. Cha, B. Kim, in *IEEE International Microwave Symposium Digest*, Honolulu, 2007, pp. 81–84

6.23. Y. Chung, S. Cai, W. Lee, Y. Lin, C. Wen, K. Wang, T. Itoh, Electron. Lett. **37**, 1199 (2001)

6.24. Y. Chung, C. Hang, S. Cai, Y. Qian, C. Wen, K. Wang, T. Itoh, in *IEEE International Microwave Symposium Digest*, Seattle, 2002, pp. 433–436

6.25. P. Colantonio, F. Giannini, R. Giofre, E. Limiti, A. Serino, M. Peroni, P. Romanini, C. Proietti, in *Proceedings of European GaAs and Related Compounds Application Symposium GAAS*, Paris, 2005, pp. 673–676

6.26. Compound Semiconductor, Comp. Semicond. **6**, 15 (2000)

6.27. J. Conlon, N. Zhang, M. Poulton, J. Shealy, R. Vetury, D. Green, J. Brown, S. Gibb, in *Compound Semiconductor IC Symposium Technical Digest*, San Antonio, 2006, pp. 85–88

6.28. A. Couturier, S. Heckmann, V. Serru, T. Huet, P. Chaumas, J. Fontecave, M. Camiade, J.P. Viaud, S. Piotrowicz, in *IEEE International Microwave Symposium Digest*, Honolulu, 2007, pp. 813–816

6.29. S. Cripps, in *IEEE International Microwave Symposium Digest*, Boston, 1983, pp. 221–223

6.30. S. Cripps, *RF Power Amplifiers for Wireless Communications* (Artech House, Boston London, 1999)

6.31. A. Darwish, K. Boutros, B. Luo, B. Huebschman, E. Viveiros, H. Hung, IEEE Trans. Microw. Theory Tech. **54**, 4456 (2006)

6.32. D. Dawson, in *IMS Workshop WMG: Solid-State Power Invades the Tube Realm*, Honolulu, 2007

6.33. A. de Hek, P. Hunneman, M. Demmler, A. Hülsmann, in *Proceedings of European GaAs and Related Compounds Application Symposium GAAS*, Munich, 1999, pp. 276–280

6.34. I. Dettmann, E. Chigaeva, L. Wu, M. Berroth, in *EEEfCOM Workshop*, Ulm, June 2005

6.35. M. Drinkwine, T. Winslow, D. Miller, D. Conway, B. Raymond, in *Proceedings of International Conference on GaAs Manufacturing Technology*, Vancouver, 2006, pp. 187–190

6.36. W. Eckl, B. Friedel, G. Fischer, H. Schenkel, in *IMS Workshop Advances in GaN-based Device and Circuit Technology: Modeling and Applications*, Forth Worth, 2004

6.37. M. Engels, R. Jansen, in *IEEE International Microwave Symposium Digest*, Atlanta, 1993, pp. 757–760

6.38. D. Fanning, L. Witkowski, C. Lee, D. Dumka, H. Tserng, P. Saunier, W. Gaiewski, E. Piner, K. Linthicum, J. Johnson, in *Proceedings of International Conference on GaAs Manufacturing Technology*, New Orleans, 2005, p. 8.3

6.39. H. Finlay, R. Jansen, J. Jenkins, I. Eddison, IEEE Trans. Microw. Theory Tech. **36**, 961 (1988)

6.40. G. Fischer, in *IEEE Wireless and Microwave Technology Conference*, Clearwater, 2004, p. FD-1

6.41. R. Freitag, H. Henry, E. Lee, M. Pingor, H. Salvo, 160 W MODAR Wind Shear Detection/Weather System. IEEE Trans. Microw. Theory Tech. **43**, 1703 (1995)

6.42. J. Gajadharsing, O. Bosma, P. van Westen, in *IEEE International Microwave Symposium Digest*, Fort Worth, 2004, pp. 529–532

6.43. N.L. Gallou, J. Villemazet, B. Cogo, J. Cazaux, A. Mallet, L. Lapierre, in *Proceedings of European Microwave Conference*, Munich, 2003, pp. 273–276

6.44. S. Gao, S. Sanabria, H. Xu, S. Long, S. Heikman, U. Mishra, R. York, IEE Proc. Microw. Ant. Propag. **153**, 259 (2006)

6.45. S. Gao, H. Xu, S. Heikman, U. Mishra, R. York, in *Asia Pacific Microwave Conference*, Suzhou, 2005

6.46. S. Gao, H. Xu, S. Heikman, U. Mishra, R. York, IEEE Microw. Wireless Compon. Lett. **16**, 28 (2006)

6.47. S. Gao, H. Xu, U. Mishra, R. York, in *Compound Semiconductor IC Symposium Technical Digest*, San Antonio, 2006, pp. 259–262

6.48. J. Gassmann, P. Watson, L. Kehias, G. Henry, in *IEEE International Microwave Symposium Digest*, Honolulu, 2007, pp. 615–618

6.49. E. Griffin, in *IEEE International Microwave Symposium Digest*, Boston, 2000, pp. 709–712

6.50. G. Soubercaze-Pun, J.G. Tartarin, L. Bary, J. Rayssac, E. Morvan, B. Grimbert, S. Delage, J. DeJaeger, J. Graffeuil, in *IEEE International Microwave Symposium Digest*, San Francisco, 2006, pp. 747–750

6.51. K. Gupta, R. Garg, I. Bahl, P. Bhartia, *Microstrip Lines and Slotlines.* (Artech House, Boston London, 1996)

6.52. U. Gustavsson, Master thesis, Orebro University, 2006

6.53. C. Hang, Y. Qian, T. Itoh, in *IEEE International Microwave Symposium Digest*, Phoenix, 2001, pp. 1079–1082

6.54. G. Hanington, P. Chen, P. Asbeck, L. Larson, IEEE Trans. Microw. Theory Tech. **47**, 1471 (1999)

6.55. W. Haydl, A. Tessmann, K. Züfle, H. Massler, T. Krems, L. Verweyen, J. Schneider, in *Proceedings of European Microwave Conference* Prague, 1996, pp. 996–1000

6.56. W. Heinrich, IEEE Trans. Microw. Theory Tech. **41**, 45 (1993)

6.57. M. Heins, J. Carroll, M. Kao, J. Delaney, C. Campbell, in *IEEE International Microwave Symposium Digest*, Forth Worth, 2004, pp. 149–152

6.58. S. Hong, Y. Woo, I. Kim, J. Kim, J. Moon, H. Kim, J. Lee, B. Kim, in *IEEE International Microwave Symposium Digest*, Honolulu, 2007, pp. 1247–1250

6.59. S. Hsu, P. Valizadeh, D. Pavlidis, J. Moon, M. Micovic, D. Wong, T. Hussain, in *GaAs IC Symposium Technical Digest*, Monterey, 2002, pp. 85–88

6.60. T. Hussain, A. Kurdoghlian, P. Hashimoto, W. Wong, M. Wetzel, J. Moon, L. McCray, M. Micovic, in *IEDM Technical Digest*, Washington DC, 2001, pp. 581–584

6.61. T. Hussain, M. Micovic, T. Tsen, M. Delaney, D. Chow, A. Schmitz, P. Hashimoto, D. Wong, J. Moon, M. Hu, J. Duvall, D. McLaughlin, in *Proceedings of International Conference on GaAs Manufacturing Technology*, Miami, 2004, pp. 25–27

6.62. A. Huttunen, R. Kaunisto, in *IEEE International Microwave Symposium Digest*, Honolulu, 2007, pp. 1437–1440

6.63. A. Inoue, A. Ohta, S. Goto, T. Ishikawa, Y. Matsuda, in *IEEE International Microwave Symposium Digest*, Fort Worth, 2004, pp. 1947–1950

6.64. M. Iwamoto, A. Jayaraman, G. Hanington, P. Chen, A. Bellora, W. Thornton, L. Larson, P. Asbeck, Electron. Lett. **36**, 1010 (2000)

6.65. K. Iyomasa, K. Yamanaka, K. Mori, H. Noto, H. Ohtsuka, M.N.S. Yoneda, Y. Kamo, Y. Isota, in *IEEE International Microwave Symposium Digest*, Honolulu, 2007, pp. 1255–1258

6.66. B. Jacobs, B. van Straaten, M. Kramer, F. Karouta, P. de Hek, E. Suijker, R. van Dijk, Proc. MRS **693**, 629 (2001)

6.67. A. Jararaman, P. Chen, G. Hanington, L. Larson, P. Asbeck, IEEE Microw. Guided Wave Lett. **8**, 121 (1998)

6.68. A. John, R. Jansen, in *IEEE International Microwave Symposium Digest*, San Francisco, 1996, pp. 745–748

6.69. Y. Kamo, T. Kunii, H. Takeuchi, Y. Yamamoto, M. Totsuka, T. Shiga, H. Minami, T. Kitano, S. Miyakuni, T. Oku, A. Inoue, T. Nanjo, H. Chiba, M. Suita, T. Oishi, Y. Abe, Y. Tsuyama, R. Shirahana, H. Ohtsuka, K. Iyomasa, K. Yamanaka, M. Hieda, M. Nakayama, T. Ishikawa, T. Takagi, K. Marumoto, Y. Matsuda, in *IEEE International Microwave Symposium Digest*, Long Beach, 2005, pp. 495–498

6.70. K. Kong, B. Nguyen, S. Nayak, M. Kao, in *Compound Semiconductor IC Symposium Technical Digest*, Palm Springs, 2005, pp. 232–235

6.71. V. Kaper, R. Thompson, T. Prunty, J. Shealy, in *IEEE International Microwave Symposium Digest*, Fort Worth, 2004, pp. 1145–1148

6.72. V. Kaper, V. Tilak, H. Kim, R. Thompson, T. Prunty, J. Smart, L. Eastman, J. Shealy, in *GaAs IC Symposium Technical Digest*, Monterey, 2002, pp. 251–254

6.73. O. Kappeler, R. Quay, F. van Raay, R. Kiefer, R. Reiner, H. Walcher, S. Müller, M. Mikulla, M. Schlechtweg, G. Weimann, D. Wiegrer, U. Seyfried, W. Templ, in *IEDM Technical Digest*, Washington DC, 2005, pp. 385–386

6.74. A. Katz, M. Kubak, G. DeSalvo, in *IEEE International Microwave Symposium Digest*, San Francisco, 2006, pp. 1364–1367

6.75. A. Kawano, N. Adachi, Y. Tateno, S. Mizuno, N. Ui, J. Nikaido, S. Sano, in *Asia Pacific Microwave Conference*, Suzhou, 2005

6.76. D. Keogh, J. Li, A. Conway, D. Qiao, S. Raychaudhuri, P. Asbeck, Int. J. High Speed Electron. Systems **14**, 831 (2004)

6.77. L. Khan, Proc. IRE **40**, 803 (1952)

6.78. T. Kikkawa, in *Compound Semiconductor IC Symposium Technical Digest*, Monterey, 2004, pp. 17–20

6.79. D. Kimball, P. Draxler, J. Jeong, C. Hsia, S. Lanfranco, W. Nagy, K. Linthicum, L. Larson, P. Asbeck, in *Compound Semiconductor IC Symposium Technical Digest*, Palm Springs, 2005, pp. 89–92

6.80. D. Kimball, J. Jeong, C. Hsia, P. Draxler, P. Asbeck, D. Choi, W. Pribble, R. Pengelly, in *IEEE Radio and Wireless Symposium*, San Diego, 2006, pp. n.a.

6.81. H. Klockenhoff, R. Behtash, J. Würfl, W. Heinrich, G. Tränkle, in *IEEE International Microwave Symposium Digest*, San Francisco, 2006, pp. 1846–1849

6.82. K. Kobayashi, Y.C. Chen, I. Smorchkova, R. Tsai, M. Wojtowicz, A. Oki, in *Radio Frequency Integrated Circuit Symposium Digest*, Honululu, 2007, pp. 585–588

6.83. K. Kobayashi, Y. Chen, I. Smorchkova, R. Tsai, M. Wojtowicz, A. Oki, in *IEEE International Microwave Symposium Digest*, Honululu, 2007, pp. 619–622

6.84. J. Komiak, W. Kong, P. Chao, K. Nicols, in *GaAs IC Symposium Technical Digest*, Orlando, 1996, pp. 111–114

6.85. J. Komiak, W. Kong, K. Nicols, in *IEEE International Microwave Symposium Digest*, Seattle, 2002, pp. 905–907

6.86. D. Krausse, R. Quay, R. Kiefer, A. Tessmann, H. Massler, A. Leuther, T. Merkle, S. Müller, C. Schwörer, M. Mikulla, M. Schlechtweg, G. Weimann, in *Proceedings of European GaAs and Related Compounds Application Symposium GAAS*, Amsterdam, 2004, pp. 71–74

6.87. K. Krishnamurthy, S. Keller, U. Mishra, M. Rodwell, S. Long, in *GaAs IC Symposium Technical Digest*, Seattle, 2000, pp. 33–36

6.88. B. Kruger, in *IEEE International Radar Conference*, Annapolis, 1998, pp. 227–232

6.89. P. Ladbrooke, *MMIC Design: GaAs FETs and HEMTs* (Artech House, Boston London, 1989)

6.90. A. Lan, M. Wojtowicz, I. Smorchkova, R. Coffie, R. Tsai, B. Heying, M. Truong, F. Fong, M. Kintis, C. Namba, A. Oki, T. Wong, IEEE Microw. Wireless Compon. Lett. **16**, 425 (2006)

6.91. L. Larson, P. Asbeck, D. Kimball, in *Compound Semiconductor IC Symposium Technical Digest*, Palm Springs, 2005, pp. 1–4

6.92. J. Lee, L. Eastman, K. Webb, IEEE Trans. Microw. Theory Tech. **51**, 2243 (2003)

6.93. S. Lee, B. Green, K. Chu, K. Webb, L. Eastman, in *IEEE International Microwave Symposium Digest*, Boston, 2000, pp. 549–552

6.94. S. Lee, B. Cetiner, H. Torpi, S. Cai, J. Li, K. Alt, Y. Chen, C. Wen, K. Wang, T. Itoh, IEEE Trans. Electron Devices **48**, 495 (2001)

6.95. Y.S. Lee, Y.H. Jeong, IEEE Microw. Wireless Compon. Lett. **17**, 622 (2007)

6.96. J. Lees, J. Benedikt, K. Hilton, J. Powell, R. Balmer, M. Uren, T. Martin, P. Tasker, Electron. Lett. **41**, 1284 (2005)

6.97. B. Levush, in *IMS Workshop WMG: Solid-State Power Invades the Tube Realm*, Honululu, 2007

6.98. C. Lin, H. Liu, C. Chu, H. Huang, Y. Wang, C. Liu, C. Chang, C. Wu, C. Chang, in *Compound Semiconductor IC Symposium Technical Digest*, San Antonio, 2006, pp. 165–168

6.99. G. Ma, Q. Chen, O. Tornblad, T. Wei, C. Ahrens, R. Gerlach, in *IEDM Technical Digest*, Washington DC, 2005, pp. 361–364

6.100. A. Maekawa, T. Yamamoto, E. Mitani, S. Sano, in *IEEE International Microwave Symposium Digest*, San Francisco, 2006, pp. 722–725

6.101. L. Maloratsky, Microwave & RF **39**, 79 (2000)

6.102. E. Martinez, in *GaAs IC Symposium Technical Digest*, Monterey, 2002, pp. 7–10

6.103. J. Martins, P. Cabral, N. Carvalho, J. Pedro, IEEE Trans. Microw. Theory Tech. **54**, 4432 (2006)

6.104. P. McGovern, D. Williams, P. Tasker, J. Benedikt, J. Powell, K. Hilton, R. Balmer, T. Martin, M. Uren, in *IEEE International Microwave Symposium Digest*, Fort Worth, 2004, pp. 825–828

6.105. D. Meharry, R. Lendner, K. Chu, L.G.K. Beech, in *IEEE International Microwave Symposium Digest*, Honolulu, 2007, pp. 631–634

6.106. M. Micovic, A. Kurdhoglian, P. Hashimoto, M. Hu, M. Antcliffe, P. Willadsen, W. Wong, R. Bowen, I. Milosavljevic, A. Schmitz, M. Wetzel, D. Chow, in *IEDM Technical Digest*, San Francisco, 2006, pp. 425–428

6.107. M. Micovic, A. Kurdoghlian, A. Janke, P. Hashimoto, P. Wong, D. Moon, J. McCray, C. Nguyen, IEEE Trans. Electron Devices **48**, 591 (2001)

6.108. E. Mitani, in *IMS Workshop WSG: Solid-State Power Invades the Tube Realm*, Honolulu, 2007

6.109. E. Mitani, H. Haematsu, S. Yokogawa, J. Nikaido, Y. Tateno, in *Proceedings of International Conference on GaAs Manufacturing Technology*, Vancouver, 2006, pp. 183–186

6.110. W. Nagy, J. Brown, R. Borges, S. Singhal, IEEE Trans. Microw. Theory Tech. **51**, 660 (2003)

6.111. Y. Nanishi, H. Miyamoto, A. Suzuki, H. Okumura, N. Shibata, in *Proceedings of International Conference on GaAs Manufacturing Technology*, Vancouver, 2006, pp. 45–48

6.112. R. Negra, T. Chu, M. Helaoui, S. Boumaizs, G. Hegazi, F. Ghannouchi, in *IEEE International Microwave Symposium Digest*, Honolulu, 2007, pp. 795–798

6.113. M. Neuburger, M. Kunze, I. Daumiller, T. Zimmermann, A. Dadgar, A. Krost, S. Hettich, F. Gruson, H. Schumacher, E. Kohn, in *Proceedings of International Conference on GaAs Manufacturing Technology*, Miami, 2004, pp. 111–114

6.114. M. Neuburger, M. Kunze, I. Daumiller, T. Zimmermann, A. Dadgar, A. Krost, S. Hettich, F. Gruson, H. Schumacher, E. Kohn, in *Proceedings of European GaAs and Related Compounds Application Symposium GAAS*, Paris, 2005, pp. 225–228

6.115. C. Nguyen, M. Micovic, D. Wong, A. Kurdoghlian, P. Hashi noto, P. Janke, L. McCray, J. Moon, in *GaAs IC Symposium Technical Digest*, Seattle, 2000, pp. 11–14

6.116. M. Nishijima, T. Murata, Y. Hirose, M. Hikita, N. Negoro, H. Sakai, Y. Uemoto, K. Inoue, T. Tanaka, D. Ueda, in *IEEE International Microwave Symposium Digest*, Long Beach, 2005, pp. 299–302

6.117. T. Ogawa, T. Iwasaki, H. Maruyama, K. Horiguchi, M. Nakayama, Y. Ikeda, H. Kurebayashi, in *IEEE International Microwave Symposium Digest*, Fort Worth, 2004, pp. 537–540

6.118. M. Ohtomo, IEEE Trans. Microw. Theory Tech. **41**(6–7), 495 (1993)

6.119. Y. Okamoto, A. Wakejima, K. Matsunaga, Y. Ando, T. Nakayama, K. Kasahara, K. Ota, Y. Murase, K. Yamanoguchi, T. Inoue, H. Miyamoto, in *IEEE International Microwave Symposium Digest*, Long Beach, 2005, pp. 491–494

6.120. V. Paidi, S. Xie, R. Coffie, B. Moran, S. Heikman, S. Keller, A. Chini, S. DenBaars, U. Mishra, S. Long, M. Rodwell, IEEE Trans. Microw. Theory Tech. **51**, 643 (2003)

6.121. J. Palmour, S. Sheppard, R. Smith, S. Allen, W. Pribble, T. Smith, Z. Ring, J. Sumakeris, A. Saxler, J. Milligan, in *IEDM Technical Digest*, Washington DC, 2001, pp. 385–388

6.122. R. Pantocha, IEEE Trans. Microw. Theory Tech. **37**, 1675 (1989)

6.123. P. Parikh, Y. Wu, M. Moore, P. Chavarkar, U. Mishra, B. Neidhard, L. Kehias, T. Jenkins, in *Lester Eastman Conference High Performance Devices*, Newark, 2002, pp. 56–57

6.124. R. Pengelly, in *Proceedings of Wireless and Microwave Technology Conference*, Clearwater, 2004, pp. RC-5

6.125. P. Piel, M. Miller, B. Green, in *IEEE International Microwave Symposium Digest*, Fort Worth, 2004, pp. 1363–1366

6.126. S. Piotrowicz, E. Chartier, J. Jaquet, D. Floriot, J. Coupat, C. Framery, P. Eudeline, P. Auxemery, in *IEEE International Microwave Symposium Digest*, Fort Worth, 2004, pp. 1527–1530

6.127. G. Ponchak, in *Proceedings of European GaAs and Related Compounds Application Symposium GAAS*, Munich, 1999, pp. 414–417

6.128. G. Ponchak, Z. Schwartz, S. Alterovitz, A. Downey, in *Proceedings of European GaAs and Related Compounds Application Symposium GAAS*, Amsterdam, 2004, pp. 41–44

6.129. M.J. Poulton, W. Leverich, J. Shealy, R. Vetury, J. Brown, D. Green, S. Gibb, in *IEEE International Microwave Symposium Digest*, San Francisco, 2006, pp. 1327–1330

6.130. W. Pribble, J. Palmour, S. Sheppard, R. Smith, S. Allen, T. Smith, Z. Ring, J. Sumakeris, A. Saxler, J. Milligan, in *IEEE International Microwave Symposium Digest*, Seattle, 2002, pp. 1819–1822

6.131. T. Quach, R. Fitch, J. Gillespie, T. Jenkins, R. Neidhard, E. Nykiel, G. Via, P. Watson, J. Wiedemann, in *Proceedings of International Conference on GaAs Manufacturing Technology*, Miami, 2004, pp. 103–106

6.132. R. Quay, R. Kiefer, F. van Raay, H. Massler, S. Ramberger, S. Müller, M. Dammann, M. Mikulla, M. Schlechtweg, G. Weimann, in *IEDM Technical Digest*, San Francisco, 2002, pp. 673–676

6.133. R. Quay, F. van Raay, A. Tessmann, R. Kiefer, M. Dammann, M. Mikulla, M. Schlechtweg, G. Weimann, in *Proceedings WOCSDICE*, Venice, 2007, pp. 349–352

6.134. F. Raab, IEEE Trans. Microw. Theory Tech. **45**, 2007 (1997)

6.135. F. Raab, P. Asbeck, S. Cripps, P. Kenington, Z. Popovic, N. Pothecary, J. Sevic, N. Sokal, IEEE Trans. Microw. Theory Tech. **50**, 814 (2002)

6.136. F. Raab, P. Asbeck, S. Cripps, P. Kenington, Z. Popovic, N. Pothecary, J. Sevic, N. Sokal, High Freq. Electron. **2**, 22 (2003)

6.137. V. Radisic, C. Pobanz, H. Ming, M. Micovic, M. Wetzel, P. Janke, M. Yu, C. Ngo, D. Dawson, M. Matloubian, in *Radio Frequency Integrated Circuits Symposium Digest*, Boston, 2000, pp. 43–46

6.138. M. Riaziat, R. Majidi-Ahy, I. Feng, IEEE Trans. Microw. Theory Tech. **38**, 245 (1990)

6.139. M. Rosker, H. Dietrich, C. Bozada, A. Hung, G. Via, in *Proceedings of International Conference on GaAs Manufacturing Technology*, Vancouver, 2006, pp. 41–45

6.140. M. Rosker, in *IEEE RF IC Symposium Digest*, Honolulu, 2007, pp. 159–162

6.141. A. Royet, B. Cabon, O. Rozeau, T. Ouisse, T. Billon, IEE Proc. Microw. Ant. Propag., **149**, 253 (2002)

436   References of Chapter 6

6.142. M. Rudolph, R. Behtash, K. Hirche, J. Würfl, W. Heinrich, G. Tränkle, in *IEEE International Microwave Symposium Digest*, San Francisco, 2006, pp. 1899–1902

6.143. C. Sanabria, X. Hongtao, S. Heikman, U. Mishra, R. York, IEEE Microw. Wireless Compon. Lett. **15**, 463 (2005)

6.144. H. Sano, K. Otobe, Y. Tateno, N. Adachi, S. Mizuno, A. Kawano, J. Nikaido, S. Sano, in *Asia Pacific Microwave Conference*, Suzhou, 2005

6.145. M. Schlechtweg, W. Haydl, A. Bangert, J. Braunstein, P. Tasker, L. Verweyen, H. Massler, W. Bronner, A. Hülsmann, K. Köhler, IEEE J. Solid-State Circuits **31**, 1426 (1996)

6.146. D. Schmelzer, S. Long, in *Compound Semiconductor IC Symposium Technical Digest*, San Antonio, 2006, pp. 96–99

6.147. C. Schöllhorn, W. Zhao, M. Morschbach, E. Kasper, IEEE Trans. Electron Devices **50**, 740 (2003)

6.148. P. Schuh, R. Leberer, H. Sledzik, M. Oppermann, B. Adelseck, H. Brugger, R. Behtash, H. Leier, R. Quay, R. Kiefer, in *IEEE International Microwave Symposium Digest*, San Francisco, 2006, pp. 726–729

6.149. P. Schuh, R. Leberer, H. Sledzik, M. Oppermann, B. Adelseck, H. Brugger, R. Quay, M. Mikulla, G. Weimann, in *Proceedings of European Microwave Integrated Circuits Conference*, Manchester, 2006, pp. 241–244

6.150. P. Schuh, R. Leberer, H. Sledzik, D. Schmidt, M. Oppermann, B. Adelseck, H. Brugger, R. Quay, F. van Raay, M. Seelmann-Eggebert, R. Kiefer, W. Bronner, in *IEEE International Microwave Symposium Digest*, San Francisco, 2006, pp. 1324–1327

6.151. R. Schwindt, V. Kumar, O. Aktas, J. Lee, I. Adesida, in *Compound Semiconductor IC Symposium Technical Digest*, Monterey, 2004, pp. 201–203

6.152. K. Seemann, S. Ramberger, A. Tessmann, R. Quay, J. Schneider, M. Riessle, H. Walcher, M. Kuri, R. Kiefer, M. Schlechtweg, in *Proceedings of European Microwave Conference*, Munich, 2003, pp. 383–386

6.153. S. Sheppard, B. Pribble, R. Smith, A. Saxler, S. Allen, J. Milligan, R. Pengelly, in *Proceedings of International Conference on GaAs Manufacturing Technology*, Vancouver, 2006, pp. 175–178

6.154. S. Sheppard, W. Pribble, D. Emerson, Z. Ring, R. Smith, S. Allen, J. Palmour, in *Device Research Conference*, Denver, 2000, pp. 37–38

6.155. H. Shimawaki, H. Miyamoto, in *Proceedings of European GaAs and Related Compounds Application Symposium GAAS*, Paris, 2005, pp. 377–380

6.156. K. Shiojima, T. Makimura, T. Kosugi, T. Suemitsu, N. Shigekawa, M. Hiroki, H. Yokoyama, in *IEEE International Microwave Symposium Digest*, San Francisco, 2006, pp. 1331–1334

6.157. K. Shiojima, T. Makimura, T. Kosugi, S. Sugitani, N. Shigekawa, H. Ishikawa, T. Egawa, Electron. Lett. **40**, 775 (2004)

6.158. D. Snider, IEEE Trans. Electron Devices **14**, 851 (1967)

6.159. M. Südow, K. Andersson, N. Billström, J. Grahn, H. Hjelmgren, J. Nilsson, P. Nilsson, J. Stahl, H. Zirath, N. Rorsman, IEEE Trans. Microw. Theory Tech. **54**, 4072 (2006)

6.160. Y. Takada, H. Sakurai, K. Matsushita, K. Masuda, S. Takatsuka, M. Kuraguchi, T. Suzuki, M. Hirose, H. Kawasaki, K. Takagi, K. Tsuda, IEICE Tech. Rep. **105**, 39 (2005)

6.161. K. Takagi, K. Masuda, Y. Kashiwabara, H. Sakurai, K. Matsushita, S. Takatsuka, H. Kawasaki, Y. Takada, K. Tsuda, in *Compound Semiconductor IC Symposium Technical Digest*, San Antonio, 2006, pp. 265–268

6.162. T. Takagi, M. Mochizuki, Y. Tarui, Y. Itoh, S. Tsuji, Y. Mitsui, IEICE Trans. Electron. **E78-C**, 936 (1995)

6.163. R. Therrien, A. Chaudhari, S. Singhal, C. Snow, A. Edwards, C. Park, W. Nagy, J. Johnson, A. Hanson, K. Linticum, I. Kizilyalli, in *IEEE International Microwave Symposium Digest*, Honolulu, 2007, pp. 635–638

6.164. R. Therrien, W. Nagy, I. Kizilyalli, RF Design, 27 (2007)

6.165. N. Ui, H. Sano, S. Sano, in *IEEE International Microwave Symposium Digest*, Honolulu, 2007, pp. 1259–1262

6.166. N. Ui, S. Sano, in *IEEE International Microwave Symposium Digest*, San Francisco, 2006, pp. 718–721

6.167. F. van den Bogaart, A. de Hek, in *IEE Tutorial Colloquium on Design of RFICs and MMICs*, London, 1997, pp. 8/1–8/6

6.168. M. van Heiningen, F. van Vliet, R. Quay, F. van Raay, M. Seelmann-Eggebert, in *Proceedings of European Microwave Integrated Circuits Conference*, Manchester, 2006, pp. 75–78

6.169. F. van Raay, R. Quay, R. Kiefer, F. Benkhelifa, B. Raynor, W. Pletschen, M. Kuri, H. Massler, S. Müller, M. Dammann, M. Mikulla, M. Schlechtweg, G. Weimann, IEEE Microw. Wireless Compon. Lett. **15**, 460 (2005)

6.170. F. van Raay, R. Quay, R. Kiefer, W. Bronner, M. Seelmann-Eggebert, M. Schlechtweg, M. Mikulla, G. Weimann, in *IEEE International Microwave Symposium Digest*, San Francisco, 2006, pp. 1368–1371

6.171. F. van Raay, R. Quay, R. Kiefer, W. Fehrenbach, W. Bronner, M. Kuri, F. Benkhelifa, H. Massler, S. Müller, M. Mikulla, M. Schlechtweg, G. Weimann, in *Proceedings of European GaAs and Related Compounds Application Symposium GAAS*, Paris, 2005, pp. 233–236

6.172. F. van Rijs, S. Theeuwen, in *IEDM Technical Digest*, San Francisco, 2006, pp. 205–208

6.173. B. Vassilakis, A. Cova, in *Asia Pacific Microwave Conference*, Suzhou, 2005

6.174. B. Vassilakis, A. Cova, W. Veitschegger, in *Compound Semiconductor IC Symposium Technical Digest*, Monterey, 2004, pp. 3–7

6.175. N. Vellas, C. Gaquiere, Y. Guhel, M. Werquin, F. Bue, R. Aubry, S. Delage, F. Semond, J. Jaeger, IEEE Electron Device Lett. **23**, 461 (2002)

6.176. G. Vendelin, *Design of Amplifiers ans Oscillators by the S-Parameter Method* (Wiley, New York, 1982)

6.177. R. Vetury, J. Shealy, D. Green, J. McKenna, J. Brown, S. Gibb, K. Leverich, P. Garber, M. Poulton, in *IEEE International Microwave Symposium Digest*, San Francisco, 2006, pp. 714–717

6.178. J. Vincent, D. van der Merve, in *IEEE AFRICON*, Stellenbosch, 1996, pp. 749–752

6.179. A. Wakejima, T. Asano, T. Hirano, M. Funabashi, K. Matsunaga, IEEE J. Solid-State Circuits **40**, 2054 (2005)

6.180. A. Wakejima, K. Matsunaga, Y. Okamoto, Y. Ando, T. Nakayama, H. Miyamoto, Electron. Lett. **41**, 1371 (2005)

6.181. A. Wakejima, K. Matsunaga, Y. Okamoto, K. Ota, Y. Ando, T. Nakayama, H. Miyamoto, in *IEEE International Microwave Symposium Digest*, San Francisco, 2006, pp. 1360–1363

6.182. F. Wang, A. Yang, D. Kimball, K. Larson, P. Asbeck, IEEE Trans. Microw. Theory Tech. **53**, 1244 (2005)

6.183. R. Wang, M. Cole, in *GaAs IC Symposium Technical Digest*, Orlando, 1996, pp. 111–114

6.184. M. Werquin, C. Gaquiere, Y. Gubel, N. Vellas, D. Theron, B. Boudart, V. Hoel, M. Germain, J.D. Jaeger, S. Delage, Electron. Lett. **41**, 46 (2005)

6.185. P. White, T. O'Leary, in *GaAs IC Symposium Technical Digest*, San Diego, 1995, pp. 277–280

6.186. W. Wu, B. Thibeault, J. Xu, R. York, S. Keller, B. Keller, U. Mishra, in *Device Research Conference*, Santa Barbara, 1999, pp. 50–51

6.187. Y. Wu, P. Charvarkar, M. Moore, P. Parikh, B. Keller, U. Mishra, in *IEDM Technical Digest*, San Francisco, 2000, pp. 375–376

6.188. Y. Wu, D. Kapolnek, J. Ibbetson, P. Parikh, B. Keller, U. Mishra, in *IEEE International Microwave Symposium Digest*, Boston, 2000, pp. 963–965

6.189. Y. Wu, M. Moore, A. Saxler, T. Wisleder, P. Parikh, in *Device Research Conference*, State College, PA, 2006, pp. 151–152

6.190. Y. Wu, A. Chin, K.H. Shih, C.C. Wu, C. Liao, S. Pai, C. Chi, IEEE Electron Device Lett. **21**, 442 (2000)

6.191. S. Xie, V. Paidi, R.C.A. Heikman, B. Moran, A. Chini, S. DenBaars, U. Mishra, S. Long, M. Rodwell, IEEE Microw. Wireless Compon. Lett. **13**, 284 (2003)

6.192. S. Xie, V. Paidi, S. Heikman, L. Shen, A. Chini, U. Mishra, M. Rodwell, S. Long, Int. J. High Speed Electron. Systems **14**, 847 (2004)

6.193. H. Xu, S. Sanabria, N. Pervez, S. Keller, U. Mishra, R. York, in *IEEE International Microwave Symposium Digest*, Fort Worth, 2002, pp. 1509–1512

6.194. H. Xu, S. Sanabria, N. Pervez, S. Keller, U. Mishra, R. York, in *IEEE International Microwave Symposium Digest*, Long Beach, 2005, pp. 1345–1348

6.195. J. Xu, S. Keller, G. Parish, S. Heikman, U. Mishra, R. York, in *IEEE International Microwave Symposium Digest*, Boston, 2000, pp. 959–962

6.196. S. Xu, K. Gan, G. Samudra, Y. Liang, J. Sin, IEEE Trans. Electron Devices **47**, 1980 (2000)

6.197. T. Yamamoto, T. Kitahara, S. Hiura, in *IEEE International Microwave Symposium Digest*, Honolulu, 2007, pp. 1263–1266

6.198. T. Yamamoto, E. Mitani, K. Inoue, M. Nishi, S. Sano, in *Proceedings of the European Microwave Integrated Circuits Conference* Munich, 2007, pp. 173–176

6.199. K. Yamanaka, K. Iyomasa, H. Ohtsuka, M. Nakayama, Y. Tsuyama, T. Kunii, Y. Kamo, T. Takagi, in *Proceedings of European GaAs and Related Compounds Application Symposium GAAS*, Paris, 2005, pp. 241–244

6.200. K. Yamanaka, K. Mori, K. Iyomasa, H. Ohtsuka, M. Nakayama, Y. Kamo, Y. Isota, in *IEEE International Microwave Symposium Digest*, Honolulu, 2007, pp. 1251–1254

6.201. M. Yu, M. Matloubian, P. Petre, L.R. Hamilton, R. Bowen, M. Lui, H. Sun, C. Ngo, P. Janke, D. Baker, R. Robertson, IEEE J. Solid-State Circuits **34**, 1212 (1999)

6.202. J. Zolper, Wide Bandgap Semiconductor RF Electronics Technology, MTO Industry Briefing, Sept. 2001, http://www.compoundsemiconductor.net/cws/article/magazine/11332

References of Chapter 7

7.1. N. Adachi, Y. Tateno, S. Mizuno, A. Kawano, J. Nikaido, S. Sano, in *IEEE International Microwave Symposium Digest*, Long Beach, 2005, pp. 507–510

7.2. M. Adlerstein, J. Gering, IEEE Trans. Electron Devices **47**, 434 (2000)

7.3. M. Akita, S. Kishimoto, T. Mizutani, IEEE Electron Device Lett. **22**, 376 (2001)

7.4. O. Aktas, A. Kuliev, V. Kumar, R. Schwindt, S. Toshkov, D. Costescu, J. Stubbins, I. Adesida, Solid-State Electron. **48**, 471 (2004)

7.5. N. Ashcroft, N. Mermin, *Solid-State Physics* (Saunders College, Forth Worth, 1976)

7.6. A. Barnes, A. Boetti, L. Marchand, J. Hopkins, in *Proceedings of European GaAs and Related Compounds Application Symposium GAAS*, Paris, 2005, pp. 5–12

7.7. J. Bernat, M. Wolter, P. Javorka, A. Fox, M. Marso, P. Kordos, Solid-State Electron. **48**, 1825 (2004)

7.8. R. Blanchard, A. Cornet, J. del Alamo, IEEE Electron Device Lett. **21**, 424 (2001)

7.9. R. Blanchard, M. Sommerville, J.D. Alamo, K. Duh, P. Chao, IEEE Trans. Electron Devices **47**, 1560 (2000)

7.10. J.D. Blauwe, D. Wellekens, G. Groeseneken, L. Haspeslagh, J.V. Houdt, L. Deferm, H. Maes, IEEE Trans. Electron Devices **45**, 2466 (1998)

7.11. M. Borgarino, R. Menozzi, Y. Baeyens, P. Cova, F. Fantini, IEEE Trans. Electron Devices **45**, 366 (1998)

7.12. K. Boutros, W. Luo, B. Brar, in *Reliability of Compound Semiconductors Workshop*, San Antonio, 2006, pp. n.a.

7.13. K. Boutros, P. Rowell, B. Brar, in *Proceedings of the International Reliability Physics Symposium*, Phoenix, 2004, pp. 577–578

7.14. N. Bovolon, R. Schultheis, J. Müller, P. Zwicknagl, E. Zanoni, IEEE Electron Device Lett. **19**, 469 (1998)

7.15. N. Braga, R. Mickevicius, V. Rao, W. Fichtner, R. Gaska, M. Shur, in *Compound Semiconductor IC Symposium Technical Digest*, Palm Springs, 2005, pp. 149–152

7.16. H. Brech, W. Brakensiek, D. Burdeaux, W. Burger, C. Dragon, G. Formicone, B. Pryor, D. Rice, in *IEDM Technical Digest*, Washington DC, 2003, pp. 359–362

7.17. J. Brown, S. Lee, D. Lieu, J. Martin, R. Vetury, M. Poulton, J. Shealy, in *IEEE International Microwave Symposium Digest*, Honolulu, 2007, pp. 303–306

7.18. S. Cai, Y. Tang, R. Li, Y. Wei, L. Wong, Y. Chen, K. Wang, M. Chen, Y. Zhao, R. Schrimpf, J. Keay, K. Galloway, IEEE Trans. Electron Devices **47**, 304 (2000)

7.19. C. Canali, F. Magistral, M. Sangalli, C. Tedesco, E. Zanoni, G. Castellaneta, F. Marchetti, in *Proceedings of the International Reliability Physics Symposium*, Cardiff, 1991, pp. 206–213

7.20. P. Cappelletti, R. Bez, A. Modelli, A. Visconti, in *IEDM Technical Digest*, San Francisco, 2004, pp. 489–492

7.21. P. Chao, W. Hu, H. DeOrio, A. Swanson, W. Hoffmann, W. Taft, IEEE Electron Device Lett. **18**, 441 (1997)

7.22. H. Chiu, T. Yeh, S. Yang, M. Hwu, Y. Chan, IEEE Trans. Electron Devices **50**, 1532 (2003)

7.23. Y. Chou, D. Leung, R. Lai, R. Grundbacher, M. Barsky, Q. Kan, R. Tsai, M. Wojtowicz, D. Eng, L. Tran, T. Block, P. Liu, M. Nishimoto, A. Oki, IEEE Trans. Electron Devices **24**, 378 (2003)

7.24. Y. Chou, D. Leung, I. Smorchkova, M. Wojtoicz, R. Grundbacher, L. Callejo, Q. Kan, R. Lai, P. Liu, D. Eng, A. Oki, Microelectron. Reliab. **44**, 1033 (2004)

7.25. A. Christou, P. Tang, J.M. Hu, IEEE Trans. Electron Devices **39**, 2229 (1992)

7.26. K. Chu, P. Chao, M. Pizzella, R. Actis, D. Meharry, K. Nichols, R. Vaudo, X. Xu, J. Flynn, J. Dion, G. Brandes, IEEE Electron Device Lett. **25**, 596 (2004)

7.27. R. Coffie, Y. Chen, P. Smorchkova, Y.C. Chou, M. Wojtowicz, A. Oki, in *Proceedings International Reliability Physics Symp.* , San Jose, 2006, pp 99–102

7.28. R. Coffie, Y. Chen, P. Smorchkova, B. Heying, W. Sutton, Y.C. Chou, W.B. Luo, M. Wojtowicz, A. Oki, in *Proceedings International Reliability Physics Symp.* , Phoenix, 2007, pp 568–571

7.29. A. Conway, M. Chen, P. Hashimoto, P. Willadsen, M. Micovic, in *Proceedings of the International Conference on GaAs Manufacturing Technology*, Austin, 2007, pp. 99–102

7.30. J. Cooper Jr., M. Melloch, R. Singh, A. Agarwal, J. Palmour, IEEE Trans. Electron Devices **49**, 658 (2002)

7.31. A. Dabiran, A. Osinsky, P. Chow, R. Fitch, J. Gillespie, N. Moser, T. Jenkins, J. Sewell, D. Via, A. Crespo, J. LaRoche, F. Ren, S. Pearton, in *Proceedings of High Temperature Electronics Conference*, Santa Fe, 2006, pp. 329–333

7.32. M. Dammann, A. Leuther, R. Quay, M. Meng, H. Konstanzer, W. Jantz, M. Mikulla, Microelectron. Reliab. **44**, 939 (2004)

7.33. M. Dammann, Personal Communication, 2007

7.34. A.M. Darwish, A. Bayba, H. Hung, in *IEEE International Microwave Symposium Digest*, Honolulu, 2007, pp. 311–314

7.35. A. Darwish, B. Hübschman, R.D. Rosario, E. Viveiros, H. Hung, in *Compound Semiconductor IC Symposium Technical Digest*, Palm Springs, 2005, pp. 145–148

7.36. I. Daumiller, C. Kirchner, M. Kamp, K. Ebeling, E. Kohn, IEEE Electron Device Lett. **20**, 448 (1999)

7.37. P. de Jong, G. Meijer, A. van Roermund, IEEE J. Solid-State Circuits **33**, 1999 (1998)

7.38. J. del Alamo, A. Villanueva, in *IEDM Technical Digest*, San Francisco, 2004, pp. 1019–1022

7.39. M. die Forte-Poisson, S. Delage, S. Cassette, Mater. Sci. Semicond. Process. **4**, 503 (2001)

7.40. D. Dieci, G. Sozzi, R. Menozzi, E. Tediosi, C. Lanzieri, C. Canali, IEEE Trans. Electron Devices **48**, 1929 (2001)

7.41. D. DiSanto, C. Bolognesi, Electron. Lett. **41**, 503 (2005)

7.42. Y. Dora, A. Chakraborty, S. Heikman, L. McCarthy, S. Keller, S. DenBaars, U. Mishra, IEEE Electron Device Lett. **7**, 529 (2006)

7.43. D. Dumka, C. Lee, H. Tserng, P. Saunier, Electron. Lett. **40**, 1554 (2004)

7.44. R. Esfandiari, T. O'Neill, T. Lin, R. Kono, IEEE Trans. Electron Devices **37**, 1174 (1990)

7.45. M. Faqir, A. Chini, G. Verzellesi, F. Fantini, F. Rampazzo, G. Meneghesso, E. Zanoni, J. Bernat, P. Kordos, in *Reliability of Compound Semiconductors Workshop*, San Antonio, 2007, pp. 25–31

7.46. M. Fresina, M. Logue, J. Fendrich, T. Rogers, in *Proceedings of the International Symposium on GaAs and Related Compounds*, Baltimore, 2001, pp. 203–222

7.47. R. Gaska, Q. Chen, J. Yang, A. Osinsky, M. Khan, M. Shur, IEEE Electron Device Lett. **18**, 494 (1997)

7.48. A. Geissberger, M. Fresina, L. Kapitan, C. Baratt, K. Tan, M. Hoppe, D. Streit, T. Block, M. Lammert, A. Oki, in *Proceedings of the International Conference on GaAs Manufacturing Technology*, Vancouver, 1999, p. 5.7

7.49. D. Gogl, H. Fiedler, M. Spitz, B. Parmentier,

7.50. S. Goodmann, F. Auret, F. Koschnick, J. Spaeth, B. Beaumont, P. Gibart, MRS Internet J. Nitride Semicond. Res. **4S1**, G6.12 (1999)

7.51. D. Green, S. Gibb, B. Hosse, R. Vetury, D. Grider, J. Smart, J. Cryst. Growth **272**, 285 (2004)

7.52. C. Hatfield, G. Bilbro, S. Allen, J.W. Palmour,

7.53. S. Heikman, S. Keller, S. DenBaars, U. Mishra, Appl. Phys. Lett. **81**, 439 (2002)

7.54. J. Hilsenbeck, E. Neubauer, J. Würfl, G. Tränkle, H. Obloh, Electron. Lett. **36**, 980 (2000)

7.55. T. Hisaka, Y. Nogami, H. Sasaki, A. Hasuike, N. Yoschida, K. Hayashi, T. Sonada, A. Villanueva, J. del Alamo, in *GaAs IC Symposium Technical Digest*, San Diego, 2003, pp. 67–70

7.56. S. Hsu, P. Valizadeh, D. Pavlidis, in *GaAs IC Symposium Technical Digest*, Monterey, 2002, pp. 85–88

7.57. A. Immorlica, in *IMS Workshop WFI: GaN Device and Circuit Reliability*, Honolulu, 2007

7.58. Y. Inoue, S. Masuda, M. Kanamura, T. Ohki, K. Makiyama, N. Okamoto, K. Imanishi, T. Kikkawa, N. Hara, H. Shigematsu, K. Joshin, in *IEEE International Microwave Symposium Digest*, Honolulu, 2007, pp. 639–642

7.59. JEDEC, *Guidelines for GaAs MMIC and FET Life Testing* (Electronic Industries Association, Washington DC, 1993), JEP118

7.60. J. Jimenez, U. Chowdhury, M. Kao, A. Balistreri, C. Lee, P. Saunier, P. Chao, W. Hu, K. Chu, A. Immorlica, J. del Alamo, J. Joh, M. Shur, in *Reliability of Compound Semiconductors Workshop*, San Antonio, 2006, pp. n.a.

7.61. J. Joh, J. del Alamo, in *IEDM Technical Digest*, San Francisco, 2006, pp. 415–418

7.62. J. Joh, L. Xia, J.A. del Alamo, in *IEDM Technical Digest*, Washington DC, 2007, pp. 385–388

7.63. J. Johnson, J. Gao, K. Lucht, J. Williamson, C. Strautin, J. Riddle, R. Therrien, P. Rajagopal, J. Roberts, A. Vescan, J. Brown, A. Hanson, S. Singhal, R. Borges, E. Piner, K. Linthicum, Proc. Electrochem. Soc. **7**, 405 (2004)

7.64. A. Jordan, Microelectron. Reliab. **18**, 267 (1978)

7.65. D. Keogh, P. Asbeck, T. Chung, J. Limb, D. Yoo, J. Ryou, W. Lee, W. Lee, S. Chen, R. Dupuis, Electron. Lett. **42**, 661 (2006)

7.66. S. Khanna, J. Webb, A. Houdayer, C. Carlone, IEEE Trans. Nucl. Sci. **47**, 2322 (2000)

7.67. T. Kikkawa, K. Imanishi, M. Kanamura, K. Joshin, in *Proceedings of the International Conference on GaAs Manufacturing Technology*, San Diego, 2002, pp. 171–174

7.68. T. Kikkawa, M. Kanamura, T. Ohki, K. Imanishi, K. Makiyama, N. Okamoto, N. Hara, K. Joshin, in *Proceedings of the International Conference on GaAs Manufacturing Technology*, Austin, 2007, pp. 91–94

7.69. T. Kikkawa, T. Maniwa, H. Hayashi, M. Kanamura, S. Yokokawa, M. Nishi, N. Adachi, M. Yokoyama, Y. Tateno, K. Joshin, in *IEEE International Microwave Symposium Digest*, Fort Worth, 2004, pp. 1347–1350

7.70. H. Kim, V. Tilak, B. Green, H. Cha, J. Smart, J. Shealy, L. Eastman, in *Proceedings of the International Reliability Physics Symposium*, Nara, 2001, pp. 214–218

7.71. G. Koley, H. Kim, L. Eastman, M. Spencer, Electron. Lett. **39**, 1217 (2003)

7.72. B. Kopp, T. Axness, C. Moore, in *IEEE International Microwave Symposium Digest*, Denver, 1997, pp. 583–586

7.73. T. Kunii, M. Totsuka, Y. Kamo, Y. Yamamoto, H. Takeuchi, Y. Shimada, T. Shiga, H. Minami, T. Kitano, S. Miyakuni, S. Nakatsuka, A. Inoue, T. Oku, T. Nanjo, T. Oishi, T. Ishikawa, Y. Matsuda, IEEE JEDEC, 137 (2003)

7.74. T. Kunii, M. Totsuka, Y. Kamo, Y. Yamamoto, H. Takeuchi, Y. Shimada, T. Shiga, H. Minami, T. Kitano, S. Miyakuni, S. Nakatsuka, A. Inoue, T. Oku, T. Nanjo, T. Oishi, T. Ishikawa, Y. Matsuda, in *Compound Semiconductor IC Symposium Technical Digest*, Monterey, 2004, pp. 197–200

7.75. K. Kurishima, S. Yamahata, H. Nakajima, H. Ito, N. Watanabe, IEEE Electron Device Lett. **19**, 303 (1998)

7.76. J. Kuzmik, D. Pogany, E. Gornik, P. Javorka, P. Kordos, Solid-State Electron. **48**, 271 (2004)

7.77. C. Lee, H. Tserng, L. Witkowski, P. Saunier, S. Guo, B. Albert, R. Birkhan, G. Munns, Electron. Lett. **40**, 1547 (2004)

7.78. C. Lee, L. Witkowski, M. Muir, H. Tserng, P. Saunier, H. Wang, J. Yang, M. Khan, in *Proceedings of IEEE Lester Eastman Conference on High Performance Devices*, Newark, 2002, pp. 436–442

7.79. C. Lee, L. Wittkowski, H. Tserng, P. Saunier, R. Birkhahn, D. Olson, D. Olson, G. Munns, S. Guo, B. Albert, Electron. Lett. **41**, 155 (2005)

7.80. M. Levinshtein, S. Rumyantsev, M. Shur (eds.), *Properties of Advanced Semiconductor Materials: GaN, AlN, InN, BN, SiC, SiGe* (Wiley, New York, 2001)

7.81. C. Li, J. Duster, K.T. Kornegay, IEEE Electron Device Lett. **24**, 72 (2003)

7.82. W. Liu, IEEE Trans. Electron Devices **43**, 220 (1996)

7.83. D. Look, Phys. Stat. Sol. B **228**, 293 (2001)

7.84. P. Maaskant, M. Akhter, J. Lambkin, L. Considine, IEEE Trans. Electron Devices **48**, 1822 (2001)

7.85. K. Matsushita, S. Teramoto, H. Sakurai, Y. Takada, J. Shim, H. Kawasaki, K. Tsuda, K. Takagi, in *Proceedings of the International Conference on GaAs Manufacturing Technology*, Austin, 2007, pp. 87–90

7.86. G. Meneghesso, E.D. Bortoli, A. Paccagnella, E. Zanoni, C. Canali, IEEE Electron Device Lett. **16**, 336 (1995)

7.87. G. Meneghesso, T. Grave, M. Manfredi, M. Pavesi, C. Canali, E. Zanoni, IEEE Trans. Electron Devices **47**, 2 (2000)

7.88. G. Meneghesso, F. Rampazzo, P. Kordos, G. Verzellesi, E. Zanoni, IEEE Trans. Electron Devices **53**, 2932 (2006)

7.89. M. Meneghini, L. Trevisanello, S. Levada, G. Meneghesso, G. Tamiazzo, E. Zanoni, T. Zahner, U. Zehnder, V. Härle, U. Strauß, in *IEDM Technical Digest*, Washington DC, 2005, pp. 1009–1012

7.90. S. Mertens, J. del Alamo, IEEE Trans. Electron Devices **49**, 1849 (2002)

7.91. J. Mittereder, S. Binari, P. Klein, J. Roussos, D. Katzer, D. Storm, D. Koleske, A. Wickenden, R. Henry, Appl. Phys. Lett. **83**, 1650 (2003)

7.92. G. Moore, in *IEEE International Solid-State Circuits Conference*, vol. XLVI, San Francisco, 2003, pp. 20–23

7.93. F. Mu, C. Tan, M. Xu, IEEE Trans. Electron Devices **48**, 2740 (2001)

7.94. S. Nakamura, G. Fasol, *The Blue Laser Diode* (Springer, Berlin Heidelberg New York, 1997)

7.95. R. Narasimhan, L. Sadwick, R. Hwu, IEEE Trans. Electron Devices **46**, 24 (1999)

7.96. P. Neudeck, R. Okojie, L. Chen, Proc. IEEE **90**, 1065 (2002)

7.97. J. Nikaido, T. Kikkawa, E. Mitani, S. Yokokawa, Y. Tateno, in *Proceedings of the International Conference on GaAs Manufacturing Technology*, New Orleans, 2005, pp. 97–100

7.98. A. Oki, in *IMS Workshop WSG: Solid-State Power Invades the Tube Realm*, Honolulu, 2007

7.99. J. Osvald, J. Kuzmik, G. Konstantinidis, P. Lobotka and A. Georgakilas Microelectronic Engineering, **81**, 181, (2005)

7.100. S. Oyama, T. Hashizume, H. Hasegawa Applied Surface Science **190**, 322 (2002)

7.101. V. Palankovski, S. Vitanov, R. Quay, in *Compound Semiconductor IC Symposium Technical Digest*, San Antonio, 2006, pp. 107–110

7.102. J. Palmour, J. Milligan, J. Henning, S. Allen, A. Ward, P. Parikh, R. Smith, A. Saxler, M. Moore, Y. Wu, in *Proceedings of European GaAs and Related Compounds Application Symposium GAAS*, Amsterdam, 2004, pp. 555–558

7.103. N. Pan, J. Elliott, M. Knowles, D. Vu, K. Kishimoto, J. Twynam, H. Sato, M. Fresina, G. Stillman, IEEE Electron Device Lett. **19**, 115 (1998)

7.104. M. Papastamatiou, N. Arpatzanis, G. Papaioannou, C. Papastergiou, A. Christou, IEEE Trans. Electron Devices **44**, 364 (1997)

7.105. E. Piner, S. Singhal, P. Rajagopal, R. Therrien, J. Roberts, T. Li, A. Hanson, J. Johnson, I. Kizilyalli, K. Linthicum, in *IEDM Technical Digest*, San Francisco, 2006, pp. 411–414

7.106. J. Piprek (ed.), *Nitride Semiconductor Devices: Principles and Simulation* (Wiley-VCH, Weinheim, 2007)

7.107. G. Ponchak, S.K.H. Ho-Chung, in *Digest of IEEE Microwave Millimeter-Wave Monolithic Circuits Symposium*, San Diego, 1994, pp. 69–72

444   References of Chapter 7

7.108. W. Pribble, S. Sheppard, R. Smith, S. Allen, J. Palmour, T. Smith, Z. Ring, J. Sumakeris, A. Saxler, J. Milligan, in *IMS Workshop on Wide Bandgap Technologies*, Seattle, 2002

7.109. F. Rampazzo, R. Pierobon, D. Pacetta, C. Gaquiere, D. Theron, B. Boudart, G. Meneghesso, E. Zanoni, Microelectron. Reliab. **44**, 1375 (2004)

7.110. P. Roblin, H. Rohdin, *High-Speed Heterostructure Devices* (Cambridge University Press, Cambridge, 2002)

7.111. P. Saunier, C. Lee, A. Balistreri, D. Dumka, J. Jimenez, H. Tserng, Y. Kao, P. Chao, K. Chu, A. Souzis, I. Eliashevich, S. Guo, J. del Alamo, J. Joh, M. Shur, in *Device Research Conference*, South Bend, 2007, pp. 35–36

7.112. P. Schmid, K.M. Lipka, J. Ibbetson, N. Nguyen, U. Mishra, L. Pond, C. Weitzel, E. Kohn, IEEE Electron Device Lett. **19**, 225 (1998)

7.113. R. Shaw, D. Sanderlin, J. DeJulio, in *Reliability of Compound Semiconductors Workshop*, San Antonio, 2007, pp. 3–20

7.114. J. Shealy, in *CSIC-IC Symposium Short Course: Emerging Technologies from Defense to Commercial*, Palm Springs, 2005

7.115. S. Sheppard, B. Pribble, R. Smith, A. Saxler, S. Allen, J. Milligan, R. Pengelly, in *Proceedings of the International Conference on GaAs Manufacturing Technology*, Vancouver, 2006, pp. 175–178

7.116. N. Shmidt, D. Davydov, V. Emtsev, I. Krestnikov, A. Lebedev, W. Lundin, D. Poloskin, A. Sakharov, A. Usikov, A. Osinsky, Phys. Stat. Sol. B **216**, 533 (1999)

7.117. G. Simin, V. Adivarahan, J. Yang, A. Koudymov, S. Rai, M. Khan, Electron. Lett. **41**, 774 (2005)

7.118. R. Singh, D. Capell, M. Das, L. Lipkin, J.W. Palmour, IEEE Trans. Electron Devices **50**, 471 (2003)

7.119. S. Singhal, A. Chaudhari, A. Hanson, J. Johnson, R.T.P. Rajagopal, T. Li, C. Park, A. Edwards, E. Piner, I. Kizilyalli, K. Linthicum, in *Reliability of Compound Semiconductors Workshop*, San Antonio, 2006, pp. 21–24

7.120. S. Singhal, A. Hanson, A. Chaudhari, P. Rajagopal, T. Li, J. Johnson, W. Nagy, R. Therrien, C. Park, A. Edwards, E. Piner, K. Linthicum, I. Kizilyalli, in *Reliability of Compound Semiconductors Workshop*, San Antonio, 2006

7.121. S. Singhal, T. Li, A. Chaudhari, A. Hanson, R. Therrien, J. Johnson, W. Nagy, J. Marquart, P. Rajagopal, J. Roberts, E. Piner, I. Kizilyalli, K. Linthicum, Microelectron. Reliab. **46**, 1247 (2006)

7.122. M.H. Sommerville, J.D. Alamo, P. Saunier, IEEE Trans. Electron Devices **45**, 1883 (1998)

7.123. A. Sozza, C. Dua, E. Morvan, M. diForte Poisson, S. Delage, F. Rampazzo, A. Tazzoli, F. Danesin, G. Meneghesso, E. Zanoni, A. Curutchet, N. Malbert, N. Labat, B. Grimbert, J.D. Jaeger, in *IEDM Technical Digest*, Washington DC, 2005, pp. 590–593

7.124. A. Sozza, C. Dua, E. Morvan, B. Grimber, S. Delage, Microelectron. Reliab. **45**, 1617 (2005)

7.125. A. Sozza, C. Dua, E. Morvan, B. Grimbert, V. Hoel, S. Delage, N. Chatuverdi, R. Lossy, J. Wuerfl, Microelectron. Reliab. **44**, 1369 (2004)

7.126. A. Stopel, A. Khramtsov, O. Katz, S. Solodky, T. Baksht, Y. Knafo, M. Leibovitch, Y. Shapira, in *Proceedings of the International Conference on GaAs Manufacturing Technology*, New Orleans, 2005, p. 14.19

7.127. T. Sudarshan, G. Gradinaru, J. Yang, M. Khan, Electron. Lett. **34**, 927 (1998)

7.128. T. Suemitsu, Y. Fukai, H. Sugiyama, K. Watanabe, H. Yokoyama, Microelectron. Reliab. **42**, 47 (2002)

7.129. B. Surridge, J. Law, B. Oliver, W. Pakulski, H. Strackholder, M. Abou-Khalil, G. Bonneville, in *Proceedings of the International Conference on GaAs Manufacturing Technology*, San Diego, 2002, p. 3A

7.130. H. Tan, J. Williams, C. Yuan, S. Pearton, in *MRS Symposium*, Boston, 1996, pp. 807–811

7.131. R. Therrien, S. Singhal, J. Johnson, W. Nagy, R. Borges, A. Chaudhari, A. Hanson, A. Edwards, J. Marquart, P. Rajagopal, C. Park, I. Kizilyalli, K. Linthicum, in *IEDM Technical Digest*, Washington DC, 2005, pp. 568–571

7.132. V. Tilak, M. Ali, V. Cimalla, V. Manivannan, P. Sandvik, J. Fedison, O. Ambacher, D. Merfeld, in *Proceedings of Materials Research Society Symposium*, Warrendale, 2004, pp. 593–597

7.133. R. Trew, Y. Liu, W. Kuang, G. Bilbro, in *Compound Semiconductor IC Symposium Technical Digest*, San Antonio, 2006, pp. 103–106

7.134. G. Umana-Membreno, J. Dell, G. Parish, B. Nener, L. Faraone, U. Mishra, IEEE Trans. Electron Devices **50**, 2326 (2003)

7.135. P. Valizadeh, D. Pavlidis, in *GaAs IC Symposium Technical Digest*, San Diego, 2003, pp. 78–81

7.136. P. Valizadeh, D. Pavlidis, IEEE Trans. Electron Devices **52**, 1933 (2005)

7.137. R. Versari, A. Pieracci, IEEE Trans. Electron Devices **46**, 1228 (1999)

7.138. A. Villanueva, J.A. del Alamo, T. Hisaka, T. Ishida in *IEDM Technical Digest*, Washington DC, 2007, pp. 393–396

7.139. C. Wang, J. Vac. Sci. Technol. B **20**, 1821 (2002)

7.140. X.W. Wang, W. Zhu, X. Guo, T. Ma, J. Tucker, M. Rao, in *IEDM Technical Digest*, Washington DC, 1999, pp. 209–212

7.141. A. Ward, in *IMS Workshop WFI GaN Device and Circuit Reliability*, Honolulu, 2007

7.142. C. Wilson, A. O'Neill, S. Baier, J. Nohava, IEEE Trans. Electron Devices **43**, 201 (1996)

7.143. J. Würfl, J. Hilsenbeck, E. Nebauer, G. Tränkle, H. Obloh, W. Österle, Microelectron. Reliab. **40**, 1689 (2000)

7.144. H. Xing, P. Chavarkar, S. Keller, S. DenBaars, U. Mishra, IEEE Electron Device Lett. **24**, 141 (2003)

7.145. F. Yamaki, K. Ishii, M. Nishi, H. Haematsu, Y. Tateno, H. Kawata, in *Proceedings of the International Conference on GaAs Manufacturing Technology*, Austin, 2007, pp. 95–98

7.146. B. Yang, W. Li, E. Casterline, in *Proceedings of the International Symposium on GaAs and Related Compounds*, Seattle, 2000, pp. 53–61

7.147. S. Yoshida, J. Suzuki, J. Appl. Phys. **85**, 7931 (1999)

7.148. L. Yu, Q. Liu, Q. Quao, S. Lau, J. Redwing, J. Appl. Phys. **84**, 2099 (1998)

7.149. E. Zanoni, G. Meneghesso, G. Verzellesi, F. Danesin, M. Meneghini, F. Rampazzo, A. Tazzoli, F. Zanon, in *IEDM Technical Digest*, Washington DC, 2007, pp. 381–384

References of Chapter 8

8.1. Z. Aboush, J. Benedikt, J. Priday, P. Tasker, in *10th High Frequency Postgraduate Student Colloquium*, University of Leeds, 2005, p. WE4B4

8.2. N. Adachi, Y. Tateno, S. Mizuno, A. Kawano, J. Nikaido, S. Sano, in *IEEE International Microwave Symposium Digest*, Long Beach, 2005, pp. 507–510

8.3. A. Agrawal, R. Clark, J. Komiak, in *IEEE International Microwave Symposium Digest*, San Francisco, 1996, pp. 995–999

8.4. Y. Ando, Y. Okamoto, H. Miyamoto, N. Hayama, T. Nakayama, K. Kasahara, M. Kuzuhura, in *IEDM Technical Digest*, Washington DC, 2001, pp. 381–384

8.5. A. Angelini, M. Furno, F. Cappelluti, F. Bonani, M. Pirola, G. Ghione, in *Proceedings of European GaAs and Related Compounds Application Symposium GAAS*, Paris, 2005, pp. 145–148

8.6. I. Anjoh, A. Nishimura, S. Eguchi, IEEE Trans. Electron Devices **45**, 743 (1998)

8.7. A. Barnes, A. Boetti, L. Marchand, J. Hopkins, in *Proceedings of European GaAs and Related Compounds Application Symposium GAAS*, Paris, 2005, pp. 5–12

8.8. K. Beilenhoff, P. Quentin, S. Tranchant, O. Vaudescal, M. Parisot, H. Daembkes, in *Proceedings of European GaAs and Related Compounds Application Symposium GAAS*, Milano, 2002, pp. 331–334

8.9. F. Benkhelifa, R. Kiefer, S. Müller, F. van Raay, R. Quay, R. Sah, M. Mikulla, G. Weimann, in *Proceedings of the International Conference on GaAs Manufacturing Technology*, New Orleans, 2005, p. 8.4

8.10. K. Bertilsson, C. Harris, H. Nilsson, Solid-State Electron. **48**, 2103 (2004)

8.11. R. Bolam, V. Ramachandran, D. Coolbaugh, K. Watson, IEEE Trans. Electron Devices **50**, 941 (2003)

8.12. M. Brophy, A. Torrejon, S. Petersen, K. Avala, L. Liu, in *Proceedings of the International Conference on GaAs Manufacturing Technology*, Scottsdale, 2003, p. 4.1

8.13. E. Burgemeister, W. von Muench, E. Pettenpaul, J. Appl. Phys. **50**, 5790 (1979)

8.14. A. Cetronico, M. Cicolani, M. Comparini, U.D. Marcantonio, R. Giordani, L. Marescialli, in *Proceedings of European Microwave Conference*, Paris, 2005, p. 4

8.15. A. Chandrasekhar, S. Stoukatch, S. Brebels, J. Balachandran, E. Beyne, W. de Raedt, B. Nauwelaers, A. Poddar, in *Proceedings of European GaAs and Related Compounds Application Symposium GAAS*, Munich, 2003, pp. 427–430

8.16. E. Chang, R. Dean, J. Proctor, R. Elmer, K. Pande, IEEE Trans. Semiconductor Manufacturing **4**, 66 (1991)

8.17. L. Chen, G. Hunter, P. Neudeck, in *MRS Symposium*, San Francisco, 2000, pp. T8.10.1–6

8.18. H. Chiu, S. Yang, Y. Chan, IEEE Trans. Microw. Theory Tech. **48**, 2210 (2001)

8.19. M. Comparini, C. Leone, P. Montanucci, M. Tursini, in *Proceedings of European Microwave Conference*, Milano, 2002, pp. 57–60

8.20. J. Conlon, N. Zhang, M. Poulton, J. Shealy, R. Vetury, D. Green, J. Brown, S. Gibb, in *IEEE Compound Semiconductor IC Symposium Technical Digest*, San Antonio, 2006, pp. 85–88

8.21. T. Daly, J. Fender, B. Duffin, M. Kottke, in *Proceedings of the International Conference on GaAs Manufacturing Technology*, Scottsdale, 2003, p. 10.6

8.22. A. Darwish, A. Bayba, H. Hung, in *IEEE International Microwave Symposium Digest*, Fort Worth, 2004, pp. 2039–2042

8.23. A. Darwish, A. Bayba, H. Hung, IEEE Trans. Microw. Theory Tech. **52**, 2611 (2004)

8.24. A. David, T. Fujii, B. Moran, S. Nakamura, S. DenBaars, C. Weisbuch, H. Benisty Appl. Phys. Lett. 88, 133514 (2006)

8.25. H. Davidson, N. Coletta, J. Kerns, D. Makowiecki, in *Proceedings of the Electronic Components and Technology Conference*, Las Vegas, 1995, pp. 538–541

8.26. R. Dietrich, A. Wieszt, A. Vescan, H. Leier, R. Stenzel, W. Klix, Solid-State Electron. **47**, 123 (2003)

8.27. C. Drevon, in *IEE Seminar of Packaging and Interconnects at Microwave and mm-Wave Frequencies*, London, 2000, pp. 8/1–8/4

8.28. L. Eastman, V. Tilak, J. Smart, B. Green, E. Chumbes, R. Dimitrov, H. Kim, O. Ambacher, N. Weimann, T. Prunty, M. Murphy, W. Schaff, J. Shealy, IEEE Trans. Electron Devices **48**, 479 (2001)

8.29. G. Ellis, J. Moon, D. Wong, M. Micovic, A. Kurdoghlian, P. Hashimoto, M. Hu, in *IEEE International Microwave Symposium Digest*, Fort Worth, 2004, pp. 153–156

8.30. M. Elta, A. Chu, L. Mahoney, R. Cerretani, W. Courtney, IEEE Electron Device Lett. **3**, 127 (1982)

8.31. J. Evans, in *IEEE Compound Semiconductor IC Symposium Technical Digest*, San Antonio, 2006, pp. 211–214

8.32. D. Fanning, L. Witkowski, C. Lee, D. Dumka, H. Tserng, P. Saunier, W. Gaiewski, E. Piner, K. Linthicum, J. Johnson, in *Proceedings of the International Conference on GaAs Manufacturing Technology*, New Orleans, 2005, p. 8.3

8.33. K. Filippov, A. Balandin, MRS Internet J. Nitride Semicond. Res. **8**, (2003)

8.34. J. Fleurial, A. Borshchevsky, M. Ryan, W. Phillips, E. Kolawa, T. Kacisch, R. Ewell, in *Proceedings of the International Conference on Thermoelectrics*, Dresden, 1997, pp. 641–645

8.35. D. Floriot, J. Jacquet, E. Chartier, J. Coupat, P. Eudeline, P. Auxemery, H. Blanck, in *Proceedings of European GaAs and Related Compounds Application Symposium GAAS*, Paris, 2005, pp. 541–544

8.36. J. Freeman, in *IEEE International Microwave Symposium Digest*, Fort Worth, 2004, pp. 2039–2042

8.37. J. Galiere, J. Valard, E. Estebe, in *Proceedings of European GaAs and Related Compounds Application Symposium GAAS*, Amsterdam, 2004, pp. 591–594

8.38. R. Gaska, A. Osinsky, J. Yang, M. Shur, IEEE Electron Device Lett. **19**, 89 (1998)

8.39. J. Gassmann, P. Watson, L. Kehias, G. Henry, in *IEEE International Microwave Symposium Digest*, Honolulu, 2007, pp. 615–618

8.40. S. Green, D. Perrottet, B. Richerzhagen, in *Proceedings of the International Conference on GaAs Manufacturing Technology*, Vancouver, 2006, pp. 145–148

8.41. E. Gu, C. Jeon, H. Choi, G. Rice, M. Dawson, E. Illy, M. Knowles, Thin Solid Films, **453/454**, 26 (2004)

8.42. H. Hommel, H. Feldle, in *Proceedings of European Radar Conference*, Amsterdam, 2004, pp. 121–124

8.43. H. Hu, C. Zhu, Y. Lu, M. Li, B. Cho, W. Choi, IEEE Electron Device Lett. **23**, 514 (2002)

8.44. H. Hu, C. Zhu, X. Yu, A. Chin, M. Li, B.J. Cho, D. Kwong, P. Foo, M.B. Yu, X. Liu, J. Winkler, IEEE Electron Device Lett. **24**, 60 (2003)

8.45. A. Jezowski, B. Danilchenko, M. Bockowski, I. Grzegory, S. Krukowski, T. Suski, T. Paszkiewicz, Solid-State Commun. **128**, 69 (2003)

8.46. R. Joshi, P. Neudeck, C. Fazi, J. Appl. Phys. **88**, 265 (2000)

8.47. B. Kang, S. Kim, F. Ren, J. Johnson, R. Therrien, P. Rajagopal, J. Roberts, E. Piner, K. Linthicum, S. Chu, K. Baik, B. Gila, C. Abernathy, S. Pearton, Appl. Phys. Lett. **85**, 2962 (2004)

8.48. D. Katsis, J. vanWyk, IEEE Trans. Compon. Pack. Technol. **29**, 127 (2006)

8.49. K. Kobayashi, Y. Chen, I. Smorchkova, R. Tsai, M. Wojtowicz, A. Oki, in *IEEE International Microwave Symposium Digest*, Honululu, 2007, pp. 619–622

8.50. J. Komiak, W. Kong, K. Nicols, in *IEEE International Microwave Symposium Digest*, Seattle, 2002, pp. 905–907

8.51. A. Kopp, M. Borkowski, G. Jerinic, IEEE Trans. Microw. Theory Tech. **50**, 827 (2002)

8.52. B. Kopp, in *IEEE International Microwave Symposium Digest*, Boston, 2000, pp. 705–708

8.53. A. Kowalczyk, V. Alder, C. Amir, F. Chiu, C.P. Chng, W.J.D. Lange, Y. Ge, S. Ghosh, T.C. Hoang, B. Huang, S. Kant, Y.S. Kao, C. Khieu, S. Kumar, L. Lee, A. Liebermensch, X. Liu, N.G. Malur, A.A. Martin, H. Ngo, S. Oh, I. Orginos, L. Shih, B. Sur, M. Tremblay, A. Tzeng, D. Vo, S. Zambare, J. Zong, IEEE J. Solid-State Circuits **36**, 1609 (2001)

8.54. S. Krukowski, A. Witek, J. Adamczyk, J. Jun, M. Bockowski, I. Grzegory, B. Lucznik, G. Nowak, M. Wroblewski, A. Presz, S. Gierlotka, S. Stelmach, B. Palosz, S. Porowski, P. Zinn, J. Phys. Chem. Solids **59**, 289 (1998)

8.55. M. Kuball, J. Hayes, M. Uren, T. Martin, J. Birbeck, R. Balmer, B. Hughes, IEEE Electron Device Lett. **23**, 7 (2002)

8.56. M. Kuball, A. Sarua, H. Ji, M. Uren, R. Balmer, T. Martin, in *IEEE International Microwave Symposium Digest*, San Francisco, 2006, pp. 1339–1342

8.57. J. Kuzmik, S. Bychikhin, M. Neuburger, A. Dadgar, A. Krost, E. Kohn, D. Pogany, IEEE Trans. Electron Devices **52**, 1698 (2005)

References of Chapter 8

8.58. J. Kuzmik, P. Javorka, A. Alam, M. Marso, M. Heuken, P. Kordos, IEEE Trans. Electron Devices **49**, 1496 (2002)
8.59. F.D. la Hidalga, M. Deen, E. Gutiérrez, IEEE Trans. Electron Devices **47**, 1098 (2000)
8.60. P. Ladbrooke, *MMIC Design: GaAs FETs and HEMTs* (Artech House, Boston London, 1989)
8.61. J. Laskar, S. Nuttinck, S. Pinel, in *Proceedings of the International Conference on GaAs Manufacturing Technology*, Miami, 2004, pp. 217–220
8.62. L. Lecheminoux, N. Gosselin, in *IEEE International Electronics Manufacturing Technology Symposium*, San Jose, 2003, pp. 255–260
8.63. D. Leung, Y. Chou, C. Wu, R. Kono, J. Scarpulla, R. Lai, M. Hoppe, D. Streit, in *Radio Frequency Integrated Circuits Symposium Digest*, Anaheim, 1999, pp. 153–156
8.64. W. Liu, A. Balandin, Appl. Phys. Lett. **85**, 5230 (2004)
8.65. M. Ludwig, C. Buck, F. Coromina, M. Suess, in *IEEE International Microwave Symposium Digest*, Long Beach, 2005, pp. 1619–1622
8.66. M. Mahalingam, E. Mares, in *IEEE International Microwave Symposium Digest*, Phoenix, 2001, pp. 2199–2202
8.67. Y. Mancuso, P. Gremillet, P. Lacomme, in *Proceedings of European Microwave Conference*, Paris, 2005, pp. 817–820
8.68. C. Martin, T. Kerr, W. Stepko, T. Anderson, in *Proceedings of the International Conference on GaAs Manufacturing Technology*, Miami, 2004, pp. 291–294
8.69. S. Marsh, J. Clifton, K. Vanner, J. Cockrill, I. Davies, in *Proceedings International Symp. Elec. Dev. for Microwave and Optoelectronic Applications EDMO*, London, 1997, pp. 169–174
8.70. S. McCarthy, P. Smith, J. Walker, N. Padfield, in *IEEE International Microwave Symposium Digest*, Long Beach, 2005, pp. 1023–1026
8.71. D. McQuiddy, R. Gassner, P. Hull, J. Mason, J. Bedinger, Proc. IEEE **79**, 308 (1991)
8.72. K. Meyyappan, P. McCluskey, L. Chen, IEEE Trans. Devices Mater. Rel. **3**, 152 (2003)
8.73. E. Mitani, M. Aojima, A. Maekawa, S. Sano, in *Proceedings of the International Conference on GaAs Manufacturing Technology*, Austin, 2007, pp. 213–216
8.74. M. Micovic, A. Kurdoghlian, H. Moyer, P. Hashimoto, A. Schmitz, I. Milosavjevic, P. Willadesn, W. Wong, J. Duvall, M. Hu, M. Delaney, D. Chow, in *IEEE International Microwave Symposium Digest*, Fort Worth, 2004, pp. 1653–1656
8.75. J. Mizoe, S. Amano, T. Kuwabara, T. Kaneko, K. Wada, A. Kato, K. Sato, M. Fujise, in *IEEE International Microwave Symposium Digest*, Anaheim, 1999, pp. 475–478
8.76. J. Moore, A. Smith, D. Nguyen, S. Kulkarni, in *Proceedings of the International Conference on GaAs Manufacturing Technology*, Miami, 2004, pp. 175–178
8.77. W. Nagy, S. Singhal, R. Borges, W. Johnson, J. Brown, R. Therrien, A. Chaudhari, A. Hanson, J. Riddle, S. Booth, P. Rajagopal, E. Piner, K. Linthicum, in *IEEE International Microwave Symposium Digest*, Long Beach, 2005, pp. 483–486

8.78. T. Nakatsuka, J. Itoh, T. Yoshida, M. Nishitsuji, T. Uda, O. Ishikawa, IEEE J. Solid-State Circuits **33**, 1284 (1998)

8.79. P. Neudeck, R. Okojie, L. Chen, Proc. IEEE **90**, 1065 (2002)

8.80. C. Ng, K. Chew, S. Chu, IEEE Electron Device Lett. **24**, 506 (2003)

8.81. S. Nuttinck, B. Banerjee, S. Venkataraman, J. Laskar, M. Harris, in *IEEE International Microwave Symposium Digest*, Philadelphia, 2003, pp. 221–223

8.82. S. Nuttinck, E. Gebara, J. Laskar, B. Wagner, M. Harris, in *IEEE International Microwave Symposium Digest*, Seattle, 2002, pp. 921–924

8.83. S. Nuttinck, R. Mukhopadhyay, C. Loper, S. Singhal, M. Harris, J. Laskar, in *Proceedings of European GaAs and Related Compounds Application Symposium GAAS*, Amsterdam, 2004, pp. 79–82

8.84. S. Nuttinck, B. Wagner, B. Banerjee, S. Venkataraman, E. Gebara, J. Laskar, H. Harris, IEEE Trans. Microw. Theory Tech. **51**, 2445 (2003)

8.85. Y. Ohno, M. Kuzuhara, IEEE Trans. Electron Devices **48**, 517 (2001)

8.86. V. Palankovski, S. Selberherr, in *Proceedings of European Conference on High Temperature Electronics Conference*, Berlin, 1999, pp. 25–28

8.87. J. Park, P. Sercel, Comp. Semicond. **48**, 517 (2002)

8.88. D. Perrottet, A. Spiegel, S. Amorosi, B. Richerzhagen, in *Proceedings of the International Conference on GaAs Manufacturing Technology*, New Orleans, 2005, p. 14.16

8.89. N. Pham, P. Sarro, K. Ng, J.N. Burghartz, IEEE Trans. Electron Devices **48**, 1756 (2001)

8.90. S. Pinel, S. Chakraborty, M. Roellig, R. Kunze, S. Mandal, H. Liang, C. Lee, R. Li, K. Lim, G. White, M. Tentzeris, J. Laskar, in *IEEE International Microwave Symposium Digest*, Seattle, 2002, pp. 1553–1556

8.91. S. Pinel, A. Marty, J. Tasselli, J. Bailbe, E. Beyne, R.V. Hoof, S. Marco, J. Morante, O. Vendier, M. Huan, IEEE Trans. Compon. Pack. Technol. **25**, 244 (2002)

8.92. K. Pipe, R. Ram, A. Shakouri, IEEE Photonics Technol. Lett. **14**, 433 (2002)

8.93. W. Pribble, S. Sheppard, R. Smith, S. Allen, J. Palmour, T. Smith, Z. Ring, J. Sumakeris, A. Saxler, J. Milligan, in *IMS Workshop Wide Bandgap Technologies*, Seattle, 2002

8.94. P. Regoliosi, A.R.A.D. Carlo, P. Romanini, M. Peroni, C. Lanzieri, A. Angelini, M. Pirola, G. Ghione, IEEE Trans. Electron Devices **53**, 182 (2006)

8.95. L. Sadwick, J. Chern, R. Nelson, R. Hwu, in *Proceedings of the High Temperature Electronics Conference*, Santa Fe, 2006, pp. 377–379

8.96. A. Sarua, H. Ji, M. Kuball, M. Uren, T. Martin, K. Hilton, R. Balmer, in *Proceedings of the International Conference on GaAs Manufacturing Technology*, Vancouver, 2006, pp. 179–182

8.97. A. Sarua, H. Ji, M. Kuball, M. Uren, T. Martin, K. Hilton, R. Balmer, IEEE Trans. Electron Devices **53**, 2438 (2006)

8.98. A. Sarua, J. Hangfeng, K. Hilton, D. Wallis, M. Uren, T. Martin, M. Kuball, IEEE Trans. Electron Devices **54**, 3152 (2007)

8.99. C. Schaffauser, O. Vendier, S. Forestier, F. Michard, D. Geffroy, C. Drevon, J. Villemazet, J. Cazaux, S. Delage, J. Roux, in *Proceedings of European GaAs and Related Compounds Application Symposium GAAS*, Paris, 2005, pp. 537–540

8.100. F. Schmückle, A. Jentzsch, W. Heinrich, J. Butz, M. Spinnler, in *IEEE International Microwave Symposium Digest*, Phoenix, 2001, pp. 1903–1906

8.101. P. Schuh, R. Leberer, H. Sledzik, D. Schmidt, M. Oppermann, B. Adelseck, H. Brugger, R. Quay, F. van Raay, M. Seelmann-Eggebert, R. Kiefer, W. Bronner, in *IEEE International Microwave Symposium Digest*, San Francisco, 2006, pp. 1324–1327

8.102. M. Seelmann-Eggebert, P. Meisen, F. Schaudel, P. Koidl, A. Vescan, H. Leier, Diamond Relat. Mater. **10**, 744 (2001)

8.103. K. Seemann, S. Ramberger, A. Tessmann, R. Quay, J. Schneider, M. Riessle, H. Walcher, M. Kuri, R. Kiefer, M. Schlechtweg, in *Proceedings of European Microwave Conference*, Munich, 2003, pp. 383–386

8.104. S. Selberherr, *Analysis and Simulation of Semiconductor Devices* (Springer, Wien New York, 1984)

8.105. J. Shealy, in *GaAs-IC Symposium Short Course: Emerging Technologies from Defense to Commercial*, San Diego, 2003

8.106. J. Shealy, in *CSIC-IC Symposium Short Course: Emerging Technologies from Defense to Commercial*, Palm Springs, 2005

8.107. S. Sheppard, W. Pribble, D. Emerson, Z. Ring, R. Smith, S. Allen, J. Milligan, J. Palmour, in *Proceedings of IEEE/Cornell Conference on High Performance Devices*, Ithaca, 2000, pp. 232–236

8.108. C. Shih, J. Sjöström, R. Bagger, P. Anderson, Y. Yu, G. Ma, Q. Chen, T. Aberg, in *IEEE International Microwave Symposium Digest*, San Franscisco, 2006, pp. 889–892

8.109. S. Singhal, J. Brown, R. Borges, E. Piner, W. Nagy, A. Vescan, in *Proceedings of European GaAs and Related Compounds Application Symposium GAAS*, Milano, 2002, pp. 37–40

8.110. G. Slack, R. Tanzilli, R. Pohl, J. Vandersande, J. Phys. Chem. Solids **48**, 641 (1987)

8.111. M. Spencer, J. Palmour, C. Carter, IEEE Trans. Electron Devices **49**, 940 (2002)

8.112. N. Stath, V. Hïle and J. Wagner Materials Science and Engineering B **80**, 224 (2001)

8.113. M. Südow, K. Andersson, N. Billström, J. Grahn, H. Hjelmgren, J. Nilsson, P. Nilsson, J. Stahl, H. Zirath, N. Rorsman, IEEE Trans. Microw. Theory Tech. **54**, 4072 (2006)

8.114. Y. Suh, D. Richardson, A. Dardello, S. Mahon, J. Harvey, in *Proceedings of European Microwave Conference*, Paris, 2005, pp. 545–548

8.115. J. Sun, H. Fatima, A. Koudymov, A. Chitnis, X. Hu, H. Wang, J. Zhang, G. Simin, J. Yang, M. Khan, IEEE Electron Device Lett. **24**, 375 (2003)

8.116. R. Therrien, A. Chaudhari, S. Singhal, C. Snow, A. Edwards, C. Park, W. Nagy, J. Johnson, A. Hanson, K. Linticum, I. Kizilyalli, in *IEEE International Microwave Symposium Digest*, Honolulu, 2007, pp. 635–638

8.117. B. Thibeault, B. Keller, Y. Wu, P. Fini, U. Mishra, C. Nguyen, N. Nguyen, M. Le, in *IEDM Technical Digest*, Washington DC, 1997, pp. 569–572

8.118. V. Turin, A. Balandin, Electron. Lett. **40**, 81 (2004)

8.119. M. van Heijningen, J. Priday, in *Proceedings of European Microwave Conference*, Amsterdam, 2004, pp. 357–360

8.120. F. van Raay, R. Quay, R. Kiefer, W. Bronner, M. Seelmann-Eggebert, M. Schlechtweg, M. Mikulla, G. Weimann, in *IEEE International Microwave Symposium Digest*, San Francisco, 2006, pp. 1368–1371

8.121. R. Vandersmissen, J. Das, W. Ruythooren, J. Derluyn, M. Germain, D. Xiao, D. Schreurs, G. Borghs, in *Proceedings of the International Conference on GaAs Manufacturing Technology*, New Orleans, 2005, p. 13.3

8.122. C. Varmazis, G. D'Urso, H. Hendricks, Semiconduct. Int. **23**, 87 (2000)

8.123. F. Villard, J. Prigent, E. Morvan, C. Dua, C. Brylinski, F. Temcamani, IEEE Trans. Microw. Theory Tech. **51**, 1129 (2003)

8.124. A. Ward, in *IMS Workshop WFI GaN Device and Circuit Reliability*, Honolulu, 2007

8.125. M. Watts, in *Proceedings of High Temperature Electronics Conference*, Santa Fe, 2006, pp. 371–376

8.126. M. Werner, W. Fahmer, IEEE Trans. Ind. Electron. **48**, 249 (2001)

8.127. J. Wilson, P. Raad, Int. J. Heat Mass Transf. **47**, 3707 (2004)

8.128. M. Winser, in *Proceedings of European Microwave Conference*, Paris, 2005, pp. 493–496

8.129. Y Wu, D. Kapolnek, J. Ibbetson, P. Parikh, B. Keller, U. Mishra, IEEE Trans. Electron Devices **48**, 586 (2001)

8.130. Y. Wu, B. Keller, S. Keller, D. Kapolnek, S. Denbaars, U. Mishra, IEEE Electron Device Lett. **17**, 455 (1996)

8.131. J. Xu, W. Yin, J. Mao, IEEE Microw. Wireless Compon. Lett., **17**, 55 (2007)

8.132. L. Xue, C. Liu, H. Kim, S. Kim, S. Tiwari, IEEE Trans. Electron Devices **50**, 601 (2003)

8.133. M. Yang, C. Huang, A. Chin, C. Zhu, M. Li, D. Kwong, IEEE Electron Device Lett. **24**, 306 (2003)

8.134. B. Yeats, IEEE Trans. Electron Devices **45**, 939 (1998)

8.135. T. Young, Surface and Coatings Technology **202**, 1208 (2007)

8.136. J. Zimmer, G. Chandler, in *Proceedings of the International Conference on GaAs Manufacturing Technology*, Austin, 2007, pp. 129–132

8.137. A. Ziroff, M. Nalezinski, W. Menzel, in *Proceedings of European GaAs and Related Compounds Application Symposium GAAS*, Amsterdam, 2004, pp. 491–494

8.138. J. Zolper, in *Proceedings of the International Conference on GaAs Manufacturing Technology*, Scottsdale, 2003, p. 1.2

Index

acceptor
  passivation 316
activation energy 327
adhesion 169
adhesive 340
AESA 88
AFM 100, 107, 128
airbridge 337
AlGaN/GaN 33
alignment
  gate 56
alloy scattering 135
ambient 316
amplifier
  bandwidth 88
  C-band 285
  Chireix 278
  class-A 215, 216, 240, 259, 275, 288
  class-A/B 275
  class-B 275, 288
  class-C 240, 275
  class-D 275
  class-E 75, 275, 284
  class-F 275, 278
  Class-S 276
  delay-matched 297
  distributed 297
  Doherty 279, 281, 283, 285
  driver 282
  envelope elimination restoration 278, 283
  envelope-tracking 283
  hybrid 78, 288
  Ku-band 79
  linearization 284
  low-noise 271, 300
  mm-wave 298
  multiband 280
  multicarrier 280
  push-pull 75, 283, 288
  single-ended 282
  switch-mode 281, 284
  X-band 287
analysis
  atomic force microscopy 107
  failure 324
  photoluminescence 107
  sensitivity 297
  surface potential 179
  thermal 343
  X-ray diffraction 107
annealing 145
  device 156
  full wafer 192
  high temperature 145
  rapid thermal 156
application
  DC-DC conversion 89
  electronic warfare 89
  radar 88
  space 89
atomic force microscope 111
attenuator 308

backgating 216
backside

metallization   140
processing   56, 142, 192, 341
sapphire   193
SiC   193
band structure
  AlN   16
  BN   20
  GaN   11
  InN   18
bandgap
  discontinuity   135
  energy   13
bandwidth
  3 dB   288
  relative   295
barrier
  layer   318
    thickness   77, 132
  thinning   153
  tunneling   153
base
  transport factor   209
base-station   74, 76, 279, 283
  GaAs   280
baseband   257
BEEM   129
binary   4
bipolar   209
BN   5, 18
bond wire   289
bonding
  atomic   121
boron nitride   5
breakdown
  RF   239
  soft   251
  voltage   52
    maximum   207
    MIM   338
    open collector   210
    three-terminal   208
    two-terminal   207
breakdown voltage   315
buffer
  codoping
    carbon   116
    Fe   99, 329
  degradation   329
  GaN

relaxed   36
  growth   111
  HVPE   62
  interlayer   99
  isolation   98, 329
  layer
    intermediate   181
    isolation   98
    thickness   99
bulk modulus   7
burn-in   182, 314
burnout   311

C-AFM   118
C-V measurement   30
CAIBE   166
calculation
  pseudopotential   14
  variational principle   17
cap layer
  doped   172, 220
  GaN
    n-doped   134, 137
    p-doped   134, 173
capacitance   280
  MIM   337, 338
capture time
  constant   130
carrier
  compensation   96, 100, 169
  concentration
    intrinsic   333
    origin   218
  transport
    AlN   15
    GaN   8
    InN   17
    limitation   219
  velocity   8
    quasi-saturation   201
cascode   227, 297
CAVET   54
CDMA   74, 75, 250, 283
channel   137
  current density   220
  formation   33
characteristic
  velocity-field   10, 46, 47
characterization

linearity   197
charge
  conservation   246
  control   47, 170, 218
  polarization-induced   32
  relaxation   246
circuit   3, 86, 271
cleaning   182
cleaving   358
codoping
  buffer   99
  compensation   100
  indium   95
  MOCVD
    carbon   100
    silicon   100
coefficient
  absorption   7
  thermal expansion   6, 55, 347
cointegration   297
combiner
  Wilkinson   297
concentration
  background   16
conductivity
  thermal   7, 22, 347
  surface   357
confinement   35, 81
contact   45, 144
  degradation   315
  e-beam pattern definition   147
  edge definition   147
  gate
    copper   155
    formation   142
    Ir   155
    Mo   155
    Ni   155
    Pd   155
    WSi   155
    WSi/p-   155
  morphology   147
  ohmic   140, 144, 147, 327
    degradation   317
    formation   142
    high temperature   191
    implanted   45, 149
    n-type   45, 146
    p-type   45, 146

recess   45
regrown   151
resistance   147
Schottky   45, 151, 180, 324, 328
  gas sensitivity   156
  mechanical stress   154
  n-   154
  p-   154
  thermal   94
conversion   66
converter
  DC–DC   67
cooling
  active   365
  backside   352
  flip-chip   355
  micro channel   365
  thermoelectric   366
coplanar   337
  transmission-line   287
copper diamond   347
copper molybdenum   347
copper tungsten   347, 349
cost
  per area   55
Cripps
  impedance   240
  load   240, 295
criterion
  failure   312
critical field
  dielectric   19, 51, 52
  impact ionization   21
  saturation   10, 24, 47, 199
cross section
  TEM   325
  trap   179
crossover
  direct-indirect   23, 26
current
  amplification   209
  collapse   54, 188, 212
    cryogenic temperatures   181
  density   219
  dispersion   54, 179, 212
  gain   209
    cut-off frequency   51, 202, 207, 221
    HBT   85, 210
  minimum   202

pinch-off　324
slump　54
surface　143
time dependence　54
cut-off
　frequency　202
　　current gain　210
　　stability　221

damage
　crystal　324
　etch　85, 317
　radiation　316
defect　42, 127
　grain boundary　59
　reduction　220
　residual　55
　screw　57
　sub-surface　59
degradation
　contacts　327
　DC-　318
　dislocation-induced　324
　graceful　287
　RF　322
design
　neural network　297
　techniques　295
device　3
　access region　179
　bipolar　84
　characteristics　197
　characteriszation　197
　degradation　55, 311
　double heterojunction　35, 78
　hour　314
　In-based　85
　isolation　142
　large-signal　202
　lateral　54
　leakage　220
　linearity　204
　MISFET　190
　mm-wave　81
　modeling　197
　passivation　174
　　ungated region　174
　power　66
　statistics　314

vertical　54
DHBT　37, 78, 235
　GaN/InGaN/GaN　37
DHHEMT　33, 35, 36, 78, 124
　AlGaN/GaN　79
　AlGaN/InGaN　78
　mm-wave　79
diamond　5, 349
　abrasive　91
　CVD　65
　substrate　121
dicing　316, 358
die-attach　359
dielectric
　deposition　184
　gate　189
　high-k　189
diffusion
　barrier　147
　metal　312
diode　40
　GaN　65
　laser　40
　SiC　65
dislocation　59, 129
　edge　118
　MBE　117
　screw　118
　threading　9, 10, 54
dispersion　52, 99, 197, 211
　analysis　214
　characterization
　　time-domain　214
　current　54
　frequency　52
　low-frequency　212
　model　217
　　virtual gate　217
　origin　54
　reduction　140
　removal
　　field plate　180
　　surface　179
　temperature dependence　216
　time-domain　212
distortion　74, 256, 259
　second order　204
distribution
　lognormal　314

DLTFS   181
DLTS   128, 129, 179
  Fourier   181
  temperature-dependent   181
domain
  frequency-   216, 243
  time   211
donor   127
  passivation   316
  shallow   92
doping   127, 134
  activation   37
  diffusion   329
  Fe
    dislocation   100
    migration   100
  isoelectronic   135, 137
  magnesium   93
  n-   38, 40
  p-
    degenerate   37
  passivation   184
  polarization   40, 134
  selective   143
drain current   199
droop
  phase   358
  power   358
DX center   181

e-beam
  edge definition   148
  pattern recognition   145, 148
EDGE   258
EER   278
effect
  environmental   324
  hot carrier   324
  radiation   331
  thermal
    dynamic   357
effective mass   12
efficiency
  drain   203
  power-added   203
electroluminescence   180
electromigration   328
electronics
  harsh environments   89

ELO   63, 101, 122
  HBT   122
ELOG   see ELO
energy
  activation   37, 45, 312
  alignment   41
    valence band   13
  conservation   246
  valence band
    offset   41
engineering
  bandgap   37
  polarization   4, 17
  HBT   37
epitaxial lateral overgrowth   101, 122
epitaxy   91
ESD   308, 313
etch
  CAIBE   166
  contamination   168
  damage   168, 169, 317
    plasma induced   166
  delay   167, 172
  depth control   167
  ECR   166
  nonselective   172
  photo accelerated   170
  product   194
  RF-source power   167
  RIE   166
  selectivity   171
  surface morphology   169
etching
  additive gas   167
  contamination   168
  dry   166
  ECR   168
  hydrogen   92
  ICP   166
  ion channeling   168
  mesa   166
  methodologies   166
  RIE   166, 167
  selectivity
    AlN   169
  wet   169
    AlN   169
    GaN   169
    InN   170

photo-electrochemically enhanced
170
photo-enhanced   170
EVM   76, 278, 282

failure
analysis   325
catastropic   311
parametric   311
Fermi-level
pinning   29, 42, 100, 190
AlGaN   44
buffer   99
GaN   44
FET   68, 70, 71
doped channel   83
GaN/InGaN   83
indium   82
RF-switch   67
FIB   325, 327, 328
field
distribution   152
piezoelectric   36
plate   71, 74, 160, 183
gate   76
gate-connected   164
source-connected   165
polarization   50, 326
shaping   50, 140
figure-of-merit   48, 51
Baliga   51
Eastman   51
Johnson   51
Nguyen   52
Shenai   52
filter   243, 275
SAW   48
FIT rate   313
flip-chip   288, 354
underfill   356
fluorine   316
FOM   48, 51, 67
fracture toughness   5
frequency
dispersion   see dispersion
frequency band   89
C-   76, 89, 279, 285, 308
K-   79
Ka-   79

Ku-   79
L-   74, 279
mm-wave   79
Q-   89
S-   74, 89, 279
V-   81
X-   77, 89, 287, 288, 308

gain
associated   52
compression   133, 152
premature   55, 212
soft   285
conversion   307
MAG   221
maximum stable   221
MSG   221
unilateral   221
gain-bandwidth product   297
gain-breakdown product   170
galvanic   339
high-power   195
GaN
bulk   46
cap   318
junction FET   69
non-polar   5
p-   84, 153
p-doping
implant   150
refractive index   7
gate   140
adhesion   153, 154
annealing   156
barrier   140
extension   160
finger
length   352
lag   183, 216
leakage   151, 188
defect correlation   92
pitch   352
pre-linearization   285
recess   74, 165
sinking   315, 324, 328
virtual   55, 230
gate oxide
thermal oxidation   189
generation   20

dynamic   177
grinding   192
growth
   Ga-face   30, 33, 35, 37
   InN
      two-step   126
   kinetics   93
   MBE   29
   MOCVD   29
   molecular beam   112
   N-face   30, 33, 35
   pressure   96
   selective area   85
   temperature   94, 96
      channel   95
      InN   124
      sapphire   94
   V/III ratio   95, 96
   window   107
GSM   258, 279
Gunn effect   47

hardness
   Vickers   5, 15
harmonic
   control   284
   termination   242, 277, 278
HBT   37, 84, 92, 209, 210, 234, 277, 311
   AlGaN/GaN   37, 84
   base resistance   84
   base transit time   85
   current amplification   85
   degradation   316
   fabrication   84
   InGaN/GaN   85
   leakage   85
   npn   84
   pnp   84
HCI   320
heat capacity   22, 358
heat dissipation per area   366
heat spreader   347
HEMT   71, 92
   AlGaN/InGaN   16
   double heterostructure
      AlGaN/GaN/AlGaN   35
   dual-gate   81
   enhancement-mode   133, 173, 198
   GaN   71

GaN/InGaN   33
InAlN/GaN   16, 35
PAE   80
unpassivated   134
heterojunction
   field-effect transistor
      double   262
      single   262
HFET
   GaN   71
   InAlN/GaN   17, 126
   InGaN   83
   inverted   35
   prematching   80
homogeneity   107, 118
Hooge's law   264
Hooge-parameter   264
HTCC   362
HTCVD   57
HVPE   55, 60, 63, 122
   sapphire   123
hydride vapor phase epitaxy   see
      HVPE
hydrogen   92, 316
   pasivation   128
   sensitivity   156

ideality factor   152, 207, 315
imaging
   thermal   344
impact ionization   20, 216, 251, 315
   rate   49
IMPATT   65
impedance   280
   level   74, 88
implantation   143
   isolation   143
incorporation
   oxygen   100
indium   85, 123
inductance   337
InGaN
   p-channel   17
injection efficiency   209
InN   123
   MBE   17
   MOCVD   16
input characteristics   206
integration   341

hybrid   363
MMIC   337
interaction
   phonon
      optical   46
interface   40
   polarity   30
interlayer
   AlN   15, 64, 135, 137
      InAlN/GaN   126
   growth temperature   95
intermodulation   256, 304
   behaviour   280
   distortion   74
   nulling   256
interpolation
   cubic   23
   quadratic   23
intervalley separation   13
ion
   channeling   331
   implantation
      activation   150
ionization
   incomplete   84
irradiation
   ion   331
   neutron   331
   proton   331
isolation   35, 55, 140, 317
   doping   143
   implantation   142

Kelvin probe   83, 118
kink effect   181
knee
   voltage   203, 315, 324
   walkout   239

lag
   drain   98, 179
   gate   179
large-signal
   characterization   235
   modeling   235
      dispersion   248
   thermal modeling   252
large-signal model   241
   Angelov   247

behavioural approach   255
   Curtice   247
   Materka   247
   small-signal approaches   243
lattice
   constant   3, 32
   mismatch   54, 329, 359
layer
   barrier   117
   buffer   94
   cap   137
      MOCVD   133
   channel   117
      doped   69
   nucleation   94
LDMOS   279, 280, 355, 360
   cointegration   337
   integration   363
leakage   140, 145
   collector   37
   drain   140
   gate   140, 183, 315, 330
   pn-junction   333
   thermionic   333
LEEN   180
LEO   see ELO
lifetime
   carrier   37, 181, 210
limit
   electrical   3, 48
   thermal   3
linearity   88, 259, 279
   baseband   258
   bias dependence   257
   device   255
   harmonic impedance   258
   input   269, 300
   sweet spot   257, 259
lithography   157
   contact   157
   e-beam   158
   optical   157
LNA   300
   C-band   301
   MMIC   82
   overdrive   82
   survivability   87, 301
   X-band   301
load impedance   237

loadline
  dynamic   239
loadpull   211, 235, 236
  bias dependence   240
  pulsed
    intermodulation   237
low-noise   71, 300
LTCC   362
luminescence   179

management
  thermal   61, 291, 337
mask
  erosion   143
  Ni   150
  phase shift   157
mass
  density   6
matching
  harmonic   295
material
  high-k   307
maximum frequency of oscillation   221
MBE   1, 91, 92
  AlGaN growth   114, 115
  ammonia   112, 121
  buffer growth   115
  Beryllium   116
  codoping   116
  defects   131
  doping
    n-   130
    p-   130
  growth temperature   114
  lithium gallate   122
  MOMBE   121
  nucleation layer   115
  plasma source   112, 118
  reactive   115
  SiC   115, 118
  SICOI   121
  silicon   115, 121
  Smart Cut©   121
  source
    oxygen   117
measurements
  carrier velocity   47
  phototransient   180
mechanism

degradation   314
memory
  effect   74, 258, 259, 280, 281
  electrical   258
  reactor   40
  thermal   257, 258
MEMS   297
mesa
  etching   166, 167
  isolation   142
  ridge   143
  structure   143
MESFET   68
  GaN   69
  SiC   68, 333
metal
  diffusion   315, 324
  galvanic   140, 142
mHEMT   300
MIC   86, 271
microcracking   327
micropipe   60, 339
  SiC   57
microscopy
  electron-beam current   118
  polarized light   59
microstrip   337
  transmission-line   78, 340
microwave
  generation   89
migration   327
MIM   86, 272, 338
MISFET   70, 190
  SiC   333
  SiN   70
MISFIT   315
misorientation   97
mixer   307
  dual gate   308
  resistive   308
mm-wave   79
MMIC   86, 271
  coplanar   86, 292
  LNA   87
    C-band   87
    microstrip   87
    temperature   301
  microstrip   86
  sapphire substrate   298

X-band   289
mobility
  defect dependence   102
  drift   103
  impurity scattering   9, 102
  lattice scattering   102
  low-field   9
  model   8
  peak   103
MOCVD   91, 92, 104
  AlGaN/GaN   92
  defects   129
  doping
    p-   128
  gas flow   95
  heterostructure   103
  InAlN/GaN   124
  InN   124
  sapphire   111
  Si   111
  SiC   109
model
  backgating   218
  charge control   218
  compact   197
  drain-lag   249
  integrability   245
  large-signal
    diode   251
    dispersion   247
  noise   269
  Pospieszalski   269
  small-signal   232
  substrate effects   230
modeling
  analytical   244
  charge
    equation   245
  current equation   246
  distributed   352
  impact ionization   252
  large-signal   197
    thermal   254
  linearity   260
  material   49
  passive   271
  small-signal   197
    HBT   234
  soft-breakdown   252

table-based   245
MODFET   *see* HEMT
module
  multichip   364
  TRX   77, 78, 342
morphology   94, 96, 121
MOSDHFET   70
MOSFET   70
  GaN   70
  InGaN   83
  intermodulation   70
MOSHFETs   133
mounting   337
MTBF   313
MTTF   313
multiband   88, 280, 364
multicarrier   280, 282

NBTI   320
nitridation   96
noise   52, 71
  $1/f$ 264
  analysis   262
  channel   265
  equivalent circuit   266
  low-frequency   262
    substrate   265
  modeling   197
  RF-   81, 87, 265
normally-off HEMT   *see* HEMT
    enhancement-mode
nucleation
  AlGaN   95
  AlN   113, 114
  GaN   94, 109
  layer   98
  MBE
    SiC   114
  MOCVD   95, 98
  sapphire   94
  SiC   95
  temperature   94, 114

OFDM   76, 258, 261, 283, 284
ONO   190
operation
  high-temperature   66, 311, 332
operation voltage   40, 66, 81, 240
  increase   285

optical stepper   56, 142
  resolution   158
oscillation
  GUNN   65
  odd-mode   286
oscillator   307
  Colpitts   307
  DC- to RF-efficiency   307
  differential   307
  ring   307
output
  conductance   54, 78, 201, 205
  power   236
    density   71
    linear   88
    saturated   71, 202
overdoping   220
oxide   187
  gate   189
  native   4, 156

package
  air cavity   364
  ceramic   363
  efficiency   361
  leadless   364
  selection   361
packaging   337, 358
PAE   203, 240, 279
  system   88
PAMBE   83, 115, 117
PAR   257, 258, 278, 284
PAS   128
passivation   140, 174
  Cat   173
  cryogenic temperatures   181
  degradation   317
  dielectric   187
  epitaxy   188
  high-field effects   186
  in-situ   137
  low-k   187
  MgO   187
  oxide   187
  polyimide   187
  reliability   186
  SiN   187
    Cat   186
    CVD   185

PECVD   185
  stress   185
  surface-related   179
  thickness   185
PCDE   262
peak-to-average ratio   see PAR
PEC   170
phase-noise   307
photo-enhanced chemical etching   101
photodetector
  ultraviolet   26
photoionization   180
photoluminescence   128
photonic bandgap   362
photoreflectance spectroscopy   134
PIC   49
piezoelectric effect
  inverse   326
pinch-off   78
  cold   214
  hot   318
pinning   see Fermi-level pinning
pits
  hexagonal   329
plasma etch
  SF_6 194
pn-diodes   151
polarity intermixing   29
polarization
  effects   28
  piezoelectric   30
power
  density per area   341
  slump   54, 212
power amplifier   271, 279
  broadband   297
  high-voltage   274
  highly-linear   242, 258
  Ku-band   79
  L-band   74
  mm-wave   79
  SiC   69
  X-band   77
power-added efficiency   51, 279, 280, 287
precursor   93, 99, 123
predistortion   258, 284, 285
  analog   285, 298
  digital   282, 285

GaN HFET   261
  memoryless   284
process
  flow   142
  planar   143
processing
  backside   142
  high temperature   191
  surface   140
product
  intermodulation   204
properties
  mechanical
    AlGaN   23
    AlN   15
    BN   19
    GaN   5
    InN   17
  optical
    AlGaN   23
    AlN   15
    BN   19
    GaN   5
    InN   17
PVT   57
pyrolysis   124

quaternary   23

radar
  active electronically scanned arrays
      88
  airborne
    phased array   88
  ground based   88
  naval   88
  phased array   77
  spaceborne   88
radiation
  damage   316
  space   89
radiation hardness
  neutron   331
reactor
  concept   93, 96
  memory   329
  MOCVD   96
  susceptor   96
real-space transfer   47, 259

recess   139, 170
  etching
    RIE   173
  gate   142, 171
  ohmic   45
recombination   20, 109, 177
  direct   15
  dynamic   177
  factor   209
  Schottky-Read-Hall   180
refractive index   15
relaxation
  time
    hot electron   11
    thermal   254
reliability   139, 285
  buffer   329
  case-study   327
  device   311
  drain-lag   321
  field plate   318
  gate lag   321
  interface   330
  LDMOS   322
  polarization effect   317
  pulsed-DC   323
  testing
    pinch-off   318
repeated node architecture   297
reproducibility   69, 91, 139
  reliability   317
resistance
  access   135, 170
  base   209
  channel   201
  contact   140, 146
  drain   201
  gate   164
  negative differential   65
  NiCr   339
  on-   200
  series   177
  sheet   38, 106
  source   201
  thermal   351
  thermal boundary   351
resistor   272, 273, 337, 339
RIE   166
rock salt   5

roughness
   interface   82, 103
   RMS   111
   surface   29, 57, 111
RTA   100, 147, 153

saturation velocity   27
   AlN   16
   effective   200
   GaN   10
   heterostructure   47
   InN   18
sawing   358
scaling
   gate length   49, 111
   wafer   107, 139
scattering
   acoustic   102
   alloy   103
   Coulomb   103
   dipole   103
   dislocation   103
   interface   103
   optical   102
   phonon   9
   surface donor   103
   threading dislocations   103
Schottky
   interface
      roughness   154
Schottky contact   see contact
scribing   358
seal
   hermetic   358, 361
selectivity
   etch   194
self-biasing   236
self-heating   191, 215, 256, 351
sensing
   pressure   90
sensor   89
   gas   89
sheet
   carrier concentration   104, 109
   carrier density   52, 135, 197, 262
   resistance   107, 340
SHFET   33
SHHEMT   33, 35
SiC

MESFET   327
   polytype   57
   Si segregation   57
SiCopSiC   122
SiGaNtic   112
signal
   envelope   242
silane   93, 185
   di   93
SIMS   130, 184
simulation
   EM   296
   Monte Carlo   10, 46
SiN   5
   passivation   182
SiO_2 70, 189
SIP   see system-in-package
slump
   RF-current   54
SOI   121, 280, 332
solder   345, 359
   eutectic   364
SopSiC   122
spark plug   89
SPDT   308
spectroscopy
   deep-level transient   179
   photoionization   179, 180
   Raman   344
spiking   315
SRPES   45
SSPA   79, 287
stability
   amplifier   294
   thermal   192
STEM   327, 328
strain   36, 42, 329
   compressive   33, 36
   engineering   112
stress   329
   mechanical   56, 154, 316
submount   365
substrate   55, 91
   AlN   62, 63, 272
   bulk growth   64
   backside deposition   92
   codoping   60
   criteria   55
   diamond   65

free-standing
  GaN   63
GaN   62
generic   274
hetero   54
homo   63
isolation   56
native   62
prebake   92
preparation
  in-situ   92
quasi   123
sapphire   61, 74, 354
semi-insulating   60
Si   75
  isolation   62
  thermal expansion   62
SiC   56, 353
  HPSI   60
silicon   62, 193, 354
surface   91
thinning   74
ZnO   64
superlattice   189
surface   40
  breakdown   316
  charge
    controlled structure   74
  cleaning   183
  contamination   182
  corrosion   316
  engineering   174
  morphology   324
    etching   143
  polish
    ECMP   91
  preparation   182
  pretreatment   183
  roughness   107, 111, 152, 158
  state   33, 180
    creation   186
susceptor   93
switch   308
  SPDT   67
  TX/RX   67
system
  PAE   77
system-in-package   362

TDD   10, 54, 98, 100, 101
TEM   118, 128, 315
temperature
  annealing   147
  coefficient   349
  cryogenic   216
  Debye   332
  growth   95
  junction   311
  substrate   205, 321
ternary   23, 27
test
  accelerated   311
  step   319
thermal
  conductivity   55, 345
    dislocations   22
    intrinsic   22
    SiC   56
  effect   205
  expansion   7, 15, 357
  management   271, 343
  modeling   345
  resistance
    extraction   253
  stability
    ohmic contact   148
  transport   329
thickness
  critical   43
thinning   140, 192, 340
threading
  dislocation   100, 101, 181
threshold voltage   198
  shift   54, 212, 315, 321
TMAl   93
TMGa   93
TMIn   93
transconductance   140, 311, 315
  DC   202
  maximum   198
  RF   202
transformation
  impedance   295
transient   211
  analysis   178
  transport   18
transmission-line   337
  losses   62

microstrip   273
transport
  high-field   16, 17
  minority carrier   84
  transient   11
trap   42, 177, 312
  bulk   177
    buffer   177
  characterization   178
  charging   315
  interface   177–179
  location   177
  mechanisms   177
  nucleation layer   177
  parameters   42
  surface   177
tunneling   50, 118, 153, 180, 320
  Fowler-Nordheim   191

uniformity   106, 139
  substrate   107
  temperature   107
unilateral gain   221
UWB   88, 308

VCO   307
velocity
  heterostructure   47
  peak   8, 16
  saturation   10, 24
velocity-field characteristics   18
viahole   340
  blow-out   361
  drilling   195

etching   55, 140, 193, 340
  SiC   193
  silicon   194
  source   288
  thermal   254
void control   358
Volterra series   256

wafer
  bow   56, 111
  bowing   55, 56, 61, 90, 158, 359
  breakage
    sapphire   61, 193
  dicing   358
  etching   340
  four-inch   61, 107, 113, 142, 193
  full annealing   192
  mounting   340
  sawing   358
  scaling   61
  scribing   358
  thinning   55, 61, 192, 316, 340
    sapphire   61
  three-inch   106
  two-inch   106
  warp   55, 56
WCDMA   74, 258, 259
  two-carrier   75
WiMAX   76, 280, 282, 317
wurtzite   5, 10, 130

XRD   107, 128

zincblende   5, 6, 10, 69

Springer Series in
MATERIALS SCIENCE

Editors: R. Hull R. M. Osgood, Jr. J. Parisi H. Warlimont

50 **High-Resolution Imaging
and Spectrometry of Materials**
Editors: F. Ernst and M. Rühle

51 **Point Defects in Semiconductors
and Insulators**
Determination of Atomic
and Electronic Structure
from Paramagnetic Hyperfine
Interactions
By J.-M. Spaeth and H. Overhof

52 **Polymer Films
with Embedded Metal Nanoparticles**
By A. Heilmann

53 **Nanocrystalline Ceramics**
Synthesis and Structure
By M. Winterer

54 **Electronic Structure and Magnetism
of Complex Materials**
Editors: D.J. Singh and
D. A. Papaconstantopoulos

55 **Quasicrystals**
An Introduction to Structure,
Physical Properties and Applications
Editors: J.-B. Suck, M. Schreiber,
and P. Häussler

56 **SiO₂ in Si Microdevices**
By M. Itsumi

57 **Radiation Effects
in Advanced Semiconductor Materials
and Devices**
By C. Claeys and E. Simoen

58 **Functional Thin Films
and Functional Materials**
New Concepts and Technologies
Editor: D. Shi

59 **Dielectric Properties of Porous Media**
By S.O. Gladkov

60 **Organic Photovoltaics**
Concepts and Realization
Editors: C. Brabec, V. Dyakonov, J. Parisi
and N. Sariciftci

61 **Fatigue in Ferroelectric Ceramics
and Related Issues**
By D.C. Lupascu

62 **Epitaxy**
Physical Principles
and Technical Implementation
By M.A. Herman, W. Richter, and H. Sitter

63 **Fundamentals
of Ion-Irradiated Polymers**
By D. Fink

64 **Morphology Control of Materials
and Nanoparticles**
Advanced Materials Processing
and Characterization
Editors: Y. Waseda and A. Muramatsu

65 **Transport Processes
in Ion-Irradiated Polymers**
By D. Fink

66 **Multiphased Ceramic Materials**
Processing and Potential
Editors: W.-H. Tuan and J.-K. Guo

67 **Nondestructive
Materials Characterization**
With Applications to Aerospace Materials
Editors: N.G.H. Meyendorf, P.B. Nagy,
and S.I. Rokhlin

68 **Diffraction Analysis
of the Microstructure of Materials**
Editors: E.J. Mittemeijer and P. Scardi

69 **Chemical–Mechanical Planarization
of Semiconductor Materials**
Editor: M.R. Oliver

70 **Applications of the Isotopic Effect
in Solids**
By V.G. Plekhanov

71 **Dissipative Phenomena
in Condensed Matter**
Some Applications
By S. Dattagupta and S. Puri

72 **Predictive Simulation
of Semiconductor Processing**
Status and Challenges
Editors: J. Dabrowski and E.R. Weber

73 **SiC Power Materials**
Devices and Applications
Editor: Z.C. Feng

Springer Series in
MATERIALS SCIENCE

Editors: R. Hull   R. M. Osgood, Jr.   J. Parisi   H. Warlimont

74 **Plastic Deformation
in Nanocrystalline Materials**
By M.Yu. Gutkin and I.A. Ovid'ko

75 **Wafer Bonding**
Applications and Technology
Editors: M. Alexe and U. Gösele

76 **Spirally Anisotropic Composites**
By G.E. Freger, V.N. Kestelman,
and D.G. Freger

77 **Impurities Confined
in Quantum Structures**
By P.O. Holtz and Q.X. Zhao

78 **Macromolecular Nanostructured
Materials**
Editors: N. Ueyama and A. Harada

79 **Magnetism and Structure
in Functional Materials**
Editors: A. Planes, L. Mañosa,
and A. Saxena

80 **Micro- and Macro-Properties of Solids**
Thermal, Mechanical
and Dielectric Properties
By D.B. Sirdeshmukh, L. Sirdeshmukh,
and K.G. Subhadra

81 **Metallopolymer Nanocomposites**
By A.D. Pomogailo and V.N. Kestelman

82 **Plastics for Corrosion Inhibition**
By V.A. Goldade, L.S. Pinchuk,
A.V. Makarevich and V.N. Kestelman

83 **Spectroscopic Properties of Rare Earths
in Optical Materials**
Editors: G. Liu and B. Jacquier

84 **Hartree–Fock–Slater Method
for Materials Science**
The DV–X Alpha Method for Design
and Characterization of Materials
Editors: H. Adachi, T. Mukoyama,
and J. Kawai

85 **Lifetime Spectroscopy**
A Method of Defect Characterization
in Silicon for Photovoltaic Applications
By S. Rein

86 **Wide-Gap Chalcopyrites**
Editors: S. Siebentritt and U. Rau

87 **Micro- and Nanostructured Glasses**
By D. Hülsenberg and A. Harnisch

88 **Introduction
to Wave Scattering, Localization
and Mesoscopic Phenomena**
By P. Sheng

89 **Magneto-Science**
Magnetic Field Effects on Materials:
Fundamentals and Applications
Editors: M. Yamaguchi and Y. Tanimoto

90 **Internal Friction in Metallic Materials**
A Handbook
By M.S. Blanter, I.S. Golovin,
H. Neuhäuser, and H.-R. Sinning

91 **Time-dependent Mechanical Properties
of Solid Bodies**
By W. Gräfe

92 **Solder Joint Technology**
Materials, Properties, and Reliability
By K.-N. Tu

93 **Materials for Tomorrow**
Theory, Experiments and Modelling
Editors: S. Gemming, M. Schreiber
and J.-B. Suck

94 **Magnetic Nanostructures**
Editors: B. Aktas, L. Tagirov,
and F. Mikailov

95 **Nanocrystals
and Their Mesoscopic Organization**
By C.N.R. Rao, P.J. Thomas
and G.U. Kulkarni

96 **Gallium Nitride Electronics**
By R. Quay

97 **Multifunctional Barriers
for Flexible Structure**
Textile, Leather and Paper
Editors: S. Duquesne, C. Magniez,
and G. Camino

98 **Physics of Negative Refraction
and Negative Index Materials**
Optical and Electronic Aspects
and Diversified Approaches
Editors: C.M. Krowne and Y. Zhang